Advanced Micro & Nanosystems
Volume 2
CMOS – MEMS

Related Wiley-VCH titles:

Baltes, H., Brand, O., Fedeer, G. K., Hierold, C. Korvink, J. G., Tabata, O. (eds.)
Enabling Technology for MEMS and Nanodevices
2004
ISBN 3-527-30746-X

Baltes, H., Fedder, G. K., Korvink, J. G. (eds.)
Sensors Update Vol. 13
2004
ISBN 3-527-30745-1

Baltes, H., Fedder, G. K., Korvink, J. G. (eds.)
Sensors Update Vol. 12
2003
ISBN 3-527-30602-1

Hesse, J., Gardner, J. W., Göpel, W. eds.)
Sensors Applications:

Öberg, P. Å., Togawa, T., Spelman, F. A. (eds.)
Sensors in Medicine and Health Care
2004
ISBN 3-527-29556-9

Marek, J., Trah, H.-P., Suzuki, Y., Yokomori, I. (eds.)
Sensors for Automotive Technology
2003
ISBN 3-527-29553-4

Tschulena, G., Lahrmann, A. (eds.)
Sensors in Household Appliances
2003
ISBN 3-527-30362-6

Gassmann, O., Meixner, H. (eds.)
Sensors in Intelligent Buildings
2001
ISBN 3-527-29557-7

Tönshoff, H. K., Inasaki, I. (eds.)
Sensors in Manufacturing
2001
ISBN 3-527-29558-5

Pearce, T. C., Schiffman, S. S., Nagle, H. T., Gardner, J. W. (eds.)
Handbook of Machine Olfaction
Electronic Nose Technology
2003
ISBN 3-527-30358-8

Ajayan, P., Schadler, L. S., Braun, P. V.
Nanocomposite Science and Technology
2003
ISBN 3-527-30359-6

Caruso, F. (ed.)
Colloids and Collid Assemblies
2003
ISBN 3-527-30660-9

Decher, G., Schlenoff, J. B. (eds.)
Multilayer Thin Films
Sequential Assembly of Nanocomposite Materials
2003
ISBN 3-527-30440-1

Gómez-Romero, P., Sanchez, C. (eds.)
Functional Hybrid Materials
2003
ISBN 3-527-30484-3

Komiyama, M., Takeuchi, T., Mukawa, T., Asanuma, H.
Molecular Imprinting
From Fundamentals to Applications
2003
ISBN 3-527-30569-6

Advanced Micro & Nanosystems
Volume 2

CMOS – MEMS

Volume Editors
O. Brand and G. K. Fedder

WILEY-VCH Verlag GmbH & Co. KGaA

Series Editors

Prof. Dr. Henry Baltes
Physical Electronics
Laboratory
ETH Zürich
Hönggerberg, HPT-H6
8093 Zürich
Switzerland
baltes@iqe.phys.ethz.ch

Prof. Dr. Oliver Brand
School of Electrical and
Computer Engineering
Georgia Institute of
Technology
Atlanta, GA 30332-0250
USA
oliver.brand@ece.gatech.edu

Prof. Dr. Gary K. Fedder
ECE Department &
The Robotics Institute
Carnegie Mellon
University
Pittsburgh,
PA 15213-3890
USA
fedder@ece.cmu.edu

Volume Editors
Prof. Dr. Oliver Brand
Prof. Dr. Gary K. Fedder

Prof. Dr. Christofer Hierold
Chair of Micro-
and Nanosystems
ETH Zürich
ETH-Zentrum, CLA H9
Tannenstr. 3
8092 Zürich
Switzerland
christofer.hierold@micro.mavt.ethz.ch

Prof. Dr. Jan G. Korvink
IMTEK Institute for
Microsystem Technology
University of Freiburg
Georges-Köhler-Allee
103/03.033
79110 Freiburg
Germany
korvink@imtek.de

Prof. Dr. Osamu Tabata
Department
of Mechanical
Engineering
Faculty of Engineering
Kyoto University
Yoshida Honmachi,
Sakyo-ku
Kyoto 606-8501
Japan
tabata@mech.kyoto-u.ac.jp

Cover picture
Top left: 100-pixel infrared sensor array by Phyiscal Electronics Laboratory (PEL), ETH Zurich, Switzerland.
Top right: Dual-axis acceleration sensor ADXL-203 by Analog Devices Inc., USA.
Bottom right: Packaged pressure sensor KP-100 by Infineon Technologies AG, Germany.

■ All books published by Wiley-VCH are carefully produced. Nevertheless, authors, editors, and publisher do not warrant the information contained in these books, including this book, to be free of errors. Readers are advised to keep in mind that statements, data, illustrations, procedural details or other items may inadvertently be inaccurate.

Library of Congress Card No.: applied for

British Library Cataloguing-in-Publication Data
A catalogue record for this book is available from the British Library.

**Bibliographic information published
by Die Deutsche Bibliothek**
Die Deutsche Bibliothek lists this publication in the Deutsche Nationalbibliografie; detailed bibliographic data is available in the Internet at <http://dnb.ddb.de>

© 2005 WILEY-VCH Verlag GmbH & Co. KGaA, Weinheim, Germany

All rights reserved (including those of translation in other languages). No part of this book may be reproduced in any form – by photoprinting, microfilm, or any other means – nor transmitted or translated into machine language without written permission from the publishers. Registered names, trademarks, etc. used in this book, even when not specifically marked as such, are not to be considered unprotected by law.

Printed in Singapore
Printed on acid-free paper

Composition   K+V Fotosatz GmbH, Beerfelden
Printing and Bookbinding   Markono Print Media Pte Ltd, Singapore

ISBN-13:   978-3-527-31080-7
ISBN-10:   3-527-31080-0

Preface

We, the *CMOS-MEMS* volume editors, welcome you to this second installment of *Advanced Micro & Nanosystems*. Today's microelectromechanical systems (MEMS) are built much the same way as silicon integrated circuits (ICs) are, borrowing a variety of materials and processes from the IC industry. It is thus not surprising that from the early days of MEMS more than three decades ago, researchers have tried to co-integrate microelectromechanical devices with bipolar or CMOS circuitry. The challenges, achievements and prospects of the research and development work in the area of CMOS-integrated MEMS, or for short CMOS-MEMS, are the topics of the present book. CMOS technology and micromachining techniques are combined to fabricate a wide spectrum of microsystems ranging from physical sensors, e.g., pressure and inertial sensors, to chemical and biochemical sensing systems.

The field of CMOS-MEMS dates back to the mid 1980s, but has grown substantially in the past decade. An early example of CMOS-integrated MEMS is the CMOS-based piezoresistive pressure sensing system developed by NEC and published 1987 in the *IEEE Journal of Solid-State Circuits*. Integrated pressure sensors using bipolar instead of CMOS technologies were developed already in the late 1970s, e.g., at the University of Michigan and Case Western University. In 1993, Analog Devices introduced the first CMOS-integrated surface-micromachined accelerometer, the ADXL-50, to the commercial market. Recently, the same company has released the first CMOS-integrated gyroscope with on-chip circuitry capable of detecting capacitance changes in the zepto-F (10^{-21} F) range. Another exemplary integrated microsystem is the Texas Instruments Digital Micromirror Device (DMD^{TM}) featuring up to 1.3 million individually addressable micromirrors on a single chip, which would hardly be feasible without the underlying CMOS addressing electronics. Established fabrication processes, co-integration of powerful analog and digital circuitry, and the possibility of large sensor arrays are the clear benefits of using CMOS technologies to develop MEMS. Besides the above mentioned, by now almost classical applications, CMOS-MEMS have in recent years also found their way into new areas, including chemical and biochemical sensing, biomimetics, acoustics, and RF components.

The present volume of *Advanced Micro & Nanosystems* has been divided into eleven chapters, providing a broad overview on current and past activities in the area

Advanced Micro and Nanosystems. Vol. 2. CMOS – MEMS.
Edited by H. Baltes, O. Brand, G. K. Fedder, C. Hierold, J. Korvink, O. Tabata
Copyright © 2005 WILEY-VCH Verlag GmbH & Co. KGaA, Weinheim
ISBN: 3-527-31080-0

of CMOS-MEMS. To our best knowledge, it is the first book dedicated to CMOS-MEMS and hopefully will serve as a valuable reference for you, our reader.

The volume starts off with Oliver Brand's overview chapter on CMOS-MEMS fabrication approaches, organizing the different process flows according to where the micromachining process steps are added to the CMOS base line, i.e., pre-CMOS, intermediate-CMOS, and post-CMOS approaches. Design of CMOS-MEMS requires not only the knowledge of electrical, but also thermal and mechanical material properties.

In the second chapter, Patrick Ruther and Oliver Paul present a detailed overview on measurement techniques used to extract relevant properties of mainly thin film materials and have assembled a large amount of material property data in a valuable tabular format.

Chapter 3 by Gary Fedder, Junseok Chae, Haluk Kulah, Khalil Najafi, Tim Denison and Steve Lewis takes a close look at CMOS-integrated accelerometers and gyroscopes, including some of the commercially most successful CMOS-MEMS products. Miniaturized acoustic devices, namely loudspeakers and microphones, have become ubiquitous in our daily life; we find them in our phones, computers and our children's toys. Recent advances in the area of CMOS-based acoustic devices are discussed by John Neumann and Kaigham Gabriel in Chapter 4.

With CMOS transistors reaching ever higher frequencies with each new technology cycle, the use of CMOS-MEMS approaches to build RF MEMS devices, such as switches or high-frequency variable capacitors, has potential to create wholly on-chip wireless systems. Chapter 5 by Tamal Mukherjee and Gary Fedder gives an overview on recently developed RF CMOS-MEMS components and circuits.

Pressure sensors were the first commercially available high-volume MEMS product, and thus it is not surprising that a large number of CMOS-integrated pressure sensors have been demonstrated over the past 20 years. In Chapter 6, Hans-Jörg Timme reviews the theoretical background for micromachined pressure sensors and compares different CMOS-based implementation strategies.

The following three chapters concentrate on chemical and biological applications of CMOS-MEMS. Keeping in mind the possibility of sensor arrays and on-chip analog and digital electronics, CMOS-based microsystems are especially suited for small, hand-held devices in applications ranging from chemical safety, security and access control to biomedical diagnostics. In Chapter 7, Andreas Hierlemann discusses basic chemical sensor concepts and reviews chemical sensors based on CMOS technology. Access control is an important aspect in our daily life and there is hope that secure biomimetic access control systems will at some point diminish the need to memorize dozens of passwords. In Chapter 8, Christofer Hierold, Gerd Hribernig, and Thomas Scheiter take a close look at capacitive fingerprint sensor systems, discussing not only the actual CMOS-based fingerprint sensors, but all the system aspects that need to be addressed while developing a biometric authentication system. Finally, Jan Lichtenberg and Henry Baltes discuss in Chapter 9 the relatively young field of CMOS-based biochemical sensing systems, including biosensor arrays and cell-based assays.

Micromachining techniques are used to machine, e.g., the silicon substrate and can provide microstructures with excellent thermal isolation. Besides this, thermal sensors often require only resistive elements and can be readily integrated with CMOS technology. Chapter 10 by Tayfun Akin discusses new developments in the areas of CMOS-based thermal radiation sensors, thermal flow sensors and thermal converters. The present AMN volume closes with a chapter by Christoph Hagleitner and Kay-Uwe Kirstein on circuit and system integration, taking an in-depth look at all the relevant aspects, including common analog front-end circuitry architectures for integrated microsensors, and important building blocks for CMOS-MEMS, such as filters, analog-to-digital converters, current/voltage references and calibration/data interfaces.

Last but not least, this is the time and place to thank our authors for their hard work and timely chapter contributions. The editors are grateful to the publisher, Wiley-VCH, for the support of the book series. In particular, we thank the publishing editor, Dr. Martin Ottmar, for the management of *AMN* and the project editor, Dr. Waltraud Wüst, for her never-ending support in carrying this volume from the early concept phase all the way to the actual printing.

This book will be the last in the series for which we benefit from having Henry Baltes as a co-editor. We are deeply grateful to Henry for his many contributions to the new *AMN* series and its precursor, *Sensors Update*. As one of the leaders in the CMOS-MEMS field, it is fitting that this volume be on the topic that Henry helped to pioneer. Thank you, Henry!

Oliver Brand and Gary K. Fedder, Volume Editors

November 2004
Atlanta and Pittsburgh

Foreword

We are proud to present the second volume of *Advanced Micro & Nanosystems* (*AMN*), entitled *CMOS-MEMS* and entirely dedicated to this topic. It examines and illustrates the recent, significant advances in the field of CMOS-integrated microelectromechanical systems. With all of the series editors having a background in using CMOS processes for microsystem development, a book like this has been on our minds for quite some time. Oliver Brand and Gary Fedder as volume editors have now assembled an international team of experts from academia and industry as chapter authors to cover all the facets of CMOS-MEMS, and also act as chapter contributors themselves in this work that we believe you will find useful and exciting.

Covering recent advances from the world of micro and nanosystems, future *AMN* issues will either focus on a particular subject, such as the present *CMOS-MEMS*, or be a carefully chosen set of cutting-edge overview and review articles like the first *AMN* volume on *Enabling Techniques for MEMS and Nanodevices*.

Looking ahead, we hope to welcome you back, dear reader, to the upcoming third member of the *AMN* series, in which we leave the area of silicon-based micro- and nanosystems to take a close look at the fascinating field of *Microengineering of Metals and Ceramics*. The articles will range from the design, tooling, advanced replication techniques, automation and quality assurance all the way to the resulting properties of materials and microcomponents. To cover such a wide spectrum, we are very glad to have the support of two well-known experts in the field, Prof. Detlef Löhe from the University of Karlsruhe, Germany, and Prof. Jürgen Haußelt from the University of Freiburg, Germany, who will edit this volume together.

Henry Baltes, Oliver Brand, Gary K. Fedder,
Christofer Hierold, Jan G. Korvink, and Osamu Tabata, Series Editors

November 2004
Zurich, Atlanta, Pittsburgh, Freiburg and Kyoto

Advanced Micro and Nanosystems. Vol. 2. CMOS – MEMS.
Edited by H. Baltes, O. Brand, G. K. Fedder, C. Hierold, J. Korvink, O. Tabata
Copyright © 2005 WILEY-VCH Verlag GmbH & Co. KGaA, Weinheim
ISBN: 3-527-31080-0

Contents

Preface V

Foreword IX

List of Contributors XIII

1 **Fabrication Technology** 1
  O. Brand

2 **Material Characterization** 69
  J. O. Paul and P. Ruther

3 **Monolithically Integrated Inertial Sensors** 137
  G. K. Fedder, J. Chae, H. Kulah, K. Najafi, T. Denison, J. Kuang, and S. Lewis

4 **CMOS–MEMS Acoustic Devices** 193
  J. J. Neumann and K. J. Gabriel

5 **RF CMOS MEMS** 225
  T. Mukherjee and G. K. Fedder

6 **CMOS-based Pressure Sensors** 257
  H.-J. Timme

7 **CMOS-based Chemical Sensors** 335
  A. Hierlemann

8 **Biometric Capacitive CMOS Fingerprint Sensor Systems** 391
  C. Hierold, G. Hribernig, and T. Scheiter

9 **CMOS-based Biochemical Sensing Systems** 447
  J. Lichtenberg, H. Baltes

Advanced Micro and Nanosystems. Vol. 2. CMOS – MEMS.
Edited by H. Baltes, O. Brand, G. K. Fedder, C. Hierold, J. Korvink, O. Tabata
Copyright © 2005 WILEY-VCH Verlag GmbH & Co. KGaA, Weinheim
ISBN: 3-527-31080-0

10   **CMOS-based Thermal Sensors**   *479*
     T. Akin

11   **Circuit and System Integration**   *513*
     C. Hagleitner and K.-U. Kirstein

Subject Index   *579*

List of Contributors

Prof. Dr. Tayfun Akin
Department of Electrical & Electronics
Engineering
Middle East Technical University
06531 Ankara
Turkey

Prof. Dr. Henry Baltes
Physical Electronics Laboratory
ETH Zurich
Hoenggerberg, HPT-H6
8093 Zurich
Switzerland

Prof. Dr. Oliver Brand
School of Electrical and Computer
Engineering
Georgia Institute of Technology
777 Atlantic Drive
Atlanta, GA 30332-0250
USA

Dr. Junseok Chae
Solid State Electronics Laboratory
Center for Wireless Integrated
Microsystems (WIMS)
Department of Electrical Engineering
and Computer Science (EECS)
The University of Michigan
1301 Beal Ave.
Ann Arbor, MI 48109-2122
USA

Dr. Tim Denison
Analog Devices Inc (ADI)
21 Osborn Street
Cambridge, MA 02139
USA

Prof. Dr. Gary K. Fedder
Department of Electrical and Computer
Engineering & The Robotics Institute
Carnegie Mellon University
Pittsburgh, PA 15213-3890
USA

Prof. Dr. Kaigham J. Gabriel
Department of Electrical and Computer
Engineering & The Robotics Institute
Carnegie Mellon University
5000 Forbes Ave.
Pittsburgh PA 15213-3890
USA

Dr. Christoph Hagleitner
IBM Zurich Research Laboratory
Saumerstrasse 4
8803 Ruschlikon
Switzerland

Prof. Dr. Andreas Hierlemann
Physical Electronics Laboratory
ETH Zurich
Hoenggerberg, HPT-H 4.2
8093 Zurich
Switzerland

Prof. Dr. Christofer Hierold
Chair of Micro- and Nanotechnology
ETH Zurich
ETH Zentrum, CLA H 9
8092 Zurich
Switzerland

Dipl.-Ing. Gerd Hribernig
Straßgangerstraße 315
8054 Graz
Austria

Advanced Micro and Nanosystems. Vol. 2. CMOS – MEMS.
Edited by H. Baltes, O. Brand, G. K. Fedder, C. Hierold, J. Korvink, O. Tabata
Copyright © 2005 WILEY-VCH Verlag GmbH & Co. KGaA, Weinheim
ISBN: 3-527-31080-0

Dr. Kay-Uwe Kirstein
Physical Electronics Laboratory
ETH Zurich
Hoenggerberg, HPT H4.1
8093 Zurich
Switzerland

Dr. Jinbo Kuang
Analog Devices Inc (ADI)
21 Osborn Street
Cambridge, MA 02139
USA

Dr. Haluk Kulah
Department of Electrical & Electronics Engineering
Middle East Technical University
06531 Ankara
Turkey

Steve Lewis
Analog Devices Inc (ADI)
21 Osborn Street
Cambridge, MA 02139
USA

Dr. Jan Lichtenberg
Physical Electronics Laboratory
ETH Zurich
Hoenggerberg, HPT-H4.2
8093 Zurich
Switzerland

Prof. Dr. Tamal Mukherjee
Department of Electrical and Computer Engineering
Carnegie Mellon University
Pittsburgh, PA 15213-3890
USA

Prof. Dr. Khalil Najafi
Solid State Electronics Laboratory
Center for Wireless Integrated Microsystems (WIMS)
Department of Electrical Engineering and Computer Science (EECS)
The University of Michigan
1301 Beal Ave.
Ann Arbor, MI 48109-2122
USA

Dr. John J. Neumann Jr.
Intel Corp.
2413 NW 229th Ave.
Mailstop: RA3-410
Hillsboro, OR 97124
USA

Prof. Dr. Oliver Paul
Institute for Microsystem Technology
IMTEK
Microsystem Materials Laboratory
University of Freiburg
Georges-Koehler-Allee 103
79110 Freiburg
Germany

Dr. Patrick Ruther
Institute for Microsystem Technology
IMTEK
Microsystem Materials Laboratory
University of Freiburg
Georges-Koehler-Allee 103
79110 Freiburg
Germany

Dr. Thomas Scheiter
Infineon Technologies AG
St. Martin Str. 76
81541 München
Germany

Dr. Hans-Jörg Timme
Infineon Technologies AG
Balanstraße 73
81541 München
Germany

1
Fabrication Technology

*O. Brand, School of Electrical and Computer Engineering,
Georgia Institute of Technology, Atlanta, GA, USA*

Abstract

This chapter provides an overview on fabrication technologies for CMOS-based microelectromechanical systems (MEMS). The first part briefly introduces the basic microfabrication steps, highlights a CMOS process sequence and how CMOS materials can be used for microsystems design. While a number of microsystems can be fabricated within the regular CMOS process sequence, the focus of the chapter is on combining CMOS technology with micromachining process modules. CMOS-compatible bulk and surface micromachining techniques are introduced in the second part of the chapter together with an overview of the design challenges faced when combining mechanical microstructures and electronics on the same substrate. The micromachining modules can either precede (pre-CMOS), follow (post-CMOS) or be performed in between (intra-CMOS) the regular CMOS process steps. The last part of the chapter provides an extensive overview on the different CMOS-based MEMS approaches found in the literature.

Keywords

Micromachining; CMOS-based MEMS; MEMS fabrication; microsystem fabrication

1.1	CMOS Technology	2
1.1.1	Basic Microfabrication Steps	4
1.1.1.1	Thin Film Deposition	5
1.1.1.2	Patterning	6
1.1.1.3	Etching	8
1.1.1.4	Doping	9

1.1.2	CMOS Process Sequence	9
1.1.3	CMOS Materials for Micro- and Nanosystems	11
1.1.4	CMOS Microsystems	14
1.2	**CMOS-compatible Micromachining Process Modules**	**17**
1.2.1	Bulk Micromachining	18
1.2.2	Surface Micromachining	22
1.3	**CMOS-compatible Design of MEMS and NEMS**	**23**
1.3.1	Tolerable Process Modifications	24
1.3.2	Design Rule Modifications	26
1.3.3	Simulation of Circuitry and MEMS	27
1.4	**CMOS and Micromachining**	**28**
1.4.1	Pre-CMOS Micromachining	33
1.4.2	Intra-CMOS Micromachining	37
1.4.3	Post-CMOS Micromachining	43
1.4.3.1	Post-CMOS Micromachining of Add-on Layers	43
1.4.3.2	Post-CMOS Micromachining of CMOS Layers	49
1.5	**Conclusion**	**56**
1.6	**References**	**57**

1.1
CMOS Technology

State-of-the-art CMOS processes, such as IBM's 9S2 process based on SOI (silicon-on-insulator) technology on 300 mm wafers, feature a minimal physical gate length of less than 100 nm and up to eight (copper) metallization levels (see Fig. 1.1, [1]). Such advanced CMOS processes are required for the fabrication of today's and tomorrow's microprocessors comprising tens of millions of transistors on a single chip. An example is Apple Computer's 64-bit PowerPC-G5 processor with more than 58 million transistors [2], manufactured using IBM's 90 nm CMOS technology.

Researchers at IBM's T.J. Watson Research Center have recently used the copper-based interconnect technology of such modern CMOS processes to fabricate microelectromechanical devices, namely r.f. switches and resonators [3, 4]. Up to now, however, most commercially available microsystems combining (micromachined) transducer elements and integrated electronics on a single chip rely on CMOS or BiCMOS processes with minimum feature sizes typically between 0.5 and 3 µm and 4 or 6 in wafer sizes. While the underlying CMOS technologies are between 10 and 15 years old, their capabilities are sufficient for most microsystem applications. An example is the pressure sensor KP100 by Infineon Technologies, a surface micromachined pressure sensor array with on-chip circuitry for signal conditioning, A/D conversion, calibration and system diagnostic, which is based on a 0.8 µm BiCMOS technology on 6 in wafers [5].

A typical cross-section of a sub-µm (0.5–1.0 µm) CMOS technology used for CMOS-based microelectromechanical systems (MEMS) is shown in Fig. 1.2 [6].

1.1 CMOS Technology

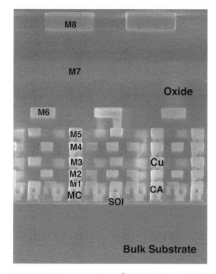

Fig. 1.1 Cross-section of IBM's 90 nm CMOS technology 9S2 with 8-level copper metallization (labeled M1–M8) with close-up of three metal–oxide–semiconductor field effect transistors (MOSFETs). Images courtesy of International Business Machines Corporation; unauthorized use not permitted

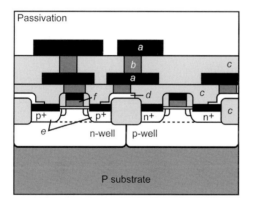

a Aluminum metallization
b Tungsten via
c Oxide
d TiN local interconnect
e LDD source/drain
f Polysilicon gate

Fig. 1.2 Schematic cross-section of typical sub-µm (0.5–1.0 µm) CMOS technology with two-level aluminum metallization and TiN local interconnects. Adapted from [6]

The twin-well technology is based on 6 in p-type wafers and uses a polysilicon/silicide gate, low-doped drain (LDD) technology for source and drain formation, silicide source/drain contacts and a two-level metallization based on tungsten plugs and aluminum interconnects. A thermal oxide separates adjacent transistors, chemical vapor deposition (CVD) silicon dioxide layers are used as dielectric layers between the metallization levels and a PECVD (plasma enhanced CVD) silicon nitride layer or a silicon dioxide, silicon nitride sandwich are employed as passivation layer. The CMOS fabrication sequence is briefly highlighted in Section 1.1.2. More detailed process descriptions can be found in a number of microelectronics textbooks, e.g. [6–8].

When designing CMOS-based MEMS or microsystems, the designer must adhere, to a great extent, to the chosen CMOS process sequence in order not to sacrifice the functionality of the on-chip electronics. This limits the available 'design space' for the integrated microsystems, as e.g. materials, material properties and layer thicknesses are determined by the CMOS process. In the following, a brief introduction into integrated circuit fabrication will be given: the basic fabrication steps are highlighted (Section 1.1.1) and a CMOS process sequence is summarized (Section 1.1.2). Section 1.1.3 discusses how the different CMOS materials and layers can be used in micro- and nanosystems and Section 1.1.4 depicts a few microsystems that can be completely formed within a regular CMOS sequence.

1.1.1
Basic Microfabrication Steps

The fabrication of integrated circuits (ICs) using CMOS or BiCMOS technology is based on four basic microfabrication techniques: deposition, patterning, doping and etching. Fig. 1.3 illustrates how these techniques are combined to build up an IC layer by layer: a thin film, such as an insulating silicon dioxide film, is deposited on the substrate, a silicon wafer. A light-sensitive photoresist layer is then deposited on top and patterned using photolithography. Finally, the pattern is transferred from the photoresist layer to the silicon dioxide layer by an etching process. After removing the remaining photoresist, the next layer is deposited and struc-

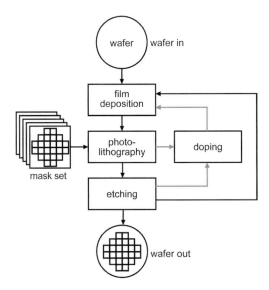

Fig. 1.3 Flow diagram of IC fabrication process using the four basic microfabrication techniques: deposition, photolithography, etching and doping. Adapted from [8]

tured, and so on. Doping of a semiconductor material by ion implantation, the key step for the fabrication of diodes and transistors, can be performed directly after photolithography, i.e. using a photoresist layer as mask, or after patterning an implantation mask (e.g. a silicon dioxide layer).

Silicon is the standard substrate material for IC fabrication and, hence, the most common substrate material in microfabrication in general. It is supplied as single-crystal wafers with diameters between 100 and 300 mm. In addition to its favorable electrical properties, single-crystal silicon also has excellent mechanical properties [9], which enable the design of micromechanical structures. CMOS processes for digital electronics typically use low-doped (doping concentration in the 10^{16} cm^{-3} range) silicon wafers, whereas processes for mixed-signal or analog electronics are often based on high-doped (doping concentration in the 10^{19} cm^{-3} range) wafers with a low-doped epitaxial layer to minimize latch-up. The choice of the substrate material might already require a compromise between the requirements for the MEMS part and the on-chip electronics: the fabrication of membrane structures for, e.g., pressure sensors is typically based on anisotropic silicon etching in a potassium hydroxide (KOH) solution (see Section 1.2). High p-type doping ($N_A \geq 10^{19}$ cm^{-3}) substantially reduces the silicon etch rates in KOH solutions, thus preventing the use of highly p-doped CMOS substrates in combination with KOH etching.

In the following, a brief overview on the four basic microfabrication steps will be given. More details can be found in textbooks and reference books on semiconductor processing [6–8, 10, 11].

1.1.1.1 Thin-film Deposition

The two most common thin-film deposition methods in microfabrication are *chemical vapor deposition* (CVD), performed at low pressure (LPCVD), atmospheric pressure (APCVD) or plasma-enhanced (PECVD), and *physical vapor deposition* (PVD), such as sputtering and evaporating. Typical CVD and PVD film thicknesses are in the range of tenths of nanometers up to a few micrometers. Other film deposition techniques include electroplating of metal films (e.g. the copper metallization in state-of-the-art CMOS processes) and spin- or spray-coating of polymeric films such as photoresist. Both processes can yield film thicknesses from less than 1 μm up to several hundreds of micrometers.

Dielectric layers, predominantly silicon dioxide, SiO_2, and silicon nitride, SiN_x, are used as insulating material, as mask material and for device passivation. Silicon dioxide is either thermally grown on top of a silicon surface (thermal oxide) at high temperatures (900–1200 °C) in an oxidation furnace or it is deposited in a CVD system (CVD oxide). CVD oxides can be deposited at temperatures between 300 and 900 °C, with the high-temperature depositions usually yielding better film properties. Low-temperature CVD oxide films are typically deposited in PECVD systems and high-temperature CVD oxide films in LPCVD equipment. Silicon nitride layers deposited in LPCVD furnaces are commonly used as masking material during local oxidation of silicon (LOCOS process), while PECVD silicon nitride films are used for e.g. device passivation.

Highly doped *polycrystalline silicon* (polysilicon) is used as gate material for metal oxide semiconductor field effect transistors (MOSFETs), as electrode and resistor materials, for piezoresistive sensing structures, as thermoelectric material, and for thermistors. Polysilicon microstructures released by sacrificial layer etching are also widely used in sensor applications (see Section 1.4). Polysilicon is usually deposited in an LPCVD furnace using silane (SiH_4) as gaseous precursor.

Metal layers are used, e.g., for electrical interconnects, as electrode material, for resistive temperature sensors (thermistors) or as mirror surfaces. Metals, which are widely used in the microelectronics industry, such as aluminum, titanium and tungsten, are routinely deposited by sputtering. Depending on the application, a large number of other metals, including gold, palladium, platinum, silver or alloys, can be deposited with PVD methods. A number of metals and metal compounds, such as Cu, WSi_2, $TiSi_2$, TiN and W, can be deposited by CVD. Metal CVD processes are less common, but can provide improved step coverage or local deposition of metals. Whereas aluminum has been the standard metallization in IC fabrication for many years, the state-of-the-art sub-0.25 µm CMOS technologies often feature copper as interconnect material, owing to its lower resistivity and higher electromigration resistance as compared with aluminum. An example is IBM's interconnect metallizations based on the so-called damascene process [12], which employ copper films electroplated in a dielectric mold. After each metallization step, planarization is achieved with a chemical–mechanical polishing (CMP) step.

Polymers such as photoresist are commonly deposited by spin- or spray-coating. Polymers can be used as dielectric materials, passivation layers, and as chemically sensitive layers for chemical and biosensors ([13]; see also Chapter 7).

1.1.1.2 Patterning

Photolithography is the standard process to transfer a pattern, which has been designed with computer-aided-engineering (CAE) software packages, on to a certain material. The process sequence is illustrated in Fig. 1.4. A mask with the desired pattern is created. The mask is a glass plate with a patterned opaque layer (typically chromium) on the surface. Electron-beam lithography is used to write the mask pattern from the CAE data. In the photolithographic process, a photoresist layer (photostructurable polymer) is spin-coated on to the material to be patterned. Next, the photoresist layer is exposed to ultraviolet (UV) light through the mask. This step is done in a mask aligner, in which mask and wafer are aligned with each other before the subsequent exposure step is performed. Depending on the mask aligner generation, mask and substrate are brought in contact or close proximity (contact and proximity printing) or the image of the mask is projected (projection printing) on to the photoresist-coated substrate. Depending on whether positive or negative photoresist was used, the exposed or the unexposed photoresist areas, respectively, are removed during the resist development process.

Fig. 1.4 Schematic of a photolithographic process sequence to structure a thin-film layer

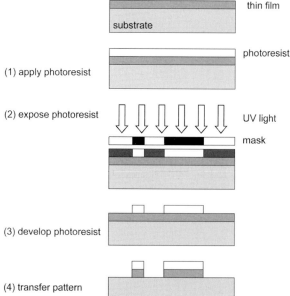

The remaining photoresist acts as a protective mask during the subsequent etching process, which transfers the pattern onto the underlying material. Alternatively, the patterned photoresist can be used as a mask for a subsequent ion implantation. After the etching or ion implantation step, the remaining photoresist is removed, and the next layer can be deposited and patterned.

The so-called *lift-off technique* is used to structure a thin-film material, which would be difficult to etch. Here, the thin-film material is deposited on top of the patterned photoresist layer. In order to avoid a continuous film, the thickness of the deposited film must be less than the resist thickness. By removing the underneath photoresist, the thin-film material on top is also removed by 'lifting it off', leaving a structured thin film on the substrate.

Thick photostructurable polymer layers, such as SU-8 [14], can be used as a mold for electroplating metal structures. A thick polymer layer is deposited on top of a metallic seed layer and photostructured. During the subsequent electroplating process, the metal is only deposited in the areas where the seed layer is exposed to the plating solution, i.e. the polymer layer acts as a plating mold.

Recently, microcontact printing or soft lithography [15] has been introduced as an additional method for pattern transfer. A soft polymeric stamp is used to reproduce a desired pattern directly on a substrate. Routinely, feature sizes on the order of 1 μm can be achieved with this technique. The polymer stamp, often made from poly(dimethylsiloxane) (PDMS), is formed by a molding process using a master fabricated with conventional microfabrication techniques. After 'inking' the stamp with the material to be printed, the stamp is brought in contact with the substrate material, and the pattern of the stamp is reproduced. Surface properties of the substrate can therefore be modified to, e.g., locally promote or prevent molecule adhesion. Soft lithography has been specifically developed for biological applications such as patterning cells or proteins with the help of, e.g., self-assembled monolayers (SAMs) [15].

1.1.1.3 Etching

The two different categories of etching processes include wet etching using liquid chemicals and dry etching using gas-phase chemistry. Both methods can be either isotropic, i.e. provide the same etch rate in all directions, or anisotropic, i.e. provide different etch rates in different directions (see Fig. 1.5). The important criteria for selecting a particular etching process encompass the material etch rate, the selectivity for the material to be etched, and the isotropy/anisotropy of the etching process. An overview on various etching chemistries used in microfabrication can be found in [16].

Wet etching is usually isotropic with the important exception of anisotropic silicon wet etching in, e.g., alkaline solutions, such as potassium hydroxide (see Section 1.2). Moreover, wet etching typically provides a better etch selectivity for the material to be etched in comparison with neighboring other materials. An example includes wet etching of silicon dioxide using hydrofluoric acid-based chemistries. SiO_2 is isotropically etched in dilute hydrofluoric acid (HF–H_2O) or buffered oxide etch, BOE (HF–NH_4F). Typical etch rates for high-quality (thermally grown) silicon dioxide films are 0.1 µm/min in BOE.

Dry etching, on the other hand, is often anisotropic, resulting in a better pattern transfer, as mask underetching is avoided (see Fig. 1.5). Therefore, anisotropic dry etching processes, such as *reactive ion etching* (RIE), of thin-film materials are very common in the microelectronics industry. In an RIE system, reactive ions are generated in a plasma and are accelerated towards the surface to be etched, thus providing directional etching characteristics. Higher ion energies typically result in more anisotropic etching characteristics, but also in reduced etching selectivity.

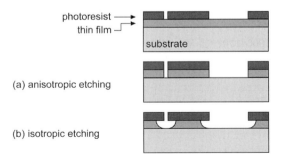

Fig. 1.5 Schematic of isotropic and anisotropic thin-film etching

1.1.1.4 Doping

Doping is used to modify the electrical conductivity of semiconducting materials such as silicon or gallium arsenide. It is hence the key process step for fabricating semiconductor devices such as diodes and transistors. In the case of silicon, doping with phosphorus or arsenic yields n-type silicon, whereas p-type silicon results from boron doping. By varying the dopant concentration of n-type silicon from 10^{14} to 10^{20} cm^{-3}, the resistivity at room temperature can be tuned from approximately 40 to 7×10^{-4} Ω cm.

Dopant atoms are introduced by either ion implantation or diffusion from a gaseous, liquid or solid source. Ion implantation has become the key process to introduce precisely defined quantities of dopants in the microelectronics industry. The substrate material, i.e. a silicon wafer, is bombarded with accelerated ionized dopant atoms in an ion implanter. The result is approximately a Gaussian distribution of the dopant atoms in the substrate wafer with a mean penetration depth controlled by the acceleration voltage. A high-temperature diffusion process can then be used to additionally 'drive-in' the dopant until a desired doping profile has been achieved.

1.1.2
CMOS Process Sequence

To be able to integrate microelectromechanical devices with CMOS circuitry, the designer must have an excellent understanding of the underlying CMOS process sequence. The particular process flow is, of course, strongly dependent on the chosen CMOS technology and a detailed description of a CMOS technology goes way beyond the scope of this chapter. Nevertheless, we briefly summarize a typical CMOS process sequence in the following, highlighting the main process steps and their importance for co-integration of CMOS and MEMS. We thereby follow the CMOS process sequence described in detail in [6] (see schematic cross-section in Fig. 1.2), which is typical for a sub-µm technology with minimal feature sizes between 0.5 and 1 µm.

The starting wafer material is a lightly p-doped (100) wafer with a typical doping concentration of $N_A \approx 10^{15}$ cm^{-3}. The first step is the definition of the active areas by local oxidation of silicon (LOCOS), thus growing a thick (~ 0.5 µm) field oxide in the areas between the individual transistors. Next, the p-wells for the n-channel MOSFETs and the n-wells for the p-channel MOSFETs are implanted. A joint drive-in for both wells establishes the desired junction depth of 2–3 µm. Typical drive-in times are 4–6 h at 1000–1100 °C. We will see later (Section 1.2) that the n-well diffused in the p-substrate can be used to define accurately the thickness of a silicon membrane. Such membranes are commonly released by anisotropic wet etching from the back of the wafer using an electrochemical etch-stop technique at the p–n junction between n-well and p-substrate [17, 18].

After n- and p-well formation, the MOSFET gate and channel regions are engineered. First, channel implants for the n- and the p-channel transistors are implanted to adjust their threshold voltages to the desired values. After removing the implantation oxides in the active area, the gate oxide with a thickness ≤10 nm in modern CMOS processes is thermally grown in the active areas. Next, a 0.3–0.5 µm thick polysilicon layer for the gate electrodes is deposited across the wafer in an LPCVD furnace operating at about 600 °C and doped by ion implantation. Finally, the polysilicon layer is patterned to define the actual gate regions. In MEMS, the gate polysilicon can also be used for resistors, piezoresistors, thermopiles, electrodes and as structural materials. The last application often requires a high-temperature anneal of the polysilicon to reduce its residual stress to values acceptable for the microstructures. Such a high-temperature step can be critical at this stage in the CMOS process, as it might effect previous doping distributions and, hence, the CMOS device characteristics.

After gate formation, the source/drain regions are implanted. In typical sub-µm CMOS technologies, this is done using a LDD (lightly doped drain) process. It provides a gradient in the doping of the source/drain regions towards the channel region, reducing the peak value of the electric field close to a channel and, hence, increasing device reliability. First, phosphorus (or arsenic as alternative n-type dopant) is implanted in the source/drain of the NMOS transistors to form n^- regions, followed by a boron implantation of the source/drain of the PMOS transistors to form p^- regions. Next, a conformal spacer dielectric layer is deposited on the wafer and anisotropically etched back, leaving sidewall spacers along the edges of the polysilicon gates. After growing a thin screen oxide for the following implantation, the source/drain regions of the NMOS and PMOS transistors not protected by the sidewall spacer are successively implanted to form n^+ and p^+ regions, respectively. The final step of the source/drain engineering is a furnace anneal, typically at ∼ 900 °C for 30 min, to activate the implants, anneal implant damage and drive the junctions to their final depth. Alternatively to the furnace anneal, a much shorter rapid thermal anneal at higher temperatures can be performed (e.g. 1 min at 1000–1050 °C). The fabrication of the active devices is now completed. Any subsequent high-temperature step (above 700–800 °C) necessary for the MEMS fabrication must be carefully qualified, as it might affect the doping distributions in the active devices, thus potentially changing the device characteristics.

In the back end of the process, the individual active devices are interconnected on the wafer to form circuits and pads for input/output connections off the chip are created. Although a large number of back-end metallization process flows with up to eight metallization levels exist, the exemplary CMOS process described in [6] uses three metallization levels with a local interconnect level based on titanium nitride and two wiring levels based on aluminum. The contacts to the source/drain regions and to the gate polysilicon are based on titanium silicide ($TiSi_2$). To this end, a thin titanium layer (50–100 nm) is sputtered on the wafer after removal of the implantation oxide. During an annealing step at about 600 °C in N_2, the titanium reacts with Si where they are in contact (e.g. source, drain and gate polysilicon) to form $TiSi_2$ and with N_2 to form TiN elsewhere. The resulting TiN layer is patterned to create a local interconnect. Subsequently, the wafer surface is typically planarized using a PSG (phosphosilicate glass) or BPSG (borophosphosilicate glass) layer reflown at 800–900 °C. Modern CMOS processes often use chemical mechanical polishing (CMP) for interconnect and interconnect dielectric planarization. In the process described in [6], each of the following wiring levels uses CVD tungsten vias with a TiN adhesion/barrier layer and an aluminum (with a small percentage of Si and Cu) interconnect layer. Finally, the passivation layer is deposited (typically by PECVD) and patterned to form the pad openings necessary to contact the device from the outside. The composition of the passivation layer and especially its residual stress can be adapted according to the needs of the microstructures (see Section 1.3.1). After passivation, the wafers are annealed at low temperatures (400–450 °C) for about 30 min in forming gas (10% H_2 in N_2) to alloy the metal contacts.

The CMOS process presented in [6] and briefly described here requires 16 masks. A schematic device cross-section is shown in Fig. 1.2.

1.1.3
CMOS Materials for Micro- and Nanosystems

The particular CMOS technology chosen for the implementation of a micro- or nanoelectromechanical system (MEMS or NEMS) dictates the overall process sequence, the doping profiles and junction depths of doped silicon regions, and the material properties and thicknesses of the different thin-film layers. In general, only minimal adaptations can be made in order not to compromise the performance of the CMOS circuits (see Section 1.3). However, the different layers of the CMOS process can be used for the fabrication of the microstructures themselves. Tab. 1.1 summarizes the different doping regions and layers of a typical CMOS process and their use in MEMS and NEMS.

Two examples, namely a CMOS-based mass-sensitive chemical sensor [19–21] and a CMOS-based thermal imager [22, 23], will be discussed in the following. The mass-sensitive chemical sensor (see Fig. 1.6) is based on a 150 µm long and 140 µm wide cantilever beam consisting of the n-well of the CMOS process covered by the CMOS dielectrics [21]. Thus, the n-well and the CMOS dielectrics are used as structural materials. The cantilever is released after completion of the CMOS process by three post-CMOS micromachining steps: first, a silicon membrane is formed by anisotropic wet etching from the back of the wafer in combination with an electrochemical etch-stop technique at the p–n junction between p-substrate and n-well; thereafter, the cantilever is released by two reactive ion etching (RIE) steps. The two aluminum metallization layers are used to form a planar coil on top of the cantilever, enabling the generation of transverse vibrations in the presence of an external DC magnetic field parallel to the cantilever length. The transverse vibration are detected with stress-sensitive diode-connected PMOS transistors, arranged in a Wheatstone bridge configuration at the cantilever's clamped edge. Alternatively, piezoresistors can be formed using either the p^+-source/drain implantation of a PMOS transistor or the n^+-doped gate polysilicon. The cantilever beam is coated with a chemically sensitive polymer layer. Upon absorption of analyte in the polymer layer, the cantilever's mass increases and, hence, its resonance frequency decreases. The change of resonance frequency is

Tab. 1.1 Common CMOS materials and their use in micro- and nanoelectromechanical systems (MEMS and NEMS)

CMOS layer/structure	Use in MEMS and NEMS
n-well/p-well	Structural material
	Thermal conductor/mass
Source/drain implantation	Resistor
	Piezoresistor
	Thermopile
	Electrode
Field oxide	Structural material
	Thermal insulator
	Sacrificial material
Gate polysilicon	Resistor
(and optional 2nd polysilicon)	Piezoresistor
	Thermopile
	Electrode
	Structural material
	Sacrificial material
Contact and intermetal oxides	Structural material
	Thermal insulator
	Sacrificial material
Metallization	Conductor
(and optional multi-level metallizations)	Mirror
	Thermal conductor
	Electrode
	Structural material
	Sacrificial material
Passivation	Structural material
	Thermal insulator
	Stress compensation
	Infrared radiation absorber

sensed by incorporating the resonant cantilever into an amplifying feedback loop [20, 21].

The thermal imager shown in Fig. 1.7 is based on a $\sim 3\times 3$ mm^2 membrane consisting of the dielectric layers of the CMOS process [22, 23]. The membrane is released by wet anisotropic silicon etching from the back of the wafer after completion of the regular CMOS process sequence. The thick field oxide is used as an intrinsic etch-stop layer. The CMOS dielectrics, i.e. the field oxide, the contact oxide, the intermetal oxide and the passivation, are used as structural materials. A grid of electroplated gold lines provides additional structural support to the membrane and divides it into 100 pixels. The gold lines are electroplated after the CMOS process in a standard process step normally preparing the wafers for TAB (tape automated bonding). Sandwiched in between the CMOS dielectrics on each pixel is a polysilicon/aluminum thermopile and a polysilicon heating resistor. The

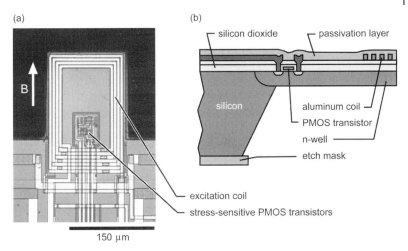

Fig. 1.6 (a) Photograph and (b) schematic cross-section of a cantilever-based mass-sensitive gas sensor. The cantilever structure features an integrated planar coil for electromagnetic excitation of transverse vibrations in the presence of a DC magnetic field and PMOS transistors in a Wheatstone bridge arrangement for deflection detection [21]. Photograph courtesy of C. Vancura, ETH Zurich, Switzerland

incoming infrared (IR) radiation is absorbed in the CMOS thin-film sandwich

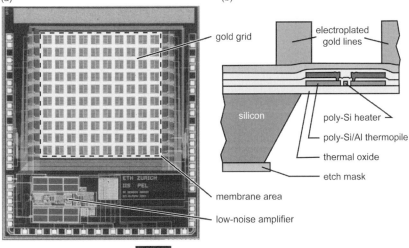

Fig. 1.7 (a) Photograph and (b) schematic cross-section of a CMOS-based infrared radiation sensor array. The sensor array is located on a micromachined membrane consisting of the dielectric layers of the CMOS process. An electroplated gold grid divides the membrane in a 10×10 array of pixels, each incorporating a thermopile with 16 polysilicon/aluminum thermocouples for temperature sensing. On-chip circuitry includes a multiplexer and a low-noise chopper amplifier [22, 23]. Photograph courtesy of Prof. H. Baltes, ETH Zurich, Switzerland

(including the passivation), resulting in a measurable temperature elevation of the individual pixels. All structures necessary for IR radiation sensing are completely formed within the regular CMOS process sequence [22, 23].

1.1.4
CMOS Microsystems

A number of microsensors can be completely formed within the regular CMOS process sequence, typically not requiring any additional process steps. Well-known examples include temperature sensors [24, 25], magnetic field sensors (especially Hall sensors) [26] and CMOS imagers [27, 28]. An additional subset of CMOS-based microsystems only requires either the modification of a CMOS layer or the deposition and patterning of additional layers, but no micromachining steps. A few selected examples will be given in the following.

Chemical sensors and biosensors relying on an electrochemical sensing principle require an electrode in contact with the sample to be sensed. Examples include amperometric sensors, palladium-gate FET and ISFET (ion-sensitive field effect transistor) structures, and also chemoresistors and chemocapacitors. A number of these electrochemical sensors have been co-integrated with CMOS circuitry (see Chapter 7), typically requiring deposition and patterning of special metal electrodes and/or passivation layers in addition to the regular CMOS process sequence.

Examples are the CMOS-based biosensor arrays developed recently for DNA analysis [29, 30] and recording of neural activity [31]. The sensor arrays are based on a standard 0.5 µm CMOS process optimized for analog applications [30]. After deposition and patterning of the second aluminum layer, a silicon dioxide layer is deposited, followed by a planarization step using CMP and the deposition of a silicon nitride passivation. The actual sensor electrodes are fabricated on top of the nitride passivation. First, vias are etched to enable contacts to the aluminum metallization and are filled with a Ti/TiN barrier layer and CVD tungsten [30]. In the case of the DNA arrays, the final interdigitated gold electrodes are deposited by evaporation of a Ti/Pt/Au electrode stack, which is patterned using a lift-off technique ([30]; see Fig. 1.8). In the case of the sensor arrays for neural activity recording, the sensor electrodes and the contact pads are defined by depositing and lift-off patterning of a Ti/Pt layer. Subsequently, a dielectric layer sandwich consisting of different TiO_2 and ZrO_2 layers is deposited and opened at the location of the bond pads. Neural activity is recorded capacitively with the sensor electrodes covered by the protective dielectric layer sandwich. Finally, a gold layer is deposited on the Pt pads and structured using a lift-off process [31].

Researchers at ETH Zurich have recently reported a CMOS-integrated microelectrode array for stimulation and recording of natural neural networks [32]. The microsystem is fabricated using a 0.6 µm CMOS process in combination with a two-mask post-CMOS process sequence to deposit and pattern biocompatible platinum electrodes. The post-CMOS process sequence starts with the deposition and patterning of 50 nm TiW and 270 nm Pt. The metal layer sandwich is structured

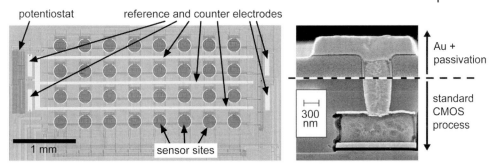

Fig. 1.8 (a) Photograph of an 8×4 element DNA sensor array with a single sensor diameter of 200 μm and a sensor pitch of 400 μm; (b) SEM photograph of sensor cross-section showing the standard CMOS metallization, the tungsten vias and the gold sensor electrodes. Adapted from [30]

using a lift-off technique. Finally, a 1 μm silicon nitride sealing layer is deposited by PECVD and patterned with RIE.

A CMOS-based biochemical multisensor microsystem requiring no micromachining has been developed by IMEC, K. U. Leuven and Siemens [33]. The microsystem combines, on a single CMOS chip, an array of ISFETs, an amperometric oxygen sensor and a conductometric cell. The biochemical analysis system is based on double-metal, 1.2 μm CMOS technology. A special ISFET module has been integrated into the regular CMOS process sequence to form a protective LPCVD nitride layer on top of the ISFET gates. In addition, Ti/Pt electrodes for the amperometric sensor are deposited and patterned and Ag/AgCl reference electrodes are formed by electroplating and electrochloridation [33, 34]. ISFET structures requiring no modifications to the CMOS process sequence have been presented in [35].

Capacitive chemical microsensors based on interdigitated metal electrodes can be fabricated completely within the regular CMOS process sequence. Examples include microsensors for detection of humidity [36, 37] and volatile organic compounds in air [38, 39]. Typically, the interdigitated electrode structure is formed by the metallization layers of the CMOS process. The capacitive sensor structure detects changes in the dielectric constant of a sensing layer deposited on top of it upon absorption of analyte molecules. CMOS-integrated capacitive humidity sensors are produced by Sensirion, Switzerland [40, 177].

Hall plates can be formed completely within a regular (Bi)CMOS process sequence and Hall sensor systems with on-chip circuitry are commercially available, e.g. from Micronas [41], Infineon Technologies [42], Allegro Microsystems [43] and Melexis [44]. Regular Hall plates are arranged parallel to the chip surface and are sensitive to magnetic fields perpendicular to the chip surface. Using spinning-current methods for offset reduction, commercially available CMOS integrated Hall sensors have offsets as low as 0.5 mT [45]. To improve sensor performance, i.e. sensitivity and offset, the Hall sensors have been combined with integrated magnetic flux concentrators by bonding and patterning thin high-permeability, low-coercivity ferromagnetic layers to the chip surface [46]. In addition to concentrating the magnetic flux at the location of the Hall sensors, the flux concentrators allow the measurement of magnetic fields in the chip plane with standard lateral Hall sensors [46]. Alternatively, magnetic fields parallel to the chip surface can be sensed by vertical Hall sensors, rotated 90° to the chip surface [47, 238]. Recently, CMOS-based vertical Hall sensors have been fabricated by developing a pre-CMOS trench etching technology to define the geometry of the Hall plates ([47]; see Section 1.4.1).

Ferromagnetic films not only are used in combination with Hall sensors, but are also essential for highly sensitive fluxgate sensors. The operation of a fluxgate sensor requires a ferromagnetic core which needs to be saturated periodically by the control circuitry. CMOS-based fluxgate sensors with minimal detectable magnetic fields in the nanotesla range (typical noise levels in the range 5–100 nT/$\sqrt{Hz}$) have been demonstrated at the Fraunhofer Institute IMS [48, 49], ETH Zurich [50, 51] and EPF Lausanne [52]. In [49], a ferromagnetic $Ni_{81}Fe_{19}$ core is embedded in the intermetal dielectrics between the two metallization layers of a CMOS process. In this way, the required excitation and pick-up coils consisting of metal-1 and metal-2 lines can be wound around the core. The electron beam-evaporated nickel–iron cores are sandwiched between tantalum layers, serving as adhesion layers and diffusion barriers. The metal sandwich is patterned using lift-off techniques. In [50, 51], two 1 μm thick ferromagnetic NiFeMo cores are electroplated on top of the CMOS chip. Finally, the approach presented in [52] uses a soft-magnetic amorphous alloy (Metglas 2714A, Honeywell), which is mounted on top of the CMOS die and structured using a photolithographic process.

The final two examples both require a direct contact with the surface of a CMOS chip during sensing. Fingerprint sensors are used for access control and authentication and are covered in detail in Chapter 8. In the case of a capacitive sensor, a two-dimensional electrode array measures the capacitance between the chip surface and the finger's surface touching the chip with a resolution of typically 500 dpi. The fingerprint sensor developed by Siemens is based on a double-metal, 0.8 μm CMOS process and features, on a single chip, a 256×256 pixel sensor array with a pitch of 50 μm, the necessary data acquisition circuitry, A/D conversion and a parallel interface [53]. The sensor is protected against electrostatic air discharge, caused by touching the sensor with a charged finger, using a grounded refractory metal grid (see Chapter 8).

Wire bonding remains the predominant method for providing electrical interconnections between chip and substrate. Increasing bonding speed paired with decreasing pad-pitch requires careful optimization of the wire bonding process and a profound understanding of the physical processes occurring during the actual bonding process. Recently, CMOS-based force sensors have been developed for in situ investigation of the forces acting on the bond pad during thermosonic ball-wedge wire bonding [54, 55]. The test chips comprise an array of xyz-force sensors connected to a multiplexer (see Fig. 1.9). Each xyz-sensor features three Wheatstone bridges with piezoresistors to measure the x, y and z-components of the force acting on the bond pad during the wire bonding process. The p^+ and n^+

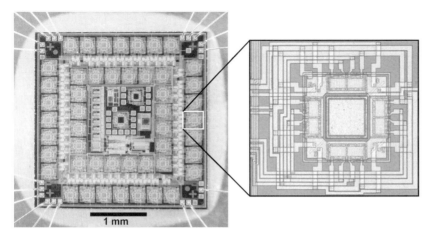

Fig. 1.9 Test chip with 48 xyz-force sensors connected to a multiplexed bus; the close-up of one xyz-force sensor shows the test pad with a size of 65 µm and the surrounding piezoresistors for x, y and z-force sensing [54]

source/drain implantations of a double-metal, 0.8 µm CMOS process are used to form the piezoresistors surrounding the bond pad. Hence, the force sensors can be completely formed within the regular CMOS process sequence, potentially allowing the implementation of bonding test structures into regular CMOS designs.

In all of the above cases, no micromachining steps are involved. The focus of the remainder of this chapter (and the main focus of this book) is on CMOS-based micro- and nanosystems requiring either bulk or surface micromachining to release micromechanical structures.

1.2
CMOS-compatible Micromachining Process Modules

The basic microfabrication processes described earlier are often combined with special micromachining steps to produce (three-dimensional) microstructures, such as cantilevers, bridges and membranes. In the following, the fundamental micromachining techniques are reviewed. More details on micromachining techniques can be found in dedicated books on microsystem technology [56–59].

The micromachining techniques are categorized into bulk micromachining [60] and surface micromachining processes [61] (see Fig. 1.10). In the case of bulk micromachining, the microstructure is formed by machining the relatively thick bulk substrate material, whereas in the case of surface micromachining, the microstructure comprises thin-film layers, which are deposited on top of the substrate and selectively removed in a defined sequence to release the MEMS structure.

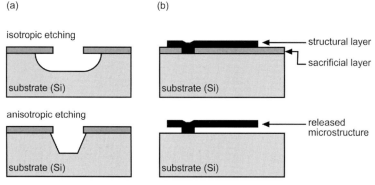

Fig. 1.10 Schematic of (a) bulk and (b) surface micromachining

1.2.1
Bulk Micromachining

Bulk micromachining techniques [60], i.e. etching techniques to machine the (silicon) substrate, can be classified into isotropic and anisotropic, and into wet and dry etching techniques, as can be seen in Tab. 1.2.

The most common *isotropic wet* silicon etchant is HNA, a mixture of hydrofluoric acid (HF), nitric acid (HNO_3), and acetic acid (CH_3COOH). In this etching system, nitric acid oxidizes the silicon surface and hydrofluoric acid etches the grown silicon dioxide layer. The acetic acid controls the dissociation of HNO_3, which provides the oxidation of the silicon. The etch rates and the resulting surface quality strongly depend on the chemical composition [58].

Anisotropic wet etching of silicon is the most common micromachining technique and is used to release, e.g. membrane and beam structures. Anisotropic wet etchants etch single-crystalline silicon with different etch rates along different crystal directions. The resulting etch grooves are bound by crystal planes, along which etching proceeds at slowest speed, i.e. the (111) planes of silicon. In case of the commonly used (100) silicon wafers, the (111) planes are intersecting the wa-

Tab. 1.2 Examples of etching techniques for machining the silicon substrate

Type	Wet etching	Dry etching
Isotropic	HNA system HF–HNO_3–CH_3COOH	Vapor-phase etching XeF_2
Anisotropic	Alkali metal hydroxide solutions KOH, NaOH Ammonium hydroxide solutions $(CH_3)_4NOH$ (TMAH), NH_4OH EDP solutions Other solutions, e.g. hydrazine	Plasma etching RIE, deep-RIE

fer surface at an angle of 54.7°, yielding the typical pyramid-shaped etch grooves shown in Fig. 1.11. Masking materials for anisotropic silicon etchants are silicon dioxide and silicon nitride. It is important to note that 'convex' corners of the etch mask (as shown in Fig. 1.11) are underetched in the case of (100) silicon substrates, leading to, e.g., completely underetched cantilever structures. The etch rates in preferentially etched crystal directions such as the ⟨100⟩ and the ⟨110⟩ directions, and the ratio of the etching rates in different crystal directions depend strongly on the exact chemical composition of the etching solution and the process temperature [57, 58, 60, 62].

The most common anisotropic silicon etching solution is potassium hydroxide, KOH. As an example, a 6-M KOH solution at 95 °C provides a ⟨100⟩ etch rate of 150 µm/h and an anisotropy, i.e. etch rate ratio, between the ⟨100⟩ and ⟨111⟩ directions of 30–100:1 [63]. Since the etch rate of silicon dioxide in KOH solutions is fairly high (for thermal oxide ~1 µm/h in 6 M KOH solution [58]), silicon nitride films are often used as etching mask. KOH solutions are very stable, yield reproducible etching results and are relatively inexpensive. KOH is, therefore, the most common anisotropic wet etching chemical in industrial manufacturing. The disadvantages of KOH include the relatively high SiO_2 and Al etch rates, which require protection of IC structures during etching. Etching with KOH is typically performed from the back of the wafer, with the front side protected by a mechanical cover and/or a protective film [63]. Another issue is the detrimental impact of

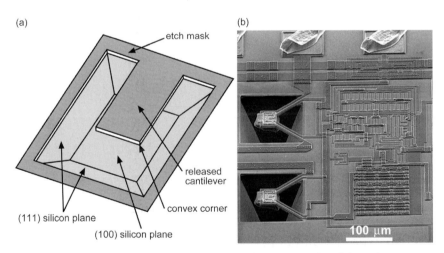

Fig. 1.11 (a) Schematic of a cantilever beam released by anisotropic silicon etching from the front side of the wafer. The etching mask defines the cantilever shape. The underetching of the cantilever structure starts at convex corners. The resulting etch groove is bound by characteristic (111) side walls and a (100) bottom surface. (b) SEM photograph showing two bulk-micromachined thermal converters cointegrated with CMOS circuitry. The devices are released from the front side of a CMOS wafer by combining anisotropic silicon etching using TMAH with an electrochemical etch-stop technique. SEM photograph courtesy of Prof. G.T.A. Kovacs, Stanford University, USA

alkali metal ions on the characteristics of MOSFET structures. Investigation of MOSFET characteristics after KOH etching from the back of CMOS wafers, however, did not reveal any etching-related damage [63].

Alternative silicon etchants are ammonium hydroxide compounds, such as tetramethyl ammoniumhydroxide (TMAH), and ethylenediamine–pyrocatechol (EDP) solutions. Certain EDP formulations, such as EDP type S, exhibit relatively low Al and SiO_2 etch rates, which make them suitable for releasing microstructures from the front side of CMOS wafers [232]. However, EDP solutions age rapidly, are potentially carcinogenic and are very difficult to dispose of. TMAH solutions exhibit similar etching characteristics to EDP, but are easier to handle. By controlling the pH by, e.g., dissolving silicon in the etching solution, the etch rate for aluminum metallizations can be reduced [60, 64], making TMAH also a candidate etchant for releasing microstructures from the front side of CMOS wafers. More detailed discussions of wet etching of silicon can be found, e.g., in [57, 58].

Reliable etch stop techniques are very important for achieving reproducible etching results. As already mentioned, wet anisotropic silicon etchants 'stop' etching, i.e. the etch rate is reduced by at least 1–2 orders of magnitude, as soon as a (111) silicon plane or a silicon dioxide (or silicon nitride) layer is reached. In addition, the etch rate is greatly reduced in highly boron doped regions (doping concentration $\geq 10^{19}$ cm^{-3}). The etching can also be stopped at a p–n junction using a so-called electrochemical etch stop technique (ECE) [56, 65]. This method has been extensively used to release silicon membranes and n-well structures (see Fig. 1.12). ECE relies on the passivation of silicon surfaces when an anodic potential is applied that is sufficiently high with respect to the potential of the etching solution.

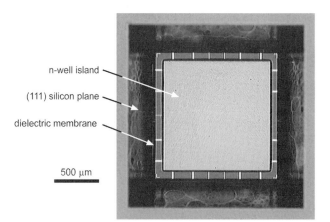

Fig. 1.12 Micrograph of an anisotropically etched cavity of a capacitive chemical microsystem (see Chapter 7, Fig. 7.32). At the bottom of the cavity, an n-well island structure carrying a thermally stabilized capacitive sensor [67] is visible. The n-well is suspended by a membrane consisting of the CMOS dielectric layers (the embedded metal interconnects connecting the sensor are clearly visible)

Isotropic dry etching of silicon can be done using xenon difluoride, XeF_2. This vapor-phase etching method exhibits excellent etch selectivity with respect to aluminum, silicon dioxide, silicon nitride and photoresist, all of which can be used as etch masks. However, the resulting etched silicon surfaces are fairly rough. The XeF_2 silicon etch rates depend on the loading (size of the overall silicon surface exposed to the etchant) with typical values of ~ 1 μm/min. XeF_2 etching systems are commercially available from XACTIX [66]. (Alternatively, reactive ion etching (RIE) can be used for isotropic dry etching see also *anisotropic dry etching* below).

Anisotropic dry etching of silicon is usually performed by reactive ion etching (RIE) in plasma-assisted etching systems. By controlling the process parameters, such as process gases and process pressure, the etching can be rendered either isotropic or anisotropic. The dry-etching anisotropy mainly originates from the direction of ion bombardment, and is, therefore, independent of the crystal orientation of the substrate material. Most bulk etching of silicon is accomplished using fluorine free radicals with SF_6 as a typical process gas. Adding chlorofluorocarbons results in polymer deposition in parallel with etching, which leads to enhanced anisotropy.

Very high aspect ratio microstructures can be achieved with deep (D)RIE, a method which has gained importance during recent years. DRIE systems rely on high-density plasma sources and an alternation of etching and polymer-assisted sidewall protection steps. In a process known as the Bosch process [68], a mixture of trifluoromethane and argon is used for polymer deposition. Owing to the ion

Tab. 1.3 Comparison of characteristics of common bulk silicon etchants; the etch rates given are typical numbers, the actual etch rates depend on the process parameters (sources for etching rates and selectivities: HNA [9, 58], KOH [58], TMAH [57, 72], XeF_2 [66], DRIE [57])

	HNA	KOH 6 M	TMAH 22%	XeF_2	DRIE
Etch type	Wet	Wet	Wet	Dry	Dry
Anisotropic?	No	Yes	Yes	No	Yes
Si (100) etch Rate [a] (μm/min)	typically > 3 at 25 °C	2.5 at 95 °C	0.6 at 80 °C	typically 1	2–3
SiO_2 etch rate (nm/min)	30–70 at 25 °C	15 at 95 °C	0.1 at 80 °C	1:10000 [c]	1:120–200 [c]
SiN_x etch rate (nm/min)	No data available [b]	≤0.01 at 95 °C	1:3600 [c] at 95 °C	1:100 [c]	No data available
ECE etch stop?	Yes	Yes	Yes	No	No
Cost	Low	Low	Low	High	High

[a] For isotropic etchants, the etch rate is independent of the crystal orientation; for anisotropic dry etching, the etch rate given is in direction of the ion bombardment.
[b] SiN_x etch rate in HNA is smaller than SiO_2 etch rate.
[c] Selectivities between etch rates SiO_2:Si and SiN_x:Si are given rather than etch rate itself

bombardment, the polymer deposition on the horizontal surfaces can almost be prevented, while the sidewalls are passivated with a Teflon-like polymer. In the second process step, SF_6-based etching chemistry provides silicon etching in the non-passivated regions, i.e. the horizontal surfaces. Both process steps are alternated, resulting in typical silicon etch rates of 1–3 µm/min with an anisotropy of the order of 30:1 [60]. Silicon dioxide and photoresist layers can be used as etch masks. The DRIE system achieves exceptional anisotropy, which is independent of the crystal orientation, but is far more expensive than e.g. a simple wet-etching setup, and can process only one wafer at a time. Commercial etchers of this type are available from, e.g., Surface Technology Systems (STS) [69], Unaxis Semiconductor [70] and Alcatel [71].

The characteristics of the most common bulk silicon etchants are summarized in Tab. 1.3. In addition to the described 'basic' micromachining processes, a large number of specific silicon-based micromachining processes have been developed. The ones relevant for the fabrication of CMOS-based microsystems will be discussed in Section 1.4.

1.2.2
Surface Micromachining

The most commonly used surface micromachining process is sacrificial-layer etching [61]. In this process, a microstructure, such as a cantilever beam or a suspended plate, is released by removing a sacrificial thin-film material, which was previously deposited underneath the microstructure. The release of polysilicon microstructures by removing a sacrificial silicon dioxide film is the most popular surface micromachining technique [61]. Sacrificial aluminum etching (SALE) has been developed to release dielectric microstructures with embedded metal layers [73]. Metallic microstructures deposited by low-temperature PVD processes can use polymer films as sacrificial layers, which are removed using, e.g., an oxygen plasma [74, 75].

A prominent example of a device based on surface micromachined microstructures is the digital micromirror device (DMD) developed by Texas Instruments ([74], see Section 1.4.3). The DMD consists of an array of micromirrors (see Fig. 1.13a), fabricated on top of a CMOS substrate by deposition and patterning of four metal and two polymer layers. The micromirror array with a pitch of 17 µm is released by removing the polymer sacrificial layers. Fig. 1.13b shows a surface-micromachined bolometer structure made from polycrystalline $Si_{57}Ge_{43}$ at IMEC (Leuven, Belgium) and Fig. 1.13c gives details of a 4 µm thick, released polysilicon microstructure fabricated at Analog Devices (Norwood, MA, USA).

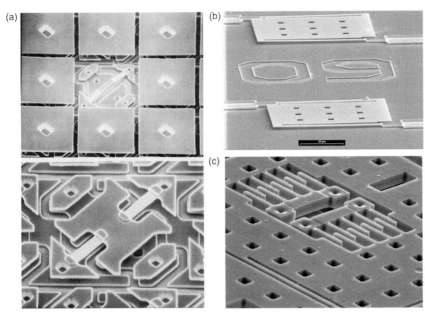

Fig. 1.13 (a) Top: SEM photograph of 3×3 array of pixels of Digital Micromirror Device (DMD^{TM}) by Texas Instruments; the mirror of the center pixel has been removed to show the underlying metal structures. Bottom: SEM photograph shows details of the micromirror yoke and hinges (images from Texas Instruments DLP image library: *http://www.dlp.com*). (b) SEM image of surface-micromachined poly-Si57Ge43 bolometer structures. Courtesy of IMEC, Leuven, Belgium. (c) Detail of surface-micromachined 4 µm thick polysilicon microstructure. Courtesy of Steve Lewis, Analog Devices, Norwood, MA, USA

1.3
CMOS-compatible Design of MEMS and NEMS

Can I modify the CMOS process sequence to co-integrate microelectromechanical systems with CMOS circuitry? Can I modify the CMOS process design rules to implement my microstructure? Maybe the questions should be less 'Can I?' but rather 'Where and how much can I?'. In the following section, we would like to give the reader an idea of possible process and design modifications. First and foremost, any modification on the CMOS process sequence and the established design rules for a particular process must not compromise the characteristics and yield of the circuitry components. Therefore, every process modification has to be properly qualified. Even if a process modification is not affecting the circuit characteristics, it might be difficult to implement it in a process run done at a commercial CMOS foundry: it can be very challenging to persuade a CMOS foundry to use pre-processed wafers as starting material or to interrupt the regular process sequence and have additional process steps performed (probably even outside the CMOS foundry) before resuming the 'standard' process sequence. We will see in Section 1.4 that 'substantial' process modifications, as required for pre-CMOS and intra-CMOS approaches, most often require in-house CMOS capabilities. In the following, we will concentrate on 'small' process modifications that might be tolerated by a majority of independent CMOS foundries. Since the author's background is especially in the area of post-CMOS microsystem approaches, most of the discussed process modifications will enable different post-CMOS micromachining modules.

1.3.1
Tolerable Process Modifications

If the microstructures are to be released by wet anisotropic silicon etching (especially from the back of the wafer), the wafer starting material for the CMOS process must be considered carefully. Modern CMOS processes often use epitaxial wafers with a weakly p-doped epitaxial layer on top of heavily p-doped substrate as starting material in order to improve latch-up stability. If the substrate p-type doping is above 10^{19} cm^{-3}, the silicon etch rates in common anisotropic etchants, such as KOH and TMAH, are drastically reduced. In addition, the starting material has typically a rather broad specification range for the substrate doping, which, in case of highly p-doped substrates, can result in substantial etch rate variations from wafer to wafer. To ensure compatibility with anisotropic silicon etching, either epi-wafers with reduced substrate doping ($\leq 5 \times 10^{18}$ cm^{-3}) or low p-doped non-epi wafers can be used as a starting material [18].

A second challenge for the bulk-micromachining using anisotropic wet etchants is the relatively high interstitial oxygen concentration in the wafer starting material, as required for internal gettering in the CMOS process. With an interstitial oxygen concentration larger than its solid solubility, the oxygen precipitates during annealing steps in the form of oxide particles. Defects caused by oxygen precipitation are commonly used for internal gettering of transition metal impurities during CMOS processing. The oxygen precipitates and the associated crystal defects in CMOS-processed wafers deteriorate the quality of etched cavities, resulting in uneven (111) sidewalls (with crater-like depressions) and large, locally varying underetching of the silicon nitride etch mask, yielding membranes with poor geometric definition [18, 76]. It should be noted that membranes with well-defined lateral dimensions can always be achieved by appropriate design, e.g. using either a p^{++}-doped 'etch-stop' ring surrounding the membrane [77] or an electroplated metal ring [22, 63], defining the mechanical edge of the membrane. A reduction of the interstitial oxygen concentration in the starting material from $\sim 8 \times 10^{17}$ to $(6.0–6.9) \times 10^{17}$ cm^{-3} resulted in a strongly improved quality of the released microstructures (see Fig. 1.14; [18, 76]). However, the wafer material with reduced defect density also has reduced internal gettering capability and external gettering using, e.g., hard mechanical damage on the wafer back, must be employed. The introduced crystal defects on the wafer back have to be removed at the end of the CMOS process, prior to the deposition of the etch mask.

Any additional (high-temperature) process step performed during or after the regular CMOS process sequence must be considered carefully in terms of the

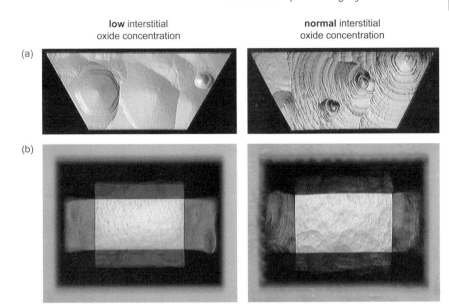

Fig. 1.14 Quality of etch cavities released by wet anisotropic etching using a 27 wt% (6 M) KOH solution at 90 °C; photographs of (a) (111) sidewalls and (b) (100) etch fronts are shown for test wafers with a normal interstitial oxygen concentration of $\sim 8 \times 10^{17}$ cm^{-3} and a low interstitial oxygen concentration of $(6.0–6.9) \times 10^{17}$ cm^{-3}; prior to KOH etching, the wafers were exposed to a thermal simulation of a CMOS process. Adapted from [18]

overall thermal budget of the process. The overall thermal budget critically influences the various doping profiles and thus the resulting device characteristics. Prolonged additional high-temperature process steps with peak temperatures ≥800 °C are likely only possible prior to the channel and source/drain implantations. Medium-temperature processes, such as LPCVD deposition of polysilicon at about 600 °C, have been performed after the source/drain implantations [78], but prior to the back-end aluminum metallization. High-temperature annealing steps required, e.g., for stress relief in the deposited polysilicon layers have to be carefully evaluated, as their thermal budget might influence shallow junction profiles. Of course, the initial doping profiles can be adapted so that additional thermal process steps are taken into account, but this generally requires substantial re-qualification of the CMOS process. The standard aluminum metallization employed in most CMOS processes with minimal feature sizes above 0.25 μm is known to withstand maximum process temperatures of about 450 °C (recent work indicates compatibility with temperatures up to 525 °C [79]), strongly limiting the range of process steps that can be performed after completion of the CMOS process sequence (see Section 1.4.3).

Deposition and patterning of the passivation layer are typically the last process steps of the regular CMOS process sequence. Hence, the passivation composition can often be adjusted to the customer's needs. If the passivation is part of the released microstructure, its residual stress can be used to tune the stress of the overall microstructure. An example is the thermal imager shown in Fig. 1.7. The membrane with embedded infrared sensor array has a layered structure comprising the different dielectric layers of the CMOS process with polysilicon and metal structures sandwiched in between them. The overall stress of the layer sandwich without the CMOS passivation is compressive in this example, which could result in membrane buckling. To reduce the overall compressive stress in the membrane, a passivation layer with tensile stress is deposited. The stress of a PECVD silicon oxynitride passivation could be controlled in the range from −300 to +300 MPa by choosing an appropriate low frequency (400 kHz) to high frequency (13.56 MHz) power ratio and chamber pressure in the used PECVD system [63]. It should be noted that the stress in the passivation can influence the characteristics of the previous layers, e.g. the electromigration behavior of the underlying metal lines.

1.3.2
Design Rule Modifications

Typically, CMOS foundries provide rule-files for their CMOS processes for a number of supported design environments in order to perform design rule checks (DRCs) and extraction of the layout for layout-versus-schematic (LVS) check. The enforced design rules ensure a high yield of the fabricated circuit components within the given process specifications, but might be problematic for the MEMS part. Two examples are given in the following: in order to release microstructures, such as the thermal converters shown in Fig. 1.11, from the front of the wafer, the silicon substrate must be exposed to the etchant in certain areas on the wafer. This can be achieved in a CMOS process by superimposing an active area (i.e. no field oxide), a contact (i.e. no contact oxide), a via (i.e. no intermetal oxide) and a pad opening (i.e. no passivation), thus locally removing all dielectric layers of the CMOS process and exposing the silicon substrate to the environment [80]. The standard design rules of the used CMOS process will, e.g., not allow a via without metal below and on top of it, because a via in a CMOS circuit only makes sense as an interconnect between two metallization levels. Thus, the automated design-rule checker (DRC) will give error messages. In another area one might want to use a non-connected aluminum area as a mirror surface. The DRC will again give an error message because of a non-connected conducting area. Both design examples make no sense in a circuitry environment, but are useful for the MEMS part and, very important, do not compromise the integrity of most CMOS processes (for completeness, it should be mentioned, that sub-0.5 μm CMOS technologies with plated vias might not allow a contact layer without metal overlap in order not to compromise via plug plating and CMP). So, how can one allow such design rule violations and still use the extremely helpful DRC? The ideal case is to write a complete set of design rules for the MEMS areas, having the circuitry checked by the foundry-supplied 'standard' design rule set and the MEMS by an extended design rule set. This approach might be initially more time consuming, but will, in the long run, prevent design errors in the complex MEMS designs.

Modern sub-μm CMOS processes use lithography based on wafer steppers, providing an array of step fields on the wafer with no mutual connection. Post-CMOS micromachining based on anisotropic wet etching in combination with an electrochemical etch-stop technique (see Section 1.4.3) requires the application of etching potentials to (structural) n-well and substrate contacts across the wafer [17, 18]. To supply these etching potentials, a contact network implemented in the metal-1 and metal-2 level of the CMOS process routes the etching potentials from large contact pads to the individual etch contacts. To achieve this, each metal mask step field is surrounded by a predefined frame (see Fig. 1.26b), routing one etching potential on metal-1 level and the second etching potential on metal-2 level to reduce the risk of short-cuts between the etch networks [18]. The frames are connected to each other at the corners of the step field by metal structures placed in the scribe channel, i.e. the individual metal-level step fields actually overlap during the stepper-based lithography. The construction of these metal bridges must not affect the regular test structures in the scribe channel. Within the reticle, the individual etch contacts are connected to the metal frame. In order to have large contact pads for applying the etch potentials with spring-loaded contacts, a dedicated 'contact' step field (see Fig. 1.26a, top of the wafer) is printed which has the same size as the other step fields. This is done on the second metal mask level using a special 'contact' reticle. In addition, a 'blank' reticle is used to remove the metal around the edges of the wafer in order to avoid short circuits in the etch network at the wafer edge. The described method requires three additional masks ('blank' reticle, 'contact' reticle and 'contact opening' reticle), which can, however, be reused if the reticle size from design to design is not changing. Besides exposing the wafer on the metal mask level with different reticles, no modification in the process flow is required.

1.3.3
Simulation of Circuitry and MEMS

Traditionally, MEMS and IC designers have used very different design tools. While IC designers rely on schematic-driven circuit simulators offered by the major electronic design automation (EDA) companies, such as Cadence [81], Mentor Graphics [82] and Synopsys [83], the MEMS designer typically relies on finite-element modeling (FEM) software, such as ANSYS [84], FEMLAB [85], CoventorWare [86], or IntelliSuite [87], for multi-domain analysis of their microstructures. In order to simulate and design integrated circuits based on a particular CMOS process, the CMOS foundries supply process-specific design kits, including design rules, process specifications, transistor-level models and analog and digital cell libraries, to support the major EDA tools. In order to simulate CMOS-based microsystems including micromechanical transducers and analog and digital circuitry, behavioral models for the transducer elements are required. To be compatible with the standard mixed-signal simulators delivered with common EDA packages (e.g. SPECTRE [81], ADVance MS [82], SABER [83]), these behavioral models must be expressed in an analog hardware description language (HDL), such as

Verilog-A or VHDL-A. The generation of such models for the transducers involving multiple signal domains from either the layout or the results of the FEM simulation is not straightforward. Simple lumped-element circuit models of the transducers might be developed manually on a case-by-case basis. For certain categories of microstructures (e.g. comb-drive resonators), the generation of macromodels is supported by academic [88, 89] and commercial tools [86, 90]. INTEGRATOR, developed by Coventor [86], is able to generate reduced-order macromodels of dynamic mechanical systems, consisting of spring, mass and damping elements, from detailed 3-D finite element (FEM) or boundary element (BEM) simulations for export in standard circuit simulators. NODAS, developed at Carnegie Mellon [88, 91, 92] is a library of parameterized components, including beams, plate masses, anchors, electrostatic comb drives and gaps, to simulate surface-micromachined MEMS structures using the SABER and SPECTRE simulators. Complex microstructures are build by interconnecting individual library elements. NODAS also has the ability to generate automatically the layout from the developed schematic.

Once the top-level layout of the integrated microsystem is completed, a design rule check (DRC) and a layout-versus-schematic (LVS) check are performed. In order to account for different design rules in the circuitry and the MEMS part, the standard design rule files supplied by the CMOS foundry might need to be extended (see Section 1.3.2). In addition, the standard extraction rules can be adapted in order to recognize and extract at least the electrical features of the transducer elements [93]. This allows the verification of the top-level design by comparison with the simulated top-level schematic and avoids, e.g., wiring errors. Some of the available tools also permit extraction of non-electric features [94].

More details on MEMS modeling in general and the extension of circuit simulation to include micromachined devices can be found in several books and overview articles [94–98].

1.4
CMOS and Micromachining

The integration of micromachining processes with CMOS technology can be accomplished in different ways. The additional process steps (or process modules) can either precede the standard CMOS process sequence (pre-CMOS) or they can be performed in between the regular CMOS steps (intra-CMOS) or after the completion of the CMOS process (post-CMOS) [99, 100]. In the case of post-CMOS micromachining, the microstructures are built from either the CMOS layers themselves or from additional layers deposited on top of the CMOS wafer. Tab. 1.4 summarizes various CMOS-based microsystem approaches found in the literature. Some of these approaches require several additional process modules, e.g. a pre-CMOS and a post-CMOS module; in these cases, we have categorized them by their first non-standard process sequence. The cited publications are exemplaric and the list provided is by no means considered to be all-inclusive.

Tab. 1.4 CMOS-based microsystems using pre-, intra- or post-CMOS process modules to implement the microstructures

	Surface micromachining	Bulk micromachining
Pre-CMOS process modules	Sandia National Laboratories (M^3EMS) [101] – Inertial sensors [102] UC Berkeley, Analog Devices, Raytheon and Sandia National Laboratories – Accelerometer [103] University Michigan – Accelerometer [104] Analog Devices, Palo Alto Research Center and UC Berkeley (*ModMEMS*) [105] – Inertial sensors [106] Analog Devices and UC Berkeley (*SOI-MEMS*) – Accelerometer [107, 108] Analog Devices and Clare (*Optical iMEMS*) – Optical switches [109] VTT Information Technologies and Micro Analog Systems [110]	MIT – Pressure sensor [111, 112] Univ. Michigan – Pressure sensor [113] Tohoku Univ. – Accelerometer [114] – Pressure sensor [115] Fraunhofer Institute – Infrared radiation sensor [116] ETH Zurich – Vertical Hall sensor [47]
Intra-CMOS process modules	Analog Devices (*iMEMS*) [78, 117] – Accelerometer [107, 118] – Gyroscope [107, 119] Infineon Technologies (Siemens) – Pressure sensor [5, 120, 121] – Accelerometer [122] – Ultrasound transducers [123] Motorola (now Freescale) – Pressure sensor [124, 125] Fraunhofer Institute IMS – Pressure sensor [126–128] RWTH Aachen and Fraunhofer Institute IMS – Pressure sensor [129, 130] Univ. Duisburg, EPOS and Fraunhofer Institute IMS – Tactile sensor [131] Bosch – Accelerometer [132] Toyota – Pressure sensor [133, 134]	Univ. Michigan – Pressure sensor [77, 136] – Mass flow [77] – Bioprobes [137, 138] – Thermal converter [139] – Infrared imager [140] LG Electronics Inst. of Technology and Seoul National Univ. – Accelerometer [141]

Tab. 1.4 (cont.)

	Surface micromachining	Bulk micromachining
	NEC – Infrared imager [135] IBM – Resonators [3] – RF switches [3, 4]	
Post-CMOS micromachining of add-on layers	UC Berkeley [142, 143] – Inertial sensors – Resonators IMEC – Si/Ge MEMS [144] Stanford Univ. – Variable capacitors [75, 145] Texas Instruments – DMD [74, 146] Univ. Bremen and Infineon – Acceleration switch [147, 148] Delphi, General Motors and Univ. Michigan – Gyroscope [149–151] Honeywell – Thermal imager [152]	austriamicrosystems – Accelerometer [153] ETH Zurich and Micronas – Pressure sensor [154]
Post-CMOS micromachining of CMOS layers	ETH Zurich – Pressure sensor [155, 156] – Fluid density sensor [157] TU Denmark – Resonators [158, 159]	NEC – Pressure sensor [160] Bosch – Pressure sensor [161, 162] Motorola (now Freescale) – Pressure sensor [163] Silicon Microstructures – Pressure sensor [164] Toyota, Toyoda and Tohoku Univ. – Pressure [165] Toyota and Ritsumeikan Univ. – Infrared imager [166] Fraunhofer Institute and European Silicon Structures – Pressure sensor [167] Fraunhofer Institute, TU Berlin and Univ. Stuttgart – Pressure sensor [168]

Tab. 1.4 (cont.)

	Surface micromachining	Bulk micromachining
		ETH Zurich [17, 80]
		– Chemical sensors [19, 169, 170]
		– Thermal imager [22, 23, 171]
		– Tactile sensor [172]
		– Thermal converter [173]
		– Proximity sensors [174]
		– Flow sensor [175]
		– Force sensors [176]
		Sensirion [177]
		– Flow sensor [178]
		Stanford Univ. [179]
		– Thermal converter [180]
		– Vacuum sensor [181]
		– Neural probes [182]
		– Bandgap reference [183]
		– Tactile sensor [184]
		Middle East TU
		– Infrared bolometer [185, 186]
		– Thermopile [187]
		NIST and George Washington Univ. [188]
		– Gas sensor [189, 190]
		– Accelerometer [190]
		– Power sensor [191]
		UC Berkeley, George Washington Univ. and NIST
		– Accelerometer [192]
		Carnegie Mellon Univ. [193, 194]
		– Accelerometer [195]
		– Gyroscope [196, 197]
		– Microphone [198]
		– Loudspeaker [199]
		– Infrared imager pixel [200]
		– Variable capacitor [201]
		Akustica [202]
		– Acoustic devices
		George Washington Univ., Naval Research Labs and Carnegie Mellon Univ.
		– Chemical sensor [203]
		MEMSIC [204]
		– Accelerometer [205]

Tab. 1.4 (cont.)

Surface micromachining	Bulk micromachining
	Univ. Michigan
	– Infrared sensor [206]
	– Accelerometer [207]
	TU Berlin
	– Inkjet printhead [208]
	Toyohashi Univ.
	– Accelerometer [209]
	Univ. Twente
	– Microphone [210]
	Warwick Univ. and Univ. Cambridge
	– Gas sensor [211, 212]
	Seoul National Univ.
	– Pressure sensor [213]
	Daimler-Benz and Dialog Semiconductor
	– Accelerometer [214]

Depending on the chosen integration path, a number of fabrication constraints are imposed on the micromachining steps in order not to deteriorate the performance of the CMOS electronics. An important example is the thermal budget allowed for the micromachining process steps. Polysilicon microstructures are deposited at temperatures between 575 and 625 °C in an LPCVD furnace and typically require thermal annealing at temperatures ≥900 °C to reduce residual stresses [215, 216]. However, after deposition of the aluminum metallization of a CMOS process, the maximum process temperature is limited to ≤450 °C in order not to degrade the aluminum-silicon contacts. Therefore, polysilicon cannot be deposited after the completion of a CMOS process with standard aluminum metallization. In order to enable the deposition of polysilicon microstructures *after* the completion of the CMOS process sequence, an alternative high-temperature stable metallization, such as tungsten, must be used for the CMOS process [61, 142]. Considering that IC manufacturers have invested enormous resources into the development of reliable, multi-level aluminum interconnect technologies, and further considering the inferior resistivity of tungsten versus aluminum, it seems unlikely that such a process would be adopted in industry. Alternatively, the standard polysilicon gate material of the CMOS process is used as well for the microstructures or an additional structural polysilicon layer is deposited and structured *before* the standard CMOS metallization is applied. In this approach, the regular CMOS process sequence is interrupted before the metal deposition, a dedicated micromachining module is inserted and then the CMOS process sequence resumes with the back-end aluminum interconnect technology. This intra-CMOS approach minimizes performance degradations for both electronic and mechanical components, but requires interruption of the CMOS process sequence and, more critical, the need to return CMOS wafers into a CMOS line after performing non-standard process steps. As a result, this fabrication approach for CMOS-integrated polysilicon microstructures has been commercialized by companies with in-house CMOS or BiCMOS fabrication facilities (e.g. Analog Devices [78, 107, 118] and Infineon [5, 120]).

1.4.1
Pre-CMOS Micromachining

Pre-CMOS micromachining or 'MEMS-first' fabrication approaches avoid thermal budget constraints during the MEMS fabrication. In this way, e.g. thick polysilicon microstructures requiring stress relief anneals at temperatures up to 1100 °C can be co-integrated with CMOS circuitry. Typically, the MEMS structures are buried and sealed during the initial process module. After the wafer surface is planarized, the pre-processed wafers with embedded MEMS structures are used as starting material for the subsequent CMOS process. Challenges include the surface planarization required for the subsequent CMOS process and the interconnections between MEMS and circuitry areas.

The M^3EMS (Modular, Monolithic MicroElectroMechanical Systems) technology developed at Sandia National Laboratories was one of the first demonstrations of the MEMS-first integration concept [61, 101]. In this approach, the multi-layer polysilicon microstructure is built in a trench, which has been etched into the bulk silicon using an anisotropic wet silicon etchant. After formation of the polysilicon microstructures, the trench is refilled with LPCVD oxide and planarized with a CMP (chemical mechanical polishing) step. Subsequently, the wafers with embedded microstructures are used as starting material in an unmodified CMOS process, fabricating CMOS circuitry in areas adjacent to the MEMS areas. The CMOS metallization is used to interconnect circuitry and MEMS areas. The back-end of the process requires additional masks to open the protective silicon nitride cap over the MEMS areas prior to the release of the polysilicon structures by silicon oxide sacrificial layer etching. A cross-section of the M^3EMS technology used for the fabrication of inertial sensors [102] is shown in Fig. 1.15a. Theoretically, the planarized wafer with embedded MEMS structures can serve as starting material for any microelectronics foundry service, since the technology does not require significant modifications of the CMOS process sequence [102]. Of course, the pre-processed starting material requires stringent qualification by the CMOS foundry in order not to compromise their process yield. A resonant accelerometer fabricated with Sandia's M^3EMS technology has been reported in [103]. Researchers at the University of Michigan have developed a similar trench-based MEMS-first technology to co-integrate polysilicon microstructures with a 3 μm CMOS technology [104].

Recently, an alternative pre-CMOS MEMS process called *Mod MEMS* has been demonstrated by Analog Devices, Palo Alto Research Center and UC Berkeley

(a)

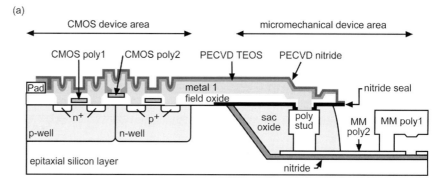

(b)

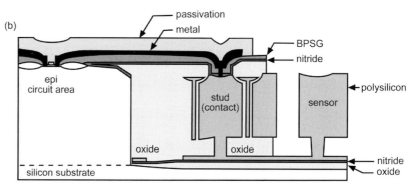

Fig. 1.15 Schematic cross-section of two pre-CMOS MEMS processes for fabrication of monolithically integrated polysilicon microstructures: (a) M³EMS technology by Sandia National Laboratories. Adapted from [101]. (b) Mod MEMS technology by Analog Devices, Palo Alto Research Center and UC Berkeley. Adapted from [105]

[105]. Mod MEMS enables the integration of 5–10 µm thick polysilicon MEMS devices with sub-µm CMOS circuitry. In contrast to the Sandia approach, the thick polysilicon structures are build on top of the silicon substrate and not in an etched trench (see Fig. 1.15b). An 1100 °C anneal ensures nearly stress-free polysilicon layers with very small stress gradients [105], which is especially important for thick polysilicon microstructures. The intra-CMOS approach used by Analog Devices for the fabrication of their ADXL and ADXRS series inertial sensors (see Section 1.4.2) does not allow such high annealing temperatures and thus limits the thickness of the structural polysilicon layer. After forming isolation trenches to provide electrical isolation between MEMS regions at different potential, a 2 µm capping oxide/nitride sandwich is deposited on the polysilicon, the MEMS structural regions are defined by a polysilicon etch step and the sidewalls of the MEMS regions are passivated by a thermal oxidation. Next, a selective epitaxial silicon growth process is used to provide planarization around the thick MEMS

structures. With the polysilicon structures encapsulated with a silicon oxide layer, the epitaxial silicon layer only grows in the wafer regions surrounding the MEMS regions. After the selective epi process, the wafer surface is planarized with a CMP process and a seal layer sandwich is deposited and patterned to protect the MEMS areas during the CMOS process. At this point, the wafers are ready for standard CMOS or BiCMOS processing. The CMOS circuitry is formed in the grown epi regions with only a 12 µm exclusion zone between MEMS and active circuitry. Similar to the Sandia M^{3}EMS process, the CMOS metallization connects the polysilicon microstructures with the circuitry. After completion of the IC process, the MEMS areas are opened up, the thick polysilicon layer is structured with an anisotropic dry etching step and the microstructures are released by sacrificial oxide etching. The feasibility of this pre-CMOS MEMS approach has been demonstrated by successfully fabricating integrated accelerometers and gyroscopes [105, 106]. Similar to the M^{3}EMS process, the Mod MEMS process can be used in conjunction with various IC foundry processes as long as the starting material can be qualified for the particular IC process.

Single-crystalline silicon microstructures can be implemented in a pre-CMOS fabrication approach using either SOI (silicon-on-insulator) wafers as substrate material [107–110] or by incorporating sealed cavities using wafer bonding [111, 112].

Originally demonstrated at UC Berkeley [108], the SOIMEMS technology has been further developed by Analog Devices as a next-generation process for the monolithic integration of inertial sensors. Compared with Analog Devices' current high-volume iMEMS technology (see Section 1.4.2), SOIMEMS offers thicker structural layers (10 µm instead of 4 µm), yielding more robust sensor structures, and a more advanced BiCMOS technology (0.6 µm instead of 3.0 µm minimal feature sizes) enabling more on-chip functionality. A cross-section of the SOIMEMS technology is depicted in Fig. 1.16a [107, 108]. The fabrication process comprises both a pre-CMOS (trench isolation) and a post-CMOS (microstructure definition and release) fabrication module, but has the advantage that all of the circuit processing is done in one process module [107]. The fabrication process starts with etching trenches in the SOI wafers (having a 10 µm device layer) to establish isolated areas on the wafer [108, 217]. The DRIE trench etching stops on the buried oxide layer of the SOI substrates. After trench refill and surface planarization, the regular 0.6 µm BiCMOS process sequence is executed. Interconnects between circuitry and microstructures are established with the standard IC metallization. After completion of the BiCMOS process sequence, the structural regions are cleared from all dielectrics and the microstructures are defined using a DRIE trench etching step. Finally, the microstructures are released by etching the buried oxide layer underneath them using a hydrofluoric acid based etch. Temporary photoresist pedestals prevent the structures from collapsing during drying [107]. First commercial acceleration sensors (ADXL40) based on Analog Devices' SOI-MEMS technology are expected to be launched in 2004 [107].

The basic SOIMEMS technology developed by Analog Devices has been extended to co-integrate electrostatic optical switches with on-chip electronics [109]. The re-

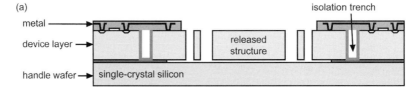

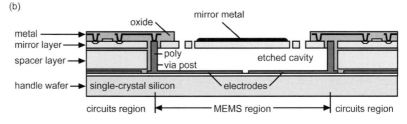

Fig. 1.16 Cross-sections of SOI-based integrated MEMS technologies by Analog Devices: (a) SOIMEMS with 10 μm device layer for fabrication of single-crystalline silicon inertial sensors. Adapted from [107]. (b) Optical iMEMS based on triple-stack substrate for fabrication of integrated optical switches. Adapted from [109]

sulting 'Optical iMEMS' technology uses a three-layer silicon stack as substrate material (see Fig. 1.16 b). The custom-made triple-stack substrate consists of a 10 μm thick mirror layer on top of a sacrificial spacer layer (10–80 μm thick), which is connected to the handle wafer. A patterned polysilicon layer embedded between two dielectric layers forms the electrode structures for mirror actuation between the silicon spacer layer and handle wafer. Trenches are etched using DRIE to contact the buried polysilicon layer and to achieve device isolation. After surface planarization, the preprocessed wafers enter a 3 μm high-voltage CMOS process provided by Clare for on-chip circuit fabrication. The circuitry provides 200 V high-voltage transistors for mirror actuation and 10 V CMOS for the position sense electronics [109]. After completion of the circuit process, the MEMS area is cleared from all circuit dielectrics and the mirror structures are defined in the mirror layer by DRIE. The mirrors are finally released by etching the spacer silicon layer underneath the mirror with xenon difluoride, XeF_2. During this etch, the mirrors themselves are protected using photoresist and silicon oxide layers. Process details can be found in [109].

An alternative SOI-based technology for the monolithic integration of CMOS electronics with MEMS has been demonstrated by VTT Information Technology and Micro Analog Systems [110]. In a pre-CMOS fabrication module, vacuum cavities are formed in defined regions of the buried oxide. The so-called 'plug-up' process sequence consists of (i) DRIE trench etching of micron-sized access holes into the device layer, (ii) deposition of a semipermeable polysilicon layer, (iii) local removal of the buried oxide through the pinholes of the polysilicon layer, thus releasing the microstructures, (iv) plugging of the access holes by conformal LPCVD polysilicon deposition at reduced pressure and finally (v) surface planarization to prepare the wafers for IC processing [110]. The pre-CMOS process sequence can be complemented by isolation trenches and substrate contacts and is followed by a 1 µm CMOS process based on molybdenum gates.

A pre-CMOS approach using wafer bonding to incorporate sealed cavities has been developed at MIT [111, 112]. The cavities are etched into a handle wafer and sealed by silicon fusion bonding the device wafer on to the handle wafer. The device wafer is subsequently thinned to the desired thickness. In [111], the device wafer is an epi-wafer and a combination of grinding, polishing and anisotropic wet etching with electrochemical etch-stop is used to thin the wafer down to the epitaxial layer. In this way, sealed cavities bound by membranes with uniform thickness are formed. Subsequently, the pre-processed wafers are run through a 1.75 µm twin-well CMOS process. After completion of the CMOS process sequence, post-CMOS micromachining steps can be used to access the sealed cavities by dry etching from the wafer front or anisotropic wet etching from the wafer back [111]. The technology has been successfully demonstrated by fabricating piezoresistive pressure sensors with circular membranes [111]. The fabrication of complex microstructures by combining silicon fusion bonding and DRIE has been described in [218].

A number of CMOS-based MEMS processes fabricate the on-chip circuitry in recessed cavities anisotropically etched into the silicon wafers prior to the CMOS process sequence [113–115]. In this way, the front side of the sensor wafers with integrated transducer elements and circuitry can be (anodically) bonded to constraint (glass) wafers. The approach facilitates the fabrication of capacitive sensor structures having counter electrodes on the glass wafer and provides a zero-level packaging for the integrated microsystems.

Silicon trench etching and polysilicon refill have become a standard isolation technology in CMOS processes. A similar trench technology has been used to define the active area of a vertical Hall plate [47]. After sidewall implantation of the vertical Hall plate, the trenches are isolated with a silicon oxide layer and refilled with polysilicon. After surface planarization, the silicon wafers with embedded Hall plates are processed using an unmodified CMOS process, providing sensor contacts, interconnects and read-out circuitry. The resulting vertical Hall sensors are sensitive to magnetic fields in the chip surface. The Hall plates are geometrically well defined by the pre-CMOS trench technology and decoupled from the substrate by a reverse biased p–n junction at the bottom of the trenches. The combination of the trench technology with either SOI substrates or silicon wet etching removes the p–n junction at the bottom of the trenches, yielding dielectrically isolated vertical Hall devices [219].

1.4.2
Intra-CMOS Micromachining

Intermediate micromachining is most commonly used to integrate polysilicon microstructures in CMOS/BiCMOS process technologies. Inserting the micromachining process steps before the back-end interconnect metallization ensures process compatibility with the polysilicon deposition and anneal. The polysilicon annealing temperature is typically limited to about 900 °C in order not to affect the doping profiles of the CMOS process. Alternative post-CMOS approaches [142] require modified interconnect metallizations based on, e.g., tungsten to achieve the necessary high-temperature stability for the subsequent polysilicon deposition and anneal.

Commercially available examples of polysilicon microstructures, fabricated with CMOS/BiCMOS processes with intermediate micromachining, include Analog Devices ADXL series accelerometers and ADXRS series gyroscopes [220], Infineon Technologies' KP100 series pressure sensors [42] and Freescale's (Motorola) MPXY8000 series pressure sensors [221]. Not surprisingly, all three companies have in-house IC processing capabilities, facilitating the chosen interleaved process sequences and allowing fine-tuning of the overall process sequence to minimize degradation in both electronic and mechanical components. Business models based on complete outsourcing of the microsystem fabrication will be very unlikely able to use CMOS processes with intermediate micromachining, because a CMOS foundry will probably not accept wafers back into their line after a number of micromachining steps have been performed elsewhere.

In the following, we will briefly highlight the process technologies developed by Analog Devices and Infineon Technologies to fabricate integrated polysilicon microstructures.

Analog Devices employs a 24 V capable 3 µm BiCMOS process with trimable thin-film resistors for the fabrication of their integrated accelerometers and gyroscopes ([119], see also Chapter 3). Surface-micromachined polysilicon structures with a thickness of 2–4 µm are formed within a micromachining module inserted into the process sequence before the BiCMOS back-end interconnect metallization. An early version of the process [78, 96] for integrated acceleration sensors used 24 masks, including 13 for the electronics and 11 for the mechanical structure and the interconnects to the electronics. A cross-section illustration of Analog Devices' integrated MEMS technology is shown in Fig. 1.17. The fabrication starts with the front end of the BiCMOS technology, creating n-wells, sources, drains and polysilicon gates for the MOS transistors and bases and emitters for the bipolar transistors. In the course of the front-end circuit process, diffused n^+-runners to connect the capacitive microstructures to the on-chip electronics and n^+ ground plane regions are formed. After transistor fabrication, the circuit areas are covered with an LPCVD nitride layer and a BPSG (borophosphosilicate glass) layer and the sensor regions are cleared from all dielectrics down to the gate oxide. Subsequently another LPCVD nitride layer is deposited and patterned, which will later serve as an etch-stop layer during the sensor release etch. Next, the 1.6 µm sacrificial oxide layer and the 2 µm structural polysilicon are successively deposited and patterned. The polysilicon layer is doped by a phosphorus implantation and annealed, targeting a slightly tensile final stress in order to prevent microstructure warping or buckling. After depositing an additional thin oxide over the wafer and removing this oxide and the sacrificial oxide in the circuit areas, the BiCMOS process sequence is resumed with the back-end interconnect formation. Finally, a passivation layer sandwich consisting of a PECVD oxide and nitride film are deposited and patterned and the inertial sensors are released on a wafer level by sa-

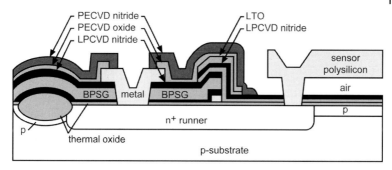

Fig. 1.17 Schematic cross-section of Analog Devices' integrated MEMS technology with an n$^+$-diffusion interconnect structure between polysilicon microstructure and on-chip electronics. Adapted from [78]

crificial oxide etching. During the release and subsequent drying step, temporary photoresist pedestals prevent the microstructures from collapsing. After sensor release, the sensors are tested and the on-chip SiCr resistors are trimmed, followed by dicing, die and wire bonding and application of an antistiction coating. During wafer sawing from the back of the wafer, the released mechanical elements are protected by two layers of tape (with recessed cavities) on the front side of the wafer. Details on these non-standard packaging process steps can be found in [107, 117]. More recent iMEMS versions use polysilicon ground planes, thicker 2 µm sacrificial oxide layers and thicker 4 µm structural polysilicon layers [107].

In contrast to Analog Devices, Infineon Technologies uses the standard capacitor polysilicon layer of a 0.8 µm BiCMOS process as mechanical layer for their pressure sensors ([5, 120]; see also Chapter 6). Obviously, the rather small pressure sensors with a diameter of 70 µm are less sensitive to residual stress within the polysilicon layer and do not require a dedicated low-stress polysilicon structural layer. A schematic cross-section of the surface-micromachined pressure sensors are shown in Fig. 1.18. The standard process sequence of the 16-mask BiCMOS process is stopped before the back-end interconnect metallization to insert a single-mask micromachining module. The basic pressure sensor structure is formed within the course of the BiCMOS process sequence. The lower electrode is made from the n-well, the 600 nm field oxide serves as sacrificial layer and the 400 nm capacitor polysilicon as structural layer and top electrode [5, 120]. Within the micromachining module, the polysilicon membranes are released by sacrificial layer etching and the cavities are sealed. After perforating the membranes in a dry etching step, the oxide sacrificial layer is etched using vapor HF through these holes. Finally, the resulting cavities are sealed with a process optimized for the vertical etch channels, yielding a typical cavity pressure of 300 mbar [120]. After completion of the micromachining module, the regular BiCMOS back-end process is employed to form the aluminum interconnects and passivate the microsystem. The final pad etch is used to open the contact pads and form the oxide boss structures on the pressure sensors.

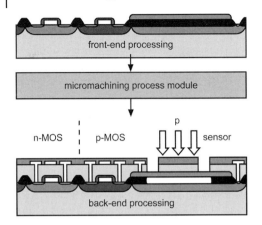

Fig. 1.18 Schematic cross-section of Infineons' integrated MEMS technology for the fabrication of pressure sensors. Adapted from [120]

In addition to the processes by Analog Devices and Infineon, substantial work has been invested at the Technical University of Aachen and the Fraunhofer Institute IMS in the development of a monolithically integrated capacitive pressure sensor technology [126]. The underlying CMOS process remains unchanged until the source/drain implantation. The n^+ implantation is used as one electrode for the capacitive pressure sensors. After drain/source implantation, the pressure sensors are formed by depositing a thin silicon nitride insulation layer, depositing and patterning a sacrificial oxide layer, and finally the polysilicon structural layer. After releasing the polysilicon membranes by a hydrofluoric acid etch, the cavities are sealed with an LPCVD oxide layer, which also serves as the contact oxide for the metallization. Subsequently, the CMOS process is completed with the metallization and passivation. The pressure sensor consists of an array of polysilicon membranes connected in parallel. Different pressure ranges can be covered by adjusting the membrane diameter from 25 to 125 µm [127]. Recently, the capacitive sensor technology has been combined with a CMOS technology on SIMOX (Separation by IMplantation of OXygen) substrates for high-temperature applications up to 250 °C [130]. The pressure sensor is insulated towards the substrate by a thin silicon nitride layer deposited before the sacrificial oxide. The developed capacitive pressure sensor technology has been employed for, e.g., catheter flow sensors [129], intraocular pressure sensor implants [128] and tactile sensors [131].

In order to minimize the influence of the structural polysilicon deposition and anneal on the doping profiles of the CMOS process, the structural polysilicon should be deposited as early as possible in the process sequence (see, e.g., MEMS-first approaches in Section 1.4.1). A unique approach to integrate thick polysilicon layers in a BiCMOS technology has been proposed by Bosch and the Fraunhofer Institute ISiT [132] for the fabrication of a surface-micromachined accelerometer. After the definition of the buried layer, a sacrificial oxide sandwich consisting of an oxide and a LPCVD polysilicon layer is deposited and patterned to define the sensor areas. Subsequently, the BiCMOS process is continued with the deposition of the epitaxial silicon layer in the circuit areas. At the same, a thick polycrystalline silicon layer (epi-poly) is grown in the sensor areas with the LPCVD polysilicon layer acting as nucleation layer. The 10 μm thick epi-poly is grown at high temperatures and exhibits almost ideal mechanical properties [132]. In a later process stage, the epi-poly layer is doped together with the gate polysilicon of the BiCMOS process. After completion of the BiCMOS process metallization and passivation, the epi-poly is patterned and the sacrificial oxide layer etched with vapor HF to release the accelerometer structure. The accelerometer structure is electrically contacted through the buried layer.

All processes described so far relied on a polysilicon structural layer and sacrificial oxide etching. A 1K-element pressure sensor array based on a sacrificial polysilicon layer has been developed by Toyota Central R&D Laboratories [133, 134]. The micromachining module for the sensor fabrication is inserted into the CMOS process flow after source/drain implantation. The MEMS module consists of deposition and patterning of a silicon nitride base layer to protect the silicon substrate, a polysilicon sacrificial layer, followed by the membrane sandwich consisting of silicon nitride top and bottom layer with a polysilicon sensing layer in between. The sacrificial layer is etched using a KOH solution through etch holes located at the edges of the 50×50 μm membranes [133, 134]. Finally, the back-end of the CMOS process is performed and the released cavities are sealed using a PECVD silicon nitride film. The pressure sensor array features piezoresistive pressure sensing using the polysilicon piezoresistors embedded in the silicon nitride membrane (see Fig. 1.19). The array has been used for high-resolution tactile imaging [133, 134].

Using a sacrificial polysilicon layer as well, a monolithic infrared focal plane array has been developed at NEC [135]. The sensor array with a pixel pitch of 50 μm is based on titanium bolometer structures embedded in silicon dioxide suspended plates. The devices are fabricated by inserting a micromachining module consist-

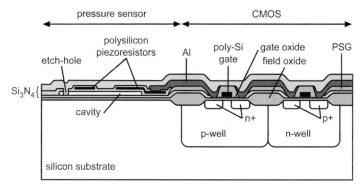

Fig. 1.19 Schematic cross-section of surface-micromachined pressure sensor element developed at Toyota Central R&D Laboratories for use in tactile imagers. Adapted from [133]. The silicon nitride membrane structure with embedded polysilicon piezoresistors is formed in an intra-CMOS process module and released by etching a sacrificial polysilicon layer

ing of several silicon dioxide and metal deposition and patterning steps after the source/drain implantation of the CMOS process [135].

Researchers at IBM's T.J. Watson Research Center have recently incorporated r.f. switches and resonators into the copper-based interconnect technology of state-of-the-art CMOS processes [3, 4]. Certain adaptations had to be made to the standard copper (dual)-damascene process sequence [222] used by IBM for interconnect formation in order to (i) encapsulate the copper and, thus, prevent its oxidation, (ii) introduce suitable contact materials for the switches and (iii) provide a sacrificial layer for copper microstructure release. The microstructures are released by removing an organic sacrificial layer in an oxygen plasma. All required dielectric films are produced using PECVD at temperatures of 400 °C or less and all metal films are deposited by sputtering or a combination of sputtering and electroplating [3].

Bulk micromachining using wet anisotropic silicon etchants in combination with p^{++} etch-stop techniques have been used by Wise and co-workers at the University of Michigan for the fabrication of CMOS-based mass-flow sensors [77], pressure sensors [77, 136], microelectrode recording arrays [137, 138], thermal converters [139] and infrared sensors [140]. The microsystems are fabricated using a modified p-well CMOS process. The highly p-doped regions are diffused into the silicon substrate wafer after the CMOS p-well implantation. The p-well implant dose had to be modified from the baseline CMOS process to account for the additional p-well oxidation and boron segregation into the masking oxide in the merged process [77, 137]. The p- well drive-in is accomplished simultaneously with the p^{++} diffusion. Using a 16 h

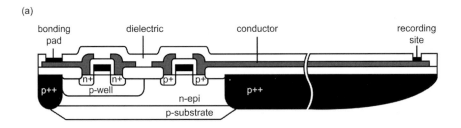

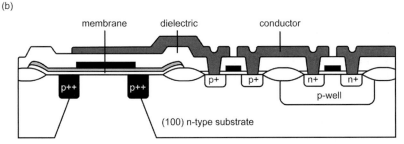

Fig. 1.20 Schematic cross-section of (a) needle probe (adapted from [137]) and (b) mass flow sensor (adapted from [77]) developed at the University of Michigan; diffused p^{++}-regions are used as etch-stop layers during the microstructure release by anisotropic wet etching

diffusion at 1175 °C, rim and p-well depths of 15 and 5 μm are obtained, respectively. If required, an additional shallow p^{++} diffusion is implemented. The p^{++} regions define (a) the lateral dimensions of dielectric membranes by providing a non-etched p^{++} rim around them and (b) the thickness of silicon microstructures, e.g. membranes for pressure sensors or shafts for needle probes, by the depth of the diffusion (see Fig. 1.20). Depending on the actual microsystem, additional process modifications to the CMOS baseline, such as special dielectric layers [77] or metallizations [137], have been incorporated. The microstructures are released after completion of the CMOS process sequence with the p^{++} regions providing an intrinsic etch-stop. In the early 1990s, a number of impressive CMOS-based microsystems were realized based on this process, including a multi-sensor chip comprising six different sensor types and their electronics [77].

1.4.3
Post-CMOS Micromachining

Probably the greatest advantage of post-CMOS micromachining approaches is that the fabrication can be completely outsourced. After completion of the regular CMOS process sequence, which can, in principle, be performed at any CMOS foundry, the post-CMOS micromachining steps can be done at a dedicated MEMS foundry. The price to pay for this fabrication flexibility is the stringent thermal budget for all process steps following standard CMOS technologies with aluminum metallizations. A maximum process temperature of $\sim 450 °C$ excludes high-temperature deposition and annealing steps, such as polysilicon deposition in an LPCVD furnace. PECVD processes, sputtering, electroplating and most wet and dry bulk and surface micromachining processes are, however, well suited for the post-CMOS approach. During the micromachining etching/release step the CMOS electronics might require special protection.

Basically, one can distinguish two general post-CMOS micromachining approaches: the microstructures either are formed by machining the CMOS layers themselves or by building the complete microstructures on top of the CMOS substrate. In the first approach, most of the microstructure is already formed within the regular process sequence. In this case, the post-CMOS process module typically requires very few process steps, such as an etching step to release the microstructure or an additional deposition step. Building the complete MEMS on top of the CMOS substrate might require more process steps but can save valuable real estate, because the MEMS part can be build directly on top of the CMOS circuitry. In the following, we will provide examples for both fabrication approaches.

1.4.3.1 Post-CMOS Micromachining of Add-on Layers
Building the MEMS on top of a CMOS substrate, most processes in this category use surface micromachining techniques and in particular sacrificial layer etching to release the microstructures. Based on the required process temperatures, one can distinguish two basic categories of post-CMOS add-on micromachining mod-

ules: (i) low-temperature modules with process temperatures up to ~ 100–150 °C, which are typically based on PVD (physical vapor deposition) or electroplating of metal layers and use polymers or metals as sacrificial layers, and (ii) medium-temperature modules requiring process temperatures over 300 °C, which are often based on CVD (chemical vapor deposition) processes for the structural and the sacrificial layers.

Crucial for all add-on surface micromachining modules is a good planarity of the underlying CMOS substrate and both good electrical and mechanical contact between the microstructures and the CMOS circuit. The on-chip circuitry can either surround the MEMS or be located underneath the microstructures, saving valuable CMOS real estate.

Because of its desirable mechanical properties, the integration of polysilicon microstructures after the completion of a CMOS process was studied carefully in the early 1990s [142]. The LPCVD deposition and stress-relief anneal of thin polysilicon films require process temperatures of ~ 600 and ≥ 900 °C, respectively, which are not compatible with the standard aluminum (or copper) metallization used in most of today's CMOS processes. To accommodate the required high-temperature post-processing module, modifications to a baseline CMOS process were made at the metallization and passivation level only [142]. Metal–silicon contacts use a high-temperature stable titanium silicide and titanium nitride barrier metallurgy and a tungsten metallization for circuit interconnect. The passivation layer should planarize the wafer surface and protect the underlying circuitry not only from the environment but also from the hydrofluoric acid used to release the microstructures. To this end, a passivation layer consisting of a sandwich of LPCVD phosphosilicate glass (PSG, for surface planarization) and low-stress LPCVD silicon nitride (for HF protection) has been chosen. The subsequent micromachining module consisting of a ground-plane polysilicon (SP1) and two structural polysilicon layers (SP2 and SP3, see Fig. 1.21a) requires nine additional lithography steps. Polysilicon–polysilicon interconnects are used between circuitry and microstructure. The process sequence is described in detail in [142]. To minimize the effect of the high-temperature steps on the circuit characteristics, both the PSG densification and the polysilicon stress-relief anneal are done by RTA (rapid thermal anneal) steps at 900 °C. Additional high-temperature steps include the LPCVD nitride (835 °C) and the polysilicon (610 °C) depositions. A slight shift of the transistor characteristics could be observed, indicating a doping redistribution during the high-temperature post-processing steps.

To avoid doping redistribution and the need for high-temperature stable interconnect metallizations, the post-processing temperature must be reduced to below ~ 500 °C [79]. To achieve this, polycrystalline silicon–germanium films have been investigated recently [143, 144, 223] as an alternative to polysilicon films. Depending on the germanium concentration and the deposition pressure, polycrystalline Si–Ge films can be deposited at temperatures of 450 °C or even lower, making the process compatible with a standard CMOS aluminum metallization. The poly-SiGe films are deposited in either an LPCVD furnace [143, 223] or a PECVD system [144], with the latter method showing increased deposition rates. In [143], two post-CMOS micromachining approaches for the co-integration of poly-SiGe microstructures with

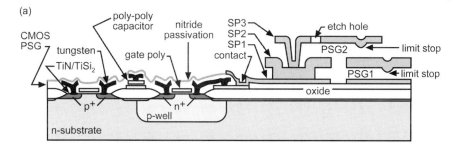

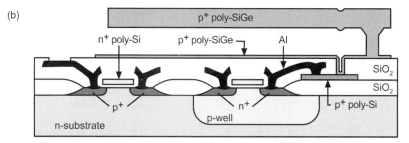

Fig. 1.21 Schematic cross-sections of (a) polysilicon and (b) polycrystalline silicon germanium (poly-SiGe) microstructures fabricated by post-CMOS surface micromachining techniques on top of a completed CMOS substrate wafer. Both fabrication processes were developed at the University of California at Berkeley. Adapted from [142] and [143]

CMOS circuitry are investigated: the first approach uses n-type poly-Ge deposited at 400 °C as structural layer and SiO_2 as sacrificial layer, the second approach p-type poly-$Si_{0.35}Ge_{0.65}$ as structural layer and poly-Ge as sacrificial layer (see schematic in Fig. 1.21 b). Whereas the first approach requires a special CMOS passivation to protect the circuitry during the microstructure release, the second approach uses hydrogen peroxide for sacrificial layer etching and no special layers are needed to protect CMOS metallization and dielectric layers. Similarly to the earlier poly-Si technology [142], a polysilicon–polysilicon contact connects the MEMS to the circuitry. Even though the deposited poly-$Si_{0.35}Ge_{0.65}$ films already exhibit a relatively low as-deposited stress of only –10 MPa, it is expected that the film properties can be improved for MEMS applications with further optimization of deposition and annealing parameters [143, 223], making poly-SiGe a promising candidate for post-CMOS integration of MEMS.

The use of PVD techniques for film deposition can further lower the processing temperatures of post-CMOS micromachining modules. Silicon films sputter-deposited at room temperature using a DC magnetron sputter system have been deposited on both polyimide and silicon dioxide sacrificial layers for MEMS applications [145]. After 3 h anneal in forming gas at 350 °C, films with a thickness of 2 and 5 µm showed residual tensile stress of the order of 70 MPa. A drawback of the sputtered silicon films are very high sheet resistances in the MΩ/sq. range before annealing and in the GΩ/sq. range after the 350 °C anneal. Cladding the structural silicon layer with two 50 nm thick TiW layers resulted in a decreased electrical resistivity of 25 Ω/sq. To prove the CMOS compatibility of this approach, integrated variable capacitors have been fabricated in a post-CMOS approach using the TiW-clad sputtered silicon process [145]. The microstructures have been released by dry etching the polyimide sacrificial layer in an oxygen plasma.

A commercial example of a metal-based MEMS fabricated on top of a CMOS substrate using low-temperature processes only is the DMD (digital micromirror device) developed by Texas Instruments [74, 146]. The DMD, an array of electrostatically actuated torsional micromirrors (used as light switches), creates the image in DLP-based (digital light processing) projection displays. The mechanical structure of a DMD pixel consists of alternating layers of patterned aluminum and air gaps and is built on top of a CMOS static random-access memory (SRAM) cell using surface-micromachining techniques. After completion of the 0.8 μm double-metal CMOS process, the final dielectric layer is planarized using CMP and vias are generated for interconnecting the mirror with the underlying circuitry [146]. As can be imagined, the surface planarization throughout the MEMS process is crucial, as non-uniformities in the mirror surface would result in non-tolerable contrast changes on the final picture. The formation of the 16×16 μm micromirror superstructure requires six additional photolithographic steps to define four aluminum layers and two sacrificial photoresist layers (see schematic in Fig. 1.22). The four aluminum layers are used to fabricate (1) the yoke address electrodes and the bias/reset bus, (2) the torsional hinges, (3) the

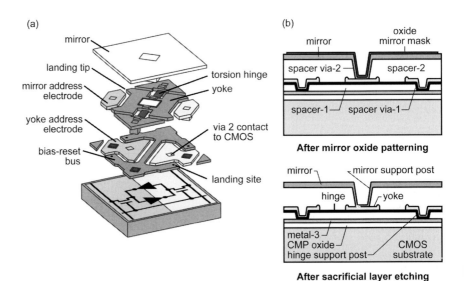

Fig. 1.22 (a) Schematic of the DMD superstructure consisting of four aluminum layers and two sacrificial polymer layers on top of a CMOS SRAM cell. Adapted from [146]. (b) Cross-section of DMD superstructure before and after release by polymer sacrificial layer etching. Adapted from [74]

mirror address electrodes, yoke and hinge support posts and (4) the actual mirror. The aluminum layers are sputter-deposited and the final mirror superstructures are released by etching the polymer sacrificial layer in a plasma etcher. Finally, an antistiction coating is applied to prevent stiction of the micromirrors to the landing pads during operation. The fabrication and packaging process have been described in detail in [74, 146].

Thicker metal structures can be achieved by electroplating techniques. Examples of electroplated microstructures on top of CMOS substrates include a gold acceleration threshold switch developed by Infineon and the University of Bremen ([147, 148], see Fig. 1.23a) and a nickel ring gyroscope developed by Delphi-Delco Electronics, General Motors and the University of Michigan at Ann Arbor ([149–151], see Fig. 1.23b). In the following, the fabrication process of the acceleration switch is briefly highlighted. After completion of the CMOS process, a thin sacrificial photoresist layer is spun-on and patterned for electrical and mechanical contacts between microstructure and circuitry. Subsequently a titanium/gold seed layer is sputtered on the wafer and a thick photoresist plating mold is spun-on and patterned. Finally, gold is electroplated into the photoresist plating mold and the microstructures are released by removing plating mold, seed layer and sacrificial photoresist layer. The post-CMOS fabrication process has been described in detail in [147].

The electroplated ring gyroscope is fabricated using a similar post-CMOS fabrication sequence [149, 150]. An early post-CMOS process sequence [224, 225] is summarized in the following (see cross-section in Fig. 1.24). After completion of the CMOS process, the aluminum metallization is passivated using a LTO (low-temperature oxide)–SOG (spin-on-glass)–LTO sandwich to protect the metallization and CMOS circuitry from the subsequent etching steps in the sensor process. After formation of via holes to connect to the underlying aluminum IC metallization, Ti/W and Au are deposited and patterned. The Ti/W layer serves as an adhesion layer for the gold and as a diffusion barrier to keep the gold from interacting with the aluminum. Next, a conductive sacrificial spacer (Cr/Al/Cr) is deposited and patterned. This layer defines the movable portions of the sensor element and also serves as a plating base layer for electroforming the microstructure. Then, the electroforming mold is defined and the sensor element is formed by selective electroplating in the open mold areas. Finally, the mold and sacrificial layer are removed, completing the sensor process. Fig. 1.23b shows close-up photographs of the photoresist plating mold and the fabricated metal ring and electrodes.

Both additive electroplating technologies are room temperature processes and do not affect the performance of the underlying CMOS circuits. As with most surface-micromachined structures, control of the thin-film stress and stress gradients is a major challenge.

If the monolithic integration of MEMS and CMOS circuitry is either technologically or economically not feasible, a hybrid integration based on wafer bonding of a MEMS wafer to a CMOS circuit wafer might be a viable alternative. Examples of this approach include an acceleration sensor developed by austriamicrosystems

(a)

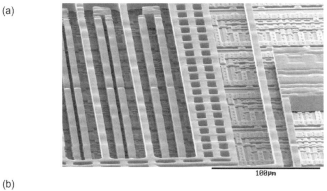

(b)

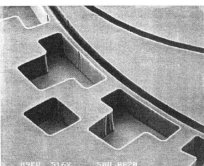 

Fig. 1.23 (a) Detail of acceleration threshold switch by Infineon Technologies and the University of Bremen, fabricated using gold electroplating on top of a CMOS substrate. Reprinted with permission from [148]. (b) Detail of photoresist mold (left) and electroplated nickel structure (right) of vibrating ring gyroscope developed by General Motors, the University of Michigan and Delphi Electronics [148, 149]. SEM photographs courtesy of Prof. K. Najafi, University of Michigan, USA

[153] and a pressure sensor developed at ETH Zurich and Micronas [154]. The acceleration sensor is based on a polysilicon cantilever beam fabricated on the MEMS wafer and released by wet anisotropic silicon etching [153]. The cantilever beam forms the movable electrode of the capacitive microsystem. Static actuation and sensing electrodes and also the readout electronics are implemented on the CMOS wafer. MEMS and CMOS wafer are finally hermetically bonded to each other using an Au–Si eutectic bonding technique, resulting in a controlled spacing between cantilever beam and static electrodes. The capacitive pressure sensor is based on a silicon membrane micromachined on the MEMS wafer using wet anisotropic etching from the back of the wafer [154]. The MEMS wafer is solder bonded on to the substrate wafer with the counter electrode. In both approaches, CMOS and MEMS wafers can be processed independently in a CMOS and MEMS foundry, respectively, and are joined in the final wafer bonding step.

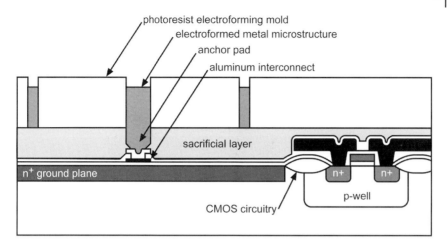

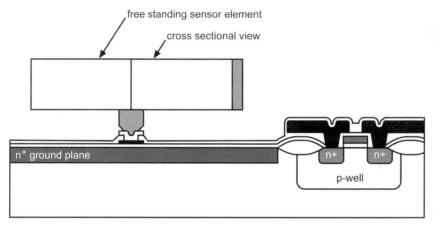

Fig. 1.24 Schematic cross-section of electroplated ring gyroscope developed by General Motors, Delphi-Delco Electronics and the University of Michigan (a) before and (b) after removal of the photoresist plating mold and the conductive sacrificial layer. Adapted from [149]

1.4.3.2 Post-CMOS Micromachining of CMOS Layers

In this approach, microstructures are released by micromachining the CMOS substrate wafer itself after the completion of the regular CMOS process sequence. By far the majority of demonstrated devices rely on bulk micromachining processes, such as wet and dry anisotropic and isotropic silicon etching, but surface-micromachining approaches have also been proposed.

Piezoresistive pressure sensors based on bulk-micromachined silicon membranes were the earliest commercially successful application of silicon micromachining, with the first complete silicon pressure transducer catalog distributed by National Semiconductor in 1974 (an excellent overview of the early MEMS efforts was published in 1982 by Petersen [9]). It is therefore not surprising that early attempts to co-integrate transducers and electronics also targeted bulk-micromachined pressure sensors. Fully integrated and temperature-compensated pressure sensors using bipolar technology were demonstrated in 1979 [226, 227]. Shortly thereafter, the first integrated capacitive pressure sensor with bipolar circuitry was demonstrated [228]. The first CMOS-integrated piezoresistive silicon pressure sensor was developed by NEC in the mid-1980s [160]: the sensor consists of a thin, square silicon diaphragm with four piezoresistors in a Wheatstone bridge configuration located along the clamped edges. p-Type wafers with an n-type epitaxial layer were used as starting material for the p-well CMOS process. While an additional implantation step for the piezoresistors was implemented in the CMOS process sequence, the actual silicon micromachining steps were performed after the completion of the CMOS process sequence. The membranes were released by anisotropic etching from the back of the wafer using a hydrazine–water etchant [160]. An electrochemical etch-stop technique was applied to stop the etching automatically at the p–n junction between p-substrate and n-epitaxial layer. Finally, the sensor wafer was anodically bonded to a glass constraint wafer. The early NEC pressure sensor featured on-chip circuitry for a stable Wheatstone bridge biasing, temperature compensation and signal amplification.

Today, CMOS-integrated piezoresistive pressure sensors are commercially available from several companies, including Bosch [229], Freescale (Motorola) [221] and Silicon Microstructures [230]. Although the basic transducer structure of these microsystems is still a bulk-micromachined silicon membrane with implanted piezoresistors, far more circuitry components are implemented in these modern systems. The Freescale (Motorola) design features an on-chip digital signal processor (DSP) and non-volatile memory for calibration, temperature compensation and the ability to implement customer-specific features [163]. The CMOS process is only slightly modified to provide the optimal doping profile for the piezoresistors and to deposit the etch mask for the membrane release on the back of the wafer. The membrane is released using a timed wet etching step, yielding membranes with a ± 2 μm thickness tolerance across the wafer [163]. Finally, the sensor wafer is anodically bonded to a glass wafer in vacuum.

Bulk micromachining from the back of the wafer using silicon anisotropic etching has become one of the standard post-CMOS micromachining modules, releasing not only membranes but also cantilevers and suspended microstructures. Substantial work in this area has been done e.g. at ETH Zurich [19, 22, 169–171, 174–176, 231], the University of Michigan [77, 104, 137, 138, 206], Tohoku University [114, 115], and the Fraunhofer Institutes [167, 168]. A typical device cross-section after post-CMOS bulk micromachining is shown in Fig. 1.25. In addition to the pressure sensors mentioned earlier, e.g. accelerometers, flow sensors, ultrasound proximity sensors, thermal converters, infrared radiation sensors, and chemical sensors have been fabricated using this approach (see Tab. 1.4).

While potassium hydroxide (KOH) solutions have become the most common wet etchant used in bulk micromachining from the back of the wafer, various etch-stop techniques are employed to control the etch result. In addition to a

1.4 CMOS and Micromachining

(a) etch-stop on thermal oxide

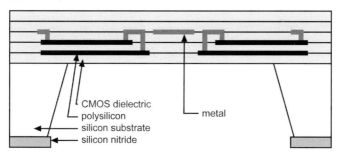

(b) electrochemical etch-stop

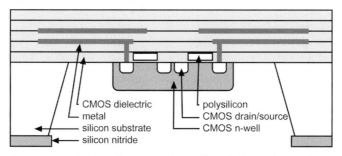

Fig. 1.25 Post-CMOS bulk micromachining from the back of the wafer

timed etch (e.g. [163]), silicon dioxide layers, highly p^{++}-doped silicon regions and p–n junctions are commonly used as etch-stop layers. Membranes consisting of the dielectric layers of the CMOS process are released by using the field oxide on top of the silicon substrate as an intrinsic etch-stop layer [232]. The use of SOI-based CMOS processes offers the possibility of employing the buried oxide layer of the SOI substrates as an etch-stop layer in order to release single-crystalline silicon structures. Alternatively, a buried etch-stop oxide can be locally fabricated before the CMOS process sequence (i.e. as a pre-CMOS sequence) by high-dose oxygen implantation with subsequent high-temperature annealing [116].

Early CMOS-based pressure sensor designs [160, 161] relied on electrochemical etch-stop (ECE) techniques [65] to cease the silicon etching at the p–n junction between a p-type substrate and an n-type epitaxial layer. In this implementation, a special wafer starting material had to be introduced, as commercial CMOS processes are typically not based on p-doped substrates with n-type epitaxial layers. Fortunately, the ECE technique can be also applied at the p–n junction between the n-well of a CMOS process and the p-doped substrate [17, 179, 214]. In this case, however, the electrochemical potential provided by a potentiostat has to be connected to each individual 'mechanical' n-well on the wafer. A scheme to modify commercial CMOS device technologies for application of ECE in a four-electrode configuration has been presented in [17] (see also Section 1.3.2): to supply the ECE potentials to the sensor structures, a contact field and a wafer-wide contact network are generated (see Fig. 1.26). The contact network routes the n-well and p-substrate potential to the respective contacts on the sensor structures. The chosen process sequence comprises additional standard photolithography steps at the metallization and passivation mask levels performed exclusively on wafer steppers. Hence, there are no restrictions on the minimum feature size of the process or the wafer diameter.

The p^{++} etch-stop techniques have been used extensively by Wise and co-workers for the fabrication of CMOS-based mass-flow sensors [77], microelectrode recording arrays [137] and IR sensors [205]. Even though the microstructure release is done after the completion of the CMOS process sequence, the p^{++} etch-stop requires a preparatory intra-CMOS process module to diffuse highly p-doped regions into the silicon substrate wafer after the p-well implantation (see Section 1.4.2).

Bulk micromachining from the back of the wafer requires the deposition and patterning of a hard mask on the back of the wafers. Prior to the deposition of the hard mask, any processing residues and damage caused by the CMOS process have to be removed from the back side of the wafer. Damage in the wafer surface can lead to an intolerable large mask underetching during the MEMS release. The preparation of the wafer back can be achieved, e.g., using a spin etcher [18]. The hard mask typically consists of a PECVD silicon nitride layer, if necessary with a pad of oxide underneath. Care must be taken to minimize the pinhole density for the subsequent anisotropic wet etching step. The etch mask is patterned using a double-side mask aligner, aligning the patterns on the wafer back to front-side structures. Subsequently, the CMOS wafer is typically mounted in a wafer holder to protect the wafer front with the CMOS circuits from the wet etchant. In the case of wet anisotropic etching with an electrochemical etch-stop technique, the

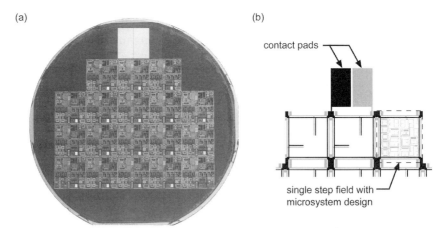

Fig. 1.26 (a) CMOS wafer with large contact pads to enable anisotropic wet etching from the back of the wafer with an electrochemical etch-stop technique. (b) Schematic of etch network routing the etching potentials for the structural n-well and substrate contacts from the contact pads to the individual microstructures. Adapted from [17, 18]

wafer holder supplies the etching potentials to the wafer (see, e.g., [233]). The need for mounting every wafer in a mechanical holder might be cumbersome, but is currently still the most reliable method to protect the front-side structures and the wafer edges. Recently, alternative protection schemes based on metal films [234] or polymer layers [18] have been proposed.

Alternative to wet etching, bulk micromachining from the back of the wafer can be performed with DRIE systems, resulting in almost vertical sidewalls independent of the silicon crystal orientation. DRIE techniques have gained significant momentum over the past few years, but the equipment required is expensive, only single wafers can be processed at a time and no ECE or p^{++} etch-stop can be used. On the other hand, DRIE can achieve structures, e.g. narrow support bars, that cannot be achieved by KOH etching of (100) wafers.

The post-CMOS release of microstructures from the front of the wafer using anisotropic silicon etchants was pioneered by Baltes and co-workers [80, 235]. By superimposing active area, contact, via and pad opening in the design of the integrated microstructures, silicon areas on the wafer surface are exposed to the ambient at the end of the regular CMOS process sequence. Anisotropic wet etchants, such as EDP and TMAH, will etch the silicon substrate in these areas (see Section 1.2.1). Alternatively, isotropic dry etching techniques based on, e.g., XeF_2 can be used for microstructure release [188]. In the case of anisotropic wet etchants, the etch mask consisting of the dielectric layers of the CMOS process will be underetched at convex corners, allowing the release of dielectric microstructures, such as cantilever beams and bridges [80, 232]. Polysilicon and metal layers can be embedded in these microstructures, enabling the fabrication of a variety of microsensors, including flow sensors [236], power sensors [173] and IR sensors [185, 186]. Care must be taken to protect the aluminum contact pads of the CMOS process during the etching step. Certain EDP and TMAH formulations have relatively small aluminum etch rates, allowing a maskless wet release of the microstructures if the etching step is not too long. KOH etching solutions cannot be used for front-side release because of the relatively large silicon dioxide etch rate. In order to permit active electronics on the microstructures, the bulk micromachining from the front of the CMOS wafer has been successfully combined with an electrochemical etch-stop technique on CMOS n-wells [179] (see Fig. 1.11b). For example, CMOS-based thermal converters [180] and focal plane arrays [185, 186] have been fabricated in this way.

An alternative bulk micromachining technology from the front side of CMOS wafers using dry etching steps has been developed at Carnegie Mellon University [193]. The post-CMOS micromachining module uses the top metal interconnect layer as etch mask for the microstructure definition. In this way, the minimum feature sizes, such as minimum beam widths and gaps, are defined by the CMOS design rules and can be scaled with the CMOS technology. The actual laminated microstructures consisting of the CMOS dielectrics with polysilicon and metal layers sandwiched in between are released using two dry etching steps (see schematic in Fig. 1.27a): in the first anisotropic etching step using CHF_3–O_2 etch chemistry, the oxide areas not protected by the metal mask are etched to the sili-

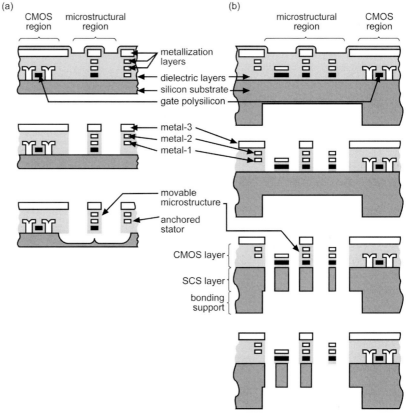

Fig. 1.27 Cross-section of post-CMOS process sequences developed at Carnegie Mellon University to release (a) dielectric (adapted from [193]) and (b) crystalline silicon microstructures (adapted from [194]). Both processes are based on a series of anisotropic and isotropic dry etching steps

con substrate; in the second isotropic etching step using SF_6–O_2 chemistry, the oxide beams are underetched, releasing the microstructures. The isotropic release etch has a vertical to lateral etch rate of about 2:1, safely undercutting 16 µm wide structures for a typical etch depth of 25 µm. The process technology has been used to fabricate integrated accelerometers [195], gyroscopes [197], IR sensors [200] and acoustic devices [198, 199]. To construct a speaker or microphone, a mesh-type membrane is released with the described post-CMOS dry etching sequence. The released mesh is conformally coated with polymer in a CVD process, yielding a continuous, airtight membrane [199] (for more detail, see Chapter 4).

Depending on the CMOS process, the released dielectric layer sandwich with embedded polysilicon and metallization lines can be subject to large residual stress and stress gradients, causing large microstructures to curl. To be able to release single-crystalline silicon microstructures, the maskless post-CMOS micromachining process developed at Carnegie Mellon University has been combined with

DRIE and an anisotropic etch step from the back of the wafer ([194]; see schematic in Fig. 1.27b). The new post-CMOS process sequence starts with a deep anisotropic backside etch, leaving a 10–100 μm thick single-crystal silicon membrane. The membrane thickness ultimately controls the thickness of the microstructure. Next, an anisotropic dry etching step is performed from the front of the wafer, removing silicon dioxide where it is not protected by the top metal layer. Then, in contrast to the earlier technology [193], an anisotropic instead of isotropic silicon etching step is used to release the microstructures, which now consist of a thick silicon layer in addition to the dielectric layer sandwich. With an optional lateral underetch, the silicon under small beams can be removed to achieve, e.g., electrical isolation of certain silicon areas. The process has been used, e.g., for the fabrication of a z-axis accelerometer [194] and a gyroscope [196].

A unique post-CMOS bulk-micromachining process based on silicon-on-glass (SOG) bonding has been demonstrated recently at the University of Michigan [207]. A schematic of the SOG monolithic integration process is shown in Fig. 1.28. First, conventional IC fabrication is performed on a silicon wafer. A glass substrate with recessed cavities and a shielding metal is also prepared. The metal is deposited and patterned on the glass substrate not only to avoid the micro-loading effect from the following DRIE, but also to protect the IC from the large electric field during the bonding process. Next, the fabricated silicon wafer is anodically bonded to the glass substrate and is thinned to the desired thickness using CMP. Finally, metal contacts are formed, and DRIE is used to define the MEMS structure, silicon islands and dielectric bridges.

So far, post-CMOS bulk-micromachining modules have been discussed to release microstructures consisting of the layers of the CMOS process. In addition, surface-micromachining techniques can be used to remove thin-film layers of the CMOS process selectively, thus releasing different types of microstructures. An example is the sacrificial aluminum etching (SALE) technique developed at ETH Zurich [237]. In this post-CMOS micromachining module, the first metal layer of the CMOS process is selectively removed to release microstructures comprised of the intermetal dielectric, the upper metallization layer and the passivation. During the sacrificial aluminum etching using standard wet etchants [237], the electrical

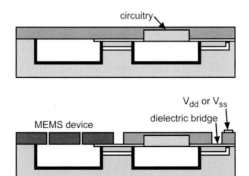

Fig. 1.28 Schematic cross-section of monolithic silicon-on-glass microsystems developed at the University of Michigan using a post-CMOS micromachining module [207]. Courtesy of Dr. J. Chae, University of Michigan

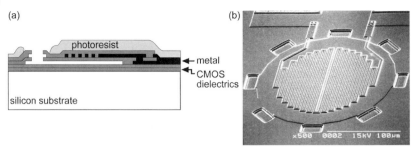

Fig. 1.29 (a) Schematic cross-section and (b) SEM photograph of a Pirani-type pressure sensor released in a post-CMOS module by sacrificial aluminum etching (SALE). Adapted from [155]

contact pads have to be protected either using a photoresist layer or electroplated, etch-resistant contact bumps. Fig. 1.29 shows a cross-section and an SEM photograph of a thermal pressure sensor realized with the SALE process [155]. The circular pressure sensing membrane is released by etching the sacrificial aluminum layer through a number of etch holes around its perimeter. The second aluminum layer is used to form a resistor inside the membrane, acting as both heating element and temperature sensor. A complete thermal pressure-sensing system based on the surface-micromachined sensor cells has been presented [156].

A second example is a mass-sensitive nanocantilever fabricated recently using a post-CMOS surface-micromachining technique [158]. The approach uses an innovative laser beam direct lithography process to structure an aluminum etch mask. The aluminum mask is used to transfer the cantilever structure to the gate polysilicon layer of the CMOS process. Finally, the polysilicon cantilever having submicron width and height is released by locally removing the field oxide layer underneath the cantilever using BOE (buffered oxide etch). All these process steps are performed after the completion of the regular CMOS process sequence. To do this, the passivation and upper polysilicon layer are removed in the sensor area. A thin aluminum layer acting as etch mask is deposited on top of the thin oxide separating the two polysilicon layers [158].

1.5
Conclusion

A large number of process technologies for the co-integration of MEMS with CMOS circuitry has been demonstrated over the past 20 years. Even though the material and process choices are limited to ensure compatibility with the underlying CMOS process, this fabrication constraint is, for many applications, outweighted by the possibility of on-chip electronics and the proven reliability of well-established CMOS technologies. High-volume applications (e.g., inertial and pressure sensors) and applications requiring sensor arrays (e.g., DMD and infrared imagers) are predestined for CMOS integration. As examples, Analog Devices is producing about 1 million inertial sensors per week [107] and Texas Instruments' DMD chips consist of up to 1.3 million, individually addressable, CMOS-integrated micromirrors [74]. While physical sensors currently clearly dominate the CMOS-based sensor market, chemical and biological microsensors are likely to benefit from the CMOS-integration as well, enabling small-size sensing arrays for, e.g., hand-held, battery-operated (bio)chemical sensing systems.

While many of today's CMOS-based microsystems still require the insertion of dedicated (micromachining) modules in between the regular CMOS process steps, a clear trend towards not interrupting the CMOS sequence can be observed. This can be achieved by adding the required process modules either before (pre-CMOS) or after (post-CMOS) the CMOS process steps. From a manufacturing point of view, especially the post-CMOS approach is appealing: the CMOS process can be completed in a dedicated CMOS foundry, followed by a micromachining module processed in a dedicated MEMS foundry. This approach allows especially small and medium size MEMS companies without in-house CMOS capability to design and produce CMOS-based MEMS.

1.6
References

1   IBM Microelectronics, *http://www.ibm.com/chips/*.
2   Apple Computer, G5 Processor, *http://www.apple.com/g5processor/*.
3   C.V. JAHNES, J. COTTE, J.L. LUND, H. DELIGIANNI, A. CHINTHAKINDI, L.P. BUCHWALTER, P. FRYER, J.A. TORNELLO, N. HOIVIK, J.H. MAGERLEIN, D. SEEGER, "Simultaneous fabrication of RF MEMS switches and resonators using copper-based CMOS interconnect manufacturing methods." In: *Proc. IEEE Microelectromechanical Systems Conference (MEMS 2004)*; **2004**, pp. 789–792.
4   N. HOIVIK, C.V. JAHNES, J. COTTE, J.L. LUND, D. SEEGER, J.H. MAGERLEIN, "RF MEMS switches using copper-based CMOS interconnect manufacturing technology." In: *Proc. Solid-State Sensor, Actuator and Microsystem Workshop, Hilton Head Island*; **2004**, pp. 93–94.
5   C. HIEROLD, "Intelligent CMOS sensors." In: *Proc. IEEE Int. Conf. on Micro Electro Mechanical Systems (MEMS 2000)*; **2000**, pp. 1–6.
6   J.D. PLUMMER, M.D. DEAL, P.B. GRIFFIN, *Silicon VLSI Technology: Fundamentals, Practice and Modeling*; Prentice Hall: Englewood Cliffs, NJ, **2000**.
7   S.A. CAMPBELL, *The Science and Engineering of Microelectronic Fabrication*, 2nd edn; Oxford University Press: Oxford, **2001**.
8   S.M. SZE, *Semiconductor Devices: Physics and Technology*, 2nd edn; Wiley: New York, **2002**.
9   K.E. PETERSEN, "Silicon as a mechanical material." *Proc. IEEE* **1982**, *70*, 420–457.
10 S.M. SZE (ed.), *VLSI Technology*, 2nd edition; McGraw-Hill: New York, **1988**.
11 C.Y. CHANG, S.M. SZE (eds.), *ULSI Technology*, McGraw-Hill: New York, **1996**.
12 D. EDELSTEIN, J. HEIDENREICH, R. GOLDBLATT, W. COTE, C. UZOH, N. LUSTIG, P. ROPER, T. MCDEVITT, W. MOTSIFF, A. SIMON, J. DUKOVIC, R. WACHNIK, H. RATHORE, R. SCHULZ, L. SU, S. LUCE, J. SLATTERY, "Full copper wiring in a sub-0.25 micro-m CMOS ULSI technology." In: *Technical Digest, IEEE International Electron Devices Meeting*; **1997**, p. 773–776.
13 A. HIERLEMANN, D. LANGE, C. HAGLEITNER, N. KERNESS, A. KOLL, O. BRAND, H. BALTES," Application-specific sensor sys-

tems based on CMOS chemical microsensors," *Sens. Actuators B* **2000**, *70*, 2–11.

14 H. LORENZ, M. DESPONT, N. FAHRNI, J. BRUGGER, P. VETTIGER, P. RENAUD, "High-aspect-ratio, ultrathick, negative-tone near-UV photoresist and its applications for MEMS," *Sens. Actuators A* **1998**, *64*, 33–39.

15 G. M. WHITESIDES, E. OSTUNI, S. TAKAYAMA, X. JIANG, D. E. INGBER, "Soft lithography in biology and biochemistry." *Annu. Rev. Biomed. Eng.* **2001**, *3*, 335–373.

16 K. R. WILLIAMS, R. S. MULLER, "Etch rates for micromachining processing," In: *IEEE J. Microelectromechan. Syst.* **1996**, *5*, 256–269; K. R. WILLIAMS II, K. GUPTA, M. WASILIK, "Etch rates for micromachining processing – Part II." *IEEE J. Microelectromechan. Syst.* **2003**, *12*, 761–778.

17 T. MÜLLER, M. BRANDL, O. BRAND, H. BALTES, "An industrial CMOS process family adapted for the fabrication of smart silicon sensors," *Sens. Actuators A* **2000**, *84*, 126–133.

18 T. MÜLLER, "An Industrial CMOS Process Family for Integrated Silicon Sensors," *PhD Thesis*, no. 13463; ETH Zurich, **1999**.

19 D. LANGE, C. HAGLEITNER, A. HIERLEMANN, O. BRAND, H. BALTES, "Complementary metal oxide semiconductor cantilever array on a single chip: mass-sensitive detection of volatile organic compounds." *Anal. Chem.* **2002**, *74*, 3084–3095.

20 D. LANGE, C. HAGLEITNER, C. HERZOG, O. BRAND, H. BALTES, "Electromagnetic actuation and MOS-transistor sensing for CMOS-integrated micromechanical resonators." *Sens. Actuators A* **2003**, *103*, 150–155.

21 Y. LI, C. VANCURA, C. HAGLEITNER, J. LICHTENBERG, O. BRAND, H. BALTES, "Very high Q-factor in water achieved by monolithic, resonant cantilever sensor with fully integrated feedback." In: *Proc. IEEE Sensors Conference*; **2003**, pp. 809–813.

22 A. SCHAUFELBUHL, N. SCHNEEBERGER, U. MUNCH, M. WAELTI, O. PAUL, O. BRAND, H. BALTES, C. MENOLFI, Q. HUANG, E. DOERING, M. LOEPFE, "Uncooled low-cost thermal imager based on micromachined CMOS integrated sensor array." *IEEE J. Microelectromechan. Syst.* **2001**, *10*, 503–510.

23 A. SCHAUFELBUEHL, U. MUNCH, O. BRAND, H. BALTES, "256-pixel thermoelectric infrared imager in CMOS technology." *Sens. Actuators A* to be published.

24 A. BAKKER, "CMOS smart temperature sensors – An overview." In: *Proc. IEEE Sensors Conference 2002*; **2002**, pp. 1423–1427.

25 A. BAKKER, J. HUIJSING, *High-Accuracy CMOS Smart Temperature Sensors*; Kluwer: Dordrecht, **2000**.

26 R. S. POPOVIC, *Hall Effect Devices*, 2nd edn. IOP: Bristol, **2003**.

27 O. YADID-PECHT, R. ETIENNE-CUMMINGS, *CMOS Imagers: From Phototransduction to Image Processing*; Kluwer: Dordrecht, **2004**.

28 E. R. FOSSUM, "CMOS image sensors: electronic camera-on-a-chip." *IEEE Trans. Electron Devices* **1997**, *44*, 1689–1698.

29 F. HOFMANN, A. FREY, B. HOLZAPFL, M. SCHIENLE, C. PAULUS, P. SCHINDLER-BAUER, D. KUHLMEIER, J. KRAUSE, R. HINTSCHE, E. NEBLING, J. ALBERS, W. GUMBRECHT, K. PLEHNERT, G. ECKSTEIN, R. THEWES, "Fully electronic DNA detection on a CMOS chip: device and process issues." In: *Proc. IEEE International Electron Devices Meeting (IEDM 2002)*; **2002**, pp. 488–491.

30 R. THEWES, F. HOFMANN, A. FREY, M. SCHIENLE, C. PAULUS, P. SCHINDLER-BAUER, B. HOLZAPFL, R. BREDERLOW, "CMOS-based DNA sensor arrays." In: H. BALTES, O. BRAND, G. K. FEDDER, C. HIEROLD, J. G. KORVINK, O. TABATA (eds). *Advanced Micro- and Nanosystems*; Wiley-VCH: Weinheim, **2004**, vol. 1, pp. 383–412.

31 B. EVERSMANN, M. JENKNER, F. HOFMANN, C. PAULUS, R. BREDERLOW, B. HOLZAPFL, P. FROMHERZ, M. MERZ, M. BRENNER, M. SCHREITER, R. GABL, K. PLEHNERT, M. STEINHAUSER, G. ECKSTEIN, D. SCHMITT-LANDSIEDEL, R. THEWES, "A 128 * 128 CMOS biosensor array for extracellular recording of neural activity." *IEEE J. Solid-State Circuits* **2003**, *38*, 2306–2317.

32 W. FRANKS, F. HEER, I. MCKAY, S. TASCHINI, R. SUNIER, C. HAGLEITNER, A. HIERLEMANN, H. BALTES, "CMOS mono-

lithic microelectrode array for stimulation and recording of natural neural networks." In: *Proc. Transducers '03*; **2003**, pp. 963–966.

33 E. LAUWERS, J. SULS, W. GUMBRECHT, D. MAES, G. GIELEN, W. SANSEN, "A CMOS multiparameter biochemical microsensor with temperature control and signal interfacing." *IEEE J. Solid-State Circuits* **2001**, *36*, 2030–2038.

34 F. VAN STEENKISTE, E. LAUWERS, J. SULS, D. MAES, K. BAERT, W. GUMBRECHT, P. ARQUINT, R. MERTENS, G. GIELEN, W. SANSEN, K. ABRAHAM-FUCHS, "A biochemical CMOS integrated multi-parameter microsensor." In: *Proc. Transducers '99*; **1999**, pp. 1188–1190.

35 J. BAUSELLS, J. CARRABINA, A. ERRACHID, A. MERLOS, "Ion-sensitive field-effect transistors fabricated in a commercial CMOS technology." *Sens. Actuators B*, **1999**, *57*, 56–62.

36 T. BOLTSHAUSER, C. A. LEME, H. BALTES, "High sensitivity CMOS humidity sensors with on-chip absolute capacitance measurement system." *Sens. Actuators B* **1993**, *15*, 75–80.

37 T. BOLTSHAUSER, "CMOS Humidity Sensors", *PhD Thesis*, No. 10320; ETH Zurich, **1993**.

38 A. KOLL, S. KAWAHITO, F. MAYER, C. HAGLEITNER, D. SCHEIWILLER, O. BRAND, H. BALTES, "Flip-chip packaged CMOS chemical microsystem for detection of volatile organic compounds." In: *Proc. SPIE* **1998**, *3328*, 223–232.

39 A. KOLL, "CMOS Capacitive Chemical Microsystems for Volatile Organic Compounds", *PhD Thesis*, No. 13460; ETH Zurich, **1999**.

40 S. CHRISTIAN, "New generation of humidity sensors." *Sens. Rev.*, **2002**, *22*, 300–302.

41 Micronas, Zurich, Switzerland, http://www.micronas.com/products/overview/sensors/index.php.

42 Infineon Technologies, Munich, Germany, http://www.infineon.com/sensors/.

43 Allegro Microsystems, Worcester, MA, http://www.allegromicro.com/hall/.

44 Melexis Microelectronic Systems, Belgium, http://www.melexis.com/prod_hall.asp.

45 R. S. POPOVIC, Z. RANDJELOVIC, D. MANIC, "Integrated Hall-effect magnetic sensors." *Sens. Actuators A* **2001**, *91*, 46–50.

46 R. S. POPOVIC, P. M. DRLJACA, C. SCHOTT, "Bridging the gap between AMR, GMR, and Hall magnetic sensors." In: *Proc. IEEE Int. Conf. on Microelectronics (MIEL 2002)*; **2002**, pp. 55–58.

47 R. STEINER-VANHA, F. KROENER, T. OLBRICH, R. BARESCH, H. BALTES, "Trench-Hall devices." *J. Microelectromech. Syst.* **2000**, *9*, 82–87.

48 H. GRUGER, R. GOTTFRIED-GOTTFRIED, "Performance and applications of a two axis fluxgate magnetic field sensor fabricated by a CMOS process." *Sens. Actuators A* **2001**, *91*, 61–64.

49 R. GOTTFRIED-GOTTFRIED, W. BUDDE, R. JAHNE, H. KUCK, B. SAUER, S. ULBRICHT, U. WENDE, "A miniaturized magnetic-field sensor system consisting of a planar fluxgate sensor and a CMOS readout circuitry." *Sens. Actuators A* **1996**, *54*, 443–447.

50 M. SCHNEIDER, S. KAWAHITO, Y. TADOKORO, H. BALTES, "High sensitivity CMOS microfluxgate sensor." In: *Technical Digest of Int. Electron Device Meeting (IEDM '97)*; **1997**, pp. 907–910.

51 M. SCHNEIDER, "CMOS Magnetotransistors and Fluxgate Vector Sensors," *PhD Thesis*, No. 12746; ETH Zurich, **1999**.

52 P. M. DRLJACA, P. KEJIK, F. VINCENT, R. S. POPOVIC, "Low noise CMOS microfluxgate magnetometer." In: *Proc. Transducers '03*; **2003**, pp. 304–307.

53 S. JUNG, R. THEWES, T. SCHEITER, K. F. GOSER, W. WEBER, "A low-power and high-performance CMOS fingerprint sensing and encoding architecture." *IEEE J. Solid-State Circuits* **1999**, *34*, 978–984.

54 J. SCHWIZER, M. MAYER, O. BRAND, *Force Sensors for Microelectronic Packaging Applications*; Springer: Berlin, **2004**.

55 J. SCHWIZER, W. H. SONG, M. MAYER, O. BRAND, H. BALTES, "Packaging test chip for flip-chip and wire bonding process characterization." In: *Transducers 2003, Digest of Technical Papers*; **2003**, pp. 440–443.

56 W. MENZ, J. MOHR, O. PAUL, *Microsystems Technology*, Wiley-VCH: Weinheim, **2001**.

57 G. T. A. Kovacs, *Micromachined Transducers Sourcebook*; McGraw-Hill: New York, **1998**.
58 A. Heuberger, *Mikromechanik*; Springer: Berlin, **1991**.
59 M. Madou, *Fundamentals of Microfabrication*; CRC Press: Boca Raton, FL, **1997**.
60 G. T. A. Kovacs, N. I. Maluf, K. E. Petersen, "Bulk micromachining of silicon," *Proc. IEEE* **1998**, *86*, 1536–1551.
61 J. M. Bustillo, R. T. Howe, R. S. Muller, "Surface micromachining for microelectromechanical systems." *Proc. IEEE*, **1998**, *86*, 1552–1574.
62 M. Shikida, K. Sato, K. Tokoro, D. Uchikawa, "Differences in anisotropic etching properties of KOH and TMAH solutions." *Sens. Actuators A* **2000**, *80*, 179–188.
63 U. Münch, "Industrial CMOS Technology for Thermal Imagers." PhD Thesis, No. 13801; ETH Zurich, **2000**.
64 O. Tabata, "pH-controlled TMAH etchants for silicon micromachining." *Sens. Actuators A* **1996**, *53*, 335–339.
65 B. Kloeck, S. D. Collins, N. F. de Rooij, R. L. Smith, "Study of electrochemical etch-stop for high-precision thickness control of silicon membranes." *IEEE Trans. Electron Devices* **1989**, *36*, 663–669.
66 XACTIX, Pittsburgh, PA, http://www.xactix.com/.
67 C. Hagleitner, A. Koll, R. Vogt, O. Brand, H. Baltes, "CMOS capacitive chemical microsystem with active temperature control for discrimination of organic vapors." In: *Transducers 99 Digest of Technical Papers*; **1999**, pp. 1012–1015.
68 F. Lärmer, A. Schilp, "Method of anisotropically etching silicon," German Patent DE 4241045, US Patent 5501893; **1994**.
69 STS Surface Technology Systems, Newport, UK, http://www.stsystems.com/.
70 UNAXIS Semiconductor, Balzers, FL, http://www.semiconductors.unaxis.com/.
71 Alcatel Vacuum Technology, France, http://www.adixen.com/adixen_avt/.
72 O. Tabata, R. Asahi, H. Funabashi, K. Shimaoka, S. Sugiyama, "Anisotropic etching of silicon in TMAHsolutions." *Sens. Actuators A* **1992**, *34*, 51–57.
73 D. Westberg, O. Paul, G. I. Andersson, H. Baltes, "Surface micromachining by sacrificial aluminum etching." *J. Micromech. Microeng.* **1996**, *6*, 376–384.
74 P. F. van Kessel, L. J. Hornbeck, R. E. Meier, M. R. Douglass, "A MEMS-based projection display." *Proc. IEEE* **1998**, *86*, 1687–1704.
75 C. W. Storment, D. A. Borkholder, V. Westerlind, J. W. Suh, N. I. Maluf, G. T. A. Kovacs, "Flexible, dry-released process for aluminum electrostatic actuators." *J. Microelectromechan. Syst.* **1994**, *3*, 90–96.
76 T. Müller, G. Kissinger, A. C. Benkitsch, O. Brand, H. Baltes, "Assessment of silicon wafer material for the fabrication of integrated circuit sensors." *J. Electrochem. Soc.* **2000**, *147*, 1604–1611.
77 E. Yoon, K. D. Wise, "An integrated mass flow sensor with on-chip CMOS interface circuitry." *IEEE Trans. Electron Devices* **1992**, *39*, 1376–1386.
78 T. A. Core, W. K. Tsang, S. J. Sherman, "Fabrication technology for an integrated surface micromachined sensor." *Solid State Technol.* **1993**, 39–47.
79 S. Sedky, A. Witvrouw, H. Bender, K. Baert, "Experimental determination of the maximum post-process annealing temperature for standard CMOS wafers." *IEEE Trans. Electron Devices* **2001**, *48*, 377–385.
80 D. Moser, M. Parameswaran, H. Baltes, "Field oxide microbridges, cantilever beams, coils and suspended membranes in SACMOS technology." *Sens. Actuators A* **1990**, *23*, 1019–1022.
81 Cadence Design Systems, San Jose, CA, http://www.cadence.com/.
82 Mentor Graphics, Wilsonville, OR, http://www.mentor.com/.
83 Synopsys, Mountain View, CA, http://www.synopsys.com/.
84 ANSYS, Canonsburg, PA, http://www.ansys.com/.
85 Comsol, Stockholm, Sweden, http://www.comsol.com/.
86 Coventor, Cary, NC, http://www.coventor.com/.
87 Intellisense Software, Woburn, MA, http://www.intellisensesoftware.com/.
88 G. K. Fedder, Q. Jing, "A hierarchical circuit-level design methodology for microelectromechanical systems." *IEEE Trans.*

89 SUGAR, http://www-bsac.eecs.berkeley.edu/cadtools/sugar/sugar/.
90 MEMSCAP, Crolles, France, http://www.memscap.com/.
91 G. K. FEDDER, "Structured design of integrated MEMS." In: *Proc. IEEE Int. Micro Electro Mechanical Systems Conference (MEMS 1999)*; **1999**, pp. 1–8.
92 T. MUKHERJEE, G. K. FEDDER, R. D. BLANTON, "Hierarchical design and test of integrated microsystems." *IEEE Design Test Comput.* **1999**, *16(4)*, 18–27.
93 C. HAGLEITNER, A. HIERLEMANN, H. BALTES, "CMOS single-chip gas detection systems: Part II." In: *Sensors Update*, H. BALTES, G. K. FEDDER, J. G. KORVINK (eds.); Wiley-VCH: Weinheim, **2003**, *12*, 51–120.
94 T. MUKHERJEE, G. K. FEDDER, D. RAMASWAMY, J. WHITE, "Emerging simulation approaches for micromachined devices." In: *IEEE Trans. Computer-Aided Design Integrated Circuits Syst.* **2000**, *19*, 1572–1589.
95 A. NATHAN, H. BALTES, *Microtransducer CAD: Physical and Computational Aspects*; Springer: Berlin, **1999**.
96 S. D. SENTURIA, *Microsystem Design*; Kluwer: Dordrecht, **2001**.
97 S. D. SENTURIA, "CAD challenges for microsensors, microactuators, and microsystems." *Proc. IEEE* **1998**, *86*, 1611–1626.
98 J. G. KORVINK, H. BALTES, "Microsystem Modeling." In: *Sensors Update*; VCH: Weinheim, **1996**, *2*, 181–209.
99 O. BRAND, H. BALTES, "CMOS-based microsensors and packaging," *Sens. Actuators A* **2001**, *92*, 1–9.
100 H. BALTES, O. BRAND, A. HIERLEMANN, D. LANGE, C. HAGLEITNER, "CMOS MEMS – Presence and future." In: *Proc. IEEE Micro Electro Mechanical Systems 2002 (MEMS 2002)*; **2002**, pp. 459–466.
101 J. SMITH, S. MONTAGUE, J. SNIEGOWSKI, J. MURRAY, P. MCWHORTER, "Embedded micromechanical devices for the monolithic integration of MEMS with CMOS." In: *Proc. IEEE IEDM '95* **1995**, pp. 609–612.
102 J. J. ALLEN, R. D. KINNEY, J. SARSFIELD, M. R. DAILY, J. R. ELLIS, J. H. SMITH, S. MONTAGUE, R. T. HOWE, B. E. BOSER, R. HOROWITZ, A. P. PISANO, M. A. LEMKIN, W. A. CLARK, T. JUNEAU, "Integrated micro-electro-mechanical sensor development for inertial applications." *IEEE Aerospace Electron. Syst. Mag.* **1998**, *13*, 36–40.
103 A. A. SESHIA, M. PALANIAPAN, T. A. ROESSIG, R. T. HOWE, R. W. GOOCH, T. R. SCHIMERT, S. MONTAGUE, "A vacuum packaged surface micromachined resonant accelerometer." *J. Microelectromechan. Syst.* **2002**, *11*, 784–793.
104 Y. B. GIANCHANDANI, H. KIM, M. SHINN, B. HA, B. LEE, K. NAJAFI, C. SONG, "A fabrication process for integrating polysilicon microstructures with post-processed CMOS circuits." *J. Micromech. Microeng.* **2000**, *10*, 380–386.
105 J. YASAITIS, M. JUDY, T. BROSNIHAN, P. GARONE, N. POKROVSKIY, D. SNIDERMAN, S. LIMB, R. HOWE, B. BOSER, M. PALANIAPAN, X. JIANG, S. BHAVE, "A modular process for integrating thick polysilicon MEMS devices with sub-micron CMOS." In: *Proc. SPIE* **2003**, *4979*, 145–154.
106 S. A. BHAVE, J. I. SEEGER, X. JIANG, B. E. BOSER, R. T. HOWE, J. YASAITIS, "An integrated, vertical-drive, in-plane-sense microgyroscope." In: *Proc. Transducers '03*; **2003**, pp. 171–174.
107 M. W. JUDY, "Evolution of integrated inertial MEMS technology." In: *Proc. Solid-State Sensor, Actuator and Microsystem Workshop, Hilton Head Island*; **2004**, pp. 27–32.
108 M. A. LEMKIN, T. N. JUNEAU, W. A. CLARK, T. A. ROESSIG, T. J. BROSNIHAN, "A low-noise digital accelerometer using integrated SOI-MEMS technology." In: *Proc. Transducers '99*; **1999**, pp. 1294–1297.
109 T. J. BROSNIHAN, S. A. BROWN, A. BROGAN, C. S. GORMLEY, D. J. COLLINS, S. J. SHERMAN, M. LEMKIN, N. A. POLCE, M. S. DAVIS, "Optical iMEMS – A fabrication process for MEMS optical switches with integrated on-chip electronics." In: *Proc. Transducers '03*; **2003**, pp. 1638–1642.
110 J. KIIHAMAEKI, H. RONKAINEN, P. PEKKO, H. KATTELUS, K. THEQVIST, "Modular integration of CMOS and SOI-MEMS using plug-up concept." In: *Proc. Transducers '03*; **2003**, pp. 1647–1650.

111 L. Parameswaran, C. Hsu, M. A. Schmidt, "A merged MEMS-CMOS process using silicon wafer bonding." In: *Proc. IEEE IEDM '95*; **1995**, pp. 613–616.

112 L. Parameswaran, C. Hsu, M. A. Schmidt, "IC process compatibility of sealed cavity sensors." In: *Transducers 97, Digest of Technical Papers*; **1997**, pp. 625–628.

113 A. V. Chavan, K. D. Wise, "A monolithic fully-integrated vacuum-sealed CMOS pressure sensor." *IEEE Trans. Electron Devices* **2002**, *49*, 164–169.

114 Y. Matsumoto, M. Esashi, "Integrated silicon capacitive accelerometer with PLL servo technique." *Sens. Actuators A* **1993**, *39*, 209–217.

115 T. Kudoh, S. Shoji, M. Esashi, "An integrated miniature capacitive pressure sensor." *Sens. Actuators A* **1991**, *29*, 185–193.

116 M. Muller, W. Budde, R. Gottfried-Gottfried, A. Hubel, R. Jahne, H. Kuck, "A thermoelectric infrared radiation sensor with monolithically integrated amplifier stage and temperature sensor." *Sens. Actuators A* **1996**, *54*, 601–605.

117 K. H.-L. Chau, R. E. Sulouff, "Technology for the high-volume manufacturing of integrated surface-micromachined accelerometer products." *Microelectron. J.* **1998**, *29*, 579–586.

118 K. H.-L. Chau, S. R. Lewis, Y. Zhao, R. T. Howe, S. F. Bart, R. G. Marcheselli, "An integrated force-balanced capacitive accelerometer for low-g applications." *Sens. Actuators A* **1996**, *54*, 472–476.

119 J. A. Geen, S. J. Sherman, J. F. Chang, S. R. Lewis, "Single-chip surface micromachined integrated gyroscope with 50 degrees/h Allan deviation." *IEEE J. Solid-State Circuits* **2002**, *37*, 1860–1866.

120 T. Scheiter, H. Kapels, K.-G. Oppermann, M. Steger, C. Hierold, W. M. Werner, H.-J. Timme, "Full integration of a pressure-sensor system into a standard BiCMOS process." *Sens. Actuators A* **1998**, *67*, 211–214.

121 C. Hierold, B. Clasbrummel, D. Behrend, T. Scheiter, M. Steger, K. Oppermann, H. Kapels, E. Landgraf, D. Wenzel, D. Etzrodt, "Low power integrated pressure sensor system for medical applications." *Sens. Actuators A* **1999**, *73*, 58–67.

122 C. Hierold, A. Hildebrandt, U. Naeher, T. Scheiter, B. Mesching, M. Steger, R. Tiefert, "A pure CMOS surface-micromachined integrated accelerometer." *Sens. Actuators A* **1996**, *57*, 111–116.

123 P.-C. Eccardt, K. Niederer, T. Scheiter, C. Hierold, "Surface micromachined ultrasound transducers in CMOS technology." In: *Proc. IEEE Ultrasonics Symposium*; **1996**, pp. 959–962.

124 B. P. Gogoi, D. Mladenovic, "Integration technology for MEMS automotive sensors." In: *Proc. IEEE IECON '02*; **2002**, *4*, 2712–2717.

125 B. P. Gogoi, S. Jo, R. August, A. McNeil, M. Fuhrmann, J. Torres, T. F. Miller, A. Reodique, M. Shaw, K. Neumann, D. Hughes, D. J. Monk, "A 0.8 µm CMOS integrated surface micromachined capacitive pressure sensor with EEPROM trimming and digital output for a tire pressure monitoring system." In: *Proc. 2002 Solid-State Sensor, Actuator and Microsystems Workshop, Hilton Head*; **2002**, pp. 181–184.

126 H. Dudaicevs, M. Kandler, Y. Manoli, W. Mokwa, E. Spiegel, "Surface micromachined pressure sensors with integrated CMOS read-out electronics." *Sens. Actuators A* **1994**, *43*, 157–163.

127 H. K. Trieu, M. Knier, O. Koster, H. Kappert, M. Schmidt, W. Mokwa, "Monolithic integrated surface micromachined pressure sensors with analog on-chip linearization and temperature compensation." In: *Proc. IEEE Int. Conf. on Micro Electro Mechanical Systems (MEMS 2000)*; **2000**, pp. 547–550.

128 K. Stangel, S. Kolnsberg, D. Hammerschmidt, B. J. Hosticka, H. K. Trieu, W. Mokwa, "A programmable intraocular CMOS pressure sensor system implant." *IEEE J. Solid-State Circuits* **2001**, *36*, 1094–1100.

129 R. Kersjes, F. Liebscher, E. Spiegel, Y. Manoli, W. Mokwa, "An invasive catheter flow sensor with on-chip CMOS readout electronics for the on-line determination of blood flow." *Sens. Actuators A* **1996**, *54*, 563–567.

130 K. Kasten, N. Kordas, H. Kappert, W. Mokwa, "Capacitive pressure sensor with monolithically integrated CMOS readout circuit for high temperature applications." *Sens. Actuators A* **2002**, *97/98*, 83–87.

131 M. Leineweber, G. Pelz, M. Schmidt, H. Kappert, G. Zimmer, "New tactile sensor chip with silicone rubber cover." *Sens. Actuators A* **2000**, *84*, 236–245.

132 M. Offenberg, F. Lamer, B. Elsner, H. Munzel, W. Riethmuller, "Novel process for a monolithic integrated accelerometer." In: *Proc. Transducers '95*; **1995**, pp. 589–592.

133 S. Sugiyama, K. Kawahata, H. Funabashi, M. Takigawa, I. Igarashi, "A 32*32 (1k)-element silicon pressure-sensor array with CMOS processing circuits." *Electron. Commun. Jpn, Part 2 (Electronics)* **1992**, *75*, 64–76.

134 S. Sugiyama, K. Kawahata, M. Yoneda, I. Igarashi, "Tactile image detection using a 1k-element silicon pressure sensor array." *Sens. Actuators A* **1990**, *22*, 397–400.

135 A. Tanaka, S. Matsumoto, N. Tsukamoto, S. Itoh, K. Chiba, T. Endoh, A. Nakazato, K. Okuyama, Y. Kumazawa, M. Hijikawa, H. Gotoh, T. Tanaka, N. Teranishi, "Infrared focal plane array incorporating silicon IC process compatible bolometer." *IEEE Trans. Electron Devices* **1996**, *43*, 1844–1850.

136 A. DeHennis, K. D. Wise, "A fully-integrated multi-site pressure sensor for wireless arterial flow characterization." In: *Proc. Solid-State Sensor, Actuator and Microsystems Workshop, Hilton Head Island*; **2004**, pp. 168–171.

137 J. Ji, K. D. Wise, "An implantable CMOS circuit interface for multiplexed microelectrode recording arrays." *IEEE J. Solid-State Circuits* **1992**, *27*, 433–443.

138 Y. Yao, M. N. Gulari, J. F. Hetke, K. D. Wise, "A self-testing multiplexed CMOS stimulating probe for a 1024-site neural prosthesis." In: *Proc. Transducers '03*; **2003**, pp. 1213–1216.

139 E. Yoon, K. D. Wise, "A wideband monolithic RMS-DC converter using micromachined diaphragm structures." *IEEE Trans. Electron Devices* **1994**, *41*, 1666–1668.

140 A. D. Oliver, W. G. Baer, K. D. Wise, "A bulk-micromachined 1024-element uncooled infrared imager." In: *Proc. Transducers '95*; **1995**, *2*, 636–639.

141 Y. Yee, J. U. Bu, K. Chun, J.-W. Lee, "An integrated digital silicon micro-accelerometer with MOSFET-type sensing elements." *J. Micromech. Microeng.* **2000**, *10*, 350–358.

142 J. M. Bustillo, G. K. Fedder, C. T.-C. Nguyen, R. T. Howe, "Process technology for the modular integration of CMOS and polysilicon microstructures." *Microsys. Technol.* **1994**, *1*, 130–141.

143 A. E. Franke, J. M. Heck, T.-J. King, R. T. Howe, "Polycrystalline silicon-germanium films for integrated microstructures." *J. Microelectromechan. Syst.* **2003**, *12*, 160–171.

144 C. Rusu, S. Sedky, B. Parmentier, A. Verbist, O. Richard, B. Brijs, L. Geenen, A. Witvrouw, F. Laermer, S. Kronmueller, V. Lea, B. Otter, "New low-stress PECVD poly-SiGe layers for MEMS." *J. Microelectromechan. Syst.* **2003**, *12*, 816–825.

145 K. A. Honer, G. T. A. Kovacs, "Integration of sputtered silicon microstructures with pre-fabricated CMOS circuitry." *Sens. Actuators A* **2001**, *91*, 386–397.

146 M. A. Mignardi, "From IC's to DMD's." *Texas Instrum. Tech. J.* **1998**, *15*(3), 56–63.

147 M. Wycisk, T. Tonnesen, J. Binder, S. Michaelis, H.-J. Timme, "Low-cost post-CMOS integration of electroplated microstructures for inertial sensing." *Sens. Actuators A* **2000**, *83*, 93–100.

148 S. Michaelis, H.-J. Timme, M. Wycisk, J. Binder, "Additive electroplating technology as post-CMOS process for the production of MEMS acceleration-threshold switches for transportation applications." *J. Micromech. Microeng.* **2000**, *10*, 120–123.

149 S. Chang, M. Chia, P. Castillo-Borelley, W. Higdon, Q. Jiang, J. Johnson, L. Obedier, M. Putty, Q. Shi, D. Sparks, S. Zarabadi, "An electroformed CMOS integrated angular rate sensor." *Sens. Actuators A* **1998**, *66*, 138–143.

150 D. R. Sparks, S. R. Zarabadi, J. D. Johnson, Q. Jiang, M. Chia, O. Larsen, W. Higdon, P. Castillo-Borelley, "A CMOS integrated surface micromachined

angular rate sensor: its automotive applications." In: *Proc. Transducers '97*; **1997**, pp. 851–854.

151 D. R. Sparks, X. Huang, W. Higdon, J. D. Johnson, "Angular rate sensor and accelerometer combined on the same micromachined CMOS chip." *Microsyst. Technol.* **1998**, *4*, 139–142.

152 B. E. Cole, R. E. Higashi, R. A. Wood, "Monolithic two-dimensional arrays of micromachined microstructures for infrared applications." *Proc. IEEE* **1998**, *86*, 1679–1686.

153 M. Brandl, V. Kempe, "High performance accelerometer based on CMOS technologies with low cost add-ons." In: *Proc. IEEE Int. Conf. on Micro Electro Mechanical Systems (MEMS 2001)*; **2001**, pp. 6–9.

154 B. Rogge, D. Moser, H. Oppermann, O. Paul, H. Baltes, "Solder-bonded micromachined capacitive pressure sensors." In: *Proc. SPIE* **1998**, *3514*, 307–315.

155 O. Paul, H. Baltes, "Novel fully CMOS-compatible vacuum sensor." *Sens. Actuators A* **1995**, *46*, 143–146.

156 A. Haberli, O. Paul, P. Malcovati, M. Faccio, F. Maloberti, H. Baltes, "CMOS integration of a thermal pressure sensor system." In: *Proc. 1996 IEEE Int. Symp. on Circuits and Systems (ISCAS '96)*; **1996**, pp. 377–380.

157 D. Westberg, O. Paul, G. I. Andersson, H. Baltes, "A CMOS-compatible device for fluid density measurements." In: *Proc. IEEE Int. Workshop on Micro Electro Mechanical Systems (MEMS '97)*; **1997**, pp. 278–283.

158 Z. J. Davis, G. Abadal, B. Helbo, O. Hamsen, F. Campabadal, F. Perez-Murano, J. Esteve, E. Figueras, J. Verd, N. Barniol, A. Boisen, "Monolithic integration of mass sensing nano-cantilevers with CMOS circuitry." *Sens. Actuators A* **2003**, *105*, 311–319.

159 Z. J. Davis, G. Abadal, E. Forsen, O. Hansen, F. Campabadal, E. Figueras, J. Esteve, J. Verd, F. Perez-Murano, X. Borrise, S. G. Nilsson, I. Maximov, L. Montelius, N. Barniol, A. Boisen, "Nanocantilever based mass sensor integrated with CMOS circuitry." In: *Proc. Transducers '03*; **2003**, pp. 496–499.

160 T. Ishihara, K. Suzuki, S. Suwazono, M. Hirata, H. Tanigawa, "CMOS integrated silicon pressure sensor." *J. Solid-State Circuits* **1987**, *22*, 151–155.

161 H.-J. Kress, F. Bantien, J. Marek, M. Willmann, "Silicon pressure sensor with integrated CMOS signal-conditioning circuit and compensation of temperature coefficient." *Sens. Actuators A* **1996**, *25*, 21–26.

162 J. Marek, "Microsystems for automotive applications." In: *Proc. Eurosensors XIII*; **1999**, pp. 1–8.

163 X. Ding, W. Czarnocki, J. P. Schuster, B. Roeckner, "DSP-based CMOS monolithic pressure sensor for high-volume manufacturing." In: *Proc. Transducers '99*; **1999**, pp. 362–365.

164 J. G. Markle, M. L. Dunbar, H. V. Allen, R. Bornefeld, W. Schreiber-Prillwitz, O. Stoever, "A single-chip pressure sensor." *Sens. Mag.* **2004**, *21*, 27–31.

165 T. Nagata, H. Terabe, S. Kuwahara, S. Sakurai, O. Tabata, S. Sugiyama, M. Esashi, "Digital compensated capacitive pressure sensor using CMOS technology for low-pressure measurements." *Sens. Actuators A* **1992**, *34*, 173–177.

166 N. Fujitsuka, J. Sakata, Y. Miyachi, K. Mizuno, K. Ohtsuka, Y. Taga, O. Tabata, "Monolithic pyroelectric infrared image sensor using PVDF thin-film." *Sens. Actuators A* **1998**, *66*, 237–243.

167 J. Weber, S. Seitz, U. Steger, B. Folkmer, U. Schaber, A. Plettner, H. L. Offereins, H. Sandmeier, E. Lindner, "A monolithically integrated sensor system using sensor-specific CMOS cells." *Sens. Actuators A* **1995**, *46*, 137–142.

168 F. V. Schnatz, U. Schoneberg, W. Brockherde, P. Kopystynski, T. Mehlhorn, E. Obermeier, H. Benzel, "Smart CMOS capacitive pressure transducer with on-chip calibration capability." *Sens. Actuators A* **1992**, *34*, 77–83.

169 C. Hagleitner, A. Hierlemann, D. Lange, A. Kummer, N. Kerness, O. Brand, H. Baltes, "Smart single-chip gas sensor microsystem." *Nature* **2001**, *414*, 293–296.

170 M. Graf, D. Barrettino, M. Zimmermann, A. Hierlemann, H. Baltes, S. Hahn, N. Barsan, U. Weimar, "CMOS

monolithic metal-oxide sensor system comprising a microhotplate and associated circuitry." *IEEE Sens. J.* 2004, 4, 9–16.

171 U. MUNCH, D. JAEGGI, K. SCHNEEBERGER, A. SCHAUFELBUHL, O. PAUL, H. BALTES, J. JASPER, "Industrial fabrication technology for CMOS infrared sensor arrays." In: *Proc. Transducers '97;* 1997, pp. 205–208.

172 T. SALO, T. VANCURA, O. BRAND, H. BALTES, "CMOS-based sealed membranes for medical tactile sensor arrays." In: *Proc. IEEE Micro Electro Mechanical Systems 2003 (MEMS 2003);* 2003, pp. 590–593.

173 D. JAEGGI, H. BALTES, D. MOSER, "Thermoelectric AC power sensor by CMOS technology." *IEEE Electron Device Lett.* 1992, 13, 366–368.

174 M. R. HORNUNG, O. BRAND, *Micromachined Ultrasound-Based Proximity Sensors;* Kluwer: Dordrecht, 1999.

175 F. MAYER, A. HABERLI, H. JACOBS, G. OFNER, O. PAUL, H. BALTES, "Single-chip CMOS anemometer." In: *IEEE IEDM 1997, Technical Digest;* 1997, pp. 895–898.

176 D. LANGE, M. ZIMMERMANN, C. HAGLEITNER, O. BRAND, H. BALTES, "CMOS 10-cantilever array for constant-force parallel scanning AFM." In: *Proc. Transducers '01/Eurosensors XV;* 2001, pp. 1074–1077.

177 Sensirion AG, Zurich, *Switzerland, http://www.sensirion.com/*.

178 D. MATTER, T. KLEINER, B. KRAMER, B. SABBATTINI, "Microsensor-based gas flow meter wins innovation prize." *ABB Rev.* 2003, 3, 49–50.

179 R. J. REAY, E. H. KLAASSEN, G. T. A. KOVACS, "Thermally and electrically isolated single crystal silicon structures in CMOS technology." *IEEE Electron Device Let.* 1994, 15, 399–401.

180 E. H. KLAASSEN, R. J. REAY, G. T. A. KOVACS, "Diode-based thermal RMS converter with on-chip circuitry fabricated using standard CMOS technology." In: *Proc. Transducers '95;* 1995, pp. 154–157.

181 E. H. KLAASSEN, G. T. A. KOVACS, "Integrated thermal-conductivity vacuum sensor." *Sens. Actuators A* 1997, 58, 37–42.

182 M. D. HILLS, D. T. KEWLEY, J. M. BOWER, G. T. A. KOVACS, "Active SOI-based neural probes." In: *Proc. Solid-State Sensor, Actuator and Microsystems Workshop, Hilton Head Island;* 2002, pp. 193–197.

183 R. J. REAY, E. H. KLAASSEN, G. T. A. KOVACS, "A micromachined low-power temperature-regulated bandgap voltage reference." *IEEE J. Solid-State Circuits* 1995, 30, 1374–1381.

184 B. J. KANE, M. R. CUTKOSKY, G. T. A. KOVACS, "A traction stress sensor array for use in high-resolution robotic tactile imaging." *IEEE J. Microelectromechan. Syst.* 2000, 9, 425–434.

185 D. SABUNCUOGLU, S. EMINOGLU, T. AKIN, "A low-cost uncooled infrared microbolometer detector in standard CMOS technology." *IEEE Trans. Electron Devices* 2003, 50, 494–502.

186 S. EMINOGLU, M. YUSUF TANRIKULU, T. AKIN, "A low-cost 64×64 uncooled infrared detector array in standard CMOS." In: *Proc. Transducers '03;* 2003, pp. 316–319.

187 T. AKIN, Z. OLGUN, O. AKAR, H. KULAH, "An integrated thermopile structure with high responsivity using any standard CMOS process." *Sens. Actuators A* 1998, 66, 218–224.

188 N. H. TEA, V. MILANOVIC, C. A. ZINCKE, J. S. SUEHLE, M. GAITAN, M. E. ZAGHLOUL, J. GEIST, "Hybrid postprocessing etching for CMOS-compatible MEMS." *IEEE J. Microelectromechan. Syst.* 1997, 6, 363–372.

189 J. S. SUEHLE, R. E. CAVICCHI, M. GAITAN, S. SEMANCIK, "Tin oxide gas sensor fabricated using CMOS micro-hotplates and in-situ processing." *IEEE Electron Device Lett.* 1993, 14, 118–120.

190 M. Y. AFRIDI, J. S. SUEHLE, M. E. ZAGHLOUL, D. W. BERNING, A. R. HEFNER, R. E. CAVICCHI, S. SEMANCIK, C. B. MONTGOMERY, C. J. TAYLOR, "A monolithic CMOS microhotplate-based gas sensor system." *IEEE Sens. J.* 2002, 2, 644–655.

191 M. GAITAN, J. SUEHLE, J. R. KINARD, D. X. HUANG, "Multijunction thermal converters by commercial CMOS fabrication, In: *IEEE Instrumentation and Measurement Technology Conference (IMTC/93);* 1993, pp. 243–244.

192 V. MILANOVIC, E. BOWEN, M. E. ZAGHLOUL, N. H. TEA, J. S. SUEHLE, B. PAYNE,

M. Gaitan, "Micromachined convective accelerometers in standard integrated circuits technology." *Appl. Phys. Lett.* **2000**, *76*, 508–510.

193 G. K. Fedder, S. Santhanam, M. L. Reed, S. C. Eagle, D. F. Guilou, M. S.-C. Lu, L. R. Carley, "Laminated high-aspect-ratio microstructures in a conventional CMOS process." *Sens. Actuators A* **1996**, *57*, 103–110.

194 H. Xie, L. Erdmann, X. Zhu, K. J. Gabriel, G. K. Fedder, "Post-CMOS processing for high-aspect-ratio integrated silicon microstructures." *J. Microelectromechan. Syst.* **2002**, *11*, 93–101.

195 H. Luo, G. Zhang, L. R. Carley, G. K. Fedder, "A post-CMOS micromachined lateral accelerometer." *IEEE J. Microelectromechan. Syst.* **2002**, *11*, 188–195.

196 H. Xie, G. K. Fedder, "Fabrication, characterization, and analysis of a DRIE CMOS-MEMS gyroscope." *IEEE Sens. J.* **2003**, *3*, 622–631.

197 H. Luo, X. Zhu, H. Lakdawala, L. R. Carley, G. K. Fedder, "A copper CMOS-MEMS z-axis gyroscope." In: *Proc. IEEE Int. Conf. on Micro Electro Mechanical Systems (MEMS 2002)*; **2002**, pp. 631–634.

198 J. J. Neumann, K. J. Gabriel, "A fully-integrated CMOS-MEMS audio microphone." In: *Proc. Transducers '03*; **2003**, pp. 230–233.

199 J. J. Neumann, K. J. Gabriel, "CMOS-MEMS membrane for audio-frequency acoustic actuation." *Sens. Actuators A* **2002**, *95*, 175–182.

200 H. Lakdawala, G. K. Fedder, "CMOS micromachined infrared imager pixel." In: *Proc. Transducers 2001*; **2001**, pp. 556–559.

201 A. Oz, G. K. Fedder, "RF CMOS-MEMS capacitor having large tuning range." In: *Proc. Transducers '03*; **2003**, pp. 851–854.

202 Akustica Inc., Pittsburgh, PA, http://www.akustica.com/.

203 I. Voiculescu, M. Zaghloul, R. A. McGill, E. J. Houser, S. Stepnowski, E. Sokolovski, J. Stepnowski, J. Vignola, G. K. Fedder, "Resonant microcantilever gas sensor fabricated in CMOS technology for the detection of chemical agents." In: *Proc. Solid-State Sensor, Actuator and Microsystem Workshop, Hilton Head Island*; **2004**, pp. 57–58.

204 MEMSIC, North Andover, MA, http://www.memsic.com/.

205 M. Bugnacki, J. Pyle, P. Emerald, "A micromachined thermal accelerometer for motion, inclination and vibration measurement." *Sensors* **2001**, *18*, 98–104.

206 A. D. Oliver, K. D. Wise, "A 1024-element bulk-micromachined thermopile infrared imaging array." *Sens. Actuators A* **1999**, *73*, 222–231.

207 J. Chae, "High-Sensitivity, Low-Noise, Multi-Axis Capacitive Micro-Accelerometers," PhD Thesis, University of Michigan, Ann Arbor, MI; **2003**.

208 P. Krause, E. Obermeier, W. Wehl, "A micromachined single-chip inkjet printhead." *Sens. Actuators A* **1996**, *53*, 405–409.

209 H. Takao, H. Fukumoto, M. Ishida, "A CMOS integrated three-axis accelerometer fabricated with commercial submicrometer CMOS technology and bulk-micromachining." *IEEE Trans. Electron Devices* **2001**, *48*, 1961–1968.

210 M. Pedersen, W. Olthuis, P. Bergveld, "High-performance condenser microphone with fully integrated CMOS amplifier and DC-DC voltage converter." *IEEE J. Microelectromechan. Syst.* **1997**, *7*, 387–394.

211 J. W. Gardner, J. A. Covington, F. Udrea, T. Dogaru, C.-C. Lu, W. Milne, "SOI-based micro-hotplate microcalorimeter gas sensor with integrated BiCMOS transducer." In: *Proc. Transducers '01/Eurosensors XV*; **2001**, pp. 1688–1691.

212 F. Udrea, J. W. Gardner, "SOI CMOS gas sensors." In: *Proc. IEEE Sensors 2002*; **2002**, pp. 1379–1384.

213 K. Chun, H. Kim, "Monolithic integration of the digitized pressure sensor and the micromechanical switch for the application of a pressure transponder." *Microelectron. J.* **1998**, *29*, 621–626.

214 H. Seidel, U. Fritsch, R. Gottinger, J. Schalk, J. Walter, K. Ambaum, "A piezoresistive silicon accelerometer with monolithically integrated CMOS-circuitry." In: *Proc. Transducers '95*; **1995**, pp. 597–600.

215 P. J. FRENCH, "Polysilicon: a versatile material for microsystems." *Sens. Actuators A* **2002**, *99*, 3–12.

216 P. J. FRENCH, B. P. VAN DRIEENHUIZEN, D. POENAR, J. F. L. GOOSEN, R. MALLEE, P. M. SARRO, R. F. WOLFFENBUTTEL, "The development of a low-stress polysilicon process compatible with standard device processing." *IEEE J. Microelectromech. Syst.* **1996**, *5*, 187–196.

217 S. LEWIS, S. ALIE, T. BROSNIHAN, C. CORE, T. CORE, R. HOWE, J. GEEN, D. HOLLOCHER, M. JUDY, J. MEMISHIAN, K. NUNAN, R. PAYNE, S. SHERMAN, B. TSANG, B. WACHTMANN, "Integrated sensor and electronics processing for >10^8 iMEMS inertial measurement unit components." In: *Proc. IEEE Int. Electron Device Meeting (IEDM 2003)*; **2003**, pp. 39.1.1–39.1.4.

218 E. H. KLAASSEN, K. PETERSEN, J. M. NOWOROLSKI, J. LOGAN, N. I. MALUF, J. BROWN, C. STORMENT, W. McCULLEY, G. T. A. KOVACS, "Silicon fusion bonding and deep reactive ion etching: a new technology for microstructures." *Sens. Actuators A* **1996**, *52*, 132–139.

219 R. SUNIER, P. MONAJEMI, F. AYAZI, T. VANCURA, H. BALTES, O. BRAND, "Precise release and insulation technology for vertical Hall sensors and trench-defined MEMS." In: *Proc. IEEE Sensors Conference*; **2004**, to be published.

220 Analog Devices, Norwood, MA, http://www.analog.com/imems/.

221 Freescale Semiconductor (prior Motorola Semiconductor), Austin, TX, http://www.freescale.com/.

222 C. ANDRICACOS, C. UZOH, J. O. DUKOVIC, J. HORKANS, H. DELIGIANNI, "Damascene copper electroplating for chip interconnections." *IBM J. Res. Devel.* **1998**, *42*, 567–574.

223 T. J. KING, R. T. HOWE, S. SEDKY, G. LIU, B. C.-Y. LIN, M. WASILIK, C. DUENN, "Recent progress in modularly integrated MEMS technologies." *Proc. IEEE Int. Electron Decive Meeting (IEDM 2002)*; **2002**, pp. 199–202.

224 M. PUTTY, K. NAJAFI, "A micromachined vibrating ring gyroscope." In: *Proc. Solid-State Sensors and Actuators Workshop, Hilton Head*; **1994**, pp. 213–220.

225 M. PUTTY, A Micromachined Vibrating Ring Gyroscope, *PhD Thesis*, University of Michigan, Ann Arbor, MI; **1995**.

226 J. M. BORKY, K. D. WISE, "Integrated signal conditioning for silicon pressure sensors." *IEEE Trans. Electron Devices* **1979**, *26*, 1906–1910.

227 W. H. KO, J. HYNECEK, S. F. BOETTCHER, "Development of a miniature pressure transducer for biomedical applications." *IEEE Trans. Electron Devices* **1979**, *26*, 1896–1905.

228 C. S. SANDER, J. W. KNUTTI, J. D. MEINDL, "Monolithic capacitive pressure sensor with pulse-period output." *IEEE Trans. Electron Devices* **1980**, *27*, 927–930.

229 Bosch, Germany, http://www.bosch.com/.

230 Silicon Microstructures, Milpitas, CA, http://www.si-micro.com/.

231 M. GRAF, D. BARRETTINO, P. KAESER, J. CERDA, A. HIERLEMANN, H. BALTES, "Smart single-chip CMOS microhotplate array for metal-oxide-based gas sensors." In: *Proc. Transducers '03*; **2003**, pp. 123–126.

232 H. BALTES, O. PAUL, O. BRAND, "Micromachined thermally based CMOS microsensors." *Proc. IEEE* **1998**, *86*, 1660–1678.

233 AMMT – Advanced Micromachining Tools, Frankenthal, Germany, http://www.ammt.de/.

234 U. MUNCH, O. BRAND, O. PAUL, H. BALTES, M. BOSSEL, "Metal film protection of CMOS wafers against KOH." In: *Proc. IEEE Int. Conf. on Micro Electro Mechanical Systems*; **2000**, pp. 608–613.

235 M. PARAMESWARAN, H. P. BALTES, L. RISTIC, A. C. DHADED, A. M. ROBINSON, "A new approach for the fabrication of micromechanical structures." *Sens. Actuators* **1989**, *19*, 289–307.

236 D. MOSER, R. LENGGENHAGER, H. BALTES, "Silicon gas flow sensors using industrial CMOS and bipolar IC technology." *Sens. Actuators A* **1991**, *27*, 577–581.

237 O. PAUL, D. WESTBERG, M. HORNUNG, V. ZIEBART, H. BALTES, "Sacrificial aluminum etching for CMOS microstructures." In: *Proc. IEEE MEMS '97*; **1997**, pp. 523–528.

238 E. SCHURIG, M. DEMIERRE, C. SCHOTT, R. S. POPOVIC, "A vertical Hall device in CMOS high-voltage technology." *Sens. Actuators A* **2002**, *97–98*, 47–53.

2
Material Characterization

O. Paul and P. Ruther, IMTEK Institute of Microsystem Technology, University of Freiburg, Germany

Abstract

The characterization of materials for microelectromechanical systems (MEMS), in particular for those based on CMOS technology, is a challenging task. The relevant materials are mostly thin films that are difficult to handle experimentally using the classical macroscopic characterization techniques. Numerous new methods have therefore been developed to measure the physical properties of samples of dielectric, conducting, and semiconducting thin film materials in the combined electrical, thermal, and mechanical energy domains. Explicitly, the properties discussed in this chapter include the electrical resistivity, thermopower, thermal conductivity, heat capacity, elastic modulus, Poisson's ratio, thermal expansion coefficient, fracture strength, and piezoresistive coefficients. The chapter provides a review of various methods applicable to the extraction of these properties and on experimental data acquired using them. Many commercial CMOS processes have at least partially been analyzed with respect to material parameters relevant for MEMS applications. In this chapter, data from such processes are compiled in numerous tables and are complemented with values measured in academia and research institutes. Since other valuable sources of information are available concerning micromachining properties of MEMS materials, these are not considered here.

Keywords

Material property, material parameter, complementary metal oxide semiconductor, thin film, microelectromechanical system, microsystem, characterization method, measurement technique, process dependence, electrical resistivity, thermopower, Seebeck coefficient, thermal conductivity, heat capacity, specific heat, elastic modulus, Young's modulus, Poisson's ratio, thermal expansion coefficient, piezoresistive coefficient, fracture strength, fracture toughness.

Advanced Micro and Nanosystems. Vol. 2. CMOS – MEMS.
Edited by H. Baltes, O. Brand, G. K. Fedder, C. Hierold, J. Korvink, O. Tabata
Copyright © 2005 WILEY-VCH Verlag GmbH & Co. KGaA, Weinheim
ISBN: 3-527-31080-0

2 Material Characterization

2.1	**Introduction** *71*	
2.1.1	Transduction Effects in CMOS Materials *73*	
2.1.1.1	Electrical-Electrical *74*	
2.1.1.2	Electrical–Thermal and Thermal–Electrical *74*	
2.1.1.3	Thermal–Thermal *75*	
2.1.1.4	Mechanical–Mechanical *76*	
2.1.1.5	Thermal–Mechanical and Mechanical–Thermal *77*	
2.1.1.6	Mechanical–Electrical *77*	
2.1.2	Test Structures *78*	
2.2	**Electrical and Thermoelectric Properties** *80*	
2.2.1	Electrical Resistivity and Sheet Resistance *80*	
2.2.2	Electrical Material Data *82*	
2.2.3	Thermopower *82*	
2.2.4	Thermoelectric Material Data *86*	
2.3	**Thermal Properties** *86*	
2.3.1	Thermal Conductivity *86*	
2.3.1.1	Planar Test Structures *88*	
2.3.1.2	Surface Micromachined Test Structures *89*	
2.3.1.3	Membrane-based Test Structures *89*	
2.3.1.4	Cantilever-based Structures *90*	
2.3.1.5	Thermal van der Pauw Test Structure *92*	
2.3.2	Heat Capacity *95*	
2.4	**Mechanical Properties** *100*	
2.4.1	Bulge Test *100*	
2.4.1.1	Theory *102*	
2.4.1.2	Reduction to Dimensionless Form *103*	
2.4.1.3	Membranes *105*	
2.4.1.4	Determining Load-deflection Models *105*	
2.4.1.5	Square Diaphragms *106*	
2.4.1.6	Long Diaphragms *109*	
2.4.1.7	Thermal Expansion *114*	
2.4.1.8	Blister Test *114*	
2.4.2	Wafer Curvature *114*	
2.4.3	Passive Test Structures *115*	
2.4.4	Electrostatically Actuated Beams *117*	
2.4.5	Microtensile Test *119*	
2.4.6	Nanoindentation *121*	
2.4.7	Bridge Bending *122*	
2.4.8	Mechanical Material Data *123*	
2.5	**Conclusion** *130*	
2.6	**References** *130*	

2.1
Introduction

This chapter gives an overview of properties of materials used in integrated MEMS and of methods to measure them. Acquisition of such data started in the 1980s and has expanded significantly since then. A considerable number of general MEMS material property data are now spread across the archival literature and conference proceedings, and are available in more concentrated form on dedicated websites [1]. Nevertheless, determining such properties for IC technology-based MEMS remains a necessity and a challenge.

Unlike fundamental constants, material properties vary strongly from case to case. This tendency is inherent to the notion of *material* which approaches only in the rarest cases an ideal state with uniquely defined properties. High-purity monocrystalline silicon is such an example, where material properties attain ideal values with, e.g., carrier concentration defined by temperature only. Rather, the properties of real materials result from a wide range of imperfections from the atomic to the macroscopic level. These imperfections turn a single piece of ideal matter into a broad class of technically interesting materials able to solve a range of engineering problems. Imperfections include non-stoichiometric composition, impurities, distorted chemical bonds, absence of long-range order, precipitates, voids, grain boundaries, twinning and texture, among others. Material science and technology is the art of combining these aspects, by fundamental insight or experience, using appropriate processes.

The properties of materials fabricated using CMOS (or more generally IC) technology are of particular interest. With highest priority, CMOS materials have been designed for controlled and reliable electrical conduction and insulation and functionalities related to these. Only in recent years have mechanical and thermal considerations begun to receive a similar level of attention. Further, the range of traditional CMOS conductors such as diffusion, polysilicon and aluminum based metallizations has been extended by copper metallizations, tungsten plugs, silicides and diffusion barrier metallizations. Conventional intermetal insulators such as silicon dioxides produced using plasma-enhanced chemical vapor deposition (PECVD) or spin-on-glass (SOG) coating are being replaced by porous or organic low-k dielectrics. Ensuring the thermomechanical integrity of the CMOS back-end structures built by combining these novel materials represents a significant challenge for IC process engineers [2].

The perspective of MEMS engineers on CMOS materials is usually different. It is the possibility of benefitting from a broader range of effects beyond the electrical signal domain that constitutes the driving force behind many CMOS MEMS developments. Transduction mechanisms including mechanical, thermal, optical, electrochemical and other effects enable nonelectrical inputs to be converted into the final electrical signal extracted from the microstructure. Conversely, electrical inputs are transduced into non-electrical signals in CMOS actuators. In strong contrast to IC technology, electrical conduction/insulation of respective materials is often taken for granted.

As a consequence of these different perspectives, MEMS engineers wishing to take advantage of IC materials for their product are rarely able to build on a complete and reliable material data base. However, such data are a central ingredient for any attempt to

- predict the producibility of a device and its performance,
- optimize a device within the limits dictated by the IC technology,
- ascertain the statistical distribution of the device performance,
- ensure the long-term stability of the device performance and
- guarantee the reliability of the product.

The principal goal of this chapter is to provide an overview of the methods available for the acquisition of such data and to summarize the data already available in the literature. The focus is on materials fabricated using CMOS technology and by MEMS foundries and on phenomenological properties rather than fundamental physical theories behind them.

Materials for integrated MEMS with properties documented in the literature include those from the following industrial sources:

- Austriamicrosystems AG (AMS), Austria, represented by the CMOS technologies CYE (0.8 µm), CAE (1.2 µm) and CBT (2 µm).
- EM Microelectronic-Marin SA (EM), Switzerland, represented by the 2 µm low- and medium-voltage CMOS processes ALP2LV and ALP2MV, respectively.
- European Silicon Structures (ES2), now ATMEL, France, represented by the ECPD10 process.
- Infineon Technologies (Infineon), Germany, 6 in standard BiCMOS technology on (100) 675 µm thick p-type wafers.
- International Business Machines (IBM), USA.
- Lucent Technologies (Lucent), USA.
- Micronas GmbH (Micronas), Germany, 0.8 µm process.
- Robert-Bosch GmbH (Bosch), Germany, publicly accessible epipoly technology.
- Siemens, Germany.
- ST Microelectronics, Italy, epipoly technology THELMA.
- Taiwan Semiconductor Manufacturing Company (TSMC), Taiwan, 0.6 µm CMOS technology.
- Texas Instruments (TI), USA.
- X-Fab, Germany, represented by the XC10 process.

Published work quoted in this chapter relying on technologies of IBM, Lucent, Siemens and TI does not reveal the exact IC process designation. The abbreviations defined in this list are used throughout the chapter and in particular in Tables 2.1 to 2.14 listing material property data.

The next section recalls transduction effects usable in CMOS MEMS and the concept of diagnostic test structures to extract the corresponding material parameters. Sections 2.2, 2.3 and 2.4 describe methods to extract electrical and thermo-electric, thermal and mechanical properties of CMOS materials, respectively. Compilations of data available in the literature are presented in these respective locations. Finally, conclusions are drawn and directions for further work are sketched.

2.1.1
Transduction Effects in CMOS Materials

Tab. 2.1 summarizes the most important electrothermomechanical material properties and transduction effects used in CMOS MEMS. Each intersection of a column and a row lists material coefficients relating a cause (input signal) to a measurable effect (output signal). In order to set the stage for the following sections, the individual intersection fields including the respective coefficients contained therein are discussed within due brevity. Before this is done, however, let us note that all coefficients are temperature dependent. This is not explicitly mentioned in the thermal input line but has important consequences for transducer operation. It implies that in practice most CMOS microtransducers are to some extent also temperature sensors if no counter-measures are adopted. Measuring temperature coefficients of the various properties is thus almost as important as measuring the property values themselves. More generally, each effect is a potential source of undesired cross-sensitivities, further substantiating the need for material property characterization.

Tab. 2.1 Electrothermomechanical transduction effects available for CMOS technology-based MEMS: signals from three main signal domains accessible with standard CMOS materials are related to outputs in the same domains via material constants; the individual coefficients are commented in the text in further detail

		Output signal domain		
		Electrical	Thermal	Mechanical
Input signal domain	Electrical	Resistivity ρ Sheet resistance R_{sq} Dielectric constant ε_r	Peltier coefficient Π_P Thomson coefficient γ	Capacitive forces
	Thermal	Seebeck coefficient a_S	Heat capacities c and ρc Thermal conductivity κ Thermal diffusivity a	Thermal expansion coefficient a_{th}
	Mechanical	Piezoresistance coefficients (PC) π_{ijkl}, π_{11}, π_{12}, π_{44} Inversion layer PC Π_{11}, Π_{12}, Π_{44}	Thermoelastic friction	Residual stress σ_0 Residual strain ε_0 Compliance coefficients S_{ij} Elastic stiffness coefficients C_{ij} Young's modulus E Poisson's ratio ν Yield stress σ_y Fracture strength σ_f Fracture toughness K_C

2.1.1.1 Electrical–Electrical

Charge carrier transport in conductors is governed by the generalized Ohm's law $E = \rho j$, or equivalently $j = \sigma E$, where E, j, ρ and $\sigma = \rho^{-1}$ denote the electrical field vector, the current density vector, the second rank resistivity tensor and its inverse, the electrical conductivity tensor [3]. In unstressed silicon, ρ and σ are isotropic and usually represented by the scalar resistivity $\rho = [q(\mu_n n + \mu_p p)]^{-1}$ and conductivity $\sigma = \rho^{-1}$, with the elementary charge q, the electron and hole densities n and p, and the corresponding mobilities μ_n and μ_p. Mobilities and charge densities are functions of the temperature T and of the acceptor and donor doping densities, N_A and N_D, respectively [4, 5]. Since the temperature dependence of ρ is often exploited in CMOS MEMS or has to be compensated as an undesired effect, its linear temperature coefficient (TCR) β_R, defined as $\beta_R = \rho^{-1} \partial \rho / \partial T$ is an important quantity. In view of its definition, β_R enables small resistance changes ΔR to be translated into temperature changes ΔT via the linear approximation $\Delta R = R_0 \beta_R \Delta T$. Resistor materials available in CMOS IC technology are provided by various types of diffusions in the monocrystalline substrate material, variants of polysilicon layers with different doping concentrations and the metal layers.

In addition to its obvious use in the resistors of integrated circuits, ρ is also utilized in CMOS MEMS in the form of heater elements. In these structures, Joule heat is intentionally generated at a rate per volume of $g_{th} = \rho j^2$.

The dielectric constant ε_r describes the tendency of a material to be polarized by an external electrical field. In CMOS MEMS it takes effect in capacitive sensors in which part of the electrical field energy resides in the thin dielectric films insulating capacitor electrodes from each other. Commonly applied low-frequency values of ε_r are 11.7 [5] and 11.9 [4] for silicon, 3.9 for silicon dioxide [5] and 7 for LPCVD silicon nitride, Si_3N_4 [5].

2.1.1.2 Electrical–Thermal and Thermal–Electrical

Transport in the combined electrical and thermal energy domains in homogeneous materials with cubic symmetry or isotropic structure is governed by the relations [6]

$$E = \rho j - a_S \nabla T \tag{1}$$

$$j_h = -\Pi_P j - \kappa \nabla T \tag{2}$$

where a_S, ∇T, j_h, Π_P and κ denote the Seebeck coefficient, the temperature gradient, the heat flux density, the Peltier coefficient and the thermal conductivity, respectively. In other terms, the Seebeck coefficient, also termed thermoelectric coefficient, relates a temperature gradient to a resulting electrical field. Temperature differences are thus translated into electrical potential differences, a possibility exploited in numerous thermoelectric CMOS microsensors [6, 7]. Conversely, in an isothermal sample an electrical current entrains a heat current. The Peltier coefficient Π_P and the Seebeck coefficient a_S are related through the Kelvin relation $\Pi_P = T\alpha_S$, a consequence of the theory of irreversible thermodynamics due to Onsager [8]. Material pairs for the realization of thermoelectric devices and Peltier elements in CMOS technology are provided by diffusions against metal, polysilicon against metal, or pairs of semiconducting materials with different and preferably opposite doping.

In addition, the Thomson coefficient γ is related to the temperature dependent α_S by $\gamma = Td\alpha_S/dT$. This is the coefficient describing the small heat generation or absorption density g_{th} in conducting samples simultaneously subjected to a temperature gradient ∇T and an electrical current density j, i.e. $g_{th} = \gamma j \cdot \nabla T$.

2.1.1.3 Thermal–Thermal

Thermal material properties are important parameters for all thermal microsystems [6, 7]. In the field of CMOS MEMS these include infrared radiation detectors [9], gas flow sensors [10], pressure sensors [11], thermal converters [12, 13] and radiation sources [14, 15]. The two most important thermal properties are the thermal conductivity κ introduced in Eq. (2) and the heat capacity c.

The thermal conductivity of a material relates the temperature gradient ∇T in the material to the heat flow j_h it causes. In CMOS MEMS, the substrate material, the metallization layers and, to a lesser extent, the polysilicon layers are favored as high thermal conductivity materials useful for thermally short-circuiting parts of thermal structures, i.e. to homogenize temperature profiles or to prevent undesired temperature changes. In contrast, the dielectric layers, including various silicon oxides and nitrides, are the preferred thermal insulators in view of their low κ values.

Loosely speaking, the heat capacity is the energy required to change the temperature of a material sample, per unit temperature change and per unit of mass. In contrast to the thermodynamics of gases where the specific heats c_v and c_p are distinguished, for temperature changes at fixed volume and pressure, respectively, the corresponding values for solid materials lie closely within each other and are usually designated by the unique constant c. Often it is more convenient to use the specific heat referred to the volume, i.e. ρc, where ρ denotes the density of the material. This parameter affects the thermal response of microstructures through the heat transport equation

$$\rho c \frac{\partial}{\partial t} T + \nabla \cdot j_h = g_h \qquad (3)$$

with g_h being the heat generation/absorption rate per unit volume. Inserting $j_h = -\kappa \nabla T$ into Eq. (3), dividing by ρc and assuming a negligible temperature dependence of κ reveals the thermal diffusivity $a = \kappa/\rho c$ to be the central parameter for the time-dependent response of thermal microstructures.

For first rough estimates, the value 24.93 J K^{-1} from the Dulong–Petit law provides a useful approximation of the specific heat per amount of material containing a number of atoms equal to Avogadro's number $N_A = 6.022 \times 10^{23}$. This is a mole for elemental solids and an integer fraction of a mole for non-elemental materials.

2.1.1.4 Mechanical–Mechanical

Materials used in CMOS MEMS have amorphous structure or cubic crystal symmetry. In isotropic materials such as thermal oxide and CVD dielectric thin films, only two coefficients are required to describe the linearly elastic response to a mechanical load, i.e. the elastic modulus or Young's modulus E and Poisson's ratio v. Both parameters are conveniently defined using a situation with uniaxial load, namely by $\sigma_{ii}=E\varepsilon_{ii}$ and $\varepsilon_{jj}=-v\varepsilon_{ii}$, where i and j denote the load direction and any direction perpendicular to the load, and σ_{ii} and ε_{ii} (ε_{jj}) denote stress and strain tensor components, respectively. In other terms, E is the longitudinal strain per applied uniaxial stress, while v is the ratio of the transverse to longitudinal strains under uniaxial load. The resistance against shear loads is written as $\sigma_{ij}=G\varepsilon_{ij}$, with orthogonal directions i and j and the shear modulus $G=E/2(1+v)$ [5] or $G=E/(1+v)$ [16], depending on the mathematical definition of the shear strain and stress components. In materials with cubic symmetry such as silicon, the role of E, v and G is taken over by the elastic stiffness coefficients C_{11}, C_{12} and C_{44} or equivalently by the compliance coefficients S_{11}, S_{12} and S_{44}. It has to be kept in mind, however, that this identification is only applicable to coordinate systems aligned with the crystal axes of the cubic material. If another coordinate system is more suitable to model the particular case, the corresponding components of the tensors of elastic stiffness or compliance coefficients C_{ijkl} and S_{ijkl}, respectively, have to be computed using standard tensor transformation rules [3].

Also of great importance to mechanical structures built from thin films is the residual stress σ_0 in the component layers. This material parameter is related to the residual strain ε_0 via $\sigma_0=\varepsilon_0 E/(1+v)$ in the case of biaxial residual strain. Residual strain has two main causes: thermal mismatch between the thin-film and its substrate, leading to so-called extrinsic residual stress, and internal residual stress due to processing conditions, e.g., nucleation conditions and compactification under particle bombardment during film deposition.

The yield stress σ_y (e.g. $\sigma_{0.02}$) is defined as the stress value beyond which an irreversible plastic deformation strain (e.g. 0.02%) is left in the sample after unloading. Silicon and the CMOS dielectric layers show elastic behavior up to their brittle fracture [17], thus σ_y cannot be determined. In contrast, metal thin films typical of CMOS technology have shown the expected plastic behavior, e.g. in bulge and microtensile tests [18, 19]. The value of σ_y of metal alloys depends strongly on texture, grain dimensions and dislocation and defect densities. To the authors' knowledge, this value has not been determined systematically for CMOS metallizations.

The fracture stress and toughness parameters are essential to ascertain material strength and the reliability of mechanical structures. The strength σ_f is the tensile stress at which a material fails. The fracture toughness parameter K_C is related to the stress needed for crack propagation in the material [20, 21]. Depending on the loading mode, different toughness values are distinguished. The toughness K_C accessible to controlled measurements on the microscale is K_{IC} corresponding to the normal stress mode in which the two surfaces of a crack are perpendicularly pulled apart.

2.1.1.5 Thermal–Mechanical and Mechanical–Thermal

Thermal expansion of micromaterials is as unavoidable as temperature changes. Leading to undesired deformations of the sensor and thus unstable sensor output, thermal expansion may be a source of concern. On the other hand, differences in the thermal expansion of different materials are taken advantage of in bimorph and multimorph elements [22]. This effect is used in CMOS ultrasound emitters elegantly excited into resonance by thermomechanical actuation [23–25].

The thermal expansion coefficient a_{th} is defined as the relative length change of a material per temperature change. In terms of the temperature dependent material strain ε it is expressed as

$$a_{th} = \varepsilon^{-1} d\varepsilon/dT \tag{4}$$

The inverse effects, e.g. the heat generation due to mechanical strain, is relevant in mechanical devices operating at high frequencies. Irreversible heat generation due to thermoelastic friction has been shown to play a role in limiting the quality factor Q of microelectromechanical resonators [26, 27].

2.1.1.6 Mechanical–Electrical

Piezoresistivity is applied in numerous CMOS MEMS devices to translate loads, stress and the resulting strains into a modulated electrical signal. Whereas in metals the piezoresistance effect is mainly due to geometric effects, in semiconductors it is dominated by changes in the band structure [28, 29]. Piezoresistive gauge factors of silicon are roughly two orders of magnitude larger than those of typical CMOS metals [30]. This beneficial property has found wide applications in silicon-based sensors probing mechanical signals such as pressure [30], force [31–35] and acceleration [30]. For a detailed description of the piezoresistance effect we refer to the literature [5, 36] and Chapter 3.

On the phenomenological level, in monocrystalline silicon the piezoresistance effect is described using the three piezoresistive coefficients π_{11}, π_{12} and π_{44}. The first coefficient is the relative resistance change of a resistor collinear with an applied uniaxial stress, per unit of stress, the second is the analogous quantity for a resistor perpendicular to the load and π_{44} describes the occurrence of off-diagonal elements in the electrical resistivity tensor under shear loads [3, 5, 30]. All three definitions require the mechanical situation to refer to the cubic crystal axes. In situations where rotated coordinate systems are more appropriate for modeling a device, the analysis will involve the fourth-rank tensor of piezoresistive coefficients π, the coefficients π_{ijkl} of which are obtained from π_{11}, π_{12} and π_{44} using standard tensor transformation rules [3]. The piezoresistive coefficients of semiconductors depend on temperature (roughly as T^{-1}), and on the doping level. Higher doping reduces both the absolute value of the piezoresistive coefficients and their temperature dependence [5, 36].

In addition to diffused resistors, polysilicon resistors [25] and inversion layers [33, 34] have been used as stress sensors in CMOS MEMS. The piezoresistive coefficients of the carriers in inversion layers differ from those in diffused resistors and have been denoted Π_{11}, Π_{12} and Π_{44} [33, 34, 37].

2.1.2
Test Structures

Characterizing transduction effects in CMOS materials is a challenging task. A particular difficulty resides in the small dimensions of the material samples to be characterized, with thicknesses of a few hundred nanometers to a few micrometers in the direction perpendicular to material plane, and in-plane lengths of a few micrometers to a few hundred micrometers. Such small dimensions make sample handling delicate. In addition, sufficient coupling to a measurement setup is hard to guarantee. As an example, for the determination of mechanical properties based on the traditional tensile test, the sample structure has to be grabbed in a controlled way to load it with an adequate force. Even slight misalignments alter the stress distribution. This in turn influences the mechanical response and hence the material property values extracted from the experiments. Similarly, in thermal measurements, the thermal coupling to heat sources and sinks and to temperature sensors is known to be critical. Insufficient thermal coupling results in temperature drops between the setup and the material under test. The corresponding thermal resistances translate into an uncertainty in the measured thermal properties.

In addition, with the small sample sizes, surface effects, errors due to inaccurate geometries, and edge effects play an important role. Heat transfer along thin-films in thermal conductivity measurements is strongly influenced by radiative heat loss. As an example, a 1 µm thick PECVD silicon nitride film with an assumed thermal conductivity of 1.5 W m^{-1} K^{-1} is considered. It is clamped between a heat source and a heat sink at 300 K parallel to each other. Assume the distance d spanned by the film to be $d=1$ mm, the average spectral emissivity of the material in the wavelength range between 2 and 15 µm to be $\varepsilon=0.3$ and the experiment to be performed in vacuum. Under these conservative conditions, more than 50% of the heat flux injected into the sample at the heat source are lost to the ambient by thermal radiation. With $d=300$ µm, this fraction drops to roughly 8%, and with $d=100$ µm to a mere 1%. Extraction of the thermal conductivity undisturbed by thermal radiation thus requires smaller samples. If d can be defined with an accuracy of 1%, the corresponding accuracy in the extracted thermal conductivity will at best be 1% as well. Needless to say that clamping a thin film sample with the required micrometer precision between heat source and sink is a challenge.

The method of choice to avoid these difficulties is based on process-compatible diagnostic structures, that is, process monitors. Rather than extracting a piece of material from the CMOS MEMS wafer and introducing it into a dedicated test setup, a diagnostic device comprising all the necessary functions for successful property extraction is constructed and fabricated. In a sense, the result is a microsensor optimized for sensing one of its own properties. This test vehicle, robustly supported on a chip or wafer, is more easily transferred into an experimental setup than an isolated material piece. In turn, the setup has to maintain controlled pressure and background temperature only.

The partial transfer of functionality from the setup to the test monitor results in an increased complexity in design and processing. Test structures for thermal properties require the integration of components for heat dissipation and temperature sensing. Mechanical test structures may benefit from independent on-chip stress sensors. In the case of CMOS MEMS material characterization, however, this may be achieved with minimal additional effort. Full advantage is taken of the standard processing and structuring capabilities of the CMOS MEMS process to be characterized. Hence the goal of characterizing materials as-processed can be reached.

The use of process monitors is firmly established in IC technology, e.g. for photolithography control, line width and sheet and contact resistance measurements, and transistor parameter evaluation [38, 39]. These monitors of geometric and electrical properties are inspected optically and operated electrically, respectively. The main difference of MEMS diagnostic structures with these standard monitors lies in the more complex functionality of the test structures and the know-how required for reliable property extraction.

A catalog of requirements for test structures of the described kind may contain the following criteria, among others:

- *Simple geometry*. This enables simple models to be applied for the translation of the raw measured data into a material property. Test structures with complex geometries are nevertheless possible and a certain level of geometric complexity is often unavoidable owing to process design rules. Finite-element simulations may in such cases provide the link between measured data and material properties. However, numerical uncertainties have to be carefully assessed.
- *Robustness*. During fabrication, in particular during micromachining, test structures are exposed to high mechanical loads, due to internal mechanical stress and fabrication steps such as rinsing and drying. These may cause microcracking and fracture of the structure. Further, the integrity of the structures should be guaranteed even under process variation, e.g. when the same structure is used to characterize different processes.
- *Minimal micromechanical structuring*. Even though wet chemical etching may negligibly contribute to the thermal budget of a material, such processing often represents an undesired complication in the overall process, especially if the micromachining is not a standard step of the MEMS fabrication flow under consideration. When micromachining is unavoidable and has to be performed as a post-CMOS process, the materials characterization is delayed to the very end of the process chain. No in-process monitoring is then possible.
- *Simple measurement method*. Ideally, the method should enable the desired material parameter to be extracted directly from easily accessible raw signals. Electrical signals are suitable candidates, since process engineers may be more familiar with electrical and optical testing than with the possibly more complex physics involved in the inner workings of MEMS test structures. Current, voltage, power, resistance, frequency and phase are excellent candidates as measurement signals.

2.2
Electrical and Thermoelectric Properties

2.2.1
Electrical Resistivity and Sheet Resistance

Electrical resistivity measurements of CMOS thin-films are either performed at the wafer level prior to structuring of the thin films or with structured samples. Data are usually documented by the foundries in the form of specification lists tabulating expectation values and variances for the sheet resistance R_{sq} and the temperature coefficient β_R of the conducting layers including diffusions, gate and capacitor layers and metallizations. These data are usually made accessible after signature of a non-disclosure agreement, if they are not publicly available anyway [40].

The most common wafer-level technique to measure the resistivity ρ is the four-point probe technique illustrated in Fig. 2.1a [38]. The sample surface is contacted with four collinear probes with spacing s in between. A small constant current I_{14} is injected and extracted through the outer two contacts 1 and 4, respectively. A voltage V_{23} is measured between the inner two probes 2 and 3 with a high-impedance voltmeter. Thus, a negligible current is drawn from the voltage probes. Consequently, parasitic resistances such as (i) contact resistance R_c, (ii) probe resistance R_p and (iii) spreading resistance R_{sp} associated with the current-carrying probes contribute negligibly to the resistivity measurement. For thin-film samples of finite size, ρ is given by

$$\rho = 2\pi s \left(\frac{V_{23}}{I_{14}}\right) CF \tag{5}$$

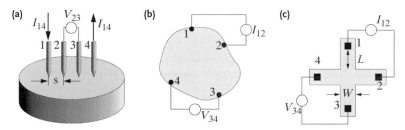

Fig. 2.1 (a) Collinear four-point probe with equidistant probe spacing s. (b) Arbitrarily shaped sample with four contacts at its circumference (van der Pauw method). (c) Cross bridge test structure for sheet resistance measurement.

where CF denotes a well documented correction factor which takes into account sample size, thickness and shape as well as the orientation of the probes with respect to the sample borders [38, 39].

For samples of arbitrary shape as shown in Fig. 2.1b, van der Pauw developed a method based on conformal mapping to extract the resistivity of thin films [41]. The resistivity is obtained by solving

$$\exp\left(-\frac{\pi t R_{12,34}}{\rho}\right) + \exp\left(-\frac{\pi t R_{23,41}}{\rho}\right) = 1 \qquad (6)$$

where t denotes the sample thickness. The resistances $R_{ij,kl}$ in Eq. (6) are defined as $R_{ij,kl} = V_{kl}/I_{ij}$, where I_{ij} and $V_{kl} = V_k - V_l$ denote the current applied between contacts i and j and the voltage drop measured between contacts k and l, respectively. Van der Pauw's theory applies to samples of (i) uniform thickness, with (ii) sufficiently small contacts at their periphery and (iii) simple connectivity, i.e containing no holes [41]. In practice, van der Pauw's method is applied to symmetrically shaped structures such as the Greek-cross device shown in Fig. 2.1c. In this specific case, with $R_{ij,kl} = R_{jk,li}$, the resistivity ρ is obtained from

$$\rho = \frac{\pi t R_{12,34}}{\ln 2} \qquad (7)$$

It was shown that contact size effects can be neglected for $L > 1.02\,W$, where L and W denote the length and width of the Greek-cross arms as shown in Fig. 2.1c [42].

With thin layers, sheet resistance R_{sq} rather than bulk resistivity ρ is measured, particularly if the resistivity of the sample varies throughout the thickness of the film. The sheet resistance is defined as the resistance between two opposite edges of a square-shaped sample of any sidelength L. For layers with homogeneous resistivity ρ the sheet resistance is $R_{sq} = \rho/t$. In view of Eq. (7), R_{sq} is equal to $\pi R_{12,34}/\ln 2$. In the more general case where ρ varies across the layer, such as in diffusions or sandwiches of conducting layers, R_{sq} reads

$$R_{sq}^{-1} = \int \rho(z)^{-1} dz \qquad (8)$$

where the integral is performed across the thickness of the layer.

It is recommended to combine measurements with opposite currents $\pm I_{12}$. This serves the purpose of cancelling possible voltmeter offsets and parasitic thermoelectric voltages. Then, the sheet resistance is obtained as

$$R_{sq} = \frac{\pi}{\ln 2} \frac{V_{34}(I_{12}) - V_{34}(-I_{12})}{2I_{12}} \qquad (9)$$

2.2.2
Electrical Material Data

The sheet resistance R_{sq} and resistivity ρ of conductive CMOS thin films are routinely monitored by the IC fabs and foundries using the four-point or van der Pauw methods. In addition to the data selectively disclosed by the CMOS foundry to their customers, sheet resistances and resistivities of CMOS conducting materials have been measured independently by users. As an example, Fig. 2.2 shows the resistance R and its temperature coefficient β_R as a function of temperature for three samples of the metal 1 and two samples of the polysilicon layer of a 1.2 μm process of Austriamicrosystems (AMS) [43]. Tab. 2.2 lists values for R_{sq} and ρ for CMOS thin-film materials obtained from various processes as well as bulk materials. Despite their fabrication with different processes, the experimental results reveal correlations between the electronic properties. Both the electrical conductivity $\sigma = \rho^{-1}$ and the temperature coefficient of resistance (TCR) increase with dopant concentration.

2.2.3
Thermopower

Methods for measuring the thermoelectric coefficient a_S of CMOS materials have been directly derived from Eq. (1) and the possibility of fabricating thermocouples. Instead of using a thermocouple to measure a temperature difference, a well-defined temperature difference is applied in order to extract the Seebeck coefficient from the thermovoltage.

Fig. 2.3 illustrates the principle on the basis of a planar test structure [43] to extract the Seebeck coefficient of polysilicon/CMOS metal thermocouples. A polysilicon sample (material 1) is connected at both ends to the lower CMOS metal layer (material 2). A resistive heater close to one contact enables the temperature T_{hot}

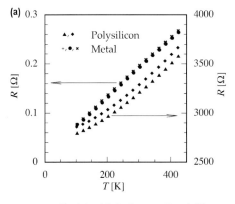
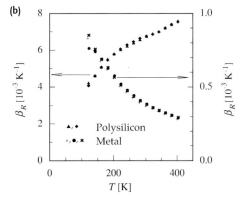

Fig. 2.2 (a) Resistance R and (b) temperature coefficient β_R as a function of temperature for three samples of the metal 1 layer and two samples of polysilicon layer of a 1.2 μm process of AMS (reproduced by permission of IEEE [43])

Tab. 2.2 Values of resistivity ρ, sheet resistance R_{sq} and the respective TCR β_R for conducting thin films of various CMOS processes

Material	ρ (10^{-3} Ω cm)	R_{sq} (Ω)	TCR β_R (10^{-3} K^{-1})	Comments[a]
Polysilicon				
n-poly	0.76±0.01		0.84±0.01	AMS, gate poly [43]
	0.85		0.86	AMS, poly1, n$^+$ [45]
	1.03		0.89	EM, poly1, n$^+$ [45]
	96		−4.4	EM, poly2, n [45]
	1.22		0.83	EM, poly3, n$^+$ [45]
	1.58		0.54	ES2, poly1, n$^+$ [45]
	2.25		0.49	ES2, poly2, n$^+$ [45]
	0.813±0.001			Infineon [46]
	2.13		0.431	[47]
	0.945		0.685	[47]
		27	0.75	n$^+$, poly1 resistor [40]
p-poly	5.8		−0.14	AMS, poly2, p$^+$ [45]
	16.2		−0.59	EM, poly4, p$^+$ [45]
	2.214±0.004			Infineon [46]
		20×10^3	−5.0	High-resistance poly2 [40]
		355	−0.2	p$^+$, poly2 resistor [40]
Metal				
Metal 1		(65±0.7)×10^{-3}	2.96±0.1	AMS [45]
		(44±0.2)×10^{-3}	4.38±0.09	EM [45]
		(59±0.5)×10^{-3}	3.29±0.1	ES2 [45]
Metal 2	3.6×10^{-3}		3.02±0.08	AMS [43]
		(36±0.4)×10^{-3}	3.01±0.1	AMS [45]
		(29±0.2)×10^{-3}	4.28±0.09	EM [45]
		(30±0.3)×10^{-3}	3.23±0.1	ES2 [45]
Bulk materials				
Al	2.733×10^{-3}			[48]
Cu	1.725×10^{-3}			[48]
Ni	7.2×10^{-3}			[48]
Pt	10.8×10^{-3}			[48]
W	5.44×10^{-3}			[48]
Ti	39×10^{-3}			273 K [48]

a) For abbreviations of process and company names, see Section 2.1.

of this contact to be increased by ΔT above the temperature T_{cold} of the second contact. According to Eq. (1), under open-circuit conditions, i.e. with $j=0$, the temperature gradients ∇T_i in the two materials cause electrical fields E_i given by

$$E_i(x) = -a_{S,i}(T_i(x))\nabla T_i(x), \quad i = 1, 2 \tag{10}$$

The thermoelectric voltage V_{te} measured at the ends of the two material 1 lines is therefore

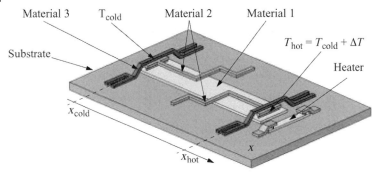

Fig. 2.3 Operating principle of planar CMOS test structure to determine the Seebeck coefficient between polysilicon (material 1) and the lower metal (metal 1 = material 2). Temperature difference ΔT between hot and cold contacts is obtained by power dissipation in a resistive heater. Thermovoltage V_{te} is measured at the ends of the material 2 lines. Temperatures are measured resistively using the upper metal (metal 2 = material 3). All thin-film components are sandwiched between dielectric CMOS layers, removed here for clarity. Adapted from [43]

$$V_{te} = -\int_{x_{cold}}^{x_{hot}} [E_1(x) - E_2(x)]dx = \int_{T_{cold}}^{T_{hot}} a_{S,12}(T)dT \qquad (11)$$

where x_{cold}, x_{hot}, T_{cold}, T_{hot} and $a_{S,12}$ denote the locations and temperatures of the hot and cold contacts, respectively, and the relative Seebeck coefficient, $a_{S,12} = a_{S,1} - a_{S,2}$, between the two materials. For sufficiently small ΔT, $a_{S,12}$ is well approximated by

$$a_{S,12}(T_{cold} + \Delta T/2) = \frac{V_{te}}{\Delta T} \qquad (12)$$

Hence the measurement of V_{te}, T_{cold} and $\Delta T = T_{hot} - T_{cold}$ enables the Seebeck coefficient at $T = T_{cold} + \Delta T/2$ to be determined. Whereas the measurement of V_{te} is straightforward, the accurate determination of T_{cold} and T_{hot} requires appropriate and careful design of the test structure. This was achieved in the example of Fig. 2.3 by placing metal 2 (material 3) resistors right above the polysilicon/metal 1 contacts. They are integrated in a four-point measurement configuration enabling the accurate determination of local temperature changes.

Fig. 2.4 shows two diagnostic structures to measure $a_S(T)$ of CMOS IC polysilicon thin films against the lower CMOS metallization [43]. The planar test structure in Fig. 2.4a is that schematically illustrated in Fig. 2.3. The structure was realized using standard CMOS technology. In contrast, the cantilever based diagnostic structure in Fig. 2.4b required additional post-CMOS bulk micromachining using an ethylenediamine–pyrocatechol–pyrazine (EDP) solution at 95 °C. It consists of a 100 μm wide and 260 μm long cantilever with four connection lines suspended over a cavity. The suspended structure is made of all CMOS dielectrics and con-

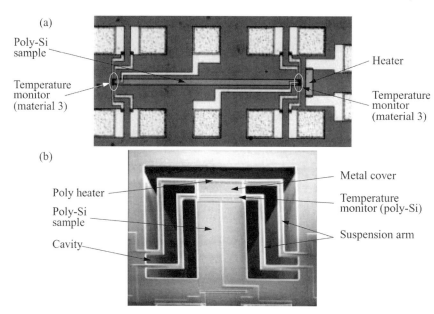

Fig. 2.4 (a) Optical micrograph of a planar test structure and (b) SEM image of a micromachined test structure based on a suspended cantilever, to determine the Seebeck coefficient of CMOS polysilicon. Both structures contain a contacted polysilicon sample, a heater and temperature monitors (reproduced by permission of IEEE [44])

tains the polysilicon sample to be characterized. Two resistors made of gate polysilicon are integrated into the tip of the cantilever. While the resistor at the edge of the beam is used as heater, the second serves as temperature monitor. The temperature distribution at the end of the beam is homogenized with an integrated rectangular cover made of both metal layers. This is merged with the hot contact of the polysilicon sample. Thus, the temperature difference between hot contact and temperature monitor is minimized. The curved geometry of the connection lines reduces heat loss and ensures effective stress relaxation. The cold contact is located above the thermally highly conductive silicon substrate, 40 μm away from the etch pit. Its temperature increase during operation is three orders of magnitude smaller than ΔT of the hot contact [43].

The Seebeck effect in diffusions in the substrate material can also be characterized by both types of structures. The first thermoelectric measurements on silicon were made on externally heated macroscopic substrate pieces [49]. The relevant temperature difference was determined by applying discrete external temperature sensors pressed against the sample. Similar experiments have been performed with polysilicon layers on bulk silicon substrates [47, 50]. In addition, the Seebeck coefficient of monocrystalline silicon covering a broad range of resistivities was measured against aluminum using cantilevers with integrated thermoelements based on metal-contacted diffused resistors [51]. The results showed a logarithmic dependence on resistivity, as expected from theoretical considerations [6].

2.2.4
Thermoelectric Material Data

As an example, Fig. 2.5 shows the temperature dependent Seebeck coefficient α_S of n-doped gate polysilicon against the lower metal, obtained with the two test structures shown in Fig. 2.4. The α_S of both layers is negative, in agreement with published data for n-doped silicon [49, 50, 52]. The agreement between the thermoelectric coefficients obtained with both methods is better than 2.1 µV at 300 K [43]. The linear dependence of α_S on temperature is well accounted for by the theory of the thermopower of degenerately doped semiconductors [53].

Tab. 2.3 lists measured values of Seebeck coefficient for polysilicon deposited under various process conditions and for monocrystalline silicon. In general, the Seebeck coefficient increases with increasing resistivity or, equivalently, decreasing doping concentration. Further, $|\alpha_S|$ of n^+-silicon is smaller than that of comparably doped p^+-material.

2.3
Thermal Properties

2.3.1
Thermal Conductivity

Like the electrical conductivity, the thermal conductivity of materials can be anisotropic and is correctly described by the second rank thermal conductivity tensor, κ. In view of the planarity of CMOS and integrated MEMS thin films and their amorphous structure or perpendicular texture with in-plane isotropy, κ is conveniently reduced to an in-plane thermal conductivity κ_l and a transverse thermal conductivity κ_t perpendicular to the film surface. Consequently, methods for in-plane and out-of-plane thermal conductivity determination are distinguished. Four cate-

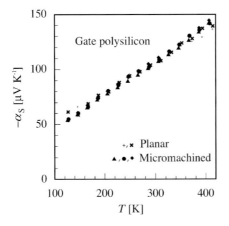

Fig. 2.5 Temperature-dependent Seebeck coefficient of n-doped CMOS gate polysilicon against lower metal layer (1.2 µm process of AMS) measured using planar and micromachined test structures. Adapted from [43]

Tab. 2.3 Values of Seebeck coefficient a_S, temperature coefficient $a_S^{-1} da_S/dT$ and resistivity of polysilicon thin films produced with various CMOS processes, and of bulk silicon and metals

Material (method)	a_S (μV K^{-1})	$a_S^{-1} da_S/dT$ (10^{-6} K^{-1})	ρ ($\mu\Omega$m)	Comments[a]
n-Polysilicon				
Micromachined test structure	−107.2 ± 1.5		7.6 ± 0.1	Gate poly, AMS, 1.2 μm [43]
	−95.7 ± 0.6			Capacitor poly, AMS, 1.2 μm [43]
	−120		8.5	Poly1, n$^+$, AMS [45]
	−108		10.3	Poly1, n$^+$, EM [45]
	−520		960	Poly2, EM [45]
	−111		12.2	Poly3, n$^+$, EM [45]
	−108		15.8	Poly1, n$^+$, ES2 [45]
	−128		22.5	Poly2, n$^+$, ES2 [45]
	−87.8 ± 1.2	3124 ± 28	6.51 ± 0.36	Gate poly, ALP2LV [54]
	−111.3 ± 1.5	2792 ± 35	6.61 ± 0.17	Gate poly, CAE, gate poly [54]
	−115.1 ± 0.5	2897 ± 33	11.54 ± 1.15	Gate poly, ECPD10, gate poly [54]
	−366.9	1650	604	High-resistance poly, ALP2LV [54]
Planar test structure	−105.2 ± 3		7.6 ± 0.1	Gate poly, AMS, 1.2 μm [43]
	−93.6 ± 1.3			Capacitor poly, AMS, 1.2 μm [43]
	−57 ± 9		0.813 ± 0.001	Infineon [46]
	−204 ± 18		64 ± 8	n-poly2, Infineon [55]
Thin film on substrate	−132		21.3	[47]
	−75		9.45	[47]
	−425			P-doped, $N_D = 2.0 \times 10^{19}$ cm^{-3} [50]
	−525			P-doped, $N_D = 1.0 \times 10^{19}$ cm^{-3} [50]
p-Polysilicon				
Micromachined test structure	190		58	Poly2, p$^+$, AMS [45]
	330		162	Poly4, p$^+$, EM [45]
Planar test structure	103 ± 17		2.214 ± 0.003	Infineon [46]
	183 ± 8		42 ± 5	p-poly2, Infineon [55]
Thin film on bulk substrate	315			$N_A = 2.0 \times 10^{19}$ cm^{-3} [50]
	380			$N_A = 1.0 \times 10^{19}$ cm^{-3} [50]
n-Silicon	−733			Arsenic-doped, $N_D = 2.2 \times 10^{18}$ cm^{-3} [49]
	−306			Arsenic-doped, $N_D = 2.7 \times 10^{19}$ cm^{-3} [49]
	−200			[51]
p-Silicon	1000			B-doped, $N_A = 1.0 \times 10^{18}$ cm^{-3} [49]
	500			B-doped, $N_A = 1.5 \times 10^{19}$ cm^{-3} [49]
	1090			$N_A = 1.7 \times 10^{17}$ cm^{-3} [51]
	860			$N_A = 1.8 \times 10^{18}$ cm^{-3} [51]
	550			$N_A = 1.7 \times 10^{19}$ cm^{-3} [51]
Al	−1.7			[6]
Cu	1.85			[6]

a) For abbreviations of process and company names, see Section 2.1.

gories of test structures are available to measure κ values, differing by fabrication technology and device geometry. These are (i) planar test structures based on thin-film resistors on thin films on substrate, (ii) test structures fabricated using surface micromachining and (iii) membrane- and (iv) cantilever-based test structures requiring bulk micromachining.

2.3.1.1 Planar Test Structures

Planar test structures with resistors integrated on a substrate have been used to measure the thermal conductivity of the bulk substrate material [56], the buried oxide layer of silicon-on-insulator (SOI) wafers [57] and thin films [58–61]. Fig. 2.6 shows a schematic cross-section and the operating principle of such structures. Three long, parallel metal lines are patterned on top of the thin film to be characterized. The inner line serves as a heater and simultaneously enables the temperature T_{in} at this location to be determined. Dissipating a heating power P_h causes a temperature increase of thin film and substrate. In contrast, the left and right resistors are biased with a small probe current causing negligible self-heating. Hence they monitor the respective temperatures T_l and T_r of the substrate underneath exploiting the temperature-dependent resistance of the metal. From T_l and T_r the substrate temperature $T_{in,S}$ below the inner resistor is obtained from a model of the temperature distribution. Assuming one-dimensional thermal conduction in the thin insulating layer, the thermal resistance R_T and consequently the transverse thermal conductivity κ_t of the thin layer is given by [58]

$$R_T = \frac{d}{\kappa_t} = Lw \frac{T_{in} - T_{in,S}}{P_h} \qquad (13)$$

where d, L and w denote the layer thickness and the length and width of the inner resistor, respectively. Heater and temperature monitors are preferably connected in four-point configurations for accurate temperature measurements. A similar direct measurement of κ_t was demonstrated, with the thin film under investigation vertically sandwiched between two resistors [60].

While measurements are conveniently carried out in the direct current (DC) mode with constant heating power [58–60], the transient response to sinusoidal heating signals has also been analyzed using the 3ω method [56, 61]. This meth-

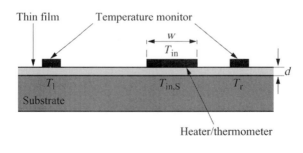

Fig. 2.6 Cross-section of a planar test structure to determine the transverse thermal conductivity of thin films on a bulk substrate. Adapted from [58]

od is described in further detail in Section 2.3.2, as it also enables one to measure the thermal diffusivity. Although planar test structures are easily implemented as test vehicles for process monitoring, their application range is limited to layers with low thermal conductivity such as silicon oxides and nitrides.

2.3.1.2 Surface Micromachined Test Structures

Surface micromachined test structures have been used to determine the thermal conductivity of polysilicon [46, 62] and LPCVD silicon nitride [63] thin films. They use simple microbridge structures [62, 63] or complex arrangements of bridges suspended over the substrate [46]. A structure of the latter type (shown schematically in Fig. 2.7) was designed to characterize CMOS polysilicon layers and was realized using sacrificial layer technology [46]. When an electric current is applied to the central heater, the generated heat dominantly flows away through the beams to the rim of the structure. An advantage of the structure is that the polysilicon layer does not need to be sandwiched between other materials such as silicon oxides.

2.3.1.3 Membrane-based Test Structures

Thermal test structures based on thin-film membranes have been used to determine the thermal conductivity of polysilicon [64] and silicon oxide/nitride layer sandwiches [65]. They apply resistors acting as heaters and temperature monitors integrated in the membrane material to be analyzed. Typically, long rectangular membranes are utilized with the resistors parallel to the long side of the membrane edges. Thus, lateral thermal conduction through the membrane layer is the dominant heat transport mode. With two integrated resistors, as shown schematically in Fig. 2.8, the thermal in-plane conductivity κ_l is approximated by [64]

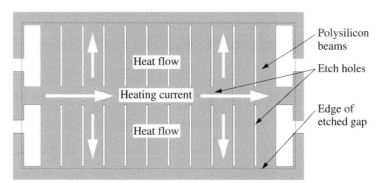

Fig. 2.7 Schematic top view of a surface micromachined test structure to determine the in-plane thermal conductivity κ_l of polysilicon layers. The structure consists of parallel polysilicon ribs connected by a central polysilicon heater spine. The surface micromachined gap underneath the structure is indicated by the dashed line. Adapted from [46]

90 | 2 Material Characterization

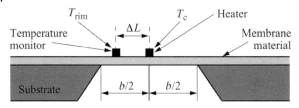

Fig. 2.8 Schematic cross-section of a membrane-based test structure to determine the in-plane thermal conductivity κ_l of thin films. Adapted from [64]

$$\kappa_l = \frac{P_h}{2A_c} \frac{\Delta L}{(T_c - T_{rim})} \tag{14}$$

where P_h, A_c, ΔL, T_c and T_{rim} denote the heat dissipated in the central resistors, the cross-sectional area for heat conduction, the distance between the two resistors, and the temperatures of the central resistor and temperature monitor, respectively.

2.3.1.4 Cantilever-based Structures

Bulk micromachined cantilever-based test structures for the measurement of thin-film thermal conductivities have been extensively used for the measurement of in-plane thermal conductivities of CMOS thin films [45, 54, 66–71]. An example of such a structure is shown in Fig. 2.9a. It consists of a 200 µm wide and 150 µm long cantilever suspended over a micromachined cavity. The cantilever is composed of CMOS thin films. Four thin arms extend from the rim of the cavity to the tip of the cantilever. The meandering geometry of the arms serves for the successful wet micromachining of the structure and stress relaxation. Two gate polysilicon resistors are integrated into the free end of the cantilever and serve as heater and temperature monitor, respectively. To homogenize the temperature distribution over the two resistors, a metal cover is integrated into the cantilever tip as shown schematically in Fig. 2.9b. Such structures have been fabricated using several standard CMOS IC processes followed by post-CMOS anisotropic silicon etching in EDP [71].

The thermal conductivity of the cantilever layer sandwich is determined by measuring its static thermal conductance. When a power P_h is dissipated in the heater, the temperature of the cantilever end is increased by ΔT with respect to the substrate temperature T_0. Under appropriate experimental conditions and with suitable test structure dimensions, heat transfer occurs dominantly to the silicon support by conduction along the cantilever and the four suspension arms. Thus, the applied power P_h and resulting temperature difference ΔT are related by $P_h/\Delta T = G_{total} = G_c + G_a$, where G_c and G_a denote the thermal conductance of the cantilever and the four arms, respectively. The conductance of the cantilever is given by

(a)

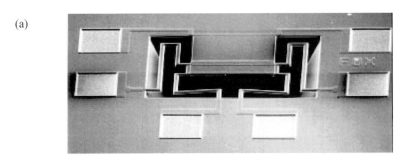

(b)

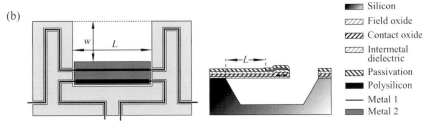

Fig. 2.9 (a) SEM image of a micromachined test structure to determine the thermal conductivity of CMOS IC thin films [68]. The cantilever width is 200 µm. (b) Schematic top view and cross-section of the thermal microstructure (reproduced by permission of American Insitute of Physicis [68])

$$G_c = \sum \frac{\kappa_i d_i w}{L} \qquad (15)$$

where κ_i, d_i, w and L denote the thermal conductivity, thickness, width, and length of the component layers (indexed with i) of the cantilever.

Measurements are performed in vacuum to minimize conductive and convective heat losses through the surrounding gas. The radiative heat transfer contribution P_{rad} is minimized by making the test structures sufficiently small. For small temperature increases ΔT above ambient temperature T_0, P_{rad} is estimated using a linear approximation of the Stefan–Boltzmann law, i.e.

$$P_{\text{rad}} = 4\varepsilon\sigma_{\text{SB}} T_0^3 \Delta T w L \qquad (16)$$

where ε and σ_{SB} denote the spectral emissivity of the structure at thermal radiation wavelengths corresponding to $T_0+\Delta T$, and the Stefan–Boltzmann constant $\sigma_{\text{SB}} = 5.67 \times 10^{-8}$ W m^{-2} K^{-4}. Thus P_{rad} roughly scales as wL whereas the conductive heat transfer across the test structure components scales as w/L. As a consequence, reducing L suppresses P_{rad} to a negligible level compared with the heat flow along the cantilever and the connection arms.

The thermal conductance G_a of the side arms is separately determined using a reference structure consisting only of the heating and temperature monitoring section of the cantilever, without the main cantilever body. Subtracting the ratios $P_h/\Delta T$

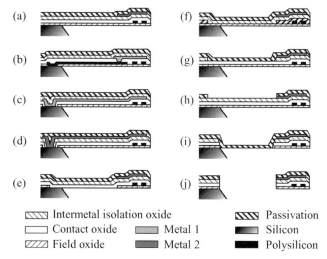

Fig. 2.10 Schematic cross-sections of cantilevers to determine the thermal conductivities of individual CMOS thin films (reproduced by permission of American Institute of Physics [68])

for test structures with and without cantilever body thus provides the thickness-weighted thermal conductivities of the cantilever sandwich, i.e. $\Sigma \kappa_i d_i$. Division by the total thickness gives the average thermal conductivity of the cantilever sandwich.

To determine the thermal conductivities κ_i of individual sandwich layers, the composition of the layer sandwich of the cantilever body is systematically varied as shown in Fig. 2.10, leading to different measured sums $\Sigma \kappa_i d_i$. By subtraction, the thermal conductivities of the individual layers distinguishing the various sandwiches are obtained.

Thermal conductivities of thin films of several CMOS processes are listed in Tabs. 2.4 and 2.5 in Section 2.3.2. Fig. 2.11 a–d shows temperature-dependent thermal conductivities of CMOS thermal oxides, CVD dielectric thin-films, polysilicon layers and metallizations of different CMOS processes.

2.3.1.5 Thermal van der Pauw Test Structure

An alternative approach to determine the temperature-dependent thermal conductivity $\kappa(T)$ of thin CMOS IC layers makes use of a thermal analogy of the van der Pauw method [41]. A thermal van der Pauw test structure is shown in Fig. 2.12 [69]. It consists of a Greek-cross structure suspended on four arms above a micromachined cavity. The folding of the arms enables the cavity to be bulk micromachined into the silicon substrate using EDP etching. At the same time, the thermal conductance contributed by the arms is decreased.

The arms are made of the complete sandwich of dielectric CMOS layers. Each contain four CMOS metal lines. The Greek cross consists of a stack of CMOS thin films. Various sandwiches were realized enabling the thermal conductivities

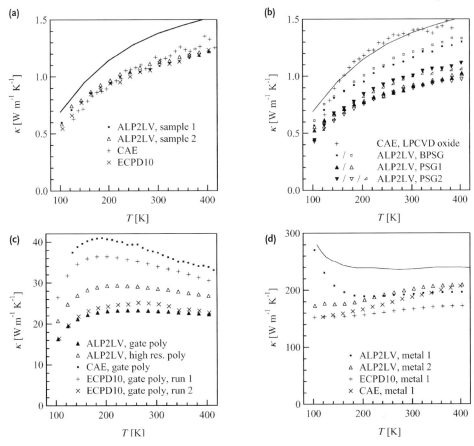

Fig. 2.11 Thermal conductivity versus temperature for (a) thermal oxides, (b) CVD dielectric thin films, (c) polysilicon layers and (d) metallizations of three CMOS processes (reproduced by permission of IEEE [71]). Solid lines indicate recommended values for (a, b) bulk fused silica [72] and (d) pure bulk aluminum [73]. For abbreviations of process names, see Section 2.1

of individual CMOS layers and thin-film sandwiches to be extracted, by a procedure similar to that of the cantilever-based structures.

Each branch of the Greek cross contains an integrated meandering polysilicon resistor, contacted in a four-point configuration to the metal lines in the corresponding suspension arm. Each resistor is used as a heater for the dissipation of controlled heating powers and, simultaneously, as a temperature monitor, exploiting the temperature-dependent resistivity of polysilicon. Additionally, each resistor is capped with a rectangular metal structure for temperature homogenization.

The principle of operation is based on the van der Pauw theory described in Section 2.2.1, translated into the thermal domain. It is based on the observation that heat conduction and electrical current conduction are governed by laws of

Fig. 2.12 SEM image of a micromachined thermal van der Pauw structure to measure in-plane thermal conductivities of thin films (a) Entire structure; (b) close-up of central Greek cross with integrated polysilicon resistors and connection lines (reproduced by permission of IEEE [69])

identical form, so that conclusions obtained in one domain can be translated into the other. For this purpose, the electrical potential is identified with the temperature field T, the electrical conductivity σ with the thermal conductivity κ and the electrical current density j_{el} with the heat flux density j_h; current injection/extraction reads injection/extraction of heat power; and voltage probes are replaced by temperature monitors. Thus, thermal resistances $R^{th}_{ij,kl}$ are defined as

$$R^{th}_{ij,kl} = \frac{T_k - T_l}{P_{ij}} \tag{17}$$

where P_{ij} denotes the heat power injected into branch i and extracted from branch j of the Greek cross and T_k and T_l denote the temperature values at the other two branches. The numbering of the branches is identical with that in Fig. 2.1c in Section 3.2.1. As in the electrical van der Pauw method, two separate thermal resistance measurements are in principle required to extract the thermal sheet resistance from

$$\exp\left(\frac{-\pi R^{th}_{12,34}}{R^{th}_{sq}}\right) + \exp\left(\frac{-\pi R^{th}_{23,41}}{R^{th}_{sq}}\right) = 1 \tag{18}$$

In analogy with the electrical sheet resistance $R_{sq} = (\Sigma \sigma_i t_i)^{-1}$ of a sandwich of layers with individual electrical conductivities σ_i and thicknesses t_i, the thermal sheet resistance R^{th}_{sq} is defined as

$$R^{th}_{sq} = \left(\sum \kappa_i t_i\right)^{-1} \tag{19}$$

with the thermal conductivities κ_i of the individual component layers of thickness t_i.

For the analogy with the electrical van der Pauw method to be complete, it is necessary for the heat flow to be constrained to the plane of the conductive film.

This is ensured by giving the micromachined Greek-cross sandwich the best possible thermal insulation, namely vacuum, and designing it with sufficiently small dimensions to suppress the radiative heat transfer to a negligible level compared with the in-plane heat flows [69].

A further challenge is the condition of vanishing heat flow into the temperature sensors during the two measurement phases. This condition corresponds to the high impedance voltage measurement in the electrical van der Pauw method. However, this condition is not easily satisfied because of the presence of the suspension arms. These conductive elements constitute a finite thermal impedance to thermal ground, that is, the substrate. It was shown, however, that the exact replication of the van der Pauw conditions is not necessary for extracting the thermal sheet resistance of the sandwich [69]. In fact, it is sufficient to measure the thermal response of the structure: the temperature changes ΔT_i of the four Greek cross branches to heating powers P_j individually dissipated in the four heating resistors have to be measured. The procedure to extract R_{sq}^{th} thus involves the following four steps:

1. Determination of the temperature coefficient of resistance (TCR) β_i of the four polysilicon resistors. Resistance changes ΔR_i can then be translated into temperature changes ΔT_i.
2. Determination of the thermal response ΔT_i of the structure to the dissipation of defined heat powers P_j. With sufficiently small powers, the ΔT_i values are well described by the linear approximation $P = R\Delta T$, with $\Delta T = (\Delta T_1, \Delta T_2, \Delta T_3, \Delta T_4)$ and $P = (P_1, P_2, P_3, P_4)$ and where R denotes the linear thermal response matrix with elements $R_{ij} = \partial P_i / \partial \Delta T_j$ extracted from these measurements.
3. Determination of the thermal resistance of the suspension arms from the elements of R and construction of the thermal response matrix R' with elements $R'_{ij} = R_{ij} - \delta_{ij}(R_{i1} + R_{i2} + R_{i3} + R_{i4})$ compensated for the heat loss through the suspension arms.
4. Extraction of R_{sq}^{th} from the elements of R'.

Technical details of the procedure have been described in [69, 70]. Using the thermal van der Pauw method, the thermal conductivities of various CMOS dielectric thin-film sandwiches and layers and aluminum-based CMOS metallization were determined. Results are listed in Tab. 2.4.

2.3.2
Heat Capacity

Measuring the heat capacity c of thin films was found by the authors to be truly a challenge, much more so than measuring the thermal conductivity or the thermopower. In principle, from the transient thermal response of test structures, the thermal diffusivity $a = \kappa / \rho c$ of materials can be determined, where ρ denotes the density of the material. It is useful to perform measurements in the frequency domain and to determine the thermal response as a function of frequency. From the

Tab. 2.4 Thermal conductivity $\kappa(T)$ and corresponding temperature coefficient $\kappa^{-1}d\kappa/dT$ of insulating thin films of several CMOS processes and other sources

Material	κ (W m^{-1} K^{-1})	$\kappa^{-1}d\kappa/dT$ (10^{-6} K^{-1})	Comments[a]
Field oxide	1.087 ± 0.016	1076 ± 40	EM, ALP2LV process [71]
	1.122 ± 0.021	966 ± 77	EM, ALP2LV process [71]
	1.145 ± 0.025	1670 ± 233	AMS, CAE process [71]
	1.09 ± 0.07	1112 ± 183	ES2, ECPD10 process [71]
	1.28 ± 0.11		AMS [45, 68]
	1.5 ± 0.1		van der Pauw [70]
Contact oxide	1.16 ± 0.06	1226 ± 69	EM, ALP2LV process [71]
	1.39 ± 0.08	920 ± 218	AMS, CAE process [71]
	1.32 ± 0.18		AMS [45, 68]
	1.5 ± 0.2		ES2, ECPD10 process [45]
Intermetal oxide	0.898 ± 0.13	1328 ± 42	EM, ALP2LV process [71]
	0.76 ± 0.07	1077 ± 669	AMS, CAE process [71]
	1.16 ± 0.24		AMS [45, 68]
	1.25 ± 0.2		ES2, ECPD10 process [45]
Passivation	3.2 ± 0.5		LPCVD, $Si_{1.0}N_{1.1}$ [63]
	0.45–0.75		PECVD SiN_x [61]
	1.3		APCVD [61]
	1.5 ± 0.25		AMS [45]
	1.26 ± 0.02		van der Pauw [70]
SiO_2	1.0		κ_t, APCVD+BPSG [60]
	1.12		κ_t, PECVD [60]
	1.08–1.34		κ_t, LPCVD oxide [58]
	1.35		Bulk sample [56]
	0.66		300 nm BESOI oxide [57]
	0.82		400 nm SIMOX oxide [57]
	0.75–1.2		PECVD oxide [61]
	0.76 ± 0.07		PECVD oxide, van der Pauw [70]
Layer sandwich	1.44		van der Pauw [69]
	0.93 ± 0.04	1500	van der Pauw [70]
	2.4		SiO_2/Si_3N_4 sandwich [65]
	1.65 ± 0.25		Si oxide/nitride [68]

a) All values are for 300 K. Abbreviations: chemical vapor deposition (CVD), low pressure CVD (LPCVD), plasma-enhanced CVD (PECVD), atmospheric pressure CVD (APCVD), bonded and etched back silicon-on-insulator (BESOI) and separation by implantation of oxygen (SIMOX). For abbreviations of process and company names, see Section 2.1.

static limit $\omega \to 0$, the thermal conductivity κ is extracted and values at frequencies $\omega > 0$ provide the diffusivity from which ρc and c, if ρ is known, are extracted.

As for thin-film thermal conductivity measurements, (i) planar [74], (ii) surface micromachined [75, 76], (iii) membrane-based [65] and (iv) bulk micromachined [45, 77–80] test structures to determine thin film c, ρc and a values have been reported. Using these structures, metal thin films [77–79], polysilicon [75, 78],

2.3 Thermal Properties

Tab. 2.5 Thermal conductivity $\kappa(T)$ and corresponding temperature coefficient $\kappa^{-1}d\kappa/dT$ of conducting thin-films of several CMOS processes and other sources, and bulk materials

Material	κ (W m^{-1} K^{-1})	$\kappa^{-1}d\kappa/dT$ (10^{-6} K^{-1})	Comments[a]
Polysilicon			
Gate poly	22.4 ± 0.7	−234 ± 22	EM, ALP2LV process [71]
	37.3 ± 1.0	−1158 ± 74	AMS, CAE process [71]
	24.8 ± 2.4	−725 ± 155	ES2, ECPD10 process [71]
n-poly	45.6		$N_A = 1.6 \times 10^{19}$ cm^{-3} [64]
	31.5 ± 3.7		$\rho = (0.813 \pm 0.001) \times 10^{-3}$ Ωcm Infineon [46]
	32		Heavily P-doped [62]
	29		$R_{sq} = 25$ Ω [66]
	29		[67]
p-poly	52		$N_D = 2.4 \times 10^{19}$ cm^{-3} [64]
	31.2 ± 3.7		$\rho = (2.214 \pm 0.004) \times 10^{-3}$ Ωcm Infineon [46]
	18		$R_{sq} = 225$ Ω [66]
Metal			
Metal 1	198.4 ± 6.1	297 ± 27	EM, ALP2LV process [71]
	187.8 ± 5.8	1041 ± 38	AMS, CAE process [54]
	167.0 ± 9.1	523 ± 16	ES2, ECPD10 process [54]
Metal 2	197 ± 5.3	682 ± 43	EM, ALP2LV process [71]
	166 ± 8		AMS, $R_{sq} = (36 \pm 0.4) \times 10^{-3}$ Ω [45]
	133 ± 6		ES2, ECPD10 process, $R_{sq} = (30 \pm 0.3) \times 10^{-3}$ Ω [45]
Bulk materials			
Al	237		[48]
Cu	401		[48]
Ni	90.7		[48]
Pt	71.6		[48]
Ti	21.9		[48]
W	174		[48]
Si	148		[48]
Si	156		[5]

a) All values are for 300 K. For abbreviations of process and company names, see Section 2.1.

CMOS IC thin films including metals and dielectrics [76–78, 80] and silicon oxide/nitride sandwiches [65] have been characterized.

In the following we will describe in further detail only the hot plate-based bulk-micromachined test structure reported in [77]. This was found by the authors to provide more accurate heat capacity values than, e.g., cantilever-based structures [78, 80]. Fig. 2.13a shows an SEM image of the device. The hot plate with lateral dimensions of 102 µm×172 µm is suspended over a EDP-micromachined cavity

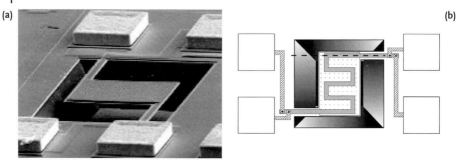

Fig. 2.13 (a) SEM image and (b) schematic view of a hot plate test structure to determine the heat capacity of thin films (reproduced by permission of MYU K. K. [77])

by two arms. Various versions of the plate are composed of different layer sandwiches of the conducting and dielectric CMOS layers. All contain an integrated polysilicon resistor as shown schematically in Fig. 2.13b serving as a heater and temperature monitor. The polysilicon heater extends through the suspension arms to the substrate, where it is connected to metal lines.

The thermal properties of the hot plate materials were extracted from the thermal response of the test structure to static and dynamic heat generation. In the static case, the resistor is powered by a constant current I_{dc}. The corresponding power $R_0 I_{dc}^2$ dissipated in the structure leads to the temperature increase $T_{dc} = R_0 I_{dc}^2 / G_{loss}$ of the hot plate, where G_{loss} denotes the overall thermal conductance due to the arms and radiative heat loss and R_0 is the resistance of the heater at small heating currents. For temperature increases of a few degrees, the resistance R of the heater is well approximated by

$$R = R_0(1 + \beta_R T_{dc}) \tag{20}$$

where β_R denotes the linear temperature coefficient of resistance (TCR). The voltage drop across the resistor is thus given by

$$U_{dc} = R_0 I_{dc} + R_0^2 \beta_R I_{dc}^3 / G_{loss} \tag{21}$$

where the last term results from the temperature increase of the structure. This term enables G_{loss} of the device to be determined.

In case of dynamic heat dissipation, an AC current $I(t) = I_{ac}\cos(\omega t)$ with amplitude I_{ac} and angular frequency ω is applied to the heating resistor. Thus, the heat power

$$P_{ac}(t) = R_0 I_{ac}^2 [1 + \cos(2\omega t)]/2 \tag{22}$$

is dissipated. It is composed of a static term establishing a stationary temperature distribution over the structure and a dynamic component of frequency 2ω. Hence the corresponding resistor temperature $T_{ac}(t)$ is

$$T_{ac}(t) = T_{ac0} + T_{ac2}(\omega)\cos[2\omega t - \phi(\omega)] \tag{23}$$

with a 2ω component with frequency-dependent amplitude $T_{ac2}(\omega)$ and phase shift $\phi(\omega)$. With the temperature-dependent resistance $R(t) = R_0[1 + \beta_R T_{ac}(t)]$, the voltage drop $U(t) = R(t)I_{ac}\cos(\omega t)$ over the resistor is then obtained. It contains a component at the third harmonic frequency, viz.

$$U_{ac3}(t) = \frac{1}{2} R_0 \beta T_{ac2}(\omega) \cos[3\omega t - \phi(\omega)] = A(\omega) \cos[3\omega t - \phi(\omega)] \tag{24}$$

Tab. 2.6 Heat capacities c and ρc and thermal diffusivity a of CMOS layers and layer sandwiches produced by various CMOS processes and of bulk materials

Material	Specific heat capacity c (J kg^{-1} K^{-1})	Volumetric heat capacity ρc (10^6 J m^{-3} K^{-1})	Thermal diffusivity a (cm^2 s^{-1})	Comments[a]
Passivation	1500±230			PECVD SiN [76]
		2.62±0.24		EM [77]
Metal				
Metal 1		2.56±0.24		EM [77]
		2.41±1.88		EM [78]
Metal 2		2.44±0.12		EM [77]
Silicon oxide		1.5±0.2		ES2, contact oxide [45]
		1.25		ES2, intermetal oxide [45]
Layer sandwich				
Dielectrics + poly		1.95±0.12		EM [77]
Si oxide + poly		1.72±0.15		EM [77]
SiO$_2$/Si$_3$N$_4$		1.68	0.0146	[65]
CMOS dielectrics		1.71±0.12		EM [78]
Oxides		1.05±0.1		AMS, field oxide, contact oxide, intermetal dielectric [45]
CMOS dielectrics + lower metal + poly		1.66±0.06		EM [80]
CMOS dielectrics		1.82±0.12		EM [80]
Polysilicon				
n-poly			0.17±0.01	[75]
Bulk materials				
Al	897			298 K [48]
Cu	385			298 K [48]
Ni	444			298 K [48]
Si	702			298 K [48]

a) Data are for room temperature. For abbreviations of process and company names, see Section 2.1.

which can be separated experimentally from the remaining DC, ω and 2ω terms. The connection of the amplitude $A(\omega)$ with the overall heat capacity $C = \sum c_i v_i$ of the central plate materials, where c_i and v_i denote the volumetric heat capacities and volumes of the component layers, is found from a simple thermal model of the structure [77] as

$$A(\omega) = \frac{R_0^2 \beta_R I_{ac}^3}{4 G_{loss}} [1 + (\tau \omega)^2]^{-1/2} \tag{25}$$

with the thermal time constant $\tau = 2C/G_{loss}$.

To determine C, the following procedure is thus followed: (i) determination of β_R of the polysilicon resistor, (ii) determination of G_{loss} from a static measurement and (iii) measurement of the frequency-dependent amplitude of U_{ac3} with the subsequent extraction of τ and C. Compared with the extraction of C from $A(\omega)$, the analogous extraction from the phase $\phi(\omega)$ was found to be less reliable.

All test structures with a resistor used simultaneously as a heater and temperature monitor fundamentally operate by a similar principle [56, 74, 77]. The distinction among the individual cases are the different thermal models, that is, the precise frequency dependence of $A(\omega)$ and the way in which loss efficiency and heat capacity are merged into a thermal time-scale of the structure. In view of the role played by the third harmonic, all these techniques are termed 3ω methods [56]. In contrast, if temperatures are measured by sensors distinct from the heater element, temperature and thus heat capacity extraction relies on 2ω terms [78, 80]. A further alternative is to carry out the measurements in time rather than frequency space [75]. The DC offset has to be compensated in this case. Heat capacity values reported in the archival literature are summarized in Tab. 2.6.

2.4
Mechanical Properties

Mechanical and piezoresistive properties of MEMS materials have been measured using several techniques. The most common methods rely on (i) static structure characterization, (ii) electrostatic structure deflection, (iii) microtensile testing, (iv) bulge testing, (v) nano-indentation and (vi) bridge bending. Each of these methods enables several mechanical parameters to be extracted, as summarized in Tab. 2.7. These various methods, including sketches of their theoretical background, experimental aspects and their merits and disadvantages are discussed in the following sections. Experimental results are summarized in Tabs. 2.9–2.14 in Section 2.4.8.

2.4.1
Bulge Test

In this method, the mechanical parameters of thin films are extracted from the load-deflection response of thin-film diaphragms under varying differential pressure p. Initial bulge tests were performed on circular metal diaphragms [81, 82]

2.4 Mechanical Properties

Tab. 2.7 Capability of different techniques to determine mechanical properties of thin-films.

Property	Method					
	Bulge test	Static structures	Electro-static beams	Micro-tensile test	Nano-indentation	Bridge bending
Residual strain ε_0	×	×	×	×		
Residual stress σ_0	×		×	×		
Residual stress gradient $d\sigma_0/dz$		×				
Elastic coefficients S_{ij}, C_{ij}	×		×	×	×	×
Young's modulus E	×		×	×		×
Poisson's ratio ν	×			×		
Plane strain modulus $E/(1-\nu^2)$	×			(×)	×	
Biaxial modulus $E/(1-\nu)$	×			(×)		
Strength σ_f	×			×	×	×
Toughness K_C				×	×	×
Thermal expansion coeff. a_{th}	×	(×)	(×)			
Piezoresistive coefficients π_{ij}				×		×

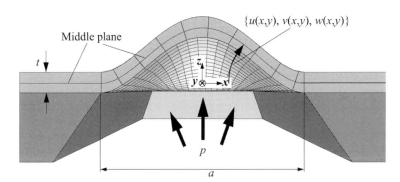

Fig. 2.14 Schematic view of the bulge test, i.e. the deflection of a thin-film diaphragm under a differential pressure load p [84]

long before the inception of MEMS. Elaborate mechanical models [83] are now available for this geometry. However, in view of MEMS fabrication techniques, square and rectangular diaphragm geometries are more commonly used. A structure of this type is shown schematically in Fig. 2.14. The thin-film diaphragm has thickness t and side lengths a and b. If it is a MEMS thin film, it is likely supported by a micromachined silicon frame. The load-deflection response of such a structure is governed by essentially three effects: first, the tendency of the perpendicular load to drive the structure away from its load-free equilibrium position; second, the counteraction by the in-plane stresses against the deformation; third, the resistance opposed by the bending stiffness of the thin film against the deformation. If the diaphragm is very thin, the influence of bending is negligible. The structure is then termed *membrane* in the language of mechanics. In contrast, if bending stiffness plays a role but the material is still thin enough to satisfy the Kirchhoff condition of negligible out-of-plane shear stresses, the structure is termed *thin plate* [85, 86]. The equilibrium conditions governing the load-deflection response of thin plates, the corresponding simpler form for membranes, and their consequences are developed in the following.

2.4.1.1 Theory

The starting assumption is that the diaphragm consists of a single isotropic material with elastic modulus E and Poisson's ratio v, and is prestrained with ε_0. The deformation of a thin plate under p is described by the vector field of the displacements $\{u(r), v(r), w(r)\}$ of the individual points $r = \{x, y\} = \{x_1, x_2\}$ of its middle plane (the neutral fiber in classical plate theory) in the two horizontal and the perpendicular directions. These displacements result in the strain tensor components ε_{xx}, ε_{yy} and ε_{xy} including the residual strain ε_0

$$\varepsilon_{xx} \equiv \varepsilon_0 + \varepsilon_{1,xx} = \varepsilon_0 + \frac{\partial}{\partial x} u(r) + \frac{1}{2} \left[\frac{\partial}{\partial x} w(r) \right]^2$$

$$\varepsilon_{yy} \equiv \varepsilon_0 + \varepsilon_{1,yy} = \varepsilon_0 + \frac{\partial}{\partial y} v(r) + \frac{1}{2} \left[\frac{\partial}{\partial y} w(r) \right]^2$$

$$\varepsilon_{xy} = \varepsilon_{yx} \equiv \varepsilon_{1,xy} = \frac{1}{2} \left[\frac{\partial}{\partial y} u(r) + \frac{\partial}{\partial x} v(r) \right] + \frac{1}{2} \left[\frac{\partial}{\partial x} w(r) \right] \left[\frac{\partial}{\partial y} w(r) \right] \quad (26)$$

which define $\varepsilon_{1,xx}$, $\varepsilon_{1,yy}$ and $\varepsilon_{1,xy} = \varepsilon_{1,yx}$ as the strain components caused by the external load in addition to the residual strain ε_0 present before the load is applied. These are translated into the in-plane stress tensor components in the mid-plane of the structure by the stress–strain relationships for thin plates [84, 87]:

$$\sigma_{xx} = \sigma_0 + \frac{E}{(1-v^2)} (\varepsilon_{1,xx} + v\varepsilon_{1,yy}) \equiv \sigma_0 + \sigma_{1,xx}$$

$$\sigma_{yy} = \sigma_0 + \frac{E}{(1-v^2)} (\varepsilon_{1,yy} + v\varepsilon_{1,xx}) \equiv \sigma_0 + \sigma_{1,yy}$$

$$\sigma_{xy} = \sigma_{yx} = \frac{E}{(1+v)} \varepsilon_{1,xy} \equiv \sigma_{1,xy} \quad (27)$$

where $\sigma_0 = \varepsilon_0 E/(1-v)$ denotes the residual stress of the thin structure. In view of the biaxial character of the residual stress, $E/(1-v)$ is termed biaxial modulus. The conditions governing the out-of-plane and in-plane equilibrium of such a structure are then [16]

$$D\Delta^2 w(r) - t\sigma_0 \Delta w(r) - t \sum_{i,j=1,2} \left[\sigma_{1,x_i x_j}(r) \frac{\partial^2}{\partial x_i \partial x_j} w(r) \right] = p \quad (28)$$

$$\sum_{j=1,2} \frac{\partial}{\partial x_j} \sigma_{1,x_i x_j}(\mathbf{r}) = 0 \qquad (29)$$

respectively, where $D = Et^3/12(1-v^2)$ denotes the flexural rigidity of the diaphragm. The first, second and third terms on the left-hand side of Eq. (28) correspond to the restoring force due to out-of-plane bending, residual membrane forces and the additional membrane forces due to the additional strain in the structure under load. The restoring forces have to match the applied pressure load p. Eqs. (28) and (29) have to be solved for u, v, and w subject to the boundary conditions

$$u(\mathbf{r}_b) = v(\mathbf{r}_b) = w(\mathbf{r}_b) = 0 \qquad (30)$$

and [24, 88]

$$D \frac{\partial^2}{\partial x_\perp^2} w(\mathbf{r}_b) = -K \frac{\partial}{\partial x_\perp} w(\mathbf{r}_b) \qquad (31)$$

where $\mathbf{r}_b$ and $\partial/\partial x_\perp$ denote the coordinates of any point of the diaphragm periphery and the derivative normal to the boundary in the outward direction. Eq. (31) implements the assumption of an elastic membrane support, relating the torque exerted by the diaphragm on the support to its slope at the boundary. The rotational spring parameter K has been estimated using finite-element (FE) simulations as $K = t^2 E_{ps}/k_r$, where $E_{ps} = E/(1-v^2)$ and k_r denote the plane-strain modulus of the diaphragm and a dimensionless geometry factor [88]. For wet etched membranes with {111} oriented cavity wall with 54.7° inclination, k_r was found to be roughly 0.75.

For elastic supports, relaxing the boundary conditions in Eq. (30) may be considered [88]. Displacements in the horizontal and vertical directions are possible and to lowest order are proportional to the in-plane and out-of-plane loads, respectively. However, for all practical purposes in bulge testing in the field of MEMS, the effects of these boundary displacements have been found by the authors to be negligible.

2.4.1.2 Reduction to Dimensionless Form

Considering Eqs. (27)–(31), one expects the load-deflection response of a rectangular membrane to depend on the geometric parameters a, b and h, on the mechanical parameters ε_0, E, v and K and on the load p. This impressive number of seemingly independent parameters is reduced, however, to a more modest number using a dimensional analysis based on the following definitions of dimensionless parameters [84, 87, 89, 90]: $\overline{x_1} = x_1/a$, $\overline{x_2} = x_2/a$, $\overline{w} = w/t$ and consequently $\overline{r} = r/a$, $\overline{r_b} = r_b/a$, $\overline{\Delta} = a^4 \Delta$, $\overline{u} = ua^2/t$, $\overline{v} = va^2/t$, $\overline{\varepsilon_i} = \varepsilon_i a^2/t^2$ and $\overline{\sigma_i} = \sigma_i a^2/E_{ps} t^2$ with ($i = 0, xx, xy, yx, yy$), $\overline{p} = pa^4/E_{ps} t^4$ and $\overline{K} = Ka/E_{ps} t^3 = a/tk_r$. With these prescriptions, Eqs. (27)–(29) and (31) are simplified to

$$\overline{\sigma}_{1,xx} = \overline{\varepsilon}_{1,xx} + \nu\overline{\varepsilon}_{1,yy}, \quad \overline{\sigma}_{1,yy} = \overline{\varepsilon}_{1,yy} + \nu\overline{\varepsilon}_{1,xx}, \quad \overline{\sigma}_{1,xy} = \overline{\sigma}_{1,yx} = (1-\nu)\overline{\varepsilon}_{1,xy} \tag{32}$$

$$\overline{\Delta}^2\overline{w}(\overline{r}) - 12\overline{\sigma}_0\overline{\Delta}\overline{w}(\overline{r}) - 12\sum_{i,j=1,2}\left[\overline{\sigma}_{1,\overline{x}_i\overline{x}_j}(\overline{r})\frac{\partial^2}{\partial\overline{x}_i\partial\overline{x}_j}\overline{w}(\overline{r})\right] = \overline{p} \tag{33}$$

$$\sum_{j=1,2,3}\frac{\partial}{\partial\overline{x}_i}\overline{\sigma}_{1,\overline{x}_i\overline{x}_j}(\overline{r}) = 0 \tag{34}$$

$$\frac{\partial^2}{\partial\overline{x}_\perp^2}\overline{w}(\overline{r}_b) = -12\overline{K}\frac{\partial}{\partial\overline{x}_\perp}\overline{w}(\overline{r}_b) \tag{35}$$

while Eqs. (26) and (30) remain of identical form except that all coordinates, displacements and strains are replaced by the corresponding dimensionless quantity. In this reduced, dimensionless form, it is evident that the load-deflection response depends only on five independent parameters: the aspect ratio $b:a$, Poisson's ratio ν and the reduced residual strain $\overline{\varepsilon}_0$, load $\overline{p}$, and rotational suspension stiffness $\overline{K}$. Therefore, for a well-defined aspect ratio, the four parameters ν, $\overline{\varepsilon}_0$, $\overline{p}$ and $\overline{K}$ completely define the load-deflection response.

If in addition the support is well modelled by a rigid clamping boundary condition, i.e. $K=\infty$, the boundary condition of Eq. (35) reduces to the usual horizontal tangent condition $\partial\overline{w}/\partial\overline{x}_\perp = 0$. The remaining independent parameters are ν, $\overline{\varepsilon}_0$, and $\overline{p}$.

The load-deflection response of clamped rectangular diaphragms is therefore compactly written in the form $\overline{w}(\overline{r}) = f_{rect}(\overline{r}, \overline{\varepsilon}_0, \overline{p}, \nu, b/a)$, irrespectively of absolute dimensions a, b and h, elastic modulus E, residual strain ε_0, and pressure p. Translated back into dimensional form, this reads

$$w(r) = t \times f_{rect}\left(\frac{r}{a}, \varepsilon_0\frac{a^2}{t^2}, p\frac{a^4}{tE_{ps}}, \nu, \frac{b}{a}\right) \tag{36}$$

The load-deflection response of square diaphragms is the special case with $a=b$. Following the same line of reasoning, circular diaphragms are found to follow a load-deflection law of the form $w_{circ}(r) = t \times f_{circ}(r/R, \varepsilon_0 R^2/t^2, pR^4/tE_{ps}, \nu)$, with the radial coordinate r and the diaphragm radius R. Performing load-deflection experiments with diaphragms of given geometry, i.e. measuring the deflection w as a function of p therefore enables one to determine the unknown parameters ε_0, E_{ps} and ν of the diaphragm material, by using them as fitting parameters. Usually the center deflection $w_0 = w(r=0)$ is taken as deflection information. If a reliable value of ν is obtained, Young's modulus can be safely extracted from E_{ps}, so that both elastic coefficients and residual strain are then known. From ε_0, E and ν, the residual stress $\sigma_0 = \varepsilon_0 E/(1-\nu)$ is finally inferred. Extracting reliable material data from the bulge test therefore requires the functions f_{rect} and f_{circ} to be determined.

2.4.1.3 Membranes

Modelling very thin bulging diaphragms as membranes is justified as long as the bending term in Eq. (28) is negligible in comparison with the other terms and in particular with the second. This is the case when $\overline{\sigma_0} = \sigma_0 a^2 / E_{ps} t^2$ is sufficiently large, i.e. when at least one of the following four conditions is satisfied: the residual stress σ_0 of the structure is large, its E value is small, its width a is large or its thickness t is small. As an example, with $\overline{\sigma_0} > 2000$, the response of long membranes at all pressure lies within 2.5% of the exact model developed in Section 2.4.1.6. In contrast, the membrane approximation fails for weakly tensile and compressive residual stresses and for small and thick diaphragms. The membrane equation derived from Eq. (28) reads

$$t\sigma_0 \Delta w(\mathbf{r}) + t \sum_{i,j=1,2} \left[\sigma_{1,ij}(\mathbf{r}) \frac{\partial^2}{\partial x_i \partial x_j} w(\mathbf{r}) \right] = -p \quad (37)$$

and has to be combined with Eqs. (26), (27), (29) and (30). A dimensional analysis shows that the rescaled out-of-plane deflection $\overline{w}(\overline{\mathbf{r}})$ of a rectangular membrane depends only on the dimensionless parameters b/a, $pa^2/\sigma_0 t^2$, $E_{ps} t^2 / \sigma_0 a^2$ and $\nu E_{ps} t^2 / \sigma_0 a^2$. The dimensional deflection function of a pressure-loaded membrane therefore takes the form

$$w(\mathbf{r}) = t \times f_{mem}\left(\frac{\mathbf{r}}{a}, \frac{pa^2}{\sigma_0 t^2}, \frac{E_{ps} t^2}{\sigma_0 a^2}, \frac{\nu E_{ps} t^2}{\sigma_0 a^2}, \frac{b}{a}\right) \quad (38)$$

Consequently, load-deflection data are analyzed using such a model by considering the unknown variables σ_0, E_{ps} and ν as fitting parameters. Since the fitting of ν is often numerically ill-defined, only a combination of E_{ps} and νE_{ps} is extracted. It is therefore essential to have at one's disposal accurate models for f_{rect} and f_{mem}. Once these are known, load-deflection data can be analyzed to extract mechanical parameters.

2.4.1.4 Determining Load-deflection Models

Three methods to obtain load deflection functions f_{rect} and f_{mem} have been used: solution of the differential equations, variational analysis and finite element (FE) analysis.

The first approach consists of explicitly constructing the deformation functions u, v and w. This is possible only in the rarest cases and at the cost of considerable simplifications. Among the explicitly solved problems are circular and rectangular membranes with tensile residual stress at small deflections [82, 91–93] and long diaphragms under plane-strain deformations [94]. Rectangular tensile membranes have been modelled by a series expansion known as the Navier solution [85, 93]. The case of long diaphragms is described in further detail in Section 2.4.1.6.

Variational analysis proceeds by constructing approximative solutions by linear superposition of base functions. The expansion coefficients are obtained by minimizing the total energy of the loaded mechanical structure [89]. For strongly deflected membranes, trigonometric functions such as

$$u_{kl}(x,y) = \sin(2k\pi x/a)\cos(l\pi y/b), \text{ with } k = 1,2,3,\ldots \text{ and } l = 1,3,5,\ldots$$
$$v_{mn}(x,y) = \cos(m\pi x/a)\sin(2n\pi y/b), \text{ with } m = 1,3,5,\ldots \text{ and } n = 1,2,3\ldots$$
$$w_{pq}(x,y) = \cos(p\pi x/a)\cos(q\pi y/b), \text{ with } p = 1,3,5,\ldots \text{ and } q = 1,3,5,\ldots \quad (39)$$

have served the purpose. In the early bulge test reports, only the terms with $k=l=m=n=p=q=1$ of this system were kept [95–97]. For membranes with large aspect ratio $b:a$, this early truncation leads to considerable errors, since the plane strain response in their middle section with steep descents towards the smaller diaphragm edges is not satisfactorily rendered. For both directions we recommend to include functions with wavelengths down to about $a/5$, where a denotes the smaller of the two side lengths.

The test functions in Eq. (39) do not satisfy clamping boundary conditions. For clamped thin plates it is better to design test functions each satisfying the required boundary conditions. For this purpose, alternative test functions were implemented and were shown to provide accurate results over a wide range of experimental conditions [84, 90, 98, 99]. Up to 2×16^2 trial functions for the in-plane deformations and 8^2 functions and for the out-of-plane deformation were included in the most advanced case [89].

Implementing the corresponding total energy and minimizing it with respect to large numbers of coefficients of the linear combinations of test functions represents a formidable task. Careful book-keeping of millions of energy terms is essential [87, 89]. For the minimization, the method of conjugated gradients was found to be efficient [100].

The third approach is finite-element analysis. After the diaphragm structure has been meshed, an approximation to the deformation functions u, v and w is obtained by computing the three-dimensional deflections of the mesh nodes. The challenge is to use well-suited finite elements and to define a mesh adequately mapping the geometric situation. If shell elements are used, the FE software has to be able to handle the geometrical nonlinearity represented by the strain-deformation relations in Eq. (26). Further, it has to be able to deal with bifurcations such as buckling and other symmetry transitions. Convergence is often found to be a delicate issue and has to be considered carefully. The software packages ANSYS [89, 101, 102] and ABAQUS and ADINA [92] have been employed for bulge test simulations.

2.4.1.5 Square Diaphragms

The load-deflection behavior of square diaphragms is illustrated in Fig. 2.15. The dimensionless central deflection amplitude $\overline{w_0}$ is shown schematically as a function of reduced residual strain $\overline{\varepsilon_0}$ and load pressure $\overline{p}$. Optical micrographs of

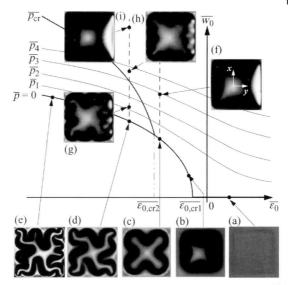

Fig. 2.15 Schematic diagram of the load-deflection response of square clamped diaphragms, i.e. reduced dimensionless center deflection $\overline{w_0}$ as a function of reduced residual strain $\overline{\varepsilon_0}$ and pressure $\overline{p}$, illustrated by optical micrographs of silicon nitride diaphragm deflection profiles. Adapted from [84]

membranes, labelled (a) to (i), illustrate the various types of deflection profiles. Vertical dashed lines, e.g. from (c) in Fig. 2.15 via (f) or from (d) via (g), (h) and (i) correspond to load-deflection experiments, i.e. they show the evolution of $\overline{w_0}$ of a diaphragm with fixed $\overline{\varepsilon_0}$ as a function of $\overline{p}$. Curves of a constant pressure value, e.g. $\overline{p_1}$, describe the deflection of diaphragms with different reduced residual strains under identical reduced loads. Diaphragms with $\overline{\varepsilon_0} > \overline{\varepsilon_{0,cr1}}$, i.e., weakly compressive or tensile residual strain, are flat in the unloaded state (Fig. 2.15a). A pressure load progressively bulges them out of the plane. The isobar $\overline{p} = 0$ to the left of $\overline{\varepsilon_{0,cr1}}$ shows the postbuckling deflection of the structures [89]. Their unloaded equilibrium position is in a bulged state (Fig. 2.15b–e), that is $\overline{w_0} \neq 0$ at $\overline{p} = 0$. Their energy is lowered by a partial out-of-plane relaxation of the in-plane stress, at the cost of a slight increase in bending energy. The value of $\overline{\varepsilon_{0,cr1}}$ is $-4.363/(1+\nu)$ corresponding to the dimensional critical residual stress $\overline{\sigma_{0,cr1}} = -4.363\, E_{ps} h^2/a^2$ [89].

At pressure higher than the $\overline{\varepsilon_0}$-dependent critical pressure $\overline{p_{cr}(\varepsilon_0)}$ and above the isobar $p=0$, the deflection profiles show all the in-plane reflection and rotation symmetries of the square (Fig. 2.15a,b,c,f,h, and i). At pressures below $\overline{p_{cr}}$, the profiles show only the in-plane symmetry under rotations by multiples of $\pi/2$ (Fig. 15d,e, and g). What happens at the pressure $\overline{p_{cr}}$ thus is a symmetry transition analogous to the buckling transition at $\overline{\varepsilon_{0,cr1}}$, where the vertical reflection symmetry of the unloaded diaphragm is broken by the buckling [101]. The critical pressure line $\overline{p_{cr}}(\overline{\varepsilon_0})$ intercepts the postbuckling line $\overline{p}=0$ at the critical reduced residual strain $\overline{\varepsilon_{0,cr2}}$. Thus, to the left of $\overline{\varepsilon_{0,cr2}}$ unloaded post-buckled profiles show the reduced symmetry, while to the right, they show the full set of symmetries of a square. Values of $\overline{\varepsilon_{0,cr2}}$ depend on v and were found, e.g., to be −212 for $v=0.125$, −206 for $v=0.25$, and −208 for $v=0.375$ [89]. Values of $\overline{p_{cr}(\varepsilon_0)}$ have proved to be hard to determine reliably, owing to convergence problems with the FE method and excessive numbers of test functions necessary to accurately model these structures using the variational method [101].

The mechanical response of square structures with $\overline{\varepsilon_0} > -200$ to pressures has been extensively analyzed using the FE method. Based on the numerical results, the following load-deflection law was found to fit the numerical results best [84]:

$$p(w_0) = E_{ps} \frac{t^4}{a^4} \left[c_1(\overline{\varepsilon_0} - \overline{\varepsilon_{0,cr1}}) \frac{w_0}{t} + c_3(v)(1 + d_3\overline{\varepsilon_0}) \frac{w_0^3}{t^3} \right] \qquad (40)$$

with $c_1 = 13.84$, $c_3(v) = 32.852 - 9.48\,v$ and $d_3 = -2.6 \times 10^{-5}$. This result was extracted from simulations covering the parameter space $0.15 < v < 0.35$ and $-200 < \overline{\varepsilon_0} < 1000$. If $\overline{\varepsilon_0}$ is extracted from load-deflection curves, its value lies within a distance of less than 5 from the true $\overline{\varepsilon_0}$. Further, relative uncertainties of the extracted E value are smaller than 1% over the entire parameter range. For circumstances where σ_0 rather than ε_0 is to be extracted, Eq. (40) is equivalent to

$$p(w_0) = \frac{c_1}{a^2}\left(t\sigma_0 + 4.363 E_{ps}\frac{t^3}{a^2}\right) w_0 + \frac{c_3(v)}{a^4}\left(tE_{ps} + d_3\sigma_0 \frac{a^2}{t}\right) w_0^3 \qquad (41)$$

As a consequence of Eq. (36), the residual strain of a compressive diaphragm material can be extracted from the central deflection of a square alone, if Poisson's ratio of the material is known. For residual strains between $\overline{\varepsilon_{0,cr1}}$ and $\overline{\varepsilon_{0,cr2}}$, Ref. [98] describes a suitable method to reach this goal. Using the variational method, the relationship between reduced central deflection and reduced residual strain was found to be well approximated by

$$\overline{w_0} = \overline{\Delta\varepsilon_0}^{-1/2} \left[c_1 + c_2 \tanh(c_3\overline{\Delta\varepsilon_0}) + \frac{c_4\overline{\Delta\varepsilon_0} + c_5\overline{\Delta\varepsilon_0}^2}{1 - c_6\overline{\Delta\varepsilon_0}^3} \right]^{1/2} \qquad (42)$$

with $\overline{\Delta\varepsilon_0} = \overline{\varepsilon_{0,cr1}}(v) - \overline{\varepsilon_0}$ and $c_1 = -0.4972 - 0.2313\,v - 0.2128\,v^2$, $c_2 = 0.0698 + 0.1625\,v + 0.2\,v^2$, $c_3 = -7.19 \times 10^{-3} - 0.0466\,v + 0.0367\,v^2$, $c_4 = -1.19 \times 10^{-3} + (5.51 \times 10^{-3})\,v^2$, $c_5 = -3.34 \times 10^{-6} - (7.43 \times 10^{-5})\,v + (1.28 \times 10^{-4})\,v^2$, and $c_6 = 3.16 \times 10^{-6} + (4.8 \times 10^{-6})\,v - (1.52 \times 10^{-5})\,v^2$.

Square diaphragms under strongly tensile residual stresses or strong loads behave as membranes. These were among the first structures to be analyzed by the bulge test. For this case, the approximate load-deflection law

$$p(w_0) = c_1 \frac{\sigma_0 t}{a^2} w_0 + c_3(v) \frac{Et}{(1-v)a^4} w_0^3 \qquad (43)$$

Tab. 2.8 Coefficients c_1 and c_3 in Eq. (43), as reported in the bulge test literature

Geometry	c_1	c_3	Type of analysis, comment [source][a]
Square	12.176	21.968	Variational, $v=0.25$ [95]
	13.64	31.7–9.36 v	FE [92]
	13.572	$16/(0.8+0.062\,v)^3$	Variational [103]
	13.8	31.9–8.65 v	Variational [104]
Rectangular	8	$64/3(1+v)$	FE [105], variational [103]
	–	$21.6(1.41-0.292\,v)$	Clamped [106]
Circular	16	$2^7/3$	Insert diameter $2r$ instead of length a in Eq. 43 [82]

a) In the references quoted, the membrane width is usually defined as $2a$. This explains the apparent discrepancy by a factor of 4 in c_1 and 16 in c_2 between published and tabulated values.

is generally accepted. It is based on FE analyses and variational calculations. The two numerical coefficients c_1 and $c_3(v)$ depend on the studies, as summarized in Tab. 2.8. Residual stresses and strains and elastic coefficients of various materials extracted using the bulge test of square membranes are listed in Tabs. 2.9–2.12 in Section 2.4.8.

2.4.1.6 Long Diaphragms

A second extremely useful geometry for bulge testing is provided by long diaphragms. These are structures with large aspect ratios $b:a$. The small edges negligibly influence the mechanical response of an extended central section of the structure. There, the load-deflection profile is invariant under continuous or discrete translations in the longitudinal direction. This part of the structure is appropriately modelled by an infinitely long diaphragm.

The mechanical response of such structures is shown schematically in Fig. 2.16. As in Fig. 2.15, the reduced center deflection amplitude $\bar{w}_0$ is shown as a function of the reduced residual stress $\bar{\varepsilon}_0$ and pressure $\bar{p}$. Selected states are illustrated by schematic three-dimensional deflection profiles labelled (a) to (f). Vertical lines correspond to load-deflection experiments, showing the progressive increase of $\bar{w}_0$ with increasing pressure, e.g. from (a) to (b) or from (d) to (e).

Structures with tensile or weakly compressive residual stress ($\bar{\varepsilon}_0 > \bar{\varepsilon}_{0,\text{cr1}}$) respond to external differential pressures by a translationally invariant plane strain deformation. Without load, the diaphragm is flat [(a) in Fig. 2.16]. Under pressure it progressively bulges out of the plane [(b) in Fig. 2.16]. At the critical reduced residual strain value $\bar{\varepsilon}_{0,\text{cr1}} = -\pi^2/3(1+v)$ corresponding to the critical residual stress $\sigma_{0,\text{cr1}} = -\pi^2 E_{\text{ps}} t^2/3a^2$, the structures undergo a buckling transition, below which they are post-buckled even for $p=0$. For $\bar{\varepsilon}_0$ values below the second critical residual strain value $\bar{\varepsilon}_{0,\text{cr2}} \approx -17.3$, the unloaded structures show a meanderlike superstructure superimposed on their plane-strain profile. Under pressure the meander-shaped profiles are progressively modified until they cross over into ripple-shaped profiles at a critical pressure level $\bar{p}_{\text{cr1}}(\bar{\varepsilon}_0, v)$ [90, 107]. With increasing pressure, the ripple amplitude

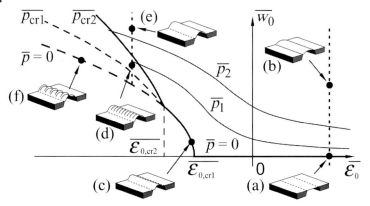

Fig. 2.16 Schematic load-deflection response diagram of long clamped diaphragms, i.e., reduced dimensionless deflection amplitude $\overline{w}_0$ and pressure p, illustrated by selected post-buckling and deflection profiles

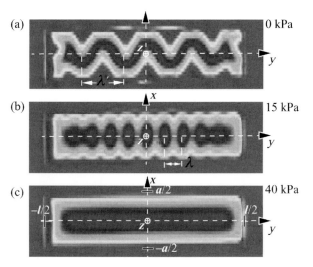

Fig. 2.17 Measured profiles of a 179 μm wide, 697 μm long and 490 nm thick PECVD silicon nitride membrane at three different pressure loads: (a) meander profile without load; (b) ripple profile at 15 kPa; (c) plane-strain profile at 40 kPa (reproduced by permission of IEEE [90])

decreases and finally vanishes completely at a second critical pressure $\overline{p}_{cr2}(\overline{\varepsilon}_0, \nu)$. This second transition offers the possibility to determine Poisson's ratio of the material [98]. An example of such a profile evolution is shown in Fig. 2.17.

With strongly tensile residual stress, the structure shows membrane behavior. Its deflection profile is

$$w(x) = \frac{p}{2t\sigma}\left[\left(\frac{a}{2}\right)^2 - x^2\right] \tag{44}$$

where x denotes the coordinate across the width a of the structure and $\sigma = \sigma_0 + 8E_{ps}w_0^2/3a^2$ is the effective stress acting on the mid-plane of the deflected diaphragm. With this, the load-defletion law of long membranes [87, 108]

$$p(w_0) = \frac{8\sigma_0 t}{a^2} w_0 + \frac{64 E_{ps} t}{3a^4} w_0^3 \tag{45}$$

is obtained. Residual stress σ_0 and plane-strain modulus E_{ps} can thus be extracted from the linear and cubic components of the membrane response.

On the buckling branch, for reduced residual strains $\bar{\varepsilon}_0$ between $\bar{\varepsilon}_{0,cr2}$ and $\bar{\varepsilon}_{0,cr1}$, the center deflection grows as

$$|w_0| = \frac{2t}{\sqrt{3}}\left(\frac{\sigma_0}{\sigma_{0,cr1}} - 1\right)^{1/2} = \frac{2t}{\sqrt{3}}\left(\frac{\varepsilon_0}{\varepsilon_{0,cr1}} - 1\right)^{1/2} \tag{46}$$

and the deflection profile is $w(x) = w_0[1 + \cos(2\pi x/a)]/2$ [108]. On this entire post-buckling branch ($\bar{p} = 0$), the mid-plane stress is $\sigma_{0,cr1}$.

In the plane-strain region, i.e., for $\bar{p} > \bar{p}_{cr2}(\varepsilon_0, v)$, Eqs. (28) and (29) simplify to a one-dimensional problem [108], viz.

$$D\frac{\partial^4}{\partial x^4}w(r) - t\sigma\frac{\partial^2}{\partial x^2}w(r) = p \tag{47}$$

with

$$\sigma = \sigma_0 + \frac{E_{ps}}{2a}\int_{-a/2}^{a/2}\left(\frac{dw}{dx}\right)^2 dx \tag{48}$$

The stress σ is the effective stress acting on the mid-plane of the structures. It is the sum of the residual stress σ_0 and the additional stress resulting from the elongation of the structures under the out-of-plane deflection. The corresponding additional strain is given by the integral in Eq. (48) divided by $2a$. Via E_{ps} this is translated into the additional stress.

Equations (47) and (48) clearly expose the non linearity of bulge test mechanics. They now have to be solved for $w(x)$ subject to the boundary conditions, i.e. Eqs. (30) and (31). For most practical purposes, the diaphragm support can be assumed as stiff, so that $dw/dx = 0$ at the boundary. The appropriate solution of Eq. (47) is

$$p = \frac{E_{ps}t^4/a^4}{\dfrac{1}{8\bar{\sigma}} - \dfrac{\cosh(\sqrt{3\bar{\sigma}}) - 1}{4\sqrt{3}\bar{\sigma}^{3/2}\sinh(\sqrt{3\bar{\sigma}})}} w_0 \quad (49)$$

where the reduced effective stress $\bar{\sigma} = \sigma a^2/E_{ps}t^2$ solves the equivalent of Eq. (48):

$$\bar{\sigma} = \frac{\sigma_0 a^2}{E_{ps}t^2} + \frac{2\left[(8+4\bar{\sigma})\sinh(\sqrt{3\bar{\sigma}})^2 - 6\bar{\sigma} - 3\sqrt{3\bar{\sigma}}\sinh(2\sqrt{3\bar{\sigma}})\right]}{\left[\sqrt{3\bar{\sigma}}\sinh(\sqrt{3\bar{\sigma}}) - 4\sinh(\sqrt{3\bar{\sigma}}/2)^2\right]^2} \frac{w_0^2}{t^2} \quad (50)$$

Equations (49) and (50) represent the exact non-linear load-deflection law $p(w_0, \sigma_0, E_{ps}, a, t)$ of long thin plates with clamped edges. One way of obtaining E_{ps} and σ_0 from a set of experimental data $\{p_i, w_{0,i}\}$ with $i=1,\ldots,N$, thus is to fit the load-deflection law to the data, with σ_0 and E_{ps} used as fitting parameters. The general case with elastic support leads to more cumbersome expressions [109]. The evolution of deflection profiles as a function of applied pressure is illustrated in Fig. 2.18. Clearly, with increasing effective stress parameter $\bar{\sigma}$, the profile undergoes a transition from plate to membrane behavior.

In situations with large effective stress, the load-deflection law implicit in Eqs. (49) and (50) merges into the membrane response of Eq. (45). This is the case with films subjected to strongly tensile stress. However, it is also true for membranes under weakly tensile or even compressive residual stress at sufficiently high pressure loads. The effective stress σ has then evolved to a highly tensile level and the membrane approximation is comfortably used.

When a long thin plate is loaded, it will ultimately break at a fracture load p_f. Fracture is most likely to initiate at the edges, where the structure experiences very large local stresses. This is due to the strong curvature in the transition from the horizontal tangent at the edge to the sloped deflection profile. The curvature causes an additional stress $\sigma_{curv}(z) = -zE_{ps}d^2w/dx^2$ on top of the mid-plane effective stress σ, where z denotes the perpendicular coordinate measured upward from the mid-plane. As an example, a 400 µm wide and 300 nm thick membrane

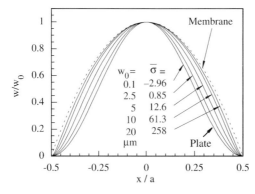

Fig. 2.18 Deflection profiles for various effective stress values $\bar{\sigma}$ calculated using Eqs. (47), (49) and (50), showing the transition from plate to membrane behavior. Adapted from [108]

with a tensile residual stress of 1 GPa and a plane-strain modulus of 300 GPa, bulging under a pressure of 100 kPa, experiences an effective stress $\sigma=1.16$ GPa, but a local maximum stress of $\sigma_{max}=3$ GPa at the bottom surface of the diaphragm at $x=\pm a/2$. Such considerations made it possible to extract fracture properties of thin dielectric films [102, 109]. Data were evaluated using Weibull statistics, well established for brittle materials [110] and are listed in Tab. 2.13 in Section 2.4.8.

The meanders and ripples in long diaphragms under strongly compressive residual stress and the corresponding hierarchy of symmetry transitions under pressure have been analyzed using the energy minimization method [90, 107]. The out-of-plane profile was decomposed into plane-strain, undulating and skew-symmetric components. The superposition of the three types of functions produces the observed meander-shaped pattern. With increasing pressure, the coefficients of the skew-symmetric components progressively decrease. At p_{cr1} they vanish and remain equal to zero beyond. Thus the remaining profile is composed of plane-strain and undulating components which combine into ripple-shaped profile, as observed. With further pressure increase, the coefficients of the undulating component progressively become smaller, and finally vanish at p_{cr2}. Beyond this point the deflection profile reduces to the plane-strain components. Experimental profiles of a diaphragm at these three stages are shown in Fig. 2.17.

The last symmetry transition, i.e. from ripples to plane-strain, has been analyzed extensively using the energy minimization method. It was shown to strongly depend on Poisson's ratio v. Qualitatively, the reason is simple: the ripple manifests the tendency of the diaphragm to relax the longitudinal residual stress component; under pressure, the additional transverse strain is translated via Poisson's ratio v into a longitudinal contraction, which ultimately compensates the longitudinal residual stress component and flattens out the structure. Beyond, the diaphragm is left in a plane-strain state. Evidently, the dependence of this transition depends on v: for $v=0.5$, the longitudinal contraction is strongest and the pressure at which compensation occurs is lowest; at the other extreme, for $v=0$ orthogonal strains are decoupled and no pressure is able to straighten the profile. The problem was quantitatively addressed by an instability analysis of plane-strain profiles with respect to the addition of undulating components [98]. The analysis showed that the dependence of p_{cr2} and corresponding deflection $w_{0,cr2}$ on v is sufficiently strong to enable v of thin-films to be stably determined. The ripple wavelength λ constitutes a further parameter serving as an independent numerical self-consistency check. A typical experiment proceeds by first determining the transition pressure and deflection, i.e. p_{cr2} and $w_{0,cr2}$, respectively, then by measuring the load-deflection response of the plane-strained diaphragm at pressures $p > p_{cr2}$ resulting in σ_0 and E_{ps}, and finally extracting Poisson's ratio. Using this method, v of a PECVD silicon nitride film with σ_0 between −57.8 and −82.3 MPa was determined as 0.253 ± 0.017 [98].

An alternative method to determine v combines bulge experiments on square and long diaphragms. The different load-deflection laws of these two geometries makes it possible to determine different combinations of elastic coefficients, from which v

is finally extracted. This was achieved with an unstressed PECVD silicon nitride with $\sigma_0 = -2.4 \pm 2.4$ MPa where $\nu = 0.235 \pm 0.04$ was found [84]. Previous attempts have possibly suffered from the lack of sufficiently accurate model for the load-deflection response of square and long diaphragms including their bending rigidity [111].

A final comment concerns the question how long is long enough. For the evaluation of parameters from profiles under plane strain, diaphragm aspect ratios $b:a$ larger than 6 provide reliable results; for the extraction of parameters from rippled and meandering profiles, aspect ratios larger than 12 are recommended. If the structure is too short, wavelength selection disturbs the profile and influences the extracted properties. Residual stresses and strains and elastic coefficients of various materials extracted using the bulge test of long membranes are listed in Tabs. 2.9–2.12 in Section 2.4.8.

2.4.1.7 Thermal Expansion

Using Eqs. (44) and (48), the residual strain ε_0 of compressively prestressed square and long diaphragms with respect to the substrate can be inferred from their post-buckling amplitude. If such measurements are performed as a function of temperature, the derivative $d\varepsilon_0/dT$ is obtained. By definition, the coefficient of thermal expansion a_f of the thin film is

$$a_f = a_s + \frac{d\varepsilon_0}{dT} \tag{51}$$

where a_s denotes the corresponding coefficient of the substrate. This procedure was demonstrated with a PECVD silicon nitride, where $a_{SiN}(T) = a_0 + a_1(T - T_0)$ with $a_0 = (1.803 \pm 0.006) \times 10^{-6}$ K^{-1}, $a_1 = (7.5 \pm 0.5) \times 10^{-9}$ K^{-2} and $T_0 = 25\,°$C was found for the temperature range between 25 and 140 °C [112].

2.4.1.8 Blister Test

In suitable cases the bulge test has also been used to determine interfacial energies. When the diaphragm material adheres weakly to the substrate, it can be forced by the applied load to delaminate from the substrate. This requires the mechanical energy stored in the deformed diaphragm and the work performed by the pressure against the membrane to be evaluated. Using these two energy values as a function of the progressing interfacial crack, the interfacial energy between membrane and substrate can be computed [113, 114].

2.4.2 Wafer Curvature

Wafer curvature characterization is one of the most widely used methods to determine the average stress in thin films rigidly attached to a substrate, e.g. a silicon wafer [115]. Use is made of the curvature induced in the substrate by the bending moment due to the stressed film. Surface profiles are measured before and after film deposition and the difference is attributed to the deposited film. The relation between the film stress and the substrate curvature is known as Stoney's equation [116]:

$$\sigma_0 = \frac{E_s t_s^2}{6(1-v_s)t}\left(\frac{1}{R} - \frac{1}{R_0}\right) \tag{52}$$

where σ_0, $E_s/(1-v_s)$, t_s, t, R and R_0 denote the film stress, the biaxial modulus of the substrate, its thickness, the film thickness and the measured radii of curvature after and before film deposition, respectively. Assumptions underlying Stoney's equation are that the substrate is isotropically elastic and free to bow, the film thickness is uniform with $t \ll t_s$, the film stress is isotropic in the film plane and has to be low enough not to induce a mechanical bifurcation of the substrate–film system [117] and gravity plays no role [118]. Extensions of Stoney's equation beyond these limits are available [119, 120]. For the commonly used {100} wafers, the biaxial modulus has the value $E_s/(1-v) = 180$ GPa. For {111} wafers, the corresponding value is 228 GPa. With films with thicknesses up to a few micrometers and substrate wafers with standard thicknesses in the range of a few hundred micrometers, the film-to-substrate thickness condition is sufficiently well satisfied. If the film shows a z-dependent stress $\sigma_0(z)$, where z denotes the coordinate perpendicular to the film plane, the term measured on the left hand side of Eq. (52) is the thickness-averaged film stress, i.e. $\int \sigma_0(z)dz/t$.

From temperature-dependent measurements, the temperature coefficient $d\sigma_0/dT = (a_f - a_s)E_f/(1-v_f)$ of the thermal stress of the thin film can be determined, by extending Eq. (51) with the thin film biaxial modulus $E_f/(1-v_f)$. Knowing the biaxial moduli of substrate and thin film, the thermal expansion coefficient a_f of the thin film can be extracted. If the film is deposited on two substrates with different coefficients of thermal expansion, both a_f and $E_f/(1-v_f)$ can be extracted from the two sets of measurements. Residual stresses of various thin-films, extracted from wafer bow are listed in Tabs. 2.9 and 2.10 in Section 2.4.8.

2.4.3
Passive Test Structures

Micromachined mechanical diagnostic structures are useful monitors for mechanical thin-film properties. The basic idea is that a micromachined structure made from a thin film reveals information about its mechanical properties. The challenge is to define appropriate geometries for optimal evaluation of these properties.

Fundamentally, if a structure is not exposed to an external force except for the effects of its residual stress, its mechanical response is governed by the plate equations, Eqs. (33) and (34), with $\bar{p}=0$. As a consequence, for a fixed geometry, the only independent parameters of the response are $\bar{\varepsilon}_0 = \varepsilon_0 a^2/t^2$ and v, as in the case of thin diaphragms. Therefore, ε_0 and v at most can be extracted from a static test structure.

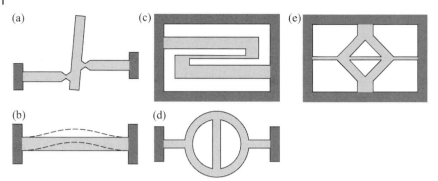

Fig. 2.19 Schematic geometries of selected surface micromachined test structures to measure residual strain of thin films: (a) pointer structures; beam structures for materials with (b) compressive and (c) tensile residual stress, (d) ring- and (e) diamond-shaped structures for materials under tensile stress

In addition to bulk micromachined diaphragms, several kinds of surface micromachined structures have been developed: rotating pointer structures, beam test structures, suspended rings and diamond-shaped structures with inner beams (see Fig. 2.19), and linear and annular overhangs. Upon surface micromachining, the residual strain of the structures relaxes and the structure is deformed. Usually, the resulting shape is characterized using optical surface profilometry and ε_0 is extracted from the deformation data. If elastic properties of the thin-film material are known, the residual stress σ_0 can be inferred.

Pointer structures have a rather intuitive operating principle [121, 122]. They amplify the strain into a directly measurable translation or rotation of one of their components. As shown schematically in Fig. 2.19a, two beams show a differential deformation after their release. Consequently, a connecting bridge experiences a rotation, which is geometrically amplified by a long pointer beam rigidly attached to the bridge. Deflection of the beam tip, and the responsible rotation are accurately measured by a Vernier gauge attached to the mobile beam end, with the counterpart fixed on the substrate [122].

Clamped-clamped beams made from a compressively prestressed material exhibit post-buckling beyond a critical length (see Fig. 2.19b). The Euler buckling amplitude w_0, i.e. the out-of-plane center deflection of a beam of length L subject to a residual strain ε_0, is

$$w_0(L, \varepsilon_0) = 2(-\varepsilon_0 L^2/\pi^2 - t^2/3)^{1/2} \tag{53}$$

The deflection shape is $w(x) = w_0[1+\cos(2\pi x/L)]$, with the beam center being at $x=0$. Note that the profile is similar to that of a buckled long membrane and that Eq. (53) is based on the assumption of beam mechanics and ideal clamping [123, 124]. Fig. 2.19c shows a variant for tensile thin films.

Since tensile stress evades determination by simple beam structures, a ring-shaped geometry with two outer attachments and a perpendicular inner beam, as

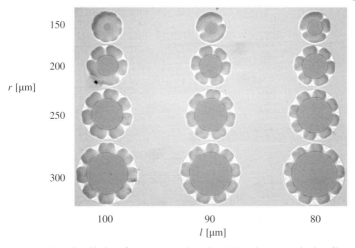

Fig. 2.20 Post-buckled surface micromachined PECVD silicon nitride thin-film annuli for residual stress measurements. Dimensions r and l denote the annulus outer radius and width, respectively (reprinted from [127], with permission from Elsevier)

shown in Fig. 2.19 d, has been proposed and successfully used to determine the residual strain of polysilicon thin films [125]. The relaxation of the longitudinal strain in the supporting beams stretches the ring into an oval which in turn compresses the transverse beam beyond its buckling point. In view of the complex geometry, it is advisable to support the extraction of strain values by FE simulations. As an alternative, similar structures with diamond-shaped geometry as shown in Fig. 2.19 e have also been used [121, 126].

A particularly aesthetic test structure is based on a thin-film annulus clamped to the substrate at its outer perimeter. If the material has compressive residual stress, it relaxes into a circular wave pattern from which the residual strain is extracted [127]. A series of such structures are shown in Fig. 2.20. The incomplete underetching of long beam structures of compressive material leads to the formation of similar linear wavy post-buckling profiles [128].

2.4.4
Electrostatically Actuated Beams

Going beyond the extraction of ε_0 and ν requires the application of external forces, as in the case of thin-film diaphragms. In the case of surface micromachined test structures, electrostatic actuation is easily implemented, in particular if the film to be characterized is polysilicon. Counter-electrodes for electrostatic actuation are handily integrated below the test element. A general schematic cross-section of such structures and their operating principle is shown in Fig. 2.21. Their testing potential is threefold [129]:

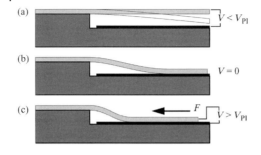

Fig. 2.21 Schematic cross-section of electrostatically actuated beams for mechanical testing of thin films. (a) Response of small actuation voltages; (b) partial release after pull-in; (c) friction measurement at large voltages. Aafter [132]

- From the response at small force levels up to the pull-in voltage, the residual stress and elastic modulus of the material can be deduced.
- Once a structure has been pulled in and the load has been removed, the response of the structure depends on the interfacial energy between the substrate and the structure: longer structures remain stuck while shorter devices with a higher stiffness are able to release themselves [130].
- If a structure is pulled in further, it is forced to move horizontally on the substrate surface. Its response then also reflects the frictional properties of the test element against the substrate surface.

A set of microelectromechanical beam test structures serving the purpose of wafer-level process monitoring and mechanical property measurement was developed at MIT [131]. This so-called M-Test, as an abbreviation of 'mechanical test', was demonstrated using MIT's dielectrically isolated wafer-bonding process. There is no evident obstacle in extending it to thin-film testing for CMOS MEMS and multi-user MEMS processes. The test structures of M-Test include doubly clamped and clamped-free beams, i.e. bridges and cantilevers, and circular clamped diaphragms. For each structure type, a closed-form expression of the pull-in voltage V_{PI} has been reported, which enables two phenomenological parameters, S and B, to be extracted, where S is more relevant in the case of cantilevers, while B strongly affects the response of the bridge and diaphragm structures [131]. These two parameters are products of material properties and geometric dimensions of the test structures including film thickness and gap width and thus provide a simple way to assess the uniformity of these products across the wafers and among wafers and batches. From S and B, based on data from geometric metrology, the local residual stress and Young's modulus are extracted. The method is particularly appealing in view of the fact that only one electrical parameter has to be measured, namely V_{PI}, and the rest is standard dimensional metrology.

Similar efforts conducted by Sandia National Laboratories are well suited for MEMS polysilicon [132]. Again, cantilevers and clamped-clamped beams of various lengths provide the experimental vehicles. A difference with the M-Test is that the deflection profile of the structures is monitored interferometrically. This enables imperfections due to, e.g., stress gradients and non-ideal supports to be recognized and taken into account.

By combining out-of-plane electrostatic actuation with a beam geometry inspired from tensile testing, the fracture strength of polysilicon was measured [133]. The test structure was fabricated using the SUMMiT process flow and features two wide electrostatically deflectable cantilevers facing each other and connected by a 1 µm wide beam section. The structure is pulled in and then progressively loaded until fracture of the gauge section occurs. Since the gauge section and parts of its two anchors rest flatly on the substrate, the gauge section experiences uniaxial load, which is a condition for reliable fracture strength extraction.

2.4.5
Microtensile Test

Several groups have developed materials characterization structures and equipment operating as downscaled versions of the classical tensile test. In this method, a long prismatic sample is subjected to a longitudinal, uniaxial load. From the load-deflection response, elastic parameters including Young's modulus E and Poisson's ratio ν, plastic properties such as σ_y, and fracture properties such as σ_f and K_C can be extracted at increasing force levels. With brittle layers, the elastic regime is directly terminated by sample fracture. Shrinking the tensile test method to microscopic dimensions presents several challenges.

The first is sample fabrication, from both design and technology points of view. Samples usually comprise a slender prismatic gauge section, i.e. the part to be uniformly loaded, with two divergent material sections at both ends. Ideally, the transitions are smooth so that stress concentrations are minimized. In contrast, sharp corners increase the probability of fracture and consequently reduce the overall mechanical strength. Typical dimensions of gauge sections are several tens to hundreds of micrometers in length, a few micrometers to a few tens of micrometers in width and fractions of micrometers to a few micrometers in thickness. A major concern in surface micromachined samples is stiction which has been prevented by laterally tethering the structures. Once etching and rinsing is completed, the tethers are delicately broken, fully releasing the structure.

The next challenge involves subjecting the structure to the load. Several methods to achieve this goal have been demonstrated. They are shown schematically in Fig. 2.22 and include:

- *Cut-frame structures.* The two ends of the gauge section of the thin film to be characterized are attached to a common silicon frame. The frame is mounted into the test setup where the sample can be loaded once the sides of the frame have been disrupted. The setup presents only the longitudinal degree of freedom and is otherwise rigid. This ensures uniaxial loading of the gauge section [134].
- *Frame-stabilized structures.* Here the substrate frame is designed to be soft in the longitudinal direction and rigid with respect to all other degrees of freedom [19, 135, 136]. The longitudinal stiffness of the frame has to be determined experimentally and subtracted from the overall stiffness of the frame–sample system in order to extract the sample properties. Measurements therefore have to go beyond sam-

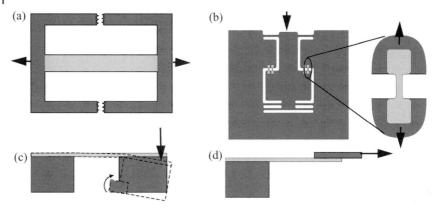

Fig. 2.22 Schematic view of methods to subject microtensile samples to a uniaxial load. (a) Cut-frame structure; frame-stabilized structure with (b) horizontal and (c) vertical load; (d) structure relying on sample gripping. Arrows indicate load location and direction

ple fracture. An interesting variant is provided by test structures where actuation of a vertical lever on the test structures is translated into the desired longitudinal load by rotation around an axis below the sample plane [136, 137]. In this case, all mechanical components have been fabricated by wet etching of silicon.

- *Sample gripping.* The sample is gripped after its tethers have been ruptured. Electrostatic force gripping has been developed to a high technical level [138–140] for both conducting and insulating layers. In the first case, the free sample end is delicately approached by a conducting probe protected by a thin-film dielectric and is electrostatically attracted against the probe. The contact area has to be large enough to avoid slipping under the longitudinal load. Alternative schemes for gripping the sample such as glueing [141, 142] and mechanical gripping mechanisms [143] have also been reported. The challenge with all these methods is to guarantee the planarity of the gauge sections during the measurement. Oblique loads lead to undesired stress components reducing the apparent fracture strength.

Measuring load-displacement curves represents a further challenge. With samples a few hundred micrometers long and exposed to strains of 10^{-3}, the relative displacement of the sample ends is in the sub-micrometer range. Whereas loads are typically in the μN to tens of mN range and can be measured using a commercial load cell, displacement measurements are more delicate. Strain measurements directly on the gauge section are preferred. An appealing approach is to use reflective patterns on the sample surfaces to characterize the surface strain interferometrically. This method has been used to measure both longitudinal and transverse strain, making it possible to extract the Poisson's ratio of thin films [144]. In this case, it has to be ensured that the reflecting structure negligibly influences the mechanical response of the sample.

Alternative relative displacement measurements at other locations of the test structure or the test setup have been performed optically [142] and capacitively

[145]. In the case of the vertically loaded frame-stabilized structures [136, 137, 146], geometric amplification by the actuation lever provides an easy access to accurate displacement data.

Recently, the external load was applied by a seismic mass accelerated under shock loads instead of being applied as a slow ramp [147]. At accelerations up to 30 000 g, the epipolysilicon gauge sections were exposed to apparent uniaxial stresses up to roughly 200 MPa. The study compared highly dynamic loading with quasi-static loading conditions. After careful analysis of the experiment, it was concluded that the dynamic fracture strength of epipolysilicon is reduced by roughly 50% from the corresponding quasistatic value.

Systematic variation of sample dimensions and subsequent careful analysis of the statistical data makes it possible to identify the location of the fracture origin. In the case of a polysilicon process, the fracture was found to originate on the edges of the samples [138]. Introducing a well-controlled imperfection such as a notch into the gauge section provides a way to evaluate the fracture toughness of the sample material [137, 148].

Various microtensile test approaches have recently been compared in a round-robin test involving seven laboratories [149]. Samples designed according to the individual groups' requirements were fabricated using the same technologies. Results showed reasonable agreement in fracture strength and Young's modulus for two types of monocrystalline silicon layers, polysilicon, nickel and titanium.

A selection of published elastic properties and fracture strength and toughness data is given in Tabs. 2.11–2.13 in Section 2.4.8.

2.4.6
Nanoindentation

Indentation is a standard technique for determining the hardness of bulk materials. The indentation process of a geometrically well-defined hard tip into the material is monitored by measuring load-displacement curves in the loading and unloading mode. During loading, plastic deformations are induced in the material, whereas in the unloading phase, the response is mainly defined by elastic properties of the material and the tester. The mechanical properties of thin films are monitored by miniaturized instruments termed nanoindenters with adapted displacement and force range [150, 151]. Again, a hardness value is measured during the loading phase, whereas from the unloading process an elastic property of the film is derived [151]. However, the compliance of the indenter with its transducers and of the substrate have to be taken into account for data evaluation. Nanoindentation is often used in combination and as validation of other measurement techniques [152, 153]. A disadvantage is the relatively large force applied to a small area of the specimen. This has caused densification or even phase transformations in the material under characterization [154]. The potential of nanoindentation is currently being extended by exploring the range of forces down to those occurring in scanning probe microscopy [155].

2.4.7
Bridge Bending

The measurement of the piezoresistive coefficients of silicon requires well-controlled mechanical stresses to be applied to a sample. A conceptually simple way of achieving this goal is to subject a silicon strip to a well-defined curvature. Its deformation is modelled using simple beam theory, from which the surface strain and stress are inferred. Diffused resistors integrated on the strip surface or thin-film resistors deposited on it can thus be characterized.

Two popular setups for beam bending are the three-point and four-point bending bridges [156]. In three-point bending the material strip to be tested is placed on top of a linear supporting wedge and pressed down by two additional wedges at equal distances from the support. The structures to be characterized is centered on the support. In contrast, in four-point bending, the strip rests on two supporting wedges and is symmetrically pressed down by two outer wedges. The advantage of four point-bending is that the central section experiences a constant bending moment. Consequently, in this section, the curvature and surface strains and stresses are constant along the beam. In contrast to three-point bending, the position of the test structure to be characterized is therefore less critical.

Ideally the surface stress has only a longitudinal component σ_{xx}, where x denotes the coordinate along the strip, while all other stress tensor components vanish. It is determined from geometrical considerations, i.e., wedge distances and vertical displacement only, without the necessity of knowing the load forces. In four-point bending, σ_{xx} is given by [156]

$$\sigma_{xx} = \frac{3tE\Delta d}{2L_o(L_o - 3L_i)} \tag{54}$$

where t, E, Δd, L_i and L_o denote the thickness and Young's modulus of the strip, the relative vertical displacement of the wedge pairs against each other, and the distance from the center to the inner supports and from the inner supports to the outer wedges, respectively.

A rarely mentioned challenge with beam bending is the proper alignment of all mechanical components. Several types of misalignment cause undesirable surface stress components altering the measurement results. These are horizontal misalignment of the wedges, out-of-plane misalignment of the wedges with respect to each other and rotation of the strip with respect to the wedge orientation. All these cause a non-symmetrical contact of the strip on the individual wedges and lead to additional in plane surface shear stress. These undesirable stress components have been numerically simulated using the FE method and measured using appropriate stress sensors [35, 157].

Piezoresistive coefficients $\pi_{11}-\pi_{12}$ and π_{44} of n- and p-silicon diffusions were determined using this method [5, 36, 158, 159]. Similarly, using CMOS field effect transistor (FET) structures, the corresponding coefficients $\Pi_{11}-\Pi_{12}$ and Π_{44} of MOSFET inversion layers were determined [33, 34, 37, 160]. The separation of π_{11}

Fig. 2.23 Multidimensional stress sensor to extract all three surface stress components, consisting of an n-well diffusion with one central and eight peripheral contacts (reproduced by permission of IEEE [35])

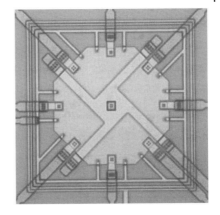

and π_{12} from measured values of $\pi_{11}-\pi_{12}$ and $\pi_{11}+\pi_{12}$ was attempted using a novel multidimensional stress sensor shown in Fig. 2.23 [35]. Experimental piezoresistive coefficients of CMOS diffusions, inversion layers and silicon samples are listed in Tab. 2.14 in Section 2.4.8.

2.4.8
Mechanical Material Data

Tabs. 2.9–2.14 list mechanical material properties of MEMS thin films, including residual strain and stress (Tabs. 2.9 and 2.10), elastic properties (Tabs. 2.11 and 2.12), fracture mechanical properties (Tab. 2.13) and piezoresistance coefficients (Tab. 2.14). A special focus is on materials produced using commercial CMOS technologies and by MEMS foundries. Companies with IC processes used for MEMS fabrication and for which material data have been compiled include those listed in Section 2.1.

Data for materials from these commerical sources were selectively complemented by references to measured properties of other MEMS thin films. The aim is to give a broader impression of the range of values that the properties can cover, depending on processing conditions. A second reason is that many methodological developments were done using films not produced as part of an IC or MEMS process. Examples are:

- The mechanical properties of silicon nitride, where a considerable amount of data was accumulated using LPCVD and PECVD equipment in university laboratories. This material forms the basis of numerous mechanical and thermal MEMS.
- Mechanical properties of polysilicon. Much of the work on the influence of annealing on the stresses in polysilicon was done in university laboratories and research institutes. The results have been transferred to MEMS foundries and may be relevant for the use of these sources.
- Fracture mechanical properties of silicon. Little is known about the fracture strength and toughness of MEMS materials besides silicon. Methods to evaluate these properties are in their infancy.

Tab. 2.9 Residual stress σ_0 of silicon oxide, polysilicon thin-films and CMOS metal layers

Material	σ_0 (MPa)	Comments[a]
Silicon oxide	−300	FOX, ALP2LV, EM, bulge test [161]
	−305	FOX, ALP2MV, EM, wafer curvature [162, 163]
	−360	FOX, CYE and CBT, AMS, wafer curvature [164]
	−300	FOX, $T_{ox}=950\,°C$ [165]
	−235	Thermal, Siemens [166]
	−485	Gate oxide, ALP2MV, EM, wafer curvature [162, 163]
	−237	Thermal, bulge test [109]
	−40 ± 10	BPSG, ALP2LV, EM, bulge test [161]
	−74	BPSG, ALP2MV, EM, wafer curvature [162, 163]
	−70 ± 5	BPSG, CYE and CBT, AMS, wafer curvature [164]
	−27 ± 3/−26 ± 3	PBSG, before/after anneal at 400 °C, Infineon [167]
	−37 ± 6	PSG, EM, ALP2LV, bulge test [161]
	−96 ± 9/−68 ± 5	PE-silane ox., before/after 400 °C anneal, Infineon [167]
	−103 ± 24/ −96 ± 21	PE-TEOS ox., before/after 400 °C anneal, Infineon [167]
	−29	IMD, ALP2MV, EM, wafer curvature [162, 163]
	−50 ± 5	IMD, CYE and CBT, AMS, wafer curvature [164]
	$\varepsilon_0 = -2.55 \times 10^{-3}$	IMD, 2 μm CMOS, EM, beam buckling [124]
	−70/−155/−190	LPCVD, as deposited/stored in air/annealed, Siemens [166]
	95/−50/−135	LPCVD, TEOS, as deposited/stored in air/annealed, Siemens [166]
	250/75/−140	LPCVD, as deposited/stored in air/annealed, Siemens [166]
Polysilicon	−260 ± 20	p⁺-Poly2, ALP2LV, EM, bulge test [161]
	−134 ± 10	n⁺-Poly1, ALP2LV, EM, bulge test [161]
	−583	n⁺-Poly, ALP2MV, EM, wafer curvature [162, 163]
	−220/−280	Undoped, as deposited (560 °C/625 °C) [135]
	330	Undoped, annealed at 650 °C [135]
	180	Undoped, annealed at 850 °C [135]
	±20	Undoped, annealed at 1050 °C [135]
	300−425	Undoped, $T_{dep}=620\,°C$ [168]
	350 ± 12	As deposited [168]
	20	Annealed at 600−1100 °C [168]
	180	As deposited, $T_{dep}=630\,°C$ [95]
	10	Phosphorus, $T_{anneal}=1050\,°C$ [169]
	19 ± 7/18 ± 7	n, before/after anneal at 400 °C, Infineon [167]
	203 ± 7/203 ± 7	p, before/after anneal at 400 °C, Infineon [167]
Metallization	73 ± 10	Metal 1, ALP2LV, EM, bulge test [161]
	30 ± 10	Metal 2, ALP2LV, EM, bulge test [161]
	232 ± 24/167 ± 17	Al, before/after anneal at 400 °C, Infineon [167]

a) Abbreviations: field oxide (FOX), borophosphosilicate glass (BPSG), phosphosilicate glass (PSG), tetraethyl orthosilicate (TEOS), inter-metal dielectric (IMD), plasma-enhanced (PE), low-pressure chemical vapor deposition (LPCVD). For abbreviations of process and company names, see Section 2.1.

Tab. 2.10 Residual strain ε_0 and stress σ_0 of silicon nitride thin-films

Material	ε_0	σ_0 (MPa)	Comments[a]
Silicon nitride		82±6	PECVD sensor passivation, ALP2LV, EM, bulge test [161]
		−224/−180	PECVD standard passivation, before/after anneal, ALP2MV, EM, wafer curvature [162, 163]
		110/171	PECVD sensor passivation, before/after anneal, ALP2MV, EM, wafer curvature [162, 163]
		−360	PECVD, passivation, CBT, AMS, wafer curv. [164]
		−120	PECVD, passivation, CYE, AMS, wafer curv. [164]
		300–10	30–60% LF power, PECVD, wafer curvature [161]
		−320±15	PECVD, bulge test [101]
		−44±5/−24±5	PECVD, before/after anneal at 400 °C, Infineon [167]
		110	PECVD, T_{dep} = 300 °C [95]
		1.3±3.8	PECVD, bulge test [108]
		−63.2±12.4	PECVD, bulge test [108]
		111.1±18	PECVD, bulge test [170]
		60.7±1	PECVD, wafer curvature [89]
		970±50	LPCVD, bulge test [102]
		1040±160	LPCVD, bulge test [102]
		1000	LPCVD, bulge test [95]
		226±2	LPCVD, T_{dep} = 835 °C [96]
	−6.26×10^{-4}		PECVD, before anneal, clamped annulus [171]
	−9.09×10^{-4}		PECVD, after anneal, clamped annulus [171]
	−3.5×10^{-4}		PECVD, post-buckling amplitude [89]
	−1.78×10^{-3}		PECVD, bulge test, symmetry transitions [90]
	−2.3×10^{-4}		PECVD, 2 µm CMOS, EM, beam buckling [124]

a) Abbreviations: plasma-enhanced chemical vapor deposition (PECVD), low-pressure chemical vapor deposition (LPCVD). For abbreviations of process and company names, see Section 2.1.

Tab. 2.11 Elastic properties of conducting materials of CMOS processes and MEMS foundries, including elastic stiffness coefficients C_{ij}, compliance coefficients S_{ij}, elastic modulus E, thermal expansion coefficient a_{th} and Poisson's ratio v.

Material	Property	Value	Comments[a]
Silicon	C_{11} (GPa)	165.8	298 K, from temperature-dependent values in [172]
	C_{12} (GPa)	63.9	
	C_{44} (GPa)	79.6	
	S_{11} (Pa^{-1})	7.74×10^{-12}	
	S_{12} (Pa^{-1})	-2.16×10^{-12}	
	S_{44} (Pa^{-1})	12.6×10^{-12}	
	E (GPa)	155–180	Epi-SOI, round-robin test, microtensile methods [149]
		130–220	Direct-bonded SOI, round-robin test, microtensile methods [149]
	a_{th} (K^{-1})	$2.362 \times 10^{-6} + (1.026 \times 10^{-8})T - (2.988 \times 10^{-11})T^2 + (3.948 \times 10^{-14})T^3$ T in °C [112]	
Polysilicon	E (GPa)	140 ± 14	p$^+$-Poly2, ALP2LV, EM, bulge test [161]
		52 ± 6	n$^+$-Poly1, ALP2LV, EM, bulge test [161]
		150–170	Undoped, $T_{dep} = 620$ °C [168]
		151 ± 6	$T_{dep} = 620$ °C [168]
		160	$T_{dep} = 630$ °C, bulge test [95]
		150 ± 30	Boron, 2–10 μm [173]
		174 ± 10	Phosphorus, $T_{anneal} = 1050$ °C [169]
		130 ± 5	Phosphorus, $T_{dep} = 610$ °C, $T_{anneal} = 1040$ °C [174]
		147 ± 6	Phosphorus, $T_{dep} = 560$ °C, $T_{anneal} = 1040$ °C [174]
		130–175	Round-robin test, microtensile methods [149]
		169	Phosphorus, n, Infineon [167]
		143	Epipoly, 15 μm thick, ST Microelectronic, microtensile test [145]
	v	0.2 ± 0.1	p$^+$-Poly2, ALP2LV, EM, bulge test [161]
		0.2 ± 0.1	n$^+$-Poly1, ALP2LV, EM, bulge test [161]
Metallization	E (GPa)	50 ± 10	Metal 1, ALP2LV, EM, bulge test [161]
		55 ± 10	Metal 2, ALP2LV, EM, bulge test [161]
		69	Aluminum, Infineon [167]
	v	0.3 ± 0.1	Metal 1, ALP2LV, EM, bulge test [161]
		0.2 ± 0.2	Metal 2, ALP2LV, EM, bulge test [161]
Al	S_{11} (Pa^{-1})	15.9×10^{-12}	[3]
	S_{12} (Pa^{-1})	-5.8×10^{-12}	[3]
	S_{44} (Pa^{-1})	35.2×10^{-12}	[3]
Cu	S_{11} (Pa^{-1})	14.9×10^{-12}	[3]
	S_{12} (Pa^{-1})	-6.3×10^{-12}	[3]
	S_{44} (Pa^{-1})	13.3×10^{-12}	[3]
W	S_{11} (Pa^{-1})	2.57×10^{-12}	[3]
	S_{12} (Pa^{-1})	-0.73×10^{-12}	[3]
	S_{44} (Pa^{-1})	6.6×10^{-12}	[3]

a) Abbreviation: silicon on insulator (SOI). For abbreviations of process and company names, see Section 2.1.

Tab. 2.12 Elastic properties of dielectric materials of CMOS processes and MEMS foundries, including elastic modulus E, Poisson's ratio v, plane-strain modulus $E/(1-v^2)$, and thermal expansion coefficient a_{th}

Material	Property	Value	Comments[a]
Silicon oxide	E (GPa)	66	Thermal [116]
		70	Thermal, bulge test [109]
		20±10	BPSG, ALP2LV, EM, bulge test [161]
		65.5±5	PSG, ALP2LV, EM, bulge test [161]
		60	PE-silane, Infineon [167]
		60	PE-TEOS, Infineon [167]
		57	BPSG, before/after 400 °C anneal, Infineon [167]
	v	0.2	Thermal [116]
		0.2±0.2	PSG, ALP2LV, EM, bulge test [161]
Silicon nitride	E (GPa)	97±6	PECVD sensor passivation, ALP2LV, EM, bulge test [161]
		150±7	PECVD, bulge test [101]
		112±5	PECVD, after anneal, clamped annulus, assumed $v=0.25$ [171]
		130	PECVD, post-buckling amplitude combined with wafer curvature [89]
		133.6	PECVD, bulge test [98]
		160	PECVD, bulge test, symmetry transitions [90]
		210	PECVD, $T_{dep}=300$ °C, bulge test [95]
		132	PECVD, Infineon [167]
		302	LPCVD, bulge test [109]
		288	LPCVD, bulge test [102]
		290	LPCVD, bulge test [95]
		260	LPCVD, $T_{dep}=850$ °C [175]
		320	LPCVD, $T_{dep}=850$ °C, $T_{anneal}=1150$ °C [175]
		276	LPCVD, $T_{dep}=835$ °C [96]
	$E/(1-v^2)$ (GPa)	134.4±3.9	PECVD, bulge test [108]
		142±2.6	PECVD, bulge test [108]
		145.5±3.8	PECVD, bulge test [170]
	v	0.253±0.017	PECVD, bulge test, ripple transition [98]
		0.235±0.04	PECVD, bulge test, combined long and square membranes [84]
		0.22±0.03	LPCVD [134]
	a_{th} (K^{-1})	$1.803\times10^{-6} + 7.5\times10^{-9}\Delta T$	PECVD, $\Delta T=(T-25$ °C), T close to 25 °C post-buckling amplitude [89]

a) Abbreviations: borophosphosilicate glass (BPSG), phosphosilicate glass (PSG), plasma-enhanced (PE), tetraethyl orthosilicate (TEOS), inter-metal dielectric (IMD), plasma-enhanced chemical vapor deposition (PECVD), low-pressure chemical vapor deposition (LPCVD). For abbreviations of process and company names, see Section 2.1.

Tab. 2.13 Tensile strength σ_f and mode I fracture toughness K_{IC} of silicon oxide, silicon nitride and polysilicon thin films, silicon and epireactor grown polysilicon (epipoly)

Material	Property	Value	Comments [a]
Silicon oxide	σ_f (GPa)	0.87–1.35	Thermal, bulge test [109]
Silicon nitride	σ_f (GPa)	9–13.2	LPCVD, bulge test [102]
		5.83±025	LPCVD, microtensile [134]
Polysilicon	σ_f (GPa)	1.81	Undoped, $T_{dep}=560C$, $T_{anneal}=650\,°C$, microtensile test [135]
		2.35	Undoped, $T_{dep}=560\,°C$, $T_{anneal}=850\,°C$, microtensile test [135]
		2.12	Undoped, $T_{dep}=560\,°C$, $T_{anneal}=1050\,°C$, microtensile test [135]
		1.7	Undoped, $T_{dep}=625\,°C$, $T_{anneal}=1050\,°C$, microtensile test [135]
		2.85–4.27	Round-robin test, various methods [134]
		1.4–2.5	Round-robin test, microtensile methods [149]
		2.3–4	4 µm thick, Infineon [176]
		2.0–2.8	Undoped, 2 µm thick, microtensile test [138]
		2.0–2.7	P-doped, 2 µm thick, microtensile test [138]
Silicon	σ_f (GPa)	7.7–2.0	Notched cantilevers, notch depth 0–500 nm [177]
		6.5–0.7	Notch depth 0–1 µm, microtensile test [137]
		17.63–11.65	Nanowires, width 200–800 nm, 295 K [178].
		1.4–2.05	Direct-bonded SOI, round-robin test, microtensile methods [149]
		1.7–2.25	Epi-SOI, round-robin test, microtensile methods [149]
		1.3–4.9	SIMOX-Si, microtensile test [140]
	K_{IC} (MPa m$^{1/2}$)	0.75–1.35	Bulk material [179]
		1.1–3.45	Microtensile test [137]
Epipoly	σ_f (GPa)	1.2	[180]
		<0.7	Bosch epipoly, highly dynamic load [147]
		4.1–4.6	Epipoly, 15 µm thick, ST Microelectronic, microtensile [145]

[a] Abbreviations: low-pressure chemical vapor deposition (LPCVD), silicon on insulator (SOI), separation by implantation of oxygen (SIMOX). For abbreviations of process and company names, see Section 2.1.

2.4 Mechanical Properties

Tab. 2.14 Piezoresistive coefficients of diffused silicon samples and MOSFET inversion layers

Material	Coefficient (10^{-11} Pa^{-1})	Value	Comments[a]
p-Si	π_{11}	6.6	$\rho = 7.8$ Ωcm [181]
	π_{12}	–1.1	$\rho = 7.8$ Ωcm [181]
	π_{44}	138.1	$\rho = 7.8$ Ωcm [181]
p$^+$-Si	π_{11}	1.1 ± 1.3	Boron, 0.6 µm CMOS, TSMC [182]
		5.7	[185]
	π_{12}	1.0 ± 0.6	Boron, 0.6 µm CMOS, TSMC [182]
		–2.3	[185]
	$\pi_{11} + \pi_{12}$	1.24 ± 1.8	Boron, 0.6 µm CMOS, TSMC [183]
	π_{44}	62	Boron, $N_S = 2 \times 10^{20}$ cm^{-3} [184]
		57.2	0.8 µm CMOS, Micronas, bridge bending [159]
		69 ± 1	Boron, 0.6 µm CMOS, TSMC [183]
		75.2 ± 2	Boron, 0.6 µm CMOS, TSMC [182]
		71	[185]
n-Si	π_{11}	–102.2	$\rho = 11.7$ Ω cm [181]
	π_{12}	53.4	$\rho = 11.7$ Ω cm [181]
	π_{44}	–13.6	$\rho = 11.7$ Ω cm [181]
n$^+$-Si	π_{11}	–36.8 ± 2.4	2 µm CMOS, EM, bridge bending + force [158]
		–34	Phosphorus, $N_S = 2 \times 10^{20}$ cm^{-3} [184]
		–28	Phosphorus, $N_S = 2 \times 10^{20}$ cm^{-3} [186]
		–30	Phosphorus, $N_S = 10^{20}$ cm^{-3} [187]
		–28.7	Phosphorus, 0.6 µm CMOS, TSMC [182]
		–38	[185]
	π_{12}	12.6 ± 1.4	Phosphorus, 0.6 µm CMOS, TSMC [182]
		19	[185]
		22.4 ± 1.8	2 µm CMOS, EM, bridge bending + force [158]
	$\pi_{11} + \pi_{12}$	–16.65 ± 1.5	Phosphorus, 0.6 µm CMOS, TSMC [183]
	π_{44}	–15	Phosphorus, $N_S = 2 \times 10^{20}$ cm^{-3} [184]
		–9.7	0.8 µm CMOS, Micronas, bridge bending [159]
		–14.2 ± 0.5	Phosphorus, 0.6 µm CMOS, TSMC [183]
		–15.5 ± 1.0	Phosphorus, 0.6 µm CMOS, TSMC [182]
		–15.9 ± 0.9	2 µm CMOS, EM, bridge bending + force [158]
		–19	[185]
MOS-channel	$\Pi_{11} - \Pi_{12}$	97–133	n-Inversion layer, AMS [160]
	$\Pi_{11} - \Pi_{12}$	–69 to –45	p-Inversion layer, AMS [160]
	$\Pi_{11} + \Pi_{12}$	80/57/85	n-Inversion layer, IBM/TI/Lucent [37]
	$\Pi_{11} + \Pi_{12}$	–5/–3/–20	p-Inversion layer, IBM/TI/Lucent [37]
	Π_{44}	18–32	n-Inversion layer, AMS [33]
	Π_{44}	–70 to –92	p-Inversion layer, AMS [33]
	Π_{44}	10/7/15	n-Inversion layer, IBM/TI/Lucent [37]
	Π_{44}	–95/–80/–100	p-Inversion layer, IBM/TI/Lucent [37]

a) For abbreviations of process and company names, see Section 2.1.

2.5
Conclusion

This chapter has demonstrated that a large body of knowledge has been accumulated concerning the measurement of electrical, thermal, and mechanical properties of MEMS materials. Based on a broad range of methods, a considerable amount of data has been gathered and is available to guide the engineer in making material-related technical choices.

Measuring material properties specifically in view of MEMS is a demanding task. Many methods require dedicated know-how in order to be applied adequately. Aspects of fabrication technology, delicate measurement techniques and complex physical analysis have to be mastered equally well to extract properties successfully from the materials. In this sense, the area of MEMS material characterization, in particular under the additional design constraints imposed by commercial CMOS or MEMS foundry technologies, still has a long way to go to reach a level of maturity comparable to that of electrical or geometric characterization in IC industry. The list of criteria in Section 2.1.2 presents a few guidelines for possible improvements and, therefore, chances for future work.

2.6
References

1 MEMS and Nanotechnology Clearinghouse, *www.memsnet.org/material*.
2 H. Ruelke, P. Huebler, C. Streck, M. Gotuaco, W. Senninger, *Solid State Technol.* **2004**, *47 (1)*, 60–64.
3 J.F. Nye, *Physical Properties of Crystals – Their Representation by Tensors and Matrices*; Oxford: Clarendon Press, **1985**.
4 S.M. Sze, *Physics of Semiconductor Devices*, 2nd edn; New York: Wiley, **1981**.
5 A. Nathan, H. Baltes, *Microtransducer CAD – Physical and Computational Aspects*; Berlin: Springer, **1999**.
6 A.W. van Herwaarden, P.M. Sarro, *Sens. Actuators* **1986**, *10*, 321–346.
7 H. Baltes, O. Paul, D. Jaeggi, *Thermal CMOS Sensors – An Overview*, in: *Sensors Update 95*; Weinheim: VCH, **1996**, pp. 121–142.
8 L. Onsager, *Phys. Rev.* **1931**, *37*, 405–426 and *Phys. Rev.* **1931**, *38*, 2265–2279.
9 A. Schaufelbühl, U. Münch, C. Menolfi, O. Brand, O. Paul, Q. Huang, H. Baltes, in: *Tech. Dig. MEMS 2001*; **2001**, pp. 200–203.
10 F. Mayer, A. Häberli, G. Ofner, H. Jacobs, O. Paul, H. Baltes, in: *Tech. Dig. IEDM '97*; **1997**, pp. 895–898.
11 O. Paul, A. Häberli, P. Malcovati, H. Baltes, in: *Tech. Dig. IEDM 94*; **1994**, pp. 131–134.
12 M. Gaitan, J. Kinard, D.X. Huang, in: *Dig. Tech. Papers Transducers '93*; **1993**, pp. 1012–1015.
13 D. Jaeggi, C. Azeredo Leme, P. O'Leary, H. Baltes, in: *Dig. Tech. Papers Transducers '93*; **1993**, pp. 462–465.
14 M. Gaitan, M. Parameswaran, R.B. Johnson, R. Chung, *Proc. SPIE* **1993**, *1969*, 363–369.
15 R.B. Johnson, R. Chung, M. Gaitan, D. Berning, *Proc. SPIE* **1994**, *2269*, 338–347.
16 L.D. Landau, E.M. Lifshitz, *Theory of Elasticity*, 3rd edn; London: Butterworth-Heinemann, **1986**.
17 K.E. Petersen, *Proc. IEEE* **1982**, *70*, 420–457.
18 D. Jaeggi, unpublished results, personal communication.

19 A. Haque, M.T.A. Saif, *Sens. Actuators A* **2002**, *97/98*, 239–245.
20 T.L. Anderson, *Fracture Mechanics – Fundamentals and Applications*, 2nd edn; Boca Raton, FL: CRC Press, **1995**.
21 D. Broek, *Elementary Engineering Fracture Mechanics*, 3rd edn; Dordrecht: Kluwer, **1986**.
22 J. Soderkvist, *J. Micromech. Microeng.* **1993**, *3*, 24–31.
23 M. Hornung, O. Brand, O. Paul, H. Baltes, C. Kuratli, Q. Huang, in: *Proc. MEMS '98*; **1998**, pp. 643–648.
24 O. Paul, H. Baltes, *J. Micromech. Microeng.* **1999**, *9*, 19–29.
25 D. Westberg, O. Paul, G.I. Andersson, H. Baltes, *Sens. Actuators A* **1999**, *73*, 243–251.
26 R. Abdolvand, G.K. Ho, A. Erbil, F. Ayazi, in: *Dig. Tech. Papers Transducers '03*; **2003**, pp. 324–327.
27 C. Zener, *Internal Friction in Solids – Part I. Theory of Internal Friction in Reeds*; Physical Review, **1937**, *52*, 230–235.
28 G. Dorda, I. Eisele, *Phys. Status Solidi* **1973**, *20*, 263–273.
29 I. Eisele, *Surf. Sci.* **1977**, *73*, 315–337.
30 B. Kloeck, N.F. de Rooij, in: *Semiconductor Sensors*, S.M. Sze (ed.), New York: Wiley-Interscience, **1994**.
31 J. Schwizer, W.H. Song, M. Mayer, O. Brand, H. Baltes, in: *Dig. Tech. Papers Transducers '03*; **2003**, pp. 440–443.
32 M. Mayer, O. Paul, D. Bolliger, H. Baltes, *IEEE Trans. Components Packag. Technol.* **2000**, *23*, 393–398.
33 M. Doelle, P. Ruther, O. Paul, in: *Proc. MEMS 2003*; **2003**, pp. 490–493.
34 M. Doelle, C. Peters, P. Gieschke, P. Ruther, O. Paul, in: *Tech. Digest IEEE MEMS 2004*; **2004**, pp. 829–832.
35 J. Bartholomeyczik, P. Ruther, O. Paul, in: *Proc. IEEE Sensors 2003 (CD-ROM), Toronto, October 2003*; **2003**, pp. 242–247; in: *Book of Abstracts, Sensors 2003, Torronto*; **2003**, pp. 50–51.
36 Y. Kanda, *IEEE Trans. Electron Devices* **1982**, *ED-29*, 64–70.
37 A.T. Bradley, R.C. Jaeger, J.C. Suhling, K.J. O'Connor, *IEEE Trans. Electron Devices* **2001**, *48*, 2009–2015.
38 D.K. Schroder, *Semiconductor Material and Device Characterization*; New York: Wiley, **1998**.
39 W.R. Runyan, T.J. Shaffner, *Semiconductor Measurements and Instrumentation*, 2nd edn; New York: McGraw-Hill, **1997**.
40 *Modular Mixed Signal Technology [XC10], process description*, rev. 3.0; **2003**, http://www.xfab.com/xfab/frontend/index.php4.
41 L.J. van der Pauw, *Philips Res. Rep.* **1958**, *13*, 1–9.
42 J.M. David, M.G. Bühler, *Solid-State Electron.* **1977**, *20*, 539–543.
43 M. von Arx, O. Paul, H. Baltes, *IEEE Trans. Semicond. Manuf.* **1997**, *10*, 201–208.
44 M. von Arx, O. Paul, H. Baltes, *Proc. IEEE ICMTS 1996*, *9*, 117–122.
45 O. Paul, M. von Arx, H. Baltes, in: *Dig. Tech. Papers, Transducers '95*; **1995**, Vol. 1, pp. 178–181.
46 M. Strasser, R. Aigner, M. Franosch, G. Wachutka, *Sens. Actuators A* **2002**, *97/98*, 535–542.
47 R.E. Jones, S.P. Wesolowski, *J. Appl. Phys.* **1984**, *56*, 1701–1706.
48 D.R. Lide (ed.) *Handbook of Chemistry and Physics*, 76th edn; Boca Raton, FL: CRC Press, **1995**.
49 T.H. Geballe, G.W. Hull, *Phys. Rev.* **1955**, *98*, 940–947.
50 F. Völklein, H. Baltes, *Sens. Mater.* **1992**, *3*, 325–334.
51 A.W. van Herwaarden, *Sens. Actuators* **1984**, *6*, 245–254.
52 A.W. Herwaarden, D.C. van Duyn, B.W. Oudheusden, P.M. Sarro, *Sens. Actuators A* **1989**, *21–23*, 621–630.
53 R.A. Smith, *Semiconductors*, 2nd edn; Cambridge: Cambridge University Press, **1978**.
54 M. von Arx, *Thermal Properties of CMOS Thin Films*, PhD-Thesis, No. 12743; ETH Zurich, **1998**.
55 M. Strasser, R. Aigner, G. Wachutka, in: *Proc. Eurosensors XIV*; **2000**, pp. 17–20.
56 D.G. Cahill, *Rev. Sci. Instrum.* **1990**, *61*, 802–808.
57 B.M. Tenbroek, R.J.T. Bunyan, G. Whiting, W. Redman-White, J. Uren, K.M. Brunson, M.S.L. Lee, C.F. Edwards, *IEEE Trans. Electron Devices* **1999**, *46*, 251–253.

58 K.E. Goodson, M.I. Flik, L.T. Su, D.A. Antoniadis, *IEEE Electron Device Lett.* **1993**, *14*, 490–492.

59 F.R. Brotzen, P.J. Loos, D.P. Brady, *Thin Solid Films* **1992**, *207*, 197–201.

60 M.B. Kleiner, S.A. Kühn, W. Weber, in: *Proc. ESSDERC '95, Erlangen-Nürnberg;* **1995**, pp. 473–476.

61 S.-M. Lee, D.G. Cahill, *J. Appl. Phys.* **1997**, *81*, 2590–2595.

62 Y.C. Tai, C.H. Mastrangelo, R.S. Muller, *J. Appl. Phys.* **1988**, *63*, 1442–1447.

63 C.H. Mastrangelo, Y.C. Tai, R.S. Muller, *Sens. Actuators A* **1990**, *21–23*, 856–860.

64 A.D. McConnel, S. Uma, K.E. Goodson, *J. Microelectromech. Syst.* **2001**, *10*, 360–369.

65 F. Völklein, *Thin Solid Films,* **1990**, *188*, 27–33.

66 O. Paul, J.G. Korvink, H. Baltes, *Sens. Actuators A* **1994**, *41/42*, 161–164.

67 F. Völklein, H. Baltes, *J. Microelectromech. Syst.* **1992**, *1*, 193–196.

68 O. Paul, M. von Arx, H. Baltes, in: *Semiconductor Characterization: Present and Future Needs*, W.M. Bullis, D.G. Seiler, A.C. Diebold (eds.); Woodbury, NY: AIP, **1996**, pp. 197–201.

69 O. Paul, P. Ruther, L. Plattner, H. Baltes, *IEEE Trans. Semicond. Manuf.* **2000**, *13*, 159–166.

70 S. Hafizovic, O. Paul, *Sens. Actuators A* **2002**, *97/98*, 246–252.

71 M. von Arx, O. Paul, H. Baltes, *J. Microelectromech. Syst.* **2000**, *9*, 136–145.

72 S. Touloukian, R.W. Powell, C.Y. Ho, P.G. Klemens, *Thermophysical Properties of Matter;* New York: Plenum, **1970**, Vol. 2, pp. 183–193.

73 S. Touloukian, R.W. Powell, C.Y. Ho, P.G. Klemens, *Thermophysical Properties of Matter;* New York: Plenum, **1970**, Vol. 1, pp. 1–9.

74 D.G. Cahill, R.O. Pohl, *Phys. Rev. B* **1987**, *35*, 4067–4073.

75 C.H. Mastrangelo, R.S. Muller, *Sens. Mater.* **1988**, *3*, 133–142.

76 P. Eriksson, J.Y. Andersson, G. Stemme, *J. Microelectromech. Syst.* **1997**, *6*, 55–61.

77 M. von Arx, L. Plattner, O. Paul, H. Baltes, *Sens. Mater.* **1998**, *10*, 503–517.

78 M. von Arx, O. Paul, H. Baltes, *IEEE Trans. Semicond. Manuf.* **1998**, *11*, 217–224.

79 D.W. Denlinger, E.N. Abarra, K. Allen, P.W. Rooney, M.T. Messer, S.K. Watson, F. Hellmann, *Rev. Sci. Instrum.* **1994**, *64*, 946–959.

80 M. von Arx, O. Paul, H. Baltes, in: *Dig. Tech. Papers Transducers '97;* **1997**, pp. 619–622.

81 J.W. Beams, in: *Structure and Properties of Thin Films*, C.A. Neugebaur, J.B. Newkirk, D.A. Vermilyea (eds.); New York: Wiley, **1959**, pp. 183–192.

82 E.I. Bromley, J.N. Randall, D.C. Flanders, R.W. Mountain, *J. Vac. Sci. Technol. B* **1983**, *1*, 1364–1366.

83 M. Sheplak, J. Dugundji, *J. Appl. Mech.* **1998**, *65*, 107–115.

84 T. Kramer, *Mechanical Properties of Compressive Silicon Nitride Thin Films*, PhD Thesis; University of Freiburg (Wissenschaftlicher Verlag, Berlin), **2003**.

85 S. Timoshenko, S. Woinowsky-Krieger, *Theory of Plates and Shells;* New York: McGraw-Hill, **1987**.

86 A.E. Love, *Treatise on the Mathematical Theory of Elasticity*, 4th edn.; New York: Dover, **1944**.

87 V. Ziebart, *Mechanical Properties of CMOS Thin Films*, PhD Thesis, No. 13457; ETH Zurich, **1999**.

88 G. Gerlach, A. Schroth, P. Pertsch, *Sens. Mater.* **1996**, *8*, 79–98.

89 V. Ziebart, O. Paul, H. Baltes, *J. Microelectromech. Syst.* **1999**, *8*, 423–432.

90 O. Paul, T. Kramer, in: *Dig. Tech. Papers Transducers 2003, Boston;* **2003**, pp. 432–435.

91 M.K. Small, J.J. Vlassak, W.D. Nix, *Mater. Res. Soc. Symp. Proc.* **1991**, *239*, 13–18.

92 J.Y. Pan, P. Lin, F. Maseeh, S.D. Senturia, in: *IEEE Solid-State Sens. Actuators Workshop, Hilton Head, SC;* **1988**, pp. 84–87.

93 S. Levy, in: *Non-Linear Problems in Mechanics of Continua, Proc. Symp. in Applied Mathematics;* **1949**, Vol. 1, pp. 197–210.

94 J.J. Vlassak, *New experimental techniques and analysis methods for the study of the mechanical properties of materials in small volumes*, PhD Dissertation; Stanford University, Palo Alto, CA, **1994**.

95 O. Tabata, K. Kawahata, S. Sugiyama, I. Igarashi, *Sens. Actuators* **1989**, *20*, 135–141.
96 R. A. Stewart, J. Kim, E. S. Kim, R. M. White, R. S. Muller, *Sens. Mater.* **1991**, *2*, 285–298.
97 M. Mehregany, M. G. Allen, S. D. Senturia, in: *IEEE Solid-State Sensors Workshop*; **1986**, pp. 58–61.
98 V. Ziebart, O. Paul, U. Münch, H. Baltes, *Mater. Res. Soc. Symp. Proc.* **1998**, *505*, 27–32.
99 D. Maier-Schneider, J. Maibach, E. Obermeier, *J. Microelectromech. Syst.* **1995**, *4*, 238–241.
100 W. H. Press, S. A. Teukolsky, W. T. Vetterling, B. P. Flannery, *Numerical Recipes in C*; Cambridge: Cambridge University Press, **1995**, pp. 420–423.
101 T. Kramer, O. Paul, *J. Micromech. Microeng.* **2002**, *12*, 475–478.
102 Jinling Yang, C. Peters, O. Paul, *Sens. Actuators A* **2002**, *97/98*, 520–526.
103 J. J. Vlassak, W. D. Nix, *J. Mater. Res.* **1992**, *7*, 3242–3249.
104 D. Maier-Schneider, J. Maibach, E. Obermeier, *J. Micromech. Microeng.* **1992**, *2*, 173–175.
105 J. Y. Pan, *A study of suspended-membrane and acoustic techniques for the determination of the mechanical properties of thin polymer films*, PhD Dissertation; MIT, **1991**.
106 E. Bonnotte, P. Delobelle, L. Bornier, *J. Mater. Res.* **1997**, *12*, 2234–2248.
107 T. Kramer, O. Paul, in: *Proc. MEMS 2003 Conference, Kyoto*; **2003**, pp. 678–681.
108 V. Ziebart, O. Paul, U. Münch, J. Schwizer, O. Paul, H. Baltes, *J. Microelectromech. Syst.* **1998**, *7*, 320–328.
109 J. Yang, unpublished results, personal communication.
110 D. G. S. Davies, *Proc. Br. Ceram. Soc.* **1973**, *22*, 429–452.
111 O. Tabata, T. Tsuchiya, N. Fujitsuka, in: *Tech. Dig. 12th Sensor Symposium*; **1994**, pp. 19–22.
112 V. Ziebart, O. Paul, H. Baltes, *Mater. Res. Soc. Symp. Proc.* **1999**, *546*, 103–108.
113 A. Doll, F. Goldschmidtboeing, P. Woias, in: *Tech. Dig. IEEE MEMS 2004 Conference, Maastricht*; **2004**, pp. 665–668.
114 J. Sizemore, D. A. Stevenson, J. Stringer, *Mater. Res. Soc. Symp. Proc.* **1993**, *308*, 165–170.
115 P. A. Flinn, *Mater. Res. Soc. Symp. Proc.* **1989**, *130*, 41–51.
116 M. Ohring, *The Material Science of Thin Films*; San Diego: Academic Press, **1992**.
117 D. E. Fahnline, *Mater. Res. Soc. Symp. Proc.* **1992**, *239*, 251–256.
118 N. A. Winfree, Y.-C. Tai, W. H. Hsieh, R. Wu, in: *Dig. Tech. Papers Transducers 1993, Yokohama*; **1993**, pp. 179–182.
119 L. B. Freund, J. A. Floro, E. Chason, *Appl. Phys. Lett.* **1999**, *74*, 1987–1989.
120 C. A. Klein, *J. Appl. Phys.* **2000**, *88*, 5487–5489.
121 J. F. L. Goosen, B. P. von Drieenhuizen, P. J. French, R. F. Wolffenbuttel, in: *Dig. Tech. Papers Transducers 93, Yokohama*; **1993**, pp. 783–786.
122 L. Lin, R. T. Howe, A. P. Pisano, in: *Proc. MEMS 93 Workshop*, Fort Lauderdale; **1993**, pp. 201–206.
123 J. R. Roark, *Roark's Formula for Stress and Strain*, 6th edn.; New York: McGraw-Hill, **1989**.
124 O. Paul, D. Westberg, M. Hornung, V. Ziebart, H. Baltes, in: *Proc. IEEE MEMS Workshop 1997, Nagoya*; **1997**, pp. 523–528.
125 H. Guckel, D. Burns, C. Rutigliano, E. Lovell, B. Choi, *J. Micromech. Microeng.* **1992**, *2*, 86–95.
126 R. Mutikainen, M. Orpana, *Mater. Res. Soc. Symp. Proc.* **1993**, *308*, 153–158.
127 T. Kramer, O. Paul, *Sens. Actuators A* **2001**, *92*, 292–298.
128 T. Y. Zhang, X. Zhang, Y. Zohar, *J. Micromech. Microeng.* **1998**, *8*, 243–249.
129 M. P. de Boer, T. M. Mayer, *Mater. Res. Soc. Bull.* **2001**, 302–304.
130 J. W. Rogers, T. J. Mackin, L. M. Phinney, *J. Microelectromech. Syst.* **2002**, *11*, 512–520.
131 P. M. Osterberg, S. D. Senturia, *J. Microelectromech. Syst.* **1997**, *6*, 107–118.
132 B. D. Jensen, M. P. de Boer, S. L. Miller, in: *Proc. MSM 99, San Juan, Puerto Rico*; **1999**, pp. 206–209.
133 M. P. de Boer, B. D. Jensen, F. Bitsie, *Proc. SPIE* **1999**, *3875*, 97–103.

134 W. N. Sharpe, J. Bagdahn, K. Jackson, G. Coles, *J. Mater. Res.* **2003**, *38*, 4075–4079.

135 S. Kamiya, J. Kuypers, A. Trautmann, P. Ruther, O. Paul, in: *Tech. Dig. IEEE MEMS 2004 Conference*; **2004**, pp. 185–188.

136 K. Sato, T. Yoshioka, T. Ando, M. Shikida, T. Kawabata, *Sens. Actuators A* **1998**, *70*, 148–152.

137 X. Li, T. Kasai, S. Nakao, H. Tanaka, T. Ando, M. Shikida, K. Sato, in: *Tech. Dig. Transducers 2003, Boston*; **2003**, pp. 444–447.

138 T. Tsuchiya, O. Tabata, J. Sakata, Y. Taga, *J. Microelectromech. Syst.* **1998**, *7*, 106–113.

139 T. Tsuchiya, A. Inoue, J. Sakata, *Sens. Actuators A* **2000**, *82*, 286–290.

140 T. Tsuchiya, M. Shikida, K. Sato, *Sens. Actuators A* **2002**, *97-/98*, 492–496.

141 Y. Higo, K. Takashima, M. Shimojo, S. Sugiura, B. Pfister, M. V. Swain, *Mater. Res. Soc. Symp. Proc.* **2000**, *605*, 241–246.

142 W. N. Sharpe, J. Bagdahn, *Mech. Mater.* **2004**, *36*, 3–11.

143 H. Ogawa, K. Suzuki, S. Kaneko, Y. Nakano, Y. Ishikawa, T. Kitahara, in: *Proc. IEEE MEMS Workshop, Nagoya*; **1997**, pp. 430–435.

144 W. N. Sharpe, K. R. Vaidyanathan, B. Yuan, R. L. Edwards, Materials for Mechanical and Optical Microsystems Symposium, *Mater. Res. Soc.* **1997**, 185–190.

145 A. Corigliano, B. DeMasi, A. Frangi, C. Comi, A. Villa, M. Marchi, *J. Microelectromech. Syst.* **2004**, *13*, 200–219.

146 T. Ando, M. Shikida, K. Sato, *Sens. Acuators A* **2001**, *93*, 70–75.

147 U. Wagner, W. Bernhard, R. Müller-Fiedler, J. Bagdahn, B. Michel, O. Paul, in: *Dig. Tech. Papers Transducers 2003, Boston*; **2003**, pp. 175–178.

148 T. Tsuchiya, J. Sakata, Y. Taga, *Mater. Res. Soc. Symp. Proc.* **1998**, *505*, 285–290.

149 T. Tsuchiya, M. Hirata, N. Chiba, R. Udo, Y. Yoshitomi, T. Ando, K. Sato, K. Takashima, Y. Higo, Y. Saotome, H. Ogawa, K. Ozaki, in: *Proc. MEMS 2003 Conference, Maastricht*; **2003**, pp. 666–669.

150 A. B. Mann, in: *Handbook of Nanotechnology*, B. Bhushan (ed.); Berlin: Springer, **2004**, pp. 687–716.

151 W. C. Oliver, G. M. Pharr, *J. Mater. Res.* **1992**, *7*, 1564–1583.

152 O. R. Shojaei, A. Karimi, *Thin Solid Films* **1998**, *332*, 202–208.

153 M. Bamber, A. Mann, B. Derby, *Mater. Res. Soc. Symp. Proc.* **2002**, *695*, 183–188.

154 W. D. Nix, *Metall. Trans.* **1989**, *20A*, 2217–2245.

155 S. P. Baker, *Thin Solid Films* **1997**, *308/309*, 289–296.

156 S. A. Gee, V. R. Akylas, W. F. van den Bogart, *IEEE Proc. Microelectron. Test Struct.* **1988**, *1*, 185–191.

157 R. E. Beaty, J. C. Suhling, C. A. Moody, D. A. Bittle, R. W. Johnson, R. D. Butler, R. C. Jaeger, in: *Proc. 40th IEEE Elect. Components and Technology Conf.*; **1990**, pp. 797–806.

158 M. Mayer, O. Paul, H. Baltes, in: *Proc. MME 1997*; **1997**, pp. 203–206.

159 U. Schiller, *Thermomechanical Offset in Integrated Hall Plates, Diploma Thesis*; IMTEK, University of Freiburg, **2001**.

160 M. Doelle, P. Ruther, O. Paul, in: *Eurosensors XVII, Guimaraes, September 2003 (CD-ROM)*; **2003**, pp. 189–192; *Book of Abstracts, Eurosensors XVII, Guimaraes*; **2003**, pp. 110–111.

161 D. Jaeggi, *Thermal Converters by CMOS Technology, PhD Dissertation*, No. 11567; ETH Zurich, **1996**.

162 U. Münch, *Industrial CMOS Technology for Thermal Imagers, PhD Dissertation*, No. 13801; ETH Zurich, **2000**.

163 U. Münch, H. Baltes, O. Paul, E. Doering, *Proc. SPIE* **1999**, *3891*, 344–351.

164 T. Müller, *An Industrial CMOS Process Family for Integrated Silicon Sensors, PhD Dissertation*, No. 13463; ETH Zurich, **1999**.

165 G. Smolinsky, T. P. H. F. Wendling, *J. Electrochem. Soc., Solid-State Sci. Technol.* **1985**, *132*, 950–954.

166 M. Stadtmueller, *J. Electrochem. Soc.* **1992**, *139*, 3669–3674.

167 H. Kapels, D. Maier-Schneider, R. Schneider, C. Hierold, in: *Proc. Eurosensors XIII, The Hague*; **1999**, pp. 393–396.

168 D. Maier-Schneider, A. Köprülülü, E. Obermeier, *J. Micromech. Microeng.* **1995**, *5*, 121.

169 R. I. Pratt, G. C. Johnson, R. T. Howe, J. C. Chang, in: *Dig. Tech. Papers Transducers 1991, San Francisco;* **1991**, pp. 205–208.

170 S. Koller, V. Ziebart, O. Paul, O. Brand, H. Baltes, P. M. Sarro, M. J. Vellekoop, *Proc. SPIE* **1998**, *3328*, 102–109.

171 T. Kramer, O. Paul, *Sens. Actuators A* **2001**, *92*, 292–298.

172 H. J. McSkimin, *J. Appl. Phys.* **1953**, *24*, 988–997.

173 H. Kahn, S. Stemmer, K. Nandakumar, A. H. Hever, R. Mullen, R. Ballarini, M. A. Huff, in: *Proc. IEEE MEMS Workshop, San Diego;* **1996**, pp. 343–348.

174 M. Biebl, G. Brandl, R. T. Howe, in: *Dig. Tech. Papers Transducers 1995, Stockholm;* **1995**, pp. 80–83.

175 D. Maier-Schneider, A. Ersoy, J. Maibach, D. Schneider, E. Obermeier, *Sens. Mater.* **1995**, *7*, 121–129.

176 H. Kapels, J. Urscher, R. Aigner, R. Sattler, G. Wachutka, J. Binder, in: *Proc. Eurosensors XIII, The Hague;* **1999**, pp. 379–382.

177 K. Minoshima, T. Terada, K. Komai, *Fatigue Fract. Mater. Struct.* **2000**, *23*, 1033–1040.

178 T. Namazu, Y. Isono, T. Tanaka, *J. Microelectromech. Syst.* **2002**, *11*, 125–135.

179 F. Ebrahimi, L. Kalwani, *Mater. Sci. Eng. A* **1999**, *268*, 116–126.

180 S. Greek, F. Ericson, S. Johanson, M. Fürtsch, A. Rump, *J. Micromech. Microeng.* **1999**, *9*, 245–251.

181 C. S. Smith, *Phys. Rev.* **1954**, *94*, 42–49.

182 B. J. Lwo, T. S. Chen, C. H. Kao, Y. L. Lin, *J. Electron. Packag.* **2002**, *124*, 197–205.

183 B. J. Lwo, T. S. Chen, C. H. Kao, Y. L. Lin, *J. Electron. Packag.* **2002**, *124*, 22–26.

184 O. N. Tufte, E. L. Stelzer, *J. Appl. Phys.* **1963**, *34*, 313–317.

185 J. N. Sweet, in: *Thermal Stress and Strain in Microelectronics Packaging,* J. H. Lau (ed.); New York: Van Nostrand Reinhold, **1993**, pp. 221–227.

186 W. Pietrenko, *Phys. Status Solidi* **1977**, *41*, 197–205.

187 S. F. Chu, *Piezoresistive Properties of Boron and Phosphorous Implanted Layers in Silicon, PhD Dissertation;* Case Western Reserve University, **1978**.

3
Monolithically Integrated Inertial Sensors

G. K. Fedder, ECE Department & The Robotics Institute, Carnegie Mellon University, Pittsburgh, PA, USA
J. Chae, K. Najafi, Department of Electrical Engineering and Computer Sciences, University of Michigan, Ann Arbor, MI, USA
T. Denison, J. Kuang, S. Lewis, Analog Devices, Inc., Cambridge, MA, USA
H. Kulah, Middle East Technical University, Ankara, Turkey

Abstract

This chapter provides a comprehensive overview of MEMS inertial sensors that are integrated with foundry electronics. A brief summary of micro-accelerometer and gyroscope applications and operation is presented, followed by descriptions of inertial sensors catalogued by the type of structural material. Integrated sensors include those made with structural layers of polysilicon, CMOS metal-dielectric stacks, plated metal and bulk silicon.

Keywords

Inertial sensors; accelerometers; gyroscopes; integrated CMOS MEMS

3.1	Introduction	139
3.1.1	Applications and Motivation	139
3.1.2	Accelerometer Operation and Metrics	140
3.1.3	Vibratory-rate Gyroscope Operation	144
3.1.4	Integrated Inertial Sensor Process Technology	145
3.2	**Integrated Polysilicon Inertial Sensors**	**146**
3.2.1	Introduction	146
3.2.2	Analog Devices, Inc. Family of Polysilicon Integrated Inertial Sensors	146
3.2.2.1	Brief History	146
3.2.2.2	XL50	146

Advanced Micro and Nanosystems. Vol. 2. CMOS – MEMS.
Edited by H. Baltes, O. Brand, G. K. Fedder, C. Hierold, J. Korvink, O. Tabata
Copyright © 2005 WILEY-VCH Verlag GmbH & Co. KGaA, Weinheim
ISBN: 3-527-31080-0

3.2.2.3	Stiction	149
3.2.2.4	Parametric Adjustments	149
3.2.2.5	Packaging	150
3.2.2.6	Testing	150
3.2.2.7	Closed-loop Versus Open-loop Architectures	150
3.2.2.8	XL76 Family Architecture	151
3.2.2.9	Voltage Mode Versus Current Mode Sensing	151
3.2.2.10	XL202	153
3.2.2.11	XL78	154
3.2.2.12	XL203	155
3.2.2.13	XRS150/300 Gyroscope	155
3.2.3	Robert Bosch Epi-polysilicon Accelerometer	160
3.2.4	Sandia National Laboratories	160
3.2.5	University of California, Berkeley (UCB)	161
3.2.6	Defence Evaluation and Research Agency in the UK	167
3.3	**Thin-film Inertial Sensors within Foundry CMOS**	**167**
3.3.1	Introduction	167
3.3.2	Siemens Gate Polysilicon Accelerometer	167
3.3.3	UCB Piezoresistive CMOS–BEOL Accelerometer	169
3.3.4	Carnegie Mellon Series of CMOS–BEOL Inertial Sensors	169
3.3.4.1	Lateral Accelerometer	169
3.3.4.2	Thermally Stabilized Accelerometer	173
3.3.4.3	Thin-film CMOS–MEMS Gyroscopes	174
3.3.5	MEMSIC Thermal Accelerometer	175
3.4	**Metal Inertial Sensors**	**176**
3.4.1	Introduction	176
3.4.2	Metal-plated Accelerometers	177
3.4.3	Vibrating Metal Ring Gyroscope	178
3.5	**Bulk Silicon Integrated MEMS Inertial Sensors**	**180**
3.5.1	Piezoresistive Bulk Silicon Accelerometers	180
3.5.2	ADI SOI MEMS	183
3.5.3	Silicon-on-Glass (SOG) Accelerometers Developed at the University of Michigan	185
3.5.4	Carnegie Mellon DRIE-Si CMOS–MEMS	186
3.6	**Future Prospects**	**187**
3.7	**References**	**188**

3.1
Introduction

3.1.1
Applications and Motivation

The first application of integrated inertial sensors was in airbag deployment for automobiles. In the past decade, applications have emerged in vehicle stability and suspension systems, car alarms, biomedical activity monitoring, active stabilization for video recorders, computer input devices, remote controls, embedded sensing in sports equipment and electronic game interfaces. Military applications include navigation and guidance of platforms, safing and arming of ordnance, impact detection for missiles and condition-based maintenance of systems. It is expected that MEMS inertial sensors will be employed increasingly as tilt, shock and vibration sensors in cell phones, hard-disk drives, appliances and other consumer electronics. Sensor fusion of MEMS inertial sensors with GPS is increasing in sophistication. All of these applications require low-cost devices made in high volume and motivate the continuing development of integrated MEMS inertial sensors. As the cost of sensors is reduced, use in disposable applications, such as package shipment shock recorders, will expand. Low-cost micro-g accelerometers will lead to new markets for navigation in consumer goods, including augmentation of GPS receivers.

Having the electronics on the same chip as the inertial sensor structure is very desirable for minimum size, minimum wire bonding, ease of assembly and the ability to place the electronics right at the structure and thus reduce parasitic loading. The latter advantage is most applicable to capacitive inertial sensors where any parasitic capacitance loading the high-impedance output nodes will lower sensitivity. This reduction in parasitic loading due to integration in turn allows the use of smaller structures to obtain the same signal-to-noise ratio, thus allowing an even smaller area die.

The difficulties of developing and supporting a process to do full integration, however, have kept all but a few groups from following this path. Therefore, most MEMS inertial sensors are not integrated on chip with electronics, but instead have their electronics interface on a separate chip. The ability to separate sensor development from circuit development, with the faster turn time this allows, and to use whatever electronics process is desired is a very powerful counter argument to integration. However, these devices are outside the scope of this chapter. Instead, the interested reader is directed to reviews of general progress with MEMS inertial sensors [1, 2]. An overview of inertial sensor technology trends, with emphasis on military applications, has been published [3].

3.1.2
Accelerometer Operation and Metrics

Microaccelerometers are second-order mass-spring-damper systems modeled by the force balance equation:

$$ma_{ext} = m\ddot{x} + b\dot{x} + kx$$

where m is the suspended proof mass, b is the damping coefficient and k is the suspension spring constant. The external acceleration, a_{ext}, of the chip substrate and package is defined in the opposite direction of the displacement, x, of the proof mass relative to the rest of the chip. This 'pendulous' accelerometer operation is illustrated in Fig. 3.1. For most applications, the bandwidth of acceleration is from DC to about 400 Hz. For example, typical airbag use requires this bandwidth. This signal band is normally much lower than the mechanical resonance, so the proof mass' displacement is approximately related to external acceleration by the DC mechanical sensitivity:

$$\frac{x}{a_{ext}} = \frac{m}{k} = \frac{1}{\omega_r^2}$$

where $\omega_r = 2\pi f_r$ is the undamped mechanical resonant frequency. Most applications, with the important exceptions of seismometry and inclinometers, require at least a 400 Hz bandwidth, which limits how far the resonant frequency can be lowered to increase sensitivity. The other primary constraints on lowering the resonant frequency are the maximum input acceleration, shock tolerance, process limitations and device area. A higher sensitivity means the proof mass will move further for a given input acceleration. The maximum input acceleration is then defined either by mechanical limit stops, by electrostatic snap-in for capacitive sensors or by saturation of the interface electronics. Contact of the proof mass with surrounding features can occur from excessive shock. The contact may lead to the mass being stuck or latched on to an adjacent surface, causing device failure. The maximum size of the proof mass and suspension is constrained by the

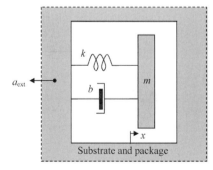

Fig. 3.1 Schematic illustrating pendulous microaccelerometer operation

3.1 Introduction

process due to residual stress and stress gradients in the materials that lead to device buckling or excessive curling. The final constraint of device area is particularly important in integrated inertial sensors, as increased transducer area translates directly into increased manufacturing cost.

Typical design values for a thin-film microaccelerometer are $m = 1\ \mu g$, $k = 2\ N/m$ and $f_r = 7.1\ kHz$. For example, the 1 μg mass may be formed from a 293 μm × 293 μm × 5 μm block of polysilicon. With this design, an external acceleration of 1 g (= 9.8 m/s^2) results in a 4.9 nm displacement of the proof mass. Average displacements of 0.049 Å must be detected to achieve a resolution of 1 milli-g, which is a common specification for commercial integrated accelerometers.

Several physical principles have been exploited for sensing proof mass displacement, including piezoresistive, capacitive and piezoelectric methods. Thermal detection has also been used to sense displacement of a controlled hot air mass for accelerometry. However, most integrated inertial sensors have employed capacitive sensing, as it has generally provided the best displacement resolution with virtually no power dissipated in the transducer element. Design of the motional sensing elements in a capacitive inertial sensor usually take the form of parallel-plate capacitors oriented in the vertical (perpendicular to substrate) direction, as in Fig. 3.2a, or of sidewall parallel-plate capacitors oriented in the lateral (in-plane to substrate) direction, as shown in Fig. 3.2b. The sidewall capacitors are usually formed from interdigitated beam fingers, called 'combs,' to increase the capacitance in a given layout area.

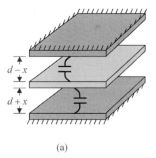

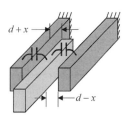

Fig. 3.2 Basic parallel-plate capacitive displacement sensors. The darker shaded structures are fixed. The lighter shaded structures are movable. (a) Vertical plate electrodes; (b) sidewall electrodes formed with interdigitated 'comb fingers'

A canonical schematic for a fully differential capacitive bridge is given in Fig. 3.3. The highest motional sensitivity will be achieved when all four capacitors in the bridge (i.e. C_1, C_2, C_3 and C_4) can change value with the proof mass displacement. Some processes place constraints on the design such that two of the capacitors remain fixed. In this case, motional sensitivity is half that of a fully differential bridge. In Fig. 3.3, a balanced AC modulation voltage, v_m, is imposed across the bridge, where the modulation may be a sinusoidal or square wave with time. The resulting sensitivity to acceleration for this configuration is

$$\frac{v_o}{a_{ext}} = \frac{1}{x\omega_r^2}\left[\frac{2v_m(C_1 - C_2)}{C_1 + C_2 + C_p}\right] \simeq \frac{2v_m}{\omega_r^2 d\left(1 + \frac{C_p}{2C_o}\right)}$$

where d is the gap between the parallel plates when the proof mass is not displaced and $C_o = C_i|_{x=0}$; $i=1, 2, 3, 4$. Of course, the output voltage, v_o, from the bridge may be amplified electronically after detection, thereby increasing the gain. The ensuing discussion refers to the transducer sensitivity, referred to the input of the interface electronics. In most integrated inertial sensors, the modulation voltage and therefore sensitivity are limited by the power rail. The gap may be decreased for higher sensitivity, although this is bounded by the minimum allowable spacing in the process. The load parasitic capacitance, C_p, reduces the sensitivity and should be kept smaller than or equal to $2C_o$. Although the parasitic capacitance is generally smaller in fully integrated devices, it is also generally true that the sense capacitor is smaller. Therefore, a comparison of sensitivity reduction due to parasitic capacitance in single-chip versus two-chip solutions must evaluate the capacitance ratio.

The resolution of an inertial sensor may be limited by thermomechanical (Brownian) noise of the proof mass or by electronic noise in the interface circuit. Damping in the micromechanical system produces the Brownian acceleration noise:

$$\sqrt{\frac{a_n^2}{\Delta f}} = \frac{\sqrt{4k_B T b}}{m}$$

where k_B is Boltzmann's constant, T is absolute temperature and Δf is the user-defined bandwidth in hertz. A rough scaling law is derived by assuming that the

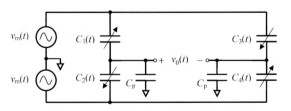

Fig. 3.3 Schematic of a fully differential capacitive bridge for detecting displacement

damping is dominated by the squeezed-film effect between the capacitor plates. Then, the damping coefficient is

$$b = \frac{\eta L_c h_c^3}{d^3} = \frac{4\eta C_o h_c^2}{\varepsilon_0 d^2}$$

where η is the viscosity of the air between the parallel plates, ε_0 is the permittivity of air, L_c is the total length of the plates in the system and h_c is the width of the plates. The damping depends on the aspect ratio of the plates and here it is assumed $h_c \ll L_c$, as would be the case in a lateral-axis capacitive sensor. The Brownian noise assuming squeezed-film damping is

$$\sqrt{\frac{a_n^2}{\Delta f}} = 4\sqrt{\frac{k_B T \eta}{\varepsilon_0}} \left(\frac{\sqrt{C_o}}{\rho A_m d}\right) \left(\frac{h_c}{h_m}\right)$$

where ρ is the proof mass density, A_m is the area of the proof mass and h_m is the height of the proof mass. Noise may be reduced by incorporating a larger proof mass and by operating in a vacuum package, which eliminates air damping. If operating in air, noise may be decreased by making the sense capacitors smaller and sense gap larger, since the squeezed-film damping is then reduced. This latter scaling effect is at some point limited by the presence of shear damping of the plate, which is neglected in the present analysis. The Brownian noise referred to the capacitive bridge output voltage is

$$\sqrt{\frac{v_{o,n}^2}{\Delta f}} \simeq \frac{8 v_m}{\rho A_m \omega_r^2 d^2} \sqrt{\frac{k_B T \eta}{\varepsilon_0}} \left(\frac{\sqrt{C_o}}{1 + \frac{C_p}{2C_o}}\right) \left(\frac{h_c}{h_m}\right)$$

Typical design values consistent with the previous thin-film microaccelerometer parameters ($m = 1$ μg, $k = \rho A_m h_m \omega_r^2 = 2$ N/m and $f_r = 7.1$ kHz) are $\eta = 1.8 \times 10^{-5}$ Pa s, $T = 294$ K, $v_m = 1$ V, $d = 2$ μm, $h_c = 5$ μm and $C_o = C_p = 25$ fF. The corresponding transducer sensitivity is 3.3 mV/g, the Brownian accelerometer noise is 15 micro-g/$\sqrt{\text{Hz}}$ and the equivalent transducer voltage noise is 48 nV/$\sqrt{\text{Hz}}$.

If the capacitive interface circuit contributes more equivalent input noise than the transducer, then the circuit limits the resolution of the accelerometer. The simplest CMOS interface circuit has the output voltage from the capacitive bridge routed to a differential transistor pair. Establishing a low-noise, stable DC bias for the input gate of the transistors is of critical importance. One of the advantages of sensing with AC modulation is the ability to frequency shift the signal band above the crossover between flicker noise and thermal noise of the interface circuit. For CMOS transistors, this crossover can be located well above 1 MHz. For such a thermal noise-limited case, the two CMOS input transistors will have a combined equivalent input voltage noise of

$$\sqrt{\frac{v_{i,n}^2}{\Delta f}} = \sqrt{8k_B T \frac{4}{3g_m}} = \sqrt{\frac{32k_B TL}{3\sqrt{2\mu_{eff} I_D C_g}}}$$

where g_m is the transconductance of the transistors, μ_{eff} is the effective mobility, I_D is the drain current, L is the gate length and C_g is the gate capacitance. The gate length is usually set to minimum dimension and the drain current is usually fixed by power constraints. The remaining design variable, gate capacitance, should be made large to achieve low noise. However, increasing the gate capacitance adds directly to the parasitic capacitance, reduces the transducer sensitivity and increases the equivalent acceleration noise contribution from the circuit. A rule of thumb to maximize resolution is to size the gate capacitance such that the equivalent input noise acceleration contributions from mechanical damping and from the circuit are equal. For a proof mass limited to 1 µg, the best resolution that can be achieved is around 10 micro-g/$\sqrt{Hz}$.

There are many capacitive interface alternatives to a balanced modulated bridge and the circuit noise analysis will change with different interface topologies. The next most popular interface technique uses switched capacitor circuits, which are reviewed in Chapter 11.

Bias (i.e. offset) and gain stability are two other important metrics for accelerometers. Most bias issues are traceable to changes in stress over time acting on the microstructure. Stress and stress gradients change from expansion over temperature intrinsic to the structure and the chip substrate. Components of extrinsic package-induced stress may be thermal based, arise from external forces acting on the package or be based on manufacturing variations in mounting, wire-bonding or encapsulation of the chip. Packaging may change its mechanical characteristics over time, producing bias drift over days, months and years.

3.1.3
Vibratory-rate Gyroscope Operation

Integrated MEMS vibratory-rate gyroscopes employ a proof mass that is excited into vibration with velocity $\vec{U}$ as measured in the reference frame of the chip. When the chip is rotated externally with rate $\vec{\Omega}$, the Coriolis acceleration of the vibratory mass in the chip frame of reference is

$$\vec{a_C} = 2\vec{\Omega} \times \vec{U}$$

The gyroscope operation is illustrated in Fig. 3.4 for the case where the velocity is driven sinusoidally in the y-axis, the rotational rate is constant around the z-axis and the Coriolis force is induced along the x-axis.

The Coriolis acceleration is detected by an accelerometer with its sensitive axis placed along the appropriate direction, which is orthogonal to the direction of vibration. In a vibratory-rate gyroscope, the velocity is sinusoidal at a fixed dither frequency and so the rotation signal band is modulated at that dither frequency.

Fig. 3.4 Schematic of vibratory-rate gyroscope operation. In this illustration, the proof mass is driven along the y axis of the chip with sinusoidal velocity U_y. A fixed external rotational rate, Ω_z, about the z-axis produces a sinusoidal Coriolis acceleration, a_C, of the proof mass in the chip frame of reference. The Coriolis acceleration induces a sinusoidal displacement along the x-axis whose amplitude is proportional to Ω_z

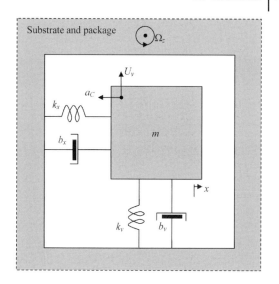

The system must demodulate the accelerometer output to provide the final gyroscope output voltage.

A useable gyroscope requires several features beyond those of an accelerometer. A stable dither velocity is needed to maintain a constant sensor gain. The dither motion along the drive axis must be balanced such that feedthrough resulting in motion on the sense axis is reduced below the sense resolution. Other external inertial force terms along the Coriolis sense axis must be rejected, both through symmetric design and using output signal processing and filtering. If these requirements are met, then the resolution of the gyroscope is directly related to the sense-axis accelerometer performance. The rate resolution is $26°/h/\sqrt{Hz}$ using an accelerometer with 10 micro-g/$\sqrt{Hz}$ resolution and a dither amplitude of 10 μm at a dither frequency of 7.1 kHz. This level of performance is considered 'rate grade' and is suitable in control applications for automobiles and stabilization for video recorders. Lower noise in the acceleration sensing is required to achieve 'inertial grade' performance of better than $1°/h/\sqrt{Hz}$.

An important point is that the vibratory-rate gyroscope allows detection of a constant rate of rotation, not just angular acceleration, which is all a simple angular accelerometer can do. It is true that integrating angular acceleration yields the rate. However, in practice, it is far too noisy an operation to yield useful rate information in all but the most undemanding applications. One must use a rate sensor instead.

3.1.4
Integrated Inertial Sensor Process Technology

Integration of inertial sensors has been accomplished within a variety of processes. The remainder of this chapter is partitioned according to the material of the micromechanical transducer structure. Integrated inertial sensors made from custom-fabricated polysilicon are reviewed first, followed by CMOS–MEMS sensors, where the microstructures are made from the CMOS interconnect stack and then metal MEMS sensors. The chapter concludes with a discussion of integrated bulk silicon sensors and predicted future directions.

3.2
Integrated Polysilicon Inertial Sensors

3.2.1
Introduction

This section is about the industrial and academic groups that have opted for integrated polysilicon sensor processing. First, the line of inertial sensors from Analog Devices, Inc. (ADI) will be presented, followed by systems designed by University of California, Berkeley (UCB) graduate students in both ADI and Sandia National Laboratories polysilicon processes. Work on integrated polysilicon for inertial sensors at Robert Bosch Corp. and the Defence Evaluation and Research Agency in the UK will be briefly covered.

3.2.2
Analog Devices, Inc. Family of Polysilicon Integrated Inertial Sensors

3.2.2.1 Brief History
In the early 1980s, Steven Sherman, an ADI design engineer, heard a presentation on the idea of forming mechanical sensors on silicon chips with IC-compatible processing that captured his imagination. Over a few years of discussions and advocacy, a small group that was led by Dr. Richard Payne was formed and funded in 1986 to try to develop an integrated sensor process initially to make accelerometers for the auto crash sensor market and then expand into other markets and sensors. The ADI team was able to demonstrate the practicality of a 'sensor middle' process using a 2 µm polysilicon sensor layer deposited and formed after all diffusions and before contact and metal deposition as a major variant of a standard 3 µm BiCMOS process in production [4] and described in more detail in Chapter 1. ADI chose a single chip process vs. the predominant competitive approach of dual sensor chip and electronics chip in the belief that integration would provide the lowest cost and smallest devices. The first product released to the market in 1993 was the XL50 [5], a 50 g full-scale voltage out accelerometer followed by a progression of devices and process improvements over the years.

3.2.2.2 XL50
The first device on the process was designed without the stability of properties of the structural polysilicon layer being fully known. What was the repeatability lot to lot of the Young's modulus? How would it perform over temperature? How would it age? How repeatable would the internal stress be? Because of these issues, it was decided

to use a force feedback design that used an electrostatic spring constant of the feedback for most of the effective spring of the device making variations over time and temperature nearly independent of the polysilicon properties as well as reducing the actual movement of the structure by the electromechanical loop gain. A conservative mechanical topology, shown in Fig. 3.5a, was chosen with a fixed-fixed beam suspension and differential capacitor comb fingers located symmetrically about the proof mass. This first design was in the original 2 μm thick polysilicon. It had approximately 1.3 μm gaps between 1.7 μm wide fingers. The spacer gap between the substrate and the polysilicon was 1.6 μm.

On early prototypes one of the first lessons in the art was learned as the spring constant was about 10× larger than calculated due to the internal tensile stress creating much stiffer springs than just the nominal mechanical Young's modulus predicted. Because of fear of folded tethers having too much vertical compliance and not having a good anti-stiction agent at the time of the XL50 development this fixed-fixed tether arrangement was maintained. It really had a good vertical restoring force. However, the lot to lot variation of this tensile stress was one of the biggest variables affecting the sensitivity of the XL50, though this was able to be adjusted for on a part by part basis.

Acceleration-induced off-center movement of the structure was detected using a differential capacitive structure and a balanced modulation carrier. Equal amplitude and opposite phase square waves at ~ 1 MHz were applied to the two fixed capacitive plates on either side of the proof mass movable plate. At precise centering, the two modulated signals cancel. When off-center, the imbalance of the differential capacitor structure causes the closer signal to become dominant. The amplitude of the detected carrier was essentially proportional to the amount of displacement from perfect center; the phasing of the detected carrier gave the direction of the movement.

Force feedback was accomplished by biasing the fixed fingers at two different, fixed, DC potentials, at 0.2 and 3.4 V, with the carrier voltages capacitively coupled to them and then feeding back an acceleration signal to the moving proof mass, which was nominally biased at 1.8 V, the mid-potential of the fixed fingers [6]. At this nominal mid-potential, the proof mass would net a zero nominal electrostatic force when physically centered between the two sets of fixed fingers.

With applied acceleration, the proof mass would move from the center position. The off-center error carrier signal caused by the movement was detected, AC amplified, synchronously demodulated and filtered with an external capacitor to set the loop crossover at ~ 400 Hz. The demodulated signal was further DC amplified and then fed back to the proof mass to force-balance it electrostatically back to the center position. This signal was level shifted, amplified and buffered by an onboard amp and presented as an analog output voltage proportional to the acceleration. The overall loop gain of the electromechanical force feedback loop was ~ 10. This made the structure about 10 times less sensitive to the polysilicon spring constant changes and reduced the amount of beam displacement by the same 10 times than without the electrostatically applied feedback. It was fortunate that the force feedback was the first electronic design as it reduced the effects of the variations of the fixed-fixed tether spring constant mentioned before.

3 Monolithically Integrated Inertial Sensors

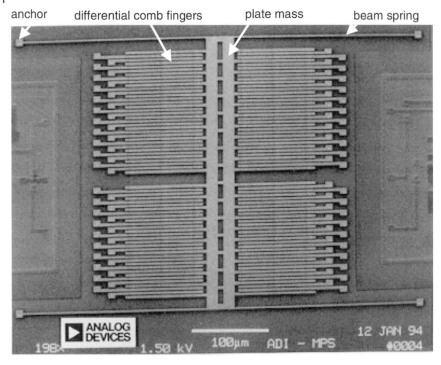

Fig. 3.5a AD XL50 - transducer

Fig. 3.5b is a simplified block diagram of the XL50.

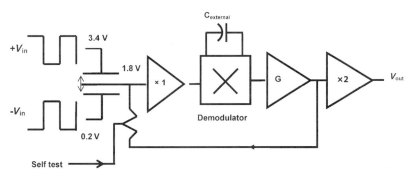

Fig. 3.5b Block diagram of the ADI XL50

To assure that the part was fully working, a self-test function was included. This digital input, when high, would cause a current to be injected into the proof mass node, which would impose a voltage change on the beam, cause it to deflect and have the deflection appear as an output of a specific delta from whatever the output was at the time of self-test actuation. This exercised the entire system, both mechanically and electrically, so that it was an excellent test of the health of the part.

The XL50 was a very successful first product and went into many applications that included close to one million cars as crash detection sensors for air bag deployment.

There were a number of problems that needed to be solved in the production of this first high-volume integrated MEMS product. The four major ones were:

1. Stiction-free release of the mechanical structure, compatible with the IC processing.
2. Adjustment of parameters without post-packaging trimming.
3. High-volume MEMS-compatible packaging.
4. High-volume electromechanical testing.

3.2.2.3 Stiction

Stiction was an issue because the 2 μm polysilicon structural layer was deposited and formed upon a 1.6 μm sacrificial oxide and, during the etching of this oxide with buffered HF, surface tension of the liquid would pull the structure down and cause it to stick. At the time of the process development, vapor-phase HF etching, commonly used now, was not a proven production process. A process was developed in which selectively etched holes were made in the sacrificial oxide under polysilicon structures. These holes, which went entirely under the polysilicon structures, were then filled with photoresist. These pedestals of photoresist would then prevent stiction when the rest of the sacrificial oxide was etched away. The pedestals would later be removed by an oxygen plasma leaving free-standing, clean, mechanical structures.

The issue with post-production stiction also had to be dealt with. Severe shocks and electrical overstress events could cause the structures to collapse and in some instances stick. Significant work went into developing a wafer-level anti-stiction agent that was able to stand up to the heat of standard IC hermetic packaging.

3.2.2.4 Parametric Adjustments

The electronics used in this integrated process did not have low-area post-packaging trimming capability. To take the approximately ±35% (or even more on the XL50) mechanical parameter variations and reduce them, to typically ±3% variation in the product, laser trimming of thin-film resistor material at the wafer stage was done to accomplish parametric adjustment. The difficulty in doing this was getting the mechanical parameters at the wafer stage. The solution was to actuate the structure electrostatically, measure certain properties, derive needed parameters and trim the electronics to compensate. In some ways this parametric adjustment method, compatible with low-cost, high-volume manufacturing, was the most impressive innovation in this technology development.

3.2.2.5 Packaging

The biggest issue with commercial MEMS in general is that of packaging. ADI uses standard hermetic IC packages for low cost and compatibility with standard IC testing equipment. The parts are assembled in a cleanroom environment to prevent particulates from causing problems. This process produces very high-quality and reliable inertial components.

The first package used was a standard 10-pin TO-100 metal can op-amp package for the XL50. Subsequent devices used cerdip, cerpac, LCC and ceramic BGA packages.

As of this writing, ADI does not cap its devices, although work is being done towards this end. It is a fairly well developed technology that has many patents from a number of different companies that have to be designed around.

3.2.2.6 Testing

ADI modified standard IC handlers to include mechanical shakers so that the parts could be tested over temperature and mechanically actuate them at the same time. Considerable time and effort went into the design of these testers. A major challenge was maintaining calibration to very accurate standards over temperature in a high-volume manufacturing environment.

The XL50 was a good start. It was shortly followed by the XL05, a $\pm 5g$ closed-loop, force-feedback design which had about a $500 \text{ micro-}g/\sqrt{Hz}$ noise floor, about 12 times lower than the XL50. The XL05 included the first folded tether design to make its spring constant insensitive to the tensile stress variations, but it did have some troubles with vertical stiction as a result. However, at this point in time, competitive price pressures dictated that lower cost, i.e. smaller area, devices had to be designed.

3.2.2.7 Closed-loop Versus Open-loop Architectures

Over time with experience from the early products, it was found that the structural polysilicon process had remarkably good mechanical properties. The aging, temperature behavior, stress properties (with proper tether design), and non-linearity issues were not things that needed to be stabilized with force feedback for the precision of applications for the markets involved. Also, the closed-loop architecture had its shortcomings. First, it required a moderate amount of electronics to create the feedback, which meant extra area and thus extra cost. Second, it required an external capacitor for the low-frequency loop compensation because the values needed were too large to produce on-chip. This extra capacitor was an applications problem and meant extra cost to users.

Another problem with the closed-loop architecture was that it was non-ratiometric to the power supply. The complexities of the force-feedback system made it difficult for the output to be ratiometric to the supply voltage without the addition of yet more circuitry. Most users wanted a power supply ratiometric voltage output since they used analog to digital converters that were on-board with their microcontrollers and these devices were ratiometric, i.e. they used the power supply voltage as their voltage reference. This meant that to use non-ratiometric parts they had to measure an on-board voltage reference. Although not a major issue, it did increase the computations and general difficulty of use.

It was also found out that the constant DC bias levels of the fixed fingers, needed to create force feedback, could cause mobile charges to move and segregate over time and create small (or not so small!) offset shifts due to the induced fields of these accumulated charge concentrations.

Therefore, in the interests of lower area and less complexity, and hence lower costs, in addition to best null stability for the next generation of accelerometer designs, it was decided to migrate to an open-loop system architecture.

3.2.2.8 XL76 Family Architecture

The XL76, or commercially available as the XL150 [5], was designed to have no external components (other than supply decoupling caps which all circuits generally need), be smaller and be ratiometric to the supplies. It also employed some improvements to the beam design that included methods to make it more immune to fabrication variations.

The suspension used was a folded-beam topology to reduce the influence of residual stress in the polysilicon on the structure. Additional anchored polysilicon beams were placed next to the suspension beams to provide a uniform etch loading that resulted in better control of the spring sidewall dimensions (Samuels patent [7]). This stabilized the signal transduction but not the resonant frequency of the structure. Through careful design centering and using these photo/etch balancing structures to make the springs and the transduction fingers etch similarly, variations in the spring constant were nominally balanced by the change in the capacitive gap transduction. As the springs got stiffer, the gaps got narrower to compensate.

The vertical stiction issue encountered with the XL05 using folded tethers was solved by a combination of being a higher g device with stiffer springs and smaller mass thus having a higher vertical restoring force with a smaller area to stick and the development of a wafer level anti-stiction agent, which took a couple of years to perfect and get into production.

3.2.2.9 Voltage Mode Versus Current Mode Sensing

Fundamentally, there are two ways to monitor the capacitive detection node. The first is just to monitor the voltage with a voltage buffer. The second is to monitor the charge or signal current, with an inverting amplifier architecture. The inverting amplifier, with its summing node at an essentially constant voltage, makes capacitive signal detection less sensitive to parasitic capacitance variations but requires considerably higher bandwidth amplifiers, depending on the gain desired, to accomplish this parasitic capacitance immunity compared with using the voltage buffer method.

Given that the sensor Brownian noise is usually less than the electronic noise, one common misconception is that the current mode or inverting amp architecture is a lower signal-to-noise ratio detection method than using a voltage buffer. Reasoning goes that since the parasitic capacitance on the sensor is at a summing node, it does not attenuate the signal as it does with the voltage buffer. It is true that the signal level is higher as a result of the summing node architecture; however, the noise level is also higher owing to the increased noise gain as a result of the parasitic capacitance. The desired figure of merit, signal-to-noise ratio, is distinctly worse with the inverting amp for low values of voltage gain than the voltage buffer and only asymptotically approaches the same signal-to-noise ratio as the voltage buffer as the inverting amplifier gain is increased.

However, one clear advantage that the inverting amplifier topology has over the voltage buffer approach is that of signal gain stability. The forward gain path for the inverting amp architecture is not affected by changes in the parasitic capacitor value whereas with the voltage buffer configuration it does. Any power supply or temperature variation of the parasitic capacitance will affect the voltage buffer signal gain and thus the volts/g at the output of the system. This can be a significant consideration in more precise applications.

For the XL76, the voltage buffer architecture was used as it was for all the earlier products. The non-linearity of the parasitically loaded capacitive divider was not an issue since at full-scale the proof mass moves only a few percent of the total gap. This yields parts with a full-scale non-linearity of well under 1%, adequate for all the envisioned uses of the part. The sensitivity change over temperature due to the parasitic capacitance variation over temperature was acceptable, although not for temperatures above about 110 °C for some applications.

A block diagram of the XL76 is shown in Fig. 3.6. Because there was no force feedback, there was no need for differing DC voltages on the fixed fingers and so the AC carriers on the fixed fingers could be switched from supply to supply. This simplified the carrier generation for lower area and made the carrier excitation increase by a factor of almost sixfold, which improved the signal-to-noise ratio of the system.

With capacitive sensors, there is always the issue of how to get the correct DC biases on the sensing node(s) without disrupting the signal detection. This is one of the main reasons for the use of carriers rather than just detecting the charge on the sensing node. The higher the carrier frequency, the lower impedance the DC potential setting network can be since the capacitive sensing impedance is inversely proportional to the frequency of the detected carrier. In the XL50 architecture, a 1 MHz carrier was used and thus the 3 MΩ feedback resistor to the capacitive sensing node of the proof mass did not load the signal appreciably.

In the XL76, the carrier was desired to be significantly lower, at about 100 kHz, since the electronics process used was far from 'fine line' at 3 µm minimum gate

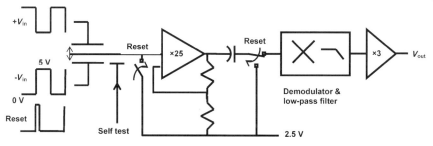

Fig. 3.6 Block diagram of the ADXL76

length, not speedy at all, and the switching architectures used needed faster operation for a given carrier frequency than the continuous time XL50 circuit architecture.

The DC bias on the proof mass sensing structure was established by a momentary switch to a reference voltage set half way between the supply rails. The switch was timed to connect and release just before the carriers switched. While connected, the demodulator would zero out the sensing amp offsets and then, when the carriers switched phase, the off-center position induced charge was impressed on the demodulation node. This acceleration signal was then filtered to 400 Hz by a switched capacitor, second-order Bessel filter. The output was a ratiometric voltage proportional to acceleration.

The first dual-axis, single-chip accelerometer, the XL276, came next. It was basically two XL76s on one chip with sensors at 90° orientation to each other, only sharing common bias and clocking circuitry. The XL105, a single-axis 5 g full-scale sensor, was developed from the basic XL76 architecture for lower g range, higher frequency and industrial and consumer applications.

The XL76 product variants are fitted in at least 30 million cars, some with multiple acceleration sensors to be able to sense the various types of crash events more accurately. It turns out that to cover lower speed, off-axis or utility pole crashes in a wide variety of car bodies, it greatly helps to have multiple sensors to discern the type of crash event and initiate the proper action, if any.

3.2.2.10 XL202

The XL202 [5], shown in Fig. 3.7, is a low-g, dual-axis accelerometer directed at consumer applications. It is a variant of the circuit architecture of the XL76 but with a single proof mass that was suspended at its four corners and free to move in both in-plane directions. The eight meander beam springs are layed out with mirror symmetry to provide an equal spring constant in the two lateral directions. Sense comb fingers are located on all four sides of the proof mass. The null accuracy of 0.5 g is limited by comb finger displacement from packaging and thermal stresses as well as x vs. y photolithography and etching effects. Unlike the XL276, the XL202 has one set of sensing electronics that were time division multiplexed

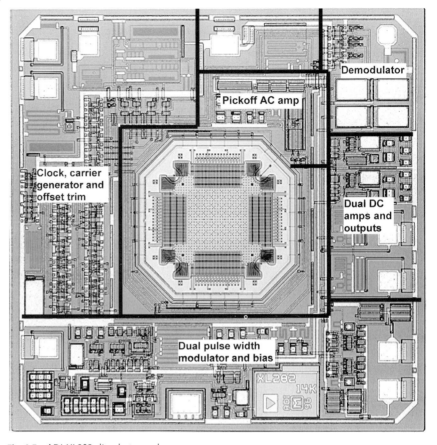

Fig. 3.7 ADI XL202 die photograph

for sensing the two axes, but have separate x- and y-axis outputs. These outputs were available as pulse width-modulated digital outputs or as analog voltages.

3.2.2.11 XL78

Price pressures again required an area reduction for the accelerometer function. The next device, the XL78, was 70% the area of the XL76. Fig. 3.8 provides a die area comparison of the XL50, XL76 and XL78 accelerometers. A combination of smaller structures and very thoughtful, minimalist circuit design accomplished this goal. One of the innovations was that of keeping the proof mass under approximately zero net electrostatic force by adjusting the position-sensing carrier signal amplitudes to be proportional to the acceleration signal amplitude in such a fashion as to reduce the amplitude of the carrier in the fixed fingers that the proof mass was moving towards. Again, a dual-axis device was also designed based on this architecture.

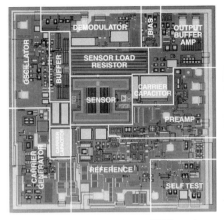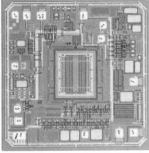

Fig. 3.8 Die area comparison of ADI accelerometers

3.2.2.12 XL203

Lessons learned in previous accelerometers in both circuit and mechanical structure design resulted in the XL203, a higher precision, higher stability, wider temperature range, low-g, dual in-plane accelerometer. In the circuit, for example, to improve the high-temperature sensitivity stability, an inverting amplifier topology was used to obviate the effect of the temperature change of parasitic capacitance value on the gain of the system. The XL203, shown in Fig. 3.9, utilized new process innovations in the structural area and a significant change in beam design to achieve typical ±50 milli-g absolute accuracy over the range −55 to 125 °C [5]. The mounting of the spring-mass and differential capacitor comb fingers was moved from the perimeter of the mechanical structure to nearly the die center. An optimal position of these anchors was obtained through extensive modeling, confirmed with prototype runs and resulted in a significant reduction of the comb-finger offset shifts from die distortion. The improved 4 μm thick, in-situ doped, structural polysilicon with an extra interconnect layer of polysilicon under the structure was used for this as well. The aspect ratio of the minimum tether width, which had been about 1:1 with the original 2 μm structural polysilicon was able to be increased to 2:1 with this process. This was important as it improved the ratio between the lateral spring constant and the vertical restoring force by a factor of almost 8, allowing compliant lateral movement with a vertical restoring force sufficient to prevent vertical stiction problems.

3.2.2.13 XRS150/300 Gyroscope

In 1994, development of a single-chip gyroscope, or angular rate to voltage converter, was started. It required many design iterations and process improvements. It was finally released to commercial production in November 2002 [5, 8].

The XRS150 is based off of a tuning-fork vibratory-rate gyroscope topology. It has an internal resonating proof mass that is attached by springs to an enclosing

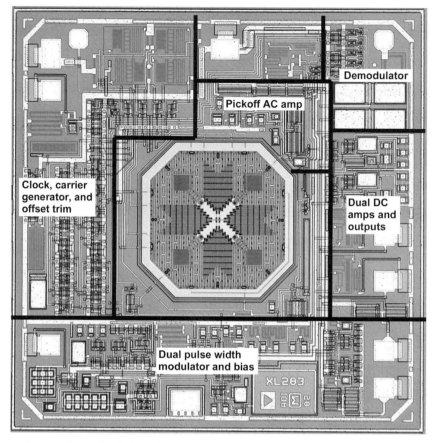

Fig. 3.9 ADI XL203 die photograph

accelerometer frame as shown in Fig. 3.10. The outside accelerometer frame is connected by springs attached to the chip's surface. The two sets of springs operate with their flexural direction at 90° from each other. The springs are designed as lever mechanisms to decrease coupling between lateral vibratory modes.

The total sensor is made of two of these structures driven anti-phase. This forms a differential sensor that better rejects vibration and any constant translational acceleration. Fig. 3.11 shows the arrangement of the structure and the various movement and sensing directions.

The resonators are driven by interdigitated comb drive structures. The Coriolis movement is sensed by conventional interdigitated differential capacitor pickoffs on the sides of the accelerometer frames. It is very tempting to increase the signal output by trying to get gain from the Q amplification of the Coriolis accelerometer structure by placing the Q peak of the accelerometer exactly on that of the resonator. However, then slight changes over temperature, time, and manufacturing can cause large shifts in sensitivity and null. For best manufacturability and stability the accel-

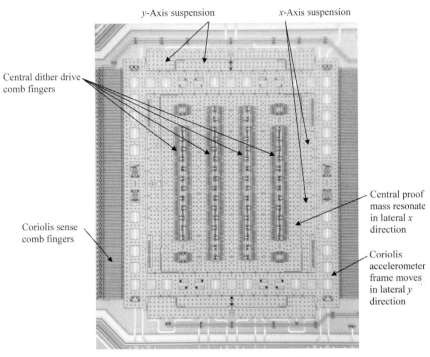

Fig. 3.10 ADI XRS150 gyroscope transducer, showing detail of one of the two resonator structures

erometer resonant frequency was placed above the resonator resonant frequency by about 16%. At this point some gain was obtained but it did not vary much over parameter variations. The signals from the two independent sensors are wired together so that common mode signals are subtracted before being presented to the processing electronics to limit the overload range that the electronics has to handle.

Extreme care was taken in the structural design to allow movement in the desired directions and reject movement in orthogonal directions. Thus mechanical movement leakage from the resonator direction into the accelerometer direction, so-called 'quadrature' movement, is rejected by a factor of nearly 100 000. This was made very difficult by having to make sure there were no mode overlaps with the structure with manufacturing variations taken into account. Considerable work was required in the photolithography and etching steps to have the part yield well in production as the requirements for sidewall angle and tether width matching were far more stringent than for accelerometers.

The device is packaged in an atmospheric ambient and not in a vacuum, as are just about all other tuning fork-type rate sensors. This non-vacuum ambient was used in order to keep costs low and to provide very high shock survival, greater than 30 000 g, as a result of the gas damping.

During the XRS150 development processing, improvements had to be made to produce a competitively specified part. Among them were adding another interconnect layer of polysilicon underneath the mechanical polysilicon and increasing

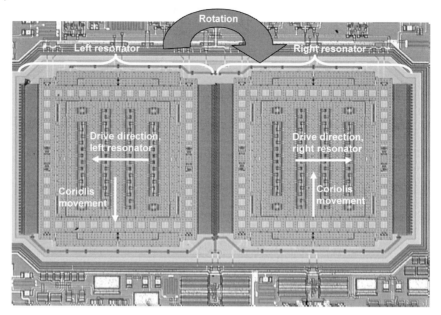

Fig. 3.11 XRS150 gyroscope transducer, showing differential lateral resonator structure

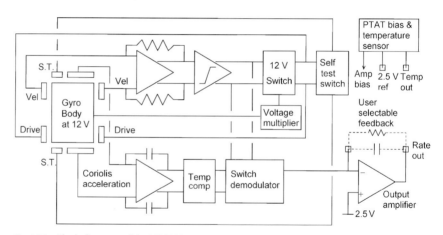

Fig. 3.12 Block diagram of the XRS150 gyroscope system

the mechanical polysilicon thickness from 2 to 4 μm. As mentioned earlier, these same process developments were used for the XL203.

The electronics is comprised of reference and biasing circuits, a voltage tripling charge pump, a resonator loop and the rate signal path. A block diagram is shown in Fig. 3.12.

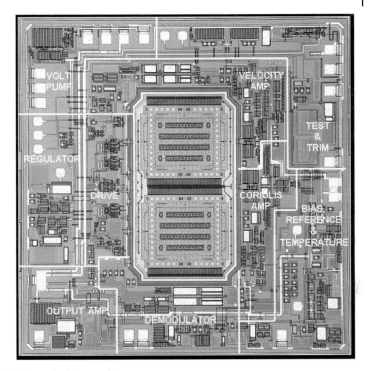

Fig. 3.13 ADI XRS150 single-chip complete gyroscope

To establish the needed velocity in the resonator to generate the Coriolis acceleration, an electromechanical oscillator is formed where the resonance of the mechanical resonators determines the oscillator frequency. This is done to obtain the mechanical amplification from the quality factor of the structures, about 45 in this design. To obtain even higher velocity, a charge pump generated, regulated 12 V supply is used to drive the structure. With this drive the resonators move about 8 µm peak to peak at about 15 kHz.

With applied constant angular rate around the normal to the surface of the chip, a force proportional to and in-phase with the velocity of the resonator is impressed upon the accelerometer frame. This force is sensed by the accelerometer frame and amplified by a low-noise differential amplifier, passed through a temperature compensation gain stage, then synchronously demodulated to produce an output voltage proportional to rate. The rate pass bandwidth is set by the user with an external capacitor.

A die photograph of the XRS150 is shown in Fig. 3.13 and some of the specifications of the part are given in Tab. 3.1. The detectable movement and capacitance detection limit are obtained at the root Allan deviation asymptote using a measurement integration of about 10 s. This $1/f$ equivalent noise floor is believed to be the result of the charge injection of the beam DC bias mechanism.

Tab. 3.1 Selected specifications for the XRS150 single-chip gyroscope

Sensitivity	12.5 mV/°/s
Full-scale range	150°/s
Spot noise	$0.05°/\sqrt{s}$
Full-scale capacitance change	$\sim$120 aF ($\sim$1.5 Å peak movement)
Detectable capacitance change	$\sim$12 zF (12×10^{-21} F)
Detectable movement	$\sim$16 Fermi (16×10^{-15} m or 0.00016 Å)
Root Allan deviation	$\sim$50°/h asymptote

3.2.3
Robert Bosch Epi-polysilicon Accelerometer

Robert Bosch Corp. reported an integrated accelerometer that incorporated a 10 µm thick structural epi-polysilicon layer in a BiCMOS process [9]. The key feature was the deposition of the polysilicon at the same time as the epitaxial silicon device layer was grown (see Chapter 1). The process eliminates issues with planarity that would arise if a thick polysilicon layer alone was deposited early in the BiCMOS flow. DRIE of the polysilicon provided a sense capacitance of more than 1 pF made from 300 µm long electrode beams with 2 µm gaps. Non-linearity was better than 0.5% within a ±35 g range.

3.2.4
Sandia National Laboratories

Sandia National Laboratories is a government-owned, contract-operated research laboratory located in Albuquerque, NM and Livermore, CA, USA, that had its beginnings in the Manhattan Project. One of its research thrusts has been in the area of integrated MEMS, partly with the goal of developing 'failsafe' nuclear arming devices, which, as part of the very painstakingly developed safety procedures, has to include mechanical elements, not just electronics, as part of the arming event.

The Sandia process incorporates a 'sensors first' approach where a well area to place microstructures is etched out of the silicon. Multiple microstructural polysilicon layers are deposited, patterned and formed, all encased in an oxide sacrificial layer whose surface is planarized (see Chapter 1). At that point, the wafers are put through whatever standard processing for the electronics is desired and then the sensors/mechanical elements are released after the electronics processing is finished. The multi-layer structural polysilicon is capable of fairly complex electromechanical structures. The Sandia website is well worth visiting [10].

Sandia opened this process to the general public to use for a modest fee. The integrated MEMS process offered a single nominal 2 µm thick structural polysilicon layer with 5 V, 2 µm CMOS [11]. A second anchored layer of polysilicon was available for signal routing and electrostatic shielding of microstructures from the substrate. A number of companies and universities built prototypes on the process. Among the academic efforts is work done by the University of California, Berkeley.

3.2.5
University of California, Berkeley (UCB)

The Berkeley Sensor and Actuator Center, an NSF Industry/University Research Center that was established in 1986, has been one of the most active University-related groups in general MEMS research. In 1992, the first integrated inertial sensor from the UCB group was a vertical-axis polysilicon accelerometer made in the MICS (Modular Integrated CMOS and microStructures) technology, based on an in-house 3 µm CMOS process at UCB [12]. The MICS process flow placed the microstructural polysilicon deposition after the entire CMOS process, where tungsten metallization for the CMOS was used to withstand the high-temperature polysilicon deposition at 600 °C and rapid thermal annealing of the polysilicon microstructures between 950 and 1050 °C [13]. In the accelerometer system, one capacitive bridge was formed with two sense capacitors made from two moving plates with capacitor gaps to the substrate, along with two fixed reference capacitors. A second capacitive bridge was almost identical, except that the two sense capacitors were made from moving plate structures with half the mass. Common-mode disturbances, such as offsets from some forms of stress, could then be cancelled by differencing the outputs of the two bridge circuits. The on-chip detection circuit included a preamplifier, variable-gain amplifier, demodulator, low-pass filter and a comparator to implement sigma–delta electrostatic force feedback. The accelerometer system worked in open-loop self-test characterization; however, full testing of the inertial performance was not reported.

In 1994, UCB teamed up with ADI for a research grant from the Technology Reinvestment Program and later DARPA, from the US government, to develop integrated MEMS devices on the ADI integrated 'iMEMS' process, based on a 3 µm BiCMOS technology. Students from UCB were the main participants in a number of project chip arrays. Students from MIT and Stanford also participated by submitting chips for testing polysilicon properties such as fatigue, ultimate tensile strength and Young's modulus and also a few accelerometer designs.

The Berkeley students submitted multiple accelerometer designs, both resonant and capacitive sensed displacement types, for both translational and angular acceleration sensing in addition to some early gyroscope designs. One of the first UCB projects in the iMEMS process was a vertical-axis accelerometer made with a 400 µm×400 µm×2 µm polysilicon proof mass [14]. The device employed a charge integrator circuit to measure the sense capacitance (nominally 500 fF) and sigma–delta electrostatic force feedback to achieve a digital output. The measured noise was 1.5 milli-g/$\sqrt{Hz}$.

Shortly after the vertical accelerometer, lateral accelerometers were reported with a differential switched-capacitor interface amplifier and sigma–delta feedback circuitry [15]. The sense capacitor comb fingers and the force-feedback drive fingers were interdigitated in a repeating cycle in each quadrant of the device and interconnected in a common-centroid arrangement to reject cross-axis accelerations better. The kT/C charge injection noise at the input nodes, having sense capacitance of 42 fF, was cancelled to first order by an auxiliary amplifier stage implementing a sample-and-hold function. The measured noise floor was 0.5 milli-g/$\sqrt{\text{Hz}}$ and measured cross-axis rejection was 45 dB for vertical-axis and 43 dB for orthogonal lateral-axis accelerations. Cross-axis measurements were believed to be limited by the shaker table, not by the device.

An early integrated vertical-axis vibratory-rate gyroscope in the iMEMS process was presented in 1996 and had a measured resolution of $1°/s/\sqrt{\text{Hz}}$ [16]. The design, shown in Fig. 3.14, used fixed-fixed flexures to restrict motion of the drive and sense modes. The drive combs are located in the middle of the device and the sense combs on two sides to form a sidewall parallel-plate differential capacitance divider. Locating the comb design on each side canceled the lateral comb-drive capacitance change arising from the drive motion coupled into the sense fingers. An on-chip transresistance amplifier detected the motional current from the capacitor sense divider. DC bias voltages on the differential sense fingers were set to generate electrostatic forces that canceled any sense-axis vibration created by manufacturing asymmetries in the drive.

The UCB group worked also with the Sandia process to produce inertial sensors that were more digitally intensive than could be handled with the ADI iMEMS process. The noise floor in most of these devices was limited by Brownian noise. For example, a single-chip tri-axial accelerometer system with three separate polysilicon transducers and sigma–delta electromechanical feedback had measured noise of 110 and 160 micro-g/$\sqrt{\text{Hz}}$ in the lateral axes and 990 micro-g/$\sqrt{\text{Hz}}$ at sampling rates

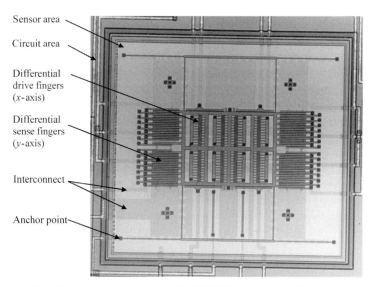

Fig. 3.14 A vertical-axis gyroscope in the ADI iMEMS process [16]

above 500 kHz [17]. Quantization noise increased above Brownian noise for lower sampling rates.

The canonical finger arrangement for a single-axis lateral accelerometer and the switched capacitor interface circuit from [17] is shown in Fig. 3.15. In this circuit topology, the proof mass is connected to the switched modulation voltage, while the stator fingers are the high-impedance nodes connected to the differential input amplifier. An input common-mode feedback (ICMFB) amplifier maintains a constant common-mode voltage on the high-impedance nodes as the modulation is switched. Charge is conserved on these nodes, resulting in the difference voltage across the integrating capacitors, C_{i1} and C_{i2}, being proportional to the difference in the sense capacitance, $C_{s2}-C_{s1}$. Drift in the high-impedance nodes is suppressed by switching the nodes to ground after each sense cycle. Correlated double sampling at two modulation voltage levels, V_{s+} and V_{s-}, eliminates any op-amp flicker noise, charge injection mismatch in switches and op-amp offset.

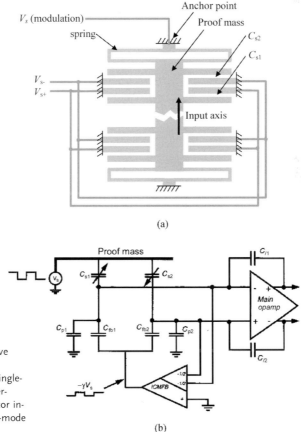

Fig. 3.15 Canonical capacitive sense topology from UCB: (a) capacitive divider for a single-axis polysilicon lateral accelerometer; (b) switched capacitor interface circuit with common-mode feedback (after [17])

A next generation of the Berkeley lateral-axis accelerometer, having a 920 µm×880 µm transducer element with 900 fF sense capacitance, was reported in 2002 [18]. The device, made in the ADI iMEMS process, was operated open loop with the change in sense capacitance detected by a fully differential switched-capacitor interface inspired by Lemkin and Boser's earlier work [17]. The measured resolution of 32 micro-g/$\sqrt{Hz}$ in air was Brownian noise limited, while the measured noise floor for operation in vacuum was pushed down to 2 micro-g/$\sqrt{Hz}$.

A resonant accelerometer design was explored in both the iMEMS and Sandia processes [19]. The transducer was designed with a lever mechanism to transfer the inertial force of the proof mass to alter the axial stress within two fixed-fixed-beam tuning fork resonators. For a given external acceleration, one resonator's frequency was lowered by induced compressive stress in the fixed-fixed flexure, while the second resonator's frequency was raised by induced tensile stress. For the iMEMS design, the measured sensitivity was 2.4 Hz/g with a root Allan variance floor of 38 mHz, which translated to an ~16 milli-g resolution for sample times above 0.4 s. A second-generation design in the Sandia process had a higher sensitivity of 45 Hz/g. Oscillation in the resonators was sustained with an on-chip feedback circuit. A large PMOS resistor in the gain-controlled transimpedance stage was the primary noise contribution in the system.

In 1997, a dual-lateral-axis gyroscope in the Sandia process employed oscillatory motion of a circular plate mass around the vertical axis and capacitive detection of the induced Coriolis acceleration around the lateral axes [20]. The device achieved 0.24°/s/$\sqrt{Hz}$ noise floor and 2°/$\sqrt{h}$ random walk, with worst-case cross-sensitivity of 16% under open-loop operation. The 300 µm diameter circular plate was suspended by four beams that provide torsional compliance about all three axes. Radial electrostatic comb drives generated the torsional oscillation. Electrostatic tuning was used to match resonance frequencies of the three rotational modes. Separate sets of differential parallel-plate capacitive bridges measured the output vibration of the plate around the x- and y-axes. The signals for both axes shared the same signal node, since the top rotor plate was a single polysilicon electrode. Different modulation frequencies for each sense axis separated the output signals in the frequency domain.

A sigma–delta force-feedback vertical-axis gyroscope made in the Sandia process had a resolution of 3°/s/$\sqrt{Hz}$ in air limited by quantization noise [21, 22]. On-chip electronics for the capacitive displacement detection were similar to those in earlier accelerometers from the same group [17, 18]. The lateral dither of the 0.3 µg proof mass was driven open loop with a 1 µm amplitude, while the sigma–delta loop nulled motion in the lateral sense axis orthogonal to the drive.

Another vertical-axis gyroscope design in the Sandia process exploited the idea from Roessig et al. [19] of coupling inertial force to fixed-fixed tuning-fork resonant sensors [23]. The device, shown in Fig. 3.16, fit into an area of about 800 µm×1.2 mm. This design produced a measured noise floor of 0.3°/s/$\sqrt{Hz}$, limited by noise in the resonant Pierce oscillator circuits. A significant bias offset was generated by direct mechanical feedthrough of the drive to the sense axis, where the feedthrough was in phase with the Coriolis acceleration.

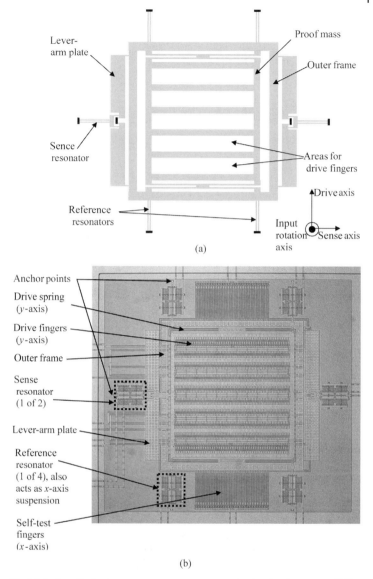

Fig. 3.16 A vertical-axis resonant output gyroscope in the Sandia process [23]: (a) mechanical schematic; (b) photograph of the device

These MEMS transducers and circuits developed by the UCB team using the Sandia process produced prototypes of all three axes of accelerometers and gyroscopes to constitute the first 6-DOF (degrees of freedom) inertial sensor suite on a single piece of silicon, albeit in separate pad frames [24].

Further funding from DARPA supported the development of the ModMems [25] process at ADI which was a different 'sensors-first' process from the Sandia process. It involved building the structures first, encasing them in sacrificial layers, then growing selective epitaxial silicon around them for the electronics, then planarizing the wafer (see Chapter 1). At that point the wafer was sent to a standard IC processing facility where the electronics of whatever type were added. As a final processing step, the sensors were released and a fully functional integrated sensor resulted. The advantage over the Sandia process was that less die area was taken up by the sensor. The Sandia process required significant space between the sensors and the electronics. Unfortunately, even though it worked very well, the process was deemed too expensive for high-volume, low-cost inertial products so it never saw commercial application.

A vertical-axis vibratory-rate gyroscope developed on the ModMems process and incorporating 6 μm thick structural polysilicon with 0.8 μm CMOS was reported in 2002 [26]. A central mass was suspended from a rigid frame and driven laterally with about 2 μm amplitude. The Coriolis acceleration acted on the central mass to move both it and the frame in the lateral direction orthogonal to the drive mode. Output motion was coupled to a differential sidewall parallel-plate capacitor connected to an interface amplifier with integrating feedback capacitors. The measured noise floor was $0.05°/s/\sqrt{Hz}$ and was limited by electronic noise.

A lateral-axis vibratory-rate gyroscope in the ModMems process followed in 2003 [27]. The transducer is shown in Fig. 3.17. The central plate proof mass was driven above its resonance, at the sense-mode resonance, to about 150 nm amplitude along the vertical (out-of-plane) axis with voltage applied to an electrode underneath the plate. The measured resolution of $8°/s/\sqrt{Hz}$ in air was believed to be affected by the sensitivity reduction caused from mismatch of the drive and sense resonance and by relatively high damping in the lateral mode.

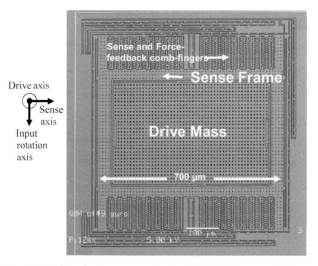

Fig. 3.17 A lateral-axis gyroscope in the ModMEMS process [27]

3.2.6
Defence Evaluation and Research Agency in the UK

As a final note on polysilicon as a material integrated within a CMOS process flow, researchers from the Defence Evaluation and Research Agency the UK performed work in 1998, similar to earlier work at ADI and UCB [28], to verify that deposition of 2 µm thick polysilicon at 600 °C and annealed at 1050 °C resulted in flat structures with less than 6 MPa compressive stress [29]. Their process flow, with structural deposition before metallization, was used to make vertical-axis plate accelerometers alongside 3 µm CMOS test transistors. The high-temperature step shifted threshold voltages by 0.03–0.06 V and the gate length was reduced by more than 1 µm.

3.3
Thin-film Inertial Sensors within Foundry CMOS

3.3.1
Introduction

As an alternative to a custom-fabricated polysilicon microstructure for the inertial transducer, the microstructure can be made directly out of the materials and layers that already exist in foundry CMOS. A handful of groups have explored various approaches to implement such devices, mostly accelerometers. In particular, foundry-CMOS microstructures have been made from the polysilicon normally used for the transistor gates and from the metal–dielectric stack that normally forms the CMOS interconnect. Integrated bulk silicon inertial sensors will be discussed later in Section 3.5.

3.3.2
Siemens Gate Polysilicon Accelerometer

In 1996, researchers from Siemens and the University of Kaiserslautern reported the first capacitive accelerometer integrated directly in a CMOS process with no special modification of the steps [30]. A cross-section and a block diagram of the analog part of the system are shown in Fig. 3.18. The micromechanical plates for the proof mass and a fixed reference capacitor were made from 350 nm thick phosphorus-doped gate polysilicon available in the 0.8 µm CMOS process, while the 600 nm-thick field oxide was the sacrificial layer for release. The fixed electrodes of the sense and reference capacitors were made from the CMOS n-well diffusion layer in the substrate. The etch stop for the bond pad openings was adjusted to stop on the top of the gate polysilicon layer. Photoresist was then patterned to expose just the polysilicon structural region. The surrounding lip of photoresist covered the intermetal dielectric layers and thus protected circuit and pad areas of the chip that were not to be micromachined. Etch holes in the movable plate of

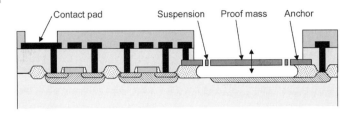

(a)

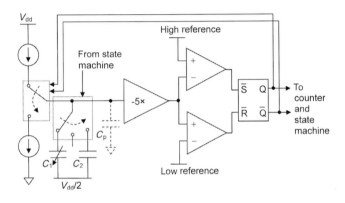

(b)

Fig. 3.18 Siemens CMOS accelerometer: (a) process cross-section; (b) block diagram of the sensing circuit (after [30])

the sense capacitor, C_s, were only 1 µm square and spaced on a 3 µm grid, which reduced the required undercut to 1 µm plus overetch. After the buffered HF release etch and photoresist removal, CO_2 critical point drying was used to inhibit sticking. The resulting 150 µm square plate mass exhibited mild compressive stress and a low residual stress gradient. A symmetric suspension of eight folded beams (estimated to be 2 µm wide and 70 µm long) absorbed the expansion from the residual stress. The reference capacitor, C_r, was made with a top plate identical with the sense capacitor, but without selected release holes so it would remain fixed after the release etch. Voltage imposed across the sense capacitor was limited to under 0.2 V to avoid electrostatic snap down. The sense, reference and parasitic capacitances were measured in sequential fashion with an on-chip dual slope integrator that output an oscillation signal, as shown in Fig. 3.18b. An on-chip 400 MHz clock counted the time for the oscillation signal corresponding to each capacitance value and subtracted counts to compensate for the parasitic capacitance. Final calculation of $(C_s-C_r)/(C_s+C_r)$ was performed with a simple arithmetic logic unit on chip. The accelerometer had a designed range of 50 g, linearity of ±1% and a measured resolution of 0.6 g in a 250 Hz signal bandwidth, corresponding to a capacitance change of 0.1 fF.

3.3.3
UCB Piezoresistive CMOS–BEOL Accelerometer

In 1995, a piezoresistive accelerometer was reported by researchers from UCB [31]. Devices were made in a commercial 2 µm double-poly double-metal CMOS process. A xenon difluoride (XeF$_2$) gas-phase etchant was used successfully to undercut the silicon isotropically without affecting the CMOS microstructures. Piezoresistors were formed from the gate polysilicon layer and aluminum flexural hinges were made with beams from the second metal layer. Three axes of accelerometers were included on a single chip by rotating one of the devices out-of-plane using a design with the metal hinges and a flip-up locking mechanism. The proof masses were perforated plates of the order of 1 mm on a side made of all the BEOL metal and dielectric layers, providing a mechanical resonance at 600 Hz. The reference resistors of the half bridge were placed in a dummy micromechanical structure (essentially a low-mass accelerometer) in order to compensate for the resistive change due to residual and thermal stresses in the beams. A 17.5 µV/g signal referred to the voltage across the piezoresistor was detected, as computed from a resistance change of 0.05 Ω/g multiplied by the bridge current of 350 µA. The measured noise of 3 g in a 500 Hz bandwidth was attributed to flicker noise of the on-chip amplifiers, which had a gain of 1000.

3.3.4
Carnegie Mellon Series of CMOS–BEOL Inertial Sensors

3.3.4.1 Lateral Accelerometer

In 1995, researchers from Carnegie Mellon introduced a process to create microstructures from the CMOS dielectric and metal stack with the capability to make beam widths and sidewall electrode gaps less than 2 µm [32]. Microstructural sidewalls are formed from a CHF$_3$–O$_2$ reactive-ion etch that is masked with the metal layers available in the selected CMOS process. The structures are undercut by first performing a timed deep reactive-ion etch of the silicon substrate followed by a timed isotropic plasma etch of the silicon. A core motivation for research with this process lies in its potential for making sensitive lateral capacitive detectors and electrostatic actuators.

Lateral-axis capacitive CMOS–MEMS accelerometers from this group have employed a symmetric design topology similar to that of the ADI single-axis polysilicon accelerometers. An example design in a three-metal 0.5 µm foundry CMOS process is shown in Fig. 3.19 [33]. The 4 µg proof mass is a 160 µm × 350 µm × 5 µm thick perforated plate made from the metal and dielectric CMOS-interconnect stack. Stiction after release was not a problem with these devices since only dry etching was used. The two-turn meandered suspension springs at each corner provided a spring constant of 0.1 N/m and a mechanical sensitivity of 36 nm/g. As is the case with the lateral polysilicon accelerometers, motion sensing was accomplished with interdigitated comb finger arrays, which maximize sidewall capacitance per unit layout area. Fig. 3.20 a illustrates how a single electrode in each finger was formed by electrically

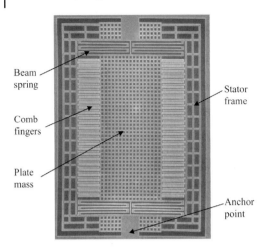

Fig. 3.19 A low-g CMOS–MEMS accelerometer [33]

connecting the three metal layers together. A capacitive bridge was formed by designing the two balanced modulation electrodes within the same mechanical finger.

Vertical-axis accelerometers were made in the same CMOS technology and same basic topology as the lateral accelerometer [34]. The multiple conductors within the comb fingers were arranged to form two sidewall capacitors, C_1 and C_2, as shown in Fig. 3.20 b. All three metal layers in one set of fingers were connected together, while metal-1 and metal-3 in adjacent sets of fingers were connected to the modulation voltages. One set of fingers had the modulation fingers assigned as rotor fingers, and a second set was assigned as stator fingers, as illustrated. This arrangement provided a balanced capacitive bridge that sensed vertical displacement.

Air damping in gaps between comb fingers in the lateral accelerometer contributed to a quality factor of ~ 10, corresponding to Brownian noise of 27 micro-g/$\sqrt{Hz}$. The measured noise floor of 1 milli-g/$\sqrt{Hz}$ was higher than expected and was attributed to additional noise from other circuits on-chip. Similar noise levels were measured in the vertical accelerometers. An improved sense circuit with modulation at 1 MHz, above the CMOS flicker noise knee, achieved a noise floor of 50 micro-g/$\sqrt{Hz}$ [35].

The CMOS–MEMS accelerometers have several differences in wiring and stator mechanics, compared with devices made from a homogeneous material such as polysilicon. The wiring schematic in Fig. 3.21 illustrates these differences for the lateral accelerometer. The first distinguishing feature of the accelerometer topology is the fully differential, common-centroid capacitive sense layout. This topology approximately doubles the sensitivity of a half-bridge topology having the same value of sensing capacitance. The two modulation signals, V_{m+} and V_{m-}, were routed to the movable fingers through electrically isolated metal layers in the suspension springs. The high-impedance sense nodes, V_{s+} and V_{s-}, were routed through the suspended frame to the on-chip preamplifier. This arrangement minimized the sense node interconnect.

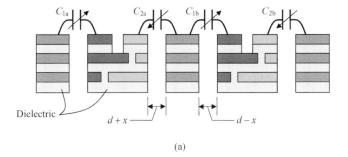

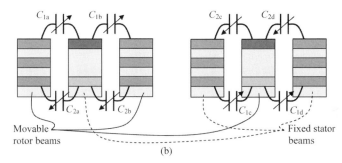

Fig. 3.20 Cross-sections of CMOS–MEMS comb-finger arrangements. Like-shaded metal electrodes are connected to the same electrical node. The finger pattern repeats to build up the desired sense capacitance. (a) Lateral accelerometer comb fingers. C_{1a} and C_{1b} are connected in parallel, as are C_{2a} and C_{2b}. (b) Vertical-axis accelerometer comb fingers. C_{1a}, C_{1b}, C_{1c} and C_{1d} are connected in parallel, as are C_{2a}, C_{2b}, C_{2c} and C_{2d}

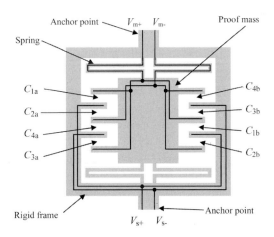

Fig. 3.21 A simplified layout view of the post-CMOS micromachined lateral accelerometer topology, illustrating a common-centroid and fully differential capacitive bridge design

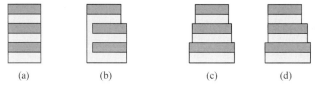

Fig. 3.22 Cross-sections of CMOS–MEMS beams to illustrate lateral curl issue. (a) Beam as layed out with metal-1, -2 and -3 layers. (b) Same beam, but with metal-3 misaligned to the left, creating an oxide sidewall and resulting in a lateral bending moment. (c) Beam as layed out with tiered widths of metal-1, -2 and -3. (d) Same beam, but with metal-3 misaligned to the left. However, in this case no oxide sidewall forms and the lateral bending moment is reduced

The parasitic capacitance of ∼120 pF loading the sense nodes was dominated by the gate capacitance of the input transistors of the interface amplifier. Sensing fingers were split into four groups of differential capacitor pairs. The resulting bridge sensitivity was 0.6 mV/g. In-plane cross-axis translation and rotation were rejected by the common-centroid differential layout. Out-of-plane translation affects sensitivity and must be compensated externally.

The second distinctive feature of the CMOS–MEMS accelerometer is that the stator fingers were connected to a stiff cantilevered frame released from the substrate to compensate for any structural curl out of the substrate plane. The typical radius of curvature of a released CMOS–MEMS metal–dielectric beam can be relatively small, between 1 and 18 mm, corresponding to tip displacement from 11 down to 0.6 µm for a 150 µm long structure. This is compared with a radius of curvature on the order of hundreds of millimeters in an optimized polysilicon technology. Without the stator frame, movable fingers and stator fingers would curl in opposing directions and the sidewall sensing capacitance would be significantly reduced. Moreover, the sidewall capacitance would change dramatically with temperature and from run to run.

A critical aspect of the design is the match of the beam layers and widths between the frame and the proof mass and comb fingers. Ideally, the layer composition and width of beams in the frame and the plate mass should be identical. In practice, there must be some deviation to account for electrical interconnect. Initial accelerometer designs had frames that were not matched perfectly to the proof mass and resulted in a vertical misalignment of the comb fingers of 2.5 µm, which reduced the sense capacitance. Later designs had comb fingers matched to within 150 nm.

Another design issue is lateral (in-plane) curl of narrow suspension beams that arises from misalignment of the inner metal layers in a multi-layer beam, as exemplified in Fig. 3.22. After the dielectric RIE micromachining step, the layer misalignment gives rise to a thin oxide sidewall on one side of the beam, making the structure asymmetric. Residual stress difference in the oxide and metal layers causes the beam to bend laterally upon release and the difference in values of the temperature coefficient of expansion causes the beam to bend with temperature changes. Both lateral curl effects are unique to CMOS–MEMS metal–dielectric structures. Initial suspension designs with beams narrower than 2.1 µm suffered from lateral curl that led to a large bias (offset) and to bias drift. Later design generations used wider spring beams and wider gaps to reduce lateral curl. It was found that designing a cut-in of the upper metal layers in the beam, as shown in Fig. 3.22c, eliminated the oxide sidewall formation even when misalignment between metal layers occurs [34].

3.3.4.2 Thermally Stabilized Accelerometer

The change in structural curl in CMOS–MEMS inertial sensors may lead to variations in sensitivity and offset, as the overlap area of the sense capacitance may change with curl. One approach taken to combat these effects is to control the local temperature of the structures to set and stabilize both vertical and lateral curl.

The vertical-axis accelerometer shown in Fig. 3.23 acted as a test bed for verifying this temperature compensation approach [36]. The topology incorporated a custom meander-spring suspension with comb 'trees' connected to the central proof mass plate. The combs for vertical motion sensing exhibited a vertical offset between the stator and movable fingers. All three conductors in both movable and stator fingers were electrically connected together.

The capacitance of the device was sensed on-chip by a switched-reset, constant-gain circuit using chopper stabilization as shown in Fig. 3.24. The total circuit gain was set to 72 by a ratio of resistors maintained at the same temperature. Temperature-independent biasing was designed with constant-g_m current sources and use of an on-chip band-gap voltage reference. For the final design iteration, the sensitivity of the transducer before circuit gain was 0.13 mV/g/V (for a 1 V_{p-p} modulation input voltage) and 9.3 mV/g after on-chip amplification. The noise floor was 2 milli-g/$\sqrt{Hz}$, limited by circuit noise.

Thermal stabilization was implemented by controlling the device temperature through joule heating of the entire device by embedded polysilicon resistors in

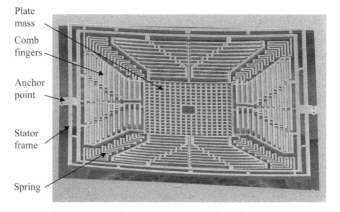

Fig. 3.23 Thermally stabilized vertical-axis accelerometer. Heating resistors are placed in the stator frame, the rotor fingers and the central plate

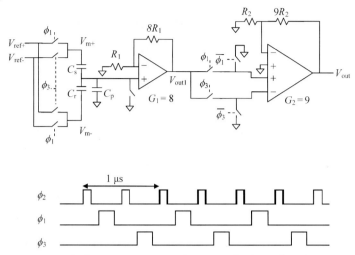

Fig. 3.24 Switched reset capacitive interface circuit for thermally stabilized accelerometer

the proof mass, movable comb and stator frame. The width of the heaters within the proof mass was designed using a finite difference modeling technique and modified Levenberg algorithm in an attempt to optimize uniformity of the temperature distribution. The device temperature was measured *in situ* by monitoring the resistance of the polysilicon and set by an external digital controller. The DC bias stability of the accelerometer improved from $1.7\,g/°C$ without temperature control to -42 milli-$g/°C$ over a $70\,°C$ temperature range by using thermal control requiring 97.5 mW of heater power.

3.3.4.3 Thin-film CMOS–MEMS Gyroscopes

The first vertical-axis gyroscope in a post-CMOS micromachined process was reported in 1997 [37]. The topology had two-axis symmetry to match the sense and drive modes. The rigid frame was meant to self-align the rotor and stator fingers in the dither combs and the accelerometer combs. However, this $600\,\mu m \times 500\,\mu m$ device suffered from vertical curl mismatch and required significant external circuitry for operation. The curl mismatch arose from the two-dimensional curl of the plates and frames. The alignment technique of using cantilevered frames compensated for only one dimension of the curl.

Later generations of vertical-axis CMOS–MEMS gyroscopes switched to a topology composed of a lateral accelerometer nested in a movable rigid frame, as shown in Fig. 3.25 [38, 39]. In this design, a lateral comb drive actuated the frame surrounding the accelerometer. The accelerometer detected the Coriolis force arising from external rotation around the vertical axis. The elastically gimbaled structure decoupled the Coriolis sense mode from the vibration drive mode, except for manufacturing-induced imbalance. The low parasitic capacitance in the CMOS–

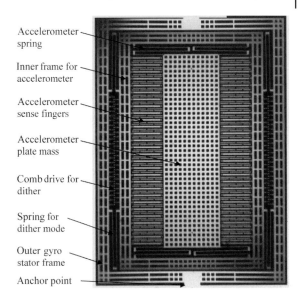

Fig. 3.25 Thin-film CMOS–MEMS vertical-axis gyroscope [38]

MEMS design produced a relatively high sensitivity of 4.5 mV/°/s at atmospheric pressure. The measured noise floor was 0.5°/s/$\sqrt{Hz}$.

Later, a lateral-axis gyroscope was developed, where the lateral-axis accelerometer was dithered out-of-plane instead of in-plane with the substrate [40]. The proof mass was vibrated in the z-direction by the vertical comb drive, using the electrode arrangement outlined in Fig. 3.20b. The minimum detectable rotation was $\sim 10°$/s at atmospheric pressure and the scale factor was 0.12 mV/°/s. Shortcomings identified during characterization motivated a redesign that focused on gyroscope non-ideality compensation schemes. The features of this design included orthogonal placement of drive spring and sense spring for reduced mode coupling, a four-part individually controllable z-drive for compensation of unbalanced driving force, integrated x- and y-actuators for compensation of off-axis motion and integrated polysilicon heaters to control the mismatch of the comb fingers on the outer frame.

3.3.5
MEMSIC Thermal Accelerometer

MEMSIC has commercialized single- and dual-axis accelerometers based on thermal convection and integrated directly in foundry CMOS [41, 42]. The accelerometer works by locally heating a small volume of gas sealed inside a chamber, then detecting any motion of the hot gas bubble from natural convection. The device is similar schematically to a CMOS thermal shear-flow sensor, which is intended to detect forced convection. The dual-axis chip is shown in Fig. 3.26. A micromachined beam includes a silicon resistor that heats the surrounding gas. The heated gas moves upon application of an inertial force and is detected by thermo-

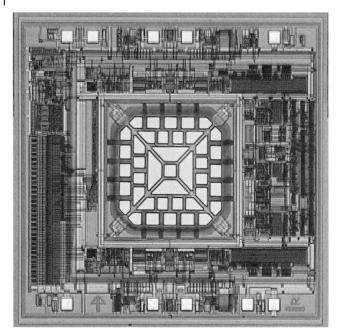

Fig. 3.26 MEMSIC dual-axis thermal accelerometer (photograph courtesy of MEMSIC)

piles embedded in microstructures on either side of the heater beam. The sensors have on-chip signal conditioning, 5 V regulators to reduce external power supply tolerance and programmability to define sensitivity, offset and select an output that is absolute or ratiometric with the power supply. Temperature stability of the signal is maintained by measuring the chip temperature and compensating with a lookup table implemented using an on-chip microcontroller with EEPROM. The output resolution is nominally 1 milli-g/$\sqrt{\text{Hz}}$. The thermal time constant of the system has a 3 dB cutoff of 30 Hz. With external circuit compensation, the usable frequency response can be extended to 160 Hz, which satisfies many consumer applications.

3.4
Metal Inertial Sensors

3.4.1
Introduction

Using metal as the structural material in inertial sensors is attractive because it allows easy integration with CMOS circuits. Metallic MEMS can be fabricated on an already processed CMOS wafer using process steps that do not require high temperatures (a maximum temperature of 120 °C) and therefore do not adversely impact the CMOS circuit performance. Metal microstructures can also be fairly thick (several tens of microns) and of high aspect ratio (as high as 10–30), both of which are important features for inertial sensors, especially when one needs lower thermal noise and higher resolution. There are a few limitations that need to be considered when using metal as the structural material, including the difference in thermal expansion coefficient and the potential for long-term fatigue. However, the simplicity of the process and the fact that all MEMS processing can be performed post-circuit fabrication with minimal disturbance to either the electronic devices or to their fabrication processes make metal microstructures appealing. This section discusses accelerometers and gyroscopes fabricated utilizing metal sensing elements.

3.4.2
Metal-plated Accelerometers

Electroplating metallic microstructures on CMOS-processed wafers is one way of fabricating integrated accelerometers. A good example of such an accelerometer is that developed by Cole and Braun [43, 44]. This design made use of an electroplated asymmetric torsional capacitor plate and two underlying sensing plates on a substrate, located symmetrically on each side of the torsion bar axis. The torsional capacitor plate and the two sensing plates form two air-gap variable capacitors with a common connection. When the sense element rotates around the torsional bar axis, one of the variable capacitances increases, whereas the other decreases.

This device was fabricated using selective electroforming where nickel was electroplated on to a conductive substrate through a patterned photoresist layer. To form the suspended sense element, a sacrificial spacer material was used. The thickness of this sacrificial layer determines the gap size for air-gap capacitors. The sense element size is 1 mm×0.6 mm×5 µm and the sense capacitances are ~150 fF. The sensitivity and dynamic range of the device can be controlled by varying the torsion bar dimensions. These process steps can be performed on a CMOS wafer.

Wycisk and co-workers reported electroplated acceleration threshold switches for inertial sensing [45–48]. These switches are fabricated by the deposition of electroplated gold into molds formed using a standard photoresist. A passivation layer is used for surface planarization and protection of the CMOS circuit. The mold is removed and the structures are then released by etching away a sacrificial layer, resulting in laterally movable metallic microstructures. This mechanical sensor design is basically a crab-leg structure operating as a plain mass-spring system.

Fig. 3.27 shows the SEM of an integrated sensor structure on top of a CMOS signal processing circuit. Device size is 3.0×3.0×1.0 mm and thickness of the mechanical structure is 16 µm. The fabricated switches have threshold values ranging from 2 to 10 g with 300–800 Hz resonance frequency.

178 | *3 Monolithically Integrated Inertial Sensors*

Fig. 3.27 Electroplated structure on the substrate (top and bottom left) and single die with cover-chip (bottom right) [48]

By combining the LIGA process with a sacrificial layer technique, the height of the structural mass can be increased significantly. Although it is expensive, LIGA provides large proof-mass, high-resolution accelerometers that can be monolithically integrated with standard CMOS process [49]. For post-CMOS LIGA processing, the electronics has to be protected by appropriate masking. Burbaum et al. [50] fabricated capacitive acceleration sensors using a movable mass suspended on a cantilever formed by LIGA. The sensing structure has a height of more than 100 µm with a capacitive gap of 3 µm.

3.4.3
Vibrating Metal Ring Gyroscope

In order to increase the sensitivity of a vibratory gyroscope, the difference between the resonant frequencies of drive and sense modes should be reduced as much as possible. Frequency splitting of two modes on the order of 0.01% is practically important in a vibratory gyroscope, especially for a vacuum-packaged high-Q (~ 10000) gyroscope. However, most vibrating gyroscopes require mechanical trimming for tuning and show strong temperature drift due to the mismatch of mode frequencies. The vibrating ring gyroscope, working at two degenerate flexural modes, has intrinsic mode matching. The first metal ring gyroscope was reported in 1994 [51]. This vibrating gyroscope utilized a high aspect ratio structure made of electroformed nickel formed on a silicon wafer that may contain on-chip circuitry and had a resolution of 0.5°/s in a 10 Hz bandwidth.

The ring gyroscope, shown in Fig. 3.28, consists of three main elements: the ring itself, the support springs and the drive, pickoff and control electrodes [52]. This device uses the two primary flexural vibration modes of the ring to sense rotation. In these modes, the ring vibrates in the plane of the substrate in an elliptically shaped pattern that has two nodal diameters, which means that the radial amplitude of the ring motion varies around the ring.

The ring gyroscope driven into vibration using electrostatic drive forces and its vibration is sensed using capacitive detection [53]. In this scheme, the ring is set at a polarization voltage to provide the bias for the capacitive detection and to avoid frequency doubling of the drive force. Each pickoff electrode incorporates a low-input capacitance unity-gain buffer amplifier to sense the small capacitance changes between the ring and the pickoff electrode due to the ring vibration. The ring to pickoff electrode capacitance and the input capacitance of the on-chip buffer amplifier form a capacitive voltage divider that is biased by the polarization voltage applied to the ring. This arrangement converts the ring vibration at the pickoff electrode to a buffered output voltage, for use by the sensor control and readout electronics, which is proportional to the ring vibration.

The ring gyroscope utilizes an electroformed metal for its structure. This electroforming process is based on fabricating a mold which defines the shapes of the sensor element and then electroplating or electroforming metal into this mold to form the desired structure. Although other micromachining techniques exist for fabricating high aspect ratio structures, such as LIGA [54–56], electroforming metal is used because of the ease of integration and the potential low cost of this pro-

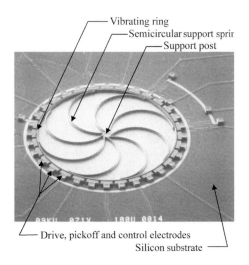

Fig. 3.28 Electroformed nickel ring gyroscope [51, 52]

cess. This method allows the use of standard CMOS technology and addition of the ring gyroscope in a series of low-temperature post-CMOS fabrication steps without any disturbance to the underlying circuitry.

The fabrication of the gyroscope started with a standard CMOS process for on-chip circuitry [57]. CMOS metallization was passivated using an LTO (low temperature oxide)–SOG (spin on glass)–LTO sandwich to protect it from the subsequent MEMS etching steps. Then, gold vias to the CMOS were fabricated, with an intermediate Ti–W adhesion and diffusion barrier. Next, a conductive Cr–Al–Cr sacrificial spacer was deposited and patterned to define the movable portions of the microstructure. Electroplating molds were defined with nearly vertical walls with aspect ratios of ~ 7. Selective electroplating in the open mold areas followed by removal of the mold and sacrificial layer completed the process. The resulting electrode to ring gap was ~ 7 µm.

Delphi reported an electroplated metal ring gyroscope of similar design on top of CMOS circuits in 1999 [58]. A separate ASIC chip implemented the electronics for four control loops to maintain the ring in resonance at a constant amplitude, to obtain the rate signal and to correct for mechanical imbalance in the ring. The measured noise floor was $0.1°/s/\sqrt{Hz}$ in a bandwidth of 25 Hz.

A concern with the use of electroformed nickel as the structural material is the difference in the thermal expansion coefficients between the ring structure and the silicon substrate. This difference in thermal expansion leads to a change in the electrode gap spacing over temperature, which in turn causes a change in the sensitivity of the sensor over temperature. To eliminate this sensitivity variation, three alternative methods may be explored: (1) The drive level may be changed linearly with measured temperature so that the sensitivity of the sensor remains nearly constant, assuming a linear relation between gap displacement and temperature. (2) One may place differential drive and sense electrodes on the inside and outside of the ring largely to eliminate the sensitivity variation. (3) The sense element may be fabricated out of silicon rather than nickel. The substrate and the sensor element would then have the same coefficient of thermal expansion. This would greatly reduce the variation in the gap spacing and also lead to lower offsets, owing to the higher Q of silicon.

3.5
Bulk Silicon Integrated MEMS Inertial Sensors

3.5.1
Piezoresistive Bulk Silicon Accelerometers

In 1992, researchers at the Fraunhofer Institute in Berlin developed a bulk silicon piezoresistive accelerometer with full on-chip electronics, based on a commercial 3 µm CMOS process [59]. The sensor was formed through backside KOH wet etching with an electrochemical etch stop at a p–n junction located on the front side. In order to use a standard n-type substrate CMOS process, p-type wafers with a 10 µm thick epitaxial n-type layer were prepared as the starting substrate. Definition of the sensor on the frontside, where details were not specified, was embedded within the CMOS process prior to contact etch and aluminum metallization. A second metal layer, made of electroplated gold, was deposited above the CMOS passivation to provide electrical contact to the p–n junction during the electrochemical etch. Aluminum could not be used as it will etch in KOH. The proof mass was etched first from the backside with the electrochemical etch stop to the 10 µm thick n-type silicon membrane, followed by an anisotropic frontside wet etch to complete release of the structure. The piezoresistors, made from the p-well layer, were arranged in a Wheatstone bridge driven by 5 V. The measured sensitivity was about 3 mV/g from 0 to 20 g with non-linearity of less than 1%. Although an on-chip operational amplifier was included and successfully tested, it was not specified whether it was used in the accelerometer system. A noise measurement was not provided.

In 1995, researchers from Daimler-Benz and Dialog Semiconductor reported a piezoresistive accelerometer in a single-polysilicon single-metal n-well 3 µm CMOS process [60]. The bulk proof mass was defined from the backside using an anisotropic silicon wet etch with an electrochemical etch stop on the n-well layer. The resulting n-well silicon membrane was then defined into the final cantilevered proof mass by a frontside dry etch step. Process steps beyond the CMOS were implantation of the piezoresistors, an additional low-pressure chemical vapor deposited film for stress compensation and a plasma-deposited film to passivate against the silicon wet etch. Wafers with recessed pits were bonded to the top and bottom of the device to provide gaps for squeezed-film damping and over-range protection and to protect the device during dicing. Piezoresistors in a Wheatstone bridge configuration were located in the thin n-well cantilever beam elements that held the proof mass to the wafer. Reference resistors in a second separate Wheatstone bridge were embedded in a low-mass structure with cantilever beam springs identical with those on the proof mass. Resistive changes from residual and thermal stress effects resulted in a common-mode signal that was canceled by differencing the bridge outputs. The on-chip electronics included a temperature-compensated reference for the Wheatstone bridge, two instrumentation amplifiers with gain of 30 to sense the bridge outputs, digital offset compensation circuits for each bridge, a low-pass filter and an adjustable-gain output amplifier. The sensitivity referred to the bridge output was 0.4 mV/g/V up to the full-scale range of 20 g, sensitivity drift was −1.8%/K before compensation and offset drift was +8 µV/K/V. For some devices, the drift was reduced by as much as fivefold by compensating with the reference bridge. However, the offset drift could not be compensated for all devices and the origin of this drift was not known.

A tri-axial bulk-silicon accelerometer using stress sensitive CMOS differential amplifiers was reported in 1997 and is shown in schematic cross-section in Fig. 3.29 [61]. The post-CMOS micromachining involved a backside wet silicon anisotropic etch that was timed to end when specially etched grooves in the frontside were exposed. This arrangement set the frontside silicon membrane thickness to 10 µm, while simultaneously creating a proof mass out of the entire substrate thickness. The backside of the structural CMOS wafer was anodically bonded to a glass substrate with 10 µm recessed pits to set a lower air gap for the proof mass. The proof mass surrounded a central support and was attached by four silicon beams, defined by a frontside RIE. Twelve piezoresistive stress sensing p-MOSFETs were placed on various parts of the suspension in a common-centroid arrangement in order to detect acceleration-induced stress from all three axes. Each set of four FETs, corresponding to a sensing axis, were used as input transistors in two differential amplifiers. Stress along a given axis produces a differential output between the two amplifiers designed for that axis, while cross-axis accelerations give rise to a common-mode output. Measured sensitivity was 192 mV/g for the vertical axis and 23 mV/g for the lateral axes for a FET current of 50 µA, a supply voltage of 10 V and a gate bias of 5 V. The uncompensated temperature coefficient was −2000 ppm/°C, linearity over the ±10 g range was better than 1.5% and compensated cross-axis error was less than 5%. An improved version of this accelerometer was reported in 2001 built on 0.8 µm CMOS technology [62].

An integrated accelerometer was reported in 2000 having a bulk silicon proof mass made from the CMOS substrate and suspended by 2 µm thick polysilicon springs [63]. The thick polysilicon is merged within the 1.5 µm CMOS process prior to aluminum metallization. The bulk proof mass is formed with a backside KOH etch. As a last fabrication step, a Pyrex cap was bonded to the front side of the accelerometer using non-conductive epoxy. Motion sensing of the proof mass used metal-gate air-gap MOSFETs, dubbed MAMOS. The gate electrode was formed with patterned metal on the Pyrex cap and spaced 1.8 µm from a floating-gate electrode of n-channel MOSFETs located on the proof mass. The drain current from these transistors is inversely proportional to the gap and therefore is sensitive to acceleration of the proof mass. Five MAMOS transistors were connected with regular p-channel FETs as a ring oscillator, resulting in a direct frequency output with a 0 g output of 5.16 MHz and sensitivity of 63 kHz/g. Careful

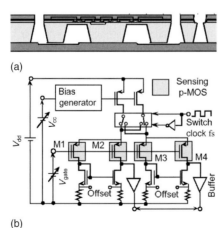

Fig. 3.29 A tri-axial piezoresistive FET accelerometer: (a) cross-section; (b) detection circuit [61]

selection of the threshold adjustment implantation (10^{12} cm^{-2} phosphorus at 70 keV) was necessary to achieve enhancement mode transistor operation and proper oscillation under all acceleration conditions. No noise level was reported.

3.5.2
ADI SOI MEMS

The polysilicon-based MEMS processing described in Section 3.2.2 suffers from three shortcomings. First, it relies on single-polysilicon single-metal (SPSM) 3 micron BiCMOS technology which limits the level of circuit functionality integrated on-chip. Second, the MEMS and circuit processing are heavily intertwined, which requires a dedicated wafer fabrication facility. Third, the structural polysilicon layer is limited in thickness to 4 µm. In order to both increase sensor functionality and reduce cost, ADI, since 2000, has been developing a next-generation MEMS technology, based primarily on the SOI MEMS process developed at UCB [64].

The main advantage of SOI MEMS over the ADI BiCMOS technology is the integration of accelerometers with much denser electronics (0.6 µm CMOS) that include in-package trim capability. Another significant advantage is that the SOI sensors are much thicker. The additional mass per unit area allows the creation of lower Brownian noise structures. The thicker structures are also patterned with higher aspect ratios, which provides not only more lateral sense capacitance per unit area, but also creates more lateral squeeze-film damping. The damping is increased to a point where a critical or an overdamped response is possible, which creates velocity preserving accelerometers in the presence of large shock events. SOI MEMS uses trench isolation to electrically isolate regions of the SOI layer. These trenches can also be released from the underlying substrate thus enabling mechanical coupled and electrical isolated structures. One application of such a structure is a differential sensor architecture in a single mechanical element and two electrical elements. In contrast, in the ADXL78 family of accelerometers a differential sensor was made from two separate sensors that were mechanically independent.

The squeeze-film damping and differential sensor architecture were exploited in two acceleration sensing applications: a high-g prototype and an automotive-grade crash accelerometer. The high-g prototype was developed under contract for DARPA [65]. It was designed to measure in-plane acceleration up to 20 000 g in a 25 kHz bandwidth, while rejecting 100 000 g cross-axis disturbances. This was accomplished by using arrays of distributed tethers in its beam design. The sensor has been successfully tested by the US Army Research Laboratory. The Army used four sensors to create an inertial "constellation" that measures in-bore vibrations of the projectile during launch as well as the in-flight spin rate. The success of this prototype demonstrated the robustness of SOI MEMS, while disrupting the existing technologies for high-g accelerometers by reducing both the size and cost of the sensor by an order of magnitude [66].

The second application of SOI MEMS is the ADXL40, an accelerometer for automobile air bag deployment applications in the mid-to-high g range (50 g to 250 g). The XL40 has a low resonant frequency of 12.5 kHz to maximize its sensitivity. Unlike all in-plane MEMS accelerometers, the ADXL40 uses only one axis of symmetry in order to minimize the die area, resulting in a 34% decrease in sensor area and 30% increase in die per wafer when compared to its BiCMOS counterpart. A view of the ADXL40 transducer is shown in Fig. 3.30. The single mechanical structure of the ADXL40 is partitioned by trench isolations into three electrical nodes: two for differential position sense and one dedicated for self-test. A unique feature of XL40 sensor is that it has two large tethers that dominate the lateral stiffness, while four smaller tethers dominate the vertical stiffness.

By having a dedicated self-test structure, XL40 makes use of the quadratic nature of the electrostatic force to achieve a larger self-test response with minimal area. In particular, the electrically isolated structure enables the application of a nearly full supply voltage across the self-test fingers independent of the sensor biasing, which is fixed at $V_{DD}/2$. This quadruples the self-test response compared to a self-test structure biased to $V_{DD}/2$, and thus substantially reduces the area for self-test.

Using 0.6-μm double-polysilicon double-metal (DPDM) CMOS allows the XL40 to have the following key improvements: first, all trims (range, bandwidth, fine-tune sensitivity and self-test) are achieved with on-chip polysilicon fuse trims, eliminating the packaging shifts found on current products using the iMEMS process. Second, on-board charge-pumps and regulators can be economically built for the sensor self-test, allowing rejection of the supply sensitivity found on current

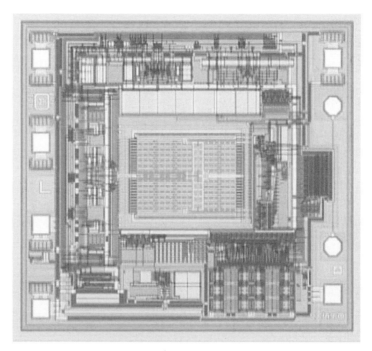

Fig. 3.30 ADXL40 SOI MEMS accelerometer

products. Finally, the number of transistors are increased by an order of magnitude, while maintaining die sizes on the order of those in iMEMS. The smaller sensor, combined with optimized anchor location and in-package trim, helps the ADXL40 achieve acceptable performance in a SOIC8 plastic package. The plastic packaging will help to maintain margins with a very low-cost accelerometer. Being the first ADI product with SOI MEMS and plastic package, the ADXL40 should help pave the way for future SOI MEMS product designs and development.

3.5.3
Silicon-on-Glass (SOG) Accelerometers Developed at the University of Michigan

A CMOS-compatible micromachined SOG *in-plane* capacitive accelerometer with CMOS interface circuitry has been developed at the University of Michigan [67, 68]. The main advantage of this structure is the post-CMOS integration technique that is fully CMOS compatible, allowing the fabrication of high-performance monolithic inertial sensors. The accelerometer is a high aspect ratio structure with a 120 µm thick single-crystal silicon proof mass and 3.4 µm lateral sense gaps and is shown in Fig. 3.31. The underlying glass substrate, which has metallized recesses fabricated, is bonded anodically to the frontside of the CMOS die. After bonding, the CMOS substrate is chem-mechanically polished on its exposed backside to thin it to the desired mechanical thickness. Then, silicon DRIE forms the bulk silicon microstructures. A CMOS switched-capacitor readout circuit and

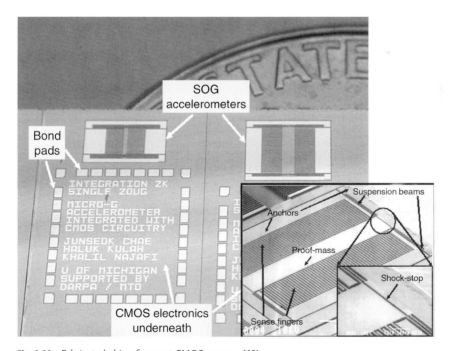

Fig. 3.31 Fabricated chip after post-CMOS process [69]

an oversampled sigma–delta modulator are used to read out capacitance changes from the accelerometer. The CMOS chip is 2.6 mm×2.4 mm in size.

3.5.4
Carnegie Mellon DRIE-Si CMOS–MEMS

In 2000, the CMOS–MEMS process used to create microstructures from the BEOL interconnect stack, as described in Section 3.3.4, was modified to incorporate bulk silicon microstructures [69]. A timed and patterned backside etch of the chip (or wafer) thins the silicon substrate to set the structural thickness to between 25 and 50 µm. An RIE through the dielectric stack is then performed from the front side to form the usual microstructures from the CMOS interconnect stack. This step is followed by a DRIE-Si etch completely through the silicon membrane. The resulting microstructures are made of bulk silicon with the CMOS–MEMS interconnect stack on top. The extra mass from the silicon substrate can be incorporated in accelerometers without increasing the area. The thick silicon substrate also greatly reduces any out-of-plane curl present from residual stress gradients in the interconnect stack. The vertical DRIE-Si CMOS accelerometer in Fig. 3.32 is based on the thin-film vertical accelerometer design, but with 25 µm of silicon added to the proof mass. A curl matching frame is not required, resulting in a simpler design compared with the thin-film CMOS–MEMS accelerometers. A vertical-axis accelerometer was later fabricated using the same comb finger arrangement as the thin-film version. Because of the larger proof mass, the quality factor was 21, about 10 times larger than that observed from the thin-film CMOS–MEMS vertical-axis accelerometer with approximately the same footprint. The accelerometer has no substrate directly under the microstructure, thereby eliminating squeeze-film air damping. However, the same circuit noise problems as in the prior thin-film CMOS accelerometers plagued this design also, limiting its resolution to 1 milli-g/$\sqrt{Hz}$.

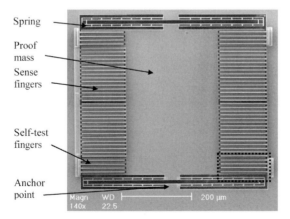

Fig. 3.32 DRIE-Si vertical-axis CMOS–MEMS accelerometer

Following on the accelerometer work, a DRIE-Si CMOS lateral-axis gyroscope was subsequently made [70]. The topology was similar to the lateral-axis thin-film gyroscope, but the proof mass included a 60 µm thick silicon layer. Output from the on-chip sense preamplifier was 0.4 mV/°/s with a noise floor of $0.02°/s/\sqrt{Hz}$ at 5 Hz. The zero-rate output drift, which was measured during an 8 h span, was around 170°/s/h. This drift was attributed to the open-loop operation, of both the drive and sense modes, and a lack of temperature stabilization. Although the process technology was demonstrated, there remain considerable circuit design improvements to complete a full on-chip system and to approach the Brownian noise limits.

3.6
Future Prospects

Most of the current integrated inertial devices offer single- or dual-axis sensing. As the fabrication, design and packaging technology matures, the cost of these devices may be expected to continue to fall. With the advent of the first commercial fully integrated gyroscope in 2002, more multi-axis inertial sensors may appear in future research and commercial efforts. As one motivation, on-chip integration provides precision alignment of the multiple transducers, lowering system packaging costs. Integrated sensor arrays may be desirable in the future for extended range operation, redundancy or to perform additional measurements for systemic error compensation.

Emerging process technologies may lead to further advances in integrated inertial sensors. For example, polysilicon–germanium (polySi–Ge) materials are deposited at low enough temperatures (400–500 °C) to allow direct fabrication of microstructures on top of metallized CMOS [71]. Poly-germanium may be used as a sacrificial layer that etches in hydrogen peroxide, also compatible with post-CMOS processing. Work remains to be done to improve the residual stress and stress gradients in the films, but this status is similar to polysilicon technology in the early 1990s. It appears likely that researchers will find ways to improve the mechanical performance.

Base work done on making inertial sensors with post-foundry CMOS processing may mature to the point of having pre-made sensor cores that can be added to any application-specific CMOS chip. Such capability would open many new markets for MEMS inertial sensing.

The current integrated inertial sensors are not accurate enough for navigation applications. The desired bias stability and resolution of around 1 micro-g for accelerometers and better than 1°/h for gyroscopes will require the larger proof masses afforded by bulk silicon, coupled with better circuit design techniques. Methods will be required to eliminate or compensate for stresses induced by packaging and by temperature excursions. Solutions may incorporate the bulk silicon processes already discussed, and also alternative processes that are no doubt being explored. Existing non-integrated processes may evolve into integrated solutions by adopting either pre-CMOS, inter-CMOS or post-CMOS micromachining approaches. Two primary factors in the approach taken are the temperatures needed in many advanced bulk silicon process flows (e.g. thermal oxidation to form precision gaps) and planarity issues after silicon etching steps.

3.7
References

1   N. Yazdi, F. Ayazi, K. Najafi, "Micromachined inertial sensors." *Proc. IEEE* **1998**, *86*, 1640–1659.
2   H. Xie, G. K. Fedder, "Integrated microelectromechanical gyroscopes." *J. Aerospace Eng.* **2003**, *16*, 65–75.
3   N. Barbour, G. Schmidt, "Inertial sensor technology trends". *IEEE Sens. J.* **2001**, 332–339.
4   S. Lewis, S. Alie, T. Brosnihan, C. Core, T. Core, R. Howe, J. Geen, D. Hollocher, M. Judy, J. Memishian, K. Nunan, P. Paine, S. Sherman, B. Tsang, B. Wachtmann, "iMEMS" Inertial Measurement Unit Components. In: *Technical Digest of the IEEE Electron Devices Meeting*; **2003**, pp. 39.1.1–39.1.4.
5   *Product Datasheets*; Analog Devices, Norwood, MA, *www.analog.com*.
6   K. H.-L. Chau, S. R. Lewis, Y. Zhao, R. T. Howe, S. F. Bart, R. G. Marcheselli, "Integrated force-balanced capacitive accelerometer for low-G applications." In: *Proceedings of the 1995 8th International Conference on Solid-State Sensors and Actuators (Transducers '95) and Eurosensors IX, Part 1, Stockholm*; **1995**, pp. 593–596.
7   H. Samuels, J. Farash, "Micromachined device with enhanced dimensional control." *US Patent 6282 960*; **2001**.
8   J. A. Geen, S. J. Sherman, J. F. Chang, S. R. Lewis, "Single-chip surface micromachined integrated gyroscope with 50 deg/hour Allan deviation." *IEEE J. Solid-State Circuits* **2002**, *37*, 1860–1866.
9   M. Offenberg, F. Laermer, B. Elsner, H. Muenzel, W. Riethmueller, "Novel process for a monolithic integrated accelerometer." In: *Proceedings of the 1995 8th International Conference on Solid-State Sensors and Actuators (Transducers '95) and Eurosensors IX, Part 1, Stockholm*; **1995**, pp. 589–592.
10 Sandia National Laboratories, Albuquerque, NM, *http://mems.sandia.gov*.
11 J. H. Smith, S. Montague, J. J. Sniegowski, J. R. Murray, P. J. J. McWhorter, "Embedded micromechanical devices for the monolithic integration of MEMS with CMOS." In: *Technical Digest of the IEEE International Electron Devices Meeting (IEDM)*; **1995**, pp. 609–612.
12 W. Yun, R. T. Howe, P. R. Gray, "Surface micromachined, digitally force-balanced accelerometer with integrated CMOS detection circuitry." In: *Technical Digest of the 5th IEEE Solid-State Sensor and Actuator Workshop, Hilton Head Island, SC*; **1992**, pp. 21–25.
13 J. M. Bustillo, G. K. Fedder, C. T.-C. Nguyen, R. T. Howe, "Process Technology for the Modular Integration of CMOS and Polysilicon Microstructures." *Microsyst. Technol.* **1994**, *1*, 30–41.
14 C. Lu, M. Lemkin, B. E. Boser, "Monolithic surface micromachined accelerometer with digital output." *IEEE J. Solid-State Circuits* **1995**, *30*, 1367–1373.
15 M. Lemkin, B. E. Boser, "Micromachined fully differential lateral accelerometer." In: *Proceedings of the 1996 IEEE Custom Integrated Circuits Conference*, San Diego, CA; **1996**, pp. 315–318.
16 W. A. Clark, R. T. Howe, R. Horowitz, "Surface micromachined z-axis vibratory rate gyroscope." In: *Technical Digest of the 7th Solid-State Sensor and Actuator Workshop, Hilton Head Island, SC*; **1996**, pp. 283–287.
17 M. Lemkin, B. E. Boser, "A three-axis micromachined accelerometer with a CMOS position-sense interface and digital offset-trim electronics." *IEEE J. Solid-State Circuits* **1999**, *34*, 456–468.
18 X. Jiang, F. Wang, M. Kraft, B. Boser, "An integrated surface micromachined capacitive lateral accelerometer with 2 mg/Hz resolution." In: *Technical Digest of the 10th Solid-State Sensor, Actuator and*

Microsystems Workshop, Hilton Head Island, SC; **2002**, pp. 202–205.
19 T. A. ROESSIG, R. T. HOWE, A. P. PISANO, J. H. SMITH, "Surface-micromachined resonant accelerometer." In: *Technical Digest of the 1997 International Conference on Solid-State Sensors and Actuators (Transducers '97), Chicago, IL*; **1997**, Vol. 2, pp. 859–862 (IEEE Cat. No. 97TH8267).
20 T. JUNEAU, A. P. PISANO, J. H. SMITH, "Dual axis operation of a micromachined rate gyroscope." In: *Technical Digest of the 1997 International Conference on Solid-State Sensors and Actuators (Transducers '97), Chicago, IL*; **1997**, Vol. 2, pp. 883–886 (IEEE Cat. No. 97TH8267).
21 X. JIANG, J. I. SEEGER, M. KRAFT, B. E. BOSER, "A Monolithic surface micromachined Z-axis gyroscope with digital output." In: *Proceedings of the 2000 Symposium on VLSI Circuits, Honolulu, HI*; **2000**, pp. 16–19.
22 J. I. SEEGER, X. JIANG, M. KRAFT, B. E. BOSER, "Sense Finger Dynamics in a SD Force-Feedback Gyroscope." In: *Technical Digest of the 9th Solid-State Sensor and Actuator Workshop, Hilton Head Island, SC*; **2000**, pp. 296–299.
23 A. A. SESHIA, R. T. HOWE, S. MONTAGUE, "An integrated microelectromechanical resonant output gyroscope." In: *Proceedings of the 15th IEEE International Conference on Micro Electro Mechanical Systems (MEMS 2002), Las Vegas, NV*; **2002**, pp. 722–726.
24 J. J. ALLEN, R. D. KINNEY, J. SARSFIELD, M. R. DAILY, J. R. ELLIS, J. H. SMITH, S. MONTAGUE, R. T. HOWE, B. E. BOSER, R. HOROWITZ, A. P. PISANO, M. A. LEMKIN, W. A. CLARK, T. JUNEAU, "Integrated microelectro-mechanical sensor development for inertial applications." *IEEE Aerospace Electron. Syst. Mag.* **1998**, *13*, 36–40.
25 J. YASAITIS, M. JUDY, T. BROSNIHAN, P. GARONE, N. POKROVSKIY, D. SNIDERMAN, S. LIMB, R. HOWE, B. BOSER, M. PALANIAPAN, X. JIANG, S. BHAVE, "A modular process for integrating thick polysilicon MEMS devices with sub-micron CMOS." *Proc. SPIE* **2003**, pp. 145–154.
26 M. PALANIAPAN, R. T. HOWE, J. YASAITIS, "Integrated surface-micromachined Z-axis frame microgyroscope." In: *Technical Digest of the IEEE International Devices Meeting (IEDM), San Francisco, CA*; **2002**, pp. 203–206.
27 S. A. BHAVE, J. I. SEEGER, X. JIANG, B. E. BOSER, R. T. HOWE, J. YASAITIS, "An integrated, vertical-drive, in-plane-sense microgyroscope." In: *Technical Digest of the 12th International Conference on Solid-State Sensors, Actuators and Microsystems (Transducers 2003)*; **2003**, Vol. 1, pp. 171–174.
28 J. M. BUSTILLO, R. T. HOWE, R. S. MULLER, "Surface micromachining for microelectromechanical systems." *IEEE Proc.* **1998**, *86*, 1552–1574.
29 D. O. KING, M. C. L. WARD, K. M. BRUNSON, D. J. HAMILTON, "Polysilicon process development for fully integrated surface-micromachined accelerometer with CMOS electronics." *Sens. Actuators A* **1998**, *68*, 238–243.
30 C. HIEROLD, A. HILDEBRANDT, U. NAEHER, T. SCHEITER, B. MENSCHING, M. STEGER, R. TIELERT, "Pure CMOS surface-micromachined integrated accelerometer," *Sens. Actuators A* **1996**, *57*, 111–116.
31 E. J. J. KRUGLICK, B. A. WARNEKE, K. S. J. PISTER, "CMOS 3-axis accelerometers with integrated amplifier." In: *Proceedings of the 1998 IEEE 11th Annual International Workshop on Micro Electro Mechanical Systems, Heidelberg*; **1998**, pp. 631–636.
32 G. K. FEDDER, S. SANTHANAM, M. REED, S. EAGLE, M. LU, L. R. CARLEY, "Laminated High-Aspect-Ratio Microstructures in a Conventional CMOS Process." *Sens. Actuators A* **1996**, *57*, 103–110.
33 H. LUO, G. ZHANG, L. R. CARLEY, G. K. FEDDER, "A post-CMOS micromachined lateral accelerometer." *J. Microelectromech. Syst.* **2002**, *11*, 188–195.
34 H. XIE, G. K. FEDDER, "Vertical comb-finger capacitive actuation and sensing for CMOS-MEMS." *Sens. Actuators A* **2002**, *95*, 212–221.
35 J. WU, G. K. FEDDER, L. R. CARLEY, "A low-noise low-offset chopper-stabilized capacitive-readout amplifier for CMOS MEMS accelerometers." In: *Technical Digest of the 2002 IEEE International Solid-State Circuits Conference (ISSCC), San Francisco, CA*; **2002**, pp. 428–478.

36 H. Lakdawala, G. K. Fedder, Temperature stabilization of CMOS capacitive accelerometers." *J. Micromech. Microeng.* **2004**, *14*, 559–566.

37 M. S. Kranz, G. K. Fedder, "Micromechanical Vibratory Rate Gyroscopes Fabricated in Conventional CMOS." In: *Proceedings of the Symposium on Gyro Technology, Stuttgart*; **1997**, pp. 3.0–3.8.

38 H. Luo, X. Zhu, H. Lakdawala, L. R. Carley, G. K. Fedder, "A copper CMOS-MEMS Z-axis gyroscope." In: *Proceedings of the 15th IEEE International Conference on Micro Electro Mechanical Systems (MEMS 2002), Las Vegas, NV*; **2002**, pp. 631–634.

39 H. Luo, G. Fedder, L. R. Carley, "Integrated multiple-device IMU system with continuous-time sensing circuitry." In: *Technical Digest of the 1997 IEEE International Solid-State Circuits Conference (ISSCC)*; **2003**, pp. 204–205.

40 H. Xie, G. K. Fedder, "A CMOS-MEMS lateral-axis gyroscope." In: *Proceedings of the 14th IEEE International Conference on Micro Electro Mechanical Systems (MEMS 2001), Interlaken*; **2001**, pp. 162–165.

41 M. Bugnacki, J. Pyle, P. Emerald, "A micromachined thermal accelerometer for motion, inclination and vibration measurement." *Sensors (Peterborough, NH)* **2001**, *18*, 98–104.

42 *MEMSIC Application Note AN-00MX-001*; MEMSIC, North Andover, MA, http://www.memsic.com/memsic/pdfs/an-00mx-001.pdf.

43 J. C. Cole, "A new sense element technology for accelerometer subsystems." In: *Proceedings of the 1991 International Conference on Solid-State Sensors and Actuators (Transducers '91), San Francisco, CA*; **1991**, pp. 93–96.

44 J. C. Cole, D. F. Braun, "Accelerometers with on-chip signal processing." *Sensors (Peterborough, NH)* **1996**, *13*, 7.

45 S. Michaelis, H.-J. Timme, M. Wycisk, J. Binder, "Acceleration threshold switches from an additive electroplating MEMS process." *Sens. Actuators A* **2000**, *85*, 418–423.

46 S. Michaelis, H.-J. Timme, M. Wycisk, J. Binder, "Additive electroplating technology as a post-CMOS process for the production of MEMS acceleration-threshold switches for transportation applications." *J. Micromech. Microeng.* **2000**, *10*, 120–123.

47 M. Wycisk, J. Binder, S. Michaelis, H.-J. Timme, "New sensor on-chip technology for micromechanical acceleration-threshold switches." *Proc. SPIE* **1999**, *3891*, 112–120.

48 M. Wycisk, T. Toennesen, J. Binder, S. Michaelis, H.-J. Timme, "Low-cost post-CMOS integration of electroplated microstructures for inertial sensing." *Sens. Actuators A* **2000**, *83*, 93–100.

49 C. Burbaum, J. Mohr, P. Bley, W. Ehrfeld, "Fabrication of capacitive acceleration sensors by the LIGA technique." *Sens. Actuators A* **1991**, *27*, 559–563.

50 J. Mohr, P. Bley, C. Burbaum, W. Menz, U. Wallrabe, "Fabrication of microsensor and microactuator elements by the LIGA-process." In: *Technical Digest of the 1991 International Conference on Solid-State Sensors and Actuators (Transducers '91), San Francisco, CA*; **1991**, pp. 607–609.

51 M. Putty, K. Najafi, "A micromachined vibrating ring gyroscope." In: *Technical Digest of the 6th Solid-State Sensors and Actuators Workshop, Hilton Head Island, SC*; **1994**, pp. 213–220.

52 M. Putty, PhD Thesis; "A micromachined vibrating ring gyroscope." University of Michigan, Ann Arbor, MI, **1995**.

53 S. Chang, M. Chia, P. Castillo-Borelley, W. Higdon, Q. Jiang, J. Johnson, L. Obedier, M. Putty, Q. Shi, D. Sparks, S. Zarabadi, "Electroformed CMOS integrated angular rate sensor." *Sens. Actuators A* **1998**, *66*, 138–143.

54 G. Engelmann, O. Ehrmann, J. Simon, H. Reichl, "Fabrication of high depth-to-width aspect ratio microstructures." In: *Proceedings of the IEEE Micro Electro Mechanical Systems Workshop, Travemuende*; **1992**, pp. 93–98.

55 A. B. Frazier, M. G. Allen, "High aspect ratio electroplated microstructures using a photosensitive polyimide process." In: *Proceedings of the IEEE Micro Electro Mechanical Systems Workshop, Travemuende*; **1992**, pp. 87–92.

56 H. Guckel, T. R. Christenson, K. J. Skrobis, D. D. Denton, B. Choi, E. G.

Lovell, J. W. Lee, S. S. Bajikar, A. T. W. Chapman, "Deep X-ray and UV Lithographies for Micromechanics." In: *Technical Digest of the 4th IEEE Solid-State Sensors and Actuators Workshop, Hilton Head Island, SC*; **1990**, pp. 38–42.

57 S. T. Cho, J. Ji, *CMOS Integrated Circuits and Silicon Micromachining, Technical Report No. 203*; University of Michigan, Ann Arbor, MI, **1992**.

58 D. Sparks, D. Slaughter, R. Beni, L. Jordan, M. Chia, D. Rich, J. Johnson, T. Vas, "Chip-scale packaging of a gyroscope using wafer bonding." *Sens. Mater.* **1999**, *11*, 97–207.

59 W. Riethmuller, W. Benecke, U. Schnakenberg, B. Wagner, "Smart accelerometer with on-chip electronics fabricated by a commercial CMOS process." *Sens. Actuators A* **1992**, *31*, 121–124.

60 H. Seidel, U. Fritsch, R. Gottinger, J. Schalk, J. Walter, K. Ambaum, "Piezoresistive silicon accelerometer with monolithically integrated CMOS-circuitry." In: *Proceedings of the 1995 8th International Conference on Solid-State Sensors and Actuators (Transducers '95) and Eurosensors IX, Part 1, Stockholm*; **1995**, pp. 597–600.

61 H. Takao, Y. Matsumoto, M. Ishida, "A monolithically integrated three-axis accelerometer using CMOS compatible stress-sensitive differential amplifiers." *IEEE Trans. Electron Devices* **1999**, *46*, 109–116.

62 H. Takao, H. Fukumoto, M. Ishida, "A CMOS integrated three-axis accelerometer fabricated with commercial submicrometer CMOS technology and bulk-micromachining." *IEEE Trans. Electron Devices* **2001**, *48*, 1961–1968.

63 Y. Yee, J. U. Bu, K. Chun, J.-W. Lee, "Integrated digital silicon micro-accelerometer with MOSFET-type sensing elements." *J. Micromech. Microeng.* **2000**, *10*, 350–358.

64 M. Lemkin, T. Juneau, W. Clark, T. Roessig, T. Brosnihan, "A low-noise digital accelerometer using integrated SOI-MEMS technology." In: *Technical Digest of the 10th International Conference Solid-State Sensors and Actuators (Transducers '99), Sendai*; **1999**, pp. 1292–1297.

65 T. J. Brosnihan, J. M. Bustillo, A. P. Pisano, R. T. Howe, "Embedded Interconnect and Electrical Isolation for High-Aspect-Ratio, SOI Inertial Instruments." In: *Technical Digest of the 9th International Conference on Solid-State Sensors and Actuators (Transducers '97)*; **1997**, pp. 637–640.

66 B. S. Davis, T. Denison, J. Kuang, "A Monolithic High-g SOI-MEMS Accelerometer For Measuring Projectile Launch and Flight Accelerations." In: *Proceedings of the Third IEEE International Conference on Sensors, Vienna*; **2004**, to be published.

67 J. Chae, H. Kulah, K. Najafi, "A hybrid silicon-on-glass (SOG) lateral micro-accelerometer with CMOS readout circuitry." In: *Proceedings of the 15th IEEE International Conference on Micro Electro Mechanical Systems (MEMS 2002), Las Vegas, NV*; **2002**, pp. 623–626.

68 J. Chae, PhD Thesis, "High-Sensitivity, Low-Noise, Multi-Axis Capacitive Micro-Accelerometers." University of Michigan, Ann Arbor, MI, **2003**.

69 H. Xie, L. Erdmann, X. Zhu, K. J. Gabriel, G. K. Fedder, "Post-CMOS processing for high-aspect-ratio integrated silicon microstructures." *IEEE J. Microelectromech. Syst.* **2002**, *11*, 93–101.

70 H. Xie, G. K. Fedder, "Fabrication, Characterization, and Analysis of a DRIE CMOS-MEMS Gyroscope." *IEEE Sens. J.* **2003**, *3*, 622–631.

71 A. E. Franke, J. M. Heck, T.-J. King, R. T. Howe, "Polycrystalline silicon-germanium films for integrated Microsystems." *IEEE J. Microelectromech. Syst.* **2003**, *12*, 160–171.

4
CMOS–MEMS Acoustic Devices

*J.J. Neumann and K.J. Gabriel, ECE Department Hamerschlag Hall,
Carnegie Mellon University, Pittsburgh, PA, USA*

Abstract

At Carnegie Mellon University, we have developed a range of acoustic devices based on CMOS–MEMS: audio-range microphones and speakers, digital sound reconstruction using arrays of speakers and ultrasonic sensors for use in liquids. We discuss the design, fabrication and performance of each these devices. We describe the mechanical design and fabrication steps to create a diaphragm in CMOS–MEMS and discuss the advantages of CMOS–MEMS membranes with regard to thermomechanical noise, sensitivity and vibration rejection. For each of the devices, the acoustics are described, with attention to the special considerations needed for MEMS-scale devices versus macroscopic devices.

Keywords

Audio; ultrasonics; acoustics; microphones; speakers; digital

4.1	Introduction	194
4.2	Microphones	195
4.2.1	Designing for Small Sizes	197
4.2.1.1	Equivalent Input Noise	197
4.2.1.2	Sensitivity and Vibration Rejection	200
4.2.2	Microphone Design and Fabrication	201
4.2.3	Acoustic Model of the CMOS–MEMS Microphone	203
4.2.4	Experimental Results	205
4.3	Speakers	208
4.3.1	Traditional Speaker versus MEMS Speaker	209
4.3.2	Fabrication	209

Advanced Micro and Nanosystems. Vol. 2. CMOS – MEMS.
Edited by H. Baltes, O. Brand, G. K. Fedder, C. Hierold, J. Korvink, O. Tabata
Copyright © 2005 WILEY-VCH Verlag GmbH & Co. KGaA, Weinheim
ISBN: 3-527-31080-0

4.3.3	Acoustics in a Closed Coupler	211
4.3.4	Results	212
4.3.5	Digital Sound Reconstruction	214
4.4	**Ultrasonics**	**217**
4.4.1	Fabrication	218
4.4.2	Acoustics in Water and Experimental Setup	219
4.4.3	Measurement and Results	220
4.4.4	Phased Array Behavior	221
4.5	**Conclusions**	**222**
4.6	**Acknowledgments**	**223**
4.7	**References**	**223**

4.1
Introduction

Microelectromechanical systems (MEMS) technology is an approach to fabrication that uses, as a basis, the materials and processes of microelectronics fabrication and conveys the advantages of *miniaturization, multiple components* and *on-chip signal processing* to the design and construction of integrated microstructures. Such systems are smarter, more reliable and more capable while also being less intrusive and less expensive than the traditional macroscopic components and systems that they replace.

MEMS products are used in applications ranging from acceleration sensors for automotive airbag safety systems to microoptical switches for telecommunications and micromirror arrays for data projectors and home theater systems. In this chapter, we explore the emerging application of MEMS technology to the design and fabrication of *acoustic transducers – microphone chips and speaker chips*. The advantages of MEMS – miniaturization, multiple components and on-chip signal processing – are all being employed to displace decades-old technology and bring the promise of new capabilities and features to audio devices and products.

Miniaturization and the relatively smaller size enabled by MEMS microphone chips open up opportunities for broader and less disruptive sampling of acoustic environments, ultrasonic devices where the wavelength of sound is on the same scale as geometries of the MEMS structures and repeatable sub-micron gaps between high-quality, reliable conductors and semiconductor materials bring the efficiencies, quality and scale of semiconductor fabrication to the manufacturing of audio devices.

Multiple components and the batch-fabricated, lithographic processes of MEMS fabrication bring the promise of arrays of acoustic elements, each element with precisely controlled geometries and material properties enabling extremely well-matched microphone pairs or clusters for the construction of multi-microphone features, including directional microphones, noise-suppression microphones and wind-immune microphones. Multiple and well-matched acoustic elements also lead to a completely new sound generation architecture, described at length later in this chapter, that relies on the direct and digital sound reconstruction enabled by the collective action of hundreds to thousands of individual and identical binary speakers.

On-chip signal processing is the final and integrating advantage that binds the miniature and multiple acoustic elements together to complete the integrated acoustic system-on-chip. Integrated analog, mixed-signal and digital circuitry not only provide the signal transduction and amplification typically needed for a sensor or actuator, but also open the door to on-chip analog-to-digital conversion and digital signal processing that will be needed to realize fully and deliver the capabilities and features of MEMS acoustic devices.

To date, most of the work reported on MEMS microphones and speakers has involved the use of piezoelectric materials and/or polymers deposited on silicon substrates [1–3]. Structures employed have included diaphragms, cantilever beams and thermally actuated domes. Work at Carnegie Mellon University (CMU) [4] has focused on building diaphragms out of the materials of standard CMOS, resulting in intrinsically integrated electronics with a simpler process flow.

A number of MEMS microphones have been built and tested using custom and captive fabrication processes, usually using silicon or silicon nitride to form the diaphragm [5–7]. The approach employed by CMU is to use the metal and oxide of CMOS fabrication to form a mesh diaphragm that acts as the skeleton for the deposition of a polymer that results in an airtight diaphragm [8]. A very different approach is taken with the 'Microflown', which uses no diaphragm, but instead directly measures the particle velocity of air moving between two heating elements [9]. An even more radical design involves a micromachined mirror attached to a diaphragm. Modulations in the strength of the laser light reflected through a fiber-optic cable are translated into an electrical signal [10].

In this chapter we discuss the design, fabrication and operation of audio-range microphones and microspeakers built using the CMOS–MEMS technology developed at CMU. We will also discuss work done on a CMOS–MEMS ultrasonic sensor array designed for use in liquids using the same technology.

4.2
Microphones

A conventional microphone is made in one of several ways. One type is a dynamic microphone, which uses a coil of wire (usually embedded in the diaphragm) which moves relative to a permanent magnet when exposed to a sound wave. These are used mainly for low-budget handheld microphones, although there are notable exceptions (e.g. Shure SM series). A much broader class of microphone is the 'condenser' or capacitive microphone. These are subdivided into two categories, electret and externally polarized. In an electret microphone, the electret material provides a permanent electric charge on either the diaphragm or backplate (which has holes to allow air flow). The change in capacitance C when sound impinges on the diaphragm creates a small voltage signal proportional to the displacement, which can be read with a high-impedance amplifier. In an externally polarized or 'phantom powered' microphone, the charge is provided by a voltage source through a very large resistance (such that $1/RC$ is smaller than the minimum frequency of interest). Electret microphones, which use a permanently charged electret material to polarize the capacitor, are typically used in lower cost consumer products, and phantom powered microphores are usually used in professional audio applications. Some microphones are also made with piezoelectric materials and are sometimes used in low-cost applications or when ruggedness is an overriding concern.

MEMS provides several advantages over conventional ways of building microphones. The most obvious are the economics of manufacturing mass quantities in the existing semiconductor fabrication and packaging infrastructure and the possibility of including integrated electronics. The smaller size also allows new applications, such as surveillance and multiple transducers in a small area. In theory, the thinner diaphragms also permit better noise, sensitivity, and vibration rejection for a given surface area than conventional microphones (see Section 4.2.1). Several approaches to MEMS microphones have already been taken. In one similar to this work, additional layers (polyimide and metal) were deposited on a CMOS chip and micromachined [5]. Polysilicon [6] or silicon [7] may also be used to form the diaphragm. It is also possible, using thermal methods, to measure particle velocity (airflow) directly in three dimensions [9], rather than pressure, which greatly simplifies sound intensity measurements when used in conjunction with a single pressure microphone.

CMOS–MEMS microphones are usually made in the condenser microphone type. In the work presented here, the traditional 'condenser' or capacitive microphone approach is used, but integrated with CMOS electronics in such a way that minimizes custom processing steps. This results in a technology which can be mass produced commercially, while maintaining design flexibility and taking advantage of advances in semiconductor fabrication as they occur in the industry at large.

At CMU, two audio-range microphone prototypes were constructed in CMOS–MEMS. In both cases, variation in the capacitance between a metal oxide–polymer diaphragm and the silicon substrate was used to transduce sound into an electrical signal (an ultrasonic 'microphone' was also made with piezoresistors, discussed in Section 4.4). In the first prototype a single diaphragm of size 700×700 µm acted as the transduction element. The capacitance between this diaphragm and the silicon substrate was part of an oscillator, and the frequency of the oscillator was modulated by the incoming sound. The carrier frequency of the oscillator was around 400 kHz, and the output was demodulated with a phase-locked loop. The second prototype was based on the same frequency-modulated concept, but the carrier frequency was designed to be around 100 MHz to facilitate demodulation by an FM radio receiver. This method was chosen to minimize the noise contribution from the demodulation section of the experimental setup, allowing us to estimate better the noise due to the transducer itself. The size of the entire CMOS chip was the same as the first prototype (2 mm square), but transduction was performed by six identical diaphragms, each 324 μm square. The smaller size was chosen to improve the high-end roll-off frequency over the first prototype. The performance of these microphones is shown in Section 4.2.4.

In this section, we will introduce the concepts necessary to understand the acoustics of MEMS microphones and present the fabrication steps and results for our microphone prototypes.

4.2.1
Designing for Small Sizes

The small size of MEMS devices is attractive, yet brings up issues of physical limits and appropriate size scales for acoustic applications. Which scenarios are realistic? What physical limitations apply? Is smaller always appropriate and/or better?

MEMS microphone/speaker design involves many of the same issues as conventional microphones/speakers, but the scale difference changes their relative importance. For example, diffraction effects are still an issue for MEMS microphones, because the packaging is still macroscopic. Because the sizes of MEMS microphone diaphragms (usually 2 mm or less) are much smaller than any audio wavelength of interest (> 17 mm), the shape of the diaphragm is not an issue; however, just like conventional microphones, diffraction effects (the increase in pressure at the face of a microphone relative to the free field) from packaging may still be the dominating effect on frequency response. Diffraction effects are ignored in this discussion. On the other hand, achieving a smooth frequency response is usually easier in the case of the MEMS microphone because of the simple mechanical structure; system resonances can be designed well above the frequency range of interest and damping can be introduced to tame the resonance of the fundamental mode of the microphone diaphragm.

Generally, one associates large-diaphragm microphones (typically 2.5 cm diameter for studio-grade microphones) with low noise floors and overall better performance. What is interesting is that although a single MEMS-scale microphone element may have worse performance than a conventional-size element, for a given total area and roll-off frequency it can be shown that an array of very thin, low-mass diaphragms will outperform a single, large diaphragm in terms of noise floor, absolute sensitivity and vibration rejection. We investigate each of these below.

4.2.1.1 Equivalent Input Noise

An important specification for microphones is equivalent input noise, usually given in terms of dB(A) SPL (the decibels are relative to a 20 μPa sound pressure level and the 'A' modifier refers to A-weighting of the noise power spectrum). For conventional microphones, larger diaphragms correlate with lower noise floors. This is due partly to the reduced thermomechanical equivalent input noise (essentially the diaphragm interacting with the Brownian motion of individual air molecules), but also to a large extent to the increased electrical capacitance facilitating the job of the preamp electronics. In the case of the micromechanical microphone, thermomechanical noise may become greater than electronic noise if the designer is not careful also about the noise of the acoustic circuit. At small size scales, the flow of individual air molecules impinging on the microphone diaphragm gives rise to an equivalent input pressure noise. This noise can be calculated in the same way as thermal (Johnson) noise in an electrical circuit.

In an electrical circuit, a voltage noise is generated across any resistance as a result of the fluctuations in potential energy of electrons due to their interactions with other electrons and scatterers in the resistor. This is called the Johnson noise. From a circuit design standpoint, this can be modeled as a white noise source in series with an 'ideal' (noiseless) resistor. The rms noise power density of the Johnson noise voltage v generated by a resistance R is $\sqrt{4k_B TR}$, where k_B is Boltzmann's constant and T is the absolute temperature [11]. Thermal noise at any point in the circuit can be calculated by substituting an ideal resistor plus a noise voltage source for every resistance in the circuit.

In analogy with electrical circuits, a 'Johnson noise' is generated by the acoustic resistances. Acoustic resistances come from any dissipative mechanism associated with the propagation of sound into or through the device. Some examples relevant to microphone design are squeeze damping, vent holes and radiation resistance. Squeeze damping is the damping due to air being squeezed outwards from between two plates moving normal to their surfaces, such as a microphone diaphragm and backplate [12]. This effect is significant at small gap sizes (around 2 µm) and increases as the inverse of the cube of the gap. This is of particular concern in MEMS microphones, relative to conventional microphones. Vent holes are an important part of microphone designs, as they serve several functions: (1) they provide a path from the back-side of the diaphragm to a 'back volume' which provides compliance, (2) they serve to damp the natural resonance of the diaphragm, providing a smoother frequency response and (3) they provide a means to equilibrate the inside of the microphone with slow changes in ambient atmospheric pressure (to maintain constant gap and sensitivity and prevent damage to the microphone). Their acoustic resistance is due to viscous (capillary) effects for very small radius tubes and depends strongly on the radius. Radiation resistance is the dissipative force that a microphone diaphragm feels as it vibrates in a medium. It should be noted that not only does the microphone diaphragm vibrate under the influence of an incoming sound wave, but also the resulting vibrations radiate new sound waves into the free field of the medium and thus take energy away from the system. The radiation resistance is proportional to frequency squared at low frequencies (where the diaphragm size is much smaller than the wavelength), but approaches a constant (the product of the air density and sound speed, divided by area) at higher frequencies. A large radiation resistance is desirable for speakers, as it determines the amount of sound energy radiated for a given displacement of the diaphragm or speaker cone. The radiation resistance generates some noise in the MEMS microphone, but by far the largest sources are the resistances of the vent(s) and squeeze damping.

An equivalent electrical circuit is shown in Fig. 4.1 for the acoustic circuit of a typical microphone. This electromechanical analogy is one in where sound pressure corresponds to voltage and volume velocity (volume of air moving past a given plane

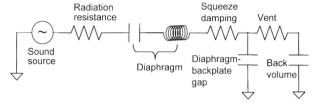

Fig. 4.1 Equivalent electrical model of a typical microphone acoustic system

during unit time) corresponds to electrical current. Other analogies are possible and are described in detail, along with formulas for acoustic components, elsewhere [13]. In any reasonable microphone design, the resistances will be kept small enough to prevent significant damping, meaning the high-frequency roll-off due to damping will be located at or above the frequency where the mass of the diaphragm (modeled by the inductor) causes a high-frequency roll-off. This allows us to make some simplifying assumptions in order to estimate the noise density in the middle of the audio range, where noise specifications are usually reported. In this limit, the total noise power is distributed uniformly across the bandwidth of the microphone [12], which we will take to be the natural resonant frequency of the diaphragm, $\omega_0 = 1/\sqrt{LC}$. This can be seen by analyzing the electrical equivalent circuit and calculating the noise voltage (pressure) across the capacitor–inductor (compliance–mass) combination, which represents the diaphragm by itself.

There is another way to analyze the noise which provides an interesting insight: in the regime described above, it turns out that the total noise (and hence the noise density, since the distribution is nearly uniform) is dependent only on the compliance of the diaphragm. What this means is that we can to a very good approximation ignore the magnitudes of the resistances and focus on the compliance to determine the overall noise. The equipartition theorem of statistical mechanics [14] states that each quadratic term in the expression for the energy (potential and kinetic) of a system has associated with it a thermal energy $k_B T/2$ when the system is in thermal equilibrium. The displacement of the diaphragm is what is transduced into the microphone signal, so the relevant expression is

$$\frac{Cp^2}{2} = \frac{k_B T}{2} \tag{1}$$

where C is the diaphragm acoustic compliance and p is the noise pressure, in analogy with the energy stored in a capacitor, $CV^2/2$.

Consider a square diaphragm [15] with thickness t, Young's modulus E and side length a. We can show that the thermomechanical equivalent input noise is

$$p^2_{\text{diaphragm}} = \frac{k_B T E t^3}{(0.61)\alpha a^6} \tag{2}$$

where $\alpha = 0.0138$. This shows clearly that larger areas are better, which seems at first to be bad news for MEMS microphones. However, we can instead consider an array of diaphragms which fill an area L^2 and choose the size of the individual diaphragms to achieve a given cutoff frequency ω_0. When the signals from the individual diaphragms are averaged together, the noise power drops by a factor $N = (L/a)^2$. Then,

$$p_{\text{array}}^2 = \frac{p_{\text{diaphragm}}^2}{N} = k_B T \frac{Et^3}{(0.61)aa^6} \frac{a^2}{L^2} = k_B T \omega_0^2 \frac{t\rho}{L^2} \tag{3}$$

This is the total noise energy; to obtain the noise density in the flat part of the passband, one should divide by $\omega_0/4$.

Now we see that the thermomechanical noise performance is determined by the areal density of the diaphragm, $t\rho$, that is, the mass per unit area. This is a quantity usually associated with the particular technology being used, rather than the particular design (especially in MEMS). Here CMOS has a great advantage: typical CMOS diaphragm thicknesses are of the order of 1–2 μm, rather than about 25 μm for a conventional microphone (the densities of the materials, usually metals, are similar).

4.2.1.2 Sensitivity and Vibration Rejection

A similar analysis can be performed that shows that the sensitivity, i.e. the change in capacitance with sound pressure, is

$$\frac{dC}{p} = \frac{0.61 L^2 \varepsilon_0 a}{g^2 \omega_0^2 t\rho} \tag{4}$$

Again, we see that it is better if the quantity $t\rho$ is kept small. It should be noted that although the ratio of the capacitance change to nominal capacitance is an obvious way to define sensitivity, making dC larger facilitates making a lower equivalent input noise for the electronics also. In MEMS technologies, the gaps can be made much smaller than in conventionally assembled microphones, so this can make up for the smaller electrode area.

The vibration rejection, which compares the diaphragm displacement due to sound pressure with displacement due to inertial effects (e.g. bumping the microphone), also improves by using diaphragms of small areal density. The following relation expresses the ratio of diaphragm displacement due to sound pressure p to the displacement due to an acceleration $\ddot{x}$:

$$\frac{x_p}{x_{\text{accel}}} = \frac{p}{\ddot{x}} \frac{1}{t\rho} \tag{5}$$

4.2.2
Microphone Design and Fabrication

The microphone diaphragms are formed from the existing CMOS layers using a variant of the CMOS-based micromachining technique developed at CMU [4, 8, 16]. A serpentine metal and oxide mesh pattern (0.9 µm wide beams and gaps) is repeated to form the diaphragm skeleton, and the underlying silicon is etched out to form a suspended mesh skeleton. A Teflon®-like conformal polymer (0.5–1.0 µm) is then deposited on the chip, covering the mesh and creating an airtight seal over a cavity. Depending on the diaphragm geometry and the gap between the diaphragm and substrate, the capacitance will typically range around 0.1–1 pF.

Vent holes are needed to allow air to flow from the sub-diaphragm cavity to a back volume (which reduces the total impedance of the system) and to provide a mechanism for controlled damping of resonant oscillations. Before the release of the CMOS structures on the front side, the back-side vent holes are deep reactive ion etched (DRIE) using either a timed etch or the oxide layer as an etch-stop. Vent hole sizes are chosen to be large enough to decrease the acoustic resistance and positioned close enough together to reduce squeeze damping in the diaphragm–substrate gap. After the vent holes have been etched, processing of the CMOS (front) side of the chip follows the steps shown in Fig. 4.2:

1. The chip comes from a CMOS fabrication foundry covered with a layer of protective glass (silicon dioxide).
2. The glass is etched anisotropically down to the silicon substrate, the metal layers acting as a mask to define the mesh structure.
3. The underlying silicon substrate is etched with a 12 min DRIE anisotropic deep etch followed by a 7 min isotropic etch. After this, the mesh structure is released from the underlying silicon and there is a gap between the diaphragm and the substrate. In the figure, we see a CMOS–MEMS beam and

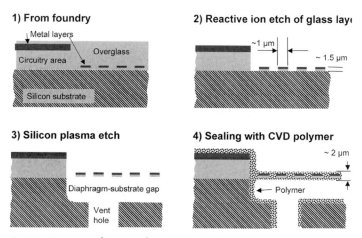

Fig. 4.2 Sequence of steps to form CMOS–MEMS diaphragms

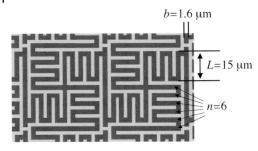

Fig. 4.3 Drawing of a typical serpentine mesh pattern. Dark areas are beams and light areas are gaps

the metal layers inside which can be used as electrodes for sensing and actuation or wires for connecting to the on-chip circuitry.

4. In the final step, the released CMOS–MEMS structure is coated with polymer in a chemical vapor deposition process. The polymer conforms to all sides of the beams, until the gaps are sealed, creating an airtight diaphragm suspended over the gap.

One difficulty in creating a large diaphragm with CMOS–MEMS is the buckling caused by stresses inherent in the oxide and metal, and also from bimorph temperature-dependent stress differences [17]. For a 50 µm long cantilever beam made of metal and oxide, the out-of-plane curl can be as much as 1 µm, increasing with the square of the length. To solve this problem, we developed the serpentine mesh design shown in Fig. 4.3. This design works as follows. Serpentine springs run in both the x and y directions, and are arranged in such a way as to cover the entire area. The x and y springs connect and cross over in the center of each unit cell. Small tabs are added near the end of the long beams (running along the boundaries of the unit cells) in order partially to close the remaining gaps, making the gaps uniform for conformal coating in a later step. The springs provide a great deal of stress relief for either compressional or tensile stress.

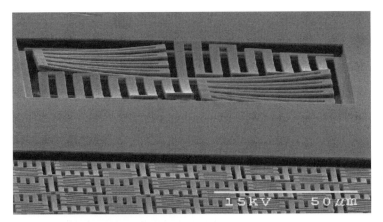

Fig. 4.4 SEM of serpentine mesh unit cell

Treating the mesh as a lumped-parameter equivalent plate, the 'effective Young's modulus' for flexural and torsional stress is about nL/b and $nL/12b$ times less, respectively, than it would be for a solid plate of the same material and thickness, where n is the number of legs in each spring section, b is the lateral width of the beams and L is the length of the spring members. The individual members which make up the springs are kept short (<50 µm) so as to limit curl. It can be seen in the scanning electron micrograph (SEM) (Fig. 4.4) that the greatest contribution to the curl in the mesh comes from the long beams, and that the mesh designs with shorter beams curl much less (bottom part of picture).

4.2.3
Acoustic Model of the CMOS–MEMS Microphone

The transduction behavior may be predicted by considering the acoustic and the electrical behavior of the chip and package. Fig. 4.5 shows a cross-section of the chip mounted on a dual in-line package, along with the corresponding equivalent electrical circuit. A hole is drilled in the DIP and a metal cover is glued over the bottom of the hole to create an isolated back volume, to which the vent holes in the chip connect. It is permissible to treat the acoustic behavior with a lumped-parameter equivalent electrical circuit because the acoustic wavelengths of interest (>17 mm) are much greater than the dimensions of the microphone and the back volume. In this analogy [13, 18], the sound pressure p (deviation from ambient atmospheric pressure) is analogous to voltage in the electrical equivalent circuit and the volume velocity U (volume of air moved per unit time) corresponds to electrical current. Thus, acoustic impedance is defined as $Z = p/U$. Acoustic compliances are modeled as capacitors, usually with one end tied to ground (the exception is modeling the diaphragm, as there is air flow on both sides). The compliance of the air in the diaphragm–substrate gap is given by $C_{gap} = (7/5)V/\rho c^2$, where V is the volume of the cavity, ρ is the density of air and c is the sound

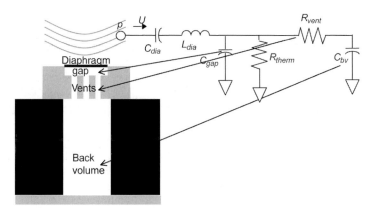

Fig. 4.5 Equivalent electrical model of microphone acoustic system

speed of the air. For macroscopic air volumes, the 7/5 factor is omitted, as the compression at audio frequencies is nearly adiabatic. At the small size scales of MEMS microphones, the area in contact with the thermally conductive silicon is high compared with the air volume, and it can be shown that the air undergoes an almost perfectly isothermal compression and expansion, which increases the compliance by the 7/5 factor (and also introduces a negligible amount of damping). Note that one end of the capacitor in the electrical equivalent circuit represents the pressure in the cavity and the other (grounded) terminal represents the wall of the cavity. The acoustic compliance of the diaphragm can be calculated based on formulas for deflection of a plate and the definition of acoustic impedance.

The mass of the air and the diaphragm are modeled as inductances ('inertance' in acoustic jargon). For a diaphragm of uniform thickness, the inductance $L = \rho' t/A$, where ρ' is the material density and A is the diaphragm area. For the vent holes, we can calculate the inductance as $L = \rho l/A$, where l is the length of the vent hole and A is the cross-sectional area. However, at MEMS scales and audio frequencies the impedance due to this inertance is negligible compared with the acoustic resistance.

There are several sources of damping (acoustic resistance). First, squeeze damping occurs between the diaphragm and the substrate. This is due to the viscosity of the air creating friction as the air is pushed sideways towards the vent holes, and is sensitive to the gap (inverse cube dependence) between the diaphragm and the cavity bottom. In the particular microphone design we investigated, this effect was dominated by the resistance of the vent holes, down to a gap of about 2 µm. The resistance of the vent holes also stems from viscosity and can be calculated from formulas for capillary resistance. Typical dimensions of the vent holes in our microphone prototypes were diameters between 32 and 75 µm and lengths from 200 to 600 µm. As mentioned earlier in this section, there is a very slight damping effect from heat loss through the silicon walls of the sub-diaphragm cavity. Another small source of damping occurs due to radiation resistance of the air in front of the microphone. This occurs because vibrations of the microphone diaphragm, due to incoming sound, radiate sound back into the air.

Because the device and packaging are both much smaller than most acoustic wavelengths of interest, we may neglect the pressure-doubling effect caused by diffraction. Material properties of the diaphragm derived in previous work [4] were used. We used an effective Young's modulus $E = 800$ MPa and a density $\rho' = 1400$ kg/m^3 for the mesh–polymer combination. It can be shown that the dissipative part of the compression is negligible in the limits of pure adiabatic and pure isothermal compression, so we set R_{therm} to infinity. C_{bv} is the compliance of the back volume, in our case several mm^3 drilled in the DIP package. This compression is approximately adiabatic because of the smaller surface-to-volume ratio. For a given sound pressure p, the change in (electrical) capacitance may be calculated by finding the volume velocity U and deriving the diaphragm displacement. The predicted response, assuming the above material properties, is shown together with the measured response in Section 4.2.4.

4.2.4
Experimental Results

The frequency response of the MEMS microphone was measured by placing it in an anechoic box [Bruel and Kjaer (B&K) 4232] and measuring its output as a loudspeaker in the box was driven over the frequency range of interest by a function generator and power amplifier. The box also contained a reference microphone (B&K 4939) placed at a symmetrical position with respect to the speakers, to measure the sound pressure level (SPL) at the microphone position. It should be noted that at higher frequencies (> 5 kHz) the sound pressure field becomes complicated owing to the wavelength being smaller than the size of the speaker, but comparable to the distance between the microphones. Hence the difference in position between the reference microphone and the device under test (DUT) may cause significant variations in the measured response. The reference microphone signal was fed through a B&K 2669 preamp and B&K 2690 Nexus conditioning amp. The electrical output of the MEMS device was connected, in the case of the prototype with 400 kHz output ('P1'), to a phase-locked loop frequency-to-voltage demodulator. In the case of the prototype with 100 MHz output ('P2'), the MEMS device was connected to the antenna input of a stereo receiver (Pioneer SX-303R) through a resistor and capacitor in series, chosen to protect the output of the MEMS device. The frequency-to-volts factor of the radio was measured as 10 µV/Hz, and a standard FM radio pre-emphasis curve was applied [19]. The signals from the MEMS device under test and the reference microphone were measured simultaneously with a signal analyzer (HP 3562A).

Fig. 4.6 shows the output of the microphones in response to the loudspeaker being driven from 20 Hz to 10 kHz, along with the predicted sensitivties based on the above-discussed acoustic models. The measured response of P1 is not flat, however for frequencies above 100 Hz it falls within a factor of two or three of the predicted sensitivity. The resonant frequency is also about a factor of two lower than predicted using the material properties estimated in previous work [4], in which the resonant peak was somewhat ambiguous because of damping and resonances from the packaging. We used other approximations which may have an effect on the acoustic parameters, such as treating the diaphragm as a flat plate (there are actually several microns of buckling), and approximating the capacitance as between two flat plates. Another feature of the response curves is the drop-off at low frequencies, which is not predicted by the acoustic model, unless a leakage path is added. We believe this may be due to cracks in the sealing polymer or holes which may be etched through the silicon but which were not sealed. A blanket mask with a grid of holes was used to pattern the vents and in some cases the timed etch reached the upper surface of the CMOS chip, creating leaks. An attempt was made to seal these holes, but we were unable (in P2) to cure this problem completely. Leaks such as these result in a response proportional to frequency, as exemplified in the P2 graph.

This effect is also visible in the P1 response below about 200 Hz. The noise level of the P2 microphone was measured by repeating the above experiment with-

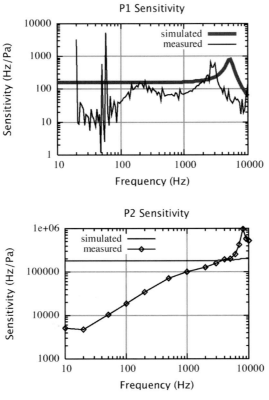

Fig. 4.6 Predicted and measured frequency response of the two prototype microphones

out driving the loudspeaker. Fig. 4.7 shows the equivalent input noise power density of the P2 MEMS microphone, unweighted and also A-weighted (A-weighting is a standard function that takes into account the frequency response of the human ear [20]). The noise power spectrum takes into account the frequency response of the microphone; however, the uncorrected (output) noise spectrum still has a $1/f$ shape, indicating that the source of the noise is electronic rather than thermomechanical. In any case, the measured noise (46 dB SPL, A-weighted) was far above the thermomechanical limit, which can be calculated from Eq. (1) to be –15 dB SPL in a 1 Hz band around 1000 Hz (or 30 dB SPL over the whole bandwidth). Possible sources of electronic noise in our device include $1/f$ noise from the transistors, Johnson noise and sensitivity to the power supply voltage (the oscillator frequency was observed to be proportional to the power supply voltage). An Agilent E3631A power supply with heavy RC filtering was used to power the device at 5 V.

The frequency response of the microphone was measured before and after exposure of the P2 microphone to various elevated temperatures (Fig. 4.8). This is an

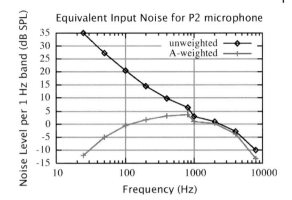

Fig. 4.7 Noise power spectrum of microphone

important issue for commercially viable microphones, as the commonly used electret microphones lose their charge and ability to function after exposure to high temperatures, e.g. sitting on a hot dashboard. Because the CMOS–MEMS microphone does not use electret to polarize the capacitor, this is not an issue. However, we wanted to make sure that the mechanical properties of the sealing polymer are not significantly affected, so we exposed the microphones to high temperatures and measured their sensitivities afterwards. For all but one measurement, the packaged chip was placed in a room temperature (20 °C) oven and the temperature was gradually increased over 30 min until the temperature noted on the graph was reached. Then the heating element was turned off and the oven was allowed to cool slowly (another 30 min) to room temperature. The chip was taken out of the oven and the frequency response was measured. The chip was raised to a number of peak temperatures in this manner. Finally, the oven was pre-heated to 250 °C and the packaged chip was placed in it for 7 min and then cooled by setting it out in the room temperature air. The measured change in frequency response relative to the unheated microphone is shown in Fig. 4.8. The behavior of the frequency response with respect to temperature changes appears small and random, after the initial heating cycle to 200 °C. This suggests an initial

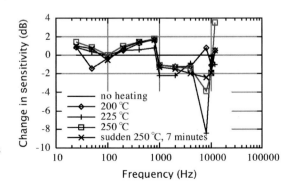

Fig. 4.8 Change in sensitivity as a function of frequency for several temperatures

'heat curing' process would guarantee repeatable performance through future heating cycles.

It appears that with some work the performance of a CMOS–MEMS microphone could reach a level sufficient for cell phones, hearing aids and other applications. Obvious improvements include covering a larger area with diaphragms (to increase capacitance) and develop low-noise circuitry that does not depend on frequency modulation. The limits of performance appear to be related to the quality of the circuit design at least as much as to physical limits of the technology.

4.3
Speakers

At CMU, microspeakers for the audible frequency range were fabricated in CMOS–MEMS. MEMS speakers face greater challenges than microphones because of their small size. As will be explained shortly, the strategy for producing a flat frequency response is different for MEMS speakers versus conventional speakers. We will see that despite the challenges of building a MEMS speaker, there are possibilities that will motivate us.

The first CMU microspeaker device, shown in Fig. 4.9, was packaged as an earphone. This allowed measurements in a standard-size 'ear simulator' measure-

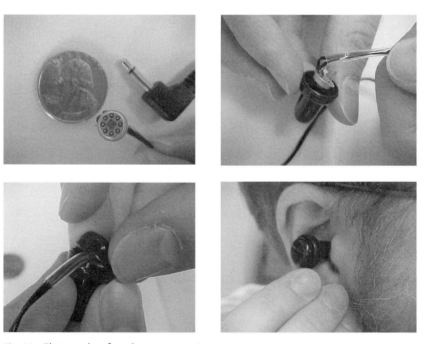

Fig. 4.9 Photographs of earphone construction

ment setup (B&K 4157), and also made a demonstration model that could be used with a portable CD player. In this section, we will discuss the acoustics of the CMOS–MEMS earphone and show the fabrication steps. The experimental and music demonstration setups will be presented. We will discuss the performance of small speakers in free space. Finally, we will consider the possibilities of arrays of microspeakers, for example, digital sound reconstruction.

4.3.1
Traditional Speaker versus MEMS Speaker

Traditional speakers work as they do owing to a fortuitous combination of speaker mechanics and acoustics. The pressure generated in the air near the cone is proportional to the velocity of the cone, and the proportionality constant is the acoustic impedance of the air, or more exactly the radiation impedance, which is a function of the ratio of the wavelength to the cone size. For calculating the sound energy radiated at a distance from the speaker, we use the real, or resistive, part of the radiation impedance, the radiation resistance. The radiation resistance approaches the characteristic impedance of the air multiplied by the area of the cone as the wavelength becomes small compared with the speaker cone. For much of the audio range, however, the wavelengths are about the size of the cone or longer and the radiation resistance is roughly proportional to the frequency [21]. Therefore, in order to have a flat response, it is necessary that the speaker be designed so that its range of operation is above the resonant frequency of the cone, so that the velocity of the cone for a given voltage is inversely proportional to the frequency (mass controlled rather than stiffness controlled).

The MEMS speaker is much smaller than the wavelengths that it is trying to excite, and in this regime the radiation resistance is proportional to the square of the frequency [21]. In addition, because its resonance frequency is designed above the audio range, the displacement is proportional to the voltage, i.e. stiffness controlled (assuming electrostatic deflection with signal plus DC bias). This results in a free-field response that goes as frequency cubed, which is not acceptable for most applications. Therefore, MEMS speakers may be useful mainly for in-ear applications such as hearing aids and portable music devices. Out-of-ear applications may still be possible using very large arrays of microspeakers and signal processing to achieve the desired frequency response.

4.3.2
Fabrication

Most of the processing steps for fabricating the CMOS–MEMS microspeaker are the same as for the CMOS–MEMS microphone. The main difference is in the packaging and the choice of geometric design parameters. The microspeaker chip contained a central diaphragm, 1442 µm square, surrounded by 27 smaller variations, structures to test the effect of varying the serpentine mesh design parameters (gaps, beam widths and number of turns). Only the large central diaphragm was electrically connected to the bond pads and was actuated [4]. Vent holes about 150 μm across were etched through the silicon substrate from the back, in a 3×3 grid behind the diaphragm, using the bottom glass layer as an etch stop.

The chips were carefully cleaned to remove photoresist and then the usual CMOS–MEMS processing [16] was performed to release the mechanical structures. Finally, a chemical vapor deposition (CVD) of C_4F_8 polymer was applied to seal the diaphragm. The gap between the microspeaker diaphragm and the silicon substrate was typically 60–80 μm deep, which is significantly greater than the gap in the microphones. This is so that the diaphragm had room enough for the large deflections necessary for generating large sound pressures. A DC bias voltage of 67 V was necessary to pull the diaphragm down to its operating position. This was chosen to be where the deflection versus voltage curve has the greatest slope, in order to maximize the sensitivity to the signal voltage (see Fig. 4.11).

The MEMS speaker device, in its housing, is placed in the ear with the diaphragm facing into the ear canal. The vent holes between the diaphragm–substrate gap and the back volume have important effects on the behavior of the device, such as reducing the acoustic impedance on the back-side of the diaphragm (allowing greater displacements for a given electric force) and adding a resistive component that damps out the unwanted resonances. The back-side of the chip needs to make contact to the outside world (or at least a significant volume of air), without creating an acoustic 'short circuit' between the back side of the diaphragm and the side facing into the ear canal. In equivalent electrical terms, the back-side of the diaphragm (capacitor–inductor combination) should have a path to ground, either directly or via a large capacitance. This can be accomplished by mounting the CMOS die on a substrate with a hole in it and sealing around the

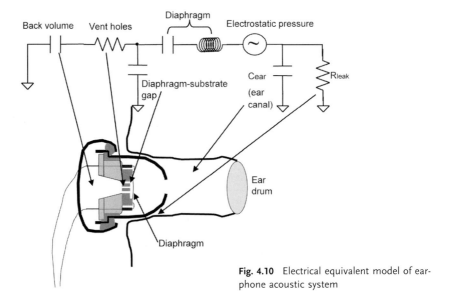

Fig. 4.10 Electrical equivalent model of earphone acoustic system

edge of the chip with epoxy, as shown in Fig. 4.10. The back-side of the substrate is then facing an enclosed, but relatively (compared to the dimensions of the CMOS–MEMS acoustic system) large air volume.

4.3.3
Acoustics in a Closed Coupler

A simplified acoustic model was developed to investigate the effects of the design parameters on the behavior of the MEMS earphone (Fig. 4.10). Like the situation with the CMOS–MEMS microphone (Section 4.2.3), most of the wavelengths of interest (audio frequencies) are much greater ($\gg$17 mm) than the size of both the device and the ear canal, so we may ignore the effects of wave propagation and model the system as discrete components (the sound pressure is assumed uniform throughout each of the air volumes). The signal voltage source in series with the diaphragm models the AC signal part of the electrostatic pressure created by the driving voltage. The voltages in the equivalent acoustic model correspond to sound pressures and currents correspond to volume velocity of air. The voltage on C_{ear} represents the sound pressure created in the ear canal. It is clear from the equivalent electrical circuit that as the volume of air that the earphone drives gets smaller, the pressure will increase. It is also clear that the acoustic impedance of the back volume should be kept smaller than or comparable to that of the diaphragm to get good performance. In our prototype CMOS–MEMS earphone, the acoustic compliance of the diaphragm was estimated to be 5.5×10^{-6} s^2 cm^4/g, which corresponds to an air volume of 7.7 cm^3. This volume is several times larger than what is available in the earphone shell, so there is some attenuation. The acoustic resistances and compliances are calculated from the device geometry and the air properties:

$$R = \frac{8l\mu}{\pi r^4} \qquad C = \frac{V}{\rho c^2} \tag{6}$$

where μ is the viscosity of air, l is the length of the vent hole, r is the radius of the vent hole, V is the volume of air trapped between the diaphragm and substrate, ρ is the density of air and c is the speed of sound in air at standard temperature and pressure. It was assumed that when the device is placed in the ear canal, there would be roughly 1 cm^3 of volume, with some leakage (modeled by R_{leak}) around the device into the outside world.

The serpentine mesh structure is too complicated to derive an exact analytical model, but with some simplifying assumptions (e.g. considering the flexure, but not the torsion of the individual beams [22]) it was possible to show that the effective stiffness is approximately 5–60 times less than that of a solid sheet of the same thickness and material, depending on whether flexural or torsional stresses dominate. Therefore, for purposes of acoustic modeling we used as a starting point the properties of polymer, as it has a density similar to the aluminum–glass beams, and is the compliant material filling the gaps between beams (in other words, we expect the polymer to set the upper limit on the stiffness). Using the results from our earphone and microphone experiments [4, 8], we were able to estimate values for the effective density and Young's modulus of the coated diaphragms.

4.3.4
Results

Optical measurements were made of static deflection as a function of voltage (Fig. 4.11) in order to determine an operating voltage, in this case 67 V. The deflection shown in the graph is given relative to the 0 V position of the diaphragm, which may be several microns above the chip surface because of the buckling or bulging mentioned above.

The CMOS–MEMS speaker chip was packaged in a conventional earphone shell for testing. First the chip was mounted on a small TO package and wire bonded. The TO package was then epoxied inside the housing from a Radio Shack 33-175B earphone in such a way as to seal off the front of the TO package from the back (Fig. 4.10). To present an acoustic load to the device similar to a human ear, we measured the acoustic behavior in a B&K 4157 Ear Simulator rather than at an arbitrary distance from a free-field reference microphone. The earphone and measurement microphone were put inside a B&K 4232 anechoic test chamber that provided about 40–50 dB isolation.

The response of the MEMS device mounted in the earphone housing was measured, driven with a 14.3 V peak signal on the top of a 67 V DC bias (Fig. 4.12). We were able to determine, by comparing results with a conventional earphone, that the resonant peaks near 4 kHz were from the earphone housing and ear simulator, and not from the earphone diaphragm itself [4]. The peaks near 12 kHz are due to the frequency response of the microphone (ear simulator) itself, which mimics the resonance of the ear canal. This peak's center frequency is sensitive to

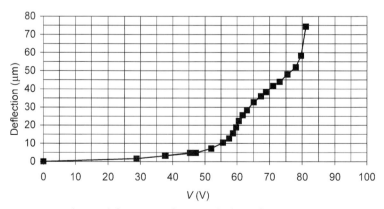

Fig. 4.11 Diaphragm deflection as a function of voltage, for a 1442 µm square diaphragm and 60 µm diaphragm–substrate gap

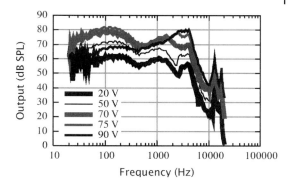

Fig. 4.12 Effect of DC bias voltage on frequency response of CMOS–MEMS earphone

the trapped air volume between the earphone diaphragm and the ear simulator's microphone, so it may not coincide with the peak in the calibration data supplied with the microphone. For this reason, we have chosen to present the data without correction for the microphone response to avoid misinterpretation.

Fig. 4.12 illustrates the effect of varying bias on the frequency response. The overall magnitude of the response increases rapidly as the bias voltage is increased from 20 to about 70 V. At around 75 V bias, a qualitative change in behavior occurs, with a broad peak forming between 2 and 4 kHz. Further increase of the bias to 80 V completes the transition, with a sudden drop in the low-frequency (20–200 Hz) response of about 15 dB, while the 3 kHz peak remains. Although we did not measure distortion, we heard a clear increase in distortion at the higher input signal levels in this region of operation, in addition to the drop-off in bass response. We believe that this change in behavior corresponds to the snapping down of the diaphragm between its two bistable states (slightly convex and slightly concave), which occurs around 75–80 V in most of the chips we studied.

Simulation results (Fig. 4.13) were compared with the experimental data to help us to estimate the polymer material properties. The set of known quantities for

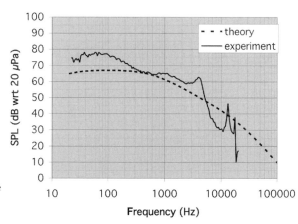

Fig. 4.13 Predicted earphone response (best fit of material property parameters)

the device include the etch depth (70 µm), diaphragm size (1442 µm^2), applied voltage (67 V bias +9.75 V rms signal), approximate diaphragm thickness (2 µm), and size (150 µm across) and length (about 500 µm) of the nine vent holes. Other knowns include the density of air and the speed of sound. It was estimated that the earphone housing fills about half of the volume of a typical ear canal or our ear simulator, so we use 1 cm^3 for the ear canal volume. Unknown quantities include the material properties of the polymer, and the leak size around the perimeter of the device. As initial trial values, we used a density of 1.4×10^3 ng/µm^3 for the polymer density and 3000 MPa for the Young's modulus. We assumed that the observed peaks near 4 and 12 kHz are artifacts of the earphone housing and microphone response (as explained earlier), so we did not try to match these. For the leak area, we started with no leak and increased the leak area. The density and Young's modulus of the diaphragm were adjusted in order to bring the simulation result in line with the experimental data.

Fig. 4.13 illustrates the results of the model calculations. Starting with the parameters described above, we adjusted the unknowns to make a reasonable fit to the data. First, the leak area around the device appears to be very small, or zero. This is to be expected since the earphone shell fits tightly into the rubber gasket supplied with the ear simulator. Second, the corner frequency of the measured earphones is ambiguous, especially given the fact that the housing itself has a significant effect on the frequency response. A range of values for both the density and the Young's modulus were simulated [4] and it was found that the earphone design was relatively insensitive to the specific material property values, probably in part because the acoustic impedance of the diaphragm was small compared with other parts of the system (back volume, ear canal and vent holes). Later work with CMOS–MEMS microphones refined our estimates to 800 MPa for the Young's modulus and 1.4×10^{-3} ng/µm^3 for the density.

4.3.5
Digital Sound Reconstruction

In Section 4.3.1 we described the strengths and weaknesses of the CMOS–MEMS speakers in relation to conventional loudspeakers. With a conventional loudspeaker, great care is taken with materials and design to create speakers with both power and linearity. We have established that any single-element MEMS-scale transducer will be relatively ineffective as a radiator at audio frequencies because of their small size (and hence acoustic coupling to air) and small displacements. At CMU, we have demonstrated a different approach to creating sound, which we call 'digital sound reconstruction' (DSR) [23]. In this approach, an array of transducers, each of limited power and linearity, together overcome both of these problems. Before MEMS, this was not a reasonable approach because of the high assembly costs and the lack of uniformity between individual transducers. However, CMOS–MEMS micromachining provides a method for easily creating large arrays and the uniformity is a result of using state of the art CMOS foundries to perform the most critical steps.

Consider a single CMOS–MEMS speaker as described in the earlier sections. The speaker size can be scaled down until the frequency response of the speaker covers the entire audio band (20 kHz). If we take an array of such speakers (call them 'speaklets'), we can then control the amplitude of the volume velocity by controlling the number of speaklets that snap up or down. Note that regardless of the linearity of an individual speaklet, as long as the speaklets have identical responses a linear overall response is achieved as the pressures add (ignoring the effect of phase cancellation due to different pathlengths). Note also that the frequency response of the individual speaklets becomes irrelevant below their resonant frequency. If one conceives of each speaklet having sequence of 'up' and 'down' states at a given sampling rate, then we have a digital speaker system.

In the past, there have been ideas put forth for building digital speakers [24] or digital speaker arrays [25], but there have been limitations with conventional technology. If one takes a single speaker and subdivides its speaker coil [24], it is difficult to achieve the exact power-of-two ratios that are necessary for linearity. Similarly, if one builds an array with conventional transducers, mismatch is also a problem, and the large overall size causes phase cancellations due to the path differences being comparable to the wavelengths of interest (not to mention the nightmare of interconnecting such a system). MEMS can solve the size and uniformity problem; CMOS–MEMS in particular facilitates interconnect and any control electronics that are necessary (for example, converting an N-bit digital word into 2^N control signals).

As proof of concept of this technology, we built [23] an array of 16×16 speaklets, each 216 µm square, on a CMOS–MEMS chip, using the AMS (Austria Micro Systems) process offered by MOSIS (Metal Oxide Semiconductor Implementa-

Fig. 4.14 Photograph of DSR chip fabricated in CMOS

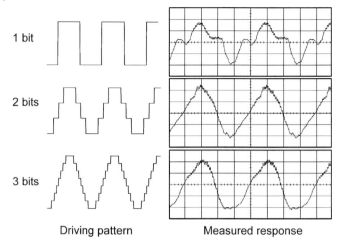

Fig. 4.15 Successive approximation to a 500 Hz sine wave by increasing the number of bits from 1 to 3

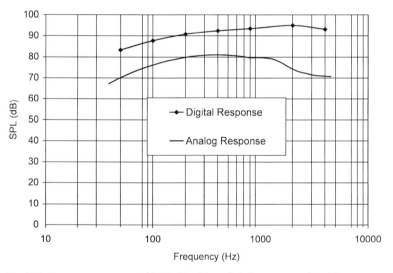

Fig. 4.16 Frequency response of DSR chip driven digitally versus analog driven

tion System, Marina del Rey, CA, USA). A photograph of the chip is shown in Fig. 4.14. Experiments were performed to verify the additive nature of the pressure pulses [26]. The impulses added as expected, but the response was slower than desired. Faster responses can be obtained with stiffer diaphragms and higher voltages and/or pulse shaping. Still, the technology works well enough that one can demonstrate a successive approximation to a low-frequency (500 Hz) sine wave, by adding in successively less significant bits (Fig. 4.15).

As mentioned earlier, the frequency response from the system driven in a digital manner should be independent (up to the cutoff frequency) of the frequency response of the individual speaklets. Fig. 4.16 shows the results of driving the array digitally (speaklets snap up and down individually in numbers proportional to targeted sound pressure) and the result of driving all the speaklets in the array together in an analog manner (voltage signal proportional to targeted sound pressure). While there is some deviation from flatness for the digital mode of driving the array, this is mainly due to the response of the ear simulator. As a final note, it should be mentioned that this array was easily audible from a distance of ~2 m in a noisy laboratory environment.

4.4
Ultrasonics

An important class of acoustic sensors is ultrasonics sensors. Ultrasonic sensors are regularly used for flaw detection and medical imaging. Both of these applications typically use piezoelectric transducers because of the large acoustic energies generated from modest voltages. Many MEMS sensors use capacitive sensing, and are called 'capacitive MEMS ultrasonic transducers' (cMUTs). Capacitive transducers suffer from parasitics in the cabling leading to read-out electronics. Integrating electronics with an on-chip amplifier, as is possible with CMOS–MEMS, is an effective way to solve this problem. Fig. 4.17 shows a waveform from a capacitive sensor on a CMOS (TSMC 0.35 µm process) sensor chip packaged for use in water and intended for gravimetric mass detection. A single diaphragm on this chip was only 130 µm square and connected to a high-impedance amplifier with a gain of about 10, the output of which was digitized. Wiring multiple diaphragms in parallel would improve the noise level, or multiple diaphragms could be placed

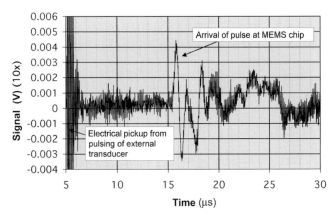

Fig. 4.17 Waveform measured with capacitive sensor on CMOS–MEMS chip with on-chip amplifier

around the chip area to perform phased array operations. The packaging of this chip was the same as the piezoresistive chip discussed here, and is discussed later.

Another approach is to use piezoresistors on chip, and provide a low-impedance output which is relatively unaffected by cable capacitance (it may also be amplified on chip if desired). It should be noted that the piezoresistors are not added to the CMOS–MEMS device through micromachining, but rather are made from the polysilicon layer normally used for transistor gates. Thus the resistors are aligned to the mechanical structures as accurately as the CMOS process would make a transistor.

In this section, we will discuss the fabrication of the sensors and how they are packaged for use in water. We will discuss the acoustic environment in which these ultrasonic sensors are used and how this differs from the environment in which audio MEMS devices are used. A description of the experimental setup will be presented, followed by results for the CMOS–MEMS sensors.

4.4.1
Fabrication

Piezoresistive sensors were fabricated using the Agilent 0.5 µm process at MOSIS (Marina del Rey, CA, USA), and the CMOS–MEMS process developed at CMU [16], described earlier (Section 4.2.2). Rather than building the diaphragms out of a serpentine mesh structure, as we did for the audio-range microphone and speaker, we used a solid plate with 0.9 µm holes 2.0 µm apart in a rectangular grid (Fig. 4.18). The holes serve to allow the underetch of the silicon during the CMOS–MEMS processing. Piezoresistors were formed by running a serpentine

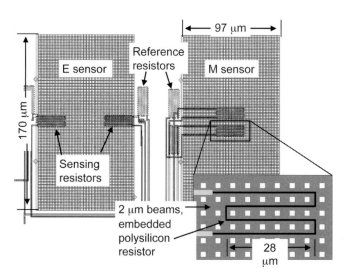

Fig. 4.18 Layout of piezoresistors on diaphragms

line of polysilicon through part of the diaphragm and performing the silicide block only on the longer sections of the polysilicon in order to maximize $\Delta R/R$ when the diaphragm flexes. The resistors were placed midway between the far ends of the diaphragm in the long direction, and so we can approximate the mechanics in the region of the resistors as being like fixed–fixed beams running in the short direction, parallel to the long members of the polysilicon resistors. For uniform loading of the beams, we expect the greatest strain at the edges (assuming a true clamped scenario), and also significant strain at the middle of the beams. Although the peak strain in the middle is only half that at the edges, there is the advantage that the strain remains significant over a longer length, so the average strain on the resistor remains large. For this reason, we made two sensor designs, one with the resistors at the edge and one with the resistors at the middle.

Our prototype chip had 16 individual sensors, each with a resistive bridge connected to an amplifier. Half of the sensors had the sensing resistors near the middle of the diaphragm ('M' sensors) and half had the resistors near the edge ('E' sensors). There were also two sensors on each chip that were wired directly to bond pads to be measured directly with external circuitry. The individual sensor diaphragms have a pitch of 150 µm.

4.4.2
Acoustics in Water and Experimental Setup

Many of the envisioned applications of ultrasonic sensing arrays take place in wet media, such as in the body, or aided by gels. After the release of the mechanical structures using the standard CMU CMOS–MEMS processing [16], we packaged the chip in a dual in-line package (DIP) and then slowly poured PDMS (polydimethylsiloxane, Dow Corning Sylgard 184) over the chip and bond wires. The thickness of the cured gel above the piezoresistive sensor was measured optically to be between 500 and 600 µm. This served to electrically insulate the wires and protect the structures on the chip. As the cured PDMS also has an acoustic impedance similar to that of water, it provides a means to couple acoustically the pressure waves in the water to the mechanical structures. The MEMS structures are designed to have an acoustic impedance significantly less than the surrounding medium (the gel), so to a good approximation their movement follows the particle velocity of the medium and provides a much wider bandwidth than conventional ultrasonic transducers such as PZT.

To hold water above the chip, plastic weighing dishes were glued to the top surface of the DIP packages with cyanoacrylate glue and then the seam was sealed with epoxy. The DIP package was inserted into a protoboard for electrical testing. Fig. 4.19 shows an equivalent electrical circuit for the sensor in our experimental setup. Designing sensors for use in liquids uses a very different set of considerations than designing for sensing in air. The acoustic impedance of a typical CMOS–MEMS structure is much greater than that of air but less than that of water. In the schematic, the water and silicone are modeled as transmission lines.

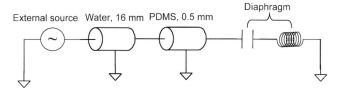

Fig. 4.19 Equivalent electrical circuit for acoustic system of ultrasonic experiments

Their length and characteristic impedance should be chosen based on the distance the waves travel through the medium and the characteristic acoustic impedance of the medium. If one compares this system with the acoustic system for the microphone (Section 4.2.3) or the speaker (Section 4.3.3), one notes that we can ignore the impedance of any air trapped in the diaphragm–substrate gap. This is reasonable as the overall scale of acoustic impedances in the system is much higher than that of the microphones and speakers. No vent holes are necessary for this reason, which simplifies the fabrication process. It is also not necessary to coat the structures with chemical vapor deposition (CVD) polymer to seal the gaps, as this is taken care of by the PDMS gel.

4.4.3
Measurement and Results

The piezoresistive transducers were characterized by exciting them with an external source, a Krautkramer PZT (lead zirconate titanate) transducer partially submerged in the water approximately 16 mm above the sensor surface. As a means to assess their sensitivity, the same measurements were taken with a MEMS capacitive transducer, for which the incoming medium displacement could be calculated from the output signal and known geometry [27]. A DC voltage of 10 V was applied across opposite corners of the bridge formed by the piezoresistors and the other two corners of the bridge, which gave signals of opposite polarity, were measured separately and algebraically subtracted later. Both types of sensors, E and M, were measured and compared. The signals from the E and M sensor elements have opposite polarities because of the opposite strains experienced in the edge versus the middle of the diaphragm. The results of the piezoresistive sensor measurements are shown in Fig. 4.20. The waveform shown is the algebraic difference of the signals from opposite corners of the resistive bridge, to reduce the effect of the electromagnetic coupling to the source transducer. It is still possible to discern the time of the initial excitation pulse applied to the external transducer. The first arrival pulse appears as a series of pulses that die out, which is consistent with a model including reflections between the water–silicone interface and the sensor chip. The acoustic impedance of water is 1.5×10^5 g/(cm² s), and we have previously estimated the acoustic impedance of the silicone gel to be 2×10^5 g/(cm² s) [28]. The acoustic impedance of the CMOS diaphragm was calcu-

Fig. 4.20 Waveform measured from piezoresistive sensor, with the source 16 mm from the surface of the sensor

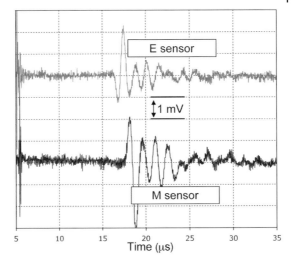

lated (based on a solid plate of the same thickness as our beams) to be 9×10^3 g/(cm² s), about an order of magnitude less than the surrounding medium.

Because the acoustic impedance of the medium (water or PDMS) dominates the system, the deflection produced in the transducer diaphragm is roughly the same over a large range of transducer acoustic impedances. Using this fact, we were able to calculate the sound pressure at the sensor due to the external source and estimate the gauge factor of the polysilicon. The deflection of the capacitive transducer, based on its geometry and voltage measurements, was approximately 7 nm at its center. Assuming the same central deflection for the piezoresistive sensor and averaging the strain along the resistors, we calculate from the measured voltage a gauge factor of –44 and –51 for the E and M resistors respectively. This is higher than the range for n-type polysilicon reported elsewhere [29, 30] of –15 to –22, but the difference may come from a different acoustic reflection pattern (pressure doubling or diffraction) in the capacitive versus piezoresistive sensors. It is possible that there is more effect of pressure doubling in the case of the piezoresistive sensor because of the greater fraction of device area that is not covered by sensors, i.e. the average acoustic impedance is closer to a hard wall.

4.4.4
Phased Array Behavior

The E and M sensors have a center-to-center distance of 150 µm. To test the phased array behavior of the sensor chip, we captured waveforms from the two sensor elements while the source transducer was positioned about 1 cm away from the sensors along the line passing through the sensor elements.

Waveforms were captured from both sensor elements of the piezoresistive chip, while positioning the source transducer at opposite ends of the water reservoir. Close up views of the detected signal are shown in Fig. 4.21, corrected for the po-

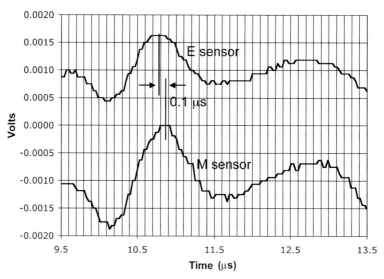

Fig. 4.21 Pulse arriving at slightly different times for the two sensor elements

larity difference between the sensor elements. In the left-hand graph, the source was on the side of the sensor chip closest to the E sensor element; we see the pulse arrival is about 0.1 μs earlier for the E sensor element than the M element, which is consistent with a 1550 m/s speed of sound in water and 150 μm element pitch. The opposite is true when the source is placed on the side of the chip closer to the M sensor element, demonstrating that an array of individual sensors may be used to determine direction. Increasing the spacing of the transducers would improve the sensitivity to direction.

4.5
Conclusions

We have explored the emerging application of CMOS–MEMS technology to the design and fabrication of both microphone chips and speaker chips: audio to ultrasonic frequencies, for use in air and in liquid. The advantages of MEMS – miniaturization, multiple components on a single chip and on-chip signal processing – have been highlighted in use for different devices in a variety of applications. MEMS technology allied with the signal processing power of standard CMOS is the key element and unique power of CMOS–MEMS, delivering new capabilities to audio products today and enabling revolutionary audio products of the future.

4.6 Acknowledgments

This material is based on work supported by the National Science Foundation under Grant No. CMS-0329880. Any opinions, findings and conclusions or recommendations expressed in this material are those of the authors and do not necessarily reflect the views of the National Science Foundation. Support was also provided by the Pennsylvania Infrastructure Technology Alliance, a partnership of Carnegie Mellon University, Lehigh University and the Commonwealth of Pennsylvania's Department of Economic and Community Development. Other support was provided by Adtranz and DARPA through the MARCO, ICESCATE and ASIMPS projects. The authors acknowledge gifts from Krautkramer Inc. and thank Dr. Lars Erdmann for advice and assistance in diaphragm processing and Professor Irving Oppenheim for his analysis of the serpentine mesh mechanics.

4.7 References

1 C. H. Han, E. S. Kim, in: *Proceedings of the 13th IEEE International Conference on Micro Electro Mechanical Systems (MEMS 2000), Miyazaki, Japan, 23–27 January 2000*; **2000**, pp. 148–152.

2 S. S. Lee, R. P. Ried, R. M. White, *J. MEMS* **1996**, 5, 238–242.

3 C. H. Han, E. S. Kim, in: *Proceedings of the 14th IEEE International Conference on Micro Electro Mechanical Systems (MEMS 2001), Interlaken, Switzerland, 21–25 January 2001*; **2001**, pp. 110–113.

4 J. J. Neumann, K. J. Gabriel, *Sens. Actuators A* **2002**, 95, 175–182.

5 M. Pedersen, W. Olthuis, P. Bergveld., *Sens. Actuators A* **1998**, 69, 267–275.

6 M. Brauer, A. Dehe, T. Bever, S. Barzen, S. Schmitt, M. Fuldner, R. Aigner, *J. Micromech. Microeng.* **2001**, 11, 319–322.

7 Y. B. Ning, A. W. Mitchell, R. N. Tait, *Sens. Actuators A* **1996**, 53, 237–242.

8 J. J. Neumann, K. J. Gabriel, in: *Proceedings of IEEE Transducers '03: the 12th International Conference on Solid-State Sensors, Actuators and Microsystems, Boston, MA, 8–12 June 2003 (www.transducers03.org)*; **2003**, pp. 230–233.

9 F. J. M. van der Eerden, H.-E. de Bree, H. Tijdeman, *Sens. Actuators A* **1998**, 69, 126–133.

10 W. W. Gibbs. *Sci. Am.* **1999**.

11 P. Horowitz, W. Hill *The Art of Electronics*; Cambridge: Cambridge University Press, **1980**.

12 T. B. Gabrielson, *J. Vibr. Acoust.* **1995**, 117, 405–410.

13 B. B. Bauer, in: L. A. Abbagnaro, *Microphones (Anthology)*; Audio Engineering Society, **1979**, New York NY., pp. 162–171.

14 H. B. Callen *Thermodynamics and an Introduction to Thermostatistics*, 2nd edn; New York: Wiley, **1985**.

15 R. J. Roark, R. G. Budynas, W. C. Young, *Roark's Formulas for Stress and Strain*; New York: McGraw-Hill, **2001**.

16 G. K. Fedder, S. Santhanam, M. L. Reed, S. C. Eagle, D. F. Guillou, M. S.-C. Lu, L. R. Carley, *Sens. Actuators A* **1996**, 57, 103–110.

17 M. S.-C. Lu, X. Zhu, G. K. Fedder, *Mater. Res. Soc. Symp. Proc.* **1998**, 518, 27.

18 H. T. Souther, in: L. A. Abbagnaro, *Microphones (Anthology)*; Audio Engineering Society, **1979**, New York NY., pp. 73–78.

19 http://www.fmsystems-inc.com/eng_desp.htm (discussion and table of pre-emphasis curve).

20 F.A. Everest, *Master Handbook of Acoustics*, 4th edn.; New York: McGraw-Hill, **2001**.
21 A.D. Pierce, *Acoustics – An Introduction to Its Physical Principles and Applications*; Woodbury, NY: Acoustical Society of America, **1991**.
22 I.J. Oppenheim, personal communication.
23 B.M. Diamond *MS Thesis*; Carnegie Mellon University, Pittsburgh, **2002**.
24 United States Patents 4,555,797, Nieuwendijk et al., November 26, 1985. Hybrid loudspeaker system for converting digital signals to acoustic signals. Inventors: Nieuwendijk, Joris A.M. (Eindhoven, NL); van Gijesel, Wilhelmus D.A.M. (Eindhoven, NL); Sanders, Georgius B.J. (Eindhoven, NL); van Niewuland, Jacob M. (Eindhoven, NL). Assignee: U.S. Philips Corporation (New York NY). United States Patent 4,360,707, Joseph et al., 23 November, 1982. Digitally driven combination coils for electrodynamic acoustic transducers. Inventors: Joseph, Joel R. (Libertyville, IL); Bleeke, William F. (Vandalia, MI). Assignee: CTS Corporation (Elkhart, IN). United States Patent 4,566,120, Nieuwendijk et al., January 21, 1986. Loudspeaker system and loudspeaker for use in a loud-speaker system for converting an n-bit digitized electric signal into an acoustic signal. Inventors: Nieuwendijk, Joris A.M. (Eindhoven, NL); Op de Beek, Franciscus J. (Eindhoven, NL); Sanders, Georgius B.J. (Eindhoven, NL); Van Gijsel, Wilhelmus D.A.M. (Eindhoven, NL); Van Nieuwland, Jacob M. (Eindhoven, NL). Assignee: U.S. Philips Corporation (New York, NY)
25 www.1limited.com.
26 B.M. Diamond, J.J. Neumann, K.J. Gabriel, in: *Proceedings of IEEE Transducers '03: the 12th International Conference on Solid-State Sensors, Actuators and Microsystems, Boston, MA, 8–12 June, 2003 (www.transducers03.org)*; **2003**, pp. 238–241.
27 J.J. Neumann, D.W. Greve, I.J. Oppenheim, in: *Proceedings of SPIE Symposium on Smart Structures and Materials/NDE 2004, San Diego, CA, 14-18 March*, vol 5391, **2004**.
28 D.W. Greve, J.J. Neumann, I.J. Oppenheim, S.P. Pessiki, D. Ozevin, in: *2003 IEEE International Ultrasonics Symposium, 5–8 October 2003 Honolulu, HI, Conference of the Ultrasonics Ferroelectrics, and Frequency Control Society*, **2003**, pp. 3F–4.
29 L. Cao, T.S. Kim, J. Zhou, S.C. Mantell, D.L. Polla, in: *Proceedings of the Thirteenth Biennial University/Government/Industry IEEE Microelectronics Symposium*; **1999**, pp. 204–210.
30 P.J. French, A.G.R. Evans, *Sens. Actuators* **1985**, *8*, 219–225.

5
RF CMOS MEMS

T. Mukherjee, G. K. Fedder, ECE Department & The Robotics Institute, Carnegie Mellon University, Pittsburgh, PA, USA

Abstract

Mobility and portability are driving miniaturization of wireless communications interfaces, leading the increased interest in single-chip radios. Achieving this goal requires integration of filtering, frequency conversion, modulation and demodulation functions, and typically involves trading off noise, power, linearity, frequency, gain and supply voltage. Higher quality factor and lower insertion loss are critical to easing these trade-offs. CMOS micromachining offers the ability to integrate a variety of components with high quality factor and low insertion loss with CMOS RF transistors, and is therefore increasingly becoming important for radio designers.

Keywords

Inductor; capacitor; varactor; resonator; filter; switch

5.1	A Brief History of RF Integration	226
5.2	**MEMS in RF Architectures**	227
5.2.1	Receiver Architectures	227
5.2.2	Multi-band Transceiver Architectures	228
5.3	**Micromachined RF Components**	229
5.3.1	RF MEMS switches	230
5.3.2	Tunable Capacitors	234
5.3.3	Suspended Inductors	238
5.3.4	Micromachined Distributed Electromagnetic Resonators	243
5.3.5	Micromechanical Resonators and Filters	244
5.4	**Circuits Based on RF CMOS MEMS**	247
5.5	**Conclusion**	250
5.6	**References**	251

Advanced Micro and Nanosystems. Vol. 2. CMOS – MEMS.
Edited by H. Baltes, O. Brand, G. K. Fedder, C. Hierold, J. Korvink, O. Tabata
Copyright © 2005 WILEY-VCH Verlag GmbH & Co. KGaA, Weinheim
ISBN: 3-527-31080-0

5.1
A Brief History of RF Integration

RF communications has been a primary technology driver since the advent of AM radio at the beginning of the 20th century. Consumer applications in radio and TV were almost mature by the time of the advent of VLSI CMOS in the 1980s. The focus switched to integrating computation instead of communication, driven by the consumer desire for portability in computation and the relative lack of desire for portability in communication. Technologically speaking, it was easier to integrate more of the same MOS transistors than to integrate the diverse types of devices needed for wireless communications.

With shrinking device feature sizes enabling computers to evolve from desktops to laptops and handhelds, mobile communication has again become the bottleneck for consumer applications. An important driver for these applications beyond the traditional performance drivers such as higher data rates and increased functionality are weight, volume and power consumption. This desire for very low weight, volume and power has driven significant technology developments aimed at monolithically integrating a radio communications system. As the GaAs, Si bipolar and Si MOS transistors crossed the 10 GHz transit frequency barrier [1] required for RF viability, they were co-opted in RF designs. By the mid-1990s, radio systems were a mishmash of GaAs (for high Q or power critical applications), discretes (primarily for tanks and filters), Si (for most of the analog portion of the front end) and CMOS (for frequency synthesis and data conversion) [2]. Within 5 years, RF CMOS became a leading contender for monolithic integration of a complete radio [3, 4]. The combination of GaAs performance in SiGe with integration into CMOS has led to its becoming a potential contender [5]. However, throughout this entire period, integrating passives into the silicon substrate has remained a bottleneck to RF performance, leading to interest in SOI [6] or SOS [7] technologies, as well as in package-level integration [8].

Today, integration has reached the point where a chip set consisting of a few RF chips can already support most required functionality. The limiting factor in further integration are the bulky expensive off-chip passive RF components, such as high-Q inductors, capacitors, varactor diodes and ceramic filters. Over the last decade, researchers in the micromachining or MEMS community have applied their technology to a variety of RF applications including switches, voltage-tunable capacitors, high-Q inductors and a variety of resonators in the electrical and mechanical domains. This variety of passive devices and the range of RF applications from 1 to 100 GHz, coupled with the potential for integration, makes RF MEMS a critical technology of interest to the wireless design community.

5.2
MEMS in RF Architectures

Several types of micromachined elements have been proposed to allow the miniaturization of communication systems by removing the off-chip components. Several recent review articles by Nguyen et al. [9, 10], Brown [11] and Tilmans et al. [12] and books by De Los Santos [13] and Rebeiz [14] have overviewed the decade of research in this area. Example RF MEMS components discussed include switches, high-Q inductors, variable capacitors and electrical and electromechanical resonators and filters. In contrast to these sources, this chapter focuses primarily on the subset of components that have been integrated with electronics.

5.2.1
Receiver Architectures

To place these components in perspective, we first consider the radio architectures of interest to wireless designers. When GSM cell phones were first introduced for the 900 MHz band, single conversion or dual conversion followed by sub-sampled data conversion or phase comparison was the norm. An example of this is the super-heterodyne receiver shown in Fig. 5.1a, which involves down-converting the radio frequency signal (at f_{IF}) to a fixed intermediate frequency of about 100 MHz (at $f_{IF}=f_{RF}-f_{LO1}$), where f_{LO1} is the first stage local oscillator frequency. Channel selection is performed using an off-chip surface acoustic wave or crystal filter. A separate filter may be needed for each band or standard, consuming premium board space on small form factor phones. The low impedance of these off-chip filters further increases the power dissipation of the on-chip interface circuits [e.g. the low-noise amplifier (LNA) and the mixer around the RF filter in Fig. 5.1a]. After channel selection, a variable gain amplifier (VGA) amplifies the signal in the de-

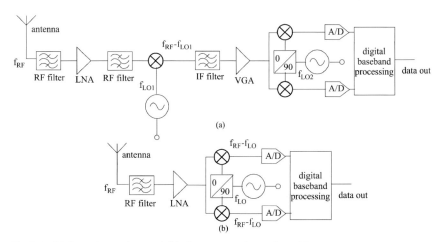

Fig. 5.1 (a) Super-heterodyne and (b) direct conversion radio architectures

sired channel and then demodulates it to baseband using in-phase (I) and quadrature (Q) signals at the second stage local oscillator frequency (f_{LO2}). The signal is then digitized, for further processing.

In contrast to the super-heterodyne architecture, the direct conversion architecture, shown in Fig. 5.1b, involves amplifying the incoming RF signal by the LNA and then directly demodulating it to baseband using I and Q signals. Channel selection and gain control are achieved after digitization by on-chip low-pass filters and a VGA in the digital baseband processor. Direct conversion of RF signals allows channel filtering to be done at baseband where implementation of power-efficient on-chip filtering is feasible. This eliminates the need for external passive filters, saving board space and cost. The reduced component count in near-zero IF architectures such as those in Fig. 5.1b gained acceptance during the 1990s and now dominates in both cell phone and wireless LAN radios.

5.2.2
Multi-band Transceiver Architectures

Fig. 5.1 only shows the receiver portion of a wireless terminal. A complete transceiver has both a transmit (Tx) and receive (Rx) path. A complete dual-band transceiver (assuming the reduced component count direct conversion architecture) is shown in Fig. 5.2. Each time a new band is added into a multi-band front end, the entire front-end architecture is duplicated. This is because band-specific filters are needed between the antenna and the LNA in the receive chain and between the power amplifier (PA) and the antenna in the transmit chain. Also, each band needs its own I and Q frequency synthesis for demodulation. This parallelization can be eliminated using micromachined tunable filters as shown in Fig. 5.3. Micromachining can also be used for performance enhancement of each of the shaded components. The diversity, transmit and receive switch can take advantage of the low insertion loss and high isolation of micromachined contact switches. The LNA and PA need tunable networks composed of a MEMS varactor and high-Q inductor for impedance matching at the multiple frequencies of interest. Similarly, the MEMS varactor and high-Q inductor can be tuned by the control signal for frequency synthesis to drive the I/Q inputs to the mixer. The receive mixers themselves can be micromechanical in nature, taking advantage of their wideband down-conversion capabilities. Therefore, even though wireless front-end architectures have moved away from the component-heavy super-heterodyne topology, the desire for multi-band radio systems is driving up the component count in modern wireless terminals, motivating the need for CMOS-based RF MEMS components described in this chapter.

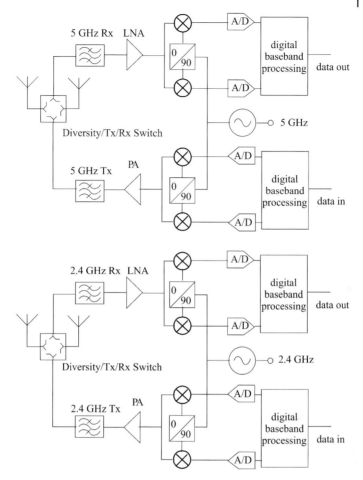

Fig. 5.2 Dual-band transceiver doubles the number of components compared with a single band

5.3
Micromachined RF Components

Following the MEMS-enhanced radio architecture shown in Fig. 5.3, we now describe the micromachined RF components including switches, variable capacitors, inductors, distributed resonators and micromechanical resonators.

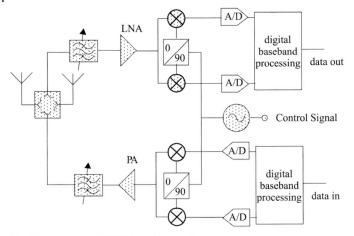

Fig. 5.3 Dual-band MEMS-based transceiver

5.3.1
RF MEMS switches

Switches are first amongst all the RF MEMS components, both in terms of the signal path of the radio architecture and in historical terms. In a single-band transceiver, an RF switch is needed for switching between the transmit and receive paths. If multiple antennas are used for diversity, an RF switch is also needed to select between the antennas. In multi-band wireless communications systems, switches are needed for routing signals to the different blocks. In phased-array antennas, RF signal routing in phase shifters requires the use of RF switches. In all such cases, the switch is right next to the antenna at the input to the receive path or the output of the transmit path.

The spectacular performance of the RF-MEMS switch (first introduced by Larson et al. [15, 16]) in comparison with PIN diodes and GaAs field effect transistors for both insertion loss when shorting and isolation when open has led to significant research in this area. In addition to the insertion loss and isolation, other important performance characteristics of RF switches include the return loss in both states, the power consumption, bandwidth, power-handling capability and linearity.

Radio frequency switching has classically been realized using PIN diodes and GaAs MESFET or JFET-based semiconductor switches. RF MEMS switching devices [11, 13, 14] also have two stable states, just like their semiconductor RF counterparts. The MEMS switch miniaturizes the conventional mechanical relay. Switching occurs through mechanical displacement of a freely movable structural member, called the armature. Just like their bulky conventional mechanical counterparts, RF MEMS switches have better isolation performance than their semiconductor competitors, especially at high frequencies (>30 GHz). Additionally, they have low loss over a wide frequency range (in particular compared with

GaAs FETs at higher frequencies). Furthermore, they have extremely low standby power consumption (in particular compared with PIN diodes). Lastly, they have excellent linearity characteristics compared with their semiconductor counterparts.

RF MEMS switches can be classified using four discriminants: the mechanical actuation mechanism, the direction of movement, the type of electrical contact and the circuit configuration. Four actuation mechanisms have been demonstrated. The highest actuation voltages have been required for the electrostatic switches. Thermally activated switches require much lower actuation voltages, but tend to consume more chip area. Magnetostatic switches also operate with low voltages, but have a large standby current. Finally, piezoelectric switches can operate with low voltage and zero current, but are slower than their electrostatic or thermal counterparts. There are two possible directions of movement: vertical and lateral. Vertical actuation typically leads to more area efficient switches. Two types of electrical configurations have been investigated: ohmic metal-to-metal contact switches that operate down to DC (and up to 60 GHz) and capacitive switches that operate in the 10–120 GHz range. The latter is similar to the tunable capacitors described in section 5.3.2. Finally, there are two types of circuit configurations: series and shunt.

Several limitations have prevented the widespread use of RF MEMS switches. In a parametric performance sense, they tend to be slow (of the order of milliseconds compared with nanoseconds for semiconductor-based switches), require extremely high control voltages (of the order of 20–80 V compared with 3–5 V in semiconductor switches) and have a low power-handling capability (<1 W compared with about 10 W). Another limitation has been their reliability, particularly for ohmic switches [17]. Raytheon Systems has demonstrated over 25 billion cycles at 10 GHz and an RF power of 100–200 mW (initial tests in [18] with the latest results in [14]). Hermetic packaging and material optimization, coupled with design for low voltage and low contact force operation, is likely to lead to over 100 billion cycles at low power levels for capacitive switches.

To integrate the switch with electronics, a batch-transfer process for low-loss microrelays was proposed by researchers at U.C. Berkeley [19]. A variety of designs were demonstrated including series and shunt switching of coplanar waveguides (CPW). Ohmic (gold–gold contact) series switches had insertion loss of <0.3 dB and isolation better than 15 dB at frequencies from 0.1 to 50 GHz. Shunt switches had better isolation (>45 dB) in the same frequency range. Actuation voltages varied from chip to chip, ranging from around 30 to 110 V. This wide variation was attributed to the variance in the height of the structures from the target chip due to variations in the transfer process. Commercial integration of a MEMS switch with a CMOS control IC in a single package has been demonstrated by Motorola [20]. The concept, shown in Fig. 5.4, leads to a switch with insertion loss under − 0.4 dB and isolation greater than −45 dB. The package includes a high-voltage charge pump and control logic chips to accommodate the low-voltage requirements in portable wireless applications. ST Microelectronics and CEA-LETI developed a thermally actuated metal–metal contact switch (with electrostatic clamping) to lower the actuation voltage for CMOS compatibility [21] as shown in Fig. 5.5. The switch can be

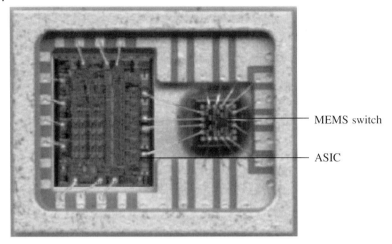

Fig. 5.4 Schematic of an integrated RF MEMS switch and CMOS control chip in a single package (source Motorola)

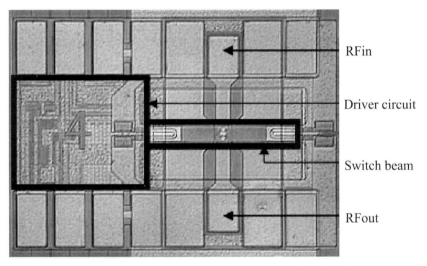

Fig. 5.5 SEM showing integration of STs above IC RF MEMS switch and its driver circuit in the ST BiCMOS7 process

integrated above any standard ST BiCMOS7 process (stopped after a metal layer, so no bonding is required). An IC driver based on the ST DriftMos components in their high-voltage CMOS process is coupled to the switch and is capable of driving 20 mA of current at 2 V for thermal actuation and delivering 10 V for electrostatic clamping. The switch was operated for more than a billion cycles and achieved 0.18 dB insertion loss and −57 dB isolation at 2 GHz on a 15 W cm low resistivity wafer [22].

Nominally, RF MEMS switches are fabricated using low-temperature processes and are therefore compatible with post-CMOS, SiGe or GaAs integration. Although over 30 different switch designs have been proposed, several of which have been monolithically integrated for switch networks, there is only one instance of integration with electronics. Rockwell Scientific has published a MEMS switch network with GaAs amplifiers [23] as shown in Fig. 5.6. The pair of MEMS switches connect the input to the two GaAs HEMTs, which provide single transistor amplification. On-chip input and output matching networks are included.

Although no connections between the RF MEMS switch and CMOS electronics have been shown, a couple of processes capable of integrating RF MEMS switches with CMOS are under development. The first wafer-level packaged RF MEMS switches fabricated in a commercial CMOS foundry were demonstrated by a team of researchers from IMEC and Alcatel Microelectronics in 2001 [24]. It is a capacitive shunt switch implemented on a coplanar waveguide using the Al layer in the CMOS process. The moveable electrode in the switch is a suspended Al bridge which is mechanically anchored and electrically connected to the ground of the CPW. This suspended bridge is formed by depositing it on top of a sacrificial photoresist that is later etched away using O_2 plasma. This effectively forms a capacitor between the metal bridge (which is at RF ground) and the signal line. In the RF-on state, the bridge is up and the switch capacitance is small and does not affect the impedance of the signal line, allowing the RF signal to pass freely through the line. When the switch is turned off, the bridge lowers and is separated from the signal line by a thin high-permittivity dielectric layer (anodized Ta_2O_3). The large capacitance causes an RF short to ground, hence no signal can flow on the signal line. The change in capacitance between the on state and off state is easily two orders of magnitude, leading to low insertion loss, combined with high isolation. Although the fabrication was carried out in a commercial CMOS manufacturing line at Alcatel, post-processing for the anodization and the sacrificial layer etch took place in a CMOS-compatible environment at IMEC. After switch fabrication, a separate silicon wafer was used to seal the switch using benzocyclobutene (BCB) as the bonding material. Some of the materials used in this process are not widely used in CMOS foundries, including the high resistivity

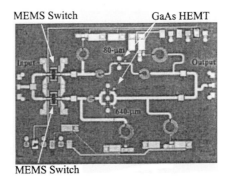

Fig. 5.6 Photograph of an X-band dual-path MEMS/GaAs amplifier [23]

(5 kW cm) Si substrate and the deposition of aluminum over a sacrificial photoresist layer. The resulting switch has an insertion loss of 0.2 dB and isolation of 5 dB at 5.2 GHz, with a switch actuation voltage of 15 V (when tested prior to packaging with the Si cap). Resistive losses in the feedthroughs arise from the need to get under the sealant to the switch. The RF performance of the package was limited owing to losses in the silicon cap.

With the emergence of damascene copper processes for CMOS interconnect, a switch structure composed of the Cu/SiO_2 back-end process was demonstrated by researchers from Singapore [25]. The electrostatic switches required the usual high voltages for actuation; however, RF characterization has not been demonstrated. IBM researchers have recently combined their expertise in copper interconnect and SiGe BiCMOS devices that are currently available through their foundry services to fabricate an RF MEMS switch [26]. Free-standing metal–MEMS structures are fabricated using a conventional back-end copper-based chemical mechanical polishing process and a dry release process. The movable structure is enclosed to prevent metal corrosion. The moving electrode is a copper bridge, with RF metallization only on one half of the device to mimic a series cantilever switch. Contact metallization is patterned on the copper moving and static electrodes. A gap of 0.2 µm is used to obtain a low actuation voltage (26 V) while ensuring adequate isolation at the design frequency of 2 GHz. Measured insertion loss is 0.3 dB and isolation is better than 30 dB from 0 to 6 GHz.

The MEMS post-processing approach following foundry CMOS has a strong potential for integrating RF MEMS switches with transistor electronics. MEMS switches have been formed out of a metal dielectric membrane using low-temperature processing for integration above the passivation in a fabricated CMOS wafer [27]. The characterized switches were fabricated on high-resistivity silicon wafers, although the authors state that the process is compatible with foundry CMOS.

In summary, the >12 years of research into RF switches has demonstrated a variety of switch topologies with excellent RF performance characteristics. The prospect for integration with CMOS or SiGe for low-power wireless front-end systems remains high. The current focus in switch research is in reliability and packaging. Once these issues have been solved, further integration with silicon is likely.

5.3.2
Tunable Capacitors

A tunable capacitor is one whose capacitance can be tuned electrically, e.g. through application of a DC (tuning) voltage. RF tunable capacitors find applications in tunable filters, in the frequency-controlling element in the LC-tank of a voltage-controlled oscillator (VCO) and for tunable load lines and matching networks. In traditional integrated circuits (ICs), tunable capacitors are implemented as p–n junction diodes (either off-chip or on-chip) and using accumulation region MOS devices. Limitations for on-chip implementation tend to include excessive series resistance (resulting in low Q) and high parasitic capacitance that limit the tuning range.

Micromachined tunable capacitors tend to be parallel-plate capacitors whose value is changed by either varying the gap or varying the overlap area between the plates. Early designs tended to use a simple gap tuning approach [28–30]. Owing to the electrostatic instability in the parallel-plate capacitor (pull-in occurs when the displacement exceeds one-third of the gap), this architecture is limited to a tunable range of 1.5:1, which is too restrictive for many wireless applications. The current focus of research is in widening the tuning range. One approach was to use a comb-based design that varied the overlap area for capacitance tuning [31]. Another approach was to use independent capacitor and tuning voltage electrodes similar to the capacitive switch architectures [32]. A third approach was to use electrothermal actuation, which does not suffer from the pull-in voltage problem [33].

From a design perspective, RF MEMS tunable capacitors should use thick (highly conductive) metal layers to minimize series resistance losses compared with diode or MOS capacitors. The potential for the RF signal in moving the capacitor plates or fingers must be minimized, implying the need for stiff springs to hold the moveable plate or fingers. Additional constraints arise when considering RF MEMS tunable capacitor implementation in CMOS processes. For example, there is only a single micromechanical layer available in MEMS etched directly from a CMOS interconnect stack with silicon undercut (using a process sequence detailed in [34] and Chapter 1). Vertical parallel-plate capacitors are not possible without additional material depositions and micromachining above the CMOS. On the other hand, lateral microactuation is relatively easy to implement.

Electrostatic actuation for RF capacitors is readily achievable, but a large tuning range requires a combination of high actuation voltage, large actuation area and a compliant suspension. All three of these requirements are detrimental for RF systems on chip. Voltages above 5 V are incompatible with conventional CMOS or BiCMOS processes. Areal efficiency for passives is important to minimize cost, especially for applications requiring multiple inductors and capacitors. Compliant suspensions produce high Brownian noise displacement, are prone to acoustic vibration and RF self-biasing and take up more chip space.

CMOS electrothermal microactuators can operate at low voltages, are area efficient and can have very stiff suspensions. One design concept for lateral electrothermal actuation exploited the ability to offset embedded layers with different temperature coefficients of expansion within beams [36]. This offset idea was first presented for use in a lateral capacitive infrared sensor [35]. Of particular importance is the ability to tailor the lateral stress gradient, and therefore lateral moment, as a function along the beam length. This ability to set an internal moment along the beam arises from different offset and width of the embedded layers.

To form CMOS MEMS electrothermal actuators, the lower metal layers embedded in the beams are laterally offset with respect to the top metal layer of the beams, as shown in the example folded-flexure design in Fig. 5.7a [36]. This particular design relieves axial residual stress. The beams have mostly dielectric along one side and aluminum on the other side. The metal offset switches sides half way down the beam length to produce a bending moment asymmetry. The resulting displacement mimics that of a guided-end beam.

A lateral residual stress gradient exists within the actuation beams at room temperature as a result of the differing residual stress. The residual stress creates an initial deflection (Fig. 5.7b), which must be accounted for in design. Electrothermal actuation up to 25 µm in a relatively compact 200×40 µm area has been achieved by passing 1–5 mA of current through polysilicon resistors embedded in the beams as shown in Fig. 5.7c. The central piston retracts laterally in response to the bending moment caused by differing thermal coefficients of expansion (TCE) of the metal and dielectric layers within the actuation beams. The actuator can be designed to self-assemble and actuate in either of two directions, depending on the ordering of the metal offset within the actuation beams.

All of the actuator beams are designed to move with no rotation at their ends. Therefore, an arbitrary number of beams may be placed in parallel to achieve high stiffness in the actuation direction without lowering the displacement capability. For example, the actuators in Fig. 5.7b and c have five parallel beams per actuator segment, thereby providing five times greater stiffness than a single beam design.

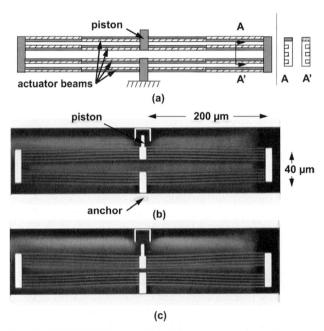

Fig. 5.7 CMOS MEMS lateral folded-flexure microactuator. (a) Layout illustrating offset metal layers embedded within the beams. A–A' is the cross section through the inner set of beams. (b) Self-actuation of 10 µm after release. (c) Electrothermal actuation demonstrating retraction of 20 µm [36]

Several CMOS MEMS tunable capacitors have been designed, fabricated and tested using the lateral electrothermal actuator technology [37, 38]. The capacitor topology shown in Fig. 5.8 has five banks of 39 comb fingers that make up the main capacitor electrodes [39]. The stator fingers are suspended from a micromachined frame that compensates for any out-of-plane curl that occurs along the finger trusses. The rotor fingers are connected to two single-ended folded-flexure electrothermal actuators located to the far right and left of the device. Upon release, the capacitor self assembles by moving the rotor to the right to engage the comb. Heating the actuators moves the rotor to the left to disengage the comb.

A separate actuator is placed adjacent to the rotor truss to operate a latching mechanism. When heated, the latch actuator retracts from the rotor, allowing it to move. Once the rotor is actuated into the desired position, the latch actuator is powered off and it clamps on to the rotor due to the self-assembly displacement. Once latched, the rotor actuators may be powered off with the capacitor remaining in its desired position. This latch sequencing provides capability to set a range of capacitances with zero input power.

Tunable capacitors with this topology fabricated in a 0.35 µm CMOS process had a 3.52 to 1 tuning range from 42 to 148 fF with a 0–12 V control voltage, 34 mW maximum heater power and an estimated Q of 50 at 1.5 GHz. For these initial prototypes, the comb fingers jammed in the gaps after about 1 µm engagement owing to manufacturing bloat of the top metal layer (which is used as a structural mask [34] for the comb fingers). The maximum capacitance was achieved by moving the rotor truss completely to the left to form a parallel-plate capacitance with the stator truss as shown in Fig. 5.9 [37].

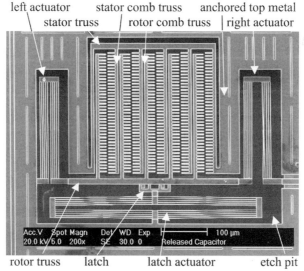

Fig. 5.8 Tunable capacitor in the Jazz semiconductor SiGe60 process. The configuration shown is latched into a low capacitance state [39]

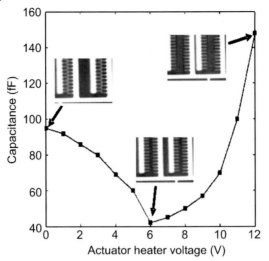

Fig. 5.9 Measured capacitance vs. actuator voltage characteristic for a capacitor in a 0.35 μm CMOS process [37]. The SEM insets indicate the finger positions at the specified control voltage. For 0 V, the fingers are partially engaged (additional engagement was not possible owing to manufacturing bloat in the metal layer used as the structural mask), for 6 V, the fingers are maximally disengaged. AT 12 V, a large capacitance is obtained by narrowing the gap between the rotor and stator comb trusses

RF MEMS tunable capacitors are still in their infancy. Capacitance values per unit area, Q-factors and tunable ranges are likely to improve in the coming years. Even so, they can already achieve comparable values to semiconductor tunable capacitors. Therefore, such devices are likely to see widespread applications in tunable filters and matching networks for next generation multi-band wireless systems.

5.3.3
Suspended Inductors

For most wireless applications, inductors in the range 0.5–10 nH with $Q > 10$ and a self-resonance frequency above 10 GHz are desirable. The lower inductance values can be obtained with a straight line of interconnect. The higher values require spiral layout topologies. Traditionally, the integration of on-chip passive inductors was of limited benefit owing to the low Q (about 5) obtained using planar spiral inductors in digital CMOS processes.

An inductor typically has multiple operating regimes. At low frequency or in a quasi-static regime, the inductor is typically dominated by the series resistance, in the interconnect and the inductance. At high enough frequencies, this model becomes invalid as the impedance of the capacitors needs to be considered. At these frequencies, the quasi-static solution of Maxwell's equation such as those obtained from FastHenry [40] has to be replaced by a full-wave solution using tools such as HFSS [41], Momentum [42] or Sonnet [43]. Of these, HFSS captures all the 3D effects and tends to be the most accurate (at the expense of simulation speed). Momentum and Sonnet are faster, as they are optimized for planar electromagnetic analysis. Lumped parameter simulation of the inductor in resonance can be accomplished by adding an ideal capacitance in parallel with the inductor. Sweeping the capacitance allows simulation of the inductor at different resonant frequencies.

The Q-factor of inductors operated in the quasi-static regime (usually at frequencies <1 GHz) is limited by series resistance of the wiring. Increasing inductor thickness can reduce this series resistance, as illustrated by the simulation in Fig. 5.10 [44]. Q increases less than linearly with geometric increase in metal thickness, indicating that increasing metal thickness has diminishing effectiveness in lowering series resistance owing to three factors. First, the resistance does not decrease as $1/t$ (where t is the interconnect thickness) owing to the high frequency skin effect [44]. Second, current crowding steers the current to one side of the metal trace [45]. Usually, the current flows on the outer sidewalls for the outer turns. As the magnetic field direction at the inner turns is opposite that of the outer turns, this current flow in the innermost turns prefers the inner sidewalls [44]. Third, for identical layouts, increasing thickness leads to reduced self-inductance owing to current crowding effectively decreasing the spiral radius. Hence the inductance drops when the thickness increases [44]. Finally, as frequency increases, the capacitance coupling to the substrate becomes increasingly important, leading to an increase in the image current flow in the substrate, and a preference for current flow in the bottom wall of the spiral. This preference for the bot-

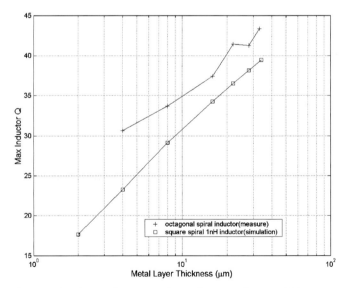

Fig. 5.10 Simulation of maximum Q as a function of copper thickness. The one-turn 1 nH inductance has a 12 μm SiO_2 spacer to a 13.5 Ω cm silicon substrate. Adapted from [44]

tom wall and the inner/outer walls due to the skin and crowding effects leads to current concentration in the corners of the metal traces and explains the less than linear improvement in Q with respect to increasing thickness shown in Fig. 5.10. Another alternative to thick metallizations is the use of low-resistivity metal traces such as copper. Thick metallizations are not specific to MEMS and are now available in many RF foundry processes.

As the frequency increases even higher, it is eventually high enough for the parasitic shunt capacitance to become appreciably low in impedance. A lumped parameter model that captures these parasitic effects appears as a π-shaped network in Fig. 5.11 around the simple series inductor (L) and resistor (R_s) between the inductor terminals (T_1 and T_2). The parasitic capacitances include the capacitance between the spiral turns (C_{t-t}) and the overlap between the spiral inductor and the return trace that crosses under the inductor (C_u). The turn-to-turn capacitance is usually less important than the overlap capacitance since the series connection of the turn-to-turn capacitances appears as a reduced effective capacitance at the terminals. Another parasitic is the overlap capacitance to the substrate (C_{ox}). These capacitances lead to image currents in the substrate that flow in the opposite direction to the main inductor, reducing its inductance. This current flow in the lossy substrate also leads to Q degradation. The final parasitics required for modeling the inductor are the substrate resistance (R_{Si}) and capacitance (C_{Si}). One solution is repositioning the inductor higher above the lossy silicon substrate. For example, the simulation results in Fig. 5.12 indicate that about 100 µm of silicon must be removed from under a 400 µm square spiral inductor to achieve the highest Q. The maximum Q is improved by about twice that of a spiral spaced from the substrate by 12 µm of SiO_2. Straight transmission-line inductors benefit from much greater depth of silicon removal, since the fields extend further.

Improving the spiral inductor quality factor was an active area of research in the late 1990s. In digital CMOS processes, a patterned ground shield [46] is often used to prevent capacitive coupling to the lossy substrate while avoiding short-circuiting of magnetic flux, at the expense of reduced self-resonance frequency. RF CMOS or SiGe BiCMOS foundries catering to the RF markets now routinely use a high-resistivity substrate [47] or thicker dielectric layers to locate the spiral in-

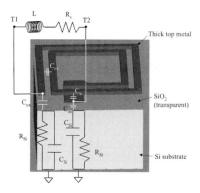

Fig. 5.11 Lumped element spiral inductor model showing parasitic elements

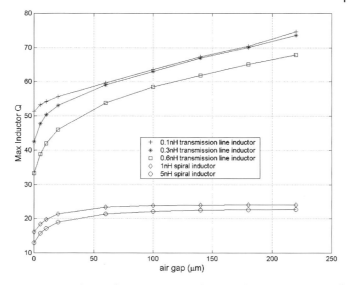

Fig. 5.12 Simulation of maximum Q as a function of air gap from the inductors to a 13.5 Ω cm silicon substrate. The inductors are made from a 4 μm thick aluminum conductor with 12 μm SiO_2 underneath. The single-turn spiral inductors are 400 μm square. After [44]

ductor further away from the lossy substrate [48] and thicker metal layers for the inductor for increased quality factor [49].

MEMS technology has also been used to improve inductor performance in digital CMOS substrates. Chang et al. [50] used front-side bulk micromachining with an anisotropic etchant (KOH or EDP) to eliminate the substrate under the spiral to create the first suspended inductor on CMOS as shown in Fig. 5.13. The membrane-supported inductor has less substrate losses and less capacitive coupling to the substrate, improving both Q and self-resonance frequency (from 0.8 to 3 GHz for a 100 nH inductor). Later work at UCLA included XeF_2-based etching of the silicon substrate [51]. The UCLA membrane inductor was on 2 μm CMOS. The same front-side bulk micromachining approach was also applied on Alcatel Mietec's double-metal single-poly 0.7 μm CMOS by IMEC [52]. In addition to removal of the silicon under the inductor, IMEC also etched away the silicon dioxide between the inductor coils.

Inspired by these initial results, several other groups demonstrated suspended inductors. Three notable approaches not yet demonstrated on CMOS have been applied: copper electroplating to create a ground plane [53], raising the inductor above the silicon substrate via self-assembly [54] and 3D micromachining for solenoid inductors [55, 56].

Returning to CMOS-based fabrication, a suspended copper inductor using copper metallization, increasingly being used in modern digital CMOS processes, was demonstrated by Carnegie Mellon researchers [57]. The post-processing [34]

Fig. 5.13 SEM of a suspended inductor formed using front-side bulk micromachining [50]

uses the same recipe as the CMOS-based capacitors described in the previouis section on tunable capacitors, viz. a dielectric RIE followed by an Si undercut (process sequence detailed in Chapter 1). Both cantilevered, as shown in Fig. 5.14, and fixed–fixed connection between the suspended inductor to the chip were shown. A comparison of quality factors prior to micromachining, after dielectric RIE, and after Si undercut, for a four-turn fixed–fixed inductor in a low-k dielectric copper process is given in Fig. 5.15. The inductor is modeled with about 5% accuracy up to the self-resonance frequency by using a lumped circuit approximation. The model is more complex than the simple π-model of Fig. 5.11. Each of the four turns is represented as an LCR network with overlap capacitance from each turn to one of the input terminals [57]. The quality factor (Q) for the inductor without micromachining tops out at 4. The dielectric RIE creates air gaps between the turns and thereby reduces the inter-turn capacitance. A modest 14% increase in Q and an 18% increase in the self-resonance frequency are the result. Undercutting the silicon results in a 179% increase in Q and a 54% increase in self-resonance frequency. Losses also occur in loop currents generated in the metal that defines the etch pit around the inductor. The eddy currents can be reduced by slotting the surrounding metal, that is shown in Fig. 5.14. Alternatively, CMOS MEMS processes that use a photoresist micromachining mask layer will eliminate this problem. The adjacent conductors in micromachined spiral inductors roughly maintain their position relative to each other after release from the substrate. Finite element simulations predict only a 70 ppm change in inductance

Fig. 5.14 Micromachined copper spiral inductor [57]

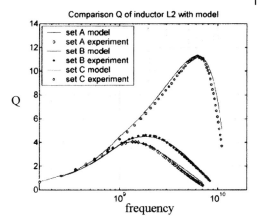

Fig. 5.15 Measured (points) and modeled (lines) quality factor as a function of frequency for 3.9 nH 336 μm diameter octagonal spiral inductor with four 20 μm wide, 20 μm pitch turns in a copper 0.18 μm CMOS process. Dataset measurements are (A) prior to micromachining, (B) after dielectric RIE and (C) after silicon undercut. After [57]

from as much as 25 μm maximum out-of-plane curl in a cantilevered design [57]. Simulated inductance change from a 100 g shock is 0.016%. The inductance in a cantilevered spiral design is about 10 times less sensitive to out-of-plane curl than a fixed–fixed design.

The focus above has been on distancing the inductor from the lossy substrate by removing it. Another potential direction where MEMS technologies have been used for higher Q inductors is in positioning the inductor high above the substrate. Researchers at KAIST have combined thick metallization (5 μm copper) to reduce series resistance with repositioning of the inductor layer about 25 μm above the substrate to minimize the substrate loss and increase self-resonance frequency [58]. A similar process is now available from MEMSCAP, which combined a thick copper process above a thick isolating polyamide separator layer for above IC inductors.

Integration of micromachined suspended inductors with tunable capacitors is essential for use in RF front-end systems. Integration of the inductor with CMOS permits tunable resonators formed from the inductor and the varactors in the foundry CMOS. Simultaneous micromachining of the inductor and tunable capacitor described in the previous section is a promising alternative for resonators with a wider tuning range and higher Q-factors and is likely to be driven by the demand for future multi-function radios.

5.3.4
Micromachined Distributed Electromagnetic Resonators

The tunable capacitors and suspended inductors are lumped parameter devices that can be combined to form resonators needed for RF filtering, oscillation or impedance matching. At even higher frequencies (K-band), microwave and millimeter-wave components such as transmission line (T-line) cavity and dielectric resonators have been integrated monolithically using microfabrication techniques.

A T-line example is the membrane supported microstripline filter from the University of Michigan [59]. This two-pole bandpass filter operating at 37 GHz was

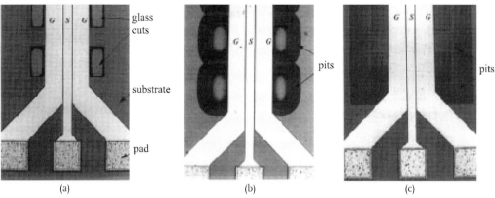

Fig. 5.16 SEM of the 50 Ω coplanar waveguide (a) after CMOS fabrication, with glass cuts from the layout glass cut masks, (b) showing the etch pits as they are isotropically etched using XeF$_2$, and (c) after anisotropic etching using EDP showing final results of combined etch

housed in a microcavity fabricated using bulk micromachining of silicon wafers and wafer bonding. Tunable resonators were formed by using a lumped-element tunable inductor and MEMS tunable capacitors along the length of the line. For higher Q, parasitic losses were kept to a minimum using microfabricated cavities [60]. By fabricating the same microstrip lines with a dielectric-filled cavity, micromachined dielectric resonators have also been demonstrated [61].

In CMOS, coplanar waveguides were fabricated through post-processing micromachining after commercial CMOS fabrication by a team from George Washington University, NIST and RF Microsystems [62]. Removal of the lossy silicon substrate using top-side micromachining reduced the losses from 38 to 4 dB/cm. Fig. 5.16 shows the 50 Ω coplanar waveguide prior to, during and after the post-foundry micromachining. The post-CMOS micromachining process used requires regions of superimposed glass cuts [63]. Superimposing glass cuts is possible for older CMOS processes, but is not compatible with deep submicron (<0.5 µm) CMOS processes.

5.3.5
Micromechanical Resonators and Filters

The inherent resistive losses in the distributed electromagnetic resonators limit the Qs compared with those of micromechanical resonators and filters, especially when operated in vacuum. Qs exceeding 80 000 were reported by Nguyen's group about a dozen years ago for low-frequency flexural mode resonators [64]. A large number of resonant filter devices have since been reported (as described in Chapter 5 of AMN Volume 1). Nguyen and co-workers have promoted the use of these filters for a variety of applications such as higher order filters for low [65] and high [66] intermediate frequency (IF) filtering.

Higher frequency versions reported recently include a 200 MHz SiC beam resonator (Q=1200) from Caltech and Case Western Reserve University [67], a 156 MHz contour-mode disk resonator ($Q=9400$) from the University of Michigan [68] and a self-aligned 1.14 GHz radial-mode (for the third vibrational mode) disk resonator with Q exceeding 1500 in both vacuum and air, also from the University of Michigan [69]. Two primary approaches have been followed to achieve the high–frequency and high-Q operation in mechanical structures. Structures with nanometer beam geometries (e.g. Roukes' resonators [67], Craighead's SCS resonators [70]) and micromechanical structures with nanometer gaps (e.g. Nguyen's beam and disk resonators [68, 71] and Ayazi's HARPSS resonators [72]) have both been demonstrated. New competitive designs and technologies are still being introduced in this arena. In many cases, the measurement technique precludes integration into wireless communication systems. For example, the Caltech approach uses Lorentz force detection [73], whereas the Cornell approach uses optical detection of motion [70].

Integration with CMOS was first demonstrated [74] through combination of planar CMOS processing with polysilicon surface micromachining. Extensions of this process for polysilicon germanium surface micromachining have been demonstrated with the potential for CMOS integration [75]. Direct integration of a filter composed of three resonators was demonstrated in the Carnegie Mellon CMOS MEMS process [76]. An SEM of the filter and the filter characteristics are shown in Fig. 5.17. Each of the resonators are comb-drive crab-leg structures. O-shaped springs couple the three individual resonators. The filter was designed using a library of lumped-element behavioral micromechanical models for a center frequency of 550 kHz and bandwidth of 92 kHz. The dimensions of the fabricated resonator were extracted and used to develop a simulation model. The simulation results matched very well with the measured results [76].

The electromechanical drive mechanism often used in micromechanical resonators and filters exerts a force that is proportional to the square of the voltage across the drive electrodes. This square law characteristic can be used for mixing [77]. Since the electromechanical force is applied on a mechanical resonator, it can be used as a filter. RF signals in the range 40–200 MHz were downconverted and filtered at a 27 MHz intermediate frequency with less than 15 dB of combined mixing conversion and filter insertion loss [77]. The mechanical filter structure in this mixer+filter device was nearly identical with a reported high-Q filter [66]. One of the concerns was the need for high voltages, both for the AC magnitudes and the DC bias levels, especially for systems with large electrode–resonator gaps. Additional concerns included interconnect parasitics in polysilicon surface micromachined structures at high (RF) frequencies. Recent extensions have reduced the electrode–resonator gap from 100 to 32.5 nm, maximizing electromechanical coupling. The combined mixer+filter (denoted 'mixler') downconverted 200 MHz signals to a 37 MHz IF with 13 dB of combined mixer conversion and filter loss [78]. Limitations include the need for a high (11 V) polarization voltage, which can be alleviated if the electromechanical gap is further reduced.

Mixler integration with CMOS is an ongoing research thrust at Carnegie Mellon University [79]. Direct integration with CMOS electronics enables drive of the mixer with an on-chip LNA. It also enables filter readout via an on-chip pre-amp. These electronics on-chip can improve impedance matching of the micromechani-

5 RF CMOS MEMS

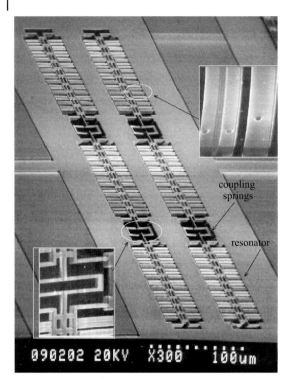

Fig. 5.17 A CMOS MEMS filter made by mechanically coupling three micropore sonators [76]

cal resonator filter with the RF signal path without the need for custom fabrication of ultra-narrow gaps. The most mature design uses two 510 kHz cantilevers as the micromechanical elements that provide their outputs into an on-chip differential amplifier, as shown in Fig. 5.18. Mixing has been demonstrated with a local oscillator frequency above 2 GHz, with the dominant design limit being the electronic mixer feedthrough. Reconfiguring the mixer for differential drive and improved physical design of the RF interconnect is expected to reduce feedthrough by three orders of magnitude, offering the potential for a CMOS MEMS-based transceiver architecture.

A cousin of the micromechanical resonator is the thin-film bulk acoustic resonator (FBAR). This is considered by the MEMS community to be a micromachined version of the conventional bulk acoustic wave resonators such as the quartz crystal. Such resonators are comprised of a parallel-plate capacitor with a piezoelectric dielectric layer. Application of an AC electrical signal at the electrodes excites bulk vibrational modes in the piezoelectric film. Acoustic resonance occurs through trapping the vibration at the reflecting electrode surfaces. Acoustic losses to the substrate have to be minimized for high Q, leading to the desire for local removal of the silicon substrate under the resonator. Such membrane-supported resonators are about a decade old [80], but have only recently become available commercially from Agilent Technologies [81]. Although they are not inte-

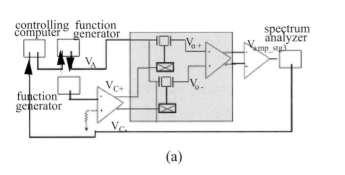 

Fig. 5.18 (a) Schematic of differential cantilever CMOS MEMS mixer filter with (b) SEM of one of the cantilevers [79]

grated on to CMOS substrates, they have been assembled with a CMOS VCO to achieve ultra-low-power performance [82].

5.4
Circuits Based on RF CMOS MEMS

Several of the micromachined RF components described in Section 5.3 have been combined with transistor electronics to form RF circuits. The earliest circuits involved multi-chip assembly with transistors.

Examples include bond-wire inductors and variable capacitors for voltage-controlled oscillators (VCOs). Researchers at U.C. Berkeley combined aluminum micromachined capacitors for frequency tuning with a discrete inductor and separately fabricated CMOS electronics on a test-board for a hybrid implementation of a VCO [83]. The capacitor had a 16% tuning range, leading to an oscillator tuning range of 14 MHz, with a phase noise of –107 dBc/Hz at 100 kHz offset from the 714 MHz carrier. A polysilicon surfaced micromachined capacitor was combined with a cross-coupled CMOS VCO core and wire-bond inductors for a VCO with a tuning range of 3.4% and a phase noise of –122 dBc/Hz at 1 MHz offset from the 2.4 GHz carrier [84]. The polysilicon capacitor [84] requires high-temperature processing, preventing integration with CMOS. The aluminum capacitor [83] can be fabricated above CMOS wafers with a maximum processing temperature of 150 °C, making it CMOS compatible. However, direct integration with CMOS has not yet been demonstrated.

The first fully integrated VCO with on-chip MEMS inductors integrated them with accumulation mode MOS capacitors and CMOS cross-coupled regenerative circuits [85]. Two VCOs were reported, at 1 and 2.6 GHz carrier frequencies. They had –124 and –117 dBc/Hz phase noise at 300 kHz from the carrier, which were some of the lowest phase noise results reported at that time. The inductor was fabricated using an above-IC thick copper micromachining process described earlier [58]. Although Qs of 20 and 27, respectively, were reported for the 5 nH (1 GHz VCO) and 1.8 nH (2.6 GHz VCO) when fabricated on a silicon test wafer, the Q of the CMOS integrated tanks were approximately 10. This is because the tank Q was being dominated by the varactor, which had been designed for excessive tuning range (in excess of 60%).

A comparison between the foundry inductor and an above-IC thick copper micromachined inductor was presented by researchers at KAIST [86]. The foundry inductor in the RF CMOS process was patterned in a top-level 2 μm thick Al/Cu layer, with typical Q-factors of about 10. The above-IC thick copper inductor had Qs in excess of 25 as described in Section 5.3. Both inductors were designed such that their Q would be maximum around 5 GHz (using Momentum software [42]). A negative-g_m (negative transconductance) cross-coupled CMOS VCO circuit with the tank comprising an inductor and two capacitors was used to compare the performances of the foundry and MEMS inductors. Two capacitors in parallel formed the tank with the inductor: a fixed MIM capacitor was used to limit the Q losses in the capacitive part of the tank; an accumulation region MOSFET varactor was used for frequency tuning (with a range of 20% around the 5 GHz center frequency). At offset frequencies of 30 kHz to 3 MHz away from the VCO's 5 GHz carrier, using the same power in the core circuits, the MEMS VCO had about 7 dB better phase noise.

The first simultaneous integration of MEMS capacitors and inductors for an RF circuit was presented in 2003 [87], using the post-CMOS micromachined capacitor [37] and inductor [57]. Two circuits have been demonstrated using this process to date. In the first circuit, reconfigurable capacitors are combined with an inductor as a π-network filter (Fig. 5.19) to switch between two desired center frequencies with the goal of achieving high Q and low insertion loss [88]. A secondary goal is to achieve similar insertion loss at both operating frequencies. This simple single-resonator filter topology was chosen as a demonstration vehicle for the first integrated MEMS LC implementation. One of the limitations of this topology is the trade-off between Q and insertion loss. For the π-network, high filter Q requires a high C_{tank}/C_{DC} ratio. On the other hand, insertion loss minimization requires a low C_{tank}/C_{DC} ratio. During design, different C_{tank}/C_{DC} ratios were chosen to compensate for the difference in inductor Q and to account for interconnect parasitic capacitances at the two frequencies. Fig. 5.20 shows the measured S_{21} response of the filter at the minimum and maximum center frequencies observed. The filter operated at 1.87 GHz (at the state where the micromechanical capacitor is engaged) to 2.36 GHz (the disengaged state). The change in center frequencies achieved is 490 MHz, showing a 23% tuning range at 2.1 GHz. The insertion loss values are −14.3 and −19.3 dB, respectively. The quality factor values are 4.4 and 9.5, respectively. Owing to incomplete release of one of the tank capacitors (C_{tank}), the measured tuning range was lower than the designed tuning range. The larger capacitance of the unreleased capacitor lowered the C_{tank}/C_{DC} ratio, leading to the increased insertion loss seen in the measurements.

The second circuit, using the same identical post-foundry micromachining process, is a voltage-controlled oscillator (VCO), targeted for use in a dual frequency-

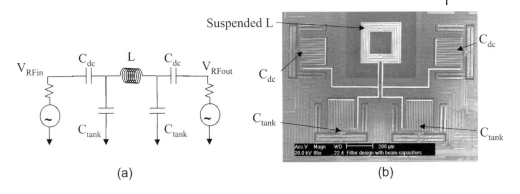

Fig. 5.19 (a) Schematic and (b) SEM of a post-foundry CMOS micromachined VCO [88]

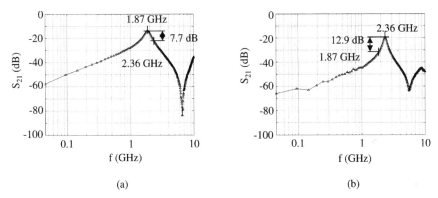

Fig. 5.20 Measured S_{21} response for RF frequency-hopping filter at (a) minimum frequency and (b) maximum frequency, showing a frequency hop of 490 MHz. The filter rejects at least 7.7 dB of the undesirable frequency component [88]

hopped receiver configuration to be used for portable applications [89]. The circuit schematic and SEM are shown in Fig. 5.21. The active circuit consists of a bipolar transistor cross-coupled pair, which has higher transconductance and lower noise than a CMOS pair. The tail current source is also an SiGe HBT for reduced noise. Emitter follower buffers to deliver up to −5 dBm to a 50 Ω load are not shown in the schematic. The LC tank consists of a differential micromachined inductor, MEMS capacitors and varactor diodes. The reconfigurable MEMS capacitors are used to hop from one frequency of interest to another. The varactors are used for fine tuning at these discrete hop frequencies. In this topology, there is a phase noise trade-off with tuning range, due to the non-linearity of the varactor diodes. To counter this, the MEMS reconfigurable capacitors are used in parallel to the varactor. The MEMS capacitors provide wide tuning over a discrete set of frequencies (here just two because of the latch design). The varactors can therefore be small in size for very narrow tuning, hence they introduce less noise.

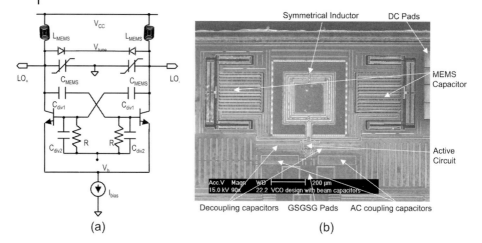

Fig. 5.21 (a) Schematic and (b) SEM of the post-foundry CMOS micromachined VCO [89]

Tab. 5.1 Comparisons with MEMS VCOs

Parameter	RFIC 2003 [86]	MTT 2000 [84]	MTT 2003 [85]
f_0 (GHz)	2.84	2.4	2.6
L (dBc/Hz)	–122	–122	–127
P (mW)	2.75	13.5	15
FoM (dB)	187	178	184
Inductance (nH)	6.25	2.8	1.8
Capacitance (pF)	0.5	1.4	2
Idea	MEMS L and C	MEMS C	MEMS L

Tab. 5.1 shows a comparison of the VCOs described above. The simultaneous integration of the MEMS inductor and capacitor in the post-foundry micromachined VCO leads to a 5-fold reduction in power compared with the other micromachined VCOs without trading off for phase noise. This leads to the highest overall figure of merit for all micromachined VCOs published to date.

5.5
Conclusion

For most MEMS devices, the goal is to interface with the physical world, with the MEMS device acting as a sensor or an actuator, where MEMS technology provides unique capabilities. For RF applications, the motivation for MEMS is to provide replacement components that are smaller and have lower loss than are available in other technologies. Therefore, the importance of single-chip integration is perhaps even more crucial than in other MEMS applications. Also, from a performance point of view, having two or more chips, one for the RF IC and others for discrete MEMS devices, is not optimal because of the importance of parasitics related to the packaging. In a multi-chip system, the signal has to leave the RF electronics chip, go through the MEMS device and return back to the RF IC. An additional benefit of single-chip integration is the reduced component count, which in turn reduces both manufacturing and test costs.

The potential for discrete RF MEMS devices is now well known. The 2003 International Technology Roadmap for Semiconductors includes the switch and the filter as potential solutions for performance breakthroughs in wireless circuits [90]. Additionally, the Roadmap also lists micromachining as a potential solution for minimizing substrate loss and for thermal isolation in silicon substrates. Integration with CMOS will hasten this potential impact of MEMS on RF devices, circuits and systems. The focus of future research in RF MEMS is likely to be in this integration. Initially, the focus is likely to be demonstration of the performance enhancements that MEMS can enable in RF circuits. The recent RF filter and VCO circuits are the beginning of this research trend. Further into the future, we are likely to see redesign of the RF system architectures to take advantage of the unique performance characteristics of MEMS. One example is the use of the reconfigurable capabilities for multi-band radios. Another example, is the use of the micromechanical beam resonator and mixer for array based RF architectures. This potential impact in RF architectures is similar to how the advent of RF CMOS led to the recent change from super-heterodyne to direct conversion architectures.

Therefore, the future for RF CMOS MEMS is extremely bright. The final outcome of RF MEMS integration with CMOS is likely to be driven by issues of system performance, manufacturing cost, miniaturization demand and the ability to open new markets such as reconfigurable radios. In this regard, the RF MEMS community needs to increase collaboration with system-level researchers in the cell phone, wireless LAN and other mobile applications to ensure fully that MEMS reaches its prospective potential in this application area.

5.6 References

1 L. николайн, L. Larson, U. Mishra, 'Ultra-high-speed modulation doped field-effect transistors: a tutorial overview,' *Proc. IEEE* **1992**, *80*, 494–518.

2 P. R. Gray, R. G. Meyer, 'Future directions in silicon ICs for RF personal communications,' in: *1995 IEEE Custom Integrated Circuits Conference, Santa Clara, CA*; **1995**, pp. 83–90.

3 T. H. Lee, *The Design of CMOS Radio-Frequency Integrated Circuits*; Cambridge: Cambridge University Press, **1998**.

4 B. Razavi, *RF Microelectronics*; Englewood Cliffs, NJ: Prentice Hall, **1998**.

5 D. L. Harame, D. C. Ahlgren, D. D. Coolbaugh, J. S. Dunn, G. G. Freeman, J. D. Gillis, R. A. Groves, G. N. Hendersen, R. A. Johnson, A. J. Joseph, S. Subbanna, A. M. Victor, K. M. Watson, C. S. Webster, P. J. Zampardi, 'Current status and future trends of SiGe BiCMOS technology,' *IEEE Trans. Electron Devices* **2001**, *48*, 2575–2594.

6 K. Joardar, 'Comparison of SOI and junction isolation for substrate cross-talk suppression in mixed-mode integrated circuits,' *Electron. Lett.* **1995**, *31*, 1230–1231.

7 R. Johnson, P. de la Houussaye, C. Chang, B. Offord, G. Imthurn, P. Asbeck, I. Lagnado, 'Silicon-on-sapphire MOSFET transmit/receive switch for L- and S-band transceiver applications,' *Electron. Lett.* **1997**, *33*, 1324–1326.

8 R. R. Tummala, G. E. White, V. Sundaram, S. Bhattacharyam, 'SoP: the microelectronics for the 21st century with integral passive integration,' *Adv. Microelectron.* **2000**, *27*, 13–19.

9 C. T.-C. Nguyen, 'Micromechanical components for miniaturized low-power communications,' in: *Proceedings of 1999 IEEE MTT-S International Microwave Symposium RF MEMS Workshop (on Microelectromechanical Devices for RF Systems: Their Construction, Reliability and Application)*, Anaheim, CA; **1999**, pp. 48–77.

10 C. T.-C. Nguyen, L. P. B. Katehi, G. M. Rebeiz, 'Micromachined devices for wireless communications,' *Proc. IEEE* **1998**, *86*, 1756–1768.

11 E. R. Brown, 'RF-MEMS switches for reconfigurable integrated circuits,' *IEEE Trans. Microwave Theory Tech.* **1998**, *46 (11)*, pp. 1868–1880.

12 H. A. C. Tilmans, W. De Raedt, E. Beyne, 'MEMS for wireless communications: from RF-MEMS components to RF-MEMS-SiP,' *J. Micromech. Microeng.* **2003**, *13*, S139–S163.

13 H. De Los Santos, *Introduction to Microelectromechanical (MEM) Microwave Systems*; Boston, MA: Artech House, **1999**.

14 G. M. Rebeiz, *RF MEMS Theory, Design and Technology*, Hoboken, NJ: Wiley, **2003**.

15 L. E. Larson, R. H. Hackett, R. F. Lohr, 'Microactuators for GaAs-based microwave integrated circuits,' in: *Transducers '91*, San Francisco, CA; **1991**, pp. 743–746.

16 L. E. Larson, R. H. Hackett, M. A Melendes, R. F. Lohr, 'Micromachined microwave actuator (MIMAC) technology – a new tuning approach for microwave integrated circuits,' in: *Microwave and Millimeter-Wave Monolothic Circuits Symposium Digest*, Boston, MA; **1991**, pp. 27–30.

17 J. DeNatale, R. Mihailovich, 'RF MEMS reliability,' in: *Proc. 12th Int. Conf. on Solid-State Sensors, Actuators and Microsystems (TRANSDUCERS '03)*; **2003**, Vol. 2, pp. 943–946.

18 B. Pillans, J. Kleber, C. Goldsmith, M. Eberly, 'RF power handling of capacitive RF MEMS devices,' in: *2002 IEEE MTT-S International Microwave Symposium Digest*, **2002**, Vol. 1. pp. 329–332.

19 V. Milanovic, M. Maharbiz, K. S. J. Pister, 'Batch transfer integration of RF microrelays,' *IEEE Microwave Guided Wave Lett.* **2000**, *10 (8)*, pp. 313–315.

20 A. P. De Silva, H. G. Hughes, 'The package integration of RF-MEMS switch and control IC for wireless applications,' *IEEE Trans. Adv. Packag.* **2003**, *26*, 255–260.

21 P. Blondy, D. Cros, P. Guillon, P. Rey, P. Charvet, B. Diem, C. Zanchi, J. B. Quoirin, 'Low voltage high isolation MEMS switches,' in: *Digest of Papers of 2001 Topical Meeting. On Silicon Monolithic Integrated Circuits in RF Systems*; **2001**, pp. 47–49.

22 D. Saias, P. Robert, S. Boret, C. Billard, G. Bouche, D. Belot, P. Ancey, 'An above IC MEMS RF switch,' *IEEE J. Solid-State Circuits* **2003**, *SC-38 (12)*, 2318–2324.

23 M. Kim, J. B. Hacker, R. E. Mihailovich, J. F. DeNatale, 'A monolithic MEMS switched dual-path power amplifier,' *IEEE Microwave Wireless Components Lett.* **2001**, *11*, 285–286.

24 H. A. C. Tilmans, H. Ziad, H. Jansen, O. Di Monaco, A. Jourdain, W. De Raedt, X. Rottenberg, E. De Backer, A. Decaussernaeker, K. Baert, 'Wafer-level packaged RF-MEMS switches fabricated in a CMOS fab,' in: *Tech. Dig. Int. Electron Devices Meeting (IEDM '01)*; **2001**, pp. 41.4.1–41.4.4.

25 C. Zhen, Y. Mingbin, G. Lihui, 'Design and fabrication of RF MEMS capacitive switch on silicon substrate with advanced IC interconnect technology,' in: *Proc. 6th Int. Conf. on Solid-State and Integrated-Circuit Technology*, **2001**, Vol. 2, pp. 739–741.

26 N. HOIVIK, C.V. JAHNES, J. COTTE, J.L. LUND, D. SEEGER, J.H. MAGERLEIN, 'RF MEMS switches using copper-based CMOS interconnect manufacturing technology,' in: *2004 Solid-State Sensors, Actuators and Microsystems Workshop, Hilton Head Island, SC*; **2004**, pp. 93–94.

27 P. ERICSSON, M. HEDSTROM, A. OLSSON, A. SCHOLES, J. SVENNEBRINK, C. VIEIDER, B. WENK, 'MEMs switches and other RF components fabricated using CMOS postprocess compatible surface micromachining,' in: *Proc. 12th Int. Conf. on Solid-State Sensors, Actuators and Microsystems (TRANSDUCERS '03), Boston, MA*; **2003**, Vol. 1, pp. 895–898.

28 D.J. YOUNG, B.E. BOSER, 'A micromachined variable capacitor for monolithic low-noise VCOs,' in: *Technical Digest, 1996 Solid-State Sensor and Actuator Workshop, Hilton Head Island, SC*; **1996**, pp. 86–89.

29 L. FAN, R.T. CHEN, A. NESPOLA, M.C. WU, 'Universal MEMS platforms for passive RF components: suspended inductors and variable capacitors,' in: *Proceedings of the Eleventh Annual International Workshop on Micro Electro Mechanical Systems (MEMS 98)*; **1998**, pp. 29–33.

30 A. DEC, K. SUYAMA, 'Micromachined electro-mechanically tunable capacitors and their applications to RF ICs', *IEEE Trans. Microwave Theory Tech.* **1998**, 46, 2587–2596.

31 J.J. YAO, S.T. PARK, J. DENATALE, 'High tuning-ratio MEMS-based tunable capacitors for RF communications applications,' in: *Technical Digest, 1998 Solid-State Sensor and Actuator Workshop, Hilton Head Island, SC*; **1998**, pp 124–127.

32 J. ZOU, C. LIU, J. SCHUTT-AINE, J. CHEN, S.-M. KANG, 'Development of a wide tuning range MEMS tunable capacitor for wireless communication systems,' in: *Technical Digest 2000 International Electron Devices Meeting (IEDM 00), San Francisco, CA*; **2000**, pp. 403–406.

33 Z. FENG, H. ZHANG, W. ZHANG, B. SU, K. GUPTA, V. BRIGHT, Y. LEE, 'MEMS-based variable capacitor for millimeter-wave applications,' in: *Technical Digest, 2000 Solid-State Sensor and Actuator Workshop, Hilton Head Island, SC*; **2000**, pp 255–258.

34 G.K. FEDDER, S. SANTHANAM, M.L. REED, S.C. EAGLE, D.F. GUILLOU, M.S.-C. LU, L.R. CARLEY, 'Laminated high-aspect-ratio microstructures in a conventional CMOS process,' *Sens. Actuators A* **1997**, 57, 103–110.

35 H. LAKDAWALA, G.K. FEDDER, 'CMOS micromachined infrared imager pixel,' in *Tech. Dig. IEEE Int. Conf. on Solid-State Sensors and Actuators (Transducers '01), Munich*; **2001**, pp. 556–559.

36 A. OZ, G.K. FEDDER, 'CMOS electrothermal lateral micromovers for actuation and self-assembly,' in *Proc. of the SEM Annual Conference on Experimental and Applied Mechanics, Charlotte, NC*; **2003**.

37 A. OZ, G.K. FEDDER, 'CMOS-compatible RF-MEMS tunable capacitors,' in: *Digest of the 2003 IEEE RFIC Symposium, Philadelphia, PA*; **2003**, pp. 611–614.

38 A. OZ, G.K. FEDDER, 'RF CMOS-MEMS capacitor having large tuning range,' in: *Tech. Dig. IEEE Int. Conf. on Solid-State Sensors and Actuators (Transducers '03), Boston, MA*; **2003**, pp. 851–854.

39 A. OZ, 'CMOS/BICMOS self-assembling and electrothermal microactuators for tunable capacitors,' *MS Thesis*, ECE Department, Carnegie Mellon University, Pittsburgh, PA; **2003**.

40 M. KAMON, M.J. TTSUK, J.K. WHITE, 'FASTHENRY: a multipole-accelerated 3-D inductance extraction program,' *IEEE Trans. Microwave Theory and Techniques* **1994**, 42, 1750–1758.

41 *High Frequency Structure Simulator*; Pittsburgh, PA: Ansoft Corporation, www.ansoft.com (version 9, 2004).

42 *EEsof EDA, Momentum Simulations*; Palo Alto, CA: Agilent Technologies, eesof.tm.agilent.com (version 2003c).

43 *Sonnet 9*; Liverpool, NY: Sonnet Software, www.sonnetusa.com (release 9, 2003).

44 X. ZHU, R. GROVES, S. SUBBANA, D. JADUS, T. MUKHERJEE, G.K. FEDDER, 'Micromachined on-chip inductor performance analysis,' in: *Proc. IEEE MEMS 2003, Kyoto*; **2003**, pp. 165–168.

45 W.B. KUHN, N.M. IBRAHIM, 'Analysis of current crowding effects in multiturn

spiral inductors,' *IEEE Trans. Microwave Theory Tech.* **2001**, *49*, 31–38.

46 C. P. Yue, S. Wong, 'On-chip spiral inductors wth patterned ground shields for Si-based RFICs,' in: *Dig. Tech. Papers of VLSI Circuits Symposium*; **1997**, pp. 85–86.

47 J. N. Burghartz, D. C. Edelstein, K. A. Jenkins, C. Jahnes, C. Uzoh, E. J. O'Sullivan, K. K. Chan, M. Soyuer, P. Roper, S. Cordes, 'Monolithic spiral inductors fabricated using a VLSI Cu-damascene interconnect technology and low-loss substrates,' in: *International Electron Devices Meeting*; **1996**, pp. 99–102.

48 D. C. Laney, L. E. Larson, P. Chan, J. Malinowski, D. Harame, S. Subbanna, R. Volant, M. Case, 'Microwave transformers, inductors and transmission lines implemented in an Si/SiGe HBT process,' *IEEE Trans. Microwave Theory Tech.* **2001**, *49*, 1507–1510.

49 R. Groves, J. Malinowski, R. Volant, D. Jadus, 'High Q inductors in a SiGe BiMOS process utilizing a thick metal process add-on module,' in: *Proceedings of the 1999 Bipolar/BiCMOS Circuits and Technology Meeting*; **1999**, pp. 149–152.

50 J. Y.-C. Chang, A. A. Abidi, M. Gaitan, 'Large suspended inductors on silicon and their use in a 2 μm CMOS RF amplifier,' *IEEE Electron Device Lett.* **1993**, *14*, 246–248.

51 J. Rael, A. Rofougaran, A. A. Abidi, 'Design methodology used in a single-chip CMOS 900 MHz spread-spectrum wireless transceiver,' in: *Proceedings 1998 Design Automation Conference*, **1998**, pp. 44–49.

52 H. A. C. Tilmans, K. Baert, A. Verbist, R. Puers, 'CMOS foundry-based micromachining,' *J. Micromech. Microeng.* **1996**, *6*, 122–127.

53 H. Jiang, W. Ye, J.-L.A. Yeh, N. C. Tien, 'On-chip spiral inductors suspended over deep copper-lined cavities,' *IEEE Trans. Microwave Theory Tech.* **2000**, *48*, 2415–2423.

54 V. M. Lubecke, B. Barber, E. Chan, D. Lopez, M. E. Gross, P. Gammel, 'Self-assembly MEMS variable and fixed RF inductors,' *IEEE Trans. Microwave Theory Tech.* **2001**, *49*, 2093–2098.

55 D. J. Young, V. Malba, J.-J. Ou, A. F. Bernhardt, B. E. Boser, 'Monolithic high-performance three-dimensional coil inductors for wireless communication applications,' in: *Tech. Dig. IEEE International Electron Devices Meeting (IEDM 97), Washington, DC*; **1997**, pp. 67–70.

56 C. H. Ahn, Y. J. Kim, M. G. Allen, 'A fully integrated micromachined toroidal inductor with nickel-iron magnetic core (the switched DC/DC boost converter application),' in: *Digest of Technical Papers, 7th International Conference on Solid-State Sensors and Actuators (Transducers '93), Yokohama*; **1993**, pp. 70–73.

57 H. Lakdawala, X. Zhu, H. Luo, S. Santhanam, L. R. Carley, G. K. Fedder, 'Micromachined high-Q inductors in a 0.18 μm Cu interconnect low-K dielectric CMOS process,' *IEEE J. Solid-State Circuits* **2002**, *37 (3)*, pp. 394–403.

58 J.-B. Yoon, Y.-S. Choi, B. I. Kim, Y. Eo, E. Yoon, 'CMOS-compatible surface-micromachined suspended-spiral inductors for multi-GHz silicon RF ICs,' *IEEE Electron Device Lett.* **2002**, *23*, 591–593.

59 P. Blondy, A. R. Brown, D. Cros, G. M. Rebeiz, 'Low-loss micromachined filters for millimeter-wave communication systems,' *IEEE Trans. Microwave Theory Tech.* **1998**, *MTT-46*, 2283–2288.

60 C. Kim I. Song, C. Song, C. Cheon Y. Kwon, S. Lee, 'A micromachined cavity resonator for mm-wave oscillator applications,' in: *TRANSDUCERS '99, Sendai*; **1999**, pp. 1268–1271.

61 B. Guillon, D. Cros, P. Pons, K. Grenier, T. Parra, J. L. Cazaux, J. C. Lalaurie, J. Graffeuil, R. Plana, 'Design and realization of high Q millimeter-wave structures through micromachining techniques,' in: *1999 IEEE MTT-S, Anaheim, CA*; **1999**, pp. 1519–1522.

62 V. Milanovic, M. Gaitan, E. D. Bowen, M. E. Zaghloul, 'Micromachined microwave transmission lines in CMOS technology,' *IEEE Trans. Microwave Theory Tech.* **1997**, *MTT-45*, 630–635.

63 N. H. Tea, V. Milanovic, C. A. Zincke, J. S. Suehle, M. Gaitan, M. E. Zaghloul, J. Geist, 'Hybrid postprocessing etching for CMOS-compatible MEMS,' *J. Microelectromech. Syst.* **1997**, *6*, 363–372.

64 C. T.-C. Nguyen, R. T. Howe, 'Quality factor control for micromechanical resonators,' in: *Technical Digest IEEE International Electron Devices Meeting (IEDM 92), San Francisco, CA*; **1992**, pp. 505–508.

65 K. Wang, C. T.-C. Nguyen, 'High-order micromechanical electronic filters,' in: *Proceedings of 1997 IEEE International Micro Electro Mechanical Systems Workshop, Nagoya*; **1997**, pp. 25–30.

66 J. R. Clark, A.-C. Wong, C. T.-C. Nguyen, 'Parallel-resonator HF micromechanical bandpass filters,' in: *Digest of Technical Papers of 1997 International Conference on Solid-State Sensors and Actuators, Chicago, IL*; **1997**, pp. 1161–1164.

67 Y. T. Yang, K. L. Ekinci, X. M. H. Huang, L. M. Schiavone, M. L. Roukes, C. A. Zorman, M. Mehregany, 'Monocrystalline silicon carbide nanoelectromechanical systems,' *Appl. Phys. Lett.* **2001**, 78 (2), pp. 162–164.

68 J. R. Clark, W.-T. Hsu, C. T.-C. Nguyen, 'High-Q VHF micromechanical contour-mode disk resonators,' in: *Technical Digest of 2000 International Electron Devices Meeting (IEDM 00), San Francisco, CA*; **2000**, pp. 493–496.

69 J. Wang, Z. Ren, C. T.-C. Nguyen, 'Self-aligned 1.14-GHz vibrating radial-mode disk resonators,' in: *Proc. 12th Int. Conf. on Solid-State Sensors, Actuators and Microsystems (TRANSDUCERS '03)*; **2003**, Vol. 2, pp. 947–950.

70 D. W. Carr, S. Evoy, L. Sekaric, J. M. Parpia, H. G. Craighead, 'Measurement of mechanical resonance and losses in nanometer scale silicon wires,' *Appl. Phys. Lett.* **1999**, 75, 920.

71 K. Wang, Y. Yu, A.-C. Wong, C. T.-C. Nguyen, 'VHF free-free beam high-Q micromechanical resonators,' in: *Technical Digest of 12th International IEEE Micro Electro Mechanical Systems Conference, Orlando, FL*; **1999**, pp. 453–458.

72 S. Pourkamali, A. Hashimura, R. Abdolvand, G. K. Ho, A. Erbil, F. Ayazi, 'High-Q single crystal silicon HARPSS capacitive beam resonators with self-aligned sub-100-nm transduction gaps,' *J. Microelectromech. Syst.* **2003**, 12, 487–496.

73 A. N. Cleland, M. L. Roukes, 'Fabrication of high frequency nanometer scale mechanical resonators from bulk Si crystals,' *Appl. Phys. Lett.* **1996**, 69, 2653.

74 C. T.-C. Nguyen, R. T. Howe, 'An integrated CMOS micromechanical resonator high-Q oscillator,' *IEEE J. Solid State Circuits* **1999**, 34, 440–445.

75 A. E. Franke, J. M. Heck, T.-J. King, R. T. Howe, 'Polycrystalline silicon-germanium films for integrated microsystems,' *J. Microelectromech. Syst.* **2003**, 12, 160–171.

76 Q. Jing, H. Luo, T. Mukherjee, L. R. Carley, G. K. Fedder, 'CMOS micromechanical bandpass filter design using a hierarchical MEMS circuit library,' in: *Thirteenth Annual International Conference on Micro Electro Mechanical Systems (MEMS 2000)*; **2000**, pp. 187–192.

77 A.-C. Wong, H. Ding, C. T.-C. Nguyen, 'Micromechanical mixer + filters,' in: *Technical Digest of 1998 IEEE International Electron Device Meeting (IEDM 98), San Francisco, CA*; **1998**, pp. 471–474.

78 A.-C. Wong, C. T.-C. Nguyen, 'Micromechanical mixer-filters ('mixlers'),' *J. Microelectromech. Syst.* **2004**, 13, 100–112.

79 J. Stillman, 'CMOS MEMS resonant mixer-filters,' *MS Thesis*; ECE Department, Carnegie Mellon University, **2003**.

80 R. Ruby, P. Merchant, 'Micromachined thin film bulk acoustic resonators,' in: *IEEE Int. Frequency Control Symp.*; **1994**, pp. 135–138.

81 R. Ruby, P. Bradley, J. Larson III, Y. Oshmyansky, D. Figueredo, 'Ultra-miniature high-Q filters and duplexers using FBAR technology,' in: *IEEE Int. Solid-State Circuits Conf. Dig. Tech. Papers*; **2001**, pp. 120–121.

82 B. P. Otis, J. M. Rabaey, 'A 300-µW 1.9-GHz CMOS oscillator utilizing micromachined resonators,' *IEEE J. Solid-State Circuits* **2003**, 38, 1271–1274.

83 D. J. Young, B. E. Boser, 'A micromachine-based RF low-noise voltage-controlled oscillator,' in: *Proc. IEEE CICC*; **1997**, pp. 431–434.

84 A. Dec, K. Suyama, 'Microwave MEMS-based voltage-controlled oscillators,' *IEEE Trans. Microwave Theory Tech.* **2000**, 48, 1943–1949.

85 E.-C. Park, Y.-S. Choi, J.-B. Yoon, S. Hong, E. Yoon, 'Fully integrated low

phase-noise VCOs with on-chip MEMS inductors,' *IEEE Trans. Microwave Theory Tech.* **2003**, *51*, 289–296.

86 E.-C. Park, S.-H. Baek, T.-S. Song, J.-B. Yoon, E. Yoon, 'Performance comparison of 5 GHz VCOs integrated by CMOS compatible high Q MEMS inductors,' in: *2003 IEEE MTT-S International Microwave Symposium Digest*; **2003**, Vol. 2, pp. 721–724.

87 A. Oz, V. K. Saraf, D. P. Ramachandran, G. K. Fedder, T. Mukherjee, 'Frequency hopping circuits based on reconfigurable MEMS capacitors,' in: *TECHCON, Dallas, TX*; **2003**.

88 D. Ramachandran, A. Oz, V. K. Saraf, G. K. Fedder, T. Mukherjee, 'MEMS-enabled reconfigurable VCO and RF filter,' in: *2004 IEEE RFIC Symposium, Fort Worth, TX*; **2004**, pp. 251–254.

89 V. K. Saraf, A. Oz, D. Ramachandran, G. K. Fedder, T. Mukherjee, 'Low-power LC-VCO using integrated MEMS passives,' in: *2004 IEEE RFIC Symposium, Fort Worth, TX*; **2004**, pp. 579–582.

90 *International Technology Roadmap for Semiconductors*; **2003**, Section on Radio Frequency and Analog/Mixed-Signal Technologies for Wireless Communications, public.itrs.net.

6
CMOS-based Pressure Sensors

H.-J. Timme, Infineon Technologies, Munich, Germany

Abstract
Silicon diaphragm pressure sensors are reviewed with respect to basic operation principles and aspects for their integration with CMOS technology. Bulk and surface micromachining provide powerful techniques for a variety of process options, including pre-CMOS, intermediate CMOS or post-CMOS integration. Piezoresistors embedded in micromachined diaphragms allow sensing of pressure-dependent stresses. Alternatively, diaphragm deflections can be sensed capacitively. Signal conditioning is mandatory for compensation of temperature and offset effects, and for calibration and linearization of transfer functions. CMOS technology not only offers stable and reproducible fabrication processes but also convenient mixed-signal circuit capabilities.

Keywords
bulk micromachining; surface micromachining; pressure sensor; diaphragm stress; diaphragm deflection; piezoresistor; capacitive sensing; switched capacitor circuits

6.1	**Introduction** *259*	
6.2	**Micromachined Pressure Sensors** *261*	
6.2.1	Hooke's Law for Elastic Solids *261*	
6.2.1.1	Stiffness and Compliance Matrices for Cubic Crystals *265*	
6.2.1.2	Definition of Elastic Constants *266*	
6.2.1.3	Hooke's Law for Cubic Crystals *268*	
6.2.1.4	Hooke's Law for Isotropic Materials *269*	
6.2.2	Plane Stress Approximation *270*	

Advanced Micro and Nanosystems. Vol. 2. CMOS – MEMS.
Edited by H. Baltes, O. Brand, G. K. Fedder, C. Hierold, J. Korvink, O. Tabata
Copyright © 2005 WILEY-VCH Verlag GmbH & Co. KGaA, Weinheim
ISBN: 3-527-31080-0

6.2.3	Deflection of Pressure-sensitive Diaphragms	271
6.2.3.1	Small Deflections of Thin Plates	271
6.2.3.2	Large Deflections of Thin Diaphragms	274
6.2.3.3	Stress Components in Circular Plates	277
6.2.3.4	Ring Plates	277
6.2.4	Piezoresistivity	279
6.2.4.1	Piezoresistivity Tensor and Piezoresistive Coefficients	279
6.2.4.2	Temperature and Doping Dependence of Piezoresistive Coefficients	281
6.2.4.3	Piezoresistive Coefficients for Diffused Layers	283
6.2.4.4	Piezoresistive Gauge Factors for Polysilicon Films	284
6.2.4.5	Piezoresistive Coefficients for Rotated Coordinate Systems	284
6.2.4.6	Piezoresistivity Coefficients for (100) Silicon Wafers	287
6.2.4.7	Piezoresistivity Coefficients for (111) Silicon Wafers	289
6.2.4.8	Two-terminal Piezoresistors	291
6.2.5	Piezoresistors in a Wheatstone Bridge	295
6.2.5.1	Diaphragm Designs with Piezoresistors	297
6.3	**CMOS-integrated Pressure Sensors**	**299**
6.3.1	Motivation	299
6.3.1.1	Bulk Versus Surface Micromachining	300
6.3.1.2	Capacitive Sensing of Diaphragm Deflection	301
6.3.2	Thickness Control for Diaphragm Formation	302
6.3.2.1	Time Etch Stop	303
6.3.2.2	Boron Etch Stop or p^{++} Etch Stop	303
6.3.2.3	Electrochemical Etch Stop (ECE)	304
6.3.3	Process Integration Strategies	304
6.3.3.1	Diaphragms with Integrated Strain Gauges	304
6.3.3.2	CMOS-integrated Bulk Micromachined Silicon Pressure Sensor	304
6.3.3.3	Bulk and Surface Micromachined Pressure Sensors with SiN Diaphragm and Polysilicon Piezoresistors	305
6.3.3.4	Solder-bonded Bulk-micromachined Capacitive Pressure Sensor	307
6.3.3.5	Absolute Capacitive Pressure Sensor with Anodically Bonded Glass Encapsulation	309
6.3.3.6	MEMS-CMOS Using Wafer Bonding	310
6.3.4	Polysilicon-based Capacitive Pressure Sensors	312
6.3.4.1	Integrated Surface Micromachined Capacitive Pressure Sensor	312
6.3.4.2	Intra-CMOS Surface Micromachining	314
6.3.4.3	Mixed Signal Processing	315
6.3.4.4	Charge Transfer in Switched Capacitor Circuits	317
6.3.4.5	Transfer Function of Delta–Sigma Modulator	321
6.3.5	Pressure Sensors for High-temperature Applications	322
6.3.5.1	Integrated Surface Micromachined Capacitive Pressure Sensor for High-temperature Applications	323
6.3.6	Micromachined Pirani Pressure Gauges	323
6.3.6.1	CMOS-integrated Thermal Pressure Sensor	324

6.3.7   Micromachined Pressure Sensors (Overview)   *325*
6.4   Conclusion   ***326***
6.5   References   ***326***

6.1
Introduction

Integrated pressure sensors combine a pressure-sensitive transducer with a signal processing circuit. Typically, the transducer itself is a micromachined diaphragm with embedded piezoresistors for stress sensing. Alternatively, the diaphragm can act as an electrode in a capacitor device with deflection-dependent capacitance. Both diaphragm stresses and deflections depend on the differential pressure applied to the diaphragm, i.e. on the pressure acting on the top surface minus the pressure acting on the bottom surface of the diaphragm. If the bottom surface of the diaphragm (for example) is part of a vacuum cavity, an absolute pressure sensor results.

Signal processing offers amplification and/or sampling of the transducer signal. In addition, compensation and calibration can be performed. Compensation addresses the reduction or elimination of temperature effects or supply voltage variations, etc. Calibration, on the other hand, adjusts the sensor's transfer function to specified parameters such as (zero-pressure) offset, sensitivity (or span) and linearity.

Silicon is the material of choice for a monolithic integration of micromachined pressure transducers and signal processing circuits. A key issue is the integration of all processing steps that are needed to manufacture both mechanical and electrical devices. Beyond the requirements of process compatibility, elastic and piezoresistive material properties of crystalline or polycrystalline silicon have to be considered, and dimensions of the micromachined structures (like diaphragm thickness) have to be controlled within tight limits.

Bulk and surface micromachining are the two principal processing techniques for thin diaphragms. In bulk micromachining, bulk material is selectively removed from a silicon wafer until a thin crystalline silicon diaphragm remains. Etch-stop techniques are used in order to control the diaphragm thickness. Surface micromachining uses thin (polycrystalline) films that have been deposited on top of sacrificial layers. After selective wet etching of the sacrificial layer, a film diaphragm remains. Both techniques feature distinct advantages, and put special demands on process integration.

The piezoresistance effect in silicon was discovered by Smith in 1954 [1] and explained by Herring and others [2–4]. Mason and Thurston [5] then described the application of the longitudinal piezoresistance effect for piezoresistive strain (or stress) gauges. The transverse and the shear piezoresistance effects were considered first by Pfann and Thurston [6].

In 1962, Tufte et al. proposed the first silicon diaphragm pressure sensor, which was based on 'integrated diaphragms' obtained by diffusion of piezoresistive strain (stress) gauges into silicon diaphragms. This approach offered considerable advantages over conventional methods with mounted or bonded strain gauges. Especially the integrated filament-type strain gauges were sensitive to both longitudinal and transverse piezoresistive effects [7].

Borky and Wise presented the first integrated silicon pressure sensor with temperature compensation in bipolar technology in 1979 [8]. One year later, a monolithically integrated capacitive pressure sensor was developed by Sander et al. [9].

In 1983, Kim and Wise first applied the electrochemical etch-stop technique for improved thickness control of silicon diaphragms in pressure sensor devices [10]. Sugiyama et al. [11] and Yamada et al. [12] reported piezoresistive pressure sensors with on-chip bipolar readout circuits. Pressure sensors with cavities that were bulk micromachined by means of a frontside undercut etch technique were introduced in 1986 [13]. This technique utilized a polysilicon sacrificial layer underneath a silicon nitride diaphragm. The anisotropic KOH wet etchants produced self-aligned V-shaped grooves inside the (100) silicon substrate. Vacuum sealing of the cavities could be performed by a wafer-level deposition process. Polysilicon strain gauges were embedded in the silicon nitride diaphragm, and offered significant advantages for high-temperature applications as compared with p–n junction isolated diffused piezoresistors in standard bulk-micromachined silicon diaphragms. The concept was further pursued and evolved into a pure surface micromachining approach [14]. Motorola introduced bulk micromachined pressure sensors based on a four-terminal transverse voltage strain gauge, the so-called X-ducer. This transducer utilized shear stress in a single piezoresistive device and eliminated the need for closely matched piezoresistors in conventional Wheatstone bridges [15, 16].

Howe and Muller fabricated free-standing polysilicon structures by means of surface micromachining in 1986. The polysilicon structures were first deposited and patterned on top of a sacrificial silicon dioxide layer, and subsequently released by etching away the sacrificial layer in hydrofluoric acid (HF) [17]. In 1987, the first stabilized (temperature and supply voltage compensated) CMOS integrated silicon pressure sensor was developed by Ishihara et al. [18].

In the early 1990s, digital interface circuits and digital compensation and calibration approaches were proposed and realized by several groups [19–24].

In 1997, Infineon Technologies introduced a monolithically integrated surface micromachined capacitive pressure sensor with digital readout and programming interface [25, 26]. This product has been in volume production since 1998, and the underlying intra-BiCMOS MEMS process has become a platform technology for other integrated pressure sensors with more advanced compensation and calibration circuits, tailored for automotive applications such as manifold air pressure sensing (MAP) [27].

This chapter is organized as follows. Section 6.2 provides a summary of basic data and techniques for understanding micromachined pressure sensors as elementary devices. After short subsections on Hooke's law for elastic solids and the plane stress approximation, small and large deflections of thin diaphragms are briefly considered. Furthermore, piezoresistivity is described on a phenomenological level, and equations are provided to calculate orientation-dependent piezoresistive coefficients. Fractional resistance changes of properly oriented and located piezoresistors can be calculated in combination with corresponding diaphragm stresses. Design considerations for Wheatstone bridge layouts are given. Section 6.3 focuses on aspects of integrating micromachined pressure transducers into bipolar or CMOS processes. Bulk and surface micromachining, capacitive sensing and diaphragm thickness control are briefly considered. Examples are provided for process integration strategies, polysilicon-based capacitive pressure sensors and pressure sensors for high-temperature applications. A subsection on micromachined Pirani pressure gauges demonstrates that the mainstream concepts for silicon pressure sensors, based on diaphragms and piezoresistive or capacitive sensing, are only a subset of the vast possibilities that CMOS-based pressure sensors have to offer. Conclusions are given in Section 6.4. Assembly and packaging issues of pressure sensors will not be addressed in this chapter.

6.2 Micromachined Pressure Sensors

6.2.1 Hooke's Law for Elastic Solids

In elastic solids, Hooke's law describes a linear relationship between the stresses σ_{ij} and the strains ε_{ij}:

$$\sigma_{ij} = c_{ijkl}\, \varepsilon_{kl}$$

where c_{ijkl} is the fourth-rank *stiffness tensor* and indices i, j, k, l run over x, y, z. Summation is over all pairs of identical indices unless stated otherwise. Owing to symmetry properties:

$$c_{ijkl} = c_{jikl} = c_{ijlk} = c_{klij}$$

the number of independent elastic constants is reduced from $3^4 = 81$ to 21. Similarly, using the fourth-rank *compliance tensor* s_{ijkl}, one has the equivalent equation

$$\varepsilon_{ij} = s_{ijkl}\, \sigma_{kl}$$

The symmetric strain tensor is defined as

$$\varepsilon_{ik} = \frac{1}{2}\left(\frac{\partial u_i}{\partial x_k} + \frac{\partial u_k}{\partial x_i}\right)$$

with displacement components u_i describing the deformation of the solid.

Note that the diagonal (dimensionless) strain tensor components – the unit elongations – are often written as ε_i, and that the non-diagonal (dimensionless) strain tensor components – the unit shear strains – are often represented as so-called engineering shear strains $\gamma_{ij} = 2\,\varepsilon_{ij}$, i.e.

$$\begin{pmatrix} \varepsilon_{xx} & \varepsilon_{xy} & \varepsilon_{xz} \\ \varepsilon_{yx} & \varepsilon_{yy} & \varepsilon_{yz} \\ \varepsilon_{zx} & \varepsilon_{zy} & \varepsilon_{zz} \end{pmatrix} = \begin{pmatrix} \varepsilon_x & \frac{1}{2}\gamma_{xy} & \frac{1}{2}\gamma_{xz} \\ \frac{1}{2}\gamma_{yx} & \varepsilon_y & \frac{1}{2}\gamma_{yz} \\ \frac{1}{2}\gamma_{zx} & \frac{1}{2}\gamma_{zy} & \varepsilon_z \end{pmatrix}$$

The diagonal stress tensor components – the normal stresses – are often referred as σ_i, and the non-diagonal stress tensor components – the shear stresses – are often given as τ_{ij} (with $i \neq j$), i.e.

$$\begin{pmatrix} \sigma_{xx} & \sigma_{xy} & \sigma_{xz} \\ \sigma_{yx} & \sigma_{yy} & \sigma_{yz} \\ \sigma_{zx} & \sigma_{zy} & \sigma_{zz} \end{pmatrix} = \begin{pmatrix} \sigma_x & \tau_{xy} & \tau_{xz} \\ \tau_{yx} & \sigma_y & \tau_{yz} \\ \tau_{zx} & \tau_{zy} & \sigma_z \end{pmatrix}$$

Normal stresses that act along the exterior normal of the reference plane element are *tensile* and have positive values. Negative normal stresses are *compressive* and act along the interior normal.

Matrix Notation

A convenient short-hand notation – called reduced index or matrix notation in contrast to the tensor notation – is

$$\sigma_\alpha = c_{\alpha\beta} \varepsilon_\beta$$

with indices $\alpha, \beta = 1, 2, ..., 6$ running over index pairs (ij), i.e.

$$1 \leftrightarrow (xx), \ 2 \leftrightarrow (yy), \ 3 \leftrightarrow (zz), \ 4 \leftrightarrow (yz), \ 5 \leftrightarrow (zx), \ 6 \leftrightarrow (xy)$$

The stress coefficients σ_α equal the stress tensor components σ_{ij} for corresponding indices $\alpha \leftrightarrow (ij)$, i.e. $\sigma_\alpha = \sigma_{ij} = \sigma_{ji}$:

$$\sigma_1 = \sigma_{xx}, \quad \sigma_2 = \sigma_{yy}, \quad \sigma_3 = \sigma_{zz} \quad \text{(normal stresses)}$$
$$\sigma_4 = \sigma_{yz}, \quad \sigma_5 = \sigma_{zx}, \quad \sigma_6 = \sigma_{xy} \quad \text{(shear stresses)}$$

Similarly, the stiffness coefficients $c_{\alpha\beta}$ equal the stiffness tensor components c_{ijkl} for corresponding indices $\alpha \leftrightarrow (ij)$ and $\beta \leftrightarrow (kl)$, i.e.

$$c_{\alpha\beta} = c_{ijkl} \quad \text{for} \quad \alpha \leftrightarrow (ij), \ \beta \leftrightarrow (kl)$$

Based on symmetry properties of the stiffness tensor, the stiffness coefficients $c_{\alpha\beta}$ form a symmetric 6×6 matrix. The strain coefficients ε_β are related to the components ε_{ij} of the symmetric strain tensor as follows:

$$\varepsilon_1 = \varepsilon_{xx}, \quad \varepsilon_2 = \varepsilon_{yy}, \quad \varepsilon_3 = \varepsilon_{zz}, \quad \varepsilon_4 = 2\varepsilon_{yz}, \quad \varepsilon_5 = 2\varepsilon_{zx}, \quad \varepsilon_6 = 2\varepsilon_{xy}$$

In the definition of the shear strains ε_4, ε_5, ε_6, the factors 2 must be introduced in order to account for the fact that the sums on the right-hand side of the tensor equation $\sigma_{ij} = c_{ijkl}\varepsilon_{kl}$ include two terms (kl) and (lk) whenever $k \neq l$.

In order to establish properly the corresponding matrix equation

$$\varepsilon_\alpha = s_{\alpha\beta}\, \sigma_\beta$$

the compliance coefficients $s_{\alpha\beta}$ must be defined in the following way:

$$s_{\alpha\beta} = \begin{cases} s_{ijkl} & \text{for } \alpha, \beta = 1, 2, 3 \\ 2\, s_{ijkl} & \text{for } \alpha = 1, 2, 3\, ,\ \beta = 4, 5, 6 \\ 2\, s_{ijkl} & \text{for } \alpha = 4, 5, 6\, ,\ \beta = 1, 2, 3 \\ 4\, s_{ijkl} & \text{for } \alpha, \beta = 4, 5, 6 \end{cases}$$

Note that the matrix of compliance coefficients $s_{\alpha\beta}$ is inverse to the matrix of stiffness coefficients $c_{\alpha\beta}$, i.e.

$$s_{\alpha\gamma}\, c_{\gamma\beta} = \delta_{\alpha\beta} = c_{\alpha\gamma}\, s_{\gamma\beta}$$

Transformation Law for Stiffness and Compliance Matrices

The components of a tensor depend on the underlying coordinate system. Under a coordinate transformation $x'_i = a_{ij} x_j$, described by an orthogonal matrix a_{ij}, one has the following transformation laws for second and fourth rank tensors, respectively:

$$\sigma'_{ij} = a_{ik}\, a_{jl}\, \sigma_{kl} \qquad c'_{ijkl} = a_{im} a_{jn} a_{kp} a_{lq}\, c_{mnpq}$$
$$\varepsilon'_{ij} = a_{ik}\, a_{jl}\, \varepsilon_{kl} \qquad s'_{ijkl} = a_{im} a_{jn} a_{kp} a_{lq}\, s_{mnpq}$$

Quantities in reduced index or matrix notation do not transform like tensors with changes of the underlying coordinate system. However, their transformation properties can easily be derived from the basic tensor equations. For this purpose, special care must be taken to handle properly summation over index pairs (ij) with $i \neq j$, i.e. whenever the corresponding index α is 4, 5 or 6.

We introduce matrices $t_{\alpha ij}$ and $r_{ij\alpha}$ such that

$$\sigma_\alpha = t_{\alpha ij}\, \sigma_{ij} \qquad \varepsilon_\alpha = r_{ij\alpha}\, \varepsilon_{ij}$$
$$\sigma_{ij} = r_{ij\alpha}\, \sigma_\alpha \qquad \varepsilon_{ij} = t_{\alpha ij}\, \varepsilon_\alpha$$

and

$$c_{\alpha\beta} = t_{\alpha ij}\, t_{\beta kl}\, c_{ijkl} \qquad s_{\alpha\beta} = r_{ij\alpha}\, r_{kl\beta}\, s_{ijkl}$$
$$c_{ijkl} = r_{ij\alpha}\, r_{kl\beta}\, c_{\alpha\beta} \qquad s_{ijkl} = t_{\alpha ij}\, t_{\beta kl}\, s_{\alpha\beta}$$

All matrix coefficients t_{aij} and r_{ija} are equal to zero except for

$$t_{1xx} = t_{2yy} = t_{3zz} = 1, \quad t_{4yz} = t_{4zy} = t_{5zx} = t_{5xz} = t_{6xy} = t_{6yx} = \frac{1}{2}$$

$$r_{xx1} = r_{yy2} = r_{zz3} = 1, \quad r_{yz4} = r_{zy4} = r_{zx5} = r_{xz5} = r_{xy6} = r_{yx6} = 1$$

These relations are determined by setting $\sigma_a = c_{a\beta}\,\varepsilon_\beta$ and $\varepsilon_a = s_{a\beta}\,\sigma_\beta$ in combination with the convention that the stiffness coefficients $c_{a\beta}$ should be identical with the stiffness tensor components c_{ijkl} for corresponding indices $a \leftrightarrow ij$.

As a consequence, the following simple identities hold:

$$t_{akl}\, r_{kl\beta} = \delta_{a\beta} \quad \text{and} \quad r_{ija}\, t_{akl} = \frac{1}{2}\left(\delta_{ik}\,\delta_{jl} + \delta_{il}\,\delta_{jk}\right)$$

Now, we can easily calculate the transformation matrix $T_{a\beta}$ for the coefficients σ_a of the symmetric stress tensor in reduced matrix notation:

$$\sigma'_a = t_{aij}\,\sigma'_{ij} = t_{aij}\,a_{ik}\,a_{jl}\,\sigma_{kl} = t_{aij}\,a_{ik}\,a_{jl}\,r_{kl\beta}\,\sigma_\beta = T_{a\beta}\,\sigma_\beta$$

and thus

$$T_{a\beta} = t_{aij}\,a_{ik}\,a_{jl}\,r_{kl\beta}$$

This transformation matrix will be given explicitly in the later section on piezoresistivity.

The coefficients ε_a transform according to

$$\varepsilon'_a = r_{ija}\,\varepsilon'_{ij} = r_{ija}\,a_{ik}\,a_{jl}\,\varepsilon_{kl} = r_{ija}\,a_{ik}\,a_{jl}\,t_{\beta kl}\,\varepsilon_\beta = R_{a\beta}\,\varepsilon_\beta$$

with the transformation matrix

$$R_{a\beta} = r_{ija}\,a_{ik}\,a_{jl}\,t_{\beta kl}$$

Next, consider the transformation of the stiffness coefficients $c_{a\beta}$ under a coordinate transformation $x'_i = a_{ij}\,x_j$. The stiffness tensor components transform according to $c'_{ijkl} = a_{im}\,a_{jn}\,a_{kp}\,a_{lq}\,c_{mnpq}$. Therefore,

$$c'_{a\beta} = t_{aij}\,t_{\beta kl}\,c'_{ijkl} = t_{aij}\,t_{\beta kl}\,a_{im}\,a_{jn}\,a_{kp}\,a_{lq}\,c_{mnpq}$$
$$= t_{aij}\,t_{\beta kl}\,a_{im}\,a_{jn}\,a_{kp}\,a_{lq}\,r_{mn\gamma}\,r_{pq\delta}\,c_{\gamma\delta}$$
$$= t_{aij}\,a_{im}\,a_{jn}\,r_{mn\gamma}\,t_{\beta kl}\,a_{kp}\,a_{lq}\,r_{pq\delta}\,c_{\gamma\delta}$$
$$= T_{a\gamma}\,T_{\beta\delta}\,c_{\gamma\delta} = T_{a\gamma}\,c_{\gamma\delta}\,T^T_{\delta\beta}$$

Now consider the transformation behaviour of the compliance coefficients $s_{a\beta}$ under a coordinate transformation $x'_i = a_{ij}\,x_j$:

$$s'_{\alpha\beta} = r_{ija}\, r_{kl\beta}\, s'_{ijkl} = r_{ija}\, r_{kl\beta}\, a_{im}\, a_{jn}\, a_{kp}\, a_{lq}\, s_{mnpq}$$
$$= r_{ija}\, r_{kl\beta}\, a_{im}\, a_{jn}\, a_{kp}\, a_{lq}\, t_{\gamma mn}\, t_{\delta pq}\, S_{\gamma\delta}$$
$$= r_{ija}\, a_{im}\, a_{jn}\, t_{\gamma mn}\, r_{kl\beta}\, a_{kp}\, a_{lq}\, t_{\delta pq}\, S_{\gamma\delta}$$
$$= R_{a\gamma}\, R_{\beta\delta}\, S_{\gamma\delta} = R_{a\gamma}\, S_{\gamma\delta}\, R^T_{\delta\beta}$$

Using the above $tr = \delta$ and $rt = \delta$ identities in combination with the orthogonality relations $a_{ij}a_{ik} = \delta_{jk}$ and $a_{ji}a_{ki} = \delta_{jk}$, one can easily show that

$$R^T_{\alpha\beta} = r_{ij\beta}\, a_{im}\, a_{jn}\, t_{amn} = T_{\alpha\beta}^{-1}$$
$$T^T_{\alpha\beta} = t_{\beta ij}\, a_{im}\, a_{jn}\, r_{mna} = R_{\alpha\beta}^{-1}$$

or, in matrix notation, $\boldsymbol{R} = (\boldsymbol{T}^{-1})^T$ and $\boldsymbol{T} = (\boldsymbol{R}^{-1})^T$, or $\boldsymbol{T}\boldsymbol{R}^T = 1 = \boldsymbol{R}\boldsymbol{T}^T$, respectively. Note that $\boldsymbol{T}^T \neq \boldsymbol{T}^{-1}$ and $\boldsymbol{R}^T \neq \boldsymbol{R}^{-1}$ in contrast to $\boldsymbol{a}^T = \boldsymbol{a}^{-1}$ as expressed in the orthogonality relations of the fundamental transformation matrix a_{ij}.

6.2.1.1 Stiffness and Compliance Matrices for Cubic Crystals

For all five crystal classes of the cubic crystal system, i.e. for O_h, O, T_d, T_h, T (in Schoenflies notation) or $m3m$, 432, $\bar{4}3m$, $m3$, 23 (in international or Hermann-Mauguin notation), respectively, the stiffness matrix has the following form (with respect to the crystallographic axes of the cubic lattice):

$$c_{\alpha\beta} = \begin{pmatrix} c_{11} & c_{12} & c_{12} & 0 & 0 & 0 \\ c_{12} & c_{11} & c_{12} & 0 & 0 & 0 \\ c_{12} & c_{12} & c_{11} & 0 & 0 & 0 \\ 0 & 0 & 0 & c_{44} & 0 & 0 \\ 0 & 0 & 0 & 0 & c_{44} & 0 \\ 0 & 0 & 0 & 0 & 0 & c_{44} \end{pmatrix}$$

The inverse compliance matrix has the same form:

$$s_{\alpha\beta} = \begin{pmatrix} s_{11} & s_{12} & s_{12} & 0 & 0 & 0 \\ s_{12} & s_{11} & s_{12} & 0 & 0 & 0 \\ s_{12} & s_{12} & s_{11} & 0 & 0 & 0 \\ 0 & 0 & 0 & s_{44} & 0 & 0 \\ 0 & 0 & 0 & 0 & s_{44} & 0 \\ 0 & 0 & 0 & 0 & 0 & s_{44} \end{pmatrix}$$

with coefficients

$$s_{11} = \frac{c_{11} + c_{12}}{(c_{11} + 2c_{12})(c_{11} - c_{12})}, \quad s_{12} = -\frac{c_{12}}{(c_{11} + 2c_{12})(c_{11} - c_{12})}, \quad s_{44} = \frac{1}{c_{44}}$$

(and vice versa for the transformation from compliance to stiffness coefficients).

Tab. 6.1 Elastic coefficients of silicon [28, 29]

Stiffness (10^{11} Pa)	c_{11}	c_{12}	c_{44}	c_A
	1.6564	0.6394	0.7951	−0.5732
Compliance (10^{-11} Pa^{-1})	s_{11}	s_{12}	s_{44}	s_A
	0.7691	−0.2142	1.2577	0.3545

Units: 10^{-11} Pa^{-1} = 10^{-12} cm^2/dyne; 1 Pa = 1 N/m^2; 1 dyne = 10^{-5} N; 1 Pa = 10^5 dyne/m^2 = 10 dyne/cm^2.

The stiffness and compliance coefficients fulfil the following conditions:

$$c_{11} > |c_{12}|, \quad c_{11} + 2c_{12} > 0, \quad c_{44} > 0$$

$$s_{11} > |s_{12}|, \quad s_{11} + 2s_{12} > 0, \quad s_{44} > 0$$

For silicon, the stiffness and compliance coefficients are given in Tab. 6.1. The anisotropy factors are given by $c_A = c_{11} - c_{12} - 2c_{44}$ and $s_A = s_{11} - s_{12} - s_{44}/2$.

For isotropic materials, one has the additional conditions

$$c_{44} = \frac{1}{2}(c_{11} - c_{12}), \quad s_{44} = 2(s_{11} - s_{12})$$

In order to avoid confusion, we will use unprimed symbols for components with reference to the crystal-axis coordinate system, i.e. with x, y, z axes aligned along the $\langle 100 \rangle$ crystal directions, and primed symbols for the components with respect to some arbitrarily oriented (rotated) rectangular coordinate system.

6.2.1.2 Definition of Elastic Constants

Using the explicit form of the compliance coefficient matrix $s_{\alpha\beta}$ for cubic crystals, Hooke's law can be written as a set of the following six equations:

$$\varepsilon_1 = s_{11}\sigma_1 + s_{12}(\sigma_2 + \sigma_3) \qquad \varepsilon_4 = s_{44}\sigma_4$$
$$\varepsilon_2 = s_{11}\sigma_2 + s_{12}(\sigma_1 + \sigma_3) \qquad \varepsilon_5 = s_{44}\sigma_5$$
$$\varepsilon_3 = s_{11}\sigma_3 + s_{12}(\sigma_1 + \sigma_2) \qquad \varepsilon_6 = s_{44}\sigma_6$$

Introducing the elastic constants E, v, and G, these equations read

$$\varepsilon_{xx} = \frac{1}{E}[\sigma_{xx} - v(\sigma_{yy} + \sigma_{zz})] \qquad 2\varepsilon_{yz} = \gamma_{yz} = \frac{1}{G}\sigma_{yz}$$

$$\varepsilon_{yy} = \frac{1}{E}[\sigma_{yy} - v(\sigma_{xx} + \sigma_{zz})] \qquad 2\varepsilon_{zx} = \gamma_{zx} = \frac{1}{G}\sigma_{zx}$$

$$\varepsilon_{zz} = \frac{1}{E}[\sigma_{zz} - v(\sigma_{xx} + \sigma_{yy})] \qquad 2\varepsilon_{xy} = \gamma_{xy} = \frac{1}{G}\sigma_{xy}$$

with

$$E = \frac{1}{s_{11}} \qquad \text{Young's modulus or modulus of elasticity}$$

$$v = -E\,s_{12} = -\frac{s_{12}}{s_{11}} \qquad \text{Poisson's ratio}$$

$$G = \frac{1}{s_{44}} \qquad \text{shear modulus or modulus of rigidity}$$

These elastic constants can be measured with help of rectangular parallelepipeds (prismatic bars) cut from a crystal along {100} planes or normal to ⟨100⟩ directions, respectively. If rectangular bars are cut along other crystal orientations, however, measured results will generally depend on the chosen orientation because of the anisotropy of cubic solids. Isotropic elastic properties can only be expected if $s_{11} - s_{12} - s_{44}/2 = 0$ or, equivalently, $c_{11} - c_{12} - 2c_{44} = 0$.

Direction-dependent elastic constants, i.e. Young's moduli E'_a, Poisson's ratios $v'_{\alpha\beta}$ and shear moduli G'_γ, can be defined as

$$E'_a = \frac{\sigma'_a}{\varepsilon'_a} = \frac{1}{s'_{aa}} \qquad \text{(no summation over } a\text{)}$$

$$v'_{\alpha\beta} = -\frac{\varepsilon'_\beta}{\varepsilon'_a} = -\frac{s'_{\beta a}}{s'_{aa}} = -\frac{s'_{a\beta}}{s'_{aa}} \qquad (a \neq \beta, \text{ no summation over } a)$$

$$G'_\gamma = \frac{\sigma'_\gamma}{\varepsilon'_\gamma} = \frac{1}{s'_{\gamma\gamma}} \qquad \sigma'_\delta = 0 \text{ for } \delta \neq \gamma \text{ (no summation over } \gamma\text{)}$$

with $a, \beta = 1', 2', 3'$ and $\gamma = 4', 5', 6'$. Note that $v'_{ij} \neq v'_{ji}$. These elastic constants can be determined by calculating the compliance coefficients $s'_{\alpha\beta}$ with respect to correspondingly rotated coordinate systems [30–32].

Example. Young's modulus is defined as the ratio of longitudinal stress to longitudinal strain, i.e. as $E'_{11} = 1/s'_{11}$. Suppose that the direction of interest is along a unit vector $\mathbf{r}$. This vector $\mathbf{r} = l_1 \mathbf{e}_x + m_1 \mathbf{e}_y + n_1 \mathbf{e}_z$ with $|\mathbf{r}| = 1$ has coordinates l_1, m_1, n_1 referred to the cubic crystal-axes coordinate system $\mathbf{e}_x, \mathbf{e}_y, \mathbf{e}_z$. If the coordinate system is rotated in such a way that the new base vector $\mathbf{e}'_x$ equals $\mathbf{r}$, then the component s'_{xxxx} of the compliance tensor is of interest. The corresponding compliance coefficient is s'_{11} and is given by [30]

$$s'_{11} = s_{11} - 2\left(s_{11} - s_{12} - \frac{1}{2}s_{44}\right)(l_1^2 m_1^2 + m_1^2 n_1^2 + n_1^2 l_1^2)$$

Volume Compressibility and Bulk Modulus

The volume compressibility of a crystal under hydrostatic pressure $\sigma_{ij} = -p\delta_{ij}$ is [30]

$$\varepsilon_{ij} = \underbrace{\left(\varepsilon_{ij} - \frac{1}{3}\delta_{ij}\varepsilon_{kk}\right)}_{\text{trace}\,=\,0} + \underbrace{\frac{1}{3}\delta_{ij}\varepsilon_{kk}}_{\text{dilatation}}$$

with the homogeneous dilatation $\varepsilon_{kk} = (dV' - dV)/dV$ (increase of unit volume element)

$$\varepsilon_{kk} = -p\,s_{kkmn}\,\delta_{mn} = -p\,s_{kkmm} = -\frac{p}{K}$$

where $1/K$ is the volume compressibility and K is the *bulk modulus* or *modulus of volume expansion*. Because the scalar s_{iikk} is invariant with respect to coordinate transformations, we have

$$s'_{iikk} = s'_{11} + s'_{22} + s'_{33} + 2(s'_{12} + s'_{23} + s'_{31}) = s_{iikk}$$

and thus for cubic systems

$$\frac{1}{K} = s_{iikk} = 3(s_{11} + 2s_{12}) = 3\,\frac{1 - 2\nu}{E} > 0$$

The linear compressibility in cubic systems is given by $s_{11} + 2s_{12}$ and is isotropic.

6.2.1.3 Hooke's Law for Cubic Crystals

According to Hooke's law, the stresses can be expressed by the strains in the following way for cubic crystals:

$$\begin{aligned}
\sigma_1 &= c_{11}\varepsilon_1 + c_{12}(\varepsilon_2 + \varepsilon_3) & \sigma_4 &= c_{44}\varepsilon_4 \\
\sigma_2 &= c_{11}\varepsilon_2 + c_{12}(\varepsilon_1 + \varepsilon_3) & \sigma_5 &= c_{44}\varepsilon_5 \\
\sigma_3 &= c_{11}\varepsilon_3 + c_{12}(\varepsilon_1 + \varepsilon_2) & \sigma_6 &= c_{44}\varepsilon_6
\end{aligned}$$

or

$$\begin{aligned}
\sigma_{xx} &= \frac{E}{(1+\nu)(1-2\nu)}\left[(1-\nu)\varepsilon_{xx} + \nu(\varepsilon_{yy} + \varepsilon_{zz})\right] & \sigma_{yz} &= 2G\,\varepsilon_{yz} \\
\sigma_{yy} &= \frac{E}{(1+\nu)(1-2\nu)}\left[(1-\nu)\varepsilon_{yy} + \nu(\varepsilon_{xx} + \varepsilon_{zz})\right] & \sigma_{zx} &= 2G\,\varepsilon_{zx} \\
\sigma_{zz} &= \frac{E}{(1+\nu)(1-2\nu)}\left[(1-\nu)\varepsilon_{zz} + \nu(\varepsilon_{xx} + \varepsilon_{yy})\right] & \sigma_{xy} &= 2G\,\varepsilon_{xy}
\end{aligned}$$

6.2.1.4 Hooke's Law for Isotropic Materials

For isotropic materials, the stiffness matrix has the same form as for cubic crystals, but the coefficients must fulfil the restriction $c_{11} - c_{12} - 2c_{44} = 0$. Then,

$$c_{11} = \lambda + 2\mu, \qquad c_{12} = \lambda c_{44} = (c_{11} - c_{12})/2 = \mu$$

where λ and μ are Lame's constants:

$$\lambda = \frac{E\nu}{(1+\nu)(1-2\nu)}, \qquad \mu = \frac{E}{2(1+\nu)}$$

In tensor notation, Hooke's law for isotropic solids is [33]

$$\sigma_{ik} = 2\mu \varepsilon_{ik} + \lambda \varepsilon_{ll} \delta_{ik} = \frac{E}{1+\nu}\left(\varepsilon_{ik} + \frac{\nu}{1-2\nu}\varepsilon_{ll}\delta_{ik}\right)$$

$$\varepsilon_{ik} = \frac{1+\nu}{E}\sigma_{ik} - \frac{\nu}{E}\sigma_{ll}\delta_{ik}$$

Hooke's Law for Isotropic Materials in Cylindrical Coordinates

In cylindrical coordinates r, ϕ, z, the strain tensor is given as [34]

$$\varepsilon_{rr} = \frac{\partial u_r}{\partial r} \qquad \varepsilon_{r\phi} = \frac{1}{2}\left(\frac{\partial u_\phi}{\partial r} - \frac{u_\phi}{r} + \frac{1}{r}\frac{\partial u_r}{\partial \phi}\right)$$

$$\varepsilon_{\phi\phi} = \frac{1}{r}\frac{\partial u_\phi}{\partial \phi} + \frac{u_r}{r} \qquad \varepsilon_{rz} = \frac{1}{2}\left(\frac{\partial u_r}{\partial z} + \frac{\partial u_z}{\partial r}\right)$$

$$\varepsilon_{zz} = \frac{\partial u_z}{\partial z} \qquad \varepsilon_{\phi z} = \frac{1}{2}\left(\frac{1}{r}\frac{\partial u_z}{\partial \phi} + \frac{\partial u_\phi}{\partial z}\right)$$

The stresses are obtained by Hooke's law (isotropic materials):

$$\sigma_{rr} = 2\mu \varepsilon_{rr} + \lambda(\varepsilon_{rr} + \varepsilon_{\phi\phi} + \varepsilon_{zz}) \qquad \sigma_{r\phi} = 2\mu \varepsilon_{r\phi}$$
$$\sigma_{\phi\phi} = 2\mu \varepsilon_{\phi\phi} + \lambda(\varepsilon_{rr} + \varepsilon_{\phi\phi} + \varepsilon_{zz}) \qquad \sigma_{\phi z} = 2\mu \varepsilon_{\phi z}$$
$$\sigma_{zz} = 2\mu \varepsilon_{zz} + \lambda(\varepsilon_{rr} + \varepsilon_{\phi\phi} + \varepsilon_{zz}) \qquad \sigma_{zr} = 2\mu \varepsilon_{zr}$$

or

$$\sigma_{rr} = \frac{E}{(1+\nu)(1-2\nu)}\left[(1-\nu)\varepsilon_{rr} + \nu(\varepsilon_{\phi\phi} + \varepsilon_{zz})\right]$$

$$\sigma_{\phi\phi} = \frac{E}{(1+\nu)(1-2\nu)}\left[(1-\nu)\varepsilon_{\phi\phi} + \nu(\varepsilon_{rr} + \varepsilon_{zz})\right]$$

$$\sigma_{zz} = \frac{E}{(1+\nu)(1-2\nu)}\left[(1-\nu)\varepsilon_{zz} + \nu(\varepsilon_{rr} + \varepsilon_{\phi\phi})\right]$$

$$\sigma_{r\phi} = \frac{E}{1+\nu}\varepsilon_{r\phi}, \qquad \sigma_{rz} = \frac{E}{1+\nu}\varepsilon_{rz}, \qquad \sigma_{\phi z} = \frac{E}{1+\nu}\varepsilon_{\phi z}$$

6.2.2
Plane Stress Approximation

For sufficiently small deflections ζ of a thin plate with thickness h, i.e. for $\zeta \ll h$, a neutral plane without stress exists in the middle of the plate [33]. The displacement vector of this neutral plane is given by

$$u_x^{(0)} = 0, \quad u_y^{(0)} = 0, \quad u_z^{(0)} = \zeta(x,y)$$

with a deflection ζ in the z direction that is a function of the lateral position on the thin plate only, i.e. $\zeta = \zeta(x,y)$. The stress components σ_{xz}, σ_{yz} and σ_{zz} are zero on both surfaces of the plate and hence can also be neglected within the plate. This gives the *plane stress* approximation with

$$\sigma_{xz} = 0, \quad \sigma_{yz} = 0, \quad \sigma_{zz} = 0$$

As a consequence of Hooke's law, we have the following requirements for the related strains:

$$\varepsilon_{xz} = 0, \quad \varepsilon_{yz} = 0, \quad \varepsilon_{zz} = -\frac{\nu}{1-\nu}(\varepsilon_{xx} + \varepsilon_{yy})$$

If these conditions are re-inserted into Hooke's law, one obtains relations between plane stresses and plane strains, a two-dimensional version of Hooke's law:

$$\sigma_{xx} = \frac{E}{1-\nu^2}(\varepsilon_{xx} + \nu\varepsilon_{yy}), \quad \sigma_{yy} = \frac{E}{1-\nu^2}(\varepsilon_{yy} + \nu\varepsilon_{xx}), \quad \sigma_{xy} = 2G\varepsilon_{xy}$$

and

$$\varepsilon_{xx} = \frac{1}{E}(\sigma_{xx} - \nu\sigma_{yy}), \quad \varepsilon_{yy} = \frac{1}{E}(\sigma_{yy} - \nu\sigma_{xx}), \quad \varepsilon_{xy} = \frac{1}{2G}\sigma_{xy}$$

Note that the strain ε_{zz} is non-zero in the plane stress approximation. By definition of the strain tensor components ε_{xz} and ε_{yz}, i.e.

$$\varepsilon_{xz} = \frac{1}{2}\left(\frac{\partial u_x}{\partial z} + \frac{\partial u_z}{\partial x}\right), \quad \varepsilon_{yz} = \frac{1}{2}\left(\frac{\partial u_y}{\partial z} + \frac{\partial u_z}{\partial y}\right)$$

the requirements $\varepsilon_{xz} = 0$ and $\varepsilon_{yz} = 0$ translate into the following conditions:

$$\frac{\partial u_x}{\partial z} = -\frac{\partial u_z}{\partial x} \simeq -\frac{\partial \zeta}{\partial x}, \quad \frac{\partial u_y}{\partial z} = -\frac{\partial u_z}{\partial y} \simeq -\frac{\partial \zeta}{\partial y}$$

where we have approximated $u_z \simeq u_z^{(0)}$. Then, as a consequence, the following expressions for the lateral displacements u_x and u_y as functions of coordinates x, y, z result:

$$u_x = -z \frac{\partial \zeta}{\partial x}, \quad u_y = -z \frac{\partial \zeta}{\partial y}, \quad u_z = \zeta(x, y)$$

The stress components are completely determined as a function of the plate deflection $\zeta(x,y)$:

$$\sigma_{xx} = -\frac{Ez}{1-v^2}\left(\frac{\partial^2 \zeta}{\partial x^2} + v\frac{\partial^2 \zeta}{\partial y^2}\right)$$

$$\sigma_{yy} = -\frac{Ez}{1-v^2}\left(\frac{\partial^2 \zeta}{\partial y^2} + v\frac{\partial^2 \zeta}{\partial x^2}\right)$$

$$\sigma_{xy} = -\frac{Ez}{1+v}\frac{\partial^2 \zeta}{\partial x \partial y}$$

The stresses on the plate (diaphragm) top and bottom surfaces are obtained by inserting $z = \mp h/2$, respectively.

In cylindrical coordinates, the plane stress approximation $\sigma_{rz} = \sigma_{\phi z} = \sigma_{zz} = 0$ yields $\varepsilon_{rz} = \varepsilon_{\phi z} = 0$, $\varepsilon_{zz} = -(\varepsilon_{rr} + \varepsilon_{\phi\phi})v/(1-v)$ and

$$\sigma_{rr} = \frac{E}{1-v^2}(\varepsilon_{rr} + v\varepsilon_{\phi\phi}), \quad \sigma_{\phi\phi} = \frac{E}{1-v^2}(\varepsilon_{\phi\phi} + v\varepsilon_{rr}), \quad \sigma_{r\phi} = 2G\varepsilon_{r\phi}$$

where the strains can be calculated from the displacements u_r, u_ϕ and the deflection ζ, respectively:

$$u_r = -z\frac{\partial u_z}{\partial r} \simeq -z\frac{\partial \zeta}{\partial r}, \quad u_\phi = -z\frac{1}{r}\frac{\partial u_z}{\partial \phi} \simeq -z\frac{1}{r}\frac{\partial \zeta}{\partial \phi}$$

6.2.3
Deflection of Pressure-sensitive Diaphragms

6.2.3.1 Small Deflections of Thin Plates

Small Deflections of Thin Circular Plates

Under the plane stress conditions for small deflections of thin plates, the deflection ζ must obey the following inhomogeneous bipotential differential equation:

$$D\Delta\Delta\zeta = D\Delta^2\zeta = p$$

where D is the *flexural rigidity* of the plate:

$$D = \frac{Eh^3}{12(1-v^2)}$$

and p represents a uniform pressure load (differential pressure between top and bottom surface of the plate). In Cartesian coordinates, the differential equation is expressed as

$$D\left(\frac{\partial^4 \zeta}{\partial x^4} + 2\frac{\partial^4 \zeta}{\partial x^2 \partial y^2} + \frac{\partial^4 \zeta}{\partial y^4}\right) = p$$

In cylindrical coordinates, using the Laplacian operator

$$\Delta = \frac{\partial^2}{\partial r^2} + \frac{1}{r}\frac{\partial}{\partial r} + \frac{1}{r^2}\frac{\partial^2}{\partial \phi^2}$$

we have

$$\Delta^2 \zeta = \frac{\partial^4 \zeta}{\partial r^4} + \frac{2}{r}\frac{\partial^3 \zeta}{\partial r^3} - \frac{1}{r^2}\left(\frac{\partial^2 \zeta}{\partial r^2} - 2\frac{\partial^4 \zeta}{\partial r^2 \partial \phi^2}\right)$$
$$+ \frac{1}{r^3}\left(\frac{\partial \zeta}{\partial r} - 2\frac{\partial^3 \zeta}{\partial r \partial \phi^2}\right) + \frac{1}{r^4}\left(4\frac{\partial^2 \zeta}{\partial \phi^2} + \frac{\partial^4 \zeta}{\partial \phi^4}\right)$$

For axial symmetry, which can be assumed for uniformly loaded circular diaphragms, we have $\zeta = \zeta(r)$, and obtain the much simpler differential equation

$$\Delta^2 \zeta = \frac{\partial^4 \zeta}{\partial r^4} + \frac{2}{r}\frac{\partial^3 \zeta}{\partial r^3} - \frac{1}{r^2}\frac{\partial^2 \zeta}{\partial r^2} + \frac{1}{r^3}\frac{\partial \zeta}{\partial r} = \frac{p}{D}$$

with general solution

$$\zeta(r) = \frac{p r^4}{64 D} + ar^2 + b + cr^2 \ln\frac{r}{R} + d \ln\frac{r}{R}$$

We consider a thin circular plate with radius R and a clamped boundary. The solution for the bending of the circular, uniformly loaded plate (of isotropic material) is

$$\zeta(r) = \frac{p}{64 D}(R^2 - r^2)^2 = p\frac{3(1-v^2)}{16 h^3 E}(R^2 - r^2)^2$$

where E is Young's modulus and v is Poisson's ratio. Note that $p = p_2 - p_1$.
As expected, the largest deflection is in the center of the plate

$$\zeta_{max} = \zeta(0) = p\frac{3(1-v^2)}{16 h^3 E} R^4$$

This equation can also be written in the following form:

$$p = \frac{16 E}{3(1-v^2)}\frac{h^3}{R^4}\zeta_{max}$$

Note that this linear relationship is valid as long as $\zeta \ll h$, i.e. for small ζ_{max}/h. For large deflections, i.e. $\zeta \gg h$, ζ is proportional to $p^{1/3}$ [33], and the load–deflection relationship results in the non-linear (cubic) equation

$$p = C_1 \frac{16E}{3(1-v^2)} \frac{h^4}{R^4} \frac{\zeta_{max}}{h} + C_2 E \frac{h^4}{R^4} \left(\frac{\zeta_{max}}{h}\right)^3$$

where the dimensionless parameters C_1 and C_2 depend on the diaphragm shape and Poisson's ratio. Note that R represents some characteristic length, such as the radius for circular diaphragms or a side length for rectangular diaphragms.

Small Deflections of Thin Rectangular Plates

For rectangular plates, the use of Cartesian coordinates is adequate in order to express the boundary conditions properly. The differential equation for the bending of plates

$$D\left(\frac{\partial^4 \zeta}{\partial x^4} + 2 \frac{\partial^4 \zeta}{\partial x^2 \partial y^2} + \frac{\partial^4 \zeta}{\partial y^4}\right) = p$$

is, however, much harder to integrate in Cartesian coordinates, and the deflections $\zeta(x,y)$ must be represented as series expansions:

$$\zeta(x,y) = \sum_{n \geq 1} a_n \Phi_n(x,y)$$

with conveniently chosen linearly independent coordinate functions $\Phi_n(x,y)$ that fulfil the (homogeneous) boundary conditions for $\zeta(x,y)$. For example, the following coordinate functions with squared cosines

$$\Phi_n(x,y) = \cos^2 \frac{(2n-1)\pi x}{a} \cos^2 \frac{(2n-1)\pi y}{b} \quad \text{with } n \geq 1$$

may be used for a rectangular plate with dimensions $-a/2 \leq x \leq a/2$ and $-b/2 \leq y \leq b/2$ and clamped edges. The expansion coefficients a_n can then be determined by Rayleigh-Ritz or Bubnov-Galerkin variational methods. If only the leading term Φ_1 of such an expansion is used, the following approximate solution for the deflection $\zeta(x,y)$ of a thin rectangular plate with clamped edges and subject to a uniform pressure load p is obtained [35]:

$$\zeta(x,y) = \frac{a^4 b^4}{D\pi^4 (3a^4 + 2a^2 b^2 + 3b^4)} p \cos^2 \frac{\pi x}{a} \cos^2 \frac{\pi y}{b}$$

In the center, the maximum deflection of a square plate ($a = b$) results as

$$\zeta_{max} = \zeta(0,0) = \frac{1}{8\pi^4} \frac{a^4}{D} p = 0.001283 \frac{a^4}{D} p$$

Another example is to use base functions:

$$\Phi_{nm}(x,y) = \left(1 - \frac{4x^2}{a^2}\right)^n \left(1 - \frac{4y^2}{b^2}\right)^m \quad \text{with } n, m \geq 2$$

Using again only the first function Φ_{22} gives the approximation [35]

$$\zeta(x,y) = \frac{49 \, a^4 b^4}{2048 \, D \, (7a^4 + 4a^2 b^2 + 7b^4)} p \left(1 - \frac{4x^2}{a^2}\right)^2 \left(1 - \frac{4y^2}{b^2}\right)^2$$

with center deflection in the special case of a square plate ($a = b$):

$$\zeta_{max} = \zeta(0,0) = \frac{49}{36864} \frac{a^4}{D} p = 0.001329 \frac{a^4}{D} p$$

6.2.3.2 Large Deflections of Thin Plates

For large diaphragm deflections, i.e. $\zeta > 0.2\,h$, the basic assumption of a neutral middle plane within the plate is no longer valid. Instead, one also has to take so-called membrane forces into account. Such membrane forces act within the plate.

The following set of two non-linear partial differential equations, the von Kármán equations, describes the large deflection behaviour of plates [33, 35]:

$$D\Delta^2 \zeta - h \left(\frac{\partial^2 \chi}{\partial y^2} \frac{\partial^2 \zeta}{\partial x^2} + \frac{\partial^2 \chi}{\partial x^2} \frac{\partial^2 \zeta}{\partial y^2} - 2 \frac{\partial^2 \chi}{\partial x \partial y} \frac{\partial^2 \zeta}{\partial x \partial y} \right) = p$$

$$\Delta^2 \chi + E \left[\frac{\partial^2 \zeta}{\partial x^2} \frac{\partial^2 \zeta}{\partial y^2} - \left(\frac{\partial^2 \zeta}{\partial x \partial y} \right)^2 \right] = 0$$

with a stress function χ that gives the membrane stresses in the midplane of the plate:

$$\sigma_{xx} = \frac{\partial^2 \chi}{\partial y^2}, \quad \sigma_{yy} = \frac{\partial^2 \chi}{\partial x^2}, \quad \sigma_{xy} = -\frac{\partial^2 \chi}{\partial x \partial y}$$

Large Axisymmetric Bending of Thin Circular Plates

The von Kármán equations can be expressed in polar coordinates r and ϕ. In the case of uniform pressure loading of circular plates, one obtains axisymmetric deflections, i.e. deflections $\zeta = \zeta(r)$ that depend only on r and not on the variable ϕ. The equations read [35]

$$D\Delta^2\zeta - h\frac{1}{r}\frac{\partial}{\partial r}\left(\frac{\partial\chi}{\partial r}\frac{\partial\zeta}{\partial r}\right) = p$$

$$\Delta^2\chi + E\frac{1}{r}\frac{\partial\zeta}{\partial r}\frac{\partial^2\zeta}{\partial r^2} = 0$$

with a stress function χ that gives the membrane stresses in the midplane of the plate:

$$\sigma_{rr}^0 = \frac{1}{r}\frac{\partial\chi}{\partial r}, \quad \sigma_{\phi\phi}^0 = \frac{\partial^2\chi}{\partial r^2}, \quad \sigma_{r\phi}^0 = 0$$

The bending moments M_r and M_ϕ are given as

$$M_{rr} = -D\left(\frac{\partial^2\zeta}{\partial r^2} + \frac{v}{r}\frac{\partial\zeta}{\partial r}\right), \quad M_{\phi\phi} = -D\left(\frac{1}{r}\frac{\partial\zeta}{\partial r} + v\frac{\partial^2\zeta}{\partial r^2}\right), \quad M_{r\phi} = 0$$

and the bending stresses are

$$\sigma_{rr} = \frac{12 M_{rr}}{h^3}z, \quad \sigma_{\phi\phi} = \frac{12 M_{\phi\phi}}{h^3}z, \quad \sigma_{r\phi} = \frac{12 M_{r\phi}}{h^3}z = 0$$

For the deflection $\zeta(r)$, we have the following boundary conditions at the contour $r = R$ of a clamped plate:

$$\zeta(R) = 0 \quad \text{and} \quad \left.\frac{\partial\zeta}{\partial r}\right|_{r=R} = 0$$

whereas the stress function $\chi(r)$ must satisfy the boundary condition [35]

$$\varepsilon_{\phi\phi}^0 = \frac{1}{E}(\sigma_{\phi\phi}^0 - v\sigma_{rr}^0) = \frac{1}{E}\left(\frac{\partial^2\chi}{\partial r^2} - \frac{v}{r}\frac{\partial\chi}{\partial r}\right) = 0 \quad \text{for} \quad r = R$$

The *ansatz*

$$\zeta(r) = f\,\Phi_1(r) \quad \text{with} \quad \Phi_1(r) = \left(1 - \frac{r^2}{R^2}\right)^2$$

where the coordinate function $\Phi_1(r)$ satisfies the homogeneous boundary conditions and the expansion coefficient f provides the maximum (center) deflection, is meaningful because it solves the problem for small deflections. For small deflections, membrane stresses can be neglected, and $\zeta(r)$ simply must fulfil the (inhomogeneous) biharmonic differential equation $D\Delta^2\zeta = p$, which can be solved analytically in the given axisymmetric case. The coefficient then results as $f = pR^4/(64D)$.

An appropriate *ansatz* for the stress function is

$$\chi(r) = f^2 \bar{\chi}(r) \quad \text{with} \quad \bar{\chi}(r) = B_1 r^2 + B_2 r^4 + B_3 r^6 + B_4 r^8$$

The coefficients B_2, B_3, and B_4 can be determined by insertion into the differential equation $\Delta^2 \chi + E\{\ldots\} = 0$ (in which f^2 drops out). Next, B_1 follows from the stress function boundary condition. One obtains

$$\chi(r) = f^2 \bar{\chi}(r) = f^2 E \left(\frac{1}{12} \frac{5 - 3v}{1 - v} \frac{r^2}{R^2} - \frac{1}{4} \frac{r^4}{R^4} + \frac{1}{9} \frac{r^6}{R^6} - \frac{1}{48} \frac{r^8}{R^8} \right)$$

Finally, the Bubnov–Galerkin method requires that the residual function of the approximation is orthogonal to the set of all linearly independent coordinate functions Φ_n, i.e. to Φ_1 in our simple approximation:

$$2\pi \int_0^R dr\, r \left[fD\Delta^2 \Phi_1(r) - f^3 h \frac{1}{r} \frac{\partial}{\partial r} \left(\frac{\partial \bar{\chi}}{\partial r} \frac{\partial \Phi_1}{\partial r} \right) - p \right] \Phi_1(r) = 0$$

where integration is over the circular plate area. This yields the following cubic equation for the expansion coefficient (maximum deflection) f:

$$\left(\frac{f}{h} \right) + \kappa \left(\frac{f}{h} \right)^3 = \eta p \quad \text{with} \quad \kappa = \frac{(1 + v)(23 - 9v)}{56}, \quad \eta = \frac{3(1 - v^2)}{16E} \frac{R^4}{h^4}$$

The (real) solution of this cubic equation is

$$\frac{f}{h} = \frac{2}{\sqrt{3\kappa}} \sinh\left[\frac{1}{3} \operatorname{arcsinh}\left(3 \frac{\sqrt{3\kappa}}{2} \eta p \right) \right]$$

For sufficiently small pressures or deflections, respectively, one obtains

$$\frac{f}{h} = \eta p - \kappa (\eta p)^3 + O(\eta^5 p^5)$$

For the specified plate type (circular plate with clamped edge, isotropic material, no stacked layers or boss structures) and a selected material (E and v), this equation relates linearity and pressure range. The design parameter is the ratio R/h between radius and thickness that enters the equation in the fourth power. The sensitivity, however, depends on the sensing method. If piezoresistive strain gauges are embedded in the diaphragm surface, then radial or circumferential stresses are measured. These stresses are proportional to f/h – and hence to the pressure p for small deflections – and a geometric design factor $(h/R)^2$. Thus the sensitivity of piezoresistive pressure sensor cells is proportional to an overall geometric factor $(R/h)^2$. Linearity requirements thus limit

$$\frac{3(1 - v^2)}{16E} \left(\frac{R}{h} \right)^4 p_{\max} = \left(\frac{f}{h} \right)_{\max} + \kappa \left(\frac{f}{h} \right)^3_{\max}$$

which leads to the following relation between sensitivity S and pressure range

$$S^2 \, p_{max} = \text{constant} = \text{limited by linearity requirements}$$

Of course, $S \propto (R/h)^2$ is also limited by the diaphragm size R and diaphragm thickness h, i.e. by layout requirements and process capabilities.

6.2.3.3 Stress Components in Circular Plates

Based on the proposed approximation, one calculates for clamped circular plates the bending stresses:

$$\sigma_{rr} = \frac{4E}{1-v^2} \frac{z}{h} \frac{f}{h} \frac{h^2}{R^2} \left[(1+v) - (3+v) \frac{r^2}{R^2} \right]$$

$$\sigma_{\phi\phi} = \frac{4E}{1-v^2} \frac{z}{h} \frac{f}{h} \frac{h^2}{R^2} \left[(1+v) - (1+3v) \frac{r^2}{R^2} \right]$$

and the membrane stresses

$$\sigma_{rr}^0 = \frac{1}{r} \frac{\partial \bar{\chi}}{\partial r} = E \frac{f^2}{h^2} \frac{h^2}{R^2} \left(\frac{1}{6} \frac{5-3v}{1-v} - \frac{r^2}{R^2} + \frac{2}{3} \frac{r^4}{R^4} - \frac{1}{6} \frac{r^6}{R^6} \right)$$

$$\sigma_{\phi\phi}^0 = \frac{\partial^2 \bar{\chi}}{\partial r^2} = E \frac{f^2}{h^2} \frac{h^2}{R^2} \left(\frac{1}{6} \frac{5-3v}{1-v} - 3 \frac{r^2}{R^2} + \frac{10}{3} \frac{r^4}{R^4} - \frac{7}{6} \frac{r^6}{R^6} \right)$$

The stress components at the upper and lower surfaces are obtained for $z = \mp h/2$. While the maximum deflection of a thin plate is proportional to the ratio R^4/h^3, the strains and stresses – and thus the sensitivity of the sensor – are proportional to the ratio R^2/h^2. As a result, the ratio of maximum plate deflection $\zeta(0)$ to plate thickness h is proportional to the square of the sensitivity design factor R^2/h^2. This limits the design of linear and sensitive low-pressure sensors, because the calculated results are only valid for sufficiently small deflections $\zeta \ll h$. Larger deflections lead to a non-linear sensitivity behavior.

6.2.3.4 Ring Plates

Consider a circular plate of radius R with a clamped edge that is stiffened in an inner circular region with radius r_0. Suppose that the inner circular part of the plate will not bend at all. The deflection of the resulting ring diaphragm can then be described as

$$\zeta(r) = \frac{pr^4}{64D} + ar^2 + b + cr^2 \ln \frac{r}{R} + d \ln \frac{r}{R} \quad \text{with} \quad r_0 \leq r \leq R$$

where the integration constants a, b, c and d are determined by boundary conditions:

6 CMOS-based Pressure Sensors

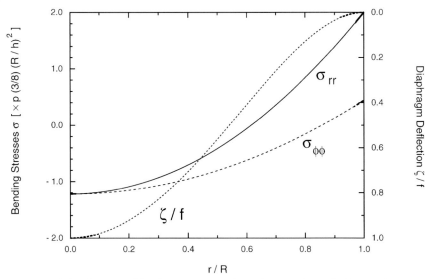

Fig. 6.1 Radial and tangential bending stresses at the surface ($z=-h/2$) of a thin circular (isotropic) plate with clamped edge and $\nu=0.22$

$$\zeta(R) = 0, \quad \frac{\partial \zeta}{\partial r}(R) = 0, \quad \frac{\partial \zeta}{\partial r}(r_0) = 0, \quad 2\pi r_0 \, Q_r(r_0) + \pi r_0^2 \, p = 0$$

For circular symmetry (i.e. symmetrically loaded circular plates), we have the following equations for the Laplacian operator Δ and the shearing force Q_r:

$$\Delta = \frac{\partial^2}{\partial r^2} + \frac{1}{r}\frac{\partial}{\partial r}, \quad Q_r = -D \frac{\partial}{\partial r} \Delta \zeta$$

The solution is

$$\zeta(r) = \frac{p}{64D}\left[(R^2 - r^2)^2 + 2r_0^2(R^2 - r^2) + 4r_0^2 R^2 \ln\frac{r}{R}\right]$$

The maximum deflection is obtained for $r = r_0$:

$$\zeta(r_0) = \frac{p}{64D}\left(R^4 - r_0^4 + 4r_0^2 R^2 \ln\frac{r_0}{R}\right)$$

For $r_0 \to R$, the maximum deflection goes to zero. If $r_0 = R/\sqrt{e} \simeq 0.6\,R$, then $\ln(r_0/R) = -1/2$ and

$$\zeta(r_0) = \frac{p}{64D}(R^2 - r_0^2)^2 = \left(1 - \frac{1}{e}\right)^2 \frac{p\,R^4}{64D} \simeq 0.4 \frac{p\,R^4}{64D}$$

Strains and stresses follow as

$$\varepsilon_{rr} = -z\frac{\partial^2 \zeta(r)}{\partial r^2} = z\frac{pR^2}{16D}\left[\left(1-\frac{r^2}{R^2}\right)\left(1+\frac{r_0^2}{r^2}\right) - 2\frac{r^2}{R^2}\left(1-\frac{r_0^2}{r^2}\right)\right]$$

$$\varepsilon_{\phi\phi} = -z\frac{1}{r}\frac{\partial \zeta(r)}{\partial r} = z\frac{pR^2}{16D}\left(1-\frac{r^2}{R^2}\right)\left(1-\frac{r_0^2}{r^2}\right)$$

and

$$\sigma_{rr} = z\frac{3pR^2}{4h^3}\left[(1+\nu)\left(1+\frac{r_0^2}{R^2}\right) - (3+\nu)\frac{r^2}{R^2} + (1-\nu)\frac{r_0^2}{r^2}\right]$$

$$\sigma_{\phi\phi} = z\frac{3pR^2}{4h^3}\left[(1+\nu)\left(1+\frac{r_0^2}{R^2}\right) - (1+3\nu)\frac{r^2}{R^2} - (1-\nu)\frac{r_0^2}{r^2}\right]$$

In the case $r_0 = 0$, the above equations reduce to the known equations for a circular plate.

6.2.4
Piezoresistivity

6.2.4.1 Piezoresistivity Tensor and Piezoresistive Coefficients

Smith discovered the piezoresistance effect in single crystal silicon in 1954 [1] and subsequently further studies on crystalline silicon were performed [2–6, 36–44]. Non-linear piezoresistive effects were investigated by [45–47] and others.

In piezoresistive materials, the resistivity tensor ρ_{ij} shows changes $\Delta\rho_{ij} = \rho_{ij} - \rho_{ij}^0$ which depend linearly on the stress components σ_{ij}, i.e.

$$\Delta\rho_{ij} = \bar{\rho}\,\pi_{ijkl}\,\sigma_{kl}$$

where π_{ijkl} are the components of the fourth-rank *piezoresistivity tensor* and $\bar{\rho}$ is the mean hydrostatic unstressed resistivity:

$$\bar{\rho} = \frac{1}{3}\left(\rho_{xx}^0 + \rho_{yy}^0 + \rho_{zz}^0\right)$$

The unstressed resistivity components are $\rho_\alpha^0 = \bar{\rho}$ for $\alpha = 1, 2, 3$ and zero for $\alpha = 4, 5, 6$ corresponding to the isotropic unstressed resistivity tensor $\rho_{ij}^0 = \bar{\rho}\delta_{ij}$ of a cubic crystal. Owing to the symmetry properties of the piezoresistivity tensor:

$$\pi_{ijkl} = \pi_{jikl} = \pi_{ijlk}$$

the following reduced index (matrix) equation is convenient as a short-hand notation:

$$\Delta\rho_\alpha = \bar{\rho}\,\pi_{\alpha\beta}\,\sigma_\beta$$

with the *piezoresistive coefficients* $\pi_{\alpha\beta}$ and indices $\alpha, \beta = 1, 2, \ldots, 6$ running over index pairs (ij) (cf. also Hooke's law, see 6.2.1). We use $\sigma_\alpha = t_{\alpha ij}\sigma_{ij}$ or

$$\sigma_1 = \sigma_{xx}, \quad \sigma_2 = \sigma_{yy}, \quad \sigma_3 = \sigma_{zz} \quad \text{(normal stresses)}$$
$$\sigma_4 = \sigma_{yz}, \quad \sigma_5 = \sigma_{zx}, \quad \sigma_6 = \sigma_{xy} \quad \text{(shear stresses)}$$

as in the matrix formulation of Hooke's law. With $\Delta\rho_\alpha = t_{\alpha ij}\Delta\rho_{ij}$ or

$$\Delta\rho_1 = \Delta\rho_{xx}, \quad \Delta\rho_2 = \Delta\rho_{yy}, \quad \Delta\rho_3 = \Delta\rho_{zz}$$
$$\Delta\rho_4 = \Delta\rho_{yz}, \quad \Delta\rho_5 = \Delta\rho_{zx}, \quad \Delta\rho_6 = \Delta\rho_{xy}$$

some factors 2 occur in relating the piezoresistive coefficients $\pi_{\alpha\beta}$ to the components π_{ijkl} of the piezoresistivity tensor, namely whenever the column index β relates to an index pair (kl) with $k \neq l$, i.e. whenever $\beta = 4, 5, 6$:

$$\pi_{11} = \pi_{xxxx}, \quad \pi_{12} = \pi_{xxyy}, \quad \pi_{21} = \pi_{yyxx}, \quad \pi_{22} = \pi_{yyyy}, \text{ etc.}$$

but

$$\pi_{16} = 2\pi_{xxxy}, \quad \pi_{26} = 2\pi_{yyxy}, \text{ etc.}$$

This can be summarized in the notation

$$\pi_{\alpha\beta} = t_{\alpha ij}\, r_{kl\beta}\, \pi_{ijkl}, \quad \pi_{ijkl} = r_{ij\alpha} t_{\beta kl} \pi_{\alpha\beta}$$

The dimensionless fourth-rank elastoresistivity tensor m_{ijkl} relates strain to resistivity changes, i.e. $\Delta\rho_{ij} = \bar{\rho}\, m_{ijkl}\, \varepsilon_{kl}$. The elastoresistivity matrix $m_{\alpha\beta}$ fulfils the relation [46]

$$m_{\alpha\beta} = \pi_{\alpha\gamma}\, c_{\gamma\beta}, \quad \Delta\rho_\alpha = \bar{\rho} m_{\alpha\beta}\varepsilon_\beta$$

For a cubic crystal that belongs to one of the three crystal classes O_h, O or T_d – such as silicon or germanium in class O_h – the matrix equation reads

$$\begin{pmatrix} \Delta\rho_1/\bar{\rho} \\ \Delta\rho_2/\bar{\rho} \\ \Delta\rho_3/\bar{\rho} \\ \Delta\rho_4/\bar{\rho} \\ \Delta\rho_5/\bar{\rho} \\ \Delta\rho_6/\bar{\rho} \end{pmatrix} = \begin{pmatrix} \pi_{11} & \pi_{12} & \pi_{12} & 0 & 0 & 0 \\ \pi_{12} & \pi_{11} & \pi_{12} & 0 & 0 & 0 \\ \pi_{12} & \pi_{12} & \pi_{11} & 0 & 0 & 0 \\ 0 & 0 & 0 & \pi_{44} & 0 & 0 \\ 0 & 0 & 0 & 0 & \pi_{44} & 0 \\ 0 & 0 & 0 & 0 & 0 & \pi_{44} \end{pmatrix} \begin{pmatrix} \sigma_1 \\ \sigma_2 \\ \sigma_3 \\ \sigma_4 \\ \sigma_5 \\ \sigma_6 \end{pmatrix}$$

if the Cartesian coordinate system is aligned with the crystallographic $\langle 100 \rangle$ axes [1]. For silicon, the piezoresistive coefficients are given in Table 6.2. Note that for general orientations of the coordinate system, the piezoresistive matrix is no long-

Tab. 6.2 Piezoresistive coefficients of Si at room temperature [1] [a]

Si material	ρ (Ω cm)	π_{11} (10^{-11} Pa^{-1})	π_{12} (10^{-11} Pa^{-1})	π_{44} (10^{-11} Pa^{-1})	π_A (10^{-11} Pa^{-1})
Monocrystalline n-type	11.7	−102.2	53.4	−13.6	−142.0
Monocrystalline p-type	7.8	6.6	−1.1	138.1	−130.4

a) 10^{-11} Pa^{-1} = 10^{-12} cm^2/dyne. Anisotropy factor $\pi_A = \pi_{11} - \pi_{12} - \pi_{44}$.

er symmetric. This is in contrast to the behavior of the stiffness and compliance matrices, which are always symmetric owing to the symmetry properties $c_{ijkl} = c_{klij}$, $s_{ijkl} = s_{klij}$ of their corresponding fourth-rank tensors. The piezoresistivity tensor π_{ijkl} does not allow permutation of the index pairs ij and kl.

For crystals belonging to the other two cubic crystal classes T_h or T, the piezoresistive coefficient matrix depends on four constants, π_{11}, π_{12}, π_{13} and π_{44} [36].

For n-type silicon, π_{11} is the dominant piezoresistive coefficient, whereas for p-type silicon the coefficient π_{44} is the most important. For the cubic crystal classes O_h, O, T_d or respectively, m3m, 432, $\bar{4}$3m, the hydrostatic pressure coefficient is given by [36]

$$-\frac{\partial}{\partial p}\left(\frac{\Delta \rho_{kk}}{\rho}\right) = \pi_{kkll} = \pi_{11} + 2\pi_{12}$$

and almost vanishes for both n- and p-type silicon.

6.2.4.2 Temperature and Doping Dependence of Piezoresistive Coefficients

Kanda [43] proposed the following model for the piezoresistive coefficients of both n- and p-type silicon:

$$\pi(N, T) = \pi_{\text{ref}}(300 \text{ K}) P(N, T)$$

with a function

$$P(N, T) = \frac{300 \text{ K}}{T} \frac{\mathcal{F}_{s-1/2}(\eta)}{\mathcal{F}_{s+1/2}(\eta)} = \frac{300 \text{ K}}{T} \frac{\mathcal{F}'_{s+1/2}(\eta)}{\mathcal{F}_{s+1/2}(\eta)}$$

that implicitly depends on temperature T and impurity concentrations N_D or N_A, respectively. The reference piezoresistive coefficients π_{ref} are taken from Smith and are valid for room temperature $T = 300$ K and lightly-doped silicon (see Tab. 6.2).

The statistical degeneracy factor is the ratio of two Fermi–Dirac integrals, and was introduced by Keyes [36] in order to consider the effect of impurity concentrations on the piezoresistance coefficients. The Fermi–Dirac integrals $\mathcal{F}_k$ are defined as [39, 48]

$$\mathcal{F}_k(\eta) = \frac{1}{\Gamma(k+1)} \int_0^\infty \frac{\varepsilon^k\, d\varepsilon}{1 + e^{\varepsilon - \eta}} \quad \text{for} \quad k > -1$$

and fulfil

$$\frac{\partial}{\partial \eta} \mathcal{F}_k(\eta) = \mathcal{F}'_k(\eta) = \mathcal{F}_{k-1}(\eta)$$

Asymptotically, i.e. for $|\eta| \gg 1$, the Fermi–Dirac integrals behave like

$$\mathcal{F}_k(\eta) \simeq e^\eta \quad \text{for } \eta \ll -1, \qquad \mathcal{F}_k(\eta) \simeq \frac{1}{\Gamma(k+2)} \eta^{k+1} \quad \text{for } \eta \gg 1$$

The parameter s describes the energy dependence of the relaxation time and depends on the scattering mechanism. One has $s = 3/2$ for impurity scattering, $s = -1/2$ for scattering by acoustic phonons, and $s = 0$ for scattering by optical phonons [46] (see also [49]). Kanda used $s = -1/2$ corresponding to lattice scattering [43] (see also [39]), which gives

$$P(N, T) = \frac{300\text{ K}}{T} \left. \frac{\mathcal{F}'_{s+1/2}(\eta)}{\mathcal{F}_{s+1/2}(\eta)} \right|_{s=-1/2} = \frac{300\text{ K}}{T} \left[(1 + e^{-\eta}) \ln(1 + e^\eta) \right]^{-1}$$

The argument η is the *reduced Fermi level* that measures the difference between the Fermi energy E_F and the conduction or valence band energies E_c or E_v, respectively, in units of $k_B T$, i.e.

$$\eta_n = \frac{E_F - E_c}{k_B T}, \qquad \eta_p = \frac{E_v - E_F}{k_B T}$$

For n-type or p-type semiconductors, the reduced Fermi levels η_n or η_p are obtained from the carrier concentrations $n \simeq N_D$ or $p \simeq N_A$ in the respective band:

$$n = N_c(T)\, \mathcal{F}_{1/2}(\eta_n) \simeq N_D \quad \text{or} \quad p = N_v(T)\, \mathcal{F}_{1/2}(\eta_p) \simeq N_A$$

where N_D and N_A are the donor or acceptor concentrations, respectively. In silicon, the effective densities of states N_c and N_v are

$$N_c(T) = 2.8 \times 10^{19} \times \left(\frac{T}{300\text{ K}} \right)^{3/2} \text{ cm}^{-3}$$

$$N_v(T) = 1.04 \times 10^{19} \times \left(\frac{T}{300\text{ K}} \right)^{3/2} \text{ cm}^{-3}$$

The Fermi–Dirac function $\mathcal{F}_{1/2}(x)$ is well tabulated [48] and different approximation equations have been proposed; see, e.g., the comparison by Wong et al.

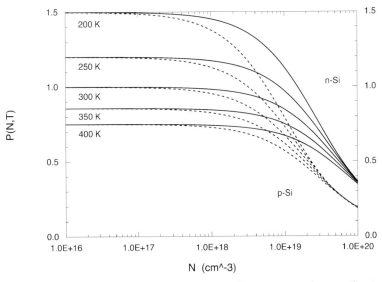

Fig. 6.2 Scaling factor $P(N, T)$ for piezoresistive coefficients in n- and p-type silicon as a function of impurity concentration and temperature (after [43])

[50]. Fig. 6.2 shows the scaling factor $P(N, T)$ for n- and p-type silicon as a function of impurity concentration and temperature.

6.2.4.3 Piezoresistive Coefficients for Diffused Layers

Kerr and Milnes [37] and Tufte and Stelzer [38] defined diffused piezoresistive coefficients for cubic semiconductors and numerically evaluated such coefficients as a function of surface impurity concentration in both p-type silicon and n-type germanium:

$$\bar{\pi}'_{\alpha\beta} = \int_0^d \pi'_{\alpha\beta}(z)\,\sigma_0(z)\,\mathrm{d}z \bigg/ \int_0^d \sigma_0(z)\,\mathrm{d}z$$

where d is the p–n junction depth or layer thickness and $\sigma_0(z)$ is the unstressed conductivity profile function. The diffused piezoresistive coefficients $\bar{\pi}'_{\alpha\beta}$ thus depend on the bulk coefficients $\pi'_{\alpha\beta}$ (and their variation with crystallographic orientation), the assumed impurity concentration profile and the surface impurity concentration N_s. Typical impurity concentrations profiles are, for example, given by the complementary error function $N_s\,\mathrm{erfc}(u)$ (for diffusion with constant surface concentration of dopant) or the Gaussian distribution $N_s\,\exp(-u^2)$ (for diffusion with constant total amount of dopant). The variable $u = z/(2\sqrt{DT})$ is a diffusion-length normalized depth. Calculations by Kerr and Milnes showed that the piezoresistive coefficients for diffused layers, $\bar{\pi}'_{\alpha\beta}$, for a surface impurity concentration N_s correspond to the respective bulk coefficients $\pi'_{\alpha\beta}$ for a (uniform) impurity concentration N that is roughly half the surface concentration N_s, i.e. $N = N_s/2.0$ or $N = N_s/2.3$ for Gaussian or complementary error function profiles, respectively [37, 46].

6.2.4.4 Piezoresistive Gauge Factors for Polysilicon Films

For a homogeneous, isotropic material, characterized by Young's modulus E and Poisson's ratio v, the gauge factor G is defined as fractional change of resistance per unit strain, i.e.

$$G = \frac{1}{\varepsilon}\frac{\Delta R}{R} = 1 + 2v + \frac{1}{\varepsilon}\frac{\Delta \rho}{\rho} = 1 + 2v + m = 1 + 2v + E\pi$$

where the elastoresistance coefficient m describes the fractional change of resistivity per unit strain and the piezoresistance coefficient π describes the fractional change of resistivity per unit stress. For silicon, the strain-induced change of resistivity is the dominant contribution to the gauge factor. The gauge factor G varies for monocrystalline silicon between –102 and 135 and for polycrystalline silicon between –30 and 40 [51].

The piezoresistance effect in polysilicon films has been widely studied [52–61]. French and Evans [55] described a new theoretical model for the piezoresistance in both n- and p-type polysilicon.

Gridchin et al. [61] presented a method for the calculation of gauge factors for textured, highly doped polysilicon films. Analytical and experimental results were given for highly doped p-type polysilicon films. Note that shear strain does not induce a change in the resistivity of textured polysilicon films (see also [58]). The piezoresistive coefficients π'_{61} and π'_{62} average out for completely random microcrystal orientation, but also for one-axis aligned microcrystals [58]. Thus uniaxial stresses never lead to transverse voltages in such polycrystalline silicon films.

6.2.4.5 Piezoresistive Coefficients for Rotated Coordinate Systems

In a rotated (primed) Cartesian coordinate systems with unit vectors e'_x, e'_y, e'_z, the (primed) coordinates r'_i of a vector r are given as $r'_i = a_{ij} r_j$. The tensor components ρ_{ij}, σ_{ij}, and π_{ijkl} transform according to

$$\rho'_{ij} = a_{ik} a_{jl} \rho_{kl}, \qquad \sigma'_{ij} = a_{ik} a_{jl} \sigma_{kl}, \qquad \pi'_{ijkl} = a_{im} a_{jn} a_{kp} a_{lq} \pi_{mnpq}$$

with the transformation (rotation) matrix a_{ij}:

$$a_{ij} = \begin{pmatrix} a_{x'x} & a_{x'y} & a_{x'z} \\ a_{y'x} & a_{y'y} & a_{y'z} \\ a_{z'x} & a_{z'y} & a_{z'z} \end{pmatrix} = \begin{pmatrix} l_1 & m_1 & n_1 \\ l_2 & m_2 & n_2 \\ l_3 & m_3 & n_3 \end{pmatrix}$$

where the index i runs over the coordinates x', y', z' with reference to the primed (new) coordinate system, while the index j runs over the coordinates x, y, z with

6.2 Micromachined Pressure Sensors

reference to the (old) coordinate system. The direction cosines are given as $a_{ij} = \cos(e'_i, e_j)$. Note that vectors r can be represented as follows with respect to coordinate systems e_x, e_y, e_z and e'_x, e'_y, e'_z:

$$r = r_x e_x + r_y e_y + r_z e_z, \qquad r = r_{x'} e'_x + r_{y'} e'_y + r_{z'} e'_z$$

The orthogonality relations

$$a_{ij} a_{ik} = \delta_{jk} \qquad \text{and} \qquad a_{ji} a_{ki} = \delta_{jk}$$

imply $a_{ij}^{-1} = a_{ji}$ (note that the transformation matrix a_{ij} is not symmetric). Only three of the nine coefficients a_{ij} are independent. According to Subsection 6.2.1, the stress and resistivity coefficients σ_α and ρ_α, respectively, transform as

$$\sigma'_\alpha = T_{\alpha\beta} \sigma_\beta, \quad \rho'_\alpha = T_{\alpha\beta} \rho_\beta$$

with a transformation matrix $T_{\alpha\beta} = t_{\alpha ij} a_{ik} a_{jl} r_{kl\beta}$. This transformation matrix $T_{\alpha\beta}$ is explicitly given as [62, 63]:

$$T_{\alpha\beta} = \begin{pmatrix} l_1^2 & m_1^2 & n_1^2 & 2m_1 n_1 & 2l_1 n_1 & 2l_1 m_1 \\ l_2^2 & m_2^2 & n_2^2 & 2m_2 n_2 & 2l_2 n_2 & 2l_2 m_2 \\ l_3^2 & m_3^2 & n_3^2 & 2m_3 n_3 & 2l_3 n_3 & 2l_3 m_3 \\ l_2 l_3 & m_2 m_3 & n_2 n_3 & (m_2 n_3 + n_2 m_3) & (n_2 l_3 + l_2 n_3) & (l_2 m_3 + m_2 l_3) \\ l_3 l_1 & m_3 m_1 & n_3 n_1 & (m_3 n_1 + n_3 m_1) & (n_3 l_1 + l_3 n_1) & (l_3 m_1 + m_3 l_1) \\ l_1 l_2 & m_1 m_2 & n_1 n_2 & (m_1 n_2 + n_1 m_2) & (n_1 l_2 + l_1 n_2) & (l_1 m_2 + m_1 l_2) \end{pmatrix}$$

A sequence of two rotations a_{ij} and $\hat{a}_{ij}$ can be expressed as $\tilde{a}_{ij} = \hat{a}_{ik} a_{kj}$. Correspondingly, we obtain $\hat{T}_{\alpha\gamma} T_{\gamma\beta} = \tilde{T}_{\alpha\beta}$ and, similarly, $T_{\gamma\beta}^{-1} \hat{T}_{\alpha\gamma}^{-1} = \tilde{T}_{\alpha\beta}^{-1}$.

The transformation law of the piezoresistive coefficients $\pi_{\alpha\beta}$ in reduced index notation follows from

$$\begin{aligned} \pi'_{\alpha\beta} &= t_{\alpha ij} \pi'_{ijkl} r_{kl\beta} = t_{\alpha ij} a_{im} a_{jn} a_{kp} a_{lq} \pi_{mnpq} r_{kl\beta} \\ &= t_{\alpha ij} a_{im} a_{jn} \pi_{mnpq} a_{kp} a_{lq} r_{kl\beta} \\ &= t_{\alpha ij} a_{im} a_{jn} r_{mn\gamma} \pi_{\gamma\delta} t_{\delta pq} a_{kp} a_{lq} r_{kl\beta} \\ &= T_{\alpha\gamma} \pi_{\gamma\delta} T_{\delta\beta}^{-1} \end{aligned}$$

with

$$T_{\delta\beta}^{-1} = t_{\delta pq} a_{kp} a_{lq} r_{kl\beta}$$

This inverse matrix equals the transpose of the matrix T except for some factors, i.e.

$$T_{\alpha\beta}^{-1} = T_{\beta\alpha} \times \begin{cases} 1 & \text{for } \alpha, \beta = 1, 2, 3 \text{ or } \alpha, \beta = 4, 5, 6 \\ 1/2 & \text{for } \alpha = 4, 5, 6 \text{ and } \beta = 1, 2, 3 \\ 2 & \text{for } \alpha = 1, 2, 3 \text{ and } \beta = 4, 5, 6 \end{cases}$$

Piezoresistivity Matrix for Arbitrarily Oriented Coordinate Systems

We calculate the matrix $\pi'_{\alpha\beta}$ for a general rotation

$$a_{ij} = \begin{pmatrix} l_1 & m_1 & n_1 \\ l_2 & m_2 & n_2 \\ l_3 & m_3 & n_3 \end{pmatrix}$$

A straightforward calculation of $\pi'_{\alpha\beta} = T_{\alpha\gamma} \pi_{\gamma\delta} T_{\delta\beta}^{-1}$ using the explicit equations for the transformation matrices $T_{\alpha\beta}$ and $T_{\alpha\beta}^{-1}$ gives the result

$$\pi'_{\alpha\beta} = \pi_{\alpha\beta} + \pi_A \begin{pmatrix} -2F_1 & F_{12} & F_{13} & 2G_{231} & 2G_{31} & 2G_{21} \\ F_{12} & -2F_2 & F_{23} & 2G_{32} & 2G_{312} & 2G_{12} \\ F_{13} & F_{23} & -2F_3 & 2G_{23} & 2G_{13} & 2G_{123} \\ G_{231} & G_{32} & G_{23} & 2F_{23} & 2G_{123} & 2G_{312} \\ G_{31} & G_{312} & G_{13} & 2G_{123} & 2F_{31} & 2G_{231} \\ G_{21} & G_{12} & G_{123} & 2G_{312} & 2G_{231} & 2F_{12} \end{pmatrix}$$

with [63]

$$F_1 = l_1^2 m_1^2 + m_1^2 n_1^2 + n_1^2 l_1^2$$
$$F_{12} = l_1^2 l_2^2 + m_1^2 m_2^2 + n_1^2 n_2^2$$
$$G_{12} = l_1 l_2^3 + m_1 m_2^3 + n_1 n_2^3$$
$$G_{123} = l_1 l_2 l_3^2 + m_1 m_2 m_3^2 + n_1 n_2 n_3^2, \quad \text{etc.}$$

In order to separate the original contributions from the new terms proportional to π_A, use has been made of the orthogonality relations $1 = a_{1i}a_{1i} = l_1^2 + m_1^2 + n_1^2$ etc., which give

$$1 = (a_{1i}a_{1i})^2 = (l_1^2 + m_1^2 + n_1^2)^2$$
$$= l_1^4 + l_1^2 m_1^2 + l_1^2 n_1^2 + m_1^2 l_1^2 + m_1^4 + m_1^2 n_1^2 + n_1^2 l_1^2 + n_1^2 m_1^2 + n_1^4$$

or

$$l_1^4 + m_1^4 + n_1^4 = 1 - \left(l_1^2 m_1^2 + l_1^2 n_1^2 + m_1^2 l_1^2 + m_1^2 n_1^2 + n_1^2 l_1^2 + n_1^2 m_1^2\right)$$
$$= 1 - 2\left(l_1^2 m_1^2 + m_1^2 n_1^2 + n_1^2 l_1^2\right), \quad \text{etc.}$$

6.2.4.6 Piezoresistivity Coefficients for (100) Silicon Wafers

Consider a (001) wafer with coordinate system $e'_x = [110]$ (perpendicular to the primary wafer flat), $e'_y = [\bar{1}10]$ (parallel to the primary wafer flat) and $e'_z = e_z = [001]$ (normal to the wafer surface) (see Fig. 6.3) [64].

The piezoresistive matrix coefficients $\pi_{\alpha\beta}$ are known with respect to the Cartesian coordinate system $e_x = [100]$, $e_y = [010]$ and $e_z = [001]$. The new (primed) coordinate system is then obtained by a rotation around the $z = z'$ axis with angle $\psi = \pi/4$. More generally, we calculate the rotation matrix a_{ij} with variable ψ: for a rotation of the coordinate system around the $e_z = [001]$ axis by an angle ψ, one has

$$a_{ij} = \begin{pmatrix} \cos\psi & \cos(\pi/2 - \psi) & 0 \\ \cos(\pi/2 + \psi) & \cos\psi & 0 \\ 0 & 0 & 1 \end{pmatrix} = \begin{pmatrix} c & s & 0 \\ -s & c & 0 \\ 0 & 0 & 1 \end{pmatrix} = \begin{pmatrix} l_1 & m_1 & 0 \\ l_2 & m_2 & 0 \\ 0 & 0 & 1 \end{pmatrix}$$

and thus $l_1 = m_2 = \cos\psi = c$, $m_1 = -l_2 = \sin\psi = s$, $l_3 = m_3 = n_1 = n_2 = 0$ and $n_3 = 1$. This gives the transformation matrix

$$T_{\alpha\beta} = \begin{pmatrix} l_1^2 & m_1^2 & 0 & 0 & 0 & 2l_1 m_1 \\ l_2^2 & m_2^2 & 0 & 0 & 0 & 2l_2 m_2 \\ 0 & 0 & 1 & 0 & 0 & 0 \\ 0 & 0 & 0 & m_2 & l_2 & 0 \\ 0 & 0 & 0 & m_1 & l_1 & 0 \\ l_1 l_2 & m_1 m_2 & 0 & 0 & 0 & (l_1 m_2 + l_2 m_1) \end{pmatrix}$$

or

$$T_{\alpha\beta} = \begin{pmatrix} c^2 & s^2 & 0 & 0 & 0 & 2sc \\ s^2 & c^2 & 0 & 0 & 0 & -2sc \\ 0 & 0 & 1 & 0 & 0 & 0 \\ 0 & 0 & 0 & c & -s & 0 \\ 0 & 0 & 0 & s & c & 0 \\ -sc & sc & 0 & 0 & 0 & (c^2 - s^2) \end{pmatrix}$$

Then, $\pi'_{\alpha\beta} = T_{\alpha\gamma}\pi_{\gamma\delta}T_{\delta\beta}^{-1}$ results in

$$\pi'_{\alpha\beta} = \begin{pmatrix} \pi_{11} - 2s^2c^2\pi_A & \pi_{12} + 2s^2c^2\pi_A & \pi_{12} & 0 & 0 & -2sc(c^2 - s^2)\pi_A \\ \pi_{12} + 2s^2c^2\pi_A & \pi_{11} - 2s^2c^2\pi_A & \pi_{12} & 0 & 0 & 2sc(c^2 - s^2)\pi_A \\ \pi_{12} & \pi_{12} & \pi_{11} & 0 & 0 & 0 \\ 0 & 0 & 0 & \pi_{44} & 0 & 0 \\ 0 & 0 & 0 & 0 & \pi_{44} & 0 \\ -sc(c^2 - s^2)\pi_A & sc(c^2 - s^2)\pi_A & 0 & 0 & 0 & \pi_{44} + 4s^2c^2\pi_A \end{pmatrix}$$

with the anistropy factor $\pi_A = \pi_{11} - \pi_{12} - \pi_{44}$. For p- and n-type (001) silicon wafers, the important piezoresistivity coefficients π'_{11}, π'_{12}, and π'_{66} are shown in Fig. 6.4.

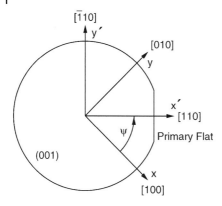

Fig. 6.3 Directions in the cubic lattice of {100} silicon wafers. The wafer surface is a (001) plane. The orientation of the directions [110], [$\bar{1}$10], [100] and [010] is indicated by the primary flat

Note: for (001) wafers with a coordinate system that is oriented with respect to the primary flat, the piezoresistive coefficient matrix is ($s = c = 1/\sqrt{2}$)

$$\pi'_{\alpha\beta} = \begin{pmatrix} \pi_{11} - \pi_A/2 & \pi_{12} + \pi_A/2 & \pi_{12} & 0 & 0 & 0 \\ \pi_{12} + \pi_A/2 & \pi_{11} - \pi_A/2 & \pi_{12} & 0 & 0 & 0 \\ \pi_{12} & \pi_{12} & \pi_{11} & 0 & 0 & 0 \\ 0 & 0 & 0 & \pi_{44} & 0 & 0 \\ 0 & 0 & 0 & 0 & \pi_{44} & 0 \\ 0 & 0 & 0 & 0 & 0 & \pi_{44} + \pi_A \end{pmatrix}$$

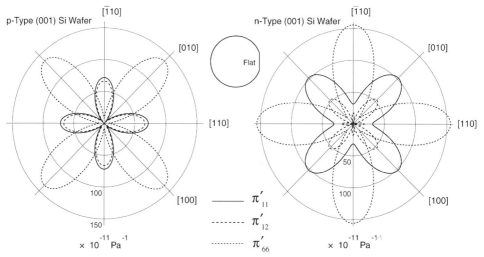

Fig. 6.4 Orientation dependence of piezoresistivity coefficients π'_{11}, π'_{12} and π'_{66} on the surface of (001) oriented *p-* and *n*-type wafers

6.2.4.7 Piezoresistivity Coefficients for (111) Silicon Wafers

Consider a (111) wafer with coordinate system $e'_x = [\bar{1}10]$ (perpendicular to the primary wafer flat), $e'_y = [\bar{1}\bar{1}2]$ (parallel to the primary wafer flat) and $e'_z = [111]$ (normal to the wafer surface) (see Fig. 6.5) [64].

The direction cosines

$$a_{ij} = \cos(e'_i, e_j) = \frac{h'_i h_j + k'_i k_j + l'_i l_j}{\sqrt{h'^2_i + k'^2_i + l'^2_i}\sqrt{h_j^2 + k_j^2 + l_j^2}}$$

give the transformation matrix

$$a_{ij} = \frac{1}{\sqrt{6}}\begin{pmatrix} -\sqrt{3} & \sqrt{3} & 0 \\ -1 & -1 & 2 \\ \sqrt{2} & \sqrt{2} & \sqrt{2} \end{pmatrix} = \begin{pmatrix} l_1 & m_1 & n_1 \\ l_2 & m_2 & n_2 \\ l_3 & m_3 & n_3 \end{pmatrix}$$

and the corresponding T matrix

$$T_{\alpha\beta} = \frac{1}{6}\begin{pmatrix} 3 & 3 & 0 & 0 & 0 & -6 \\ 1 & 1 & 4 & -4 & -4 & 2 \\ 2 & 2 & 2 & 4 & 4 & 4 \\ -\sqrt{2} & -\sqrt{2} & 2\sqrt{2} & \sqrt{2} & \sqrt{2} & -2\sqrt{2} \\ -\sqrt{6} & \sqrt{6} & 0 & \sqrt{6} & -\sqrt{6} & 0 \\ \sqrt{3} & -\sqrt{3} & 0 & 2\sqrt{3} & -2\sqrt{3} & 0 \end{pmatrix}$$

and its inverse

$$T_{\alpha\beta}^{-1} = \frac{1}{6}\begin{pmatrix} 3 & 1 & 2 & -2\sqrt{2} & -2\sqrt{6} & 2\sqrt{3} \\ 3 & 1 & 2 & -2\sqrt{2} & 2\sqrt{6} & -2\sqrt{3} \\ 0 & 4 & 2 & 4\sqrt{2} & 0 & 0 \\ 0 & -2 & 2 & \sqrt{2} & \sqrt{6} & 2\sqrt{3} \\ 0 & -2 & 2 & \sqrt{2} & -\sqrt{6} & -2\sqrt{3} \\ -3 & 1 & 2 & -2\sqrt{2} & 0 & 0 \end{pmatrix}$$

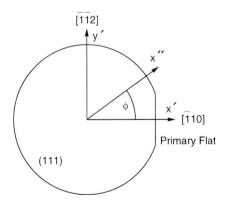

Fig. 6.5 Directions in the cubic lattice of {111} silicon wafers. The wafer surface is a (111) plane normal to the [111] direction. The orientation of the directions [$\bar{1}\bar{1}2$] and [$\bar{1}10$] is indicated by the primary flat

Using $\pi_A = \pi_{11} - \pi_{12} - \pi_{44}$, the transformed piezoresistivity matrix $\pi'_{\alpha\beta} = T_{\alpha\gamma}\,\pi_{\gamma\delta}\,T_{\delta\beta}^{-1}$ results as

$$\pi'_{\gamma\delta} = \begin{pmatrix} \pi_{11} - \frac{1}{2}\pi_A & \pi_{12} + \frac{1}{6}\pi_A & \pi_{12} + \frac{1}{3}\pi_A & -2\frac{\sqrt{2}}{6}\pi_A & 0 & 0 \\ \pi_{12} + \frac{1}{6}\pi_A & \pi_{11} - \frac{1}{2}\pi_A & \pi_{12} + \frac{1}{3}\pi_A & 2\frac{\sqrt{2}}{6}\pi_A & 0 & 0 \\ \pi_{12} + \frac{1}{3}\pi_A & \pi_{12} + \frac{1}{3}\pi_A & \pi_{11} - \frac{2}{3}\pi_A & 0 & 0 & 0 \\ -\frac{\sqrt{2}}{6}\pi_A & \frac{\sqrt{2}}{6}\pi_A & 0 & \pi_{44} + \frac{2}{3}\pi_A & 0 & 0 \\ 0 & 0 & 0 & 0 & \pi_{44} + \frac{2}{3}\pi_A & -\frac{\sqrt{2}}{3}\pi_A \\ 0 & 0 & 0 & 0 & -\frac{\sqrt{2}}{3}\pi_A & \pi_{44} + \frac{1}{3}\pi_A \end{pmatrix}$$

in the wafer-aligned coordinate system.

Rotation of Coordinate System around the [111] Axis

For (111) oriented wafers, we have expressed the piezoresistive coefficients $\pi'_{\alpha\beta}$ with respect to a coordinate system x', y', z' oriented along the directions $[\overline{1}10]$, $[\overline{1}\overline{1}2]$, and $[111]$. A rotation of the coordinate system around the z' axis ([111] direction) by an angle ϕ is then described by transformation matrices [see the discussion for (100) wafers]:

$$\hat{T}_{\alpha\beta} = \begin{pmatrix} c^2 & s^2 & 0 & 0 & 0 & 2sc \\ s^2 & c^2 & 0 & 0 & 0 & -2sc \\ 0 & 0 & 1 & 0 & 0 & 0 \\ 0 & 0 & 0 & c & -s & 0 \\ 0 & 0 & 0 & s & c & 0 \\ -sc & sc & 0 & 0 & 0 & (c^2 - s^2) \end{pmatrix}$$

with $s = \sin\phi$, $c = \cos\phi$. Calculating $\pi''_{\alpha\beta} = \hat{T}_{\alpha\gamma}\,\pi'_{\gamma\delta}\,\hat{T}_{\delta\beta}^{-1}$, we first abbreviate

$$\pi'_{\gamma\delta} = \begin{pmatrix} p_{11} & p_{12} & p_{13} & -2p_{42} & 0 & 0 \\ p_{12} & p_{11} & p_{13} & 2p_{42} & 0 & 0 \\ p_{13} & p_{13} & p_{33} & 0 & 0 & 0 \\ -p_{42} & p_{42} & 0 & p_{44} & 0 & 0 \\ 0 & 0 & 0 & 0 & p_{44} & -2p_{42} \\ 0 & 0 & 0 & 0 & -2p_{42} & p_{66} \end{pmatrix}$$

then utilize

$$p_{11} - p_{12} - p_{66} = \left(\pi_{11} - \frac{1}{2}\pi_A\right) - \left(\pi_{12} + \frac{1}{6}\pi_A\right) - \left(\pi_{44} + \frac{1}{3}\pi_A\right)$$
$$= (\pi_{11} - \pi_{12} - \pi_{44}) - \left(\frac{1}{2} + \frac{1}{6} + \frac{1}{3}\right)\pi_A = \pi_A - \pi_A = 0$$

and finally obtain

$$\pi''_{\gamma\delta} = \begin{pmatrix} \pi_{11} - \frac{1}{2}\pi_A & \pi_{12} + \frac{1}{6}\pi_A & \pi_{12} + \frac{1}{3}\pi_A & -A\frac{\sqrt{2}}{3}\pi_A & -B\frac{\sqrt{2}}{3}\pi_A & 0 \\ \pi_{12} + \frac{1}{6}\pi_A & \pi_{11} - \frac{1}{2}\pi_A & \pi_{12} + \frac{1}{3}\pi_A & A\frac{\sqrt{2}}{3}\pi_A & B\frac{\sqrt{2}}{3}\pi_A & 0 \\ \pi_{12} + \frac{1}{3}\pi_A & \pi_{12} + \frac{1}{3}\pi_A & \pi_{11} - \frac{2}{3}\pi_A & 0 & 0 & 0 \\ -A\frac{\sqrt{2}}{6}\pi_A & A\frac{\sqrt{2}}{6}\pi_A & 0 & \pi_{44} + \frac{2}{3}\pi_A & 0 & B\frac{\sqrt{2}}{3}\pi_A \\ -B\frac{\sqrt{2}}{6}\pi_A & B\frac{\sqrt{2}}{6}\pi_A & 0 & 0 & \pi_{44} + \frac{2}{3}\pi_A & -A\frac{\sqrt{2}}{3}\pi_A \\ 0 & 0 & 0 & B\frac{\sqrt{2}}{3}\pi_A & -A\frac{\sqrt{2}}{3}\pi_A & \pi_{44} + \frac{1}{3}\pi_A \end{pmatrix}$$

with abbreviations

$$A = c(c^2 - 3s^2) = c[2(c^2 - s^2) - 1]$$
$$B = s(3c^2 - s^2) = s[2(c^2 - s^2) + 1]$$

The piezoresistive effect in the (111) plane is thus isotropic, because the coefficients π''_{11}, π''_{12}, and π''_{66} do not depend on the rotation angle ϕ.

6.2.4.8 Two-terminal Piezoresistors

In stressed anisotropic ohmic conductors, one has

$$E_i = \rho_{ij} J_j = \left(\rho^0_{ij} + \bar{\rho}\pi_{ijkl}\sigma_{kl}\right) J_j$$

with the electric field vector E_i, the current density vector J_i, and the unstressed resistivity tensor ρ^0_{ij}. For cubic crystals, the unstressed resistivity is isotropic and given as $\rho^0_{ij} = \bar{\rho}\delta_{ij}$. Hence

$$E_i = \bar{\rho}\left(\delta_{ij} + \pi_{ijkl}\sigma_{kl}\right) J_j$$

Now consider a small resistor (filament) of length ℓ that is oriented along the direction $\mathbf{n}$ on the wafer surface ($|\mathbf{n}| = \sqrt{n_i n_i} = 1$). We use a wafer-aligned Cartesian coordinate system x', y', z', i.e. with direction z' normal to the wafer surface. The voltage drop V along the resistor filament is then $\mathbf{E}\mathbf{n}\,\ell = E'_i n'_i \ell$. Then, with $\mathbf{J}' = J\mathbf{n}'$ and the cross-sectional area A:

$$R = \frac{V}{I} = \frac{\ell}{JA} E'_i n'_i = \bar{\rho}\,\frac{\ell}{A}\left(\delta_{ij} + \pi'_{ijkl}\,\sigma'_{kl}\right) n_i n_j$$

With respect to the unstressed resistance $R_0 = \bar{\rho}(\ell/A)\, n'_i n'_i = \bar{\rho}(\ell/A)$, we have

$$\frac{\Delta R}{R_0} = \frac{R - R_0}{R_0} = n'_i n'_j \pi'_{ijkl}\,\sigma'_{kl}$$

Because the dyadic tensor $(\mathbf{n}' \circ \mathbf{n}')_{ij} = n'_i n'_j = N'_{ij}$ is symmetric, we define in reduced index notation $N'_\alpha = r_{\alpha ij} N'_{ij}$, $N'_{ij} = t_{\delta ij} N'_\delta$, and obtain

$$\frac{\Delta R}{R_0} = t_{\delta ij} N'_\delta\, r_{ija}\, \pi'_{\alpha\beta}\, t_{\beta kl}\, r_{kl\gamma}\, \sigma'_\gamma = N'_\alpha\, \pi'_{\alpha\beta}\, \sigma'_\beta$$

where $t_{akl}\, r_{kl\beta} = \delta_{\alpha\beta}$ has been used. Because of $n'_z = 0$, we have $N_1 = n'^2_x$, $N_2 = n'^2_y$, $N_3 = N_4 = N_5 = 0$, and $N_6 = 2 n'_x n'_y$. Hence

$$\frac{\Delta R}{R_0} = n'^2_x\, \pi'_{1\beta}\, \sigma'_\beta + n'^2_y\, \pi'_{2\beta}\, \sigma'_\beta + 2 n'_x n'_y\, \pi'_{6\beta}\, \sigma'_\beta$$

Neglecting stress components σ_{xz}, σ_{yz} and σ_{zz}, one obtains

$$\frac{\Delta R}{R_0} = n'^2_x \left(\pi'_{11}\,\sigma'_1 + \pi'_{12}\,\sigma'_2 + \pi'_{16}\,\sigma'_6\right) + n'^2_y \left(\pi'_{21}\,\sigma'_1 + \pi'_{22}\,\sigma'_2 + \pi'_{26}\,\sigma'_6\right)$$
$$+ 2 n'_x n'_y \left(\pi'_{61}\,\sigma'_1 + \pi'_{62}\,\sigma'_2 + \pi'_{66}\,\sigma'_6\right)$$

For (001) wafers with coordinate system oriented with respect to the crystallographic axes of the cubic lattice, we have

$$\pi'_{\alpha\beta} = \pi_{\alpha\beta} = \begin{pmatrix} \pi_{11} & \pi_{12} & \pi_{12} & 0 & 0 & 0 \\ \pi_{12} & \pi_{11} & \pi_{12} & 0 & 0 & 0 \\ \pi_{12} & \pi_{12} & \pi_{11} & 0 & 0 & 0 \\ 0 & 0 & 0 & \pi_{44} & 0 & 0 \\ 0 & 0 & 0 & 0 & \pi_{44} & 0 \\ 0 & 0 & 0 & 0 & 0 & \pi_{44} \end{pmatrix}$$

and thus

$$\frac{\Delta R}{R_0} = n_x'^2 \left(\pi_{11}\sigma_1' + \pi_{12}\sigma_2'\right) + n_y'^2 \left(\pi_{12}\sigma_1' + \pi_{11}\sigma_2'\right) + 2n_x'n_y'\pi_{44}\sigma_6'$$

$$= \left(n_x'^2 \pi_{11} + n_y'^2 \pi_{12}\right)\sigma_1' + \left(n_x'^2 \pi_{12} + n_y'^2 \pi_{11}\right)\sigma_2' + 2n_x'n_y'\pi_{44}\sigma_6'$$

For (001) or (111) silicon wafers with coordinate system aligned with respect to the primary flat, the piezoresistive coefficients are

$$\begin{pmatrix} \pi_{11}' & \pi_{12}' & \pi_{16}' \\ \pi_{21}' & \pi_{22}' & \pi_{26}' \\ \pi_{61}' & \pi_{62}' & \pi_{66}' \end{pmatrix} = \begin{pmatrix} \pi_{11} - \pi_A/2 & \pi_{12} + q\pi_A/2 & 0 \\ \pi_{12} + q\pi_A/2 & \pi_{11} - \pi_A/2 & 0 \\ 0 & 0 & \pi_{44} + q\pi_A \end{pmatrix}$$

with $q = 1$ for (001) wafers and $q = 1/3$ for (111) wafers. Then,

$$\frac{\Delta R}{R_0} = n_x'^2 \left(\pi_{11}'\sigma_1' + \pi_{12}'\sigma_2'\right) + n_y'^2 \left(\pi_{21}'\sigma_1' + \pi_{22}'\sigma_2'\right) + 2n_x'n_y'\pi_{66}'\sigma_6'$$

$$= \left(n_x'^2 \pi_{11}' + n_y'^2 \pi_{12}'\right)\sigma_1' + \left(n_x'^2 \pi_{12}' + n_y'^2 \pi_{11}'\right)\sigma_2' + 2n_x'n_y'\pi_{66}'\sigma_6'$$

Special Case of a Circular Diaphragm
In the special case of a circular diaphragm (plate), one can transform the stress components with respect to polar coordinates r, ϕ, i.e.

$$\sigma_1' = \sigma_{xx}' = c^2\sigma_{rr}' + s^2\sigma_{\phi\phi}' - 2sc\,\sigma_{r\phi}'$$
$$\sigma_2' = \sigma_{yy}' = s^2\sigma_{rr}' + c^2\sigma_{\phi\phi}' + 2sc\,\sigma_{r\phi}'$$
$$\sigma_6' = \sigma_{xy}' = sc(\sigma_{rr}' - \sigma_{\phi\phi}') + (c^2 - s^2)\sigma_{r\phi}'$$

with $s = \sin\phi$, $c = \cos\phi$, where ϕ denotes the angle with respect to the x' coordinate axis. For a uniform pressure load, one has $\sigma_{r\phi}' = 0$ and thus

$$\frac{\Delta R}{R_0} = \left(n_x'^2 \pi_{11}' + n_y'^2 \pi_{12}'\right)\sigma_1' + \left(n_x'^2 \pi_{12}' + n_y'^2 \pi_{11}'\right)\sigma_2' + 2n_x'n_y'\pi_{66}'\sigma_6'$$

$$= \left(n_x'^2 c^2\pi_{11}' + n_y'^2 c^2\pi_{12}' + n_x'^2 s^2\pi_{12}' + n_y'^2 s^2\pi_{11}' + 2n_x'n_y'\pi_{66}'sc\right)\sigma_{rr}'$$

$$+ \left(n_x'^2 s^2\pi_{11}' + n_y'^2 s^2\pi_{12}' + n_x'^2 c^2\pi_{12}' + n_y'^2 c^2\pi_{11}' - 2n_x'n_y'\pi_{66}'sc\right)\sigma_{\phi\phi}'$$

Example 1
Let the small piezoresistor filament at diaphragm position $(x', y') = (r, \phi)$ be oriented in radial direction, i.e. set $n_x' = \cos\phi = c$ and $n_y' = \sin\phi = s$:

$$\frac{\Delta R}{R_0} = \left(c^4 \pi'_{11} + s^2 c^2 \pi'_{12} + c^2 s^2 \pi'_{12} + s^4 \pi'_{11} + 2 s^2 c^2 \pi'_{66} \right) \sigma'_{rr}$$

$$+ \left(c^2 s^2 \pi'_{11} + s^4 \pi'_{12} + c^4 \pi'_{12} + s^2 c^2 \pi'_{11} - 2 s^2 c^2 \pi'_{66} \right) \sigma'_{\phi\phi}$$

$$= \underbrace{\left(\pi'_{11} - 2 s^2 c^2 \pi'_A \right)}_{\pi'_\ell} \sigma'_{rr} + \underbrace{\left(\pi'_{12} + 2 s^2 c^2 \pi'_A \right)}_{\pi'_t} \sigma'_{\phi\phi}$$

Example 2

Let the small piezoresistor filament at diaphragm position $(x', y') = (r, \phi)$ be oriented in tangential direction, i.e. set $n'_x = -\sin\phi = -s$ and $n'_y = \cos\phi = c$:

$$\frac{\Delta R}{R_0} = \left(c^2 s^2 \pi'_{11} + s^4 \pi'_{12} + c^4 \pi'_{12} + s^2 c^2 \pi'_{11} - 2 s^2 c^2 \pi'_{66} \right) \sigma'_{rr}$$

$$+ \left(c^4 \pi'_{11} + s^2 c^2 \pi'_{12} + c^2 s^2 \pi'_{12} + s^4 \pi'_{11} + 2 s^2 c^2 \pi'_{66} \right) \sigma'_{\phi\phi}$$

$$= \underbrace{\left(\pi'_{12} + 2 s^2 c^2 \pi'_A \right)}_{\pi'_t} \sigma'_{rr} + \underbrace{\left(\pi'_{11} - 2 s^2 c^2 \pi'_A \right)}_{\pi'_\ell} \sigma'_{\phi\phi}$$

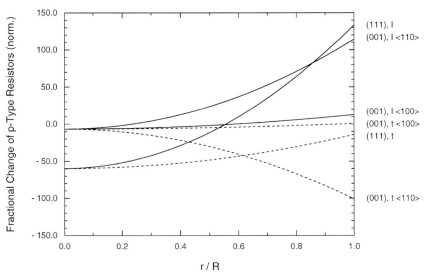

Fig. 6.6 Fractional resistance changes of two-terminal (filament) p-type piezoresistors as a function of wafer plane, piezoresistor orientation and radial location on a circular silicon diaphragm with clamped edge. As usual, small deflections $f/h \ll 1$ are assumed, and only bending stresses have been taken into account. In addition, isotropic elastic behaviour of the silicon diaphragm is assumed. The fractional resistance changes $\Delta R/R_0$ are given in dimensionless units of $(3/8)(R^2/h^2)(p/10^{11} \text{Pa})$

Now evaluate the introduced longitudinal and transverse piezoresistive coefficients π'_ℓ and π'_t:

$$\pi'_\ell = \pi'_{11} - 2s^2c^2\,\pi'_A, \qquad \pi'_t = \pi'_{12} + 2s^2c^2\,\pi'_A$$

Note that the prime and the angle ϕ refer to the wafer-aligned coordinate system. First,

$$\pi'_A = \pi'_{11} - \pi'_{12} - \pi'_{66} = (\pi_{11} - \pi_{12} - \pi_{44}) - \left(\frac{1}{2} + \frac{q}{2} + q\right)\pi_A = \frac{1-3q}{2}\pi_A$$

then

$$\pi'_\ell = \left(\pi_{11} - \frac{1}{2}\pi_A\right) - s^2c^2(1-3q)\,\pi_A = \pi_{11} - \frac{1}{2}\left[1 + 2s^2c^2(1-3q)\right]\pi_A$$

$$\pi'_t = \left(\pi_{12} + \frac{q}{2}\pi_A\right) + s^2c^2(1-3q)\,\pi_A = \pi_{12} + \frac{1}{2}\left[q + 2s^2c^2(1-3q)\right]\pi_A$$

As a consequence, π'_ℓ and π'_t are isotropic for (111) wafers ($q = 1/3$). For (001) wafers ($q = 1$), we have

$$\pi'_\ell = \pi_{11} - \frac{1}{2}\left(1 - 4s^2c^2\right)\pi_A, \qquad \pi'_t = \pi_{12} + \frac{1}{2}\left(1 - 4s^2c^2\right)\pi_A$$

With respect to the x axis of the crystallographic coordinate system, the resistor orientation is given by the angle $\psi = \phi + \pi/4$. Hence

$$1 - 4\sin^2\phi\cos^2\phi = 1 - \sin^2(2\phi) = \cos^2(2\phi) = \sin^2(2\psi) = 4\sin^2\psi\cos^2\psi$$

and

$$\pi'_\ell = \pi_{11} - 2\pi_A\sin^2\psi\cos^2\psi, \qquad \pi'_t = \pi_{12} + 2\pi_A\sin^2\psi\cos^2\psi$$

Comparison with the result for $\pi'_{\alpha\beta}$ from Subsection 6.2.4.6 shows that $\pi'_\ell = \pi'_{11}$ and $\pi'_t = \pi'_{12}$ if the primed coordinate system is oriented with its x' axis along the resistor direction $\mathbf{n}$, as expected.

6.2.5
Piezoresistors in a Wheatstone Bridge

A Wheatstone bridge, as shown in Fig. 6.7, consists of four piezoresistors R_1, R_2, R_3 and R_4, and delivers the output voltage

$$V_{out} = \frac{R_1 R_3 - R_2 R_4}{(R_1 + R_2)(R_3 + R_4)} V_{in}$$

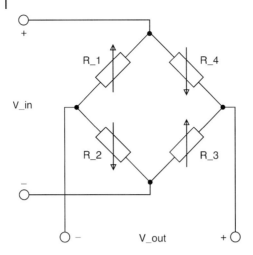

Fig. 6.7 Wheatstone bridge

where V_{in} is the input (supply) voltage. The four piezoresistors are embedded in the diaphragm surface, and are subject to their respective bending (and eventually stretching) stresses, which are a function of the applied pressure p. We can thus write ($i = 1, \ldots, 4$)

$$R_i(p) = R_i(0) + \Delta R_i(p) = R_i(0)\left[1 + \frac{\Delta R_i(p)}{R_i(0)}\right] = R_i(0)\left[1 + \rho_i(p)\right]$$

with dimensionless pressure-dependent quantities $\rho_i(p) = \Delta R_i(p)/R_i(0)$. The bridge circuit is offset-free if $R_1(0)R_3(0) - R_2(0)R_4(0) = 0$. Design and placement of the piezoresistors can be chosen in such a way that the bridge is symmetric and $R_1(p) = R_3(p)$, $R_2(p) = R_4(p)$. Then,

$$V_{out} = \frac{R_1^2 - R_2^2}{(R_1 + R_2)^2} V_{in} = \frac{R_1 - R_2}{R_1 + R_2} V_{in}$$

The offset compensation condition then reads $R_1(0) = R_2(0)$ or $R_i(0) = R_0$. As a result, an offset-compensated symmetric Wheatstone bridge delivers the output voltage

$$\begin{aligned} V_{out} &= \frac{R_1 - R_2}{R_1 + R_2} V_{in} = \frac{(1 + \rho_1) - (1 + \rho_2)}{2 + \rho_1 + \rho_2} V_{in} \\ &= \frac{\rho_1 - \rho_2}{2}\left\{1 - \frac{\rho_1 + \rho_2}{2} + O\left[\left(\frac{\rho_1 + \rho_2}{2}\right)^2\right]\right\} V_{in} \end{aligned}$$

The differential output voltage of such a well-matched bridge does not depend on the absolute value R_0 of the four piezoresistors. Thus, certain common-mode effects such as the temperature behavior $R_0(T)$ are cancelled out. In order to deliver a maximum differential signal, and to minimize non-linearity effects, the common-mode pressure signal $(p_1+p_2)/2$ should be zero. This is equivalent to pressure-independent total resistances in both branches with respect to the bridge input or output, respectively, or $R_i(p) + R_j(p) = 2R_0$ with $i = 1, 3$ and $j = 2, 4$. The orientation and location of the piezoresistors have to be selected according to this requirement $p_1(p) = -p_2(p)$.

A current-supplied Wheatstone bridge delivers the output voltage

$$V_{out} = \frac{R_1 R_3 - R_2 R_4}{R_1 + R_2 + R_3 + R_4} I_{in}$$

If we again assume symmetry, $R_3 = R_1$ and $R_4 = R_2$, and offset compensation, $R_i(0) = R_0$, we obtain

$$V_{out} = \frac{R_1^2 - R_2^2}{2(R_1 + R_2)} I_{in} = \frac{R_1 - R_2}{2} I_{in} = \frac{p_1 - p_2}{2} R_0 I_{in}$$

In contrast to conventional voltage-supplied bridges, the voltage output of current-supplied bridges does not depend on the common-mode pressure signal $(p_1+p_2)/2$, but might be affected by the temperature behavior of R_0.

6.2.5.1 Diaphragm Designs with Piezoresistors

Different layouts for piezoresistive pressure sensors with Wheatstone bridges are shown in Figs. 6.8–6.10. Fig. 6.11 shows a four-terminal piezoresistive device, as used in Motorola's X-ducer technology.

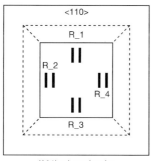
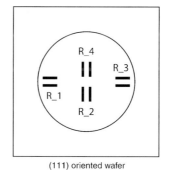

Fig. 6.8 (a) Anisotropically etched square silicon diaphragm with four p-type piezoresistors located at the rim of the diaphragm and oriented along the ⟨110⟩ directions. Each piezoresistor consists of two parallel filaments (100 μm × 12 μm). The piezoresistors R_1 and R_3 are oriented perpendicular ('radial') to their diaphragm edges, whereas the piezoresistors R_2 and R_4 are oriented parallel ('tangential') to their respective diaphragm edges. As a consequence, R_1 and R_3 behave differently under applied pressure to both R_2 and R_4 [18]. (b) Layout of silicon pressure sensor utilizing four split longitudinal (radial) piezoresistors located near the edges and the center of the diaphragm

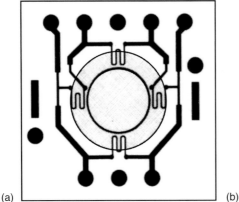 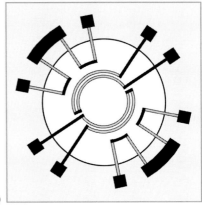

Fig. 6.9 (a) KPY60 high-pressure sensor with radially and tangentially located piezoresistors on a circular (100) Si diaphragm [65]. (b) Circular diaphragm with radial and circumferential piezoresistors [66]

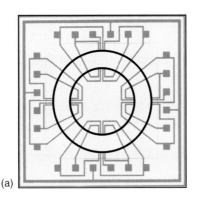 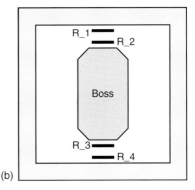

Fig. 6.10 (a) Low-pressure sensor design with circular ring diaphragm [67]. (b) Low-pressure sensor with rectangular diaphragm and center boss for improved linearity [68]

Typical configurations combine two longitudinal and two transversal piezoresistors located at the edges of (100) silicon diaphragms. On (111) silicon diaphragms, longitudinal piezoresistors may be placed near the edges and in the center. Ring diaphragms or diaphragms with a bossed center usually are equipped with tangentially (circumferentially) orientated piezoresistors at the outer and inner diameters of the ring diaphragm.

Generally, *p*-type silicon piezoresistors are chosen because they are more linear than *n*-type silicon piezoresistors. Higher doping-levels lower the temperature dependence of the piezoresistors and are thus preferred. Fig. 6.2 shows the tradeoff between temperature dependence and magnitude of the piezoresistive coefficients (gauge factors) as a function of the impurity concentration.

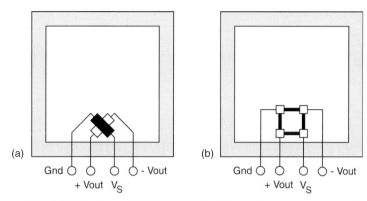

Fig. 6.11 (a) Four-terminal piezoresistive device as shear stress gauge utilizing the so-called Kanda effect [69–71]. This device is also known as Motorola's X-ducer [15]. (b) So-called picture frame bridge consisting of four p-type piezoresistors [72]

6.3
CMOS-integrated Pressure Sensors

Integrated pressure sensors combine pressure-sensitive diaphragm transducers with electronic circuits for signal amplification, processing and readout. The pressure-sensing diaphragm can be processed either in bulk or by surface micromachining. Depending on the diaphragm material, monocrystalline silicon or polysilicon, the sensing of the diaphragm deflection is typically done using implanted (diffused) piezoresistors or capacitively. The circuit is either purely analog or also provides digital conversion. CMOS or bipolar or BiCMOS base technologies are used.

6.3.1
Motivation

Monolithic integration of micromachined sensor elements with adequate CMOS circuitry for signal conditioning is an exciting and promising way to offer miniaturized sensing systems that are precisely tailored to the application needs of customers. In a first approach, this simply means to sample, to amplify, to compensate, to calibrate and to output the transducer signal. The customer is interested in functionality, a clean data interface, reliable operation and proven quality. Integration of devices, components, functional blocks, etc., takes a burden from the customer and offers value. A second approach may then further integrate such sensing systems with application processors, wherever this will help to reduce interfaces, wiring and the amount of data transfer. Data processing transforms raw data into results that are meaningful and usable, and less prone to transfer and handling errors. Integration of functionality becomes service and quality.

MEMS (microelectromechanical systems) are the key to sensor and actuator functionality. Silicon is certainly a material of choice, due not only to its well-known electrical properties, but also to its excellent mechanical properties and its remarkable stability.

6.3.1.1 Bulk Versus Surface Micromachining

Bulk micromachined silicon pressure sensors combine the excellent mechanical properties of silicon diaphragms with piezoresistive strain or stress gauges that are simply implanted or diffused into the top diaphragm surface. Locating such piezoresistors in a proper way on the diaphragm, and combining them in a Wheatstone bridge, provides a sensing element with a reasonably strong output signal (voltage). In order to add functionality, it is desirable to integrate such bulk micromachined pressure transducers with a standard CMOS circuit. Process integration, however, generally must address certain compatibility issues such as cross-contamination and thermal budgets.

Bulk micromachining in combination with different kinds of wafer bondings allows for a vast variety of possibilities in product or process definition. For example, bulk micromachining is very efficient to

- Utilize epitaxial layers for micromachined structures by means of electrochemical etch-stop techniques, especially if thicker layers (several microns) are needed.
- Create pressure ports from the wafer backside. This is important for differential pressure sensors or for pressure sensors in harsh environments. In harsh environments, the open silicon substrate cavity (diaphragm back side) can be exposed to aggressive media, whereas the chip frontside may be well covered and protected by packaging.
- Allow stacking of chips and thus micromachined structures where functional parts such as diaphragms or capacitor electrodes are provided by bonding of pre-processed wafers.
- Encapsulate chips or cavities by means of wafer bonding, as required for wafer-level packages and absolute pressure sensors.

Surface micromachining, on the other hand, offers certain advantages to pressure sensor design and fabrication as compared with bulk micromachining. This is due to the use of a sacrificial layer:

- Thin sacrificial layers allow one to fabricate capacitor electrodes, i.e. a pressure-sensitive top electrode (diaphragm) and a fixed bottom electrode (substrate), with a very small distance d between each other, important for high capacitive gauge factors (sensitivities), i.e. $\Delta C(f)/C(0) \propto f/d + O(f^2/d^2)$.
- The thickness of thin diaphragms can be precisely defined and controlled by layer deposition, avoiding the need for etch-stop techniques.
- For a given sensitivity, the lateral diaphragm dimensions scale with the diaphragm thickness, yielding small sensor devices.
- Structured sacrificial layers allow the use of isotropic etchants in combination with precise control of dimensions (no extra chip area needed, in contrast to anisotropic etching, which depends on crystallographic planes).

- Wafer front-side processing is sufficient to create shallow cavities and to release micromachined structures, thus eliminating any need for back-side processing, alignment and etching.
- Deposition processes allow one to plug small etch holes and thus to seal cavities, creating vacuum reference pressure ports without the need for wafer bonding.

Downscaling is one aspect that leads the way from bulk micromachining to surface micromachining. In the case of pressure sensors, surface micromachining allows the fabrication of diaphragms that are one or two orders of magnitude thinner than bulk micromachined diaphragms. Because, for a given mechanical sensor sensitivity, the diaphragm diameter (or linear dimension) shows the same scaling behavior as the diaphragm thickness, much smaller sensor devices can be produced in surface micromachining than in bulk micromachining. Another important aspect in favor of surface micromachining is certainly CMOS process integration, which becomes easier when no back-side processing, wafer bonding or anisotropic etching has to be performed. Nevertheless, as the requirements for pressure sensors and their features vary significantly depending on the application, there is no unique or best approach to the design, manufacture and packaging of pressure sensors. Instead, a variety of different approaches exists and will continue to exist. The art in pressure sensor product and manufacturing process design is – as for all other kinds of MEMS – to create and foster a kit of powerful and diversified techniques in order to address customer needs in the most efficient, useful and beneficial way. For this reason, it is not possible to cover the field of CMOS-based pressure sensors. However, selected examples and basic technological insights will be provided within the scope of this chapter.

6.3.1.2 Capacitive Sensing of Diaphragm Deflection

In bulk micromachining, the diaphragms are made of monocrystalline silicon and piezoresistive sensing is used. Typically, four piezoresistors embedded in the diaphragm (aligned to certain crystallographic orientations) are combined in a Wheatstone bridge. The piezoresistors and thus the bridge output are sensitive to diaphragm stress (or strain).

In surface micromachining, the diaphragm material is typically polysilicon, and the diaphragm deflections $\zeta = \zeta(r, p)$ are sensed capacitively. For rectangular diaphragms with dimensions $a \times b$, the sensor capacitance is obtained as

$$C(p) = \varepsilon_0 \int_{-a/2}^{a/2} dx \int_{-b/2}^{b/2} dy \, \frac{1}{d - \zeta(x, y, p)}$$

where d is the distance between the two electrodes if no pressure is applied. The capacitance of circular (axisymmetric) diaphragm sensors with radius R results as

$$C(p) = 2\pi\varepsilon_0 \int_0^R dr\, r\, \frac{1}{d - \zeta(r,p)}$$

The resulting capacitance signals are very small and therefore some appropriate amplification circuitry must be integrated on the chip. However, because surface micromachining shares processes used in the fabrication of standard ICs, it is a technology that offers great potential for integration into CMOS processes in order to provide the necessary circuits for signal processing.

Consider the pressure-dependent axisymmetric deflection $\zeta(r,p) = f(p)(1 - r^2/R^2)^2$ of a circular diaphragm with clamped edge. The deflection-dependent capacitance is then calculated as

$$C(f) = 2\pi\varepsilon_0 \int_0^R dr\, r\, \frac{1}{d - f(1 - r^2/R^2)^2}$$

Then, after subtraction of $C(0)$,

$$\Delta C(f) = C(f) - C(0) = \frac{\varepsilon_0 \pi R^2}{d} \int_0^R dr\, \frac{2r}{R^2}\, \frac{(f/d)(1 - r^2/R^2)^2}{1 - (f/d)(1 - r^2/R^2)^2}$$

One obtains for the fractional capacitance change

$$\frac{\Delta C(f)}{C(0)} = \frac{\operatorname{arctanh}\sqrt{f/d}}{\sqrt{f/d}} - 1 = \sum_{n=1}^{\infty} \frac{1}{2n+1}\left(\frac{f}{d}\right)^n = \frac{1}{3}\frac{f}{d} + \frac{1}{5}\left(\frac{f}{d}\right)^2 + O(f^3/d^3)$$

The simple plate capacitor model with $\zeta(r,p) = \zeta(p) = f(p)$ yields, in comparison,

$$\frac{\Delta C(f)}{C(0)} = \frac{1}{1 - f/d} - 1 = \frac{f/d}{1 - f/d} = \sum_{n=1}^{\infty}\left(\frac{f}{d}\right)^n = \frac{f}{d} + \left(\frac{f}{d}\right)^2 + O(f^3/d^3)$$

6.3.2
Thickness Control for Diaphragm Formation

The diaphragm thickness is an important parameter that enters the sensor sensitivity quadratically. In bulk micromachining, the thickness of the silicon diaphragm is determined by the etch process and etch-stop techniques in particular. Control of the diaphragm thickness is necessary to keep variations in device performance within resonable limits.

Isotropic wet etching of silicon is performed with HNA, a mixture of HF (hydrofluoric acid), HNO_3 (nitric acid) and CH_3COOH (acetic acid). The etch rates depend on the mixture ratios and range from 0.7 to 40 µm/min. Lightly p- or n-doped regions with impurity concentrations $\leq 10^{17}$ cm^{-3} exhibit etch rates reduced by a factor 150 compared with heavily doped regions [73].

The most important wet chemical anisotropic silicon etchants are KOH (potassium hydroxide), EDP (ethylenediamine pyrocatechol) and TMAH [tetramethylammonium hydroxide, $(CH_3)_4NOH$]. KOH is the most common etchant and offers etch rates in the order of 1 µm/min and a very high anisotropic {100}/{111} etch rate ratio of 400:1 [73]. The etch rates depend on concentration with a maximum around 20 wt%, and on temperature with activation energies between 0.57 eV for {100} planes and 0.70 eV for {111} planes [74]. When isopropyl alcohol (IPA) is added to KOH, the etch rate for the {110} planes is significantly reduced in contrast to the etch rate for the {100} planes. Boron doping with concentrations $\geq 10^{20}$ cm^{-3} slows the etch rate by a factor of 20 [73, 75]. Silicon dioxide is etched with etch rates of several nm/min, whereas etch rates for silicon nitride are extremely small. Disadvantages of KOH are the poor selectivity to oxide and the CMOS incompatibility due to its metal (potassium) ion content.

EDP offers smaller etch rates for {100} and {110} planes than KOH. On the other hand, the selectivity to oxide is about 100 times higher than for KOH. In addition, EDP shows lower etch rates for B^{++}-doped silicon, i.e. reduced etch rates by a factor of 50 for boron doping concentrations $\geq 7 \times 10^{19}$ cm^{-3} [75, 76]. Note that EDP is very toxic (carcinogenic), highly corrosive and difficult to handle.

TMAH is an organic etchant and thus free from metal ions and compatible with IC processing [77]. Whereas KOH and EDP etch aluminum, TMAH does not attack Al once a certain amount of Si has been dissolved in the solution [78–80]. The etch rates depend on orientation, concentration and temperature, and are typically around 1 µm/min for {100} and {110} planes. The etch anisotropy {100}/{111} is about 10:1 to 35:1. Boron doping $\geq 10^{20}$ cm^{-3} reduces the etch rate by a factor of 10 [79].

6.3.2.1 Time Etch Stop

Typical etch rates in anisotropic etching are around 1 µm/min. This allows one to control the amount of etched bulk material by the etch time if concentration- and temperature-dependent etch rates are known and monitored. The resulting diaphragm thickness is then the difference between the original wafer thickness and the estimated amount (depth) of etched material. Owing to tolerances in etch rates and wafer thickness, the time etch-stop technique is only used for manufacturing of thicker diaphragms (> 50 µm).

6.3.2.2 Boron Etch Stop or p^{++} Etch Stop

Heavily boron-doped regions (with impurity concentrations $\geq 2.2 \times 10^{19}$ cm^{-3}) show drastically reduced etch rates in KOH and EDP. For impurity concentrations of 10^{20} cm^{-3}, the etch rates are two orders of magnitude smaller than for 10^{19} cm^{-3}. For KOH, the boron etch stop works best for low concentrations. Addition of IPA improves the etch stop but also leads to large hillocks. EDP is therefore preferred for the p^{++} etch stop. Generally, a problem with the p^{++} etch stop approach is high tensile film stresses, which affect the behavior of the sensor diaphragms [81].

6.3.2.3 Electrochemical Etch Stop (ECE)

As described in Chapter 1, the ECE technique utilizes a *p–n* junction in reversed-biased mode. Typically, *p*-type substrate wafers with *n*-type epitaxial layers are used, where the *n*-epitaxial layer must be properly biased with respect to the etch bath. The etchant can be HF-based, KOH or EDP [10, 82–84].

6.3.3
Process Integration Strategies

6.3.3.1 Diaphragms with Integrated Strain Gauges

Tufte et al. described 'integrated diaphragms' in 1962 [7], where piezoresistive strain (stress) gauges were diffused into silicon diaphragms. This approach offered considerable advantages over conventional methods with mounted or bonded strain gauges. Especially the integrated (embedded) filament-type strain gauges are sensitive to both longitudinal and transverse piezoresistive effects [7].

6.3.3.2 CMOS-integrated Bulk Micromachined Silicon Pressure Sensor

The first CMOS-integrated, bulk-micromachined silicon pressure sensors were demonstrated by Ishihara et al. [18]. The sensors comprise a diaphragm with boron-implanted *p*-type piezoresistors. The standard CMOS process was enhanced by an electrochemical etch-stop (ECE) technique and electrostatic bonding of a glass wafer (see Fig. 6.12).

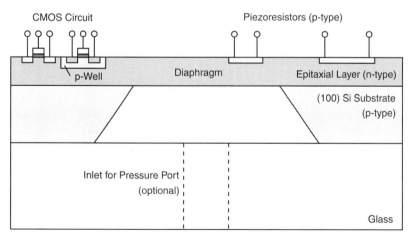

Fig. 6.12 Cross-section of CMOS-integrated bulk micromachined pressure sensor [18]

6.3.3.3 Bulk and Surface Micromachined Pressure Sensors with SiN Diaphragm and Polysilicon Piezoresistors

A special combination of surface and bulk micromachining was performed and described by Sugiyama et al. [13]. Using an alkali etchant (KOH), this method creates a cacvity under a deposited SiN layer (diaphragm) by removing the sacrificial polysilicon layer and anisotropically etching into the silicon bulk material (see Fig. 6.13a). Access to the silicon substrate is provided by four etch holes located at the corners of the diaphragm, polysilicon etch channels and the polysilicon sacrificial layer itself [13] (see Fig. 6.13b). The diaphragm has dimensions of 80 µm × 80 µm and the etch holes have diameters of 8 µm each. The method results in rectangular V-shaped cavities that serve as a vacuum chamber after the etch holes have been plugged. This concept offers the advantage of a fast undercut etching of the polysilicon sacrificial layer underneath the silicon nitride diaphragm by means of the alkali etchant KOH. As a side effect, a large pyramid-shaped cavity is etched into the silicon substrate underneath the diaphragm.

Although the entire processing can be done from the wafer front side and the cavity formation is self-aligned along the {111} crystallographic planes (cavity walls), there is a certain disadvantage to this process. Careful in situ observation of the etching process itself revealed a problem with hydrogen bubbles generated at the etching Si surface. The hydrogen bubbles may cause an oscillatory movement of the diaphragm due to accumulation inside the etched cavity and sudden exhaust through the etch holes. Such diaphragm movements can induce fractures in the diaphragms, which decrease the yield significantly [13].

An alternative process that eliminates many of the drawbacks of the anisotropic diaphragm formation is to use the surface micromachining part only. The polysilicon interlayer is no longer used as an etch channel to distribute the etchant over the area for substrate cavity formation, but instead functions as a sacrificial layer that itself precisely defines a shallow cavity underneath the diaphragm. This is possible by protecting the silicon substrate with means of an additional silicon nitride layer underneath the polysilicon layer. The volume (height) of a vacuum cavity underneath the diaphragm does not influence the operation and performance of an absolute pressure sensor, as long as the diaphragm deflection is not affected and the diaphragm strain or stress is sensed. (Capacitive pressure sensors, however, transform the diaphragm deflection into a capacitance signal that depends strongly on the distance between the diaphragm as deflected electrode and the fixed electrode at the bottom of the vacuum cavity. Hence capacitive pressure sensors demand shallow cavities of precisely defined height.) Additionally, TMAH solution can be used as etchant instead of KOH. In contrast to KOH, the organic TMAH does not contain alkali metal ions, and is fully compatible with a CMOS process [77, 78].

The polysilicon sacrificial layer is deposited (LPCVD) and structured (RIE) as a circular plate with 200 nm thickness and 100 µm diameter [14] (see Fig. 6.14). It is sandwiched between two LPCVD Si_3N_4 layers (50 nm at the bottom, 100 nm at the top). On the top SiN layer, 200 nm thick boron-doped polysilicon piezoresistors are formed with a width of 2 µm and covered by another LPCVD Si_3N_4 layer

306 | 6 CMOS-based Pressure Sensors

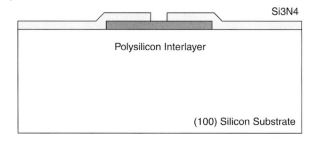

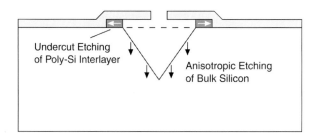

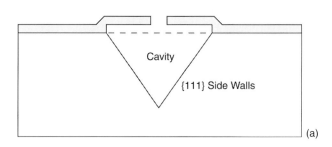

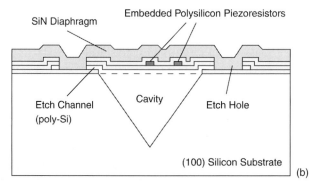

Fig. 6.13 (a) Process flow for microdiaphragm pressure sensor [13]. (b) Cross-section of microdiaphragm pressure sensor with V-shaped bulk micromachined cavity (pyramid with {111} walls) [13]

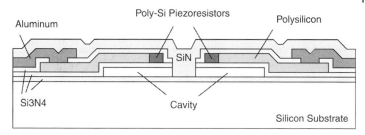

Fig. 6.14 Cross-section of surface micromachined pressure sensor with silicon nitride diaphragm and embedded polysilicon piezoresistors [14]

(300 nm thick). Contact electrodes and lead wires are provided by aluminum metallization. An etch hole with 10 μm diameter is photolithographically defined and opened at the center of the diaphragm. The polysilicon sacrificial layer is then etched by TMAH solution [77]. After the sacrificial layer etching, the etch hole is plugged by a 1 μm SiN plasma CVD process, which results in a sealed ring diaphragm that is clamped at both the perimeter and the central pillar. Finally, the silicon nitride ring diaphragm with embedded polysilicon piezoresistors features a thickness of 1.6 μm. The sealed vacuum chamber has a height of 200 nm and offers a reference pressure port with < 0.5 Torr inside pressure.

The microcavity pressure sensor provides a linearity of better than 1% over the pressure range from 0 to 300 kPa. The sensitivity is 10 μV/V/kPa. Because the piezoresistors are embedded in SiN layers and thus much better isolated than conventional implanted or diffused piezoresistors with p–n junction isolation suffering from temperature-dependent leakage currents, stable high-temperature operation is possible. The temperature coefficient of sensitivity is 0.2%/°C over a range from –50 up to 300 °C.

6.3.3.4 Solder-bonded Bulk-micromachined Capacitive Pressure Sensor

The pressure sensor described by Rogge et al. [85] consists of a CMOS signal conditioning IC with a solder-bonded pressure sensor chip (see Fig. 6.15). The pressure sensor chip offers a bulk micromachined (KOH etched with electrochemical etch-stop) diaphragm that serves as the top electrode of the resulting capacitive pressure sensor. The solder frame (Au/Sn or SnPb) that is electroplated on the sensor wafer also provides the electrical contact for the top electrode.

A reference capacitor is formed around the sensor area but within the sealed solder frame. For the read-out of the sensor, a switched capacitor amplifier circuit is used (see Fig. 6.16). The switched capacitor circuit operates with two non-overlapping clock phases 1 and 2. During phase 1, the reference (input) capacitor C_R is charged with $Q_R^{(1)} = C_R V_{in}$, while the feedback capacitor C_S is discharged, i.e. $C_S^{(1)} = 0$. Then, in phase 2, the charge $Q_R^{(1)}$ (stored in phase 1) is transferred from capacitor C_R to the capacitor C_S, as a consequence of the feedback around the operational amplifier that adjusts the potential at the inverting input terminal (−) to

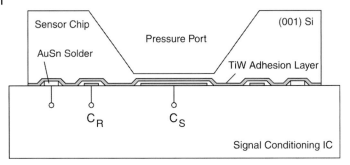

Fig. 6.15 Cross-section of capacitive pressure sensor as built from a signal-conditioning CMOS IC and a wafer-bonded bulk micromachined diaphragm chip [85]

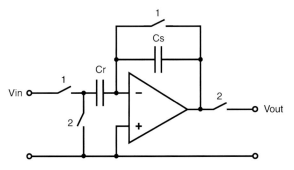

Fig. 6.16 Switched-capacitor voltage-gain amplifier for capacitive pressure sensor readout. The reference capacitor C_R is used as feed-in capacitor, whereas the pressure-dependent capacitor $C_S = C_S(p)$ is used as feedback capacitor [85]

the potential of the non-inverting input terminal (+). The output voltage V_{out} results from $Q_S^{(2)} = Q_R^{(1)}$ and is thus given as the product of the input voltage V_{in} and a voltage gain:

$$V_{out}(p) = \frac{C_R}{C_S(p)} V_{in}$$

The voltage gain equals the ratio $C_R/C_S(p)$ of reference and sensor capacitances, and thus depends on the pressure applied to the sensor element. If we assume, for simplicity, that the sensor element behaves like a plate capacitor with pressure-dependent electrode gap $d(p) = d_0 - kp$, then the voltage gain becomes a linear function of pressure p, i.e.

$$\frac{C_R}{C_S(p)} = 1 - \frac{k}{d_0} p$$

where k serves as a 'spring constant' and $C_R = C_S(0)$ has been assumed. Taking into account additional parasitic capacitances in C_S and C_R will not only lead to higher order (non-linear) terms in p, but also lower the (linear) pressure sensitivity of the voltage gain.

For pressure sensors with a diaphragm area of 1.4 mm × 1.8 mm = 2.52 mm² and a gap of 20 µm at 0 mbar between the two capacitor electrodes, a sensing capacitance $C_S = 1.5$ pF at 300 mbar (without parasitic capacitances) and a sensitivity of 0.8 fF/mbar were simulated using finite element analysis and also demonstrated experimentally [85].

6.3.3.5 Absolute Capacitive Pressure Sensor with Anodically Bonded Glass Encapsulation

A modular pressure sensor process with 20 masks that combines CMOS and bulk and surface micromachining with glass wafer bonding for hermetic encapsulation was described by Chavan and Wise [86, 87].

First, the BiCMOS circuits and the pressure sensor diaphragms are processed in recessed shallow cavities on a p-type silicon wafer (see Fig. 6.17). The transducer cavities are 8 µm deep, whereas the circuit cavities are only 2.5 µm deep. After forming the recesses, a thick n-type epitaxial layer is grown and the devices are fabricated. The cavities are surrounded by bonding anchors with a 2 µm thick second-level polysilicon on top. The second-level polysilicon is used not only for poly–poly capacitors, but also for surface-micromachined devices and for transfer leads. After a CMP step for smoothing the bonding anchor surface (roughness < 50 nm), a glass wafer is anodically bonded. This bonding process creates a vacuum-sealed cavity for the absolute pressure sensor, but also encapsulates the BiCMOS circuit and a reference capacitor (and eventually also surface-micromachined structures that were released before the bonding). Dielectrically isolated doped polysilicon leads transfer electrical signals out of the sealed cavity to bonding pads on the glass. The pressure transdu-

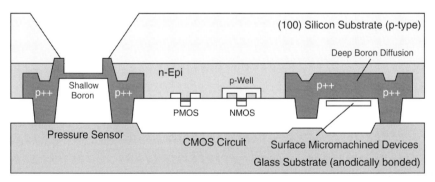

Fig. 6.17 Cross-section of CMOS-integrated pressure sensor. The bonded glass wafer provides vacuum sealing for both the absolute pressure sensor reference cavity and for the CMOS circuitry. By the same means, encapsulation of other micromechanical structures is also possible [86, 87]

cers are released by an anisotropic etch in EDP from the backside. Diffused-boron etch stops are used for thickness definition of the diaphragms with their rigid bosses in the center. Sensitivity variations of $\pm 5\,\%$ across a wafer and of $\pm 10\%$ from wafer to wafer have been obtained. The silicon diaphragms have a thickness of 4 µm and offer a Ti/TiN/Al metal electrode. The second (fixed) capacitor electrode is located on the glass substrate and consists of a Ti/Pt/Au layer.

Five pressure transducers with different diameters up to 1.1 mm are located on the chip. They cover a dynamic range from 500 to 800 Torr and use range segmentation for achieving a resolution of 25.6 mTorr, suitable for precision barometric measurements and applications (1 Torr = 101 325/760 Pa). At atmospheric pressure, the diaphragm is deflected by almost 10 µm, leaving an operating gap between the bossed center of the diaphragms and the fixed electrode of about 0.5 µm (the cavity pressure is 0.5 Torr). A sensor capacitance between 8 and 10 pF results.

Capacitance-to-voltage conversion is provided using a three-stage programmable switched-capacitor circuit. The first stage is a fully differential charge integrator. A variable-gain amplifier with differential input and single-ended output follows as second stage. Both stages feature folded-cascode operational transconductance amplifiers (OTAs). The third stage is a sample-and-hold amplifier. Correlated double sampling is used in order to eliminate the offset voltages of the operational amplifiers. The circuit can multiplex the five different pressure transducers. Offset and gain can be adjusted and calibrated by use of programmable capacitor arrays and a serial data interface.

6.3.3.6 MEMS-CMOS Using Wafer Bonding

A CMOS-integrated pressure sensor based on pre-CMOS cavities was demonstrated by Parameswaran et al. [88]. In a first step called front-end micromachining, shallow cavities are etched into a *handle* wafer and then hermetically sealed by silicon wafer bonding of a second *device* wafer (see Fig. 6.18). This device wafer is *p*-type and offers a 10 µm thick *n*-type epitaxial layer that is later used as a functional layer for MEMS structures such as pressure sensor diaphragms or beams and proof masses in inertial (acceleration or angular rate) sensors, but also for all CMOS devices.

After wafer bonding with a subsequent 1100 °C anneal, the *p*-type bulk of the device wafer is ground, polished and electrochemically etched away, leaving the *n*-type epitaxial layer in its precisely determined thickness on top of the handle wafer. Thickness control is essential for pressure sensor diaphragms with well-defined sensitivities. The second step is a standard CMOS process for fabrication of the integrated circuit. Then, so-called back-end micromachining follows as a post-CMOS process. This back-end micromachining consists of back-side bulk micromachining (KOH etch) for creating pressure ports underneath the pressure sensor diaphragms, if required. Absolute pressure sensors with pressure applied from the wafer surface do not need such a pressure port, but use the pressure inside an encapsulated cavity as reference pressure. Next, surface-micromachined structures are patterned and released from the functional layer by plasma trench

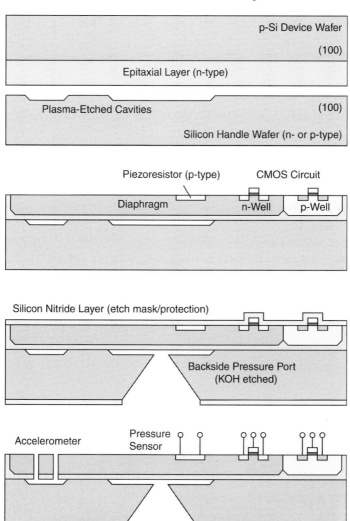

Fig. 6.18 Process flow for MEMS CMOS technology [88]. The flow starts with pre-CMOS steps. The p-type device wafer with n-type epitaxial layer is wafer-bonded to a handle wafer offering plasma-etched cavities. After grinding, polishing and etching, a substrate with epitaxial layer and buried micromachined cavities is obtained. This substrate is then CMOS processed. Finally, post-CMOS micromachining steps such as front-side trench etching, back-side anisotropic etching and some release etching can be performed

etching. Finally, a cover wafer might be bonded on top of the functional layer in order to protect moveable micromachined structures or to encapsulate a cavity above the pressure sensor diaphragm (if absolute pressure sensors are fabricated with pressure inlet from the wafer back side). Such a wafer bonding or wafer-level packaging process as part of the post-CMOS back-end micromachining process must, of course, have a low thermal budget.

The proposed combined MEMS CMOS process offers the advantage of realizing both pressure and inertial sensors together with CMOS circuitry. Moreover, it offers flexibility in pressure sensor design, as pressure ports can be realized on both the front and back-side, and reference cavities for absolute pressure sensors can be located underneath or above the diaphragm. Sensing of the stress associated with the pressure-dependent diaphragm deflection can be easily achieved by properly placed and oriented piezoresistors on top of the single-crystal diaphragm. A Wheatstone bridge delivers the voltage signal as input to the CMOS circuitry for amplification and signal conditioning. The partitioning of the MEMS CMOS process in terms of decreasing thermal budget reduces the cross effects between CMOS and micromechanical device fabrication to a minimum.

6.3.4
Polysilicon-based Capacitive Pressure Sensors

6.3.4.1 Integrated Surface Micromachined Capacitive Pressure Sensor

Monolithically integrated surface micromachined CMOS pressure sensors have been developed, demonstrated and brought into volume production by Siemens/Infineon in 1997 [25, 89, 90] (see Fig. 6.19).

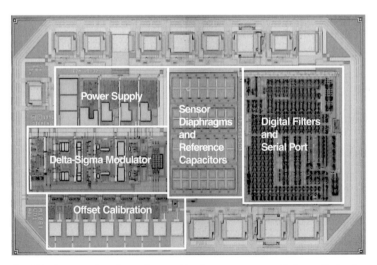

Fig. 6.19 Integrated surface micromachined pressure sensor (Infineon Technologies)

These microelectromechanical systems utilize a standard 0.8 µm BiCMOS process which is enhanced by suitable 'intra-CMOS' surface micromachining processing steps (see Fig. 6.20a). Utilizing the thin poly-2 layer from standard capacitors, pressure-sensitive capacitive devices are formed by means of a sacrificial layer etch technique. Such pressure sensor devices are combined in arrays in order to increase the total capacitance signal. A Wheatstone bridge with two sensor arrays and two reference arrays is used to subtract the zero-pressure (static) capacitance from the sensor capacitance signal. Additional calibration capacitors allow offset compensation. An oversampling delta–sigma ($\Delta\Sigma$) converter [91] performs the analog-to-digital conversion. The delta–sigma converter consists of a fully differential two-stage switched-capacitor $\Delta\Sigma$ modulator and a third-order digital FIR (finite impulse response) decimation filter. The decimation filter lowers the word rate but increases the word length from 1 to 16 bits. It removes modulation noise which dominates at high frequencies. A subsequent second-order digital IIR (infinite impulse response) low-

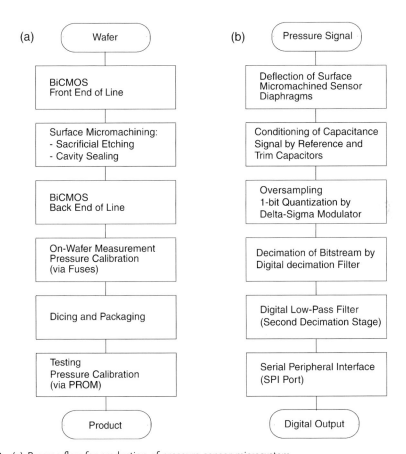

Fig. 6.20 (a) Process flow for production of pressure sensor microsystem.
(b) Signal processing in pressure sensor microsystem [25]

pass filter performs the last stage of decimation by removing out-of-band signal components and adjusting the cutoff frequency to 380 Hz. The final digital signal is available with a resolution of 16 bits at a rate of 7.8 kHz. The signal-to-noise ratio corresponds to an accuracy of 12 bits. A serial peripheral interface (SPI) allows fast communication with data transfer rates up to 500 kHz. The total signal path is shown in Fig. 6.20 b. In addition to transmitting pressure output data, this interface allows requests for information from a number of diagnosis modes. The implemented diagnosis modes perform tests of the analog and digital signal paths, and also compare the outputs from the two sensor device arrays. For calibration purposes, the interface is also used to address the trim arrays in order to try trim settings before the corresponding fuses are finally blown to freeze the programming.

6.3.4.2 Intra-CMOS Surface Micromachining

A cross-section of the surface micromachined sensor diaphragm is shown in Fig. 6.21 a. The thickness of the polysilicon-2 layer is 400 nm. In the basic (Bi)CMOS process, this layer is used for capacitors. For the formation of the pressure sensor devices, the underlying field oxide serves as a sacrificial layer. This sa-

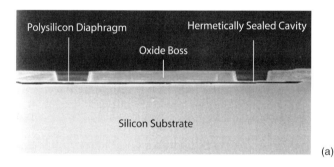

(a)

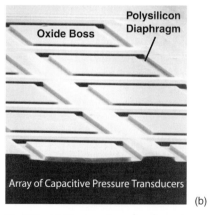

(b)

Fig. 6.21 (a) Cross-section of surface micromachined pressure sensor device. (b) Sensor array (perspective view) [25]

crificial layer is isotropically etched to create the cavity of the sensor device. During the rinse and dry cycle after the etch, sticking of the diaphragm to the cavity bottom must be avoided. Plugging of the etch holes located in the diaphragm can be achieved by means of an appropriate deposition process. The resulting internal cavity pressure at ambient temperature depends on both the pressure and the temperature during the sealing process. Sealed vacuum cavities can thus be obtained, for example, by means of low-pressure chemical vapor deposition (LPCVD). The resulting structure offers a layer stack of BPSG, intermetallic and plasma oxides, and plasma nitride on top of the polysilicon diaphragms of the pressure sensor and reference devices. In case of the pressure sensors, trenches with a width of several μm around the diaphragm perimeter determine a ring diaphragm with an oxide boss in the center (see Fig. 6.21 b). The doped polysilicon diaphragm thus behaves similarly to a moveable but rigid plate electrode of a capacitor. Each sensor array consists of 14 individual sensor devices, and offers a total capacitance of around 2 pF at a pressure of 100 kPa [25].

6.3.4.3 Mixed Signal Processing

The pressure sensitivity of an array of sensor devices is of the order of 1.5 aF/Pa or 0.15 fF per 100 Pa (1 mbar). Sampling of such tiny signals is a challenging task and requires a very noise-insensitive analog readout circuit. The chosen approach is based on a fully differential switched-capacitor delta–sigma modulator that combines a number of convenient and advantageous concepts: switched capacitor circuits are easily realized in CMOS technology and do not impose high requirements on the matching of circuit components. The fully differential architecture provides excellent power-supply and common-mode rejection rates, and eliminates switch charge injection and clock feedthrough to the first order [92]. Oversampling delta–sigma modulators perform the analog-to-digital conversion by sampling the input signal in time and by quantizing its amplitude. In contrast to non-oversampling (Nyquist rate) converters, high-precision analog anti-aliasing filters are not required by oversampling modulators. In addition, oversampling delta–sigma modulators 'shape' the quantization noise.

The noise shaping modifies the spectral density of the quantization noise by shifting quantization noise power from low to high frequencies. Once the noise has been transferred out of the signal band, it is suppressed by the digital decimation filter. The noise shaping is achieved by special feedback loops within the delta–sigma modulator. The feedback loops essentially compensate the quantization errors by a certain subtraction scheme [91, 93].

The number L of feedback loops determines the order of a delta–sigma modulator. Oversampling quantization and noise shaping by feedback loops reduces the signal-band rms quantization noise by a factor $OSR^{-(L+1/2)}$. The sampling frequency is given by $f_s = OSR\, f_{Nyquist}$, where $OSR \gg 1$ is the oversampling ratio and $f_{Nyquist}$ is the Nyquist frequency (twice the signal bandwidth). Second-order delta–sigma modulators thus offer signal-to-noise ratio improvements of 15 dB or 2.5 bits per doubling of the oversampling ratio.

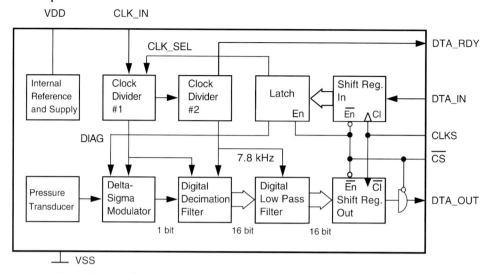

Fig. 6.22 Block diagram of pressure sensor microsystem [25]

The output of the delta–sigma modulator is a bitstream, i.e. it has a 1 bit word length. A decimation filter averages the modulator output, reduces the sampling rate by a large (decimation) factor $N \gg 1$ and increases the resolution (word length). For sinc-type decimation filters of order k, the transfer function is given as

$$H(z) = \left(\frac{1}{N} \frac{1 - z^{-N}}{1 - z^{-1}} \right)^k$$

The frequency response is obtained by substituting $z = \exp(sT)$ with $T = 1/f_s$ and the complex variable $s = \sigma + i\omega$ with $\omega = 2\pi f$. Setting $|H[\exp(i2\pi f_c/f_s)]| = 1/\sqrt{2}$, one obtains the 3 dB cutoff frequency

$$f_c \simeq \frac{f_s}{\pi N} \sqrt{6[1 - 2^{-1/(2k)}]}$$

which is inversely proportional to the decimation factor N for given filter order k and sampling frequency f_s. Frequencies above f_c are attenuated, which eliminates most of the modulation noise. Note that the cutoff frequency f_c is related to the signal delay time τ_{gr} of the sinc^k filter. One has

$$\tau_{gr} = -\frac{d\phi}{d\omega} = -\frac{d}{d\omega}\left[-\frac{1}{2}\omega k(N-1)T \right] = \frac{k}{2}(N-1)T$$

independent of frequency ω because the phase $\phi(\omega)$ is linear in ω. Hence

$$\tau_{gr} \simeq \frac{k}{2\pi f_c} \sqrt{6[1 - 2^{-1/(2k)}]}$$

which gives a signal delay time of 1 ms for a cutoff frequency of approximately 400 Hz. As a consequence, if shorter delay times are required, filter type and topology must be carefully considered and optimized for fast pressure sensors. Also note that the output data rate equals f_s/N. In the discussed example [25], the decimation filter delivers a word length of 16 bits, has the order $k = 3$ (which is considered the best choice for second-order delta–sigma modulators) and provides a cutoff frequency $f_c = 0.2575 f_s/N = 2$ kHz for an output data rate $f_s/N = 7.8$ kHz. The desired cutoff frequency of 380 Hz is finally obtained by an additional IIR digital low-pass filter. The total signal delay time of the chosen combination of FIR and IIR filters is well below 1 ms.

6.3.4.4 Charge Transfer in Switched Capacitor Circuits

A fully differential second-order delta–sigma modulator with input bridge consisting of two capacitive sensor arrays and two reference capacitor arrays is shown in Fig. 6.23. In order to minimize power noise coupling, a dedicated on-chip reference voltage generator supplies the switched capacitor circuit. The switched capacitor circuit is operated in two non-overlapping clock phases 1 and 2. In each clock phase, capacitors are charged or discharged depending on the correspondingly applied positive or negative (constant) reference voltages V_{refp} and V_{refn}, and also depending on the delta–sigma feedback mode, here called m and mq. The feedback mode is determined by the comparator output, but influences only clock phase 2. The comparator acts as a two-level (single-bit) quantizer. Each stage (integrator) of the delta–sigma modulator is equipped with a fully differential operational amplifier. A fully differential operational amplifier adjusts its balanced output voltages in such a way that the potential difference between the inverting and non-inverting input terminals vanishes. For ideal operational amplifiers, both input ports have infinite input impedance and, consequently, zero input currents. Balanced output voltages are symmetric with respect to analog ground V_h. This important property can be achieved even in non-ideal operational amplifiers by means of an internal *common-mode feedback* loop (CMFB).

The first integrator of the delta–sigma modulator takes advantage of *correlated double sampling* (CDS). Offset voltage and low-frequency noise are sampled and stored on the capacitors C_{s11} and C_{s12} in clock phase 1 and then subtracted from the signal in clock phase 2.

Properly designed switched capacitor circuits must quickly reach their equilibrium state in each clock phase. Therefore, charge transfer in switched capacitor circuits can by described in the time domain by difference equations that relate all node potentials (voltages) and all stored capacitor charges of a clock phase to those of the previous clock phase. Correspondingly, using the delay operator $z^{-1} = e^{-sT}$, the transfer function of a switched capacitor circuit can be writen in the z-domain.

318 | 6 CMOS-based Pressure Sensors

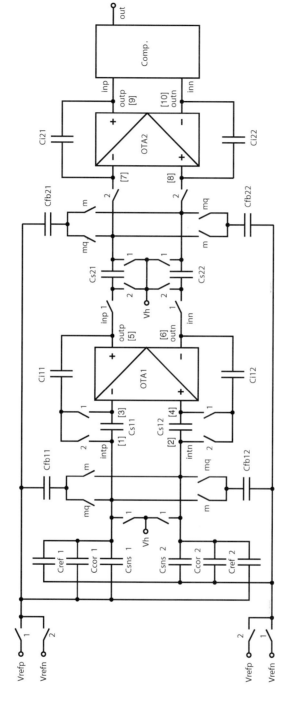

Fig. 6.23 Fully differential switched-capacitor delta–sigma modulator (without trimming capacitors and temperature compensation). The circuit is bi-phased with two non-overlapping phases 1 and 2. Switches m and mq denote different feedback modes depending on modulator (comparator) output

6.3 CMOS-integrated Pressure Sensors

In order to derive the difference equations for SC circuits, it is convenient to consider equivalent circuits for the individual phases (see Fig. 6.24). Every capacitor in the equivalent circuits is assumed to have zero initial charge in each phase. A voltage controlled charge source (VCQS) is connected in parallel to every capacitor and injects the charge stored by the capacitor in the last (previous) phase [92]. For every *inner* node of the circuit, the sum of all injected charges must be zero.

We introduce node potentials u_n (phase 1) and v_n (phase 2) for the inner nodes $n = 1, \ldots, 10$ (see Fig. 6.23). Analog ground has zero potential. Next, all node potentials (voltages) are decomposed into corresponding differential and common mode parts, i.e. $u_i = u_i^c + u_i^d$ and $v_i = v_i^c + v_i^d$ for odd indices i and $u_i = u_{i-1}^c - u_{i-1}^d$ and $v_i = v_{i-1}^c - v_{i-1}^d$ for even indices i. For simplicity, we assume *ideal* operational amplifiers with infinite gain, infinite common mode rejection and zero input voltage offset. Moreover, all integration, sample and feedback capacitors in the two inverting and non-inverting channels of the differential SC circuit are assumed to match *perfectly*, i.e. $C_{i11} = C_{i12} \equiv C_{i1}$, $C_{s11} = C_{s12} \equiv C_{s1}$, $C_{fb11} = C_{fb12} \equiv C_{fb1}$, and similarly for the second stage. However, no such matching is a priori assumed for the sensor and reference array capacitors $C_{sns1}(p)$, $C_{sns2}(p)$, and C_{ref1}, C_{ref2}. For a clock cycle $v \rightarrow u \rightarrow v'$ consisting of clock phases 1 and 2, one obtains after some calculation the following difference equations:

$$v_5^{d'} = v_5^d \mp \beta_1(V_p - V_n) + \frac{\sigma_1 + \rho_1 - \sigma_2 - \rho_2}{2 + \sigma_1 + \sigma_2 + \rho_1 + \rho_2 + 2\beta_1} u_3^c$$
$$+ \frac{(\sigma_1 - \rho_1)(1 + \sigma_2 + \rho_2 + \beta_1) + (\sigma_2 - \rho_2)(1 + \sigma_1 + \rho_1 + \beta_1)}{2 + \sigma_1 + \sigma_2 + \rho_1 + \rho_2 + 2\beta_1}(V_p - V_n)$$

$$v_7^{c'} = \frac{1}{1 + a_2 + \beta_2} v_7^c$$

$$v_9^{d'} = v_9^d + a_2 v_5^d \mp \beta_2(V_p - V_n)$$

and

$$u_3^{c'} = \frac{1}{1 + a_1} \left\{ a_1 u_3^c + \frac{2u_3^c - [(\sigma_1 - \rho_1) - (\sigma_2 - \rho_2)](V_p - V_n)}{2 + \sigma_1 + \sigma_2 + \rho_1 + \rho_2 + 2\beta_1} \right\}$$

with capacitance ratios ($\ell = 1, 2$)

$$\sigma_\ell = \frac{C_{sns\ell}}{C_{i1}}, \quad \rho_\ell = \frac{C_{ref\ell}}{C_{i1}}, \quad a_1 = \frac{C_{s1}}{C_{i1}}, \quad a_2 = \frac{C_{s2}}{C_{i2}}, \quad \beta_1 = \frac{C_{fb1}}{C_{i1}}, \quad \beta_2 = \frac{C_{fb2}}{C_{i2}}$$

After a number of clock cycles (iterations), the decoupled potential v_7^c decays to analog ground, i.e. $v_7^c \rightarrow 0$, while the decoupled potential u_3^c converges to

$$u_3^c \rightarrow -\frac{(\sigma_1 - \rho_1) - (\sigma_2 - \rho_2)}{\sigma_1 + \sigma_2 + \rho_1 + \rho_2 + 2\beta_1}(V_p - V_n)$$

320 | 6 CMOS-based Pressure Sensors

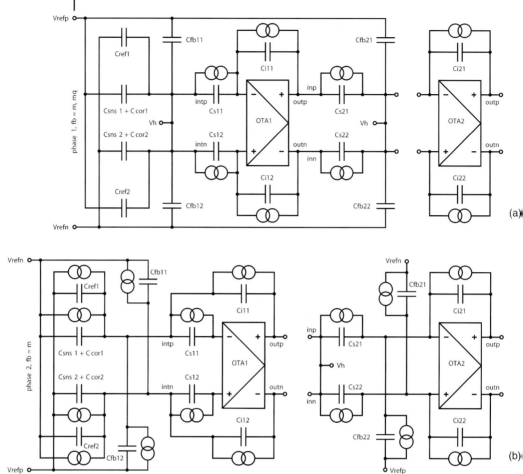

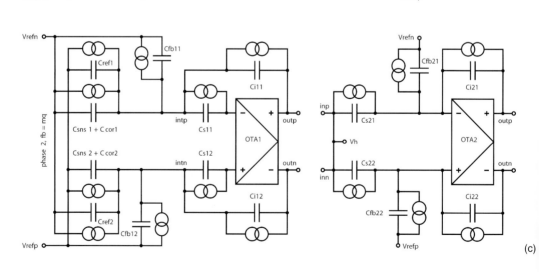

6.3.4.5 Transfer Function of Delta–Sigma Modulator

As a consequence of the above equations, the oversampling second-order delta–sigma modulator can be described by two coupled difference equations:

$$v_5^{d'} = v_5 + x_{in}\beta_1(V_p - V_n) \mp \beta_1(V_p - V_n)$$

$$v_9^{d'} = v_9 + a_2 v_5 \mp \beta_2(V_p - V_n)$$

with a *normalized* input x_{in}:

$$x_{in} = \frac{(\sigma_1 - \rho_1) + (\sigma_2 - \rho_2)}{2\beta_1} - \frac{(\sigma_1 + \rho_1) - (\sigma_2 + \rho_2)}{\sigma_1 + \sigma_2 + \rho_1 + \rho_2 + 2\beta_1} \cdot \frac{(\sigma_1 - \rho_1) - (\sigma_2 - \rho_2)}{2\beta_1}$$

Stability requires $x_{in} \in (-1, 1)$, but modulation noise performance suffers when $|x_{in}|$ gets close to 1. The second term vanishes for perfectly matched sensor and reference arrays, i.e. for $\sigma_1(p) = \sigma_2(p)$ and $\rho_1 = \rho_2$, and one obtains the 'single ended' (non-balanced) circuit result:

$$x_{in} = \frac{C_{sns1} - C_{ref1}}{C_{fb1}}$$

The quantized output of the delta–sigma modulator depends on the sign of v_9^d and determines the feedback mode:

$$Y_{\Delta\Sigma Mod} = \begin{cases} 1; v_9^d \geq 0, & \text{feedback mode } m \\ 0; v_9^d < 0, & \text{feedback mode } mq \end{cases}$$

The feedback loops for both integrators guarantee that the average modulator ouput closely tracks the input signal, i.e.

$$\text{average}(Y_{\Delta\Sigma Mod}) = \frac{1 + x_{in}}{2} + \text{modulation noise}$$

After digital decimation filtering with a final 16 bit word length, the *static* transfer function of the delta–sigma converter results as

$$Y_{\Delta\Sigma Conv} = 2^{15}(1 + x_{in})$$

For optimum performance, $C_{sns}(p)$ and C_{ref} should roughly match for some medium pressure p within the specified operating range. Offset calibration is thus mandatory in $\Delta\Sigma$-based sensor systems. Auxiliary trim capacitors are implemented in

---

Fig. 6.24 (a) Equivalent circuit for charge transfer in phase 1 (both feedback modes). (b) Equivalent circuit for charge transfer in phase 2 with feedback mode m. (c) Equivalent circuit for charge transfer in phase 2 with feedback mode mq

arrays and can be connected in parallel either to the sensor capacitance or to the reference capacitance. It was found [25] that 7 bit resolution for offset trimming is sufficient to achieve a calibration accuracy better than 1 kPa. Finally, the feedback capacitor C_{fb1} must be chosen carefully in order to maximize the output signal swing while preventing modulator overload which would adversely affect the accuracy.

6.3.5
Pressure Sensors for High-temperature Applications

The isolation of reverse-biased p–n junctions is spoiled by leakage currents that significantly increase as the operating temperature rises. This is a general problem in conventional silicon technologies, and provides a temperature limit of approximately 125 °C for applications. Silicon-on-insulator (SOI) substrates offer the possibility to replace p–n junctions by dielectric isolation (see Fig. 6.25 a). This allows one to shift the operating temperature limitation of analog silicon circuits up to 250 °C or even higher (depending on the circuit type). Piezoresistors that are implanted or diffused into silicon diaphragms also run into leakage current problems. In contrast, polysilicon piezoresistors do not require junction isolation. However, even for dielectrically isolated polysilicon piezoresistors, the temperature

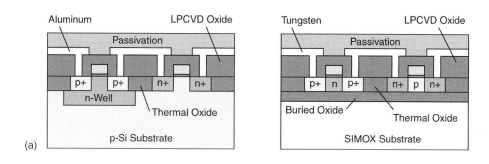

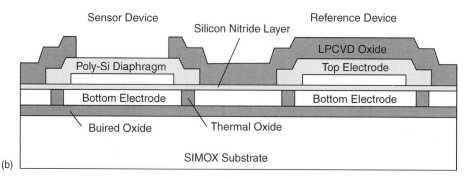

Fig. 6.25 (a) Cross-sections of standard CMOS and SOI-CMOS transistor devices [95]. (b) Comparison of surface-micromachined capacitive pressure sensor and corresponding pressure-insensitive reference devices on SIMOX substrate (cross-sections)

dependence of the piezoresistive coefficients has to be considered [14, 94]. Capacitive pressure sensors with properly isolated electrodes do not suffer from temperature dependences (unless caused by the mechanical diaphragm behavior or temperature effects of residual gas inside the sealed 'vacuum' cavity).

6.3.5.1 Integrated Surface Micromachined Capacitive Pressure Sensor for High-temperature Applications

Kasten et al. based integrated surface micromachined capacitive pressure sensors on SIMOX (separation by implantation of oxygen) substrates [95] (Fig. 6.25b). This allows for high temperature applications up to 250 °C.

Circular pressure-sensitive diaphragms are arranged in arrays in order to increase the capacitance signal. A second array is equipped with reference diaphragms that give a comparable capacitance but are not sensitive to pressure. The 120 nm thick bottom electrode is isolated from the substrate by means of the buried oxide layer of the SIMOX substrate. LOCOS (local oxidation of silicon) structures provide lateral isolation.

The cavities are defined by a 900 nm thick sacrificial oxide layer. Sacrificial etching is performed by hydrofluoric acid, and the access ports (etch holes) of the etch channels are subsequently plugged by means of an LPCVD process. The LPCVD process offers the advantage that low reference pressures of around 10^{-4} bar can be achieved inside the sealed cavities.

A switched-capacitor capacitance-to-voltage converter utilizes the capacitance ratio C_R/C_S for a first-order linearization and additionally provides an analog feedback path for a cancellation of remaining non-linearities. For this purpose, an amplification factor k must be precisely programmed (6-bit resolution) into the shift register. At room temperature, a linearity error of less than 0.4% FSO (full-scale output) results. Further stages offer offset adjustment (because of mismatch between sensor and reference capacitances), output range adjustment (0.5–4.5 V) and a sample-and-hold amplifier.

The offset temperature coefficient (TCO) has been characterized to be smaller than 0.01% FSO/°C for a pressure of 50 bar over the temperature range 25–250 °C. The TCO increases with decreasing pressure, but is still smaller than 0.09% FSO/°C for 1 bar at 250 °C.

6.3.6
Micromachined Pirani Pressure Gauges

Pirani gauges utilize the fact that the thermal conductivity of a gas is a function of pressure. The thermal conductivity can be measured by means of a cavity with a heater and a heat sink opposite to each other. A control circuit then keeps the temperature difference between heater and heat sink constant. The sensitivity is roughly proportional to the heater (and heat sink) area, whereas the dynamic pressure range becomes larger with smaller distance between heater and heat sink. Pirani-type pressure gauges can cover a pressure range of four decades. Chae et al. [96] utilized suspended p^{++} silicon as heater and as lateral heat sinks. Small micromachined gauges with integrated readout electronics were developed by Mastrangelo and Muller [97], Klaassen and Kovacs [98] and others. Micromachined Pirani gauges offer small, low-cost devices with high sensitivity which can be conveniently used for leak testing of MEMS packages [99], for example.

6.3.6.1 CMOS-integrated Thermal Pressure Sensor

Baltes and co-workers developed CMOS-compatible – i.e. post-CMOS – thermal pressure sensors utilizing the pressure-dependent heat transfer across the vertical gap of a surface-micromachined cavity [100, 101] (see Fig. 6.26).

The cavity is created by wet etching of the lower metal layer, which serves as a sacrificial layer for the Pirani device. A meandering 3 µm wide filament in the upper metal layer, sandwiched between the intermetal dielectric and the top passivation layer, serves as a heater. The heat sink is provided by the substrate, separated by a gap of 0.6 µm corresponding to the metal-1 thickness. For a 200 µm sensor diameter, the heater area has a diameter of 150 µm. During sensor operation, the voltage drop along the filament and the current through the filament are monitored and adjusted in order to achieve a constant temperature difference ΔT of 2 K. The corresponding heater current is below 10 mA in order to prevent electromigration effects. The pressure-dependent thermal conductance is then obtained as $G(p) = P/\Delta T$.

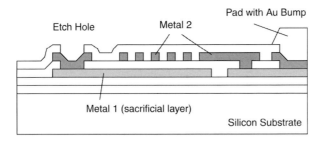

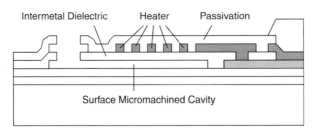

Fig. 6.26 Cross-section of thermal pressure sensor with surface micromachined cavity. Metal 1 is utilized as a sacrificial layer and metal 2 provides for the circular heating element in 3 µm wide meandering lines. The substrate underneath the cavity serves as a heat sink [100, 101]

This function has been characterized and compared with the result of a thermal model [100].

6.3.7
Micromachined Pressure Sensors (Overview)

Tab. 6.3 Bulk and surface micromachined pressure sensors

Institute or Company	Bulk micromachined	Surface micromachined	Remarks
Bosch	[102–105]		m
ETH Zurich	[85]		c, wafer bonded
FhG	[106]		c, wafer bonded
	[23, 107]		m
		[108–113]	c
		[95, 114, 115]	c, SIMOX
Hitachi	[12, 116]		m
Infineon Technologies	[67]		m
		[25–27, 90]	c
		[117, 118]	c
LETI-CEA		[119]	c, SIMOX
MIT	[88]		m, Si wafer bonding
Motorola	[15, 16, 72]		m, shear stress transducer
	[120–122]		m, shear stress transducer
		[122, 124]	c
NEC	[18, 125, 126]		m
Sandia National Laboratories		[127, 128]	p
SSI Technologies		[129]	p
Technical University of Berlin	[130]		p
	[131]		m
Tohoku	[132–134]		c, wafer bonded
Toyota	[11, 135]		m
	[24, 136]		c, wafer bonded
	[13]		p, interlayer etching
		[14, 137, 138]	p
University of Michigan	[8, 10, 32]		m
	[139]		p
	[86, 140]		c, wafer bonded
Other	[141, 142]		m, (110) wafer

Sensing: m = monocrystalline silicon piezoresistors, p = polysilicon piezoresistors, c = capacitive sensing.

Tab. 6.4 Selected and Additional Literature

Elasticity	[30, 33, 34, 143–147]
Elastic Plates	[35, 148–150]
Silicon Micromachining	[151, 152]
Surface Micromachining	[17, 153–163]
Pressure Sensor Design	[67, 68, 81, 164–177]
CMOS Integration of Sensors	[117, 178–182]

6.4 Conclusion

The past decades have shown significant developments in the field of silicon pressure sensors. Starting from basic bulk micromachined transducers with diffused piezoresistors, integration with bipolar and CMOS technology has added functionality such as compensation and calibration, programmable interfaces and diagnosis features. Today, CMOS-based pressure sensors have evolved into fully fledged microelectromechanical systems (MEMS). Cost-effective fabrication, reliability and performance are success factors that allow one to address demanding automotive, medical and industrial volume applications. Although there is a clear trend to monolithic integration and further miniaturization, a variety of possible approaches for constructing and integrating elementary pressure transducer devices continue to exist.

6.5 References

1 C. SMITH, Piezoresistance effect in germanium and silicon, *Phys. Rev.* **1954**, *94*, 42–49.
2 C. HERRING, Transport properties of a many-valley semiconductor, *Bell Syst. Tech. J.* **1955**, *34*, 237–290.
3 C. HERRING, E. VOGT, Transport and deformation-potential theory for many-valley semiconductors with anisotropic scattering, *Phys. Rev.* **1956**, *101*, 944–961.
4 F. MORIN, T. GEBALLE, C. HERRING, Temperature dependence of the piezoresistance of high-purity silicon and germanium, *Phys. Rev.* **1957**, *105*, 525–539.
5 W. MASON, R. THURSTON, Use of piezoresistive materials in the measurement of displacement, force, and torque, *J. Acoust. Soc. Am.* **1957**, *29*, 1096–1101.
6 W. PFANN, R. THURSTON, Semiconducting stress transducers utilizing the transverse and shear piezoresistance effects, *J. Appl. Phys.* **1961**, *32*, 2008–2019.
7 O. TUFTE, P. CHAPMAN, D. LONG, Silicon diffused-element piezoresistive diaphragms, *J. Appl. Phys.* **1962**, *33*, 3322–3327.
8 J. BORKY, K. WISE, Integrated signal conditioning for silicon pressure sensors, *IEEE, Trans. Electron Devices* **1979**, *ED-26*, 1906–1910.
9 C. SANDER, J. KNUTTI, J. MEINDL, A monolithic capacitive pressure sensor with pulse-period output, *IEEE Trans. Electron Devices* **1980**, *ED-17*, 927–930.
10 S. KIM, K. WISE, Temperature sensitivity in silicon piezoresistive pressure transdu-

cers, *IEEE Trans. Electron Devices* **1983**, *ED-30*, 802–810.

11 S. Sugiyama, M. Takigawa, I. Igarashi, Integrated piezoresistive pressure sensor with both voltage and frequency output, *Sens. Actuators* **1983**, *4*, 113–120.

12 K. Yamada, M. Nishihara, R. Kanzawa, R. Kobayashi, A piezoresistive integrated pressure sensor, *Sens. Actuators* **1983**, *4*, 63–69.

13 S. Sugiyama, T. Suzuki, K. Kawahata, K. Shimaoka, M. Takigawa, I. Igarashi, Micro-diaphragm pressure sensors, in *Tech. Digest International Electron Devices Meeting (IEDM '86)*; **1986**, 184–187.

14 S. Sugiyama, K. Shimaoka, O. Tabata, Surface-micromachined microdiaphragm pressure sensors, *Sens. Mater.* **1993**, *4*, 265–275.

15 J. Gragg, W. McCulley, W. Newton, C. Derrington, Compensation and calibration of a monolithic four terminal silicon pressure transducer, in *Tech. Digest IEEE Solid State Sensor Conf.*; **1984**, 21–27.

16 R. Frank, W. McCulley, An update on the integration of silicon pressure sensors, in *Proc. Wescon '85, San Francisco*; **1985**, 1–5.

17 R. Howe, R. Muller, Resonant-microbridge vapor sensor, *IEEE Trans. Electron Devices* **1986**, *ED-33*, 499–506.

18 T. Ishihara, K. Suzuki, S. Suwazono, M. Hirata, H. Tanigawa, CMOS integrated silicon pressure sensor, *J. Solid-State Circuits* **1987**, *SC-22*, 151–155.

19 N. Najafi, K. Clayton, W. Baer, K. Najafi, K. Wise, An architecture and interface for VLSI sensors, in *Tech. Digest IEEE Solid-State Sensor Workshop*; **1988**, 76–79.

20 N. Najafi, K. Wise, An organization and interface for sensor-driven semi-conductor process control systems, *IEEE Trans. Semicond. Manuf.* **1990**, *3*, 230–238.

21 A. Kjensmo, A. Hanneborg, J. Gakkestad, H. von der Lippe, A CMOS front-end circuit for a capacitive pressure sensor, *Sens. Actuators A* **1990**, *21–23*, 102–107.

22 U. Schöneberg, F. Schnatz, W. Brockherde, P. Kopystinski, T. Mehlhorn, E. Obermeier, H. Benzel, CMOS integrated capacitive pressure transducer with on-chip electronics and digital calibration capability, in *Tech. Digest Int. Conf. on Solid-State Sensors and Actuators*; **1991**, 304–307.

23 D. Hammerschmidt, F. Schnatz, W. Brockherde, B. Hosticka, E. Obermeier, A CMOS piezoresistive pressure sensor with on-chip programming and calibration, in *Digest 1993 IEEE International Solid-State Circuits Conf.*; **1993**, 128–129.

24 T. Nagata, H. Terabe, S. Kuwahara, S. Sakurai, O. Tabata, S. Sugiyama, M. Esashi, Digital compensated capacitive pressure sensor using CMOS technology for low-pressure measurements, *Sens. Actuators A* **1992**, *34*, 173–177.

25 H.-J. Timme, D. Draxelmayr, C. Hierold, S. Kolb, D. Maier-Schneider, E. Pettenpaul, T. Scheiter, M. Steger, W. Werner, Monolithic pressure sensor microsystems, in *Proc. 8th Int. Congress on Sensors, Transducers, and Systems (Sensor '97)*, Nürnberg; **1997**, 65–70.

26 T. Scheiter, H. Kapels, K.-G. Oppermann, M. Steger, C. Hierold, W. Werner, H.-J. Timme, Full integration of a pressure-sensor system into a standard BiCMOS process, *Sens. Actuators A* **1998**, *67*, 211–214.

27 C. Kolle, D. Maier-Schneider, A. Logiudice, R. Noè, E. Bodenstorfer, L. Gwehenberger, D. Draxelmayr, A monolithically integrated capacitive pressure sensor system for automotive applications, in *Proc. of 36th Int. Conf. on Microelectronics, Devices and Materials MIDEM 2000, Postojna, Slovenia*; **2000**, 123–128.

28 J. Hall, Electronic effects in the elastic constants of *n*-type silicon, *Phys. Rev.* **1967**, *161*, 756–761.

29 A. George, Elastic constants and moduli of diamond cubic Si, in *Properties of Crystalline Silicon*, R. Hull (ed.); London: INSPEC, Institution of Electrical Engineers, **1999**, 98–103.

30 J. Nye, *Physical Properties of Crystals*; Oxford: Oxford University Press, **1985**, reprinted **2004**.

31 J. Wortman, R. Evans, Young's modulus, shear modulus, and Poisson's ratio in Silicon and germanium, *J. Appl. Phys.* **1965**, *36*, 153–156.

32 S. Clark, K. Wise, Pressure sensitivity in anisotropically etched thin-diaphragm pressure sensors, *IEEE Trans. Electron Devices* **1979**, *ED-26*, 1887–1896.

33 L. Landau, E. Lifschitz, *Elastizitätstheorie*, 6th edn.; Berlin: Akademie-Verlag, **1989**.

34 A. Green, W. Zerna, *Theoretical elasticity*, 2nd edn.; New York: Dover, **1992**.

35 E. Suhir, *Structural Analysis in Microelectronic and Fiber-Optic Systems*, Volume I, New York: Van Nostrand Reinhold, **1991**.

36 R. Keyes, The effects of elastic deformation on the electrical conductivity of semiconductors, in *Solid State Physics*, Vol. 11, F. Seitz, D. Turnbull (eds.); New York: Academic Press, **1960**, 149–221.

37 D. Kerr, A. Milnes, Piezoresistance of diffused layers in cubic semiconductors, *J. Appl. Phys.* **1993**, *34*, 727–731.

38 O. Tufte, E. Stelzer, Piezoresistive properties of silicon diffused layers, *J. Appl. Phys.* **1963**, *34*, 313–318.

39 O. Tufte, E. Stelzer, Piezoresistive properties of heavily doped n-type silicon, *Phys. Rev.* **1964**, *133*, A1705–A1716.

40 F. Hock, Die Berechnung des Piezowiderstandseffektes von Silicium für meßtechnische Anwendungen, *Z. f. Angew. Physik* **1964**, *17*, 511–517.

41 J. Bretschi, A silicon integrated strain-gage transducer with high linearity, *IEEE Trans. Electron Devices* **1976**, 59–61.

42 W. Pietrenko, Einfluss von Temperatur und Störstellenkonzentration auf den Piezowiderstandseffekt in n-Silizium, *Phys. Status Solidi A* **1977**, *41*, 197–205.

43 Y. Kanda, A graphical representation of the piezoresistance coefficients in silicon, *IEEE Trans. Electron Devices* **1982**, *ED-29*, 64–70.

44 K. Yamada, M. Nishihara, S. Shimada, M. Tanabe, M. Shimazoe, Temperature dependence of the piezoresistance effects of p-type silicon diffused layers, *Electr. Eng. Jpn.*, **1983**, *103*, 8–16.

45 K. Yamada, M. Nishihara, S. Shimada, M. Tanabe, M. Shimazoe, Y. Matsuoka, Nonlinearity of the piezoresistance effect of p-type silicon diffused layers, *IEEE Trans. Electron Devices*, **1982**, *ED-29*, 71–77.

46 J. Lenkkeri, Nonlinear effects in the piezoresistivity of p-type silicon, *Phys. Status Solidi B* **1986**, *136*, 373–385.

47 K. Matsuda, K. Suzuki, K. Yamamura, Y. Kanda, Nonlinear piezoresistance effects in silicon, *J. Appl. Phys.* **1993**, *73*, 1838–1847.

48 J.S. Blakemore, *Semiconductor Statistics*, Dover Publications, Inc., Mineola, New York, **1987**, republished in 2002.

49 F. Blatt, Theory of mobility of electrons in solids, in *Solid State Physics*, Volume 4, F. Seitz, D. Turnbull (eds.); New York: Academic Press, **1957**, 199–366.

50 S. Wong, S. McAlister, Z.-M. Li, A comparison of some approximations for the Fermi-Dirac Integral of Order 1/2, *Solid-State Electron.* **1994**, *37*, 61–64.

51 M. Chaparala, B. Shivaram, Piezoresistance of c-Si, in *Properties of Crystalline Silicon*, R. Hull (ed.); London: INSPEC, Institution of Electrical Engineers, **1966**, 421–429.

52 Y. Onuma, K. Sekiya, Piezoresistive properties of polycrystalline silicon thin film, *Jpn. J. Appl. Phys.* **1972**, *11*, 20–23.

53 J. Seto Piezoresistive properties of polycrystalline silicon, *J. Appl. Phys.* **1976**, *47*, 4780–4783.

54 P. French, A. Evans, Piezoresistance in polysilicon, *Electron. Lett.* **1984**, *20*, 999.

55 P. French, A. Evans, Polycrystalline silicon strain sensors, *Sens. Actuators* **1985**, *8*, 219–225.

56 P. French, A. Evans, Polycrystalline silicon as a strain gauge material, *J. Phys. E: Sci. Instrum.* **1986**, *19*, 1055–1058.

57 D. Schubert, W. Jenschke, T. Uhlig, F. Schmidt, Piezoresistive properties of polycrystalline and crystalline silicon films, *Sens. Actuators* **1987**, *11*, 145–155.

58 Y. Kanda, K. Suzuki, Statistical model for piezoresistance in thin films, *Appl. Surf. Sci.* **1988**, *33/34*, 996–1000.

59 P. French, A. Evans, Piezoresistance in polysilicon and its applications to strain gauges, *Solid-State Electron.* **1989**, *32*, 1–10.

60 I. Obieta, F. Gracia, Sputtered silicon thin films for piezoresistive pressure microsensors, *Sens. Actuators A* **1994**, *41–42*, 685–688.

61 V. Gridchin, V. Lubimsky, M. Sarina, Piezoresistive properties of polysilicon films, *Sens. Actuators A* **1995**, *49*, 67–72.

62 C. Smith, Macroscopic symmetry and properties of crystals, in *Solid State Physics*, Volume 6, F. Seitz, D. Thurnbull (eds.); New York: Academic Press, **1958**, 175–249.

63 R. Thurston, Use of semiconductor transducers in measuring strains, accelerations, and displacements, in *Physical Acoustics*, Volume 1, Part B, W. Mason (ed.); New York: Academic Press, **1964**, 215–235.

64 D. Bittle, J. Suhling, R. Beaty, R. Jaeger, R. Johnson, Piezoresistive stress sensors for structural analysis of electronic packages, *J. Electron. Packaging* **1991**, *113*, 203–215.

65 *Semiconductor Sensors,* Infineon Technologies AG, Munich, **199**, Data Book 1999-04-01.

66 W. Eaton, *Surface micromachined pressure sensors*, PhD Dissertation, University of New Mexico, Albuquerque, NM, **1997**.

67 J. Binder, K. Becker, G. Ehrler, Silicon pressure sensors for the range 2 kPa to 40 MPa, *Siemens Components (Engl. Edn.)* **1984**, *20*, 64–67.

68 H. Sandmaier, *Untersuchung des nichtlinearen Verhaltens piezoresistiver Niederdrucksensoren auf der Basis von Silizium;* PhD Dissertation, Technical University of Munich, **1988**.

69 Y. Kanda, A. Yasukawa, Hall-effect devices as strain and pressure sensors, *Sens. Actuators* **1982**, *2*, 283–296.

70 Y. Kanda, Optimum design considerations for silicon pressure sensors using a four-terminal gauge, *Sens. Actuators* **1983**, *4*, 199–206.

71 Y. Kanda, K. Yamamura, Four-terminal-gauge quasi-circular and square diaphragm silicon pressure sensors, *Sens. Actuators* **1989**, *18*, 247–257.

72 G. Bitko, A. McNeil, R. Frank, Improving the MEMS pressure sensor, *Sensors* **2000**. http://www.sensorsmag.com/articles/0700/62/index.htm.

73 K. Petersen, Silicon as a mechanical material, *Proc. IEEE* **1982**, *70*, 420–457.

74 H. Seidel, L. Csepregi, A. Heuberger, H. Baumgärtel, Anisotropic etching of crystalline silicon in alkaline solutions: I. Orientation dependence and behavior of passivation layers, *J. Electrochem. Soc.* **1990**, *137*, 3612–3626.

75 H. Seidel, L. Csepregi, A. Heuberger, H. Baumgärtel, Anisotropic etching of crystalline silicon in alkaline solutions: II. Influence of dopants, *J. Electrochem. Soc.* **1990**, *137*, 3626–3632.

76 A. Bohg, Ethylene diamine – pyrocatechol – water mixture shows etching anomaly in boron-doped silicon, *J. Electrochem. Soc.* **1971**, *118*, 401–402.

77 O. Tabata, R. Asahi, S. Sugiyama, Anisotropic etching of silicon with quaternary ammonium hydroxide solutions, in *Tech. Digest 9th Sensor Symp.;* **1990**, 15–18.

78 O. Tabata, R. Asahi, H. Funabashi, S. Sugiyama, Anisotropic etching of silicon in $(CH_3)_4NOH$ solutions, in *Tech. Digest 6th Int. Conf. Solid-State Sensors and Actuators (Transducers '91), San Francisco*; **1991**, 811–814.

79 O. Tabata, R. Asahi, H. Funabashi, K. Shimaoka, S. Sugiyama, Anisotropic etching of silicon in TMAH solutions, *Sens. Actuators A*, **1992**, *34*, 51–57.

80 O. Tabata, pH-controlled TMAH etchants for silicon micromachining, in *Proc. Transducers '95, Stockholm;* **1995**, 83–86.

81 W. Heywang (ed.), *Sensorik*, 4th edn.; Berlin: Springer-Verlag, **1993**.

82 R. Meek, Electrochemically thinned n/n^+ epitaxial silicon – method and applications, *J. Electrochem. Soc.* **1971**, *118*, 1240–1246.

83 H. Waggener, Electrochemically controlled thinning of silicon, *Bell Syst. Tech. J.* **1970**, *50*, 473–475.

84 T. Jackson, M. Tischler, K. Wise, An electrochemical p-n junction etch-stop for the formation of silicon microstructures, *IEEE Electron Device Lett.* **1981**, *EDL-2*, 44–45.

85 B. Rogge, D. Moser, H. Oppermann, O. Paul, H. Baltes, Solder-bonded micromachined capacitive pressure sensors, *Proc. SPIE* **1998**, *3514*, 307–315.

86 A. Chavan, K. Wise, A monolithic fully-integrated vacuum-sealed CMOS pressure sensor, *IEEE Trans. Electron Devices* **2002**, *49*, 164–169.

87 A. Chavan, *An integrated high resolution barometric pressure sensing system;* PhD Dissertation, University of Michigan, Ann Arbor, MI, **2000**.

88 L. Parameswaran, C. Hsu, M. Schmidt, A merged MEMS-CMOS process using silicon wafer bonding, in *Proc. IEEE IEDM '95;* **1995**, 613–616.

89 M. Biebl, T. Scheiter, C. Hierold, H. v. Philipsborn, H. Klose, Micromechanics compatible with an 0.8 μm CMOS process, *Sens. Actuators A* **1995**, *46/47*, 593–597.

90 M. Steger, G. Nebel, J. Sauerbrey, R. Noè, T. Scheiter, C. Hierold, A monolithic μP-ready ΣΔ readout-circuit with integrated capacitive pressure sensors, in *Proc. Int. Conf. Advanced Microsystems for Automotive Applications, Berlin;* **1996**.

91 S. Norsworthy, R. Schreier, G. Themes (eds.), *Delta–Sigma data converters: theory, design, and simulation;* New York: IEEE Press, **1997**.

92 R. Unbehauen, A. Cichocki, *MOS switched-capacitor and continous-time integrated circuits and system,* 2nd edn.; Berlin: Springer, **1989**.

93 J. Candy, G. Temes, Oversampling methods for A/D and D/A conversion, in *Oversampling Delta-Sigma Converters;* New York: IEEP Press, **1992**, 1–29.

94 E. Obermeier, Polysilicon layers lead to a new generation of pressure sensors, in *IEDM Tech. Digest, Transducers '85;* **1985**, 430–433.

95 K. Kasten, N. Kordas, H. Kappert, W. Mokwa, Capacitive pressure sensor with monolithically integrated CMOS readout circuit for high temperature applications, *Sens. Actuators A* **2002**, *97/98*, 83–87.

96 J. Chae, B. Stark, K. Najafi, A micromachined Pirani gauge with dual heat sinks, in *17th IEEE Int. Conf. on Micro Electro Mechanical Systems (MEMS 2004),* Maastricht, The Netherlands, **2004**, 532–535.

97 C. Mastrangelo, R. Muller, Fabrication and performance of a fully-integrated m-Pirani pressure gauge with digital readout, in *Digest Int. Conf. on Solid-State Sensors and Actuators (Transducers '91);* **1991**, 245–248.

98 E. Klaassen, G. Kovacs, Integrated thermal-conductivity vacuum sensor, *Sens. Actuators A,* **1997**, *58,* 37–42.

99 B. Stark et al., A doubly anchored surface micromachined Pirani gauge for vacuum package characterization, in *MEMS 2003;* **2003**, 506–509.

100 O. Paul, H. Baltes, Novel fully CMOS-compatible vacuum sensor, *Sens. Actuators A* **1995**, *46/47,* 143–146.

101 A. Häberli, O. Paul, P. Malcovati, M. Faccio, F. Maloberti, H. Baltes, CMOS integration of a thermal pressure sensor system, in *Proc. 1996 IEEE Int. Symp. on Circuits and Systems (ISCAS '96),* **1996**, 377–380.

102 H.-J. Kress, F. Bantien, J. Marek, M. Willmann, Silicon pressure sensor with integrated CMOS signal-conditioning circuit and compensation of temperature coefficient, *Sens Actuators A* **1991**, *25–27,* 21–26.

103 D. Arand, J. Marek, K. Weiblen, U. Lipphardt, Integrated barometric pressure sensor with SMD packaging: example of standardized sensor packaging, in *Sensors and Actuators 1996 (SP-1133),* Society of Automotive Engineers, Warrendale, PA (SAE Technical Paper Series 960756), **1996**, 23–27.

104 H.-J. Kress, J. Marek, M. Mast, O. Schatz, J. Muchow, Integrated silicon pressure sensor for automotive application with electronic trimming, in *Sensors and Actuators 1995 (SP-1066),* Society of Automotive Engineers, Warrendale, PA, **1995**, 35–40 (SAE Technical Paper Series 950533).

105 J. Marek, Microsystems for automotive applications, in *Proc. 13th European Conference on Solid-State Transducers (Eurosensors XIII), The Hague;* **1999**, 1–8.

106 F. Schnatz, U. Schöneberg, W. Brockherde, P. Kopystynski, T. Mehlhorn, E. Obermeier, H. Benzel, Smart CMOS capacitive pressure transducer with on-chip calibration capability, *Sens. Actuators A* **1992**, *34,* 77–83.

107 J. Weber, S. Seitz, U. Steger, B. Folkmer, U. Schaber, A. Plettner, H. Offereins, H. Sandmeier, E. Lindner, A monolithically integrated sensor system

using sensor specific CMOS cells, *Sens. Actuators A*, **1995**, *46/47*, 137–142.
108 M. KANDLER, J. EICHHOLZ, Y. MANOLI, W. MOKWA, CMOS compatible capacitive pressure sensor with read-out electronics, in *Micro System Technologies 90, 1st Int. Conf. on Micro, Electro, Opto, Mechanic Systems and Components, Berlin*; **1990**, 547–580.
109 H. DUDAICEVS, M. KANDLER, Y. MANOLI, W. MOKWA, E. SPIEGEL, Surface micromachined pressure sensors with integrated CMOS read-out electronics, *Sens. Actuators A* **1994**, *43*, 157–163.
110 H. DUAICEVS, Y. MANOLI, W. MOKWA, M. SCHMIDT, E. SPIEGEL, A fully integrated surface micromachined pressure sensor with low temperature dependence, in *Proc. 8th International Conf. on Solid-State Sensors and Actuators, and Eurosensors IX, Stockholm*; **1995**, 616–619.
111 D. WEILER, O. MACHUL, D. HAMMERSCHMIDT, J. AMELUNG, B. HOSTICKA, A single-chip smart pressure sensor family with 2 dimensional calibration, in *Advanced Microsystems for Automotive Applications 2000*, S. KRÜGER, W. GESSNER (eds.); Berlin: Springer, **2000**, 289–295.
112 H. TRIEU, M. KNIER, O. KÖSTER, H. KAPPERT, M. SCHMIDT, W. MOKWA, Monolithic integrated surface micromachined pressure sensors with analog on-chip linearization and temperature compensation, in *Proc. IEEE Int. Conf. on Micro Electro Mechanical Systems (MEMS 2000)*; **2000**, 547–550.
113 K. STANGEL, S. KOLNSBERG, D. HAMMERSCHMIDT, B. HOSTICKA, H. TRIEU, M. MOKWA, A programmable intraocular CMOS pressure sensor system implant, *IEEE J. Solid-State Circuits* **2001**, *36*, 1094–1100.
114 H. TRIEU, K. KASTEN, W. MOKWA, Development of surface micromachined pressure sensors for high temperature applications, in *Fraunhofer IMS Annual Report*; **1999**, 31–33.
115 K. KASTEN, J. AMELUNG, W. MOKWA, CMOS-compatible capacitive high temperature pressure sensors, *Sens. Actuators A* **2000**, *85*, 147–152.

116 A. YASUKAWA, S. SHIMADA, Y. MATSUOKA, Y. KANDA, Design considerations for silicon circular diaphragm pressure sensors, *Jpn. J. Appl. Phys.* **1982**, *21*, 1049–1052.
117 C. HIEROLD, Intelligent CMOS Sensors, in *Proc. IEEE Int. Conf. on Micro Electro Mechanical Systems (MEMS 2000)*; **2000**, 1–6.
118 H. KAPELS, R. AIGNER, C. KOLLE, Monolithic surface-micromachined sensor system for high pressure applications, in *11th Int. Conf. on Solid-State Sensors and Actuators (Transducers '01/Eurosensors XV)*; **2001**.
119 B. DIEM, P. REY, S. RENARD, S. V. BOSSON, H. BONO, F. MICHEL, M. DELAYE, G. DELAPIERRE, SOI SIMOX; from bulk to surface micromachining, a new age for silicon sensors and actuators, *Sens. Actuators A* **1995**, *46/47*, 8–16.
120 I. BASKETT, R. FRANK, E. RAMSLAND, The design of a monolithic, signal conditioned pressure sensor, in *Proc. IEEE Custom Integrated Circuits Conference*; **1991**, 1–4.
121 R. VERMA, I. BASKETT, M. SHAH, T. MAUDIE, D. MLADENOVIC, K. SOORIAKUMAR, A monolithic integrated solution for MAP applications, in *Proc. SAE International Congress & Exposition 1997, Detroit*; **1997**, 3–9, paper 970608.
122 X. DING, W. CZARNOCKI, J. SCHUSTER, B. ROECKNER, DSP-based CMOS monolithic pressure sensor for high volume manufacturing, in *10th Int. Conf. on Solid-State Sensors and Actuators (Transducers '99), Digest of Techn. Papers, Sendai, Japan*, Volume 1; **1999**, 362–365.
123 B. GOGOI, S. JO, R. AUGUST, A. MCNEIL, M. FUHRMANN, J. TORRES, T. MILLER, A. REODIQUE, M. SHAW, K. NEUMANN, D. H. D. MONK Jr., A 0.8 µm CMOS integrated surface micromachined capacitive pressure sensor with EEPROM trimming and digital output for a tire pressure monitoring system, in *Proc. 2002 Solid-State Sensor, Actuator and Microsystems Workshop, Hilton Head Island, SC*; **2002**, 181–184.
124 B. GOGOI, D. MLADENOVIC, Integration technology for MEMS automotive sensors, in *Proc. IEEE IECON '02*; **2002**, 2712–2717.

125 T. Ishihara, M. Hirata, K. Suzuki, H. Tanigawa, CMOS integrated silicon pressure sensor, in *Proc. IEEE Custom Integrated Circuits Conf.;* **1986**, 34–37.

126 K. Suzuki, T. Ishihara, M. Hirata, H. Tanigawa, Nonlinear analyses on CMOS integrated silicon pressure sensor, in *Techn. Digest Int. Electron Devices Meeting;* **1985**, 137–140.

127 W. Eaton, J. Smith, Characterization of a surface micromachined pressure sensor array, *Proc. SPIE* **1995**, *2642*, 256–264.

128 W. Eaton, J. Smith, Planar surface-micromachined pressure sensor with a subsurface, embedded reference pressure cavity, *Proc. SPIE* **1996**, *2882*, 259–265.

129 M. Mattes, J. Seefeldt, A one chip, polysilicon, surface micromachined pressure sensor with integrated CMOS signal conditioning electronics, in *Sensors and Actuators 1996 (SP-1133)*, Society of Automotive Engineers, Warrendale, PA, **1996**, 29–34 (SAE Technical Paper Series 960757).

130 E. Obermeier, P. Kopystynski, Polysilicon as a material for microsensor applications, *Sens. Actuators A* **1992**, *30*, 149–155.

131 E. Obermeier, S. Hein, V. Schlichting, D. Hammerschmidt, F. Schnatz, B. Hosticka, A smart pressure sensor with on-chip calibration and compensation capability, *Sensors* **1995**, *20–22*, 52–53.

132 Y. Matsumoto, S. Shoji, M. Esashi, A miniature integrated capacitive pressure sensor, in *Extended Abstracts 22nd International Conf. on Solid-State Devices and Materials, Sendai;* **1990**, 701–704.

133 Y. Matsumoto, S. Shoji, M. Esashi, A miniature integrated capacitive pressure sensor, in *Techn. Digest 9th Sensor Symp.;* **1990**, 43–46.

134 T. Kudoh, S. Shoji, M. Esashi, An integrated miniature capacitive pressure sensor, *Sens. Actuators A* **1991**, *29*, 185–193.

135 S. Yamashita, K. Shimaoka, H. Funabashi, S. Sugiyama, I. Igarashi, A fully integrated pressure sensor, in *Tech. Digest of the 8th Sensor Symposium;* **1989**, 13–16.

136 H. Terabe, Y. Fukaya, S. Sakurai, A. Tabata, S. Sugiyama, M. Esahi, Capacitive pressure sensor for low pressure measurements with high overpressure tolerance, in *Tech. Digest 10th Sensor Symp.;* **1991**, 133–136.

137 S. Sugiyama, K. Kawahata, M. Yoneda, I. Igarashi, Tactile image detection using a 1k-element silicon pressure sensor array, *Sens. Actuators A* **1990**, *21–23*, 397–400.

138 S. Sugiyama, K. Kawahata, H. Funabashi, M. Takigawa, I. Igarashi, A 32×32 (1k)-element silicon pressure-sensor array with CMOS processing circuits, *Electron. Commun. Jpn., Part 2*, **1992**, *75*, 64–76.

139 E. Yoon, K. Wise, An integrated mass flow sensor with on-chip CMOS interface circuitry, *IEEE Trans. Electron Devices* **1992**, *39*, 1376–1386.

140 S. Cho, K. Najafi, K. Wise, Internal stress compensation and scaling in ultrasensitive silicon pressure sensors, *IEEE Trans. Electron Devices* **1992**, *39*, 836–842.

141 K. Kato, Y. Muramatsu, H. Fujimoto, O. Sasaki, Totally integrated semiconductor pressure sensor, in *Tech. Digest of 10th Sensor Symposium;* **1991**, 129–132.

142 T. Itoh, T. Adachi, H. Hashimoto, One-chip integrated pressure sensor, in *Tech. Digest of 11th Sensor Symposium;* **1992**, 241–244.

143 R. Hearmon, *An introduction to applied anisotropic elasticity;* London: Oxford University Press, **1961**.

144 P. Chou, N. Pagano, *Elasticity – tensor, dyadic, and engineering approaches*, 2nd edn.; New York: Dover, **1992**; unabridged and corrected republication of the work first published in 1967, Van Nostrand.

145 S. Timoshenko, J. Goodier, *Theory of Elasticity*, 3rd edn.; Singapore: McGraw-Hill, **1970**.

146 R. W. Soutas-Little, *Elasticity*, Dover Publications, Mineola, New York, **1999**.

147 P. Podio-Guidugli, *A primer in elasticity;* Dordrecht: Kluwer, **2000**.

148 S. Timoshenko, S. Woinowsky-Krieger, *Theory of plates and shells*, 2nd edn.; Singapore: McGraw-Hill, **1959**.

149 H. Reismann, *Elastic Plates, Theory and Applications;* New York: Wiley, **1988**.

150 J. Reddy, *Theory and Analysis of Elastic Plates;* Philadelphia: Taylor & Francis, **1999**.

151 M. Elwenspoek, H. Jansen, *Silicon micromachining;* Cambridge: Cambridge University Press, **1998**.

152 P. French, P. Sarro, Surface versus bulk micromachining: the contest for suitable applications, *J. Micromech. Microeng.* **1998**, *8*, 45–53.
153 R. Howe, R. Muller, Polycrystalline silicon micromechanical beams, *J. Electrochem. Soc.* **1983**, *130*, 1420.
154 H. Guckel, D. Burns, Planar-processed polysilicon sealed cavities for pressure transducer arrays, in *IEDM Tech. Digest*; **1984**, 223–225.
155 H. Guckel, D. Burns, A technology for integrated transducers, in *Proc. 3rd Int. Conf. Solid State Sensors and Actuators (Transducers '85), Philadelphia, PA*; **1985**, 90–92.
156 M. Parameswaran, H. Baltes, A. Robinson, Polysilicon microbridge fabrication using standard CMOS technology, in *Tech. Dig. Solid State Sensor and Actuator Workshop (Cat. No. 88TH0215-4), Hilton Head Island, SC*; **1988**, 148–150.
157 M. Parameswaran, H. Baltes, L. Ristic, A. Dhaded, A. Robinson, A new approach for the fabrication of micromechanical structures, *Sens. Actuators* **1989**, *19*, 289–307.
158 H. Guckel, Surface micromachined pressure transducers, *Sens. Actuators A* **1991**, *28*, 133–146.
159 C. Linder, L. Paratte, M.-A. Gretillat, V. Jaecklin, N. de Rooij, Surface micromachining, *J. Micromech. Microeng.* **1992**, *2*, 122–132.
160 R. Howe, Polysilicon integrated microsystems: technologies and applications, in *Techn. Digest, 8th Int. Conf. Solid-State Sensors and Actuators (Transducers '95/Eurosensors IX), Stockholm*; Volume 1, **1995**, 43–46.
161 L. Parameswaran, C. Hsu, M. Schmidt, IC process compatibility of sealed cavity sensors, in *1997 Int. Conf. on Solid-State Sensors and Actuators (Transducers '97), Chicago*; **1997**, 625–628.
162 W. Sharpe, B. Yuan, R. Vaidyanathan, R. Edwards, Measurements of Young's modulus, Poisson's ratio, and tensile strength of polysilicon, in *IEEE Proc. Mems '97, Amsterdam*; **1997**, 424–429.
163 P.J. French, Polysilicon: a versatile material for microsystems, Sensors and Actuators A 2002, 99, 312.
164 E. Padgett, W. Wright, Silicon piezoresistive devices, in *Semiconductor and Conventional Strain Gages*, M. Dean, R. Douglas (eds.); New York: Academic Press, **1962**, 1–19.
165 W. Mason, J. Forst, L. Tornillo, Recent developments in semiconductor strain transducers, in *Semiconductor and Conventional Strain Gages*, M. Dean, R. Douglas (eds.); New York: Academic Press, **1962**, 109–120.
166 J. Greenwood, The constraints on the design and use of silicon-diaphragm pressure sensors, *Electron. Power* **1983**, 170–174.
167 J. Voorthuyzen, P. Bergveld, The influence of tensile forces and the deflection of circular diaphragms in pressure sensors, *Sens. Actuators* **1984**, *6*, 201–213.
168 W. Ko, Solid-state capacitive pressure transducers, *Sens. Actuators* **1986**, *10*, 303–320.
169 M. Poppingr, Silicon diaphragm pressure sensors, in *Solid State Devices 1985*, P. Balk, O. Folberth (eds.); Amsterdam: Elsevier Science, **1986**, 53–70.
170 H.-L. Chau, K. Wise, Scaling limits in batch-fabricated silicon pressure sensors, *IEEE Trans. Electron Devices* **1987**, *ED-34*, 850–858.
171 O. Jäntsch, M. Poppinger, Piezowiderstandseffekte, in *Sensorik*, W. Heywang (ed.); Berlin: Springer, **1993**, 95–118.
172 J. Sweet, Die stress measurement using piezoresistive stress sensors, in *Thermal Stress and Strain in Microelectronics Packaging*, J. Lau (ed.); New York: Van Nostrand Reinhold, **1993**, 221–271.
173 L. Ristic (ed.), *Sensor Technology and Devices*; Boston: Artech House, **1994**.
174 B. Kloeck, N. de Rooij, Mechanical sensors, in *Semiconductor sensors*, S. Sze (ed.); New York: Wiley, **1994**, 153–204.
175 K. Najafi, K. Wise, N. Najafi, Integrated sensors, in *Semiconductor sensors*, S. Sze (ed.); New York: Wiley, **1994**, 473–530.
176 H. Offereins, Aufbau piezoresistiver Niederdrucksensoren aus Silicium, 2nd edn.; Aachen: Shaker **1994**, PhD Dissertation.
177 T. Lisec, Entwurf, Herstellung und Charakterisierung piezoresistiver oberflächen-mikromechanischer Drucksensoren, 2nd edn.; Aachen: Shaker, **2001**, PhD Dissertation.

178 R. Frank, Pressure sensors merge micromachining and microelectronics, *Sens. Actuators A* **1991**, *28*, 93–103.

179 R. Frank, *Understanding Smart Sensors*; Artech House, Boston, London, **1996**.

180 T. Müller, M. Brandl, O. Brand, H. Baltes, An industrial CMOS process family adapted for the fabrication of smart silicon sensors, *Sensors and Actuators A* **2000**, *84*, 126–133.

181 K. A. Honer, G. T. A. Kovacs, Integration of sputtered silicon microstructures with pre-fabricated CMOS circuitry, *Sensors and Actuators A* **2001**, *91*, 386–397.

182 H. Baltes, O. Brand, CMOS-based microsensors and packaging, *Sensors and Actuators A* **2001**, *92*, 19.

7
CMOS-based Chemical Sensors

A. Hierlemann, Physical Electronics Laboratory, ETH Zurich, Zurich, Switzerland

Abstract

This chapter provides an extensive treatment of CMOS-technology-based chemical microsensors and -systems. First, the fundamentals of the chemical sensing process itself are laid out, followed by a short discussion of the issues associated with using CMOS technology for developing chemical microsensor systems and, in particular, monolithically integrated sensor systems. Thereafter, a comprehensive overview of CMOS-based transducer structures and the corresponding integrated microsensor systems is given along with typical application examples.

Keywords

CMOS; chemical sensor; microsensor; integrated sensor; sensor system

7.1	**Chemical Sensors**	**336**
7.1.1	Simple Ad/absorption	339
7.1.2	Chemical Reaction	340
7.1.3	Charge Transfer and Electrochemical Reaction	340
7.2	**Using CMOS Technology for Developing Chemical Sensors**	**340**
7.3	**CMOS-based Chemical Sensors**	**342**
7.3.1	Chemomechanical Sensors	342
7.3.1.1	Rayleigh SAW Devices	343
7.3.1.2	Micromachined Cantilevers	345
7.3.2	Thermal Sensors	348
7.3.2.1	Catalytic Thermal Sensors (Pellistors)	349
7.3.2.2	Thermoelectric or Seebeck-effect Sensors	350
7.3.3	Optical Sensors	353

7.3.3.1	Integrated Optics	354
7.3.3.2	Fabry–Perot Microspectrometers	356
7.3.4	Electrochemical Sensors	358
7.3.4.1	Voltammetric/Amperometric Sensors	359
7.3.4.2	Potentiometric Sensors (Chemotransistors)	361
7.3.4.3	Conductometric Sensors	365
7.3.5	CMOS Multisensor Systems	376
7.3.5.1	Four-cantilever Array	377
7.3.5.2	Multiparameter Biochemical Microsensor System	378
7.3.5.3	Multitransducer Gas Sensor Microsystem	379
7.4	**Acknowledgments**	***381***
7.5	**References**	***382***

7.1
Chemical Sensors

A definition of a chemical sensor is provided by a draft from IUPAC (International Union of Pure and Applied Chemistry) [1]: 'A chemical sensor is a device that transforms chemical information, ranging from the concentration of a specific sample component to total composition analysis, into an analytically useful signal'. This rather wide definition does not include the fact that a sensor usually is a continuously operating device and that the sensing process should be reversible, in particular since intermittently operating devices exhibiting irreversible characteristics are commonly referred to as dosimeters [1, 2].

A continuously operating chemical sensor can provide *qualitative* and/or *quantitative* information on specific target compounds [3–12]. Examples of *qualitative information* include the presence or absence of certain odorants, toxic, carcinogenic or hazardous compounds. Examples of *quantitative information* include concentrations, activities or partial pressures of such specific compounds exceeding, e.g. a certain threshold-limited value (TLV) or, in the case of combustible gases, the lower explosive limit (LEL).

Chemical sensors usually consist of a sensitive layer or coating and a *transducer* [3–6] (transducer is derived from the Latin *transducere*, which means to 'transfer' or 'translate'; a device that translates energy from one domain (e.g. chemical) to another (e.g. physical) is termed a transducer). Upon interaction with a chemical species (absorption, chemical reaction, charge transfer, etc.), the physicochemical properties of the coating, such as its mass, volume, optical properties or resistance, reversibly change (Fig. 7.1). These changes in the sensitive layer are detected by a suitable transducer and are translated into an electrical signal such as a frequency, current, or voltage, which is then read out and subjected to further data treatment and processing. This is exemplified for the mass-sensitive principle in Fig. 7.1: analyte molecules are absorbed into a coating material (polymer) to an extent governed by intermolecular forces. The change in mass of the polymeric coating in turn causes a shift in the resonance frequency of the transducer, e.g., a

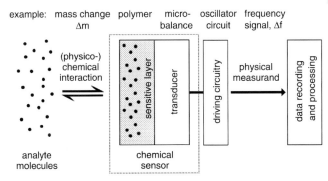

Fig. 7.1 Components of chemical sensors exemplified for the mass-sensitive principle

microbalance. This frequency shift constitutes the electrical signal to be used in subsequent data processing.

A variety of transducers based on different physical principles have been devised. Four principal categories of transducers can be distinguished following the suggestions of Janata [4, 5]:

1. chemomechanical sensors (e.g. mass changes due to bulk absorption);
2. thermal sensors (e.g. temperature changes through chemical interaction);
3. optical sensors (e.g. changes of light intensity by absorption);
4. electrochemical sensors (e.g. changes of potential or resistance through charge transfer).

The different CMOS-based chemical sensors and transducers that will be described in more detail in Section 7.3 will be organized into these four categories.

Various inorganic and organic materials serve as chemically sensitive layers that can be coated onto the different transducers [3–13]. Typical inorganic materials include metal oxides such as tin dioxide (SnO_2) for monitoring gases such as hydrogen or carbon monoxide or zirconium dioxide (ZrO_2) to detect oxygen, nitrogen oxide and ammonia. Organic layers mostly consisting of polymers such as polysiloxanes or polyurethanes are used to monitor hydrocarbons, halogenated compounds and different kinds of toxic volatile organics. A survey of typical chemically sensitive materials and their applications is given in Tab. 7.1. Further information on the coating materials will be provided in the context of the different transducers in Section 7.3.

The interaction of a chemical species with a chemical sensor either can be confined to the surface of the sensing layer or it can take place in the whole volume of the sensitive coating. Surface interaction implies that the species of interest is *adsorbed* at the surface or interface (gas/solid or liquid/solid) only, whereas volume interaction requires the *absorption* of the species and a *partitioning* between the sample phase and the bulk of the sensitive material. The different types of chemical interactions involved in a sensing process range from very weak *physisorption* (0–20 kJ/mol) through fairly strong *chemisorption* to charge transfer and chemical reactions (120–800 kJ/mol) [3–8].

Tab. 7.1 Typical sensor materials and applications

Materials	Examples	Example applications
Metals	Pt, Pd, Ni, Ag, Sb, Rh, ...	Inorganic gases such as CH_4, H_2
Ionic compounds	SnO_2, In_2O_3, $AlVO_4$, $SrTiO_3$, Ga_2O_3, ZrO_2, LaF_3, Nasicon	Inorganic gases (CO, NO_x, CH_4), exhaust gases, oxygen, ions in water
Molecular crystals	Phthalocyanines (PCs), PbPC, $LuPC_2$	Nitrogen dioxide, volatile organics
Langmuir–Blodgett films	Lipid bilayers, polydiacetylene	Organic molecules in medical applications, biosensing
Cage compounds	Zeolites, calixarenes, cyclodextrins, crown ethers	Water analysis (ions), volatile organics
Polymers	Polyurethanes, polysiloxanes, polypyrroles, polythiophenes, Nafion	Detection of volatile organics, food industry (odor and aroma), environmental monitoring in gas and liquid phase
Biological entities	Phospholipids, lipids, enzymes, receptors, proteins, cells, membranes	Medical applications, biosensing, water and blood analysis, pharma screening

High chemical selectivity and rapid reversibility place contradictory constraints on desired interactions between chemical sensor coating materials and analytes [3–6]. Low-energy, perfectly reversible (physisorptive) interactions generally lack high selectivity, whereas chemisorptive processes, the strongest of which result in the formation of new chemical bonds, offer selectivity, but are inherently less reversible. A practicable compromise has to be achieved with due regard to the specific application.

Any interaction between a coating material and an analyte is governed by chemical *thermodynamics* and *kinetics* [3–6, 14]. Thermodynamics tell us the direction of a spontaneous reaction and the composition at an equilibrium state, whereas kinetics tell us whether a kinetically viable pathway exists for that change to occur and how fast an equilibrium state will be achieved [14].

Therefore, a fundamental thermodynamic function, the Gibbs free energy, G [J], is the most important descriptor in all chemical-sensing processes: The direction of spontaneous chemical reactions is always towards lower values of G (minimization of the Gibbs energy). The interaction of an analyte, A, with a sensor coating, S, can be represented by

$$A + S \Leftrightarrow A \cdots S \tag{1}$$

Such an interaction can be described by an equilibrium constant, K, which relates the activity, a (activity denotes the effective quantity of a chemical compound that participates in a chemical reaction) of reaction products ($A \cdots S$) to those of the reactants (A and S). This constant is thus a characteristic value for the progression of the reaction ($K \leq 1$: no reaction takes place):

$$K = \frac{a_{A...S}}{a_A \cdot a_S} \quad \text{and, in general:} \quad K = \prod_i a_i^{n_i} \tag{2}$$

The index i denotes the chemical substance and n_i are the corresponding stoichiometric numbers in the chemical equation. This expression signifies that each analyte activity is raised to the power equal to its stoichiometric number and then all such terms are multiplied together. Stoichiometric numbers of the products are positive and those of the reactants are negative, i.e., reactants appear as the denominator and reaction products as the numerator.

In a thermodynamic equilibrium state ($\Delta G = 0$), the change in Gibbs energy, ΔG^0, and K are interrelated via the following equation (for details see, e.g., [14]):

$$\ln K = -\frac{\Delta G^0}{RT} \tag{3}$$

The more negative ΔG^0, the larger is K or, in other words, the larger K, the more spontaneous the reaction will occur in the case that kinetic factors will not upset such predictions. According to the Gibbs fundamental equation, ΔG^0 is composed of an enthalpy term, ΔH^0, representing the reaction heat at constant pressure, and an entropy term, ΔS^0, representing the degree of 'disorder' or, thermodynamically more precise, the number of different ways in which the energy of a system can be achieved by rearranging the atoms or molecules among the states available to them (for details, see [14]):

$$\Delta G^0 = \Delta H^0 - T \Delta S^0 \tag{4}$$

For spontaneous reactions (ΔG^0 negative), the entropy increases and/or the enthalpy term is negative, i.e., heat is released during the chemical reaction.

In the following, the thermodynamics of three prototype reactions of chemical sensors will be briefly discussed.

7.1.1
Simple Ad/absorption

A so-called partition coefficient, K_{sorption}, is used to describe the activities of free species and ad/absorbed species at thermodynamic equilibrium state:

$$K_{\text{sorption}} = \frac{a_A^{\text{sorbed}}}{a_A^{\text{free}}} \tag{5}$$

The partition coefficient is a dimensionless 'enrichment factor' that relates the activity of a compound in the sensing layer, a_A^{sorbed}, to that in the probed gas or liquid phase, a_A^{free}, and it represents a thermodynamic equilibrium constant, which is related to ΔG^0 via Eq. (3).

7.1.2
Chemical Reaction

For a simple reaction such as $n_A A + n_B B \rightleftharpoons n_C C + n_D D$, the equilibrium constant is given according to Eq. (2):

$$K = \prod_i a_i^{n_i} \text{ and in particular } K = \frac{a_C^{n_C} \cdot a_D^{n_D}}{a_A^{n_A} \cdot a_B^{n_B}} \tag{6}$$

The equilibrium constant is also related to ΔG^0 via Eq. (3). Note that interactions involving a true chemical reaction may be too strong to be reversible.

7.1.3
Charge Transfer and Electrochemical Reaction

For a reaction of type $A^+ + e^- \rightleftharpoons A$, the electrical work that is necessary to transfer the respective charges also has to be taken into account. The Nernst equation results [14]:

$$E = E_0 - \frac{RT}{zF} \ln K \tag{7}$$

where E denotes the 'electromotive force' (a spontaneous reaction with a negative ΔG according to convention always produces a positive E), E^0 is the standard cell potential, R is the molar gas constant (8.314 J/K mol) and T is the temperature in kelvin. The number of elementary charges that are transferred is denoted z and F is the Faraday constant (96485 C/mol), which is equivalent to one mole of elementary charges.

The Nernst equation now can be used to calculate the potential of any electrochemical cell or, in our case, electrochemical sensor. Electrochemical reactions can be triggered by applying currents or voltages via electrodes to the sensing layer [3–6, 15].

7.2
Using CMOS Technology for Developing Chemical Sensors

An important recent trend in developing chemical microsensors includes the miniaturization of the transducers and, increasingly, the integration of the transducers with electronics on the same chip [16–23]. Semiconductor and, in particular, CMOS technology provide excellent means to effectively realize device miniaturization and integration [16–28]. Additional benefits of using CMOS technology include favorable signal-to-noise characteristics (on-chip signal treatment), low power consumption, rapid sensor response characteristics (small sensor dimensions), the realization of smart features on the sensor chip and batch fabrication at industrial standards [23, 27, 28].

By using micromachining techniques and MEMS (microelectro-mechanical systems) technology, a variety of micromechanical structures such as membranes, cantilevers or bridges can be realized on CMOS substrates (see Chapter 1). All these micromechanical structures can be used in developing chemical microsensors and can be combined with the respective circuitry units such as driving and readout electronics on the same CMOS substrate [21–23, 27–31]. A modular approach or 'toolbox strategy' can be adopted that relies on CMOS as platform technology. The components of the toolbox include the transducers, the sensor and circuitry modules, a subset of which can be selected according to the application requirements and, subsequently, can be assembled into customized systems. Candidate components for integrated chemical sensor systems can be selected from a wealth of electronic circuits and functions that are provided by the enduring huge development efforts of the integrated-circuit (IC) industry.

A number of issues are related to realizing chemical sensor systems in CMOS-MEMS technology, the most important of which will be briefly addressed below [23].

The materials that are available for transducer realization are restricted to CMOS materials: silicon, polysilicon, silicon oxide, silicon nitride, aluminum (see also Chapter 1) and CMOS-compatible materials. There is also a limitation on available fabrication processes owing to the limited number of pre- and post-CMOS micromachining options. High-temperature steps (e.g. $>400\,°C$) are detrimental to aluminum metallization (metal oxidation, diffusion) and alter the transistor characteristics (the last high-temperature step of the CMOS process is at $\sim 400\,°C$). This aspect is of particular importance in the context of, e.g., microhotplate chemoresistors (see Section 7.3.4.3, subsection High-temperature Chemoresistors), the ceramic sensitive layer of which has to undergo a high-temperature annealing procedure.

The microsensor/system realization relies predominantly on established industrial CMOS technology to fabricate circuitry and basic sensor/transducer structures. Only little additional fabrication equipment and related technology development are needed for the sensor-specific post-processing steps.

Most chemical microsensors perform pronouncedly better in monolithic CMOS designs because the influence of parasitic capacitances and crosstalk effects can be reduced by on-chip electronics (filters, amplifiers, etc.). On-chip analog-to-digital conversion is another feature that helps to generate a stable sensor output, which can then be transferred to off-chip recording devices without losses in signal quality. The co-integration of circuitry, therefore, helps to counteract the signal reduction that is a consequence of device and transducer miniaturization.

Temperature or flow sensors can be co-integrated with the chemical sensors, which further improves sensor reliability and signal quality. Calibration, control and signal processing functions and self-test features can be also realized on the same chip.

The number of electrical connections from the chip to off-chip equipment or recording units, each of which is a potential source of device failure, can be drastically reduced by using an on-chip digital bus interface that is available in CMOS technology.

The packaging of CMOS chemical microsystems can be done using packaging methods of microelectronics. IC-based packaging techniques such as flip-chip technology or simple epoxy-based packaging methods can be modified and adapted. Here, it is important to note that the sensor package has to be defined during the initial chemical microsensor or -system conception phase, since the electronics and transducer layout of the microsystem and the component arrangement on the chip are strongly interrelated with the selected packaging strategy.

In summary, the use of CMOS technology entails a limited selection of device materials (see Chapter 1) and a predefined fabrication process for the CMOS part of the chemical microsensors. Sensor-specific or transducer-specific materials and fabrication steps have to be introduced, in most cases, as post-processing after the CMOS fabrication (holds for most of the devices presented in Section 7.3). CMOS-MEMS, on the other hand, offers unprecedented benefits in particular with regard to sensor signal quality, device performance, increased device functionality through capitalizing on the 'toolbox' approach and the availability of standard packaging solutions.

A comprehensive overview of CMOS-based and CMOS-MEMS-based chemical microsensors and -systems is given in Section 7.3.

7.3
CMOS-based Chemical Sensors

As already mentioned in Section 7.1, chemical sensors can be classified into four principal categories according to their transduction principles [4, 5]: (1) chemomechanical sensors, (2) thermal sensors, (3) optical sensors and (4) electrochemical sensors. Each of those four sensor categories will be briefly introduced, then, specific exemplary CMOS-based microtransducers and -systems will be abstracted.

7.3.1
Chemomechanical Sensors

The change in mechanical properties (e.g. mass) of a sensitive layer upon interaction with an analyte can be conveniently recorded by using micromechanical structures. Any species that can be immobilized on the sensor can, in principle, be sensed. In the simplest case such chemomechanical sensors are gravimetric sensors responding to the mass of species accumulated in a sensing layer [32–34]. Some of the sensor devices additionally respond to changes in a variety of other mechanical properties such as polymer moduli, liquid density and viscosity [32–34], which will not be further discussed here.

Gravimetric sensors provide good chemical sensitivity: mass changes in the picogram range can be detected and ppm (parts per million) to ppb (parts per billion) detection levels have been reported for gas sensors [32–34].

Most of the gravimetric sensors rely on piezoelectric materials such as quartz, aluminum nitride and zinc oxide. Using an alternating current (AC), the structures can be electrically excited into a fundamental mechanical resonance mode. The resonance frequency changes in proportion to the mass loading on the crystal or device [35]. The more mass (analyte molecules) is absorbed, e.g., into a polymer coating, the lower is the resonance frequency of the device. The following equation describes the relationship between analyte gas phase concentration changes, Δc_A [mol/L] and the invoked sensor responses, Δf_A [Hz]:

$$\Delta f_A = \Gamma \cdot M_A \cdot K \cdot \Delta c_A \tag{8}$$

where K denotes the partition coefficient (Eq. (5)), M_A [kg/mol] is the molar mass of the analyte vapor and Γ is a gravimetric constant [L/kg s] including, e.g., the frequency shift upon initial deposition of the sensitive layer, the coating density and transducer dimensions.

At low analyte concentrations (trace level), a linear correlation between the frequency shift upon analyte absorption and the corresponding analyte concentration change in the gas phase is usually observed (see Fig. 7.6), provided that the sensing film on the transducer moves synchronously with the oscillating surface [36].

The most common gravimetric devices include the *thickness-shear-mode resonator* (TSMR) or quartz microbalance (QMB) [35, 37] and the *Rayleigh surface-acoustic-wave* (SAW) device [38], both normally relying on quartz substrates. Since the TSMR is not CMOS-based and not compatible with IC technology, it will not be further detailed here.

The realization of piezoelectric transducers requires an additional layer to be patterned on the CMOS substrate since none of the CMOS materials is piezoelectric. Different materials such as aluminum nitride [39, 40] and zinc oxide (ZnO) [41–45] can be used.

In the following, two CMOS technology-compatible types of mass-sensitive devices will be described in more detail: (1) Rayleigh SAW devices on CMOS substrates with piezoelectric overlay and (2) micromachined cantilevers. CMOS-compatible flexural plate wave devices (FPWs) are treated elsewhere [28, 32, 34].

7.3.1.1 Rayleigh SAW Devices

Interdigital transducers can be used to launch and detect a surface acoustic wave on a piezoelectric substrate [38] as shown schematically in Fig. 7.2. By applying an AC voltage to a set of interdigital transducers patterned on a piezoelectric substrate with appropriate orientation of the crystal axes, one set of the fingers moves downwards and the other upwards, thereby creating an oscillating mechanical surface deformation. This surface deformation generates an acoustic wave, which propagates along the surface and is converted back into an electrical signal by de-

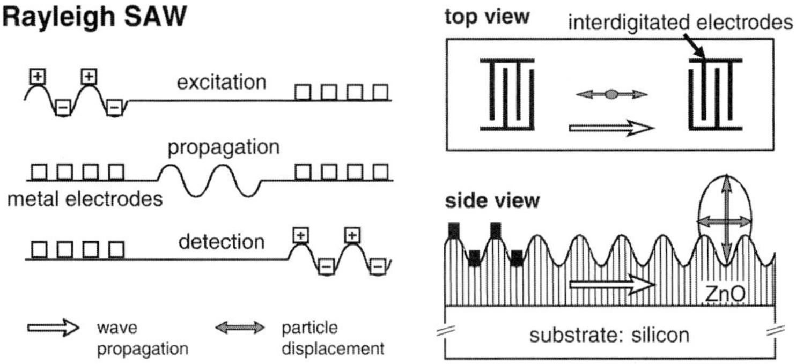

Fig. 7.2 Launching, propagation and detection of a Rayleigh-type surface acoustic wave by interdigitated transducers. The top view shows the electrode configuration and the wave propagation. The side view shows the elliptical particle displacement

forming the surface in the region of the receiving transducer. The electrical signal of the receiving transducer is recorded and represents the sensor signal.

For a given piezoelectric substrate, the acoustic wavelength and, therefore, the operating frequency of the SAW device is determined by the transducer periodicity, which is equal to the acoustic wavelength at the transducer center frequency. Typical frequencies range between 100 and 500 MHz [32–34]. Such frequencies require a sophisticated high-frequency circuit design. Therefore, a bare reference oscillator is operated together with the sensor in most cases, and the outputs are mixed to produce a difference frequency with values in the kHz range that is recorded [41, 43, 45].

The acoustic wave is confined to a surface region of approximately one acoustic wavelength thickness. The velocity and damping characteristics of the acoustic wave are hence extremely sensitive to changes at the transducer surface, such as mass loading. When used in an oscillator circuit, relative changes in the wave velocity are reflected as equivalent changes in fractional oscillation frequency. A change in mass due to, e.g., absorption of a gaseous analyte in a polymeric sensing layer thus changes the device frequency according to Eq. (8).

The acoustic (Rayleigh) wave causes an elliptical particle movement at the transducer surface (Fig. 7.2), i.e., the sensitive films deposited on top of the transducers and the piezoelectric substrate are severely deformed. Therefore, changes in the viscoelastic properties of the sensing layer can affect the sensor response [36].

The post-CMOS fabrication steps include the deposition of zinc oxide (sputtering techniques at elevated temperatures [41, 45]), the processing of the metal electrodes on top of the piezoelectric layer (aluminum, gold) and the deposition of a sensitive layer such as a polymer.

Rayleigh SAW devices cannot be used in the liquid phase, since surface-normal particle displacements occur (Fig. 7.2) so that the device radiates compressional waves into the liquid [32, 34].

Typical SAW device applications include the detection of different kinds of organic volatiles (hydrocarbons, chlorinated hydrocarbons, alcohols, etc.) by using polymeric layers [32–34, 41] or porphyrins [46] and the detection of nitrogen dioxide using phthalocyanines [43]. The interaction mechanisms involve, in most cases, fully reversible physisorption and bulk/gas-phase partitioning (see Eq. (5)).

7.3.1.2 Micromachined Cantilevers

Cantilevers are widely used in atomic force microscopy (AFM) [47, 48] and can also be used as chemical sensors [49–67]. The CMOS-based cantilever is a layered structure composed of, e.g., the dielectric layers of a standard CMOS process, silicon and the metallizations [62–67]. The cantilever base is firmly attached to the silicon support. The freestanding cantilever end is coated with a sensitive layer (Fig. 7.3).

There are two different ways to operate a cantilever as chemical sensor: (1) the *static mode*, in which the cantilever deflection upon stress changes in the sensitive layer is measured [59–61], and (2) the *dynamic mode*, in which the cantilever is excited in its fundamental mechanical resonance, and the resonance frequency change upon mass loading in the sensitive layer is recorded (Eq. (8)) [57, 58, 62–67]. These two modes impose completely different constraints on the cantilever design: (1) requires long and deformable cantilevers to achieve large deflections or large stress, whereas (2) requires short and stiff cantilevers to achieve high operation frequencies. The dynamic mode is preferable with regard to integration of electronics and simplicity of the setup (feedback loop) [50, 54, 57, 58, 62–67], whereas the static mode is easier to apply to liquid-phase measurements [48, 57].

Several cantilever excitation methods have been reported. They include electrostatic actuation using finger electrodes [68], piezoelectric actuation using materials such as zinc oxide [50], thermal actuation making use of the bimorph effect [62–64] and magnetic actuation relying on Lorentz forces [65–67]. The last two will be described here in more detail.

Thermal actuation relies on the different thermal expansion coefficients of the various layer materials forming the cantilever. This difference in material properties gives rise to a cantilever deflection upon heating. Periodic heating pulses in the cantilever base can therefore be used to excite the cantilever thermally in its resonance mode at 10–500 kHz [50, 62–64]. The area of periodic temperature variation is closely confined to the region around the heaters owing to the high excita-

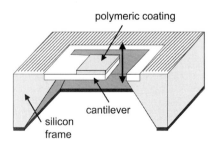

Fig. 7.3 Schematic representation of a resonating cantilever

tion frequency of ∼400 kHz (see Fig. 7.4a). At frequencies higher than the thermal actuation corner frequency of ∼1 kHz (speed of the thermal equilibration processes in the cantilever), the efficiency of the thermal actuation drops significantly (reduced cantilever oscillation amplitude and phase lag). However, the bending moment created is still sufficient to cause harmonic transverse vibrations of the cantilever with an amplitude of a few nanometers [64].

The magnetic actuation relies on Lorentz forces and requires an external magnetic field, which can be conveniently generated by including a small permanent magnet in the sensor package below the cantilever [64–66]. The resulting magnetic-field vector is in-plane and parallel to the cantilever (Fig. 7.4b).

By applying an AC current to the current loops that are patterned along the edges, cantilever oscillation is evoked in the external magnetic field by the Lorentz force. The direction of the Lorentz force is perpendicular to the cantilever (Fig. 7.4b) so that it effectuates a transverse cantilever movement.

The detection of the cantilever deformations or frequency changes can be done by embedding piezoresistors or stress-sensitive transistors in the cantilever base [57, 58, 62–67], by measuring motional capacitance changes [51], or by using optical detection by means of laser light reflection on the cantilever [52–55, 59–61].

The cantilever fabrication process includes bulk micromachining of the CMOS substrate (KOH or EDP) to achieve a membrane structure [50–67], subsequent release of the cantilevers by front-side reactive-ion etching [62–64] and the deposition of a sensitive layer by spray or drop coating. Detailed fabrication sequences for monolithic integration of the cantilevers with electronics have been described [62–67].

The 150-μm-long and 100-μm-wide cantilevers exhibit a quality factor (ratio of resonance frequency and resonator bandwidth) of ∼1000 in air at a resonance frequency of 380 kHz [64, 67] and are co-integrated with the necessary oscillator circuit, in which they constitute the frequency-determining element (Fig. 7.5). The nature of the oscillator circuit depends on the actuation principle and will be specified only for thermal actuation here [69] (for magnetic actuation, see [65–67]). The circuit is entirely integrated on the chip along with a counter to measure the

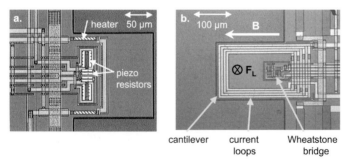

Fig. 7.4 Cantilever actuation schemes: (a) thermal actuation by using heating resistors at the cantilever base [64] and (b) magnetic actuation through Lorentz forces generated by an external magnetic field [66]

oscillation frequency. The first stage of the feedback circuit includes a low-noise, fully differential difference amplifier (DDA) with a gain of 30 dB, which amplifies the output signal of the Wheatstone bridge. The feedback additionally includes a high-pass filter followed by another amplifier, a limiter, a programmable digital delay line and a source follower to drive the heating resistor. The delay line is used to adjust the phase to achieve positive feedback at the fundamental resonance frequency. The integration of the feedback loop on chip (Fig. 7.5) greatly improves the signal-to-noise characteristics of the sensor (better frequency stability and lower electrical and thermal crosstalk). For more details on the electronics, see [69].

The mass resolution of the cantilevers is in the range of a few picograms [52–54, 59–67]. This high mass sensitivity does not necessarily imply an exceptionally high sensitivity to analytes since the area coated with the sensitive layer usually is very small (100×150 μm^2) [64]. The gas sensitivity of the cantilevers was assessed to be in the same range as that of TSMRs or Rayleigh SAW devices [64].

A typical signal of a polymer-coated (poly(ether urethane); PEUT) resonant cantilever is displayed in Fig. 7.6, showing the frequency shifts upon exposure to various concentrations of n-octane. To a first approximation, the changes in the resonance frequency of the cantilever are linearly proportional to the changes in the analyte gas concentration (see also Eq. (8)).

The sensing layer is deformed upon motion of the cantilever; therefore, modulus effects are expected to contribute to the overall signal, especially since the coating thickness may exceed the thickness of the cantilever.

Typical cantilever applications are environmental monitoring (gas and liquid phase) or biosensing in liquids. These include the detection of different kinds of organic volatiles (hydrocarbons, chlorinated hydrocarbons, alcohols, etc.; see

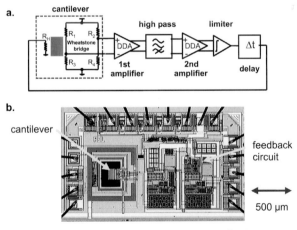

Fig. 7.5 (a) Schematic of the on-chip oscillator feedback circuitry and (b) micrograph of an integrated cantilever system including transducer (150 μm long, thermally actuated) and circuitry [64, 69]

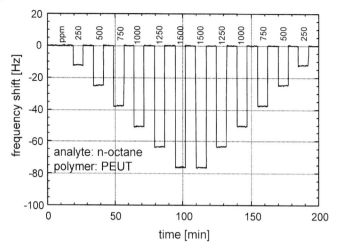

Fig. 7.6 Typical response of mass-sensitive sensors. Frequency shifts of a polymer-coated (poly(ether urethane), PEUT) cantilever upon exposure to different concentrations of an organic volatile: n-octane

Fig. 7.6) or humidity in the gas phase by using polymeric layers [53, 54, 59–67], the detection of alcohols in water [57], and the hybridization and detection of complementary strands of oligonucleotides [49].

The interaction mechanisms involve reversible physisorption and bulk/gas-phase partitioning (see Eq. (5)), in addition to receptor–ligand binding [49].

7.3.2
Thermal Sensors

Calorimetric or thermal sensors rely on the determination of the presence or concentration of a chemical species by measurement of an enthalpy change produced by the chemical to be detected [3–5, 15, 70]. Any chemical reaction (Eq. (6)) or physisorption process (Eq. (5)) releases or absorbs from its surroundings a certain quantity of heat (enthalpy term, ΔH^0, in Eq. (4)). Reactions liberating heat are termed *exothermic* and reactions abstracting heat are termed *endothermic*. Changes in the thermal budget exhibit transient behavior: Heat liberation or abstraction only occur as long as the reaction proceeds, or analyte concentrations change. There will be, however, no heat production and, hence, no measurable signal at thermodynamic equilibrium ($\Delta G = 0$) in contrast to gravimetric, optical or electrochemical sensors.

The design of an efficient thermal sensor is difficult since, on the one hand, the sensitive area must be accessible to the chemical species to be detected, albeit, on the other hand, it should be thermally as isolated as possible so that the invoked heat budget changes are detectable.

The liberation or abstraction of heat is conveniently measured as a temperature change that can be easily transduced into an electrical signal. The various types of calorimetric sensors differ in the way, in which the evolved heat is transduced.

7.3 CMOS-based Chemical Sensors

The *catalytic sensor* (often denoted 'pellistor') employs platinum resistance thermometry [71–83], whereas the thermoelectric sensor relies on the Seebeck effect [84–97]. Micromachined cantilevers can also be used as thermal sensors due to the bimorph effect [47] described in Section 7.3.1.2.

In the following, thermoelectric and catalytic calorimetric sensors will be detailed.

7.3.2.1 Catalytic Thermal Sensors (Pellistors)

A catalytic thermal sensor measures the heat evolved during the controlled combustion of flammable gaseous compounds in ambient air on the surface of a hot catalyst by means of a resistance thermometer in proximity with the catalyst. The *catalyst* is a chemical compound (often a noble metal such as platinum) enabling or accelerating a chemical reaction by provision of alternative reaction paths involving intermediates with lower activation energies than the uncatalyzed mechanism. The catalyst itself is not permanently altered by the reaction.

The heated catalyst here permits oxidation of the gas at reduced temperature and at concentrations below the *lower explosive limit* (LEL). Three elements are needed for a catalytic thermal sensor: a catalyst, a method to heat it, and a means to measure the heat of catalytic oxidation. The term 'pellistor' originally refers to a device consisting of a small platinum coil embedded in a ceramic bead or pellet (large surface area) impregnated with a noble-metal catalyst [71].

Fig. 7.7 shows a freestanding, Pt-coated polysilicon microfilament (10 μm wide, 2 μm thick) separated from the substrate by a 2 μm air gap [73, 74] to minimize heat losses to the silicon substrate. By passing an electric current through the meander, the microbridge is heated to a temperature sufficient for the Pt surface to oxidize the combustible mixture catalytically; the heat of oxidation is then measured as a resistance variation in the Pt. The combustion of methane, for example, generates 800 kJ/mol heat, which translates into a corresponding temperature change.

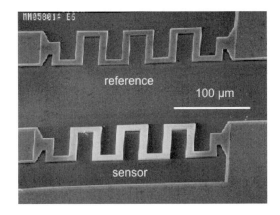

Fig. 7.7 Micrograph (SEM) of two meandered polysilicon microbridges. The lower meandered bridge is coated with a thin (∼0.1 μm) layer of platinum (CVD). In a differential gas-sensing mode, the upper uncoated filament acts to compensate changes in the ambient temperature, thermal conductivity and flow rate, while the lower filament is used to detect calorimetrically combustible gases. Reprinted from [74] with permission

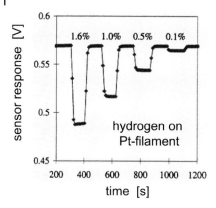

Fig. 7.8 Sensor response of a Pt-coated filament exposed to various concentrations of hydrogen in synthetic air. Reprinted from [73] with permission

The temperature change of the sensor element is proportional to the combustible concentration, when the device is operated in excess oxygen and in the mass transfer-limited regime [75, 76]. The combustion of hydrogen in dry air is exemplified in Fig. 7.8 [73, 74]. The circuit maintains a constant sensor temperature by adjusting the supplied current to keep the filament resistance at a constant value. Note that the sensor response is measured at steady state, i.e. continuous combustion. Micromachined membranes constitute an alternative to microbridges [72, 77, 79].

The fabrication steps include either surface micromachining (sacrificial-layer etching with hydrogen fluoride) for the bridges [73, 74] or bulk micromachining (KOH or EDP) to achieve membrane structures [72, 77, 79] and subsequent catalyst deposition by sputtering [77, 78] or LPCVD [73, 74]. Detailed processing sequences for the microbridges are given in [82, 83].

The main applications include the monitoring and detection of flammable gas hazards in industrial, commercial and domestic environments. Target gases include methane [75, 79, 83], hydrogen [72–76, 79], propane [72], carbon monoxide [72, 80] and organic volatiles [77, 78, 81]. The interaction process is an irreversible chemical combustion reaction at high temperature liberating the respective reaction enthalpy (Eq. (6)).

7.3.2.2 Thermoelectric or Seebeck-effect Sensors

When two different semiconductors or metals are connected at a hot junction, and a temperature difference is maintained between this hot junction and a colder point, then an open-circuit voltage is developed between the different leads at the cold point. This thermovoltage is proportional to the temperature difference itself [84]. This thermoelectric or Seebeck effect can be used to develop a thermal sensor by placing the hot junction on a thermally isolated structure such as a membrane or bridge and the cold part on the bulk chip with the thermally well-conducting silicon underneath [85–89]. To achieve large thermoelectric signals, several thermocouples are connected in series to form a thermopile. The membrane structure (hot junctions) is covered with a sensitive or chemically active layer liberating or abstracting heat upon interaction with an analyte.

Fig. 7.9a displays the cross-section of a CMOS-based thermoelectric gas sensor [89]. The sensor system relies on polysilicon/aluminum thermocouples exhibiting a Seebeck coefficient of 111 µV/K. The hot junctions are in the center of the membrane and the cold junctions on the bulk wafer material. The center part (hot junctions) of the membrane is coated with a gas-sensitive layer such as a polymer. Many thermocouples must be connected in series to achieve the desired sensitivity, i.e., to be able to measure temperature differences in the millikelvin range (Fig. 7.9b) [89].

The overall CMOS calorimetric sensor system (Fig. 7.10) includes two dielectric membranes with 300 polysilicon/aluminum thermocouples each and an on-chip amplifier. One of the membranes is coated with a gas-sensitive coating; the other one is passivated and serves as a reference [69, 87–89]. Sensor and reference are connected to the input stage of an on-chip amplifier for monitoring the temperature differences between the two membranes (Fig. 7.10). The low-noise chopper-stabilized instrumentation amplifier features a tunable gain of up to 8000 and a bandwidth of 500 Hz with an equivalent input noise of 15 nV/Hz$^{1/2}$ [69]. The anti-aliasing filter prevents downsampling of high-frequency noise into the low-frequency signal band by the A/D converter. After Sigma–Delta A/D conversion and after passing a decimation filter (13-bit word length at 800/s), the data are read out or transferred to a digital interface [69]. For more details on the circuitry, see [69].

The calorimetric detection process includes four principal steps [63]: (1) absorption and partitioning or chemical reaction, (2) generation of heat, which produces (3) temperature changes to be transformed in (4) thermovoltage changes. The final signal results from this sequence of processes, some of which are of chemical or physicochemical nature, i.e., depend on the chemical compounds involved (1, 2), and some of which are of physical nature and are device-specific (3, 4: heat sensitivity and thermovoltage generation).

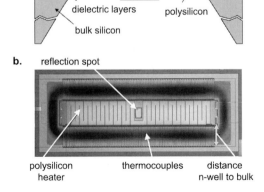

Fig. 7.9 (a) Cross-section of a CMOS calorimetric sensor showing the thermocouples (aluminum, polysilicon) and the n-well island in the center [89]. (b) Rectangular membrane (300 thermocouples) featuring a reflection spot for optical layer thickness detection and a polysilicon heater. The thermocouple length (distance from n-well to bulk) is 200 µm [89]

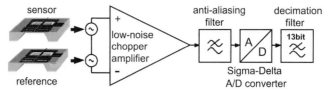

Fig. 7.10 Schematic of the calorimeter microsystem. Sensing and reference thermopiles are connected to the input stage of a low-noise chopper-stabilized instrumentation amplifier followed by an anti-aliasing filter, a Sigma–Delta A/D converter and a decimation filter [69]

The calorimetric sensor only detects changes in the heat budget at the non-equilibrium state (transients). Therefore, the sensor provides a signal, e.g. upon absorption (condensation heat) and desorption (vaporization heat) of gaseous analytes into a polymer [86–91] or during chemical reaction of an analyte with the sensing material [92–94]. The recorded thermovoltage change, ΔU, [V] is, therefore, proportional to the derivative of the analyte concentration as a function of time, dc_A/dt [mol/m³ s]:

$$\Delta U = A \cdot B \cdot V_{\text{sens}} \cdot \Delta H \cdot K \cdot \frac{dc_A}{dt} \tag{9}$$

where A [K s/J] and B [V/K] are device- and coating-specific constants describing the translation of a generated molar absorption/reaction enthalpy ΔH [J/mol] via a temperature change into a thermovoltage change. V_{sens} denotes the sensitive-layer volume and K is the partition coefficient (Eq. (5)) or reaction equilibrium constant (Eq. (6)).

Fig. 7.11 (a) and (b) show the output voltage of the microsystem and the voltage peak integrals on switching from synthetic air (nitrogen–oxygen mixture without humidity) to n-octane (900 ppm) and back to air at a temperature of 301 K [98]. The sensitive layer is poly(dimethylsiloxane) (PDMS). As already discussed above, two transient signals are produced, a positive one at the onset of the analyte exposure (liberation of predominantly condensation enthalpy) and a negative one upon terminating analyte exposure (abstraction of the enthalpy necessary for vaporizing the absorbed analyte). The net enthalpy changes can be approximated by integration over the peak area of the sensor signals [87–89, 98], which is the measurand of interest.

The post-CMOS fabrication steps include bulk micromachining (KOH or EDP) to achieve membranes [69, 85, 87, 95] and the processing of the sensitive layer by means of airbrush, dispensing, spin-coating or enzyme-immobilization methods [92–94].

Typical applications areas are environmental monitoring (gas and liquid phase) and biosensing in liquids. These include the detection of different organic volatiles in the gas phase (hydrocarbons, chlorinated hydrocarbons, alcohols, etc.; see Fig. 7.11) by using polymeric layers [69, 87–91] and the biosensing of glucose, urea and penicillin in the liquid phase by using suitable enzymes [86, 92–94].

Fig. 7.11 Calorimetric sensor signals: output voltage of the microsystem and peak integrals on switching from synthetic air to n-octane (900 ppm) and back to air at 301 K with 2 μm PDMS as sensitive layer [98]

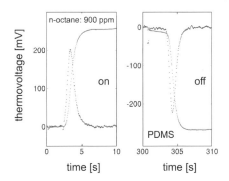

The interaction mechanisms involve reversible physisorption and bulk/gas-phase partitioning (see Eq. (5)) in addition to enzymatic chemical reactions (Eq. (6)) [86, 92–94].

7.3.3
Optical Sensors

Light can be considered as consisting either of particles (photons) or electromagnetic waves according to the *principle of duality*. Characteristic properties of the electromagnetic waves such as amplitude, frequency, phase and/or state of polarization can be used to advantage in devising chemical sensors [99–101]. The wavelength of the radiation, for example, can be tuned specifically to match the energy of a desired analyte-specific resonance or absorption process. Geometric effects (scattering) can provide additional information. In comparison with other chemical sensing methods, optical techniques, therefore, offer a great deal of selectivity already inherent in the various transduction mechanisms. Moreover, optical sensors, like any other chemical sensor, can capitalize on all the selectivity effects originating from the use of a sensitive layer [99–101].

If a sample is irradiated with visible light or electromagnetic waves, the radiation can be absorbed (intensity decrease), scattered (direction randomization, possibly frequency changes), refracted or reflected (metallic reflection, internal reflection mediated by evanescent waves) at the interface(s), or can produce phosphorescence/fluorescence (absorption–emission process) and chemiluminescence (conversion of chemical energy into light) effects [99–101].

The generation of light in CMOS or silicon devices is very difficult since there is no first-order transition from the valence band to the conduction band without the involvement of a phonon (lattice vibrations) [102]. However, direct-bandgap semiconductors (III–V semiconductors) such as gallium arsenide (GaAs) or indium phosphide (InP) show first-order radiative electron–hole recombinations with high quantum efficiency [102]. The detection of light is possible with either silicon-based devices (photodiodes) or other semiconducting materials (GaAs, InP). Consequently, integrated optical sensors and systems nowadays mostly are made of III–V semiconductors, which will not be detailed here, but offer the opportunity for fabrication and integration of lasers, waveguides, phase modulators and detectors on the same chip [103]. There is also a wealth of fiber-optical techniques and integrated optical devices, which are realized on silicon substrates without making use of CMOS technology. For details, see review articles [99–101, 103–108].

In this chapter we shall focus exclusively on CMOS-based optical sensors, examples of which include integrated optics [109–122] and Fabry–Perot-type devices [123–132]. There are also CMOS-based devices that rely on bioluminescent bacteria placed on application-specific optical integrated circuits in standard CMOS [133, 134]. The bacteria have been engineered to luminesce when a target compound such as toluene is metabolized.

7.3.3.1 Integrated Optics

Integrated optical (IO) sensors make use of guided waves or modes in planar optical waveguides. The waveguide materials usually include high-refractivity silicon dioxide or silicon nitride films on oxidized silicon wafer substrates. The guided waves or modes in planar optical waveguides include the TE (transverse electric or s-polarized, surface-normal) and the TM (transverse magnetic or p-polarized, surface-parallel) modes. Changes in the effective refractive index of a guided mode are induced by changes of the refractive index distribution in the immediate vicinity of the waveguide surface, i.e., within the penetration depth (several hundred nanometers) of the evanescent field in the sample (Fig. 7.12a) [101, 109]. The evanescent field decays exponentially with increasing distance from the wave-

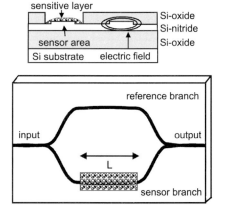

Fig. 7.12 (a) Schematic of an evanescent wave in an optical waveguide. (b) Schematic of a conventional Mach–Zehnder interferometer. The cross-section shows the separate sensor (left side, open) and reference (right side, covered) branches

guide surface. Changes in the effective refractive index can be induced by absorption of an adlayer on the surface of the waveguide from the gas or liquid phase [109–111], by interaction of an analyte molecule with a recognition structure immobilized on the waveguide surface [109, 112–120], or by changes in the refractive index of the medium adjacent to the waveguide in a flow-through configuration [109, 112, 119]. In the case of microporous waveguides, analyte molecule absorption or desorption directly into the pores of the waveguiding film itself can change the waveguide refractive index [109].

A number of different IO sensors such as grating couplers [109, 111, 112], prism couplers [121] and difference interferometers [113–120] have been developed to transform the changes in the effective refractive index into readily measurable physical quantities.

The difference interferometer realized as a *two-beam interferometer* (Mach–Zehnder interferometer) will be treated here in more detail. Mach–Zehnder IO devices (Fig. 7.12b) are monomode channel waveguides (TE or TM mode) and allow for a straightforward implementation of an interferometer structure [109, 113–120]. A waveguide is split into an open measurement path and a protected reference path and recombined after some distance. The phase difference, introduced by analyte interaction (refractive index change) in the sensing path, is detected by interference effects (Fig. 7.13b). Detection limits are in the range of several pg/mm^2 [122].

Antiresonant reflecting optical waveguides (ARROW) have recently attracted great attention owing to features such as IC-technology-compatible fabrication, high polarization selectivity, effective coupling to optical fibers and the possibility of making waveguides with very thin cladding layers [119, 120]. A typical ARROW waveguide is formed by a low-refractive-index core separated from the substrate by two cladding layers that are designed to form high-reflectivity Fabry–Perot mirrors (Fig. 7.13a). Thus, the electromagnetic field is confined by total internal reflection, as in conventional waveguides, at the core–superstrate interface and by antiresonant reflection at the core–cladding interface, which results in a leaky structure with virtual monomode characteristics. The higher order modes are filtered out by loss discrimination as a consequence of the low interface reflectivity of the cladding layers for these modes [119, 120].

Fig. 7.13b shows the signal response of an ARROW-type Mach–Zehnder interferometer to physical adsorption of a human serum albumin antibody (HSA) [120]. The upper graph shows the phase shift (measurement path and reference path) and the lower trace the intensity changes upon HSA-physisorption-induced changes. Strong intensity fluctuations can be observed at the onset of the HSA adsorption as a consequence of the phase shift induced by the changes in the refractive index in the sensor path.

The integrated-optics fabrication steps include the patterning of silicon nitride or oxide as waveguide materials (LPCVD, RIE, lithography) [109, 113–120], the deposition of silicon oxide or nitride cladding layers (PECVD) [109, 113–120] and the deposition of chemically sensitive layers by, e.g., immobilization of biological entities [113–115, 120].

a. ARROW waveguide

b. Mach-Zehnder response

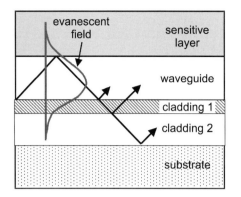
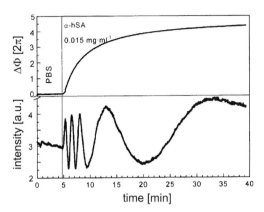

Fig. 7.13 (a) Cross-section and schematic of an ARROW waveguide: core and cladding 2 consist of silicon dioxide, cladding 1 is silicon nitride, the substrate silicon ($n_{substrate} > n_{nitride} > n_{oxide}$). (b) Signal response of an ARROW-type Mach–Zehnder interferometer to physical adsorption of a human serum albumin antibody (HSA). The upper graph shows the phase shift, $\Delta\Phi$ (measurement path and reference path) and the lower trace the intensity changes (arbitrary units) upon HSA physisorption-induced refractive index changes. Reprinted from [120] with permission

Typical applications include the detection of different organic solvents in the liquid phase (hydrocarbons, alcohols, etc.), the monitoring of sucrose and buffer solutions [109, 110, 112] and immunosensing (Fig. 7.13b) involving antibody–antigen binding experiments [109, 113–115, 120].

The interaction mechanisms include reversible physisorption (Eq. (5)) and biochemical affinity reactions (Eq. (6)) [109, 113–115, 120].

7.3.3.2 Fabry–Perot Microspectrometers

A Fabry–Perot interferometer (FPI) is an optical element consisting of two partially reflecting, low-loss, parallel mirrors separated by a gap. The optical transmission characteristics through the mirrors consist of a series of sharp resonant transmission peaks occurring when the gap is equal to multiples of half a wavelength of the incident light. These transmission peaks are caused by multiple reflections of the light in the cavity. By using highly reflective mirrors, small changes in the gap (width, absorptivity) can produce large changes in the transmission response. Even though two reflective mirrors are used, transmission through the element at the peak wavelengths approaches unity. The transmission is a function of both, the gap spacing and the radiation wavelength. The devices can be used as wavelength selector or monochromator by adjusting the gap width to achieve the desired wavelength.

A single-chip CMOS optical microspectrometer based on FPI and operating in the UV/VIS region has been reported [123, 124]. It contains an array of 16 ad-

Fig. 7.14 Cross-section of an integrated Fabry–Perot étalon. The gap width is determined by the thickness of the PECVD oxide; silver and aluminum are used as the optical coatings. A pnp-phototransistor (p+ implanted layer/n-well/p-epilayer) is located directly underneath the étalon. Adapted from [123]

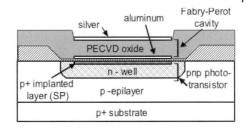

dressable Fabry–Perot étalons (500×500 µm^2) each with a different resonant cavity length realized as a PECVD silicon oxide layer sandwiched in between an aluminum and a silver layer and placed on top of an array of vertical pnp phototransistors (Fig. 7.14) [123, 125]. It additionally includes circuits for readout, multiplexing and driving a serial bus interface. The radiation source is a light bulb or light-emitting diode (LED).

A micrograph of the single-chip optical microspectrometer is shown in Fig. 7.15a, and the spectral responsivity of the 16 channels is displayed in Fig. 7.15b [123]. The FPI layer stack includes silver (45 nm)/silicon dioxide/aluminum (20 nm). The silicon dioxide thickness is used as a parameter and varies from 225 to 300 nm in 5 nm increments [123].

The fabrication steps include the deposition of the lower mirror (evaporation and lift-off of aluminum), the deposition of a silicon oxide layer of defined thickness (PECVD) and the deposition of the upper mirror layer (silver). The complete fabrication sequence of the monolithic CMOS VIS spectrometer is given in [123].

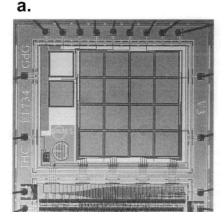
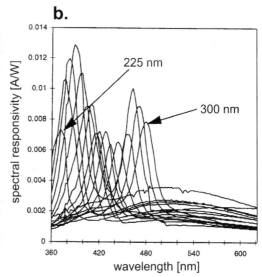

Fig. 7.15 (a) Micrograph of the single-chip optical microspectrometer and (b) spectral responsivity versus wavelength of the 16 channels. Reprinted from [123], with permission

FPI structures can be also used as gas sensors in the infrared range, where silicon becomes transparent [126–131]. The characteristic absorption wavelengths of carbon monoxide are at 4.7 µm, of carbon dioxide at 4.2 µm and of methane or hydrocarbons at 3.3 µm (IR region, molecular vibrations). Carbon dioxide sensors based on tunable FPIs (dual-wavelength measurements) are commercially available [132].

7.3.4
Electrochemical Sensors

Electrochemical sensors constitute the largest and oldest group of chemical sensors [4–10, 15]. They rely on electrochemical or charge-transfer reactions from an electrode to a solid or liquid sample phase or vice versa: $A^+ + e^- \rightleftharpoons A$ (see Eq. (7) and related text in Section 7.1).

Chemical changes take place at the electrodes or in the probed sample volume and alter the sample characteristics so that the resulting electrical signal such as charge buildup, current or resistance change can be measured. A key requirement for electrochemical sensors is a closed electrical circuit, although there may be no current flow (potentiometry). The charge transport in the sample can be ionic, electronic or mixed, whereas that in the transducer branch is always electronic.

Electrochemical sensors are usually classified according to their electro-analytical principles [4–7, 13].

Voltammetric sensors are based on the measurement of the *current–voltage* relationship in an electrochemical cell comprising electrodes in a sample phase. A potential is applied to the electrodes, and a current is measured, which is proportional to the concentration of the electroactive species of interest. *Amperometry* is a special case of voltammetry at constant potential.

Potentiometric sensors are based on the measurement of the potential at an electrode, which, in most cases, is immersed in a solution. The potential is measured at the equilibrium state, i.e., no current is allowed to flow during the measurement. According to the Nernst equation (Eq. (7)), the potential is proportional to the logarithm of the concentration of the electroactive species.

Conductometric sensors are based on the measurement of a conductance between two electrodes in a sample phase. The conductance is usually measured by applying an AC potential of small amplitude to the electrodes in order to prevent polarization. The presence of charge carriers determines the sample conductance.

Another categorization method relies on discerning the electronic components [11, 14]. There are, e.g., *chemoresistors, chemodiodes, chemocapacitors* and *chemotransistors*. The electroanalytical principles will be used as the superordinated classification scheme, and the component notation will be used within this scheme.

The following treatment will, again, be restricted to CMOS-based sensors and systems. Review articles are recommended for exploring the details of other electrochemical sensor designs (see, e.g., [135–140]).

7.3.4.1 Voltammetric/Amperometric Sensors

Amperometry is more frequently applied in chemical sensors and provides a linear current–analyte concentration relationship at a constant potential, which is predefined with regard to the target analyte.

The measurement configurations include the *two-electrode* setup [4–6, 13] with a reference electrode (RE) and a working electrode (WE), and the *three-electrode-system* [4–6, 13] in a potentiostatic configuration (Fig. 7.16): an additional auxiliary electrode (AE, sometimes denoted counter electrode, CE) is introduced for current injection into the analyte. The reference electrode in the *three-electrode* system does not become polarized and is now a true RE with a well-defined potential since there is no more current flowing through the RE as in the case of the two-electrode configuration. The three-electrode system is realized in practice with an operational amplifier (opamp) [13]. No current is flowing through the RE since the opamp features very high input impedance. The sensor signal current is measured at the working electrode.

The measured current at any given potential difference depends on the material properties, the composition and geometry of the electrodes, the concentration of the electroactive species (presumably the target analyte) and the mass transport mechanisms in the analyte phase [4–6, 13, 15].

There are two components to the measured current, a *capacitive component* resulting from the redistribution of charged and polar particles in the electrode vicinity and a component resulting from the electron exchange between the electrode and the redox species (analyte) termed *faradaic current* [4–6, 13, 15]. The faradaic component is the important measurand and is, for the case of diffusion-limited conditions, directly and linearly proportional to the target analyte concentration. The limiting current (all analyte ions are immediately charged or discharged upon arrival at the electrode) is then given by the Cottrell equation [4–6, 13, 15, 22]:

$$I_\infty = n_\text{e} \cdot F \cdot A \cdot D_\text{diff} \frac{c_\text{A}}{L_\text{diff}} \quad (10)$$

where n_e denotes the number of electrons, F is the Faraday constant, A the effective electrode area, D_diff the diffusion coefficient, c_A the target analyte concentra-

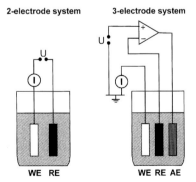

Fig. 7.16 Schematic of a two-electrode and a three-electrode configuration used for voltammetric measurements; for details, see text

tion and L_diff the diffusion length (for more mechanistic details, see [4–6, 13, 15, 22]). Correction terms to this equation for small electrodes have to be introduced to take into account the respective electrode geometry [4–6, 22]. The reference electrodes in the liquid phase are, in most cases, silver/silver chloride elements.

A picture and a schematic of a CMOS-based three-electrode amperometric sensor are shown in Fig. 7.17 [141]. The monolithic device includes the electrochemical sensor, a temperature sensor and interface circuitry. The circuitry contains an operational amplifier as potentiostat, a switched-capacitor current-to-voltage converter and a clock generator. Interface circuitry and temperature sensor are realized in 3 µm CMOS technology [141]. The circuitry needs a supply voltage of ±2.5 V, can apply voltages from +1 to –1 V to the sensor and handles current ranges from 30 nA full-scale to 1 µA full-scale. The output voltage of the temperature sensor is proportional to the absolute temperature and has a sensitivity of 125 µV/K. The total sensor dimensions are 0.75×5 mm.

Oxygen and hydrogen peroxide measurements have been carried out to characterize the sensors shown in Fig. 7.17. Signals upon dosing different concentrations of hydrogen peroxide in phosphate-buffered saline (PBS) to an amperometric sensor are shown in Fig. 7.18 [141]. The measurement was carried out at 0.7 V potential and 10 kHz clock frequency. The sensor signals (voltage values instead of current values due to the current-to-voltage converter, Fig. 7.17 b) linearly correlate with the applied hydrogen peroxide concentrations.

The fabrication steps include the deposition of additional silicon nitride as protective coating, the deposition/patterning of metal electrodes (lift-off, thermal evaporation, sputtering) [6, 13, 141–154] and of electroactive polymers, membrane materials or hydrogels (spin-casting, spraying, screen printing) [142–154]. CMOS-compatible processing sequences for amperometric sensors are given [13, 141, 150, 155].

a.

b.

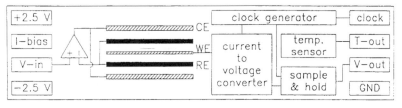

Fig. 7.17 (a) Micrograph and (b) layout of a CMOS-based three-electrode amperometric sensor. Reprinted from [141] with permission

Fig. 7.18 Amperometric sensor signal upon dosing different concentrations (1–10 mmol) of hydrogen peroxide in phosphate-buffered saline (PBS) to a two-electrode amperometric sensor. Reprinted from [141] with permission

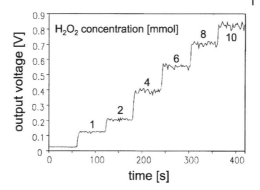

Typical applications include chemical analysis in the gas or liquid phase. Target analytes in the liquid phase comprise dissolved oxygen [148, 156], glucose [141, 143, 144], hydrogen peroxide (Fig. 7.18) [141, 145] and chlorine in drinking water [142, 147]. If the target analyte is not an electroactive species such as glucose, oxygen or carbon dioxide, then polymer electrolytes or enzymes (glucose oxidase) producing analyte-related ionic species are used as components of the sensitive electrode coatings. One of the best-known voltammetric cells is the *Clark cell* [157], which is based on a two-step-reduction of oxygen via hydrogen peroxide to hydroxyl ions in aqueous solution. The Clark cell is used to measure dissolved oxygen in blood and tissue [157].

The interaction mechanism is, in all cases, an electrochemical redox reaction (Eqs. (6) and (7)).

7.3.4.2 Potentiometric Sensors (Chemotransistors)

Potentiometry is the direct application of the Nernst equation (Eq. (7)) through measurement of the potential between non-polarized electrodes under conditions of zero current. The measurement is carried out at thermodynamic equilibrium.

The most important potentiometric sensors are based on the field-effect transistor (FET) and are among the most intensively investigated chemical sensor devices (since 1970 [158]). Field-effect based transistors, which are the most common electronic components on modern IC logic chips, rely on modulation of the charge carrier density in the semiconductor surface space–charge region through an electric field perpendicular to the device surface: The source–drain current is controlled by applying a voltage to an isolated gate electrode (see Fig. 7.19 and related text).

It is important to point out, 'that one of the most critical thrusts of the electronics industry's efforts to develop commercial field effect devices has been predicated by the need to isolate these devices from *any* variation in their chemical environment' [15], since the FET characteristics are very sensitive to such variation. Those efforts of the electronics industry are in diametric opposition to developing FET-based chemical sensors.

The *MOSFET* (metal oxide semiconductor field-effect transistor) as used for chemical gas sensing (Fig. 7.19) has a p-type silicon substrate (bulk) with two n-

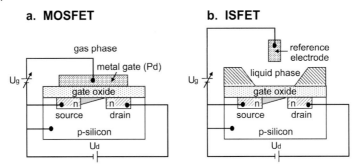

Fig. 7.19 Schematic representation of (a) a MOSFET and (b) an ISFET structure. U_g denotes the gate voltage and U_d the source–drain voltage. By replacing the metal gate of the MOSFET with an ionic solution and immersing a reference electrode in this solution, the ISFET has been developed

type diffusion regions (source and drain). The structure is covered with a silicon dioxide insulating layer, on top of which a metal gate electrode (in microelectronics typically a poly-Si gate) is deposited. When a positive voltage (with respect to the silicon) is applied to the gate electrode, electrons, which are the minority carriers in the substrate, are induced at the silicon-dioxide/semiconductor interface. Consequently, a conducting channel (n-channel) is created between the source and the drain, near the silicon dioxide/silicon interface. The conductivity of this channel can be modulated by adjusting the strength of the electric field between the gate electrode and the silicon, perpendicular to the substrate surface.

Pd-gate FET structures were demonstrated to function as a hydrogen sensor by Lundström et al. [159]. Hydrogen molecules readily absorb on the gate metal (Pt, Ir, Pd) and dissociate into hydrogen atoms. These H-atoms can diffuse rapidly through the Pd and absorb at the metal/silicon oxide interface partly on the metal and partly on the oxide side of the interface (Fig. 7.20a) [160, 161]. Owing to the absorbed species and the resulting polarization phenomena at the interface, the drain current (I_d) at constant gate voltage is altered and the threshold voltage (U_d) is shifted. The voltage shift, ΔU_d, is proportional (potentiometric sensor response is logarithmic) to the concentration or coverage of hydrogen at the oxide/metal interface (Fig. 7.20b) [162]. Sensitivity and selectivity patterns of gas-sensitive FET devices hence depend on the type and thickness of the catalytic metal used, the chemical reactions at the metal surface and the device operation temperature. CMOS-based MOSFETs detecting hydrogen have been described [163–165].

MOSFET-sensor applications also include the detection of ammonia [160, 161], amines and any kind of molecule that gives rise to polarization in a thin metal film (hydrogen sulfide, ethene, etc.) or causes charges/dipoles at the insulator surface [160, 161].

For the liquid-phase *ISFET* (ion-selective field-effect transistor), the gate metal electrode of the MOSFET is replaced by an electrolyte solution, which is contacted by a reference electrode so that the gate oxide is directly exposed to the aqueous

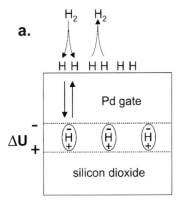
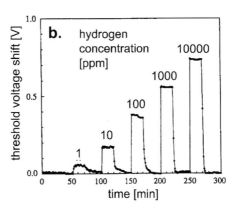

Fig. 7.20 (a) Hydrogen sensing with MOSFETs: hydrogen atoms diffuse through the Pd film and adsorb at the Pd/SiO$_2$ interface to form a dipole layer, which gives rise to voltage shifts. (b) Response (threshold voltage shifts) of a Pd-gate FET in dry air at 100 °C upon exposure to different hydrogen concentrations (1–10000 ppm). Note that the potentiometric sensor response is logarithmic. Reprinted from [162] with permission

electrolyte solution (Fig. 7.19) [158]. An external reference electrode is required for stable ISFET operation [4–6, 13]. The source–drain current is influenced by the potential at the oxide/aqueous solution interface. ISFET amplifiers with feedback keep the source–drain current constant by compensating solution-induced changes in the gate oxide potential through modulation of the gate voltage (U_g) that is applied to the reference electrode. The gate–source potential is then determined by the surface potential at the insulator/electrolyte interface. Mechanistic studies of the processes occurring at the solution/gate oxide interface (site binding model [166]) and the oxide semiconductor interface can be found in the literature [4–6, 13, 166–168]. The insulator/solution interface is assumed to represent in most cases a *polarizable* interface, i.e., there will be charge accumulation across the structure but no net charge passing through.

The fabrication of CMOS integrated field-effect-based electrochemical sensors with circuitry is described in various publications [13, 169–180]. An example of an integrated ISFET is shown in Fig. 7.21a [178]. The system includes an ISFET amplifier (source and drain follower) and two interdigitated ISFETs realized in a double-metal, 1.0 μm CMOS process (Atmel-ES2, Atmel, France). To render the fabrication CMOS compatible, the sensitive oxynitride sits on top of an electrically floating multi-conductor gate structure including the two metal layers and the gate polysilicon, all of which are electrically connected (Fig. 7.21b). In this way, the electrical charges at the surface of the oxynitride in contact with the solution directly affect the silicon surface source–drain current.

Another possibility to realize a CMOS-compatible ISFET design includes connecting a sensitive layer at the chip surface via metal lines to the polysilicon gate of the transistor at the bottom of the CMOS layer stack [179, 180]. Integrated solutions implemented in a 0.5 μm double-metal double-poly standard CMOS process

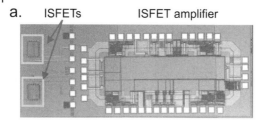

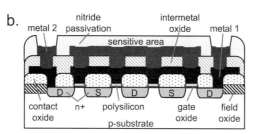

Fig. 7.21 (a) Micrograph of a CMOS chip hosting two integrated ISFETs and an ISFET amplifier. (b) Schematic and cross-section of the ISFET. For details, see text and [178]. Reprinted from [178] with permission

with temperature compensation and aluminum/tin dioxide-multilayer sensitive areas have been reported [179, 180].

Classical ISFET applications include pH-sensing (acidity or basicity) with an exposed-gate-oxide FET [158, 181]. The surface of the gate oxide contains OH-functionalities, which can be protonated and deprotonated and, therefore, when the gate oxide contacts an aqueous solution, a change of pH will change the silicon.oxide surface potential. Typical pH-sensitivities measured with silicon oxide ISFETs are

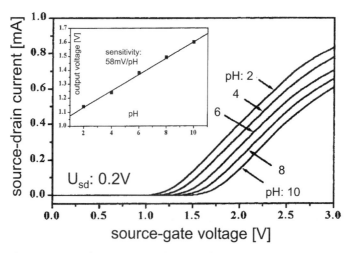

Fig. 7.22 Source–drain current versus gate voltage applied via the reference electrode upon ISFET exposure to buffer solutions of different pH. The source–drain voltage was kept constant at 0.2 V. The sensor sensitivity was 0.58 mV/pH unit (see inset). Reprinted from [179] with permission

37–40 mV per pH-unit [182]. Gate materials such as silicon nitride [169–177], oxynitride [178] and alumina [183–185] have better properties than silicon oxide with regard to pH-response (larger signals close to the maximum of 59 mV per pH-unit as given by the Nernst equation), hysteresis and drift. In practice, these layers are deposited on top of the silicon oxide by means of chemical vapor deposition (CVD). Signals recorded with an aluminum/tin dioxide multilayer on a CMOS integrated ISFET are shown in Fig. 7.22 [179, 180]. The source–drain current values are displayed as a function of the applied gate voltage upon ISFET exposure to buffer solutions of different pH. The sensor sensitivity was 58 mV/pH unit [179, 180]. pH-ISFETS are commercially available from, e.g., Honeywell [186].

ISFETs can also be covered with organic ion-selective membranes such as polyurethane, silicone rubber, polystyrene and polyacrylates containing ionophores to detect metal ions such as potassium [187, 188], sodium [189] and silver [190].

7.3.4.3 Conductometric Sensors

Conductometric techniques are a special case of AC impedance techniques. Instead of the real and imaginary components of the electrode impedance at different frequencies, only the real-valued resistive component, related to the sample (sensing material) resistance, is of interest. Since complex impedances include capacitive and inductive contributions, chemocapacitors are included here. The section on conductometric sensors is hence organized in two parts, one on chemoresistors operating at room and elevated temperature and the other on chemocapacitors.

Chemoresistors

Chemoresistors rely on changes in the electric conductivity of a film or bulk material upon interaction with an analyte. Conductance, G, is defined as the current, I [A], divided by the applied potential, U [V]. The unit of conductance is Ω^{-1} or S (siemens). The reciprocal of conductance is the resistance, R [Ω]. The resistance of a sample increases with its length, l, and decreases with its cross-sectional area, A:

$$R = \frac{1}{G} = \frac{U}{I} = \frac{1}{\kappa} \cdot \frac{l}{A} \tag{11}$$

The conductivity or specific conductance, κ [Ω^{-1} m^{-1}], is hence defined as the current density [A/m^2] divided by the electric field strength [V/m]. The reciprocal of conductivity is resistivity, ρ [Ω m]. The conductivity can be thought of as the conductance of a cube of the probed material with unit dimensions [13].

Conductometric sensors are usually arranged in a metal electrode 1/sensitive layer/metal electrode 2 configuration [4–6]. The conductance measurement is made either via a Wheatstone bridge arrangement or by recording the current at an applied voltage in a DC mode or in a low-amplitude, low-frequency AC mode to avoid electrode polarization. In Fig. 7.23a [191], a conductance cell (in this case metal oxides) and the respective equivalent electric circuits are depicted. The goal

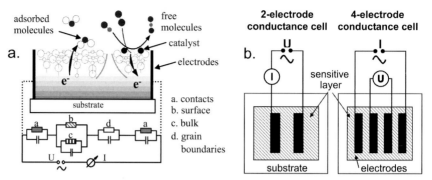

Fig. 7.23 (a) Schematic representation of a conductance cell (in this case a semiconductor sensor with tin dioxide as sensitive layer), of the different contributions (contacts, surface, bulk and grains) to the overall conductivity and of the respective equivalent circuits. Adapted from [191]. (b) Schematic representation of two- and four-electrode conductance cells. For details, see text

of conductometry is to determine the sample resistance (c). The lead wire resistances normally can be neglected. The electrode impedance (a) consists of two elements, the contact capacitance and the contact resistance. By applying an AC potential, an AC current will flow through the resistor cell. If the contact capacitance is sufficiently large, no potential will build up across the corresponding contact resistance. The contact resistance should be much lower than the sample resistance and be minimized, so that the bulk contribution dominates the measured overall conductance. If surface conductivity mechanisms differing from those in the bulk occur, this can be modeled by adding an additional surface resistance (b) to the equivalent circuit. A grain boundary in the sensing material constitutes a resistance–capacitance unit (d). The conductivity depends on the concentration of charge carriers and their mobility, either of which can be modulated by analyte exposure. In contrast to potentiometry and voltammetry, conductometric measurements monitor processes in the bulk or at the surface of the sample. Any contribution of electrode processes has to be avoided.

Therefore, in most cases, a four-electrode configuration is preferred over a simple two-electrode configuration (Fig. 7.23 b). The outer pair of electrodes is used for injecting an AC current into the sample; the potential difference is then measured at the inner pair of electrodes. The interference of electrode impedances on the measurement results is thus excluded.

Low-temperature Chemoresistors Several classes of predominantly organic materials are used for application with chemoresistors at room temperature. The chemically sensitive layer is applied over interdigitated electrodes on an insulating substrate. Electrode spacing is typically 5–100 µm, and the total electrode area is a few mm^2. The applied voltage ranges between 1 and 5 V. The most commonly applied sensitive layers include conducting polymers or carbon-loaded polymers, each of which will be briefly described here.

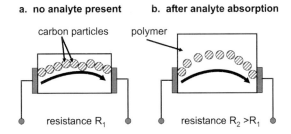

Fig. 7.24 Conductivity by particle-to-particle charge percolation in carbon-loaded polymers: (a) with no analyte present; (b) during organic volatile exposure, which causes polymer swelling

Conducting polymers such as *polypyrroles, polyaniline* and *polythiophene*, exhibit a large conjugated π-electron system, which extends over the whole polymer backbone. Partial oxidation of the polymer chain then leads to electrical conductivity, because the resulting positive charge carriers (denoted polarons or bipolarons) are mobile along the chains [11, 21]. Counter anions must be incorporated into the polymer upon oxidation to balance the charge on the polymer backbone. The conducting polymers, however, not only react with oxidizing agents, but also respond to a wide range of organic vapors and show a high cross-sensitivity to water [192–197]. Sensors are commercially available from, e.g., Osmetech and Marconi [198].

In *Carbon-black-loaded polymers*, conducting carbon black is dispersed in non-conducting polymers and deposited on an electrode structure. The conductivity is by particle-to-particle charge percolation so that if the polymer absorbs vapor molecules and swells, the particles are, on average, further apart (Fig. 7.24), and the conductivity of the film is reduced [199]. Applications of carbon-black-loaded polymers include monitoring organic solvents such as hydrocarbons, chlorinated compounds and alcohols [200–203]. Sensors are commercially available from, e.g., Cyrano Sciences [204].

The sensor-specific fabrication steps include the patterning of metal electrodes (lift-off, thermal evaporation, sputtering) [194–197, 199–203], the deposition of conducting polymers by electrochemical deposition [192–197] and the deposition of carbon-loaded polymers by spin-casting, spraying or screen printing [199–203].

CMOS-based monolithic sensor systems coated with carbon-black/polymer blends are detailed in [205, 206], and two-chip solutions with a CMOS circuitry chip and a separate sensor chip coated with conducting polymers can be found in [207, 208]. A CMOS-compatible monolithic conductivity sensor for liquid phase (Pt electrodes) has been reported [209].

High-temperature Chemoresistors The integration of heated structures with CMOS technology is particularly challenging, since operating temperatures of 250–600 °C exceed the temperature specifications of common integrated circuits (between –40 and 120 °C). The sensitive materials consequently have to be deposited on miniaturized heatable structures (microhotplates) that are thermally well isolated from the rest of the chip carrying the electronic components. Microhotplates, in most cases, consist of a thermally isolated area such as a suspended membrane structure featuring a heater, a temperature sensor and contact electrodes for the sensitive layer as shown schematically in Fig. 7.25 [210].

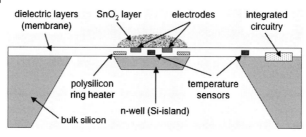

Fig. 7.25 Cross-sectional schematic of a microhotplate showing the different components and the sensitive layer [210]

The membrane, which isolates the heated area from the bulk chip, is formed by the CMOS dielectric layers (Si-oxide and Si-nitride) that exhibit low thermal conductivity. The membrane is released by etching away the bulk silicon. Depending on the micromachining procedure, it is possible to leave a silicon island underneath the heated area (Fig. 7.25) [210, 211]. Such an island can serve as a heat spreader and also mechanically stabilizes the membrane. The circuitry is arranged on the bulk chip, the temperature of which changes negligibly (temperature increase over ambient of ~1–2% of the hotplate temperature) upon hotplate heating as has been assessed by using an additional temperature sensor in the vicinity of the circuitry [210].

The microhotplate should be as symmetric as possible to achieve good temperature homogeneity over the membrane area and, as a consequence, low stress gradients. A high thermal resistance is desirable to achieve minimum power consumption, and the elevated operating temperature (250–350 °C) has to be reached with a supply voltage of 5 V. Two different designs are shown in Fig. 7.26a and b.

A front-side-etched microhotplate on a CMOS substrate is shown in Fig. 7.26a [212, 213]. The micrograph of a circular bulk-micromachined hotplate, which fea-

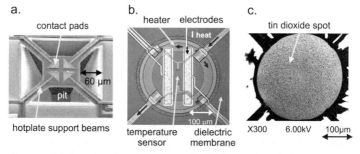

Fig. 7.26 (a) Scanning electron micrograph of a front-side-etched microhotplate suspended by beams. Reprinted from [212, 213] with permission.
(b) Circular bulk-micromachined hotplate, which features a silicon island underneath the heated area [211, 214]. (c) SEM picture of a tin-dioxide-coated bulk-micromachined hotplate. The metal oxide droplet of grainy and porous structure exclusively covers the circular heated area [215]

tures a silicon island underneath the heated area, is shown in Fig. 7.26 b [211]. The microhotplate components include in both cases a polysilicon resistive heater, a temperature sensor and noble-metal-coated (e.g. Pt) electrodes for electrical connection to the sensing film [211–215].

The circular heater and heated-area design of the device in Fig. 7.26 b also perfectly matches the shape of the sensitive-material droplet (Fig. 7.26 c) [215]: area that is not covered by the sensitive material is not heated and, consequently, the heat losses to the ambient air are minimized. The hotplate features a thermal resistance of 7.8 °C/mW.

First results with hotplates and circuitry on a single CMOS-chip have been reported recently [210, 211, 214–218]. In the following, an advanced analog/digital microsystem will be presented [211, 218]. A simplified schematic of the chip architecture is shown in Fig. 7.27 [211, 218]. The microhotplate and its components, which have already been discussed (Fig. 7.26 b), are represented as a block. The circuitry includes three major components: (a) the sensitive-layer resistance read-out, (b) the microhotplate temperature control loop and (c) the temperature read-out of the bulk chip carrying the electronics [218, 219].

The resistance of the sensitive layer is read out by using a logarithmic converter. The temperature controller is a digital programmable PID (proportional–integral–differential) controller, which precisely controls the membrane temperature (±2 °C at 350 °C) via the analog heater driving circuitry. The A/D and D/A converters that connect the analog and digital part of the circuitry are also shown in the

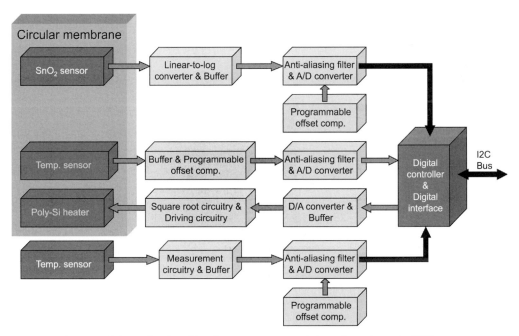

Fig. 7.27 Schematic of the analog/digital hotplate microsystem including all circuitry units [218]

block diagram. All measured data are read out via a standard serial I2C interface [220], which also allows for setting the membrane temperature and the controller parameters. The detailed circuitry implementation is published in [218, 219].

Fig. 7.28 shows the sensor chip [211, 218]. The microhotplate is located in the left part of the chip with enough free space to accommodate additional hotplates (up to three in total). The analog circuitry and the A/D and D/A converters are clearly separated and shielded from the digital circuitry for noise reasons [218]. The bulk-chip temperature sensor is located close to the analog circuitry in the center of the chip. The distance between microhotplate and circuitry is comparatively large owing to packaging requirements.

The sensitive materials used with high-temperature chemoresistors include wide-bandgap semiconducting metal oxides such as tin oxide, gallium oxide, indium oxide or zinc oxide, all of which require elevated operation temperatures (>200 °C). In general, gaseous species acting as electron donors (hydrogen) or acceptors (nitrogen oxide) adsorb on the metal oxides and form surface states, which can exchange electrons with the semiconductor. An acceptor molecule will extract electrons from the semiconductor and thus decrease its conductivity. The opposite holds true for an electron-donating surface state. A space charge layer will thus be formed. By changing the surface concentration of donors/acceptors, the conductivity of the space charge region is modulated [4–6, 191, 221–224]. In addition to the above-mentioned interaction of surface adsorbates based on *electronic* effects, the diffusion of lattice defects from the bulk of the metal oxide also occurs (*ionic* conduction) at elevated temperatures (>900 K). Lattice defects can act as donors or acceptors. Oxygen vacancies act, e.g., as intrinsic donors.

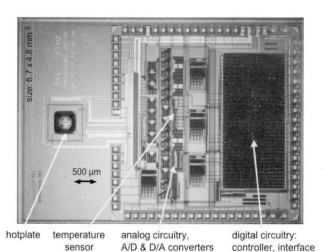

Fig. 7.28 Micrograph of the analog/digital microsystem chip showing the microhotplate (left), the analog circuitry (center) and the digital unit with serial interface (right) [211, 218]

The overall conductivity in polycrystalline samples includes contributions from the individual crystallites, the grain boundaries, insulating components such as pores and the contacts (Fig. 7.23 a). Hence the conduction mechanism in ceramic polycrystalline samples is difficult to analyze, and a variety of empirical data have been published [4–6, 191, 221–224]. The reaction between gases and the oxide surface depends on the sensor temperature, the gas involved and the sensor material [191, 221–226].

The most extensively investigated material, *tin dioxide*, is oxygen deficient and therefore an n-type semiconductor since oxygen vacancies act as electron donors. In clean air, oxygen, which traps free electrons by its electron affinity, and water are absorbed on the tin dioxide particle surface, forming a potential barrier in the grain boundaries. This potential barrier restricts the flow of electrons and thus increases the resistance. When tin dioxide is exposed to reducing gases such as carbon monoxide, the surface adsorbs the gases, and some of the oxygen is removed by reaction with water and oxygen at the surface. This lowers the potential barrier, thereby reducing the electric resistance. For details, see [4–6, 191, 221–226].

A typical sensor response as recorded with the integrated microsystem shown in Fig. 7.28 is depicted in Fig. 7.29 [211, 215]. The sensor signal output ranges from 200 to 280 digital units corresponds to a resistance range from 250 to 50 kΩ. A higher sensor signal reading represents a lower resistance of the sensitive layer so that the presence of CO, which is a reducing gas, leads to a higher sensor reading. Each analyte exposure step is 15 min at constant relative humidity of 40% (23 °C humidifier temperature) and 30 °C chip or ambient temperature. The microhotplate operating temperature was 275 °C. Low CO concentrations of 1, 3 and 5 ppm (partial pressure 0.1–0.5 Pa CO) were dosed to the sensor. As can be seen in Fig. 7.29, a concentration of 1 ppm CO is clearly detectable. The noise

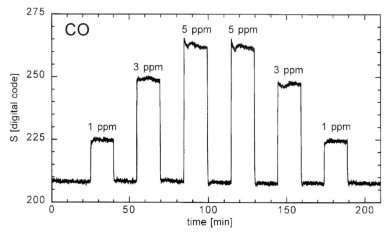

Fig. 7.29 Low-concentration signals: 1–5 ppm (0.1–0.5 Pa) CO detected with Pd-doped (0.2%) tin dioxide at 275 °C [211]

in the sensor signal is ±0.5 digits, which results in a gas concentration resolution of ±0.2 ppm [211, 215].

Semiconductor metal oxide sensors usually are not very selective, but respond to almost any analyte (carbon monoxide, nitrogen oxide, hydrogen, hydrocarbons). One method to modify the selectivity pattern includes surface doping of the metal oxide with catalytic metals such as platinum, palladium, gold and iridium [191, 221–226].

Since microhotplates have a very low thermal mass, they also allow for temperature-programmed operation modes to be applied [227–230]. By operating the device, e.g., in a cyclic thermal mode, reaction kinetics on the sensing surface are altered, producing a time-varying response signature that is characteristic for the respective analyte gas [227–230].

The post-CMOS fabrication steps include the deposition/patterning of noble-metal electrodes using lift-off, thermal evaporation or sputtering [231–235], back-side (KOH) [210, 211, 215, 231–235] or front-side (RIE, EDP, XeF_2) [212, 213] etching for membrane formation, deposition of metal oxide materials by LPCVD, sol–gel processes, sputtering or screen printing [236–240] and the subsequent sintering of the metal oxides (annealing) at elevated temperatures [241].

The fabrication of hotplates on CMOS substrates has been described [212, 213, 242–244], and reliability tests have been conducted [245]. First systems featuring hotplates and circuitry on a single CMOS chip have been presented recently [210, 211, 214–219]. Multi-chip solutions have been proposed [246, 247]. CMOS-hotplates realized on SOI (silicon-on-insulator) and SIMOX (separation by implantation of oxygen) substrates have been described [248–252].

Typical microhotplate applications include the detection of inorganic gases such as hydrogen [212, 213, 243], oxygen [212, 213], nitrogen oxide [229, 253], carbon monoxide [210–219, 247, 253] and a variety of volatile organics [212, 213, 237–243, 253] using predominantly tin dioxide as a sensitive layer.

Chemocapacitors

Chemocapacitors (dielectrometers) rely on changes in the dielectric properties of a sensing material upon analyte exposure. Interdigitated structures that are similar to those of room temperature chemoresistors are predominantly used [254–257]. In some cases, plate capacitor-type structures with the sensitive layer sandwiched between a porous thin metal film (permeable to the analyte) and an electrode patterned on a silicon support are used to increase the device sensitivity by trapping the electric field [258, 259].

The capacitances usually are measured at an AC frequency from a few up to 500 kHz. The analyte absorption in the sensitive layer on top of the electrodes induces a change in the layer dielectric properties and, consequently, a capacitance change that can be measured (Fig. 7.30). For conducting measurements at defined temperatures, sensor and reference capacitors can be placed on thermally isolated membrane structures [260–262].

The capacitors described in more detail here are fabricated exclusively with layers and processing steps available in the standard CMOS process sequence

Fig. 7.30 Schematic representation of an interdigitated capacitive sensor covered with a sensitive polymer layer. E1 and E2 denote the two sets of interdigitated electrodes and ΔC the capacitance change

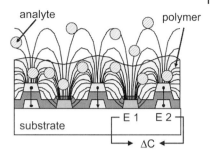

[263–268]. Electrode 1 is made from the first aluminum metal layer, and electrode 2 comprises a stack of interdigitated electrodes in the first and second metal layers, which are electrically connected on the chip, albeit being physically separated by a silicon oxide layer (Fig. 7.30). The quasi 'three-dimensional' electrode configuration was chosen to enhance the sensitivity of the coated sensor to analyte-induced capacitance changes by maximizing the sensitive-layer volume in the regions of strong electric field (see Fig. 7.30) [266–269]. The overall capacitor includes 128 electrode pairs and occupies an area of 824×814 μm². The electrode width and the interelectrode spacing are 1.6 μm [266–269].

Since the nominal capacitance of such a microstructure is of the order of 1 pF, and the expected capacitance changes (sensor signals) are in the range of a few attofarad, an integrated solution with on-chip circuitry is required [266–269]: There is no possibility of transferring minute analog signals via bond wires and cables to desktop instruments. The integrated solution includes two capacitors, a polymer-coated sensing capacitor and a silicon-nitride-passivated reference capacitor in a switched-capacitor scheme (Fig. 7.31).

The sensor response is read out as a differential signal between the polymer-coated sensing and the nitride-passivated reference capacitor. Both, sensor and reference

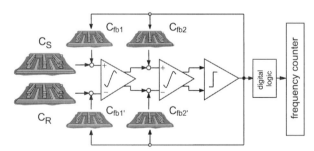

Fig. 7.31 Schematic of the fully differential second-order Sigma–Delta modulator exhibiting two switched-capacitor integrators and a subsequent comparator. Four feedback capacitors (C_{fb}) are realized as interdigital capacitors. The Sigma–Delta modulator provides a pulse-density-modulated digital output that is decimated using the frequency counter [69]

capacitors are split into two parts to improve the charge-transfer efficiency. The sensing capacitor (C_S) and reference capacitor (C_R) are incorporated in the first stage of a fully differential second-order Sigma–Delta modulator (Fig. 7.31) with two switched-capacitor integrators and a subsequent comparator [69, 270, 271]. A second-order Sigma–Delta modulator is used to achieve a shorter analog-to-digital conversion time. Since the output bit stream of the Sigma–Delta modulator is proportional to the ratio $(C_S-C_R)/(C_{fb})$, the four feedback capacitors (C_{fb} in Fig. 7.31) are realized as interdigitated capacitors with the same materials as sensing and reference capacitor in order to eliminate differences in temperature behavior and ageing. Owing to the small signal bandwidth, the output bit stream of the Sigma–Delta modulator is decimated using a frequency counter. For more details on the circuitry, see [69, 270, 271]. A micrograph of the integrated capacitive microsystem is shown in Fig. 7.32. For conducting measurements at defined temperatures, sensor and reference capacitors can be placed on thermally isolated membrane structures [260, 261].

Two effects change the capacitance of a polymeric sensitive layer upon absorption of an analyte: (1) swelling and (2) change of the dielectric constant due to incorporation of the analyte molecules into the polymer matrix [265–267].

For low analyte concentrations, the swelling is linear in the amount of absorbed analyte, expressed by the following equation [267]:

$$h_{\text{eff}} = h(1 + Q\varphi_A) \tag{12}$$

where h and h_{eff} denote the initial polymer thickness and the resulting effective thickness after analyte absorption, respectively. Q is a dimensionless non-ideality factor of the swelling and φ_A is the volume fraction of the analyte in the polymer. $Q=1$ represents ideal swelling, i.e., the total volume is given by the addition of the volume of the absorbed analyte in its liquid state to that of the polymer.

The composite dielectric constant of mixtures of non-polar liquids can be approximated for all kinds of analytes as proposed in [266, 267]. It can be generalized including Q to

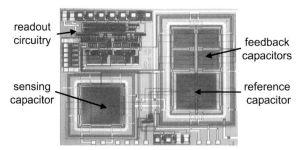

Fig. 7.32 Micrograph of a capacitive sensor system including a polymer-coated sensing capacitor, a passivated reference capacitor, four interdigitated feedback capacitors and the Sigma–Delta circuitry as detailed in the text [69]

$$\varepsilon_{\text{eff}} = \varepsilon_{\text{poly}} + \varphi_A[(\varepsilon_A - 1) - Q(\varepsilon_{\text{poly}} - 1)] \tag{13}$$

where ε_{eff} is the resulting effective dielectric constant of the polymer–analyte system and $\varepsilon_{\text{poly}}$ and ε_A are the dielectric constants of polymer and analyte (analyte in liquid state), respectively.

The presence of two relevant physical effects, ε-change and swelling, gives rise to an additional mechanism for selectivity because thin and thick sensitive layers have different sensitivity patterns [266, 267]. For a simple interdigitated structure, the space above the device containing 95% of the field lines is within a distance of half of the electrode periodicity [272]. Therefore, a thin or thick sensitive layer is defined with respect to the electrode periodicity.

For a layer thickness significantly less than half the periodicity, the region of strong electric field extends above the sensitive layer (Fig. 7.30). Upon analyte sorption, the amount of polarizable material in the sensed region of the capacitor always increases, which results in a capacitance increase regardless of the dielectric constant of the analyte. For a layer thickness greater than half the periodicity of the electrodes, almost all electric field lines are within the polymer volume. Consequently, the capacitance is determined by the composite dielectric constant of the analyte–polymer mixture. The capacity change can, therefore, be positive or negative, depending on whether analyte or polymer has a higher dielectric constant (Eq. (13)) [266, 267].

The expected differences in responses from sensors with thin and thick polymer layers have been verified experimentally. Typical sensor response profiles from capacitors coated with a thin (0.3 μm) and thick (2.3 μm) poly(etherurethane) (PEUT) layer are plotted in Fig. 7.33 [267]. The capacitor has been alter-

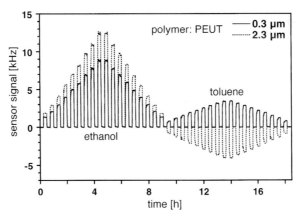

Fig. 7.33 Response profiles from two capacitors coated with PEUT layers of different thickness upon exposure to 1000–5000 ppm ethanol and 500–3000 ppm toluene as a function of time. The ratio of the dielectric constants of polymer (4.8) and analytes (toluene, 2.36; ethanol, 24.3) controls the signs of the signals from thick layers. For a thin layer, all signals are positive [267]

nately exposed to various concentrations of toluene and ethanol at 301 K and the pure carrier gas. Ethanol ($\varepsilon=24.3$) has a higher dielectric constant than PEUT ($\varepsilon=4.8$) and toluene a lower dielectric constant ($\varepsilon=2.36$). Both analytes provide positive sensor signals with thin PEUT layers, whereas with thick layers, ethanol provides a positive signal and toluene a negative signal.

The fabrication of capacitors integrated with CMOS circuitry components is described in [255–257, 263–271]. The post-CMOS steps include the optional deposition/patterning of metal electrodes other than aluminum (lift-off, thermal evaporation, sputtering), the optional etching steps to form a membrane for temperature stabilization of the capacitor [260, 261] and the deposition of sensitive layers (spin-casting, spraying, photolithography) [254–269].

Typical applications include humidity sensing with polyimide films [254, 255, 263, 264, 273–275], since water has a high dielectric constant of 78.5 (liquid state) at 25 °C, leading to large capacitance changes. CMOS-based integrated capacitive humidity sensors are commercially available from, e.g., Sensirion, Switzerland [276]. More recent applications also include the detection of organic volatiles in the gas phase using polymeric layers [265–270].

The interaction mechanisms involve reversible physisorption and bulk/gas-phase partitioning (see Eq. (5)).

7.3.5
CMOS Multisensor Systems

A significant benefit of using CMOS technology for devising chemical sensors or sensor systems is the possibility of co-integrating several identical or even different transducers along with all necessary driving circuitry on a single chip. Combinations of different chemical sensor/transducer principles and types can provide more information than using arrays of the same transducer type and varying the sensitive layer only. The single-type array, an example of which will be also shown below, has become particularly popular, as is documented in a wealth of publications [277–280]. The simultaneous application of various transducer principles is more complex (realization of different transducer and circuitry components, use of different data evaluation schemes), but has proven to be beneficial in several applications [281–289]. Multisensor and -transducer systems allow for sharing parts of the circuitry and for multiplexing signals, thus reducing the system complexity and power consumption in comparison to an array of discrete sensors or systems. Additional components such as signal-conditioning circuitry, chip memory (calibration values), an intra-chip communication system and a serial interface to off-chip microcontrollers or instruments can also be realized with such systems. These components will make the system more reliable and easier to use and will allow more functions to be offered to a potential user.

In the following, three prototype monolithic CMOS multisensor systems will be briefly presented, the transducer principles of which have been already laid out in the previous sections: (1) a four-cantilever array [64], (2) a multiparameter biochemical sensor [286], and (3) a multitransducer gas sensor microsystem [287].

7.3.5.1 Four-cantilever Array

The first system is a compact single-type array, a micrograph of which is shown in Fig. 7.34 [64]. The monolithic microsystem comprises four cantilevers, the feedback circuitry and a multiplexer to address and operate the cantilevers sequentially. The resonance frequencies of the cantilevers are in the range 380–410 kHz and differ by maximum of ±10%. The same holds for the quality factors of the resonators, which are 980±10% [64].

Three cantilevers of the array were coated with three different polymers: 0.6 μm of poly(dimethylsiloxane) (PDMS), 1.5 μm of poly(cyanopropylmethylsiloxane) (PCPMS) and 1.8 μm of poly(ether urethane) (PEUT). One cantilever was left uncoated to look into surface adsorption phenomena, which were found to be marginal. The sensors were exposed to two different n-octane concentrations (600 and 700 ppm) and, additionally, varying toluene concentrations (300–700 ppm). Since

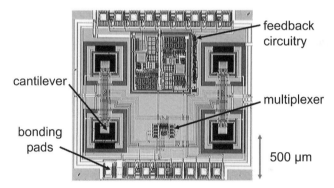

Fig. 7.34 Micrograph of an integrated cantilever array (chip size: 2×2 mm²) [64]

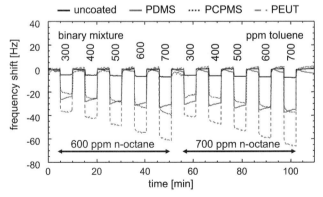

Fig. 7.35 Responses of an array of cantilevers (Fig. 7.34) coated with different polymers (PDMS, PCPMS, PEUT) upon exposure to mixtures of n-octane (600 and 700 ppm) and toluene (300–700 ppm)

the four sensors respond differently to the analyte mixtures (Fig. 7.35), the recorded data sets can be fed into multilinear regression algorithms or other multicomponent analysis tools that provide quantitative determination of both components after preceding calibration. For more details on pattern recognition and multicomponent analysis methods, see, e.g., [277–280].

7.3.5.2 Multiparameter Biochemical Microsensor System

The multiparameter biochemical microsensor system is aimed at continuous monitoring of ions, dissolved gases and biomolecules in a liquid phase such as blood (Fig. 7.36) [283, 286] and is based on an earlier design by Gumbrecht et al. [284, 285]. The eight integrated chemical sensors comprise six ion-sensitive field-effect transistors (ISFETs: 1–6 in Fig. 7.36), one oxygen sensor (7 in Fig. 7.36) and one conductometric sensor (8 a, b in Fig. 7.36), all of which can be operated in parallel [286]. An Ag/AgCl reference electrode is also integrated on the CMOS chip to get rid of external references. A flow channel (polyimide) restricts the liquid phase access to the sensor area.

The six ISFETs allow for direct contact of the electrolyte with the gate oxide. The gate oxide itself is either pH-sensitive (see ISFET Section 7.3.4.2) or the ISFET can be used as a 'Severinghaus'-type pH-FET to measure dissolved carbon dioxide (detection of carbon dioxide via dissolution in water, formation of 'carbonic

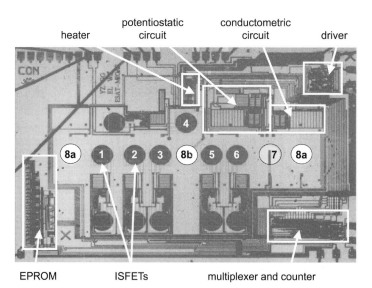

Fig. 7.36 Micrograph of the CMOS multi-parameter biochemical sensor chip, which includes six ISFETS (1–6), an (amperometric) oxygen sensor (7) and a conductometric sensor (8 a, b). The on-chip circuitry includes an EPROM, a multiplexer and counter, a driver unit, a conductometric and potentiostatic circuit and a heater. Reprinted from [286] with permission

acid' and monitoring of the pH-change). The gate oxide can also be covered with different ion-selective membranes to achieve sensitivity to target ions such as potassium. All six ISFETs or only a subset can be used. The idea was to make a standard chip to reduce manufacturing costs and then modify the chip with selective coatings according to user needs.

The integrated amperometric sensor (see also Section 7.3.4.1) can be used as a Clark-type oxygen sensor, which is based on a two-step reduction of gaseous oxygen in aqueous solution via hydrogen peroxide to hydroxyl ions.

The conductometric sensor (see Section 7.3.4.3) consists of two parallel sensors (8a), which share one common electrode (8b). A sinusoidal AC potential is applied to the electrodes, and the current, which depends on the solution composition (concentration of charged particles or ions), is recorded. The eight sensors can continuously monitor ions, dissolved gases and biomolecules via enzymatic reactions that produce charged particles.

The full system is produced in a 1.2 µm single-metal, single-poly CMOS process, and the chip size is 4.11×6.25 mm^2 [286]. The chip is operated at 5 V and hosts all driving circuitry of the sensors such as ISFET buffer amplifiers, a potentiostatic setup for the amperometric sensor and the circuitry necessary to perform a four-point conductometric measurement. In addition, the chip exhibits a temperature control unit to keep the system temperature at an adjustable value (physiological conditions). A single-bit EPROM (electrically programmable read-only memory) was implemented on-chip to make sure that the chip is only used once and then is disposed of, which is a crucial feature in medical applications. Additional on-chip electronics include units to control the chip (multiplexer, demultiplexer, 4-bit Gray counter and decoder) and units to provide the biasing and the communication to off-chip instrumentation. Owing to the high level of on-chip integration, only five external connections are needed: two for power supply, two for bidirectional communication and one for a clock signal [286].

First tests including amperometric oxygen measurements, the assessment of potassium concentrations with ISFETs (by directly connecting the ISFET buffer to a plotter) and conductometric measurements with a buffer solution have been performed [286].

7.3.5.3 Multitransducer Gas Sensor Microsystem

The multitransducer gas sensor microsystem includes three different transducers, a mass-sensitive cantilever, a capacitive sensor and a calorimetric sensor, all of which rely on polymeric coatings as sensitive layers to detect airborne volatile organic compounds (VOCs) [287–289]. The three transducers respond to fundamentally different analyte molecule properties. The thermally actuated cantilever mainly responds to the mass (volatility) of sorbed molecules (see Section 7.3.1.2), the calorimeter responds to the analyte heat of absorption/desorption (see Section 7.3.2.2) and the capacitor responds to the dielectric properties of the absorbates (see Section 7.3.4.3, subsection Chemocapacitors). The monolithic system thus simultaneously provides three different ('orthogonal') sensor responses, which can be used to classify or quantify analytes in the gas phase.

A photograph of the microsystem architecture is displayed in Fig. 7.37 [289]. The monolithic system includes the sensors, driving and signal-conditioning circuitry (sensor front ends) as already described in the previous sections dealing with the respective transducers. The system chip additionally includes a temperature sensor, since volatile absorption in polymers is strongly temperature-dependent (a 10 °C temperature increase reduces the signal by 50%). The analog-to-digital conversion is done on chip, which allows a favorable signal-to-noise ratio to be achieved, since noisy connections are avoided, and a robust digital signal is generated on-chip and then transmitted to an off-chip data port via an I2C serial interface [220]. The I2C bus interface offers the additional advantage of having only very few signal lines (essentially two) for bidirectional communication and also allows for operating multiple chips on the same bus system. An on-chip digital controller manages the sensor timing and the chip power budget. The sensors are located at one end of the chip so that the electronics can be covered with epoxy. The overall chip size is 5×7 mm^2. For more details on the system and its components, see [289].

Polymeric layers of ~ 2 µm thickness were applied to the transducers by using a drop-coating technique. The utilized polymers included ethylcellulose (EC), poly(dimethylsiloxane) (PDMS) and poly(ether urethane) (PEUT). The polymer-coated single-chip microsensor system was mounted in a gas manifold, and the sensors were then alternately exposed to air loaded with humidity or volatile or-

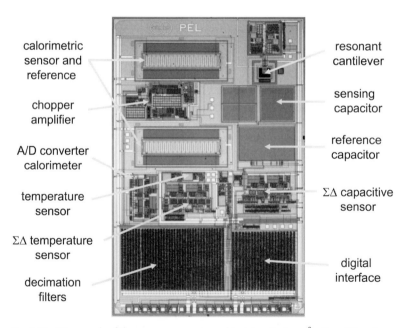

Fig. 7.37 Micrograph of the gas sensor system chip (size: 5×7 mm^2). The different components are indicated; $\Sigma\Delta$ represents Sigma–Delta converters [289]

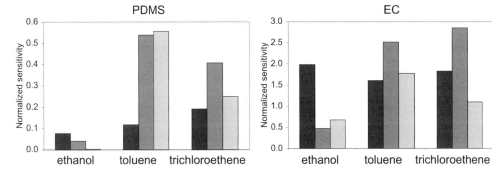

Fig. 7.38 Sensitivity as normalized to that of n-octane, which was set to unity for each transducer and polymer. The patterns of the two different polymers (PDMS and EC) differ significantly [98]

ganic compounds (ethanol, n-octane, trichloroethene and toluene) in different concentrations and pure air at 300 s intervals.

The sensitivities, i.e., the slope of sensor signal versus analyte concentration change of a polar (EC) and a non-polar (PDMS) polymer to various analytes are shown in Fig. 7.38 [98]. For comparability reasons, the analyte responses were normalized to that of n-octane, which was set to unity for each polymer and transducer, so that three bars of unity length would represent the n-octane signal in Fig. 7.38 for each polymer. Another possibility for normalizing the signals of the different transducers would rely on multiplying the signals or slopes with the analyte saturation vapor pressures [290]. It is obvious from Fig. 7.38 that polymer variation significantly changes the sensor system response pattern. The non-polar PDMS preferentially absorbs n-octane so that the response slopes of all other analytes and especially the polar ones (e.g. ethanol) are lower (values of <1). EC is polar so that the opposite holds true. The discrimination capability of a microsensor system can therefore be greatly enhanced by (a) realizing different transducer principles on a single chip and (b) additionally using various polymeric coatings.

The monolithic CMOS gas microsystem is targeted at identifying organic solvents in transport containers or providing workplace safety in, e.g., the chemical industry. It may form part of a handheld or credit-card-sized detection unit.

7.4 Acknowledgments

The author is very grateful to Professor Henry Baltes for his continuous support. He is also indebted to many highly motivated and excellent co-workers, namely Christoph Hagleitner, Dirk Lange, Markus Graf, Diego Barrettino, Cyril Vancura, Adrian Kummer, Yue Li, Kai-Uwe Kirstein, Petra Kurzawski, Wan-Ho Song, Urs Frey, Nicole Kerness, Stefano Taschini and Martin Zimmermann, who designed, fabricated, tested and packaged many of the devices described in this chapter.

7.5 References

1 R. A. Durst, R. W. Murray, K. Izutsu, K. M. Kadish, L. R. Faulkner, *Draft IUPAC Report, Commission V.5*.
2 A. Hulanicki, S. Glab, F. Ingman, *Pure Appl. Chem.* **1991**, *63*, 1247.
3 W. Göpel, J. Hesse, J. N. Zemel (eds.), *Sensors: a Comprehensive Survey*, Vol. 2/3, *Chemical and Biochemical Sensors*; Weinheim: VCH, **1991**; H. Baltes, W. Göpel, J. Hesse (eds.), *Sensors Updates*; Weinheim: Wiley-VCH.
4 J. Janata, *Principles of Chemical Sensors*; New York: Plenum Press, **1989**.
5 J. Janata, R. J. Huber, (eds.), *Solid State Chemical Sensors*, Academic Press, San Diego, **1985**.
6 M. J. Madou, S. R. Morrison, *Chemical Sensing with Solid State Devices*; Boston: Academic Press, **1989**.
7 J. Janata, M. Josowicz, P. Vanysek, M. D. DeVaney, *Anal. Chem.* **1998**, *70*, 179R–208R.
8 T. E. Edmonds (ed.), *Chemical Sensors*; New York: Chapman and Hall, **1988**.
9 E. Kress-Rodgers (ed.), *Handbook of Biosensors and Electronic Noses*; Boca Raton, FL: CRC Press, **1997**.
10 E. A. Hall, *Biosensors*; Milton Keynes: Open University Press, **1990**.
11 J. W. Gardner, P. N. Bartlett, *Electronic Noses*; Oxford: Oxford University Press, **1999**.
12 T. C. Pearce, S. S. Schiffman, H. T. Nagle, J. W. Gardner, *Handbook of Machine Olfaction*; Weinheim: Wiley-VCH, **2003**.
13 M. Lambrechts, W. Sansen, *Biosensors: Microelectrochemical Devices*; Bristol: Institute of Physics Publishing, **1992**.
14 P. W. Atkins, *Physical Chemistry*, 5th edn; Oxford: Oxford University Press, **1994**.
15 J. N. Zemel, *Rev. Sci. Instrum.* **1990**, *61*, 1579–1606.
16 S. Middelhoek, S. A. Audet, *Silicon Sensors*; London: Academic Press, **1989**.
17 S. Middelhoek, *Sens. Actuators A* **1994**, *41/42*, 1–8.
18 S. M. Sze (ed.), *Semiconductor Sensors*; New York: Wiley, **1994**.
19 R. S. Muller, R. T. Howe, S. D. Senturia, R. L. Smith, R. M. White (eds.), *Microsensors*; New York: IEEE Press, **1991**.
20 J. W. Gardner, *Microsensors*; Chichester: Wiley, **1994**.
21 J. W. Gardner, V. K. Varadan, O. O. Awadelkin, *Microsensors, MEMS and Smart Devices*; New York: Wiley, **2001**.
22 G. T. A. Kovacs, *Micromachined Transducers*; New York: MCB McGraw-Hill, **1998**.
23 A. Hierlemann, O. Brand, C. Hagleitner, H. Baltes, *Proc. IEEE Chem. Biol. Microsens.* **2003**, *91*, 839–863.
24 M. Elwenspoek, H. Hansen, *Silicon Micromachining*; Cambridge: Cambridge University Press, **1998**.
25 M. Madou, *Fundamentals of Microfabrication*; Boca Raton, FL: CRC Press, **1997**.
26 M. Gad-el-Hak, *The MEMS Handbook*; Boca Raton, FL: CRC Press, **2002**.
27 J. W. Gardner, M. Cole, F. Udrea, in: *Proc. IEEE Sensors* 2002, 721–726.
28 A. Hierlemann, H. Baltes, *Analyst* **2003**, *128*, 15–28.
29 H. L. Tuller, R. Mlcak, *Sens. Actuators B* **1996**, *35/36*, 255–261.
30 S. T. Picraux, P. J. McWorther, *IEEE Spectrum* **1998**, *12*, 24–33.
31 H. Baltes, O. Brand, *IEEE AES Syst. Mag.* **1999**, *14*, 29–34.
32 J. W. Grate, G. C. Frye, in: *Sensors Update*, H. Baltes, W. Göpel, J. Hesse (eds.); Weinheim: VCH, **1996**, Vol. 2, pp. 37–83.
33 J. W. Grate, S. J. Martin, R. M. White, *Anal. Chem.* **1993**, *65*, 940A–948A and 987A–996A.
34 D. S. Ballantine, R. M. White, S. J. Martin, A. J. Ricco, G. C. Frye, E. T. Zellers, H. Wohltjen, *Acoustic Wave Sensors: Theory, Design and Physico-Chemical Applications*; San Diego: Academic Press, **1997**.
35 G. Sauerbrey, *Z. Phys.* **1959**, *155*, 206–222.
36 S. J. Martin, G. C. Frye, S. D. Senturia, *Anal. Chem.* **1994**, *66*, 2201.
37 W. H. King, *Anal. Chem.* **1964**, *36*, 1735.
38 R. M. White, *Proc. IEEE* **1970**, *58*, 1238–1276.

39 K. M. Lakin, J. Liu, K. Wang, in: *Proc. IEEE Ultrasonics Symposium*; **1974**, 302–306.

40 J. Xia, S. Burns, M. Porter, T. Xue, G. Liu, R. Wyse, C. Thielen, in: *Proc. IEEE International Frequency Control Symposium* (Cat. No. 95CH35752); **1995**, 879–884.

41 C. T. Chuang, R. M. White, J. J. Bernstein, *IEEE Electron Device Lett.* **1982**, 6, 145–148.

42 S. J. Martin, K. S. Schweizer, A. J. Ricco, T. E. Zipperian, in: *Proc. Transducers 1985* (Cat. No. 85CH2127-9); **1985**, 445.

43 M. S. Nieuwenhuizen, A. J. Nederlof, *Sens. Actuators B* **1992**, 9, 171–176.

44 O. Brand, H. Baltes, 'Micromachined resonant sensors', in: *Sensors Update*, H. Baltes, W. Göpel, J. Hesse (eds.); Weinheim: VCH, **1998**, Vol. 4, 3–51.

45 S. Ahmadi, C. Korman, M. Zaghloul, H. Kuan Hsun, in: *Proc. of the IEEE International Symposium on Circuits and Systems, ISCAS*; **2003**, Vol. IV, 848–851.

46 C. Caliendo, P. Verardi, E. Verona, A. D'Amico, C. DiNatale, G. Saggio, M. Serafini, R. Paolesse, S. E. Huq, *Smart Mater. Struct.* **1997**, 6, 689–699.

47 J. K. Gimzewski, C. Gerber, E. Meyer, E. E. Schlittler, *Chem. Phys. Lett.* **1994**, 217, 589–594.

48 R. Berger, E. Delamarche, H. P. Lang, C. Gerber, J. K. Gimzewski, E. Meyer, H. J. Güntherodt, *Science* **1997**, 276, 2021.

49 J. Fritz, M. K. Baller, H. P. Lang, H. Rothuizen, P. Vettiger, E. Meyer, H. J. Güntherodt, C. Gerber, J. K. Gimzewski, *Science* **2000**, 288, 316–318.

50 S. S. Lee, R. M. White, *Sens. Actuators A* **1996**, 52, 41–45.

51 C. L. Britton, R. J. Warmack, S. F. Smith, P. I. Oden, R. L. Jones, T. Thundat, G. M. Brown, W. L. Bryan, J. C. DePriest, M. N. Ericson, M. S. Emery, M. R. Moore, G. W. Turner, A. L. Wintenberg, T. D. Threatt, Z. Hu, G. Clonts, J. M. Rochelle, in: *Proceedings of 20th Anniversary Conference on Advanced Research in VLSI.* Los Alamitos, CA: IEEE Comput. Soc., **1999**, 359–368.

52 E. A. Wachter, T. Thundat, *Rev. Sci. Instrum.* **1995**, 66, 3662–3667.

53 T. Thundat, G. Y. Chen, R. J. Warmack, D. P. Allison, E. A. Wachter, *Anal. Chem.* **1995**, 67, 519–521.

54 C. L. Britton, R. L. Jones, P. I. Oden, Z. Hu, R. J. Warmack, S. F. Smith, W. L. Bryan, J. M. Rochelle, *Ultramicroscopy* **2000**, 82, 17–21.

55 M. Maute, S. Raible, F. E. Prins, D. P. Kern, H. Ulmer, U. Weimar, W. Göpel, *Sens. Actuators B* **1999**, 58, 505–511.

56 B. H. Kim, F. E. Prins, D. P. Kern, S. Raible, U. Weimar, *Sens. Actuators B* **2001**, 78, 12–18.

57 A. Boisen, J. Thaysen, H. Jesenius, O. Hansen, *Ultramicroscopy* **2000**, 82, 11–16.

58 H. Jesenius, J. Thaysen, A. A. Rasmussen, L. H. Veje, O. Hansen, A. Boisen, *Appl. Phys. Lett.* **2000**, 76, 2615–2617.

59 H. P. Lang, M. K. Baller, R. Berger, C. Gerber, J. K. Gimzewski, F. Battiston, P. Fornaro, J. P. Ramseyer, E. Meyer, H. J. Güntherodt, *Anal. Chim. Acta* **1999**, 393, 59–65.

60 F. Battiston, J. P. Ramseyer, H. P. Lang, M. Baller, C. Gerber, J. K. Gimzewski, E. Meyer, H. J. Güntherodt, *Sens. Actuators B* **2001**, 77, 122–131.

61 M. Baller, H. P. Lang, J. Fritz, C. Gerber, J. K. Gimzewski, U. Drechsler, H. Rothuizen, M. Despont, P. Vettiger, F. Battiston, J. P. Ramseyer, P. Fornaro, E. Meyer, H. J. Güntherodt, *Ultramicroscopy* **2000**, 82, 1–9.

62 H. Baltes, D. Lange, A. Koll, *IEEE Spectrum* **1998**, 9, 35–38.

63 A. Hierlemann, D. Lange, C. Hagleitner, N. Kerness, A. Koll, O. Brand, H. Baltes, *Sens. Actuators B* **2000**, 70, 2–11.

64 D. Lange, C. Hagleitner, A. Hierlemann, O. Brand, H. Baltes, *Anal. Chem.* **2002**, 74, 3084–3095.

65 Y. Li, C. Vancura, C. Hagleitner, J. Lichtenberg, O. Brand, H. Baltes, in: *Proc. IEEE Sensors, Toronto*; **2003**, 244–245.

66 C. Vancura, M. Rüegg, Y. Li, D. Lange, C. Hagleitner, O. Brand, A. Hierlemann, H. Baltes, in: *Proc. IEEE Transducers 2003*, Boston, MA; **2003**, 1355–1358.

67 D. Lange, C. Hagleitner, C. Herzog, O. Brand, H. Baltes, Sens. Actuators A 2003, 103, 150–155.
68 A. Voiculescu, M. Zaghloul, R. A. McGill, in: Proc. of IEEE International Symposium on Circuits and Systems, IS-CAS; 2003, Vol. III, 922–925.
69 C. Hagleitner, A. Hierlemann, H. Baltes, 'CMOS single-chip gas detection systems, Part II,' in: Sensors Update, H. Baltes, J. Korvink, G. Fedder (eds.); Weinheim: Wiley-VCH, 2003, Vol. 12, 51–120.
70 G. C. M. Meijer, A. W. van Herwaarden, Thermal Sensors; Bristol: Institute of Physics Publishing, 1994.
71 A. R. Baker, 'Combustible gas-detecting, electrically heatable element,' UK Patent 892530; 1962.
72 M. Zanini, J. H. Visser, L. Rimai, R. E. Soltis, A. Kovalchuk, D. W. Hoffman, E. M. Logothetis, U. Bonne, L. Brewer, O. W. Bynum, M. A. Richard, Sens. Actuators A 1995, 48, 187–192.
73 R. P. Manginell, J. H. Smith, A. J. Ricco, D. J. Moreno, R. C. Hughes, R. J. Huber, S. D. Senturia, in: Technical Digest Solid State Sensor and Actuator Workshop, Hilton Head Island, SC; 1996, 23–27.
74 R. P. Manginell, J. H. Smith, A. J. Ricco, in: Proc. 4th Annual Symposium on Smart Structures and Materials; SPIE, 1997, 273–284.
75 D. W. Dabill, S. J. Gentry, P. T. Walsh, Sens. Actuators 1987, 11, 135–143.
76 M. G. Jones, T. G. Nevell, Sens. Actuators 1989, 16, 215–224.
77 R. Aigner, M. Dietl, R. Katterloher, V. Klee, Sens. Actuators B 1996, 33, 151–155.
78 R. Aigner, F. Auerbach, P. Huber, R. Mueller, G. Scheller, Sens. Actuators B 1994, 18/19, 143–147.
79 P. Krebs, A. Grisel, Sens. Actuators B 1993, 13/14, 155–158.
80 M. Gall, Sens. Actuators B 1991, 4, 533–538.
81 M. Gall, Sens. Actuators B 1993, 15/16, 260–264.
82 C. H. Mastrangelo, 'Thermal application of microbridges,' PhD Thesis; UC Berkeley, 1991.

83 A. Accorsi, G. Delapierre, C. Vauchier, D. Charlot, Sens. Actuators B 1991, 4, 539–543.
84 A. W. van Herwaarden, P. M. Sarro, Sens. Actuators 1986, 10, 321–346.
85 P. M. Sarro, A. W. van Herwaarden, W. van der Vlist, Sens. Actuators A 1994, 41/42, 666–671.
86 A. W. van Herwaarden, P. M. Sarro, J. W. Gardner, P. Bataillard, Sens. Actuators A 1994, 43, 24–30.
87 N. Kerness, A. Koll, A. Schaufelbuehl, C. Hagleitner, A. Hierlemann, O. Brand, H. Baltes, in: Proc. IEEE Workshop on Micro Electro Mechanical Systems, MEMS 2000, Myazaki, Japan; 2000, 96–101.
88 A. Koll, A. Schaufelbühl, O. Brand, H. Baltes, C. Menolfi, H. Huang, in: Proc. IEEE Workshop on Micro Electro Mechanical Systems, MEMS 99, Orlando; 1999, 547–551.
89 N. Kerness, 'CMOS-based calorimetric chemical microsensors,' PhD Thesis; ETH Zurich, ETH Diss. 14839, 2002.
90 J. Lerchner, J. Seidel, G. Wolf, E. Weber, Sens. Actuators B 1996, 32, 71–75.
91 D. Caspary, M. Schröpfer, J. Lerchner, G. Wolf, Thermochim. Acta 1999, 337, 19–26.
92 J. Lerchner, A. Wolf, G. Wolf, J. Thermal Anal. 1999, 55, 212–223.
93 P. Bataillard, E. Steffgen, S. Haemmerli, A. Manz, H. M. Widmer, Biosens. Bioelectron. 1993, 8, 89–98.
94 J. M. Köhler, E. Kessler, G. Steinhage, B. Gründig, K. Cammann, Mikrochim. Acta 1995, 120, 309–319.
95 P. M. Sarro, H. Yashiro, A. M. van Herwaarden, S. Middelhoek, Sens. Actuators 1988, 14, 191–201.
96 A. W. van Herwaarden, Meas. Sci. Technol. 1992, 3, 935–937.
97 C. A. Papadopoulos, D. S. Vlachos, J. N. Avaritsiotis, Sens. Actuators B 1996, 34, 524–527.
98 P. Kurzawski, I. Lazic, C. Hagleitner, A. Hierlemann, H. Baltes, in: Proc. IEEE Transducers 2003, Boston, MA; 2003, 1359–1362.
99 O. Wolfbeis, G. E. Boisde, G. Gauglitz, 'Optochemical sensors,' in: Sensors: a Comprehensive Survey, W. Göpel, J.

HESSE, J.N. ZEMEL (eds.); Weinheim: VCH, **1991**, Vol. 2, 573–646.
100 G. GAUGLITZ, 'Opto-chemical and opto-immuno sensors,' in: *Sensors: a Comprehensive Survey, Update*, W. GÖPEL, J. HESSE, J.N. ZEMEL (eds.); Weinheim: VCH, **1996**, Vol. 1, 1–49.
101 A. BRECHT, G. GAUGLITZ, W. GÖPEL, 'Sensors in biomolecular interaction analysis and pharmaceutical drug screening,' in: *Sensors: a Comprehensive Survey, Update*, W. GÖPEL, J. HESSE, J.N. ZEMEL (eds.); Weinheim: VCH, **1998**, Vol. 3, 573–646.
102 S.M. SZE, *Semiconductor Devices: Physics and Technology*, 2nd edn; New York: Wiley, **2002**.
103 H. ZAPPE, 'Semiconductor optical sensors,' in *Sensors: a Comprehensive Survey, Update*, W. GÖPEL, J. HESSE, J.N. ZEMEL (eds.); Weinheim: VCH, **1999**, Vol. 5, 1–45.
104 U.E. SPICHIGER-KELLER, *Chemical Sensors and Biosensors for Medical and Biological Application*; Weinheim: Wiley-VCH, **1998**.
105 O.S. WOLFBEIS, *Anal. Chem.* **2000**, *72*, 81R–90R.
106 J. DAKIN, B. CULSHAW (eds.), *Optical Fiber Sensors*; Norwood, MA: Artech House, **1997**, Vols. 3 and 4.
107 G.E. BOISDE, A. HARMER (eds.), *Chemical and Biochemical Sensing with Optical Fibers and Waveguides*; Norwood, MA: Artech House, **1996**.
108 M. SIMPSON, G. SAYLER, D. NIVENS, S. RIPP, M. PAULUS, G. JELLISON, *Trends Biotechnol.* **1998**, *16*, 332–338.
109 W. LUKOSZ, *Sens. Actuators B* **1995**, *29*, 37–50.
110 W. LUKOSZ, C. STAMM, H.R. MOSER, R. RYF, J. DÜBENDORFER, *Sens. Actuators B* **1997**, *38/39*, 316–323.
111 K. TIEFENTHALER, W. LUKOSZ, *J. Opt. Soc. Am. B* **1989**, *6*, 209–220; *Thin Solid Films* **1985**, *126*, 205–211.
112 W. LUKOSZ, P.M. NELLEN, C. STAMM, P. WEISS, *Sens. Actuators B* **1990**, *1*, 585–588.
113 F. BROSINGER, H. FREIMUTH, M. LACHER, W. EHRFELD, E. GEDIG, A. KATERKAMP, F. SPENER, K. CAMMANN, *Sens. Actuators B* **1997**, *44*, 350–355.
114 S. BUSSE, J. KÄSHAMMER, S. KRÄMER, S. MITTLER, *Sens. Actuators B* **1999**, *60*, 148–154.
115 E.F. SCHIPPER, A.M. BRUGMAN, C. DOMINGUEZ, L.M. LECHUGA, R.P. KOOYMAN, J. GREVE, *Sens. Actuators B* **1997**, *40*, 147–153.
116 R.G. HEIDEMAN, R.P. KOOYMAN, J. GREVE, *Sens. Actuators B* **1993**, *10*, 209–217.
117 R.G. HEIDEMAN, G.J. VELDHUIS, E.W.H. JAGER, P.V. LAMBECK, *Sens. Actuators B* **1996**, *35/36*, 234–240.
118 R.G. HEIDEMAN, P.V. LAMBECK, *Sens. Actuators B* **1999**, *61*, 100–127.
119 R. BERNINI, S. CAMPOPIANO, C. DE-BOER, P.M. SARRO, L. ZENI, *IEEE Sens. J.* **2003**, *3*, 652–657.
120 F. PRIETO, B. SEPULVEDA, A. CALLE, A. LLOBERA, C. DOMINGUEZ, L.M. LECHUGA, *Sens. Actuators B* **2003**, *92*, 151–158.
121 K. TIEFENTHALER, W. LUKOSZ, *Sens. Actuators* **1988**, *15*, 273–284.
122 G. GAUGLITZ, J. INGENHOFF, *Fresenius' J. Anal. Chem.* **1994**, *349*, 355–359.
123 J.H. CORREIA, G. DE GRAAF, S.H. KONG, M. BARTEK, R.F. WOLFFENBUTTEL, *Sens. Actuators A* **2000**, *82*, 191–197.
124 K. ARATANI, P.J. FRENCH, P.M. SARRO, D. POENAR, R.F. WOLFFENBUTTEL, S. MIDDELHOEK, *Sens. Actuators A* **1994**, *43*, 17–23.
125 G. DE GRAAF, R.F. WOLFFENBUTTEL, *Sens. Actuators A* **1998**, *67*, 115–119.
126 F. GRASDEPOT, H. ALAUSE, W. KNAP, J.P. MALZAC, J. SUSKI, *Sens. Actuators B* **1996**, *35/36*, 377–380.
127 H. ALAUSE, F. GRASDEPOT, J.P. MALZAC, W. KNAP, J. HERMANN, *Sens. Actuators B* **1997**, *43*, 18–23.
128 J.P. SILVEIRA, J. ANGUITA, F. BRIONES, F. GRASDEPOT, A. BAZIN, *Sens. Actuators B* **1998**, *48*, 305–307.
129 J. HAN, *Appl. Phys. Lett.* **1999**, *74*, 445–447.
130 J. HAN, D.P. NEIKIRK, M. CLEVENGER, J.T. MCDEVITT, *Proc. SPIE* **1996**, *2881*, 171–179.
131 T. CHEN, *Sens. Actuators B* **1993**, *13/14*, 284–287.
132 H. CARSON, *Sens. Rev.* **1997**, *17*, 304–306; http://www.vaisala.com.

133 M. Simpson, G. Sayler, D. Nivens, S. Ripp, M. Paulus, G. Jellison, *Trends Biotechnol.* **1998**, *16*, 332–338.

134 M. Simpson, M. Paulus, G. Jellison, G. Sayler, B. Applegate, S. Ripp, D. Nivens, in: *Technical Digest Solid-State Sensor and Actuator Workshop, Hilton Head Island, SC*; **1998**, 354–357.

135 P. Fabry, E. Siebert, 'Electrochemical sensors,' in: *The CRC Handbook of Solid State Electrochemistry*, P. J. Gellings, H. J. M. Bouwmeester (eds.); Boca Raton, FL: CRC Press, **1997**, 329–371.

136 U. Wollenberger, *Biotechnol. Eng. Rev.* **1996**, *13*, 237–266.

137 J. Kas, M. Marek, M. Stastny, R. Volf, *Bioelectrochem. Princip. Pract.* **1996**, *3*, 361–453.

138 J. M. Kauffman (ed.), *Bioelectrochem. Bioeng.* **1997**, *42*, Special Issue.

139 J. W. Schultze, V. Tsakova, *Electrochim. Acta* **1999**, *44*, 3605–3627.

140 M. R. Neuman, R. P. Buck, V. Cosofret, E. Lindner, C. C. Liu, *IEEE Eng. Med. Biol.* **1994**, *13*, 409–419.

141 W. Sansen, D. de Wachter, L. Callewaert, M. Lambrechts, A. Claes, *Sens. Actuators B* **1990**, *1*, 298–302.

142 A. van den Berg, P. D. van der Wal, B. H. van der Schoot, N. F. de Rooij, *Sens. Mater.* **1994**, *6*, 23–43.

143 R. Steinkuhl, C. Sundermeier, H. Hinkers, C. Dumschat, K. Cammann, M. Knoll, *Sens. Actuators B* **1996**, *33*, 19–24.

144 B. F. Y. Yon Hin, R. S. Sethi, C. R. Lowe, *Sens. Actuators B* **1990**, *1*, 550–554.

145 A. Schwake, B. Ross, K. Cammann, *Sens. Actuators B* **1998**, *46*, 242–248.

146 M. Dilhan, D. Estève, A. M. Gué, O. Mauvais, L. Mercier, *Sens. Actuators B* **1995**, *26/27*, 401–403.

147 A. van den Berg, A. Grisel, E. Verney-Norberg, B. H. van der Schoot, M. Koudelka-Hep, N. F. de Rooij, *Sens. Actuators B* **1993**, *13/14*, 396–399.

148 M. Wittkampf, K. Cammann, M. Amrein, R. Reichelt, *Sens. Actuators B* **1997**, *40*, 79–84.

149 M. Koudelka, *Sens. Actuators* **1986**, *9*, 249.

150 H. Hinkers, C. Sundermeier, R. Lürick, F. Walfort, K. Cammann, M. Knoll, *Sens. Actuators B* **1995**, *26/27*, 398–400.

151 H. Hinkers, T. Hermes, C. Sundermeier, M. Borchardt, C. Dumschat, S. Bücher, M. Bühner, K. Cammann, M. Knoll, *Sens. Actuators B* **1995**, *24/25*, 300–303.

152 T. Hermes, M. Bühner, S. Bücher, C. Sundermeier, C. Dumschat, M. Borchardt, K. Cammann, M. Knoll, *Sens. Actuators B* **1994**, *21*, 33–37.

153 J. Zhu, J. Wu, C. Tian, W. Wu, H. Zhang, D. Lu, G. Zhang, *Sens. Actuators B* **1994**, *20*, 17–22.

154 A. van den Berg, A. Grisel, M. Koudelka, B. H. van der Schoot, *Sens. Actuators B* **1991**, *5*, 71.

155 Z. Huixian, L. Tai Chin, L. Ralf, R. Reinhard, *Sens. Actuators B* **1998**, *46*, 155–159.

156 M. Wittkampf, G. C. Chemnitius, K. Cammann, M. Rospert, W. Mokwa, *Sens. Actuators B* **1997**, *43*, 40–44.

157 L. C. Clark, *Trans. Am. Soc. Artif. Int. Organs* **1956**, *2*, 41–48.

158 P. Bergveld, *IEEE Trans. Biomed. Eng.* **1970**, *BME-17*, 70–71.

159 I. Lundström, S. Shivaraman, C. Svensson, L. Lundkvist, *Appl. Phys. Lett.* **1975**, *26*, 55–57.

160 I. Lundström, *Sens. Actuators B* **1996**, *56*, 75–82.

161 L. G. Ekedahl, M. Eriksson, I. Lundström, *Acc. Chem. Res.* **1998**, *31*, 249–256.

162 Y. Morita, K. Nakamura, C. Kim, *Sens. Actuators B* **1996**, *33*, 96–99.

163 J. L. Rodriguez, R. C. Hughes, W. T. Corbett, P. J. McWhorter, in: *International Electron Devices Meeting 1992, San Francisco, Technical Digest* (Cat. No. 92CH3211-0); **1992**, 19.6.1–19.6.4.

164 R. C. Hughes, D. J. Moreno, M. W. Jenkins, J. L. Rodriguez, in: *Solid-State Sensor and Actuator Workshop, Hilton Head Island, SC, Technical Digest*; **1994**, 57–60.

165 A. Srivastava, N. George, J. Cherukuri, *Proc. SPIE* **1995**, *2642*, 121–129.

166 W. M. Siu, R. S. Cobbold, *IEEE Trans. Electron. Devices* **1979**, *ED-26*, 1805–1815.

167 J. Janata, *Sens. Actuators* **1983**, *4*, 255–265.

168 I. LAUKS, *Sens. Actuators* **1981**, *1*, 261–288.
169 S. ALEGRET, J. BARTOLI, C. JIMENEZ JORQUERA, M. DEL VALLE, C. DOMINGUEZ, J. ESTEVE, J. BAUSELLS, *Sens. Actuators B* **1992**, *7*, 555–560.
170 K. CHANG SOO, S. HWA IL, L. CHAE HYANG, S. BYUNG KI, in: *1997 International Conference on Solid-State Sensors and Actuators, Transducers 97, Chicago, Digest of Technical Papers* (Cat. No. 97TH8267); **1997**, 911–914.
171 I. GRACIA, C. CANE, E. LORA TAMAYO, *Sens. Actuators B* **1995**, *24*, 206–210.
172 R. L. SMITH, D. C. SCOTT, *IEEE Trans. Biomed. Eng.* **1986**, *BME-33*, 83–90.
173 S. MARTINOIA, L. LORENZELLI, G. MASSOBRIO, B. MARGESIN, A. LUI, *Sens. Mater.* **1999**, *11*, 279–295.
174 P. HEIN, P. EGGER, *Sens. Actuators B* **1993**, *14*, 655–656.
175 M. LEHMANN, W. BAUMANN, M. BRISCHWEIN, H. J. GAHLE, I. FREUND, R. EHRET, S. DRECHSLER, H. PALZER, M. KLEINTGES, U. SIEBEN, B. WOLF, *Biosens. Bioelectron.* **2001**, *16*, 195–203.
176 P. NEUZIL, *Sens. Actuators B* **1995**, *24*, 232–235.
177 L. CHAE-HYANG, S. HWA-IL, L. YOUNG-CHUL, C. BYUNG-WOOG, J. HOON, S. BYUNG-KI, *Sens. Actuators B* **2000**, *64*, 37–41.
178 J. BAUSELLS, J. CARRABINA, A. ERRACHID, A. MERLOS, *Sens. Actuators B* **1999**, *57*, 56–62.
179 C. YUAN LUNG, C. JUNG CHUAN, S. TAI PING, L. HUNG KWEI, C. WEN YAW, H. SHEN KAN, *Sens. Actuators B* **2001**, *75*, 36–42.
180 C. YUAN LUNG, C. JUNG CHUAN, S. TAI PING, C. WEN YAW, H. SHEN KAN, *Sens. Actuators B* **2001**, *76*, 582–593.
181 L. BOUSSE, P. BERGVELD, *Sens. Actuators* **1984**, *6*, 65.
182 A. VAN DEN BERG, P. BERGVELD, D. N. REINHOUDT, E. J. R. SUDHOLTER, *Sens. Actuators* **1985**, *8*, 129.
183 M. ARMGARTH, C. NYLANDER, *Appl. Phys. Lett.* **1981**, *39*, 91–92.
184 A. GRISEL, C. FRANCIS, E. VERNEY, G. MONDIN, *Sens. Actuators* **1989**, *17*, 285–295.
185 D. EWALD, A. VAN DEN BERG, A. GRISEL, *Sens. Actuators B* **1990**, *1*, 335–340.
186 *DuraFET*; http://www.honeywell.com.
187 D. N. REINHOUDT, J. F. J. ENGBERSEN, Z. BRZÓZKA, H. H. VAN DEN VLEKKERT, G. W. N. HONIG, H. A. J. HOLTERMAN, U. H. VERKERK, *Anal. Chem.* **1994**, *66*, 3618.
188 P. D. VAN DER WAL, E. J. R. SUDHOLTER, D. N. REINHOUDT, *Anal. Chim. Acta* **1991**, *245*, 159.
189 J. A. J. BRUNINK, J. R. HAAK, J. G. BOMER, D. N. REINHOUDT, M. A. MCKERVEY, S. J. HARRIS, *Anal. Chim. Acta* **1991**, *254*, 75.
190 Z. BRZÓZKA, P. L. H. M. COBBEN, D. N. REINHOUDT, J. J. H. EDEMA, J. BUTER, R. M. KELLOGG, *Anal. Chim. Acta* **1993**, *273*, 139.
191 W. GÖPEL, G. REINHARDT, 'Metal oxide sensors,' in: *Sensors Update*, H. BALTES, W. GÖPEL, J. HESSE (eds.); Weinheim: VCH, **1996**, Vol. 1, 49–120.
192 K. PERSAUD, G. H. DODD, *Nature* **1982**, *299*, 352–355.
193 K. C. PERSAUD, P. PELOSI, 'Sensor arrays using conducting polymers,' in *Sensors and Sensory Systems for an Electronic Nose*, J. W. GARDNER, P. N. BARTLETT (eds.); Dordrecht: Kluwer, **1992**.
194 P. TOPART, M. JOSOWICZ, *J. Phys. Chem.* **1992**, *96*, 7824–7830.
195 J. W. GARDNER, T. C. PEARCE, S. FRIEL, P. N. BARTLETT, N. BLAIR, *Sens. Actuators B* **1994**, *18/19*, 240–243.
196 M. COLE, J. W. GARDNER, A. W. Y. LIM, P. K. SCIVIER, J. E. BRIGNELL, *Sens. Actuators B* **1999**, *58*, 518–525.
197 J. W. GARDNER, M. VIDIC, P. INGLEBY, A. C. PIKE, J. E. BRIGNELL, P. SCIVIER, P. N. BARTLETT, A. J. DUKE, J. M. ELLIOTT, *Sens. Actuators B* **1998**, *48*, 289–295.
198 http://www.osmetech.plc.uk/; http://www.marconitech.com/.
199 M. C. LONERGAN, E. J. SEVERIN, B. J. DOLEMAN, S. A. BEABER, R. H. GRUBBS, N. S. LEWIS, *Chem. Mater.* **1996**, *8*, 2298–2312.
200 E. J. SEVERIN, B. J. DOLEMAN, N. S. LEWIS, *Anal. Chem.* **2000**, *72*, 658–668.
201 B. J. DOLEMAN, M. LONERGAN, E. J. SEVERIN, T. P. VAID, N. S. LEWIS, *Anal. Chem.* **1998**, *70*, 4177–4190.

202 S.V. Patel, M.W. Jenkins, R.C. Hughes, W.G. Yelton, A.J. Ricco, *Anal. Chem.* **2000**, *72*, 1532–1542.

203 M.P. Eastman, R.C. Hughes, W.G. Yelton, A.J. Ricco, S.V. Patel, M.W. Jenkins, *J. Electrochem. Soc.* **1999**, *146*, 3907–3913.

204 http://cyranosciences.com/.

205 J.A. Dickson, R.M. Goodman, in: *IEEE International Symposium on Circuits and Systems, Geneva, 2000, Proceedings* (Cat. No. 00CH36353); **2000**, Vol. 4, 341–344.

206 J.A. Dickson, M.S. Freund, N.S. Lewis, R.M. Goodman, in: *Solid-State Sensor and Actuator Workshop 2000, Hilton Head Island, SC, Technical Digest* (Cat. No. 00Trf-0001); **2000**, 162–164.

207 J.V. Hatfield, P. Neaves, P.J. Hicks, K. Persaud, P. Travers, *Sens. Actuators B* **1994**, *18/19*, 221–228.

208 P. Neaves, J.V. Hatfield, *Sens. Actuators B* **1995**, *26/27*, 223–231.

209 N. Kordas, Y. Manoli, W. Mokwa, M. Rospert, *Sens. Actuators A* **1994**, *43*, 31–37.

210 M. Graf, D. Barrettino, M. Zimmermann, C. Hagleitner, A. Hierlemann, H. Baltes, S. Hahn, N. Bârsan, U. Weimar, *IEEE Sens. J.* **2004**, *4*, 9–16.

211 M. Graf, D. Barrettino, S. Taschini, C. Hagleitner, A. Hierlemann, H. Baltes, *Anal. Chem.* in press.

212 J.S. Suehle, R.E. Cavicchi, M. Gaitan, S. Semancik, *IEEE Electron Device Lett.* **1993**, *14*, 118–120.

213 S. Semancik, R.E. Cavicchi, *Acc. Chem. Res.* **1998**, *31*, 279–287.

214 M. Graf, D. Barrettino, S. Taschini, C. Hagleitner, A. Hierlemann, H. Baltes, in: *Proc. IEEE MEMS, Kyoto*; **2003**, 303–306.

215 M. Graf, 'Microhotplate-based chemical sensor systems in CMOS technology,' PhD Thesis; ETH Zurich, ETH Diss. 15438, **2004**.

216 M.Y. Afridi, J.S. Suehle, M.E. Zaghloul, D.W. Berning, A.R. Hefner, R.E. Cavicchi, S. Semancik, C.B. Montgomery, C.J. Taylor, *IEEE Sens. J.* **2002**, *2*, 644–655.

217 D. Barrettino, M. Graf, M. Zimmermann, A. Hierlemann, H. Baltes, S. Hahn, N. Barsan, U. Weimar, in: *Proc. IEEE ISCAS 2002, Phoenix, AZ*; **2002**, Vol. II, 157–160.

218 D. Barrettino, M. Graf, S. Taschini, M. Zimmermann, C. Hagleitner, A. Hierlemann, H. Baltes, in: *Proc. IEEE VLSI Symposium, Kyoto*; **2003**, 157–160.

219 D. Barrettino, 'CMOS readout and control architectures for monolithic hotplate and cantilever systems,' PhD Thesis; ETH Zurich, ETH Diss. 15412, **2004**.

220 I2C; Philips, Eindhoven; http://www.semiconductors.philips.com/buses/i2c/facts.

221 G. Vandrish, *Key Eng. Mater.* **1996**, *122–124*, 185–224.

222 N. Barsan, M. Schweizer-Berberich, W. Göpel, *Fresenius' J. Anal. Chem.* **1999**, *365*, 287–304.

223 H. Geistlinger, *Sens. Actuators B* **1993**, *17*, 47–60.

224 G. Heiland, D. Kohl, in: *Chemical Sensor Technology*, T. Seiyama (ed.); Amsterdam: Elsevier, **1988**, Vol. 1, 15–38.

225 N. Barsan, U. Weimar, *J. Electroceram.* **2001**, *7*, 143–167.

226 I. Simon, N. Barsan, M. Bauer, U. Weimar, *Sens. Actuators B* **2001**, *73*, 1–26.

227 A.P. Lee, B.J. Reedy, *Sens. Actuators B* **1999**, *60*, 35–42.

228 B. Yea, T. Osaki, K. Sugahara, R. Konishi, *Sens. Actuators B* **1997**, *41*, 121–129.

229 A. Heilig, N. Barsan, U. Weimar, M. Schweizer-Berberich, J.W. Gardner, W. Göpel, *Sens. Actuators B* **1997**, *43*, 45–51.

230 P. Corcoran, P. Lowery, J. Anglesa, *Sens. Actuators B* **1998**, *48*, 448–455.

231 G. Sberveglieri, W. Hellmich, G. Müller, *Microsyst. Technol.* **1997**, *3*, 183–190.

232 V. Demarne, A. Grisel, *Sens. Actuators* **1988**, *13*, 301–313.

233 G. Faglia, E. Comini, M. Pardo, A. Taroni, G. Cardinali, S. Nicoletti, G. Sberveglieri, *Microsyst. Technol.* **1999**, *6*, 54–59.

234 H.S. Park, H.W. Shin, D.H. Yun, H.-K. Hong, C.H. Kwon, K. Lee, S-T. Kim, *Sens. Actuators B* **1995**, *24/25*, 478–481.

235 A. Götz, I. Gràcia, C. Cané, E. Lora-Tamayo, M.C. Horrillo, G. Getino, C. García, J. Gutiérrez, *Sens. Actuators B* **1997**, *44*, 483–487.

236 M. Frietsch, L.T. Dimitrakopoulos, T. Schneider, J. Goschnick, *Surf. Coat. Technol.* **1999**, *120/121*, 265–271.
237 S. Semancik, R.E. Cavicchi, K.G. Kreider, J.S. Suehle, P. Chaparala, *Sens. Actuators B* **1996**, *34*, 209–212.
238 G. Sberveglieri, G. Faglia, S. Gropelli, P. Nelli, A. Camanzi, *Semicond. Sci. Technol.* **1990**, *5*, 1231.
239 F. DiMeo, R.E. Cavicchi, S. Semancik, J.S. Suehle, N.H. Tea, J. Small, J.T. Armstrong, J.T. Kelliher, *J. Vac. Sci. Technol. A* **1998**, *16*, 131–138.
240 M. Schweizer-Berberich, J.G. Zheng, U. Weimar, W. Göpel, N. Barsan, E. Pentia, A. Tomescu, *Sens. Actuators B* **1996**, *31*, 1–5.
241 D. Briand, A. Krauss, B. van der Schoot, U. Weimar, N. Barsan, W. Göpel, N.F. de Rooij, *Sens. Actuators B* **2000**, *68*, 223–233.
242 A. Götz, I. Gracia, J.A. Plaza, C. Cane, P. Roetsch, H. Böttner, K. Seibert, *Sens. Actuators B* **2001**, *77*, 395–400.
243 S. Lie Yi, T. Zhenan, W. Jian, P.C.H. Chan, J.K.O. Sin, *Sens. Actuators B* **1998**, *49*, 81–87.
244 M. Yaowu, Y. Okawa, K. Inoue, K. Natukawa, *Sens. Actuators A* **2002**, *100*, 94–101.
245 L.Y. Sheng, C. de Tandt, W. Ranson, R. Vounckx, *Microelectron. Reliability* **2001**, *41*, 307–315.
246 N. Najafi, K.D. Wise, R. Mechant, J.W. Schwank, in: *Technical Digest. IEEE Solid-State Sensor and Actuator Workshop, Hilton Head Island, SC, 1992* (Cat. No. 92TH0403-X); **1992**, 19–22.
247 G.C. Cardinali, L. Dori, M. Fiorini, I. Sayago, G. Faglia, C. Perego, G. Sberveglieri, V. Liberali, F. Maloberti, D. Tonietto, *Analog Integrated Circuits Signal Process.* **1997**, *14*, 275–296.
248 F. Udrea, J.W. Gardner, D. Setiadi, J.A. Covington, T. Dogaru, C.C. Lu, W.I. Milne, *Sens. Actuators B* **2001**, *78*, 180–190.
249 F. Udrea, J.W. Gardner, in: *Proc. IEEE Sensors*; **2002**, 1379–1384.
250 J.A. Covington, F. Udrea, J.W. Gardner, in: *Proc. IEEE Sensors*; **2002**, 1389–1394.
251 J. Laconte, C. Dupont, D. Flandre, J.P. Raskin, in: *Proc. IEEE Sensors*; **2002**, 1395–1400.
252 J. Werno, R. Kersjes, W. Mokwa, H. Vogt, *Sens. Actuators A* **1994**, *42*, 578–581.
253 D. Vincenzi, M.A. Butturi, V. Guidi, M.C. Carotta, G. Martinelli, V. Guarnieri, S. Brida, B. Margesin, F. Giacomozzi, M. Zen, G.U. Pignatel, A.A. Vasiliev, A.V. Pisliakov, *Sens. Actuators B* **2001**, *77*, 95–99.
254 N.F. Sheppard, D.R. Day, H.L. Lee, S.D. Senturia, *Sens. Actuators* **1982**, *2*, 263–274.
255 S.D. Senturia, in: *Technical Digest Transducers 1985*; **1985**, 198–201.
256 M.C. Glenn, J.A. Schuetz, in: *Technical Digest Transducers 1985*; **1985**, 217–219.
257 D.D. Denton, S.D. Senturia, E.S. Anolick, D. Scheider, in: *Technical Digest Transducers 1985*; **1985**, 202–205.
258 G. Delapierre, H. Grange, B. Chambaz, L. Destannes, *Sens. Actuators* **1983**, *4*, 97–104.
259 H. Shibata, M. Ito, M. Asakursa, K. Watanabe, *IEEE Trans. Instrum. Meas.* **1996**, *45*, 564–569.
260 C. Hagleitner, A. Koll, R. Vogt, O. Brand, H. Baltes, in: *Technical Digest Transducers 1999, Sendai*; **1999**, Vol. 2, 1012–1015.
261 A. Koll, A. Kummer, O. Brand, H. Baltes, *Proc. SPIE* **1999**, *3673*, 308–317.
262 P. Hille, H. Strack, *Sens. Actuators A* **1992**, *32*, 321–325.
263 T. Boltshauser, H. Baltes, *Sens. Actuators A* **1991**, *25–27*, 509–512.
264 T. Boltshauser, L. Chandran, H. Baltes, F. Bose, D. Steiner, *Sens. Actuators B* **1991**, *5*, 161–164.
265 C. Cornila, A. Hierlemann, R. Lenggenhager, P. Malcovati, H. Baltes, G. Noetzel, U. Weimar, W. Göpel, *Sens. Actuators B* **1995**, *24/25*, 357–361.
266 F.P. Steiner, A. Hierlemann, C. Cornila, G. Noetzel, M. Bächtold, J.G. Korvink, W. Göpel, H. Baltes, in: *Technical Digest Transducers 1995*; **1995**, Vol. 2, 814–817.

267 A. Kummer, A. Hierlemann, H. Baltes, *Anal. Chem.* **2004**, *76*, 2470–2477.

268 A. Koll, 'CMOS capacitive chemical microsystems for volatile organic compounds,' *PhD Thesis*; ETH Zurich, ETH Diss. 13460, **1999**.

269 A. Kummer, 'Tuning sensitivity and discrimination performance of CMOS capacitive chemical microsensor systems,' *PhD Thesis*; ETH Zurich, ETH Diss. 15411, **2004**.

270 C. Hagleitner, A. Koll, R. Vogt, O. Brand, H. Baltes, in: *Proc. IEEE Transducers 1999*, Sendai; **1999**, 1012–1015.

271 S. Kawahito, A. Koll, C. Hagleitner, H. Baltes, Y. Tadokoro, *Trans. IEE Jpn.* **1999**, *119-E* (3), 138–142.

272 P. van Gerwen, W. Laureys, G. Huyberechts, M. op de Beeck, K. Baert, J. Suls, A. Varlan, W. Sansen, L. Hermans, R. Mertens, in: *Technical Digest Transducers 1997*; **1997**, Vol. 2, 907–910.

273 C. Laville, J.Y. Deletage, C. Pellet, *Sens. Actuators B* **2001**, *76*, 304–309.

274 A. Tetelin, C. Pellet, C. Laville, G. N'Kaoua, *Sens. Actuators B* **2003**, *91*, 211–218.

275 U. Kang, K.D. Wise, in: *Technical Digest Solid State Sensor and Actuator Workshop, Hilton Head Island, SC*; **1998**, 183–186.

276 http://www.sensirion.com/.

277 A.J. Ricco, R.M. Crooks, G.C. Osbourn, *Acc. Chem. Res.* **1998**, *31*, 289.

278 D.L. Massart, B.G.M. Vandeginste, S.N. Deming, Y. Michotte, L. Kaufman, *Data Handling in Science and Technology 2: Chemometrics: a Textbook*; Amsterdam: Elsevier, **1988**.

279 R.G. Brereton (ed.), *Multivariate Pattern Recognition in Chemometrics. Data Handling in Science and Technology*, Vol. 9; Amsterdam: Elsevier, **1992**.

280 A. Hierlemann, M. Schweizer-Berberich, U. Weimar, G. Kraus, A. Pfau, W. Göpel, 'Pattern recognition and multicomponent analysis,' in: *Sensors Update*, H. Baltes, W. Göpel, J. Hesse (eds.); Weinheim: VCH, **1996**, 119–180.

281 K.L. Hughes, S.L. Miller, J.L. Rodriguez, P.J. McWorther, *Sens. Actuators B* **1996**, *37*, 75–81.

282 A. van den Berg, P.D. van der Wal, B. van der Schoot, N.F. de Rooij, *Sens. Mater.* **1994**, *6*, 23–43.

283 A. Witvrouw, F. van Steenkiste, D. Maes, L. Haspeslagh, P. van Gerwen, P. de Moor, S. Sedky, C. van Hoof, A.C. de Vries, A. Verbist, A. De Caussemaeker, B. Parmentier, K. Baert, *Microsyst. Technol.* **2000**, *6*, 192–199.

284 W. Gumbrecht, D. Peters, W. Schelter, W. Erhardt, J. Henke, J. Steil, U. Sykora, *Sens. Actuators B* **1994**, *18/19*, 704–708.

285 W. Schelter, W. Gumbrecht, B. Montag, U. Sykora, W. Erhardt, *Sens. Actuators B* **1992**, *6*, 91–95.

286 E. Lauwers, J. Suls, W. Gumbrecht, D. Maes, G. Gielen, W. Sansen, *IEEE J. Solid-State Circuits* **2001**, *36*, 2030–2038.

287 C. Hagleitner, A. Hierlemann, D. Lange, A. Kummer, N. Kerness, O. Brand, H. Baltes, *Nature* **2001**, *414*, 293–296.

288 C. Hagleitner, D. Lange, A. Hierlemann, O. Brand, H. Baltes, *IEEE J. Solid-State Circuits* **2002**, *37*, 1867–1878.

289 C. Hagleitner, A. Hierlemann, H. Baltes, 'CMOS single-chip gas detection systems. Part I,' in: *Sensors Update*, H. Baltes, J. Korvink, G. Fedder (eds.); Weinheim: Wiley-VCH, **2003**, Vol. 11, 101–155.

290 A. Hierlemann, A.J. Ricco, K. Bodenhöfer, A. Dominik, W. Göpel, *Anal. Chem.* **2000**, *72*, 3696–3708.

8
Biometric Capacitive CMOS Fingerprint Sensor Systems

C. Hierold, Micro- and Nanotechnology, ETH Zurich, Zurich, Switzerland
G. Hribernig, Siemens AG Österreich, Wien, Austria
T. Scheiter, Infineon Technologies AG, Munich, Germany

Abstract

Recent market studies show that fingerprint scanning will outperform other types of biometric systems with respect to market share in a market that is forecast to double revenue every second year until 2007. This motivates efforts to provide a comprehensive review of capacitive CMOS fingerprint scanners and systems that are considered to provide unique advantages with respect to miniaturization, system integration and user benefits towards personalized biometric systems. After a general introduction to biometrics, recent research and evolution of capacitive fingerprint scanners are discussed. In Section 8.3 a particular solution for an integrated CMOS fingerprint sensor and a new methodology to approach the evaluation of sensor performance and quality are introduced. Section 8.4 provides a complete discussion of system solutions for image processing, feature extraction and verification for capacitive fingerprint sensors. Performance criteria on the system level, system integration issues and important topics for the successful deployment of fingerprint systems, including error rates and anti-fraud methods, are extensively evaluated. This chapter provides a complete overview of capacitive CMOS fingerprint sensor systems for readers to gain a quick entry into this complex area, and detailed information about relevant solutions, broad bibliographic information for further reading and an objective discussion of security issues.

Keywords

capacitive fingerprint sensor; biometric engine; verification and identification; fingerprint; biometrics

8 Biometric Capacitive CMOS Fingerprint Sensor Systems

- 8.1 **Introduction** *393*
  - 8.1.1 Types of Biometric Systems *395*
  - 8.1.2 Requirements for Biometric Systems *397*
  - 8.1.3 Outline of Section *399*
- 8.2 **CMOS Fingerprint Sensors: from the Finger Surface to the Sensor Signal** *399*
  - 8.2.1 Types of Fingerprint Sensor Systems *399*
    - 8.2.1.1 Optical transducer *399*
    - 8.2.1.2 Thermal transducers *402*
    - 8.2.1.3 Capacitive Transducers *402*
  - 8.2.2 Evolution of Capacitive Fingerprint Sensors *402*
- 8.3 **Capacitive CMOS Fingerprint Sensors: from the Sensor Signal to the Image Raw Data** *405*
  - 8.3.1 The CMOS Fingerprint Sensor Chip *405*
    - 8.3.1.1 Capacitance Measurement *406*
    - 8.3.1.2 Electromechanical Interface *407*
  - 8.3.2 Quality Requirements *409*
    - 8.3.2.1 Image Quality *410*
    - 8.3.2.2 Resistivity against ESD Air Discharge *412*
    - 8.3.2.3 Mechanical Robustness *414*
    - 8.3.2.4 Chemical Robustness *416*
- 8.4 **Image Processing, Feature Extraction and Software Solutions: from the Image Raw Data to a Person's Verification** *416*
  - 8.4.1 The Image Processing Task *416*
    - 8.4.1.1 Fingerprint Features *416*
    - 8.4.1.2 Image Processing System in General *417*
    - 8.4.1.3 Working Scheme of a Biometric System *418*
    - 8.4.1.4 Software Elements of a Fingerprint System *419*
  - 8.4.2 Performance of a Biometric System *425*
    - 8.4.2.1 Error Rates *426*
    - 8.4.2.2 Evaluation *429*
    - 8.4.2.3 Meeting the Application Requirements *432*
  - 8.4.3 System Integration *433*
    - 8.4.3.1 Technology Development *433*
    - 8.4.3.2 Sensor Device Integration *434*
    - 8.4.3.3 Software Development Kit (SDK) *434*
    - 8.4.3.4 Standard Software Architecture *435*
    - 8.4.3.5 Component-based Integration *437*
  - 8.4.4 Deployment of Fingerprint Systems *438*
    - 8.4.4.1 Verification and Identification *439*
    - 8.4.4.2 User Interaction *439*
    - 8.4.4.3 Security Issues *440*
    - 8.4.4.4 Smartcard Integration *441*
- 8.5 **Conclusions** *443*
- 8.6 **Acknowledgments** *443*

8.7	References 444
8.7.1	Journals, Conference Proceedings, Books, Reports 444
8.7.2	Links 446

8.1
Introduction

Within living memory, identifying oneself as an authorized user or a legal owner has become a standard procedure for individuals who live in social communities and interact with other individuals. For centuries people have been using passports to cross borders and keys to unlock doors. The possession of a 'device' is often enough to prove one's authorization. However, obviously this is not enough to prove a person's identity, because the device might be lost and found or it might have changed possessor intentionally or unintentionally. This basic risk of misuse of authentication by possession has been reduced by using passwords or codes to prove the user's authorization. In this case it is assumed that only the authorized person knows the code. However, this assumption might also be wrong, because the code could have been passed over to another person, again intentionally or even by force. Another problem with relying on authentication by knowledge is that the code is likely to be forgotten. In this case no proof of authorization is possible any longer. This basic risk of inconvenience of authentication by knowledge can be avoided by introducing characteristic traits of a person in the authentication process. Well-known traditional examples are passport photographs or picture IDs and a person's signature. In general, the features represented in a person's face or writing are considered to be unique and not subject to being lost or forgotten.

However, each of the above methods of authentication by knowledge (what you know), by possession (what you have) and by characteristic traits (what you are) is subject to misuse and fraud: Keys are lost or stolen, passwords are forgotten or spied out, picture IDs and signatures are faked. The resulting level of security of an authentication procedure is just a matter of the effort that is spent to protect a system, maybe by combination of different particular methods, in relation to the value that needs to be protected and the effort for an intruder to fake the system in case of misuse. Therefore, the task of an organization or a company that provides security systems for, e.g., access control, is to design the combined security level of the components such that the effort for misuse is far higher than the potential benefit for an intruder. As a consequence, those systems must be continuously improved to maintain the head start ahead of the capabilities of potential intruders.

This basic discussion about authentication or verification of a person's identity applies also and in particular to modern electronic security systems. Today's authentication by possession is provided by electronic keys, such as car keys and smartcards for, e.g., Internet access. These 'keys' are often combined with authentication by knowledge, which are passwords or PIN codes (personal identification number). These codes provide – alone or in combination with 'keys' – access to computers, networks and telephones, verify the authorized usage of bank and credit cards or provide physical access to restricted areas, just to list a few examples. The problem with these codes is that nowadays most of us have to handle, meaning memorize, at least five, or in the worst case up to 10–20 different codes for daily use in different private and corporate applications. Four-digit PIN codes are going to be substituted by six-digit codes and passwords for computer and Internet access tend to be even more complex: They are supposed to be combinations of arbitrary numbers and letters, input is case-sensitive and they are requested to be changed after a short period of time. All these constraints end up in people writing down these codes, although this is strictly forbidden for security reasons and might yield the total loss of any system provider's liability. As a consequence, the level of security of a password decreases with increasing complexity, which is considered as a password paradox.

This is the point where modern biometric systems appear. Biometrics is the science or a set of automated methods to identify individuals (i.e. authentication, verification, identification) using physical, physiological or behavioral features and characteristics [1, 2, 56, 57]. Evaluating characteristic personal traits as method for authentication was introduced above (e.g. picture IDs, signatures). The advancement in the application of biometric systems is that unbiased technical systems extract (encoding) and compare (matching) the pattern of a person's biometric features with a reference pattern at that point and time when a authentication process is requested. The reference pattern was voluntarily provided by the person to the system in advance (enrolment).

A good example of this kind of a system is demonstrated at Schiphol Airport (Amsterdam). Based on iris recognition (the system is called Privium [58]), passengers can cross the control point without showing their passport. The passenger's benefit is to save time by avoiding long queues in front of the passport checkpoint. This is simply an increase in convenience, provided by the application of a biometric system. During enrolment, the characteristic features of the iris were extracted and stored as a reference pattern on a smartcard's chip, which is in the possession of the user only. This is a very important criterion for the acceptance of such systems: the system providers must ensure that private and personal biometric data are not stored in uncontrolled databases and that these data are not subject to misuse for unauthorized identification purposes. Additionally, the authentication process must not be started without a deliberate action of the user.

One disadvantage of iris recognition is the rather bulky setup of the iris scanner. Capacitive or thermal silicon-based (full frame or swipe) fingerprint sensors instead are highly miniaturized devices. They can easily be integrated into small systems and portables such as keyboards [59], PC mouse [60], laptop PCs [61], palmtop [62] and mobile phones [63] and have already been demonstrated to fit into PC cards [61, 64] or USB sticks [65]. The integration of fingerprint sensors on smartcards has been announced [66, 67]. Fingerprint sensors are also integrated into cars as an electronic added feature to adjust a car's setup easily and conveniently for multiple drivers (seat, mirror, steering wheel, air conditioning, audio and navigation system and more) [68].

These examples show clearly that most of the applications of capacitive and thermal fingerprint sensors focus on the substitution of passwords and PIN codes for the user's convenience as an easy-to-use solution for authentication personalization. A high level of security is a prerequisite and is anticipated to be higher than that available through conventional, non-biometric methods. Fingerprint sensors are key devices for personal authentication systems that ensure that private biometric data stay under the control of the respective individual person. This is a *conditio-sine-qua-non* for the broad acceptance of those devices by the users and their widespread application in mass markets.

8.1.1
Types of Biometric Systems

According to the Biometric Market Report 2000–2005 [3] and 2003– 2007 [69] of the International Biometric Group, finger scanning is forecasted to be the most prominent technology for biometric systems (excluding automatic fingerprint identification systems, AFIS) with respect to total revenue ($\sim$34% in 2005, $\sim$52% in 2007), followed by facial scan ($\sim$15% in 2005, $\sim$11% in 2007), voice scan, signature scan and iris scan in 2005 and followed by facial scan, hand scan, iris scan and voice scan in 2007. The total biometric market (including AFIS) is estimated at <1 billion US$ in 2003, >2 billion US$ in 2005 and $\sim$4 billion US$ in 2007.

In the following section, different biometric scanning technologies are described and discussed in brief. Different types of fingerprint scanners are described in Section 8.2.1. AFIS are systems (software) to identify fingerprints automatically out of a database. These systems that have been developed in the second half of last century for forensic applications [45] are not considered in this article.

Facial scan and face recognition are probably the most user-friendly and most human-like technology for biometric systems. The task of recognizing a face is very easy for us owing to our ability to analyze even the most complex scenes almost instantaneously. Unfortunately, today's computer-based algorithms are not that advanced, which may result in high error rates for face recognition systems.

Two different main approaches are pursued: The first is based on 2D picture information that can easily be achieved by any kind of still and video imaging cameras (conventional, CCD, CMOS). As biometric features, characteristic points in the face of a person are extracted, such as the position of eye corners, and, e.g., the respective distances and angles are taken as characteristic features. For an overview, algorithms and reviews are available [4–6]. Reduced error rates are expected from recent advances in 3D facial scan technology, employing color-encoded structured light [7], which will minimize the influence of ambient light conditions.

2D face recognition is compatible with picture ID-based access systems, as used at border controls, and these picture IDs can be used as a reference for the authentication of a person. This compatibility, the chance of re-using existing camera infrastructure for, e.g., surveillance and an increased need for border security will further support the market growth of facial recognition as a widespread biometric system. However, the greatest advantage of facial recognition is its most severe drawback with respect to data security and protection of personal rights: facial recognition requires no deliberate action by a person – as a finger scan does – and can be started anywhere without permission or even knowledge of a person, and is therefore easily subject to misuse.

Like face recognition, voice scan or speaker recognition is also a biometric method that is close to the methods used by human beings to identify each other. Wherever microphones are part of an existing infrastructure such as systems for physical access control and mobile phones, speaker recognition can be implemented easily. Two different approaches are used: text-dependent and text-independent methods [1]. Whereas in the first method the system prompts the user with a predefined sentence that must be repeated, the second method analyzes the speaker based on text-independent characteristics of his/her voice. An overview of the particular methods of speaker recognition is available [8]. There are still problems associated with these methods, e.g. the voice of an individual person is subject to day-to-day variations that depend, for example, on one's health condition and the background noise.

Whereas face and speaker recognition is based on individual characteristics that are mainly influenced by genetic properties, signature scan, dynamic signature verification (DSV) or signature recognition is a biometric method that evaluates behavioral features of a person. Further examples of those not discussed here are keystroke dynamics and gait. Using signature recognition as a biometric method follows a well-accepted tradition to accept a person's signature for authentication. Technical systems will analyze not only the static image of a signature, but also the dynamics of writing that is represented by time of writing, velocity, pressure and acceleration patterns and even the angle of the pen [9, 10]. One of the major advantages of this method is that signing is an everyday procedure and simple electronic pens and panels are already available for some applications such as electronic signature for the receipt of a delivery.

The next level of accuracy in terms of how unique a feature under evaluation is for an individual person is achieved by those methods that are predominantly defined by genes and random processes in the prenatal embryonic development of a person. Examples of these features are fingerprints [11–13], the pattern of blood vessels on the retina and the individual pattern of the iris [14]. Fingerprint scanners and fingerprint recognition are discussed in detail in Sections 8.2–8.4, but retina scan will not be discussed in this chapter owing to its minor importance in the market. Iris recognition, however, has attracted increasing attention in the past and remarkable research has been reported in this field. For iris recognition, pictures are taken from a person's iris either by a multi camera setup that identifies the location of the iris first and that then takes a high-resolution image of the iris, or by a single camera that requires a cooperative, sometimes cumbersome user action to locate and focus on the small spot of the iris. Approximately 260 characteristic features (out of 400, approximately six times more than with fingerprints) are extracted, such as striations, pits, contraction furrows, collagen fibers, filaments, darkened areas in the iris, rings and freckles. Feature extraction algo-

Tab. 8.1 Qualitative comparison of biometric methods

	Comfort	Accuracy	Spread	Capital investment
Face	▫▫▫▫▫▫▫	▫▫▫▫	▫▫▫▫▫▫	▫▫▫▫
Speaker	▫▫▫▫	▫▫	▫▫▫	▫▫
Signature	▫▫▫	▫▫▫▫	▫▫▫▫	▫▫▫▫
Fingerprint	▫▫▫▫▫▫	▫▫▫▫▫▫	▫▫▫	▫▫▫
Iris	▫▫▫▫▫	▫▫▫▫▫▫▫	▫▫▫▫▫▫	▫▫▫▫▫▫▫

rithms have been proposed to analyze the characteristics of the iris [15–17] and application relevance was achieved with those of Daugman [18–20]. Iris recognition is accepted as a highly accurate biometric method. Its widespread application may be retarded, however, by the high system costs and the perceived intrusiveness to the user.

Tab. 8.1 summarizes the properties of the discussed biometric methods qualitatively. Similar analysis can be found elsewhere [1, 3].

The high accuracy of fingerprint scanning combined with its relatively low system costs and high convenience aspects will provide a good starting position for these systems in widespread consumer applications.

8.1.2
Requirements for Biometric Systems

At the very end, the requirements for a biometric system that comprises feature scanning (physical sensor) and feature extraction/matching (data processing) are defined by the requirements of the user application (market) and the capabilities of innovative technologies (supplier). In this section we will briefly list a set of first-level criteria, which are important for the acceptance of biometric systems in general and miniaturized capacitive fingerprint sensors in particular.

Reliability requirements need twofold discussions. First from the application point of view, reliability means that the correct person is accepted by the system and that the wrong person is reliably rejected. Errorless verification or authentication cannot be guaranteed by a biometric system for the following main reasons: sensor noise, variability of the data providing the biometric features (e.g. the image contrast of capacitive fingerprint sensors depends on the skin condition such as a dry or wet finger surface) and discrimination algorithms based on probability assumptions. Additionally, the presentation of the biometric features to the system, which varies with the behavior of the individual user and which may differ for different target populations and applications, is an important source of error. Authentication errors of a system are described in general by the false rejection rate (FRR) and false acceptance rate (FAR), which will be discussed in detail in Section 8.4.2.1.

Second, from the sensor device's point of view, reliability means correct functioning of the system for the specified lifetime. Fingerprint sensors are touched by the user and are therefore in close contact with a harsh environment. Sensor protection against chemicals, dirt and scratches are as important as protection against electrostatic discharge from the human body to the sensor. These issues will be discussed in Section 8.3.2.

User acceptance is a requirement that is influenced by many parameters that are partly not under the control of system designers and engineers. For example, as a consequence of an increased need for border control and security, usage of biometric systems and user acceptance are enforced by governmental and authority-driven initiatives (e.g. face and fingerprint recognition by the US Patriot Act 2002).

From a mass market point of view, those biometric systems are preferred which provide defined user advantages, such as convenience and data security, and that will substitute current authentication procedures such as passwords and PIN codes for, e.g., Internet access for computers, telephones and secure financial transactions. User acceptance comes subsequently with the non-intrusiveness of the system as perceived by the user, the ease of use and the ease of implementation in existing systems such as personal computers. Intrusiveness is often related to privacy concerns [1, 21]. The basic concerns about the application of biometric systems can be summarized in the following aspects: potential misuse of biometric data, loss of privacy and tracking. The concern about loss of privacy and tracking is, however, in general not increased by using biometric authentication, compared with other methods such as PIN codes or simply compared with using a credit card instead of cash.

Biometric data are secure if the biometric information and preferably the sensor that scans the biometric information are under the direct control of the user. This is achieved, if highly miniaturized biometric sensor systems are integrated in mobile terminals, such as telephones or PDAs, which are in the possession of the user. The biometric data are stored and processed within these devices only. The mobile terminal then communicates with another terminal, e.g. a point of sale, a teller machine, a security server or a door, and confirms the rights of the user by an encrypted connection. Even if the computing power of a fixed terminal must be used, e.g. for identification purposes (comparison of the data with a huge database), the extracted biometric features and not the original sensor data, such as the raw pixel image of a fingerprint, are transferred via an encrypted connection. The advantage is that the original sensor data such as the raw pixel image of a fingerprint cannot be reconstructed out of the extracted biometric features such as the set of minutiae of a fingerprint (e.g. finger ridge bifurcations and endings, including location and angle), stored and transferred to another system for any misuse without the approval of the user.

The third requirement for the broad use of biometric systems is low costs. Discussing cost issues includes not only fabrication costs (sensor size and technology) of the biometric system itself, but also installation, maintenance, training, support and software development and maintenance costs. Ultimately these costs must be compared with the system costs of current solutions to be substituted. For example, password-helpdesk costs are estimated [22] as US$ 300 per user and per year, which is a substantial amount for such a simple system involving the use of passwords. Miniaturized fingerprint sensors that are discussed in Section 8.2 contribute substantially to the required cost reduction for secure biometric systems.

8.1.3
Outline of Section

This section will focus on highly miniaturized fingerprint sensor systems, comprising:

- An overview of CMOS fingerprint sensors with a focus on capacitive fingerprint sensor systems (Section 8.2).
- The realization of a capacitive CMOS fingerprint system, including a discussion about sensor device quality (Section 8.3).
- Image processing, feature extraction, matching and system integration are discussed in Section 8.4. This includes also aspects of accuracy, reliability, evaluation and deployment, including a discussion of fraud and anti-fraud measures.

Hence we provide a complete system overview from scanning a fingerprint to the verification of a person, aligned with the special features of a capacitive CMOS fingerprint sensor system.

8.2
CMOS Fingerprint Sensors: from the Finger Surface to the Sensor Signal

8.2.1
Types of Fingerprint Sensor Systems

In this section, we briefly discuss different types of transducers to create images of the finger's tip: optical, thermal and capacitive transducer principles. A comparison of advantages and disadvantages is provided in Tab. 8.2.

8.2.1.1 Optical Transducer
Although there are many techniques for optical reading of a fingerprint image, the common method uses total internal reflection within a transparent prism (Fig. 8.1). Light is totally reflected at the surface of the prism, if light is travelling from the optical dense material (prism) towards an optical less dense material (air) and the condition for total reflection is fulfilled. In case a finger is pressed on the flat surface of the prism, light is not totally reflected at those areas that are covered by finger ridges. The result is a high contrast projection of the fingerprint pattern, which is focused by a lens onto a CMOS or CCD camera chip. LEDs are used as light sources.

Owing to the requirements for the optical path (focal length) for a demagnifying projection of the image to the camera chip, optical scanners are still bulky de-

Tab. 8.2 Comparison of transducer principles for fingerprint sensors

Advantages	Disadvantages
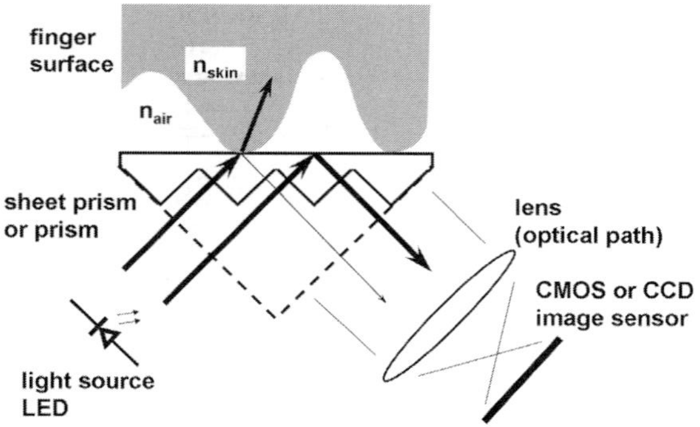 – Robust device, no risk of damage due to ESD – De-magnifying optics enables the application of low-cost CMOS image sensors – Large image capture area	– Still rather bulky (will not fit into smartcards), although improvements in miniaturization at low cost have been achieved – 2D image: No information about the depth of the valley is obtained – Risk of latent fingerprints left on the surface that can be utilized for fraud

Fig. 8.1 Optical total internal reflection transducer. n_{air} and n_{skin} are refractive indices of air and skin, respectively. Adapted from [23]

Tab. 8.2 (cont.)

	Advantages	Disadvantages
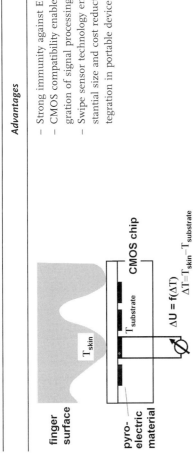	– Strong immunity against ESD – CMOS compatibility enables the integration of signal processing – Swipe sensor technology enables substantial size and cost reduction for integration in portable devices	– No image at thermal equilibrium between substrate and finger: may require heater – Image vanishes after 0.1 s: Promotes stripe sensor architecture – Requires non-CMOS process integration for pyroelectric material – 2D image: No information about the depth of the valley is obtained
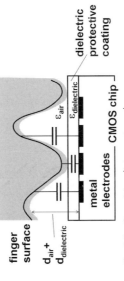	– Fully CMOS compatible, integration of signal processing easy – No temperature window without image, static image – 3D image: Information about the vertical profile of the skin surface is obtained – Swipe sensor technology enables substantial size and cost reduction for integration in portable devices	– Intrinsically sensitive to ESD (solutions known) – Thickness of dielectric coating for protection of the device limited due to its influence on the measured capacitance – Risk of latent fingerprints left on the surface that can be utilized for fraud (solution known)

Fig. 8.2 Thermal pyroelectric transducer. $T_{substrate}$, T_{skin} are temperatures of substrate and skin, respectively

Fig. 8.3 Capacitive distance transducer. ε_{air}, $\varepsilon_{dielectric}$ are dielectric constants of air and coating, respectively

vices. The use of a sheet prism (solid line in Fig. 8.1) instead of a conventional prism (dashed line) reduces the required volume. In addition, mirrors are used to fold the optical path [23]. Both measures provide optical total internal reflection transducers that can be integrated in PC keyboards and similar devices.

8.2.1.2 Thermal Transducers

Arrays of pyroelectric materials are integrated into a CMOS chip and are used as transducers that convert a temperature change into a voltage change. The sensor measures the temperature difference between the finger ridge and the substrate of the sensor (Fig. 8.2). The temperature difference between the substrate and the air in the valleys between finger ridges is negligible. Hence a dynamic image is generated that vanishes in less than 0.1 s when thermal equilibrium is reached [24]. The sensor provides no image if finger and substrate are at nearly the same temperature (e.g. body temperature) initially. For this case, controlled heating of the substrate is necessary to provide a sufficient temperature difference of 1 K minimum [25].

8.2.1.3 Capacitive Transducers

In general arrays of metal electrodes (e.g. last metal layer of a standard CMOS process) providing a spatial resolution of up to 500 dpi (50 µm capacitor array pitch) are integrated into a CMOS chip and are used as capacitor plates to measure the capacitance between the chip's surface and the finger. The individual capacitance depends on the distance of the skin surface and the dielectric constant of the medium between the capacitor plates. Therefore, capacitive fingerprint sensors are real 3D sensors providing a defined gray scale output depending on the depth of the valleys between finger ridges. On the other hand, it also reacts on, e.g., the moisture of the skin surface (wet/dry finger problem), because this will influence the dielectric constant. However, this can be automatically controlled by software adjusting the gain of the charge amplifiers in a feedback loop [23]. Furthermore, special circuits have been demonstrated to adjust and emphasize contrast automatically and to eliminate the influence of dirt and scratches on the sensor's surface [31, 32].

8.2.2
Evolution of Capacitive Fingerprint Sensors

The result of using modern VLSI CMOS processes as a platform for the integration of capacitive fingerprint sensors is twofold. First, the size of the image that should be provided by the sensor chip is between 15×15 and 20×20 mm^2. This image area pays the full area costs of the respective CMOS process, without utilizing its capability for VLSI integration in the sensor field and its potential for size reduction is limited. Therefore, the die costs of such a sensor is subject to cost reduction by yield improvement only and not by shrink and size reduction (the effect of shrinking peripheral circuit parts is neglected in this discussion).

Second, however, the available chip area underneath the sensor electrodes can be used to provide full image processing to extract the characteristic features of a person's fingerprint (minutiae encoding) on-chip without additional area consumption. This was demonstrated by Jung and co-workers [26, 28, 29] by developing a 330 dpi, 25×30 pixel, 2×2 mm^2 demonstrator chip that achieves image processing by application of hexagonal local operators implemented in pixel-parallel mixed neuron–MOS/CMOS logic circuits (0.65 µm technology generation). Additionally, the massive parallelism results in very low power dissipation of 0.25 mW at 2.5 V supply voltage and 5 kHz clock rate (encoding time < 20 ms), extrapolated to a full frame fingerprint sensor of 40 000 pixels, including nearly 5 million transistors. Another example for an integrated single-chip fingerprint sensor and identifier was demonstrated by Shigematsu and Morimura [30]. This approach stores and evaluates the data of an individual pixel in a logic circuit underneath the pixel and provides user identification by a pixel-wise comparison of the sensed fingerprint to a reference template.

Most of the proposed solutions for thermal and capacitive transducers are based on CMOS processes, which are considered as a functional advantage for the VLSI integration of signal-processing circuits. Additionally, modern CMOS processes provide low power capabilities that are important for all applications in portable devices.

However, as already mentioned, the full frame touch sensor has to pay the full area costs of the modern CMOS processes without a remarkable potential for size reduction. Therefore, the development of so-called swipe sensors is straightforward for reducing chip and sensor area substantially.

Swipe sensors are fingerprint sensors with a reduced number of sensor rows to capture the image of a fingerprint while the user is sweeping her/his finger over the sensor. More than one row or additional sensors for speed and direction are necessary to enable the software to reconstruct the picture out of individual sub-pictures.

Examples are Atmel's AT77C103A FingerChip^{TM} thermal transducer with eight sensor rows, 288 columns, 500 dpi, 0.4×14 mm^2 sensor size [24, 33], ST's Touch-Strip TCS3B capacitive transducer [70] and SINTEF's demonstrator [34] with a single sensor row and additional navigation sensors, 256 columns, 500 dpi, 7×15 mm^2 sensor size, 8 speed and 4 navigation sensors.

Sweeping a finger over a swipe sensor instead touching a full frame sensor requires a dedicated and controlled user action. If this action requires two hands for operation, e.g. in small mobile devices with non-ergonomic user interface, swipe sensors may be considered inconvenient by the user. On the other hand, the risk of latent fingerprints left on the sensor surface is obsolete for swipe sensors.

Tab. 8.3 summarizes recent research and development on capacitive integrated fingerprint sensor systems. Research effort is applied to solving the drawbacks of capacitive CMOS fingerprint sensors on the system level: ESD insensitivity is improved by grounded grid structures and diodes; protective coatings against moisture and salt are realized by dielectric layers of oxide, nitride and polyimide; pixel

Tab. 8.3 Recent research and development on integrated fingerprint sensors (without swipe sensors)

Transducer	Pixel size and density	Basic technology	Remarks	Year, reference
Capacitive with integrated driver and sensing circuit	50×50 µm², 500 dpi	0.5 µm 3.3 V, 3 metal layer digital CMOS	250 mW at 1.8 V, 60 frames per second, 90 000 pixels, 15×15 mm sensor area. Protective coating sandwich by 1 µm P-glass with 0.5 µm high density SiN on top. ESD protection provided by diode and top metal layer grid	1998 [40]
Capacitive with embedded cellular logic for image processing (encoding)	71×82 µm² hexagonal, 330 dpi 25×30 pixel demonstrator	0.65 µm double polysilicon, 3 metal layer CMOS	Low power capability: 0.25 mW at 2.5 V, 5 kHz clock rate (encoding time <20 ms), extrapolated to 40 000 pixels (including ∼5 million transistors), pixel-parallel architecture for on-chip feature extraction (encoding) by hexagonal local operators implemented in mixed neuron–MOS/CMOS logic circuits	1999 [26, 28, 29]
Capacitive with ground wall structure	50×50 µm², 500 dpi	0.5 µm standard CMOS, 3 metal layer [37, 38], gold electrode to prevent oxidation, SiN/polyimide sandwich as coating against moisture [36]	25 mW at 3.3 V, up to 50 frames per second, 57 344 pixels, integrated sensing circuit with cross-coupled differential amplifiers with charge transfer technique, ESD tolerance: ±3 kV, gold ground wall (and electrodes) [36]	2001 [36] 2002 [37] 2002 [38]
Capacitive with virtually grounded metal shield, suppressing parasitic capacitances	50×50 µm², 500 dpi	0.6 µm double polysilicon, 3 metal layer CMOS	40 mW at 5 V, 21 000 pixels, analog output, A/D converted off chip, integrated diffusion network to generate an adaptive local threshold level for pixel-level image enhancement	2002 [39]
Capacitive polysilicon TFT technology	60×60 µm², 423 dpi	Low-temperature polysilicon TFT on glass substrate, 2 metal layers, SiN coating	Low power capability: 1.2 mW at 5 V, 500 kHz, 80 000 pixels, integrated drive circuit, signal processing, A/D converter	2003 [35]

image enhancement is solved by local contrast adjustment; low power capability is improved by utilizing modern CMOS processes and intelligent circuit architectures. The cost problem with respect to total area is tackled by integrating the signal evaluation directly into the cell area (improving area efficiency) or by utilizing alternative low-cost technologies such as polysilicon TFT-based solutions or, as mentioned above, by developing swipe sensors.

8.3
Capacitive CMOS Fingerprint Sensors: from the Sensor Signal to the Image Raw Data

8.3.1
The CMOS Fingerprint Sensor Chip

In Section 8.3, the Siemens solution for a capacitive CMOS fingerprint sensor demonstrator will be discussed in detail [26, 41].

A major prerequisite for the acceptance of fingerprint scanning systems is the availability of solutions for low-cost sensors. As discussed in Section 8.2, CMOS technology offers an attractive alternative to conventional state-of-the-art optical sensors. CMOS capacitive sensor chips can be batch fabricated and no complicated system assembly with small geometric tolerances for, e.g., optical components has to be carried out. The capacitive sensor principle is illustrated in Fig. 8.3.

In order to get into the region of 5 US$ and below, the CMOS devices might even be too expensive to fabricate and novel concepts based on alternative technologies [35] or with simple passive sensor arrays on polymer materials might be

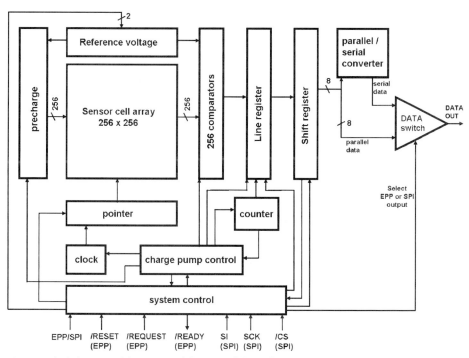

Fig. 8.4 Block diagram of the integrated fingerprint sensor. The pixel signal is converted on-chip into 8-bit digital data, which are transmitted via an asynchronous bi-directional parallel interface to a processor (e.g. PC), directly. No additional devices are necessary for sensor operation

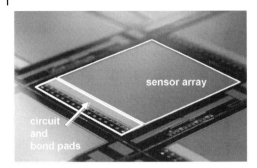

Fig. 8.5 Chip photograph of the CMOS fingerprint sensor chip (on wafer): 0.8 µm CMOS process, 256×256 sensor cells, 65 536 pixels, 50 µm pitch, 500 dpi resolution, 160 mm² active area

necessary. Passive sensor arrays are those that provide an array of sensing electrodes and wiring on a low-cost polymer substrate without active components such as switches. The sensor electrodes are wired to a hybrid integrated driver and signal evaluation chip (very small chip area) on the same substrate.

The fingerprint sensor which is used in the example was developed by Siemens Corporate Research. The chip is based on a two metal layer, 0.8 µm CMOS process. The sensor array consists of 256×256 sensor cells at a pitch of 50 µm. The lateral resolution of the sensor is then 500 dpi. The active area of the demonstrator is 160 mm², providing an image of the fingerprint with sufficient size.

In Fig. 8.4, a block diagram of the fingerprint sensor chip is shown. The chip includes data acquisition by the sensor array, direct A/D conversion of the measured signal, a controller unit, a clock generator (1 MHz) and a parallel interface which transmits the 8-bit digital data directly to a microprocessor. No additional devices are necessary to operate the fingerprint sensor system with a PC, for example. Fig. 8.5 shows a chip photograph of the sensor on wafer before dicing.

8.3.1.1 Capacitance Measurement

The basic operation principle of the sensor is explained in Fig. 8.6. The two-dimensional pixel array of the fingerprint sensor measures the capacitance between chip surface (pixel) and the finger's surface locally. A single capacitor C_p out of 65 536 is shown in Fig. 8.6. The value of this capacitor depends on the distance of the finger's surface from the chip surface. In the case of a finger ridge the distance is small and the capacitance is large. For a valley the capacitance is small. Typical capacitances which are measured for finger ridges and valley are in the range 80–30 fF. The capacitor C_p is measured by the circuit shown: during clock 1, C_p is charged. During clock 2, the charge on C_p is transferred into the capacitor C_c, which is much larger than C_p. After a number of clock cycles, the reference voltage is reached at node A and the counter is stopped by the comparator. The number of clock cycles is a measure of C_p. The number of cycles is low for large C_p, corresponding to a small distance between chip and finger surface (finger ridge=dark) and vice versa.

Based on the equivalent circuit (Fig. 8.7), C_p is given by

Fig. 8.6 Basic principle of sensor operation. The number of clock counts until C_c is charged to U_{ref} at node A is a direct measure of capacitance C_p and of the distance from the finger's surface to the chip surface

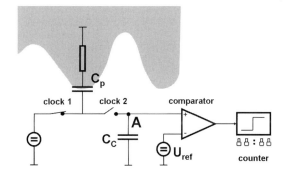

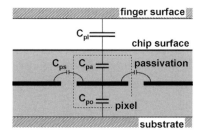

Fig. 8.7 Equivalent circuit of the pixel capacitor C_p, comprising parasitics, dielectric passivation layer and capacitance to the finger surface C_{pl}

$$C_p = C_{p0} + C_{ps} + \frac{C_{pa} C_{pl}}{C_{pa} + C_{pl}} \tag{1}$$

where C_{pl} is the variable capacitor which depends on the distance from the finger surface to the chip surface, C_{pa} is determined by the dielectric passivation layer and C_{ps} and C_{p0} are parasitic capacitances to neighbouring pixels or to the chip substrate, respectively.

8.3.1.2 Electromechanical Interface

The sensitivity S of the sensor cell is defined as dC_p/dC_{pl}:

$$S = \frac{1}{1 + 2\dfrac{C_{pl}}{C_{pa}} + \dfrac{C_{pl}^2}{C_{pa}^2}} \tag{2}$$

To optimize the sensitivity S, C_{pa} must be increased. This is achieved by reducing the thickness of the dielectric passivation layer. On the other hand, however, the passivation layer is the protective coating of the chip surfaces against mechanical and chemical influences. This protection is essential for the long-term stability of the sensor chip.

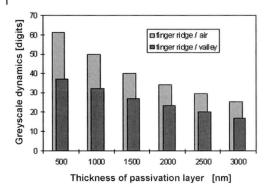

Fig. 8.8 Measured dependence of gray-scale dynamics for a given C_c between finger ridges and valleys (air) on the thickness of the passivation layer (oxide, nitride sandwich)

Therefore, there is a trade-off between sensitivity and chip protection. Fig. 8.8 shows for a given C_c the measured dependence of the gray-scale dynamics in digits depending on the thickness of the passivation layer. The passivation layer is a sandwich of silicon oxide and nitride. The application and the image-processing software define the requirements for the minimum gray-scale dynamics. Basically, a binarized image carries all the information needed for image processing. Owing to distortions, moisture and interferences, a minimum of 16 consistent distributed gray levels are required for acceptable performance (see also Section 8.4.3.2), which allows a passivation layer thickness of up to 3 µm as concluded from Fig. 8.8.

Fig. 8.9 shows an 8-bit gray-scale image of a fingerprint taken with the mentioned CMOS fingerprint demonstrator system. The image is processed and the characteristic points of the fingerprint, the so-called minutiae points, are extracted, stored as a reference (enrolment) or compared with the reference pattern (verification).

Tab. 8.4 summarizes the performance of the presented CMOS fingerprint demonstrator chip. The readout time (image capture time) is 100 ms. The high degree

Fig. 8.9 Image of a fingerprint taken with the CMOS fingerprint demonstrator chip. The features of the fingerprint, i.e. minutiae points, are marked by circles, as identified by the feature extraction software. Dark areas are finger ridges

Tab. 8.4 Performance of the fingerprint sensor demonstrator

Number of pixels	256×256
Pitch	50 µm
Lateral resolution	500 dpi
Clock (on chip)	1 MHz
Sensor area	160 mm²
Resolution	8 bits
Gray-scale dynamics (accuracy)	5–6 bits
Readout time	100 ms
V_{DD}	5 V
$U_r$ef	3.5 V

Fig. 8.10 CMOS fingerprint demonstrator chip: the high degree of miniaturization and integration is proved by assembly of the chip into a card. The sensor can be connected to the parallel port of a PC via the 16-pin interface

of system integration and miniaturization is demonstrated by the assembly of the chip into a card, as shown in Fig. 8.10. To take fingerprint scans the card is plugged directly to the parallel port of a PC.

8.3.2
Quality Requirements

To guarantee sensor operation in the field over the entire sensor lifetime, special care has to be taken when defining quality requirements for CMOS fingerprint sensors. As the CMOS sensor is, in contrast to conventional semiconductor chips, directly touched in operation, the conventional quality requirements can be applied only partly.

8.3.2.1 Image Quality

Without having a measure of image quality, which is defined with respect to the technical system that has to evaluate the captured image and which is developed to extract the characteristic features of a fingerprint, the integration of sensor hardware (=fingerprint scanner) and sensor software (=encoder and matcher; see also Section 8.4) in an optimized system is difficult. This section describes a new, reasonable approach to solving this problem.

Without having a well-defined image quality measurement procedure, the evaluation of the sensor image quality is no more than a 'good guess'. Just from a visual inspection of the two fingerprint examples in Fig. 8.11, one might conclude that the image quality of the sensor on the left is 'bad' and that on the right is 'good'. However, this example of image quality assessment can easily lead to incorrect conclusions, as judging an image may not be significant without knowing the image-capturing scenario or having a comparative setup.

In addition, it is important to evaluate the image quality 'with the eyes' of the biometric system using a particular sensor. The image (Fig. 8.11a) can easily be processed and successfully verified, even by a state-of-the-art recognition system. What a human judges as 'bad' can actually be termed 'good' by the system.

Obviously it is important to evaluate the quality of the image that is transferred by the sensor to the image processing unit. The image quality "in the eyes" of the biometric system depends either on the applied sensor principle (i.e. capacitive) or

Fig. 8.11 'Good' (left) and 'bad' (right) image quality fingerprint scans as assessed by the verification of a biometric recognition system (i.e. software for image processing and feature extraction)

8.3 Capacitive CMOS Fingerprint Sensors: from the Sensor Signal to the Image Raw Data

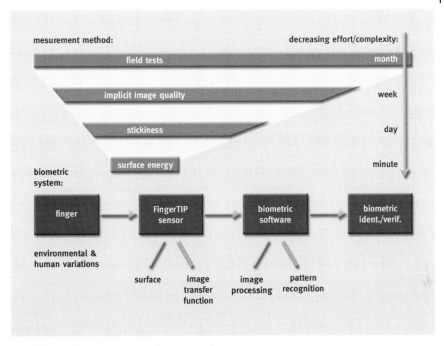

Fig. 8.12 Methodology to quantify image quality

on the performance variation per sensor device due to variations in processing parameters for example. Field tests are certainly the best way to prove image quality. But unfortunately field tests require tremendous effort in resources and time and do not provide information about image quality variations from device to device. Therefore a method is wanted that enables the fast and cheap evaluation of image quality of an individual sensor element last but not least for product quality reasons. Four different approaches are presented to measure and quantify image quality (see Fig. 8.12) are presented. Starting from the top of the figure, the methods have a decreasing complexity, as they concentrate more and more on the most important factors relevant to image quality. At the same time, the methods include progressively less parts of the complete fingerprint recognition system. This results in a remarkable improvement in evaluation and testing speed (typically from months to minutes), which allows the testing of statistically significant sensor volumes.

Field Tests

Field tests allow the evaluation of the biometric performance of a sensor system using a fixed biometric reference system. The recognition rates obtained are somehow correlated with the image quality of the sensor device, provided that environmental and human influence factors can be neglected, kept constant or averaged out and the imaging characteristics of the sensor can be compensated by image normalization. An appropriate test scenario, a carefully chosen test population and automatically adjusting normalization algorithms for image pre-processing can achieve this.

Implicit Image Quality
Implicit image quality involves sophisticated image processing algorithms that model and quantify the deviation of a real sensor image from an ideal (from the systems point of view ideal) fingerprint image. The measuring process involves acquiring a small number of fingerprint image samples from users with well-defined finger properties, such as normal, wet and dry. The resulting distribution of implicit image quality scores is typical for every type of sensor and can be compared on the basis of statistical properties.

Stickiness
Stickiness measures the adhesiveness of a sensor to moisture and finger oil, which is proposed to be the most prominent parameter for (good, bad or variations in) image quality. The stickiness measurement process involves applying fingerprints on sensors in a very accurate and reproducible way and measuring the strength of the potential latent image in a climate chamber. The whole measurement process can be automated using appropriate testing equipment.

Surface Energy
Surface energy describes how fluids are attracted by the sensor surface. This test reduces the 'image quality' of a sensor to that physical property of the surface that has been empirically identified as predominant. The surface energy is determined by dropping two different liquids on to the surface and measuring the contact angle. Since this state-of-the-art optical method is time consuming and difficult to handle, a newly developed, company proprietary method – exploiting the imaging capability of the sensor – is used, allowing very fast, robust and accurate measurements.

In conclusion, every method has specific advantages and disadvantages and can describe specific aspects of image quality. The overall sensor image quality should therefore be determined by a combination of all four methods.

8.3.2.2 Resistivity against ESD Air Discharge

The major challenge for capacitive sensors is the robustness against electrostatic air discharge resulting from the touch of a charged person. As shown in the graph in Fig. 8.13, electrostatic discharge via the air occurs at a relatively high probability (>1%) up to 8 kV if the air is very dry, which can be the case on dry winter days. Based on this kind of statistics, a certification of a sensor device with the CE certificate requires ESD air discharge robustness of at least 8 kV.

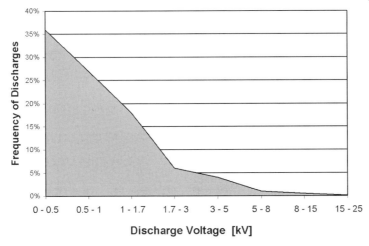

Fig. 8.13 Probability of electrostatic air discharge over discharge voltage. Adapted from [27]

In Fig. 8.14, an ESD air discharge at almost 16 kV into a sensor surface is shown. The sensor is connected to power and operates in standard mode. The ESD test is processed according to IEC 801-2 (8 kV air discharge), which is a prerequisite for CE certification. Further requirements for fingerprint sensors are (according to EMC Directive 89/33/EEC): Radio Frequency Interference (RFI) acc. EN 55022 and Radiated Fields acc. IEC 801-3. The particular sensor shown in Fig. 8.14 is protected by an integrated metal grid (Fig. 8.15b) and survived this test.

Electrostatic air discharge of as low as 1–2 kV into an unprotected CMOS sensor surface leads to severe destruction of the chip surface and subsequently a complete failure of the sensor. Fig. 8.15a shows the resulting destruction of a single electrostatic air discharge of 2 kV into an unprotected sensor surface.

Fig. 8.14 ESD air discharge at 15.7 kV into a capacitive CMOS fingerprint sensor (according to IEC 801-2). The sensor is in operative standard mode and survived this test

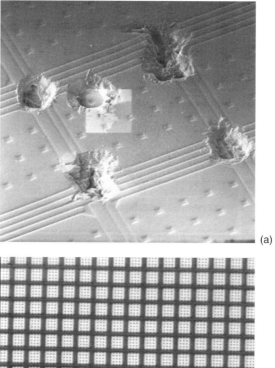

Fig. 8.15 (a) Surface of an unprotected CMOS fingerprint sensor after impact of 2 kV ESD air discharge (SEM picture). (b) Sensor with ESD protection by a grounded refractory metal grid. No damage after multiple discharges is observed

It was shown that the capacitive CMOS sensors withstand more than 12 kV discharge directly into the sensor surface when appropriate countermeasures (e.g. ESD protection by a grounded refractory metal grid, which is integrated into the sensor surface, Fig. 8.15 b) are applied.

8.3.2.3 Mechanical Robustness

Requirements concerning mechanical robustness come especially from the automotive industry. Therefore, several tests such as mechanical shock have to be implemented in the qualification test plan. Additionally, a chemical–mechanical rubbing test has been performed in order to simulate one million touches during operation. The test setup and testing conditions are shown in Fig. 8.16.

Fig. 8.16 Setup of chemical rub-off test (according to IEC 68-2-70). The testing conditions are as follows: testing force: 5 N; moving distance: 4 mm; number of tests: 10^6; test liquid: artificial sweat

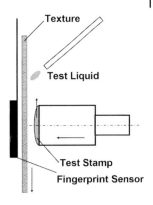

Other tests such as the pencil drop test and breakage test, which cover situations of daily use, completed the qualification procedure. The pencil drop test is illustrated in Fig. 8.17.

Finally, scratch resistivity is an important requirement for the field application of CMOS fingerprint sensors. From the automotive industry, various scratch test setups are known, for which the main evaluation criteria are based on optical inspection.

In the case of CMOS fingerprint sensors, however, a pure optical inspection after scratching at a defined force with a defined needle is not sufficient. Experimental results showed that even sensors which passed the optical evaluation after scratching failed later in the field. Therefore, the criteria were sharpened in such a way that scratched samples have to pass the active salt spray test after scratching (see Section 8.3.2.4). A typical setup of the scratch test is shown in Fig. 8.18.

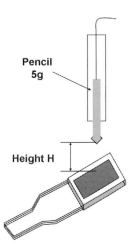

Fig. 8.17 Pencil drop test: a pencil is dropped on to the device under test. A typical requirement is that no fails are allowed to be observed after a test with a dropping height H of < 20 cm

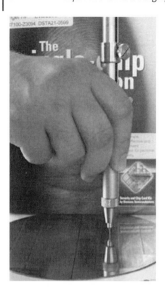

Fig. 8.18 Scratch test on CMOS fingerprint sensors (wafer level). The scratch test is according to a Siemens internal test procedure

8.3.2.4 Chemical Robustness

To guarantee chemical robustness of the sensor surface, the conventional gas and liquid tests for harsh environments can be applied (e.g. chemical resistance of a surface acc. ISO/IEC 10373, salt atmosphere test acc. MIL-STD 883, Method 1009, sweat and spittle resistance acc. DIN 53160, polluting gas test acc. IEC 68-2-60).

In addition to being resistant against liquids and gases, the resistance against sweat is a major challenge since sodium and chlorine can easily deteriorate and destroy the CMOS chip. In order to test this sweat resistance, a test can be applied which is similar to the well-known salt spray test, with the difference that the sensor is in an operative mode and images are read out continuously. During this test the sensor is covered completely with saturated salt solution. The sampled images provided by the sensor reveal the slightest leakiness of the surface as the diffusion of liquid causes failure of single pixels, complete columns and even areas.

8.4 Image Processing, Feature Extraction and Software Solutions: from the Image Raw Data to a Person's Verification

8.4.1 The Image Processing Task

8.4.1.1 Fingerprint Features

Fingerprints have been used for identification for more than 100 years; scientific studies already began in the late sixteenth century. Basic research on modern fingerprint identification was done by Galton [42] and Henry [43] at the end of the

Fig. 8.19 Basic types of minutiae: bifurcation and termination of ridges

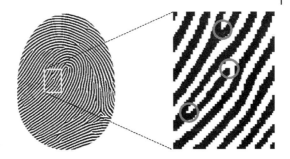

nineteenth century. They investigated the structure of fingerprints and introduced the terms *ridges* (i.e. elevations) and *valleys* and also the idea of small discriminating features, the so called *minutiae*. The minutiae are discontinuities in the ridge flow pattern. Galton defined four characteristics, which were extended over time up to eight characteristics [44]. These are ridge end, fork, island, dot, short ridge, crossover, bridge and spur. Usually automated fingerprint recognition systems (AFIS) use two basic types of minutiae only: fork (or bifurcation) and the ridge end (or termination of ridges) (Fig. 8.19).

The minutiae are classified by the following properties:

- Position: coordinates [i.e. (x, y)] or correlation (i.e. ridge count between minutiae).
- Type: fork or end.
- Angle: angle of the ridges or valleys with respect to the local ridge direction.
- Quality: technical property that evaluates the probability that the recognized minutiae is plausible (see Section 8.4.1.4).

In addition to minutiae there are additional characteristics in the global structure of a fingerprint. Henry introduced five classes of fingerprints [43] that are shown in Fig. 8.20: arch, tented arch, left loop, right loop and whorl.

Advanced fingerprint recognition systems use minutiae in addition to global structure information. In systems with restrictions on storage or computational power such as smartcards (see Section 8.4.4.4), minutiae information only is used.

8.4.1.2 Image Processing System in General

Every automated image processing system consists of five major parts:

- *Image acquisition*: Captures the raw image, triggered by a mechanical shutter or controlled by a feedback loop in the image processing system.
- *Pre-processing*: The raw image is improved by various algorithms; filters improve the image properties that are needed for interpretation and remove distortions and irrelevant information.

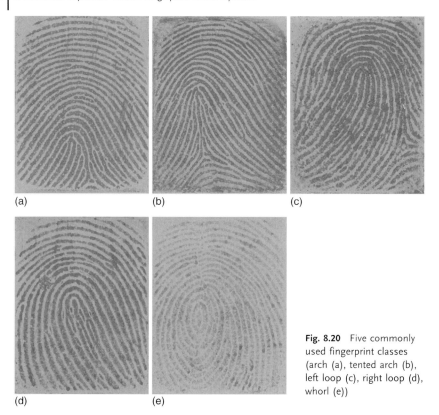

Fig. 8.20 Five commonly used fingerprint classes (arch (a), tented arch (b), left loop (c), right loop (d), whorl (e))

- *Interpretation*: Relevant features are worked out from the pixel information. The features must be unambiguous properties. Workflow continues with a description of the features.
- *Action*: Depending on the features, a conclusion is drawn and the result is presented.
- *Feedback*: If features are not clear for an action, image acquisition may be adjusted.

8.4.1.3 Working Scheme of a Biometric System

A general scheme of a biometric system is shown in Fig. 8.21. The biometric data are captured by a sensor device. The pre-processing unit produces a raw template, which is an enhanced image containing the essential information. The feature extraction component generates the feature vector including minutiae and global structure information. Pre-processing and feature extraction are called the encoder unit.

The feature vector is stored in the reference database. Some systems store an additional raw image for compatibility reasons (see Section 8.4.3.4). This procedure is called enrolment.

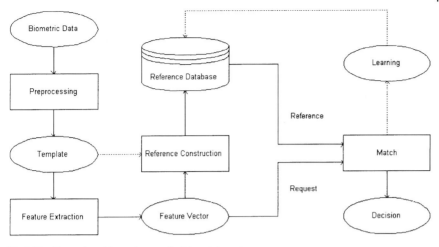

Fig. 8.21 General working scheme of a biometric system

For verification, the feature vector of the fingerprint just captured is matched with the corresponding reference in the database. The system has to take a decision if there is correspondence between inquired fingerprint and the reference. Some systems use the outcome of the matcher to improve the reference in the database.

Identification is repeated verification for the whole reference database. If there is correspondence, the identification process is stopped. For improving response time, references are presented to the matcher in sequence of usage frequency.

8.4.1.4 Software Elements of a Fingerprint System

The following description covers the basic software elements of a fingerprint system that are

- image capture of the fingerprint image
- image pre-processing
- feature extraction
- matcher.

The following sections give a brief overview of occurring problems and practical solutions. More detailed information can be found elsewhere [44].

Image Capture

The capturing process is critical for the performance of the whole system. Especially when using a capacitive sensor, the image quality depends on the capture time, as the image takes a few moments to appear and after some seconds it starts to deteriorate (see Fig. 8.22).

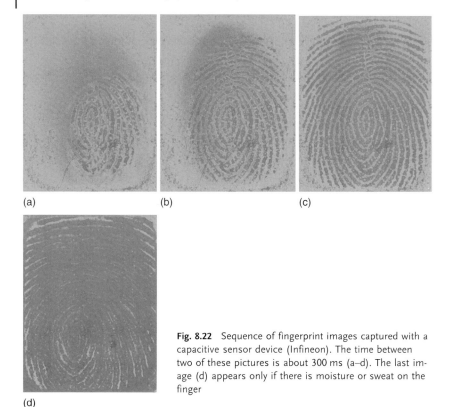

Fig. 8.22 Sequence of fingerprint images captured with a capacitive sensor device (Infineon). The time between two of these pictures is about 300 ms (a–d). The last image (d) appears only if there is moisture or sweat on the finger

A trigger is necessary to provide good image quality. The trigger can be realized in different ways:

- *Mechanical switch:* The switch underneath the sensor must have a release pressure of about 10 N. This solution does not guarantee a good image quality, as the optimal pressure depends on additional parameters such as the condition of the finger and the environment (i.e. moisture, temperature).
- *Check of image parameters:* Parameters of the captured image (i.e. contrast, modification of pixel values) are continuously checked. The image is captured when thresholds are reached. The problem is that the quality of the fingerprint image information does not necessarily correlate with the quality of the extracted features.
- *Check of fingerprint information:* Each captured image is transferred to the preprocessing and feature selection unit. The quality of the extracted features determines the moment of the final capture.
- *Prediction of match:* The captured image is processed through encoder and matcher. As long as match result improves, new images are captured. This method is most advanced and guarantees the best overall performance. The algorithms of the biometric engine have to be capable of processing more than three images per second, as the final result has to appear within 1 s.

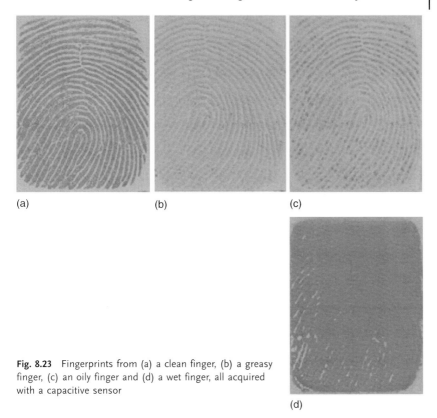

Fig. 8.23 Fingerprints from (a) a clean finger, (b) a greasy finger, (c) an oily finger and (d) a wet finger, all acquired with a capacitive sensor

Pre-processing

Depending on the sensor type the main problem of fingerprint systems is 'dirt' between the finger and sensor surface. Fig. 8.23 compares (a) fingerprint images under good conditions with fingerprint images of (b) a finger covered with grease, (c) an oily finger and (d) a wet finger. The images are typical for a CMOS capacitive sensor device.

The pre-processing unit has to improve the image quality and will remove distortions and irrelevant information. Effects of the capture (Fig. 8.23) have to be eliminated. A flow chart of typical pre-processing is shown in Fig. 8.24.

Raw Image Enhancement

In the first step, distortions have to be eliminated. A simple low-pass filtering for noise suppression is not useful as high-frequency parts of the ridge/valley information might be lost. Therefore, orientation-sensitive filters are used. First an orientation field is calculated, which holds the information on local ridge direction. Some systems use Gabor filter, which are the product of Gaussian and plane wave filters [54].

The next step is the separation of fingerprint image and sensor background. This so-called segmentation makes use of the fact that the sensor background has

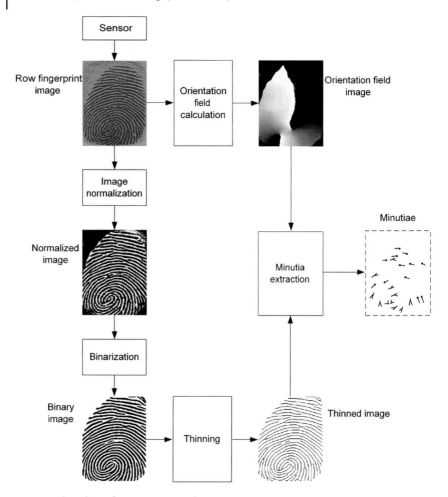

Fig. 8.24 Flow chart of pre-processing a fingerprint image

no clear orientation (i.e. outcome of the orientation field). The background is faded out.

The contrast of the segmented image can then be stretched. Owing to fingerprint pressure at capture, the contrast might change over the sensor area; therefore, contrast stretching is performed locally, on tiles of the image. For each tile, the mean value $\overline{A}$ and the standard deviation A_δ (of A that are the actual grayscale values) are calculated and two gray levels, $L\downarrow$ (Eq. 3) and $L\uparrow$ (Eq. 4), are determined as follows:

$$L\downarrow = \max(A_{\min}, \overline{A} - aA_\delta) \quad (3)$$

$$L\uparrow = \min(A_{\max}, \overline{A} + aA_\delta) \quad (4)$$

where a is a tuneable parameter and Λ_{min} and Λ_{max} are the minimal and maximal gray-scale offset values allowed. The calculation of the stretching factor φ (Eq. 5) and the gray-scale offset O (Eq. 6) is done as follows:

$$\varphi = \frac{\Lambda_{max} - \Lambda_{min}}{\max(L\uparrow - L\downarrow, \Lambda_{min})} \qquad (5)$$

$$O = \varphi\overline{A} - \frac{\Lambda_{max} - \Lambda_{min}}{2} \qquad (6)$$

Finally, the contrast stretching is done as follows:

$$A = \begin{cases} \varphi A - O & \text{for } A \in [\Lambda_{min}, \Lambda_{max}] \\ A & \text{otherwise} \end{cases} \qquad (7)$$

Binary Image Enhancement

The image binarization process is a rather complex process, although it looks very simple. Generally, the goal of the binarization process is to separate significant features from the background. In the case of fingerprint processing, binarization has to separate ridges from valleys, providing a better basis for thinning and healing processes in the next step. Various binarization algorithms have been described in the literature [46]. Results improve if, for each individual pixel and for the gray scales of its neighbourhood, a mean value and a deviation are calculated. The decision regarding a pixel being a ridge or valley depends on the mean value, the deviation and the local flow direction. A normalized fingerprint image, a thinned and an enhanced thinned fingerprint image can be seen in Fig. 8.25.

Once binarized, the image information runs through two steps of binary image enhancement:

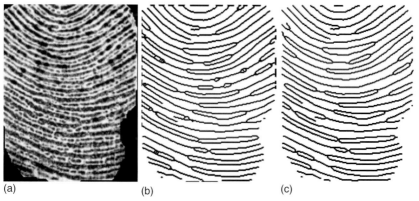

Fig. 8.25 (a) Normalized and black/white inverted fingerprint image, (b) thinned fingerprint image and (c) enhanced thinned fingerprint image

- Healing of broken lines: owing to sensor defects, dirt on the sensor and scares or cuts on the finger, ridges and valleys might be broken. So if the distance of two ends are close enough and the orientations of the end-points fit together, a straight line is added between the two end-points.
- Thinning: generally the thickness of ridges varies because of the capturing conditions. Therefore, it is necessary to reduce the thickness of the ridges until they are uniform. The thinning is an iterative process; in each step one runs through the shape pixels and tests if they can be removed without destroying the structural information. The iteration stops if no more pixels are removed. Owing to this symmetric erosion, the skeleton becomes centered within the original shape and it is of uniform width everywhere.

Feature Extraction

Most common features in automated fingerprint identification systems are the minutiae. The main difficulty for the minutiae extraction process is often the low quality of the fingerprint image. The easiest way to find minutiae is to start with the thinned image and follow the lines to a crossing or an end (see Fig. 8.25). That works well for good-quality fingerprint images as there are no artificial breaks or connections between ridges due to dirt or other disturbances on the sensor.

Advanced systems use the gray-scale image for minutiae detection. The basic idea is to track the lines following the local orientation of the ridge pattern [54]. A set of starting points is determined by superimposing a grid over the gray-scale image. For each starting point, the algorithm keeps following the ridge lines until they terminate or intersect with other lines, which defines the minutiae.

Some fingerprint systems use additional features such as global ridge structure or pores.

Matching

The matching unit determines the correspondence between a pair of fingerprints. Fingerprint matching is a difficult problem as there is a large variability in different images of the same finger. Matching algorithms have to deal with the following variations [45]:

- Shift of contact area: users never touch the sensor on exactly the same area. Especially when using a small sensor the problem increases.
- Rotation: the alignment of the finger varies up to $\pm 15°$.
- Non linear distortion: depending on the pressure on the sensor, the ridges are compressed or stretched.
- Capture conditions: condition of the skin and environmental conditions such as dirt or moisture influence the image.

Therefore, pure image matching on pixel basis does not deliver satisfactory results. Minutiae matching is the most popular method of fingerprint matching. An example of minutia matching is given in Fig. 8.26.

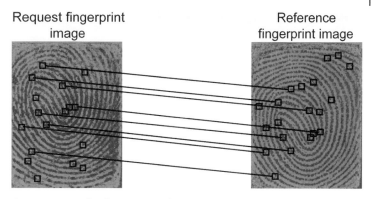

Fig. 8.26 Example of minutia matching

Minutia lists can be matched using different techniques. According to [54], the most popular are

- point set matching
- graph matching
- sub-graph isomorphism.

8.4.2
Performance of a Biometric System

The performance of a biometric system is limited by the following sources of errors:

- sensor device, capturing an image of the fingerprint (see Sections 8.3.2.1 and 8.4.1.4)
- biometric engine (i.e. biometric software; see Section 8.4.2.1)
- application scenario.

Application Scenario

Forensic systems generally produce a sequence of matching results according to the similarity of the fingerprint under investigation with the fingerprints in the database. Commercial applications usually work autonomously; they have to provide a clear decision as to whether the request fingerprint matches with a corresponding one in the database or not.

In theory, each biometry has a confidence rate for the considered characteristic features. This rate represents the ideal error rate of the biometric system. In the real world of technical realization there are a lot of technical restrictions and effects on the error rate, such as:

- scanning area limited
- imperfect image quality
- restricted gray-level resolution
- restricted accuracy in algorithmic computation
- restricted template size.

8.4.2.1 Error Rates

For the evaluation and improvement of biometric systems, different error rates are used. These error rates in general are applicable for all biometric systems. In the following, however, the definitions are applied to and clarified for fingerprint systems.

Usually the output of a biometric system is a measure coming out of the matcher unit, the so-called quantity match score. The match score describes the similarity between the matched templates (i.e. feature vectors, see Section 8.4.1.3) and is often normalized; 100% corresponds to absolute identity.

In commercial applications, the system output is always true or false, as one needs a clear decision for granting access to a system. The decision is controlled by a threshold: match scores higher than the threshold lead to a positive answer and match scores lower than the threshold result in a negative answer.

In access control systems (positive recognition system) there are two types of errors:

- Two different fingers have a match score higher than the threshold, i.e. false match.
- Two measurements of the same finger have a match score lower than the threshold, i.e. false non-match.

In general, error rates can be measured on two levels:

- Biometric engine-level rates, taking just the inputs and outputs of the engine (i.e. software).
- System-level rates, considering all the errors between user and non-biometric application.

Biometric Engine-level Error Rates

For testing biometric engine-level error rates, usually a well-known database with fingerprint images or templates as input is taken. The output is the match score. The following error rates are used for evaluation of biometric engines:
- False acceptance rate (FAR, Eq. 8): count of false match in relation to all matches of non-identical fingerprints:

$$\text{FAR}(\%) = \frac{\#\text{ false match}}{\#\text{ match}} \qquad (8)$$

- False rejection rate (FRR, Eq. 9): count of false non-match in relation to all feasible matches of identical fingers:

$$\text{FRR}(\%) = \frac{\#\text{ false non-match}}{\#\text{ match}} \qquad (9)$$

If one draws these rates for each threshold value on a chart, one obtains the typical FAR/FRR curves (Fig. 8.27). In real systems there is no threshold where FAR and FRR curves are zero, therefore other interesting points on these curves are taken to characterize the performance.

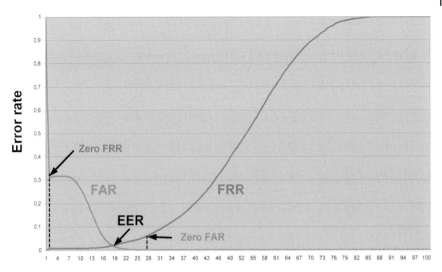

Fig. 8.27 FAR/FRR curve. Depending on the matching algorithm used, the FAR curve tends to remain at a constant level for low match score values. This in real applications irrelevant artefact is due to a non-significant random correspondence of two or three minutiae in the matched fingerprints

- Equal error rate (EER): point of intersection of FAR and FRR curve in %. EER is a fair measure if one needs to describe the performance of a biometric system by one figure only. Systems with low EER are preferred.
- Zero FAR: lowest FRR where no false match occurs.
- Zero FRR: lowest FAR where no non-false match occurs.

Zero FAR and Zero FRR are never equal, as there is no threshold value where FAR and FRR are zero in common.

These FAR/FRR values are sufficient if one wants to set up a biometric system.

Depending on the application, one sets the threshold to a higher value (e.g. more secure) or a lower value (e.g. more convenient). In the former case one accepts checking a fingerprint sometimes twice rather than accepting a false match, in the second case it is the other way round.

If one wants to compare different biometric engines, one should use a threshold-free representation. Therefore, one puts the FAR/FRR values on a logarithmic scale for each system. This receiver–operator curve (ROC, Fig. 8.28) gives a good picture of the performance of different engines. The ROC shows the limits of operation regarding the lowest possible FAR and FRR. If one compares the ROCs of different biometric engines determined on the same database, one can easily identify the better performing engine, as the curve runs more distant from the FAR axis.

More definitions and mathematical derivatives can be found in Maltoni et al.'s *Handbook of Fingerprint Recognition* [45].

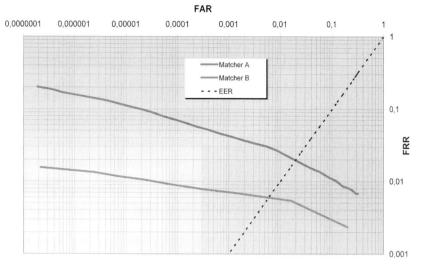

Fig. 8.28 Receiver–operator curve (ROC) from different matchers. Matcher B is the better performing biometric engine (lower EER, lower FRR for a given FAR and vice versa)

System Level Error Rates
To evaluate a biometric system, the whole workflow must be taken into account:

- user interaction
- ergonomic design of the sensor device
- sensor
- biometric engine
- application integration
- enrolment scenario
- verification/identification scenario.

Therefore, on the system level more characteristic measures (i.e. error rates) must be considered in addition to the FAR/FRR curve of the engine (Fig. 8.27). These errors are additional to the errors of the biometric engine and altogether they determine the performance of the biometric system.

Failure to Capture (FTC) Rate
Depending on the sensor and the triggering mechanism of the biometric device, image capture sometimes fails. The better the ergonomic design of the device, the lower is the FTC rate.

Failure to Enrol (FTE) Rate
This is the percentage of attempts that do not lead to a valid template and a successful user enrolment.

8.4.2.2 Evaluation

Phillips et al. [47] define three types of evaluation processes, as follows.

Technology Evaluation

Different algorithms with a given set of input/output parameters (i.e. database, type of application) are compared. Usually this scenario is used for verification contests such as the Fingerprint Verification Contest (FVC 2000, FVC 2002) [71, 72]. Participants receive a small sample of the database of fingerprints used, to tune their systems. The results of the contest are calculated with the major part of the database (e.g. 90%). Participants are evaluated by different parameters, such as EER or calculation time. In real-world applications the results do not give an appropriate answer about the performance of a whole biometric system, as there are a lot more influences on the performance (see Section 8.4.2.1). Additionally, the evaluation scenario does not take into account the individual algorithmic power that has been developed to deal with real-world scenarios, such as algorithms with user feedback or complex enrolment scenarios.

Scenario Evaluation

The overall system performance is evaluated. Each system has its own sensor device. One must ensure that all testing is carried out in the same system environment so that one can compare the results. Scenario evaluation is even more challenging if one compares different biometrics.

Operational Evaluation

The goal of operational evaluation is to find out the performance of the biometric system in a specific application environment (e.g. border control systems). These evaluations take a lot of effort, as one has to build up the whole system and use it with a specific target population for a long time. Usually one cannot compare the results of different evaluations, as the environment is critical to the results.

Databases

For developing and improving biometric algorithms, a database for continuous technical evaluation is needed. The database should meet the following requirements:

- Originate from the target biometric sensor device.
- Data must be contradiction free.
- Data should cover a wide range of environmental conditions of the device (i.e. humidity, dirt).
- Representative distribution of population (i.e. race, age, gender).
- Representative distribution in time of trial.
- A sufficiently large number of images of identical fingers and different fingers are necessary.

The results one obtains always depend on the database being used. One can establish one's own database, but must be aware of privacy legislation protecting personal data. It is a lot easier and cost effective to rely on databases provided by sensor and device manufacturers. Large databases are available at NIST (e.g. [48]). Its main area is forensic fingerprint recognition.

In practical work, databases should have at least a few hundred different fingers with some dozens of different images of each finger. For confident proof of error rates lower than FAR 10^{-4} and FRR lower than 10^{-2}, one needs:

- more than 10^5 fingerprint images different to that requested
- more than 10^3 fingerprint images of identical fingers.

For extending the proof of error rates over the size of the database, some testers extrapolate the curves of the error rates with exponential functions. Another way of effectively obtaining large databases is to generate synthetic fingerprints.

Synthetic Fingerprints
Synthetic fingerprints have a lot of advantages for evaluation reasons. One can easily generate large databases without a time-consuming gathering and one does not have privacy concerns. The task of generating synthetic fingerprints is a challenge as this generator has to model various inter-class and intra-class variations. Maltoni et al. [45] showed the problem of generating realistic impressions of the same virtual finger by simulating:

- Touching areas variation.
- Pressure of the finger on the sensor leads to non-linear distortions.
- Ridge line thickness changes with pressure.
- Cuts and scars on the finger.
- Noise.

One of the available fingerprint generators is SFINGE developed by Cappelli et al. [49]. The SFINGE method first generates a master fingerprint, which has all the characteristics of an authentic fingerprint. It starts with the shape of touched fingerprints and generates an orientation image, a frequency image and finally the ridge structure. In the second step, SFINGE generates synthetic impressions (i.e. variations in touched area on the sensor) derived from the master fingerprint. These images contain distortion, rotation, displacement, skin conditions and noise.

Evaluation Process
Depending on the target of the evaluation, an offline or online evaluation is preferred. Offline evaluation is used for biometric engine evaluation (see Section 8.4.2.1). In step one, all fingerprint images are processed through the encoder and the templates (Biometric Identification Records, BIR) are stored (see Fig. 8.29).

For FRR evaluation, each template is matched against all other templates of the identical finger. The scores of these matches are stored in a matrix for each fin-

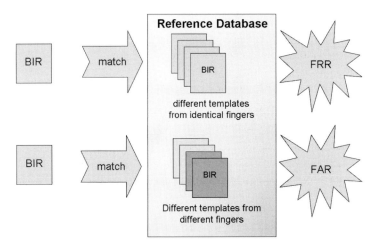

Fig. 8.29 Workflow for offline evaluation

ger. The last step is to go through the matrices, increase the threshold step by step and count the positive and negative tests for each threshold.

If a complex enrolment scenario is used (e.g. large sensor device for enrolment or image fusion techniques), one has to distinguish between templates from enrolment and templates from request. In these scenarios a lot more request templates are needed for achieving a given confidence.

For FAR evaluation, each template is matched against all templates of non-identical fingers (i.e. enrolment against request templates). All scores are stored in a matrix and the FAR values are counted for all score values.

To set up the ROC, one records the FRR values and the corresponding FAR values (see Fig. 8.28).

In technology development, automated tools are used, so that one can easily obtain FAR/FRR values after each modification of algorithms and parameters. In practical work, depending on database size and equipment, it takes some hours to some days to process the whole database. For example, for a database of 5000 fingerprint images, one has to calculate 5000 encodings (e.g. 500 ms per encoding on a PC) and 25 million matches (e.g. 100 ms on a PC including minutes writing), which take about 700 h. Especially for parameter tuning, this is a very time-consuming process and may even take weeks in extreme cases.

Online evaluation is used for scenario and operational scenarios testing (see Section 8.4.2.1). Each test is completely processed (i.e. image capture, encoding and matching). Therefore, one records each test with the corresponding outcome (i.e. score) and then calculates FAR, FRR, FTE and FTC.

Interpretation of Results
Some more interpretation of the results of the FAR/FRR evaluations is necessary for the development of a biometric application.

Stability
Slight variations in image capture conditions should not lead to severe alterations in results. A fairly good measure for the stability is the shape of the FAR and FRR curves near the EER point. The lower the gradient, the more solid is the system.

Error Distribution
During evaluation, every test has the same implications. For applications it makes a difference if the biometric system has a constant error distribution over all fingers or if there are some 'general keys' that often match with other fingers. Usually these 'general keys' occur if fingerprint images with bad image quality are accepted at enrolment. Therefore, advanced systems check the image quality carefully before accepting the fingerprint template in the database.

Enrolment Quality
As the enrolment process determines the performance of the system, one has to consider the FTE rate comparing different systems. The higher the FTE rate, the easier it is to obtain good FAR/FRR, because critical fingerprint images are already excluded during enrolment. However, in real applications users do not accept not being enrolled.

8.4.2.3 Meeting the Application Requirements
When developing a biometric (fingerprint) application, one first has to answer the following questions:

- Is the application more of a convenience type or does it need high-level security (see Section 8.4.2.1)?
- Do you have a verification or identification scenario (see Section 8.4.4.1)?
- One- or two-way authentication (i.e. combination with smartcard, see Section 8.4.4.4)?
- Small group of users using the system fairly often or large group using the system fairly rarely?
- Supervised enrolment or users' self-enrolment?
- Centralized or decentralized storage of templates?
- Computational performance requirements?

There is no one biometric fingerprint system that meets all these requirements. The more users are left alone with the system and the rarer they use the system, the more one has to take care about good usability. The biometric system must provide good feedback messages (see Section 8.4.4.2). The device interface must be of good ergonomic design.

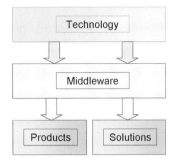

Fig. 8.30 Development chain

8.4.3
System Integration

As biometric technology matures, there are different alternatives for integrating biometrics in non-biometric applications. Depending on the software architecture and one's role in the value chain, the development challenges are different. In general, the development chain consists of product and solution development, middleware (i.e. abstraction of technology) and technology development (Fig. 8.30).

8.4.3.1 Technology Development

Technology development in biometrics is primarily image processing and pattern recognition. The starting point is the raw image provided by the sensor device driver (see Section 8.4.3.2). The encoder consists of various image processing steps; some of them are based on general algorithms and some are special fingerprint algorithms (see Section 8.4.1.4).

Technology development is an iterative task. The challenge of technology development is to achieve the right ratio of process-driven engineering and creative work. It has proved to be successful to make loops of three steps (Fig. 8.31):

1. Creative phase: collect ideas, estimate the benefits of the ideas and put them in sequence.
2. Realization phase: process-driven realization of the ideas of phase 1.
3. Evaluation phase: persistent evaluation of implemented ideas, find out strengths and weaknesses of the new algorithms.

These three steps take about 3–6 months. If the advances are high enough, algorithms could be transferred to product integration. Practical experience has shown that about five loops are needed to release the software to the market. Approximately 10 loops are needed for maturing a specific technology (e.g. Siemens biometric technology is in loop 14).

Developing a new technology or software may be necessary if one is targeting a very special hardware platform (e.g. embedded system or smartcard). For the PC platform there are a lot of mature technology platforms available on the market.

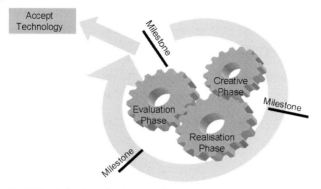

Fig. 8.31 Technology development loop

For the integration of biometrics in a specified application software, development kits or component-based technology are used in order to be fast and to save on resources.

8.4.3.2 Sensor Device Integration

For the integration of fingerprint sensor devices on *PC-based systems*, specific device drivers are provided by device manufacturers. The main tasks for these device drivers are:

- Integrate the device in the operating system.
- Convert messages from the device to the operating system or the application.
- Provide full image information out of data packages from the sensor.
- Control the low-level communication to the sensor.
- Control sensor power saving mode.

Most of the sensor devices use the universal serial bus (USB) for interfacing the host system. The sensor driver must be well integrated into the operating system. For Microsoft operating systems, there are special tests called WHQL tests (Windows Hardware Quality Labs Reference [74]). If the driver has passed these tests, the system receives a signature from Microsoft. Some manufacturers provide setup routines to install the driver and the application setup.

Low-level integration is necessary if an *embedded system* based on a microcontroller or signal processor is used for fingerprint recognition. For low-level integration, support from the sensor manufacturer is essential. Because of the absence of standards there are a lot of different interfaces proprietary to each sensor.

8.4.3.3 Software Development Kit (SDK)

Software development kits are designed for the integration of biometrics into various applications without working on biometric technology. The technology is enclosed in the SDK; the standard biometric functions such as verification, identification and enrolment are provided through simple interfaces which can be programmed by any software engineer without a deep knowledge of biometric technology.

Typically the SDK is interfaced through C/C++ or Java Code. In some cases the database is integrated into the SDK. For large-scale applications the biometric templates are stored outside the SDK.

Useful SDKs apply the following design principles [50]:

- Biometric and general processing functions (e.g. database access, set the parameters of the capturing device) are completely separated, as a consequence of a strictly modular software concept. Additionally, functions of the SDK are grouped into components. Each component reflects a feature in general.
- All dependences regarding external interfaces of the SDK functions are encapsulated. This guarantees easy handling of functions for the programmer and user of the SDK.
- Owing to the modular software concept, internal components are easy to exchange. Hence software upgrades are quickly done and are easy to handle.

Some SDKs offer an example code and user interface components (e.g. ActiveX elements) for a quick start development. A typical software architecture is shown in Fig. 8.32 and consists of the following layers:

- Control layer: management functions control the workflow of the biometric engine. Parameters such as security level or threshold (see Section 8.4.2.1) can be adjusted.
- Kernel layer: all basic image processing and biometric functions are encapsulated in the kernel. The strategy element controls the workflow of enrolment, verification and identification.
- Data layer: interfaces the hardware via sensor driver as well as the database for templates and parameters (i.e. registry in MS Windows-based systems).

Usually the functionality of the SDK is provided in a single library. For redistribution some SDK manufacturers take license fees per user or per installation.

8.4.3.4 Standard Software Architecture

Emerging technologies such as biometric technology are always vendor proprietary at the beginning. Each vendor designs their own interfaces and their own software architecture. The reasonable next step to gain a market share is to establish unified interfaces.

The main advantage of standard software architecture is that application developers write their software once and have access to various biometrics and vendor implementations. However, the advances of a specific technology create a barrier to standardization efforts, because standards always have to bring down functionality to a common platform. Functional advantages of an individual vendor cannot be considered in the standard. Applications running on non-standard hardware platforms (i.e. signal processors and microcontrollers) cannot be standard-

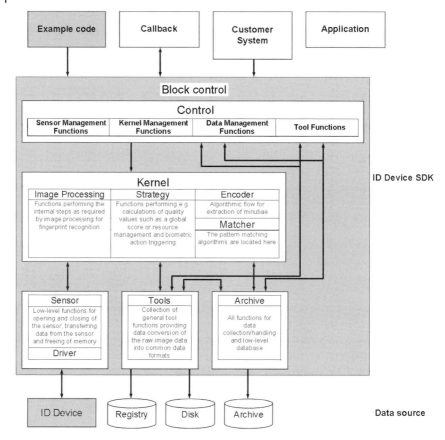

Fig. 8.32 Architecture and interfaces of Siemens ID Device SDK

ized, as the implementation of the technology has to exploit the resources of the hardware. Up to now there has been low demand on the market for standardized architecture, as customers prefer excellent functionality of complete seamless systems to open architectures.

Nevertheless, in a few years from now evolution will lead to one or more standard software architectures for biometrics. Since 1998, the BioAPI consortium has started its standardization effort and in December 1998 they published their first white paper [51].

BioAPI

The BioAPI consortium has developed a platform-independent multi-level API (application program interface) and framework (Fig. 8.33). The interfaces are designed for programming in C/C++ and Java. The multi-levels of the API should provide maximum flexibility for programmers, whether they require the use of high-level biometric function or want to thrill down to biometric technology.

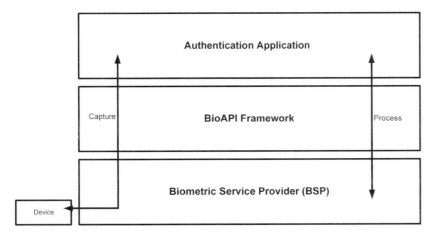

Fig. 8.33 Basic software architecture BioAPI. Adapted from [55]

Fig. 8.34 Biometric Identification Record (BIR). Adapted from [55]

Standard Biometric Templates

The minimum requirement for cooperation between different fingerprint systems is data compatibility to exchange templates between the systems (if required). The BioAPI consortium has suggested a data structure called 'biometric identification record' (BIR, see Figs. 8.29 and 8.34), which is compliant with the 'Common Biometric Exchange File Format' (CBEFF). CBEFF is described by NIST [52]. The format of the Opaque Biometric Data is in most cases vendor proprietary so that in practical use compatibility is limited to fingerprint images. The header carries a 'Format Owner' data field and the values are assigned and registered by the International Biometric Industry Association [75].

8.4.3.5 Component-based Integration

The easiest way to integrate biometrics into an application is the use of biometric authentication components. The idea is to replace the non-biometric authentication component by a biometric module. In most operating systems there is such an authentication module, which controls the access to the applications. Related to browser-based applications (i.e. inter-/intranet), the application developer has to use single Java calls or active server pages calls when using Microsoft servers. Some of these component-based systems even provide C/C++ interfaces. Hence integration of biometrics is done by just changing the authentication module or by integrating a single line of code in the application.

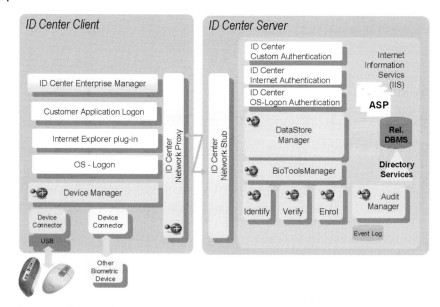

Fig. 8.35 Software architecture Siemens ID-Center

The biometric identity management system (e.g. Siemens ID-Center [73]) supplies all the functionality encapsulated in a server or a client suite:

- sensor interface for different devices
- device drivers
- biometric technology for image processing encoding and matching
- data storage on basis of conventional databases or directory services
  (e.g. Microsoft Active Directory or X.501)
- encryption and signing of biometric templates
- signing of software components
- logging and encrypted network communication
- user interface for enrolment verification and identification
- roaming templates for network use
- interfaces to biometric physical access systems.

An example of a software architecture for a biometric client/server application is shown in Fig. 8.35.

8.4.4
Deployment of Fingerprint Systems

The deployment of biometric systems depends on the motivation for applying biometrics:

- Increased security: With this motivation, usage starts in small areas with special security needs such as executive areas, computer centers or engineering depart-

ments. Typical applications are physical access control and key protection for encryption systems.
- Increased convenience: Some security applications have very complex user interfaces. Using biometric authentication is a good way to get user acceptance for these applications. For example, actual VPN systems (virtual private networks) need up to three pass phrases to gain access to the data. As these users (e.g. sales representatives) are usually not highly experienced in computer science, these applications have to be very simple. Just inserting a smartcard and pressing the finger on the sensor will have high user acceptance.
- Legal claim: In some countries personal data have to be protected by two-way authentication (e.g. HIPAA law in the USA [53]). The appropriate solution for this requirement might be biometrics in combination with smartcards. The deployment usually starts with a field trial and is followed by an organization-wide roll out.
- Cost cut: Running costs of biometric systems are lower than costs of password systems (see Section 8.1.2).

8.4.4.1 Verification and Identification

Before deploying a biometric system, the first decision is between verification or identification scenario.

In general, verification scenarios are more reliable, because the number of possible false acceptance errors increases with the size of the reference database. Identification scenarios are feasible for small groups (e.g. small office or family); groups of a thousand or even more users result in answering times of several seconds.

Identification scenarios are more convenient. Convenience in verification can be raised if one combines verification with a token or a smartcard holding a unique identifier (see Section 8.4.4.4). Another way of applying identification is to form smaller groups, such as departments in an office.

8.4.4.2 User Interaction

The user interface directly determines the false rejection rate, whereas the false acceptance rate is controlled by the biometric engine only. The interface has to be designed for intuitive usage. If there is a problem, the application has to report in clear words what to do.

Feedback

The false rejection rate can be lowered by a factor of 2–5 if user feedback is added to the application. The following messages help users to improve their personal error rates:

- *'Press harder'*: Users often just slightly tip on the sensor.
- *'Clean sensor'*: If there is dirt on the sensor, the user is asked to swipe with the finger once over the sensor.
- *'Finger is too wet'*: The user has to clean his/her finger.

8.4.4.3 Security Issues

Security engineering always starts with an attack analysis. The main attack scenarios are:

- Sensor: fakes with artificial fingerprints.
- Database: manipulation of stored fingerprint templates.
- System integrity: replacement of biometric system output.

Fraud

Sensor attacks are primarily carried out with artificial fingerprints which are accepted by the capture device as valid fingerprint images. The weak point of the sensor is the capture procedure. Depending on sensor type, there are different problems to handle:

- Full frame sensors: The image is generated by measuring the capacitance. The capacitance is influenced by electrical properties of the skin surface. Therefore, artificial fingerprints must imitate not only the 3D structure but also the electrical parameters, which is not that easy and requires good knowledge. A special problem with CMOS sensors comes from dirt left on the sensor. Under certain circumstances, this dirt can lead to a ghost image.
- Swipe sensors: Extend the measured parameters by dynamic properties (i.e. elasticity) of the skin. Therefore, fakes have to imitate the 3D structure and the electrical and dynamic features of the fingerprint. As one swipes over the sensor, there is no problem with dirt on the sensor.

Database attacks are not specific to biometrics. Therefore, there are various attack scenarios and protection measures. Actual systems are protected by symmetric or asymmetric cryptography and signed templates.

The biometric system security and integrity are part of the entire system security. Therefore, the security concept of the entire system must be extended to the biometric software and hardware components. The most important thread is exchange of components. Therefore, signature of components is state-of-the-art.

Anti-fraud Techniques

Anti-fraud techniques in biometrics concentrate on the sensor interface itself, as database and system integrity frauds are not unique to biometric systems and there exist well-established solutions.

Some biometric applications are used in a supervised mode (e.g. border control), so there is no chance of fraud in these systems as there is always someone watching users accessing the system.

- Dirt problems with CMOS sensors: As the dirt may preserve the last one to three fingerprints, a potential subsequent false acceptance or fraud attack is successfully avoided if the last images are stored in the system. In ordinary use images are always slightly different whereas dirt images are exactly identical. Therefore, before acceptance of a captured image the system checks the stored

images for identical biometric features (i.e. minutiae). If the answer is positive, the image is rejected and the user is asked for another try.
- Artificial fingerprints: Depending on the sensor, artificial fingerprints are more or less hard to create (see above). Usually one needs a cooperative user to create an artificial fingerprint. In positive scenarios (i.e. access control), security officers are not concerned about this additional threat, because cooperative users are a general threat to security and one has to make sure that there are no such risks in the organization. It is like passing on the password or a key to someone outside the organization, independent of the applied access control system. Biometric systems have to make sure that reproduced fingerprints (so-called 'latent fingerprints', e.g. fingerprints from a glass) are not accepted as valid fingerprints. Usually the latent fingerprints are not complete. There is no information about electrical properties or 3D structure. If one needs more security, one can enrol more than one finger per person and then request one or even two random fingers from the user.
- Hill climbing: If fraud attacks are answered with live pictures or score, fraud becomes easier. Alterations of the attack (e.g. improve artificial fingerprint) in combination with the system answer can lead to a successful attack. Therefore, systems with security requirements should not give advice if fingerprints are rejected. The number of tries should be limited or the pause between tries should be extended after each unsuccessful try.

In general, fraud becomes much harder if one combines the biometric system with a key or smartcard.

8.4.4.4 Smartcard Integration
The authorization process is improved if biometric systems ('what you are') are combined with some kind of key ('what you own'). In most cases, smartcards are used for that task.

Smartcards have been in use since the 1980s; one has to distinguish between the following types:

- Storage smartcards: The most common type, these carry a unique number and up to 64 kbyte of storage that can be used autonomously by different applications. Cards can be contacted through a conventional card reader or can be contact less with an RF-Interface.
- Processor and crypto smartcards: These have storage and a processing unit. They can provide keys for symmetric and asymmetric encryption. They execute an operating system optimized for security.

Smartcards and biometrics can be integrated in different ways:
- Identifier on Card (IoC)
- Template on Card (ToC)
- Matcher on Card (MoC)
- System on Card (SoC).

Identifier on Card (IoC)

This is the easiest way of combining smartcard and biometrics, especially used for improving the security of existing smartcard systems, as one does not have to replace existing smartcards. The card carries a unique identifier (e.g. employee number); this identifier is send to a centralized biometric system, where a verification is calculated between the request fingerprint image and the stored template.

Template on Card (ToC)

For decentralized application, the biometric template is stored on the card itself. The template size starts at about 200 byte for minutiae-based matcher systems; image-based matcher systems need minimum of 2 kbyte per template. In combination with an integrated smartcard fingerprint reader [54], these systems can reach high security levels.

All modern smartcards can be used and one need not to step into the operating system of the card. The security is sufficient if the values one wants to protect are within the system and not on the card.

Matcher on Card (MoC)

MoC systems use the processor of the smartcard for the matcher component. The image capture and the encoder component are executed by a fingerprint reader with a microcontroller. During the enrolment the template is stored on the secure memory of the card. For verification, the requested fingerprint is encoded in the microcontroller of the reader and the template is forwarded to the card. The card matches the requested template with the stored template and returns the credentials (e.g. digital signature) of the card.

The advantage of these systems is that the credentials on the card are highly secure as the fingerprint template does not leave the card. The disadvantage is that one has to establish a biometric enhanced card operating system and apply expensive smartcards with a powerful processor.

System on Card (SoC)

All components of the biometric system (i.e. sensor, encoder and matcher) are located on the smartcard. This would be the ideal combination of biometrics and smartcard. Because of the technical difficulties concerning power supply, processor power and sensor device, there has been no realization yet.

A major advantage of an SoC is that it would be compatible with existing infrastructure. Instead of supplying a PIN code through a pin pad, one could use a fingerprint on a conventional smartcard reader. In addition to the technical challenge, another disadvantage of a, SoC is the price of the smartcard containing the whole system.

8.5
Conclusions

Miniaturized CMOS fingerprint sensor systems provide a unique advantage for the successful deployment of such systems into mass markets: These systems have the potential to be miniature enough for integration into mobile devices, including smartcards, that are in the possession and under direct control of the user. This is an important condition for the broad acceptance of such systems with respect to data security and privacy. Those personalized mobile systems may then substitute today's cumbersome, non-secure and sometimes expensive application of multiple passwords and PIN codes for access control to security areas and networks. Furthermore, capacitive fingerprint sensors are still promising candidates to fulfil the low cost requirements for consumer applications: Further system integration, the development of swipe sensors or research on alternative materials and process technologies (e.g. thin-film technologies on polymer substrates) address this issue. Image quality, sensor reliability and anti-fraud solutions on the system level are showing reasonable progress. Especially these issues are perfect examples to illustrate the need for and importance of joint development of sensor hardware, image processing software and system integration for optimized results for advanced CMOS fingerprint sensor systems.

8.6
Acknowledgments

We would like to thank the Siemens and Infineon Fingertip Teams. In particular, special thanks go to Paul Werner von Basse, Stefan Jung, Martin Handtmann, Peter Morguet and Andreas Gaymann and to the production sites in Regensburg and Munich, Germany, both of which supported research and development of the fingerprint demonstrators. Additional thanks go to the Siemens Software Team in Graz, Austria, namely Wolfgang Marius, Josef Birchbauer, Vuk Krivec and Peter Weinzierl, who developed relevant parts of the fingerprint algorithms and application software. Many thanks are due also to Zoltan Nagy, student assistant at the Chair of Micro and Nanosystems, ETHZ, who helped with literature research and manuscript preparation.

8.7
References

8.7.1
Journals, Conference Proceedings, Books, Reports

1. R. de Luis-Garcia, C. Alberola-Lopez, O. Aghzout, J. Ruiz-Alzola, *Signal Process.* **2003**, *83*, 2539–2557.
2. G. Hribernig, D. Tukulj, M. Luetic, in: *Proceedings of the Conference Telecommunications and Mobile Computing, tcmc 2001, Workshop on Wearable Computing, 15–16 October 2001, Graz, Austria*; **2001**, Workshop 2-3-2.
3. *Biometric Market Report 2000–2005*; International Biometrics Group, One Battery Park, Plaza, New York, NY 10004, **2001**.
4. R. Chellappa, C. L. Wilson, S. Sirohey, *Proc. IEEE* **1995**, *83*, 705–740.
5. S. Akamatsu, T. Sasaki, H. Fukamachi, N. Masui, Y. Suenaga, in: *Proceedings of the 11th IAPR International Conference on Pattern Recognition, Vol. II, Conference B, Pattern Recognition Methodology and Systems*; **1992**, 217–220.
6. P. J. Phillips, A. J. O'Toole, Y. Cheng, B. Ross, H. A. Wild, *Assessing Algorithms as Computational Models for Human Face Recognition. Technical Report NISTIR 6348*; Washington, DC: National Institute of Standards and Technology, **1999**.
7. F. Forster, P. Rummel, M. Lang, B. Radig, in: *Proceedings of the 2001 International Conference on Image Processing, ICIP 2001, Vol. 2*; **2001**, pp. 598–601.
8. S. Furui, *Pattern Recognit. Lett.* **1997**, *18*, 859–872.
9. V. S. Nalwa, *Proc. IEEE* **1997**, *85*, 215–239.
10. J. G. A. Dolfing, E. H. L. Aarts, J. J. G. M. van Oosterhout, in: *Proceedings of the 14th International Conference on Pattern Recognition, August 1998, Vol. 2*; **1998**, 1309–1312.
11. A. K. Jain, S. Pankanti, S. Prabhakar, A. Ross, *Lect. Notes Comput. Sci.* **2001**, *2091*, 182–190.
12. S. Pankanti, S. Prabhakar, A. K. Jain, in: *Proceedings of the of the 2001 IEEE Computer Society Conference on Computer Vision and Pattern Recognition, CVPR 2001, Vol. 1*; **2001**, pp. 805–812.
13. A. K. Jain, S. Prabhakar, S. Pankanti, *Lect. Notes Comput. Sci.* **2001**, *2091*, 211–216.
14. J. Daugman, C. Downing, *Proc. R. Soc. London* **2001**, *268*, 1737–1740.
15. R. P. Wildes, *Proc. IEEE* **1997**, *85*, 1348–1363.
16. W. W. Boles, B. Boashash, *IEEE Trans. Signal Processing* **1998**, *46*, 1185–1188.
17. S. Lim, K. Lee, O. Byeon, T. Kim, *ETRI J.* **2001**, *23*, 61–70.
18. J. Daugman, *IEEE Trans. Pattern Anal. Machine Intell.* **1993**, *15*, 1148–1161.
19. J. Daugman, *J. Opt. Soc. Am. A* **1985**, *2*, 1160–1169.
20. J. Daugman, *IEEE Trans. Acoust. Speech Signal Process.* **1988**, *36*, 1169–1179.
21. G. Tomko, in: *Proceedings of the English Annual Conference on Computers, Austin, TX, August 1998*; **1998**, 1309–1312.
22. *Passwort Reset: Self-Service That You Will Love.* **2002**, Gartner Group, 56 Top Gallant Road, Stamford, CT 06904; Technology, *T-15-6454*.
23. X. Xia, L. O'Gorman, *Pattern Recognit.* **2003**, *36*, 361–369.
24. J.-F. Mainguet, M. Pegulu, J. B. Harris, *Future Generat. Comput. Syst.* **2000**, *16*, 403–415.
25. R. Becker, *Elektronik-Industrie* **2002**, *9*, 34–35.
26. S. Jung, C. Hierold, T. Scheiter, P. W. von Basse, R. Thewes, K. Goser, W. Weber, in: *Transducers '99, the 10th International Conference on Solid-State Sensors and Actuators, 7–10 June 1999, Sendai, Japan. Digest of Technical Papers, Vol. 2*; **1999**, 966–969.
27. Simonic R.B., in: *IEEE International Symposium on Electromagnetic Compatibility*, **1982**, 191–198.
28. S. Jung, R. Thewes, T. Scheiter, K. F. Goser, W. Weber, *IEEE J. Solid-State Circuits* **1999**, *34*, 978–984.
29. S. Jung, R. Thewes, T. Scheiter, K. Goser, W. Weber, in: *1999 Symposium on VLSI Circuits. Digest of Technical Papers*; **1999**, 161–164.

30 S. SHIGEMATSU, H. MORIMURA, *IEEE J. Solid-State Circuits* **1999**, *34*, 1852–1859.

31 H. MORIMURA, S. SHIGEMATSU, K. MACHIDA, in: *1999 Symposium on VLSI Circuits. Digest of Technical Papers*; **1999**, 157–160.

32 H. MORIMURA, S. SHIGEMATSU, T. SHIMAMURA, K. MACHIDA, H. KYURAGI, in: *2001 Symposium on VLSI Circuits. Digest of Technical Papers*; **2001**, 171–174.

33 ATMEL, *AT77C101B FingerChip^{TM}, Datasheet, 2150B-BIOM-09/03*. ATMEL, 2325 Orchard Park Way, San Jose, CA 95131, **2003**.

34 O. VERMESAN, K. H. RIISNAES, L. LE-PAILLEUR, J. B. NYSAETHER, M. BAUGE, H. RUSTAD, S. CLAUSEB, L.-C. BLYSTAD, H. GRINVOLL, R. PEDERSEN, R. PEZZANI, D. KAIRE, in: *Proceedings of the 2003 IEEE International Solid-State Circuits Conference, ISSCC 2003, San Francisco*; **2003**, 214.

35 R. HASHIDO, A. SUZUKI, A. IWATA, T. OKAMOTO, Y. SATOH, M. INOUE, *IEEE J. Solid-State Circuits* **2003**, *38*, 274–280.

36 K. MACHIDA, S. SHIGEMATSU, H. MORIMURA, Y. TANABE, N. SATO, N. SHIMOYAMA, T. KUMAZAKI, K. KUDOU, M. YANO, H. KYURAGI, *IEEE Trans. Electron. Devices* **2001**, *48*, 2273–2278.

37 H. MORIMURA, S. SHIGEMATSU, T. SHIMAMURA, N. SATO, K. MACHIDA, H. KYURAGI, *Jpn. J. Appl. Phys.* **2002**, *41*, Part 1 (4B), 2316–2321.

38 H. MORIMURA, S. SHIGEMATSU, T. SHIMAMURA, N. SATO, Y. OKAZAKI, K. MACHIDA, H. KYURAGI, *Jpn. J. Appl. Phys.* **2002**, *41*, Part 1 (10), 5951–5956.

39 K.-H. LEE, E. YOON, A 500dpi Capacitive-Type CMOS Fingerprint Sensor with Pixel-Level Adaptive Image Enhancement Scheme. *IEEE ISSCC Digest of Technical Papers* **2002**, 352–353.

40 D. INGLIS, L. MANCHANDA, R. COMIZZOLI, A. DICKINSON, E. MARTIN, S. MENDIS, P. SILVERMAN, G. WEBER, B. ACKLAND, L. O'GORMAN, in: *Proceedings of the 1998 IEEE International Solid-State Circuits Conference, ISSCC 1998, San Francisco*; **1998**, 284–285.

41 T. SCHEITER, C. HIEROLD, H.-J. TIMME, in: *Proceedings of the 1998 Microsystem Symposium, 10–11 September 1998, Delft*; **1998**, 77–85.

42 F. GALTON, *Finger Prints*; Macmillan, London, **1892**.

43 E. HENRY, *Classification and Uses of Finger Prints*; Routledge, London, **1900**.

44 A. K. HRECHAK, McHUGH, *Pattern Recognit.* **23**, 893–904, **1990**.

45 D. MALTONI, D. MAIO, A. K. JAIN, S. PRABHAKAR, *Handbook of Fingerprint Recognition*; Berlin: Springer, **2003**.

46 A. BOVIK, *Handbook of Image and Video Processing*; New York: Academic Press, **2000**.

47 P. J. PHILLIPS, A. MARTIN, C. L. WILSON, M. PRZYBOCKI, *IEEE Comput. Mag.* **2000**, 56–63.

48 C. I. WATSON, C. L. WILSON, *NIST Special Database 4, Fingerprint Database*; Washington, DC: National Institute of Standards and Technology, **1992**.

49 R. CAPPELLI, D. MAIO, D. MALTONI, in: *Proceedings of the International Conference on Pattern Recognition (16th)*; **2002**, 744–747.

50 *ID Device Software Development Kit, Programmer's Guide*; Siemens AG Austria, **2002**.

51 R. HOPKINS, BioAPI Consortium, www.bioapi.org, *Technical Whitepaper 1*; IBM.

52 *Data Format for the Interchange of Fingerprint Information*; Washington, DC: National Institute of Standards and Technology, **1991**.

53 The Privacy and Security Committee, Medical Imaging Informatics Section; *Security and Privacy: An Introduction to HIPAA. White Paper.* NEMA, www.nema.org, **2001**.

54 N. RATHA, S. CHEN, K. KARU, A. K. JAIN, *IEEE Trans. PAMI (Pattern Analysis and Machine Intelligence)*, **1996**, *18*, 799–813.

55 BioAPI Consortium, www.bioapi.org, *BioAPI Specification Version 1.1*; **2001**.

9.7.2
Links

56 The Biometrics Consortium, http//www.biometrics.org/html/introduction.html
57 Association for Biometrics, http//www.afb.org.uk
58 Schiphol Airport, http//www.schiphol.nl/schiphol/privium/privium_home.jsp
59 Cherry, http//www.cherry.de/english/advanced-line/keyboard-fingertip-id-board-g83-14000-14100.htm
60 Bromba Biometrics, http//www.bromba.com/tdidme.htm
61 Veridicom International, http//www.veridicom.com
62 Tricubes, http//www.tricubes.com/product.htm
63 NTT DoCoMo, http//www.nttdocomo.com/presscenter/pressreleases/press/pressrelease.html?param%5Bno%5D=257
64 Precise Biometrics, http//www.precisebiometrics.com/
65 Technoimagia, http//www.technoimagia.co.jp/main/e_14_product_top.htm
66 E-smart Technologies, http//www.e-smarttechnologies.com/
67 Biometric Associates, http//www.biometricassociates.com/products_access-control.html
68 Audi, http//www.audi.com/de/de/neuwagen/a8/limousine/elektronik_ausstattungen/bedienung_komfort/one_touch_memory/one_touch_memory.jsp
69 International Biometric Group, http//www.biometricgroup.com/reports/public/market_report.html
70 ST Microelectronics, http//www.st.com/stonline/products/support/touchip/products/sensor3.htm
71 The first international competition on fingerprint verification, http//bias.csr.unibo.it/fvc2000
72 The second international competition on fingerprint verification, http//bias.csr.unibo.it/fvc2002
73 Siemens ID-Center, http//siemensidcenter.com
74 Microsoft, http//www.microsoft.com/whdc
75 International Biometric Industry Association (IBIA), http//www.ibia.org

9
CMOS-based Biochemical Sensing Systems

J. Lichtenberg, H. Baltes, Physical Electronics Laboratory, ETH Zurich, Zurich, Switzerland

Abstract

Biochemical solid-state sensors are widely used for rapid and reliable compound quantification in analytical chemistry today. Monolithic integration of these sensing elements with dedicated electronic circuitry is highly desirable to improve the signal-to-noise ratio or to upscale the number of parallel sensors in a device. This has led to the development of impressive biosensing systems featuring several thousand sensor pixels for independent measurements of biochemically relevant information on single microchips.

In this chapter, the basic operating principles and design aspects of CMOS-based biochemical sensing system are presented with a focus on biosensing arrays and cell-based assays. Several device examples are complemented by a discussion of advantages and limitations of CMOS integration in this application area.

Keywords

Biosensors; biochemical sensors; microarrays; bioMEMS

9.1	**Introduction**	**448**
9.1.1	Miniaturization of Biochemical Sensing Devices	448
9.1.2	Biochemical Sensors and Sensor Systems	449
9.1.3	Integrated Electronics for Biosensing Systems	450
9.1.3.1	Advantages	451
9.1.3.2	Limitations	452
9.1.3.3	Applications for CMOS-based Biosensors	453
9.2	**Biosensor Arrays**	**454**
9.2.1	Electrochemical Read-out	454

Advanced Micro and Nanosystems. Vol. 2. CMOS – MEMS.
Edited by H. Baltes, O. Brand, G. K. Fedder, C. Hierold, J. Korvink, O. Tabata
Copyright © 2005 WILEY-VCH Verlag GmbH & Co. KGaA, Weinheim
ISBN: 3-527-31080-0

9.2.2	Optical Read-out	457
9.3	**Cell-based Assays**	**457**
9.3.1	Cell Handling	458
9.3.1.1	Flow Systems for Cells and Particles	458
9.3.1.2	Dielectrophoresis	459
9.3.2	Interfacing Electrogenic Cells and ICs	463
9.3.2.1	Special Fabrication and Design Requirements	464
9.3.2.2	Device Examples	466
9.3.2.3	Electrode Needles for *In Vivo* Electrophysiological Measurements	470
9.4	**Future Trends**	**471**
9.4.1	Scanning Probe Techniques	472
9.4.2	Cantilever-based Sensors	473
9.4.3	Biochemical Sample Pretreatment	473
9.4.4	Sensor Packaging	473
9.5	**Conclusions**	**474**
9.6	**Acknowledgments**	**474**
9.7	**References**	**474**

9.1
Introduction

Complementary-metal-oxide-semiconductor-based (CMOS) sensors successfully entered the market of physical sensors, e.g. for acceleration, flow rate and humidity. More recently, the CMOS approach has been extended to the area of biochemical analysis systems [1, 2]. The resulting devices integrate up to thousands of biochemical sensing elements together with dedicated circuitry for signal amplification, filtering, multiplexing, and conversion on a single CMOS microchip. For many applications, this monolithic integration is very beneficial as it allows for better signal-to-noise ratios, additional functionality and flexible upscaling. However, the applications and markets in the field of biochemical analysis are considerably different to those for typical, rather physical CMOS MEMS strongholds. In this chapter, the advantages and disadvantages of CMOS technology for integrated biosensing devices are discussed. Device examples are presented from the areas of specific biochemical sensors and from cell-based sensing systems. Finally, development trends expected to play an important role in the future are discussed.

9.1.1
Miniaturization of Biochemical Sensing Devices

Miniaturization by microfabrication techniques and functional integration on a single chip had a tremendous impact on the development of microsensors, including those intended for chemical and biochemical measurements. In a similar way, microfluidic devices have evolved over the last 15–20 years [3]. With the introduction of the µTAS (micro total analysis) concept in the early 1990s [4], a new paradigm was coined to develop autonomous, miniaturized instruments that provide chemical or biochemical sample-to-answer functionality in a single, fully automated device in real time. While µTAS systems are primarily associated with integrated microfluidic networks on a planar substrate, microelectronic biochemical sensors constitute an important class of building blocks for these devices. Especially if combined with suitable microfluidic components providing sample handling or calibration functionality, microelectronic sensors are an interesting alternative to conventional analytical methods.

The miniaturization and batch fabrication of biochemical sensing systems has numerous advantages:

- Virtually dead volume-free integration of several functional steps of the analysis process into one device.
- Low consumption of sample and auxiliary reagents.
- Short sample-to-answer times due to fast fluid handling and short diffusion times.
- Low energy consumption and low weight.

Of these points, functional integration is probably the most interesting feature for the majority of applications as important gains in analysis cost and time can be achieved here. In particular integrated sample pretreatment, generally a time-consuming and laborious process, adds a major customer value to a miniaturized analysis system. As a result, the cost efficiency is improved as little manual labor is needed to run the analysis and the risk of errors between subsequent processing steps is greatly reduced.

9.1.2
Biochemical Sensors and Sensor Systems

The amperometric glucose sensor represents the most successful commercial biochemical sensor to date [5]. The key component of the familiar test strips used for blood glucose monitoring, they are the driver behind the US $ 2 billion market for diabetes self-monitoring disposables. This success is due to the high customer value delivered by this combination of rapid and accurate diagnosis, small blood sample volumes and compact, user-friendly instrumentation.

Whereas typical glucose measurements are performed by a single sensor without auxiliary components, biochemical sensing systems generally integrate additional functionality around a sensor or sensor array. Examples are sample handling and pretreatment (filtering, dialysis, preconcentration), analyte labeling, sensor calibration, and sensor regeneration [6]. The i-STAT Portable Clinical Analyzer PCA (i-STAT, East Windsor, NJ, USA) is a good example of the state-of-the-art of commercially available biosensing systems for point-of-care diagnostics [7, 8]. The system, depicted in Fig. 9.1, consists of two main parts: a disposable, injection-molded analysis cartridge and a portable, battery-powered instrument. While the capital investment required for the instrument is about US $ 6,000, the disposable test cartridges sell around US $ 3 per piece. An array of electrochemical, solid-state biosensors is

9 CMOS-based Biochemical Sensing Systems

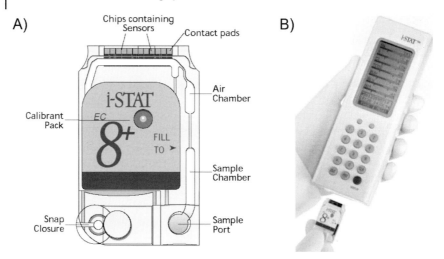

Fig. 9.1 The i-Stat system shows the typical features of a commercially available biochemical sensing system. (A) An array of solid-state biosensors is integrated in a disposable, injection-molded, polymer cartridge, which allows sample handling and metering and which contains a reservoir for calibration solutions; a row of metal pads on the top of the 2.7×4.5 cm^2 cartridge makes electrical contact with the instrument. (B) The cartridge is inserted into a hand-held instrumentation unit for operation, readout and data analysis (images by courtesy of i-Stat Inc. reprinted with permission)

integrated in this polymer cartridge, which allows sample handling and metering and which contains a reservoir for calibration solutions. To perform the measurement, the sample is injected into the inlet port of the cartridge, which is then inserted into the base instrument. In the main unit, the standard solution is pumped into the measurement chamber with the biosensors to run a calibration step. Subsequently, the chamber is filled with the patient's sample and the assay is performed. Depending on the application, different analysis cartridges are available to cover a wide range of diagnostic demands with a single instrument. Typical parameters of interest are sodium, potassium, choride, urea, glucose, lactate or hematocrit.

Although these two examples contain solid-state, electrochemical biosensors, no electronic circuitry is integrated on the sensor chip, which is the case for the majority of biosensing devices developed to date. However, the sensor fabrication using a commercial microelectronic process can be highly advantageous when it comes to the development of large sensor arrays or when the signals provided by the sensor are minute.

9.1.3
Integrated Electronics for Biosensing Systems

For a typical biochemical sensing system, the sample-to-answer path includes a conversion of chemical into electronic (analog or digital) information, e.g. the conversion of chemiluminescent light emission into a current flowing through a photodiode. As described above for the i-STAT system, most biochemical sensing instruments available today consist of two distinct parts: a complex electronic base unit with no or little sample contact and an analysis cartridge, which is replaced after one or several analysis runs. The 'cartridge' comprises here a number of possible implementations, including vials, microtiter plates, test strips, microfluidic chips and microarrays. Typically, electronic circuitry is predominantly integrated into the base unit of the instrument while the chemical analysis takes place in the low-cost cartridge.

Breaking with this traditional divide and moving part of the electronic circuitry into the measurement cartridge have a number of advantages in terms of analysis quality and system scalability. However, this undoubtedly comes at the expense of higher fabrication cost per chip area and more stringent requirements for system packaging and testing. Therefore, the technical and economic requirements for a new biochemical sensing device have to be analyzed carefully when it comes to integration of electronics on the measurement cartridge. The following sections discuss advantages and limitations of integrated CMOS circuitry for biochemical sensing systems.

9.1.3.1 Advantages
The integration of active microelectronic circuits on the sensor chip has the following major advantages.

Sensitivity
Typically, the electrical signal obtained by a miniaturized transducer decreases linearly with the active sensor area (e.g. the microelectrode). At detection limits of ng/mL, which are required for many applications in medical diagnostics, this results in signal changes of the order of nA, µV or fF, depending on the detection mode used. As a consequence, the influence of parasitics induced in the connection path between sensing site and an external instrument quickly compromises the measurement. In the case of CMOS sensors, however, a first amplifier stage can be directly integrated at the measurement site to buffer the signal for further processing. Depending on the area available, additional components can be implemented, including low-pass filters for noise suppression, high-pass filters for offset compensation and additional amplifier stages.

Scalability
A major asset of circuit integration is the possibility of creating large arrays of repeating sensing units or pixels, which can be read out in a serial fashion using on-chip multiplexers. This responds perfectly to the general trend in bioanalysis towards higher sample throughput demanded by applications such as drug screening or genetic diagnostics [9]. For instance, a commercially available microarray for the diagnosis of various cardiovascular diseases traces more than 700 target sequences in a patient's DNA in duplicate, which results in nearly 2000 parallel measurements including reference and control tests [10]. For a square array of

N elements, a row or column multiplexing scheme reduces the number of parallel data acquisition channels to $\sqrt{N}$ and a combined row–column scheme requires only a single channel. As a consequence, it is often advantageous to implement only a simple buffer amplifier at the sensing pixel, while the majority of signal-conditioning elements are placed after the multiplexer as a single instance. However, the bandwidth of the entire signal path including multiplexer as well as the pre- and post-multiplexer signal-conditioning circuitry needs to be high enough to allow for a sufficiently high scan rate.

Reduction of Interconnects
Interconnect fabrication and reliability are a major concern if larger arrays of passive sensing elements have to be connected to off-chip instrumentation. While techniques such as wire-bonding or spring-loaded contact pins work well for smaller arrays (up to 100 sensing sites), interconnect technology issues for larger arrays might raise the production costs and reduce the yield beyond what is commercially acceptable. Devices with hundreds or thousands of sites therefore require a different approach to transfer electrical signals off the chip. On-chip multiplexing not only reduces the complexity of on-chip signal-conditioning circuitry, but also reduces the number of bond wires significantly. If additionally analog-to-digital conversion and a digital serial protocol controller are integrated, even complex, multi-sensor devices require only five outside connections (data, clock, reset, GND, VDD) [11].

9.1.3.2 Limitations
Despite these interesting advantages of integrated electronics, both technical and economic issues can develop to the designer's disadvantage. The requirements for typical biosensing devices are very different from those of physical MEMS sensors, which have been commercially available with integrated CMOS electronics for a number of years. This leads to several limitations.

Fabrication Cost
The cost per area of the fully processed CMOS substrate is roughly one order of magnitude higher than for a biosensing microdevice without electronics (a typical, unprobed 150 mm CMOS wafer costs about US $ 1500 if produced in larger quantities). Additional costs for testing of the electronic features, wire bonding and increased packaging efforts have to be added to the overall bill.

Material and Process Compatibility
While the choice of substrate materials for biosensors is generally large (e.g. glass, fused silica, polymers, ceramics, silicon), CMOS-based devices are exclusively manufactured on silicon wafers. The CMOS dielectrics (Si_3N_4, SiO_2) can be used to protect the underlying circuitry to some extent from the biochemical sample and reagents. In many cases, however, these proved not to be sufficient, result-

ing in a short device life time. As an example, the dielectric breakdown of CMOS passivation layers renders electrokinetic fluid handling difficult on silicon substrates [12]. Additionally, also the biochemical analysis might be affected by an unsuitable cover layer of the microchip.

Therefore, additional, post-CMOS passivation layers, including dielectrics, metal layers or polymer coatings, may be required to suppress undesired reactions on the chip surface. However, the deposition of these layers has to be performed at low temperatures (< 400 °C) to prevent degradation of the CMOS circuitry, thereby limiting the choice of available processes significantly.

Packaging
Whereas physical sensors often measure in a non-contact mode, biochemical sensors are in direct contact with the liquid sample. The resulting constraints for the chip layout and processing and for the packaging can be challenging, as fluidic functionality, electrical interconnects and microchip protection have to be provided simultaneously.

Contamination
As a direct contact between the sample and the sensor cannot be avoided, degradation and contamination of the sensor device can result. Typical sample molecules, such as proteins, have been shown to bind unspecifically to silicon and its derivatives. During a subsequent assay on a different sample, these molecules might detach again and interfere with the analysis. As a consequence, disposable sensor concepts are preferred for many applications such as clinical diagnostics. The resulting pressure on the fabrication cost is therefore considerable.

Irreversibility
Many reactions used to detect an analyte molecule specifically through binding to a probe molecule immobilized on the sensor surface are not reversible. As a consequence, the sensor cannot be regenerated and needs to be replaced after each measurement.

9.1.3.3 Applications for CMOS-based Biosensors
Despite the number of constraints listed above, it should be stressed that CMOS integration can be favorable for certain high-volume biosensing applications and might even be imperative for others. Among the latter are techniques that require massively parallel analysis, e.g. microelectrode arrays for monitoring larger cell cultures. Also, applications suffering from extremely low signal levels can benefit from powerful on-chip signal conditioning circuitry. Having said this, one has to emphasize that there are still many areas, especially when it comes to disposable sensing devices, where CMOS is not, or not yet, competitive with current fabrication techniques.

9.2
Biosensor Arrays

CMOS technology made its first appearance in the biosensing field as a platform for multi-analyte biosensor arrays. A variety of detection techniques have been used in this context, including amperometric, potentiometric and impedance electrochemical sensors, in addition to optical labels. For reviews of biosensor principles, the reader is referred elsewhere [13–16].

9.2.1
Electrochemical Read-out

Blood gas analysis requires the simultaneous quantification of pH, pO_2 and pCO_2, which can be achieved by electrochemical sensing techniques. Arquint *et al.* combined several microsensors on a non-CMOS silicon substrate with a photopatterned silicone gasket to form a flow cell [17]. Shortly after, researchers from the Catholic University of Leuven presented a more advanced design fabricated in a 1.25 µm CMOS process, including additional sensors and interfacing electronics (Fig. 9.2) [18]. For amperometric measurements, a potentiostatic amplifier setup is integrated on the chip and a control loop with temperature sensor, amplifier and heater is provided for chip thermostating. A hybrid sensor system has been developed recently for the monitoring of *in vitro*-cultured cells, comprising a CMOS signal-conditioning chip and a sensor chip with 12 ISFETs, two temperature sensors and one conductivity sensor [19].

A two-dimensional array of 400 individually addressable platinum electrodes was used for imaging O_2, H_2O_2 and glucose distributions over a 1 cm^2 area [20]. Oxygen and hydrogen peroxide were detected amperometrically on bare electrodes, whereas glucose was determined by coating the electrodes with glucose oxidase entrapped in a conductive polypyrrole layer. The sensor chip is based on a standard CMOS process followed by deposition and patterning of a platinum and a silver layer for the sensing and reference electrodes, respectively [21]. Each sensor cell possesses a high-input-impedance (>5 GΩ) transconductance amplifier and a control unit for cell selec-

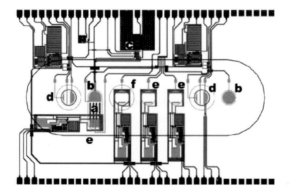

Fig. 9.2 Layout of a CMOS-integrated blood gas sensor: (a) Ag/AgCl reference electrode, (b) liquid contact electrode for biasing, (c) heater, (d) amperometric sensors, (e) and (f) ISFETs. Reprinted from [18], with permission

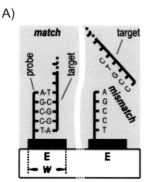 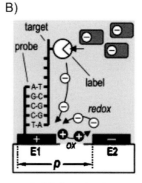

Fig. 9.3 Principle of a DNA hybridization assay. (A) A matching single-stranded DNA molecule binds specifically to a complementary strand immobilized as a probe molecule on the electrode surface, while a mismatched strand remains unbound and can be washed away. (B) The enzymatic label previously attached to the analyte molecule continuously converts substrate molecules into a redox-active species, which leads to an increase in current between the two electrodes E1 and E2. Reprinted from [1], with permission

tion. An address decoder allows the selection of individual electrodes in a serial fashion, which results in a frame rate of 0.5 per minute [20].

Probably the most evolved electrochemical biosensing system has been recently presented by Thewes and co-workers for a potentiometic DNA detection technique [22–24]. The device features 128 interdigitated microelectrode pairs in a 16×8 layout, an on-chip potentiostat for biasing of the solution, row and column decoders and analog multiplexers.

The sensing principle is based on redox cycling, which produces an electric current between two interdigitated electrodes if a matching analyte molecule has been captured. Fig. 9.3 illustrates the sensing concept, which takes advantage of the highly specific binding of a single-stranded DNA molecule to its complementary counterpart. The latter, the so-called probe, needs to be immobilized on the electrode by a suitable method [15] prior to the analysis. For each electrode, a different probe molecule can be used to allow for highly parallel analysis. Localized patterning of biomolecules is typically achieved by inkjet printing, spotting, photopatterning or micro-contact printing. If a matching analyte molecule binds to a probe molecule attached on the microelectrode site, it starts to generate redox-active compounds, i.e. electric charge carriers, by an enzymatic reaction with the surrounding solution. For this purpose, all sample molecules are labeled with an enzymatic group prior to the binding step on the microchip. As a result of the enzymatic reaction, additional redox-active compounds are produced continuously while oxidation and reduction potentials are applied to the measurement electrodes, resulting in a current increasing linearly with time. The slope $\Delta I/\Delta t$ of this current is a measure of the amount of captured analyte molecules on the particular electrode site. In practice, alkaline phosphatase is used as enzyme label, which transforms p-aminophenyl phosphate into the redox-active p-aminophenol, which results in a current flow at potentials of +300 and −100 mV compared to a pseudo-reference electrode [25].

To ensure a sufficiently large current, the electrode pairs are designed as interdigitated structures with finger width and gaps in the 1–2 µm range, a length of 100–250 µm and a circular outline. A stack of Ti/Pt/Au (50/50/500 nm) is deposited in a post-CMOS step followed by an annealing procedure [25]. To facilitate the functionalization of individual electrodes with probe molecules, compartments are fabricated by photopatterning of polybenzoxazole polymer. The pixel electronics, repeated for each sensor site, records the currents for both electrodes and

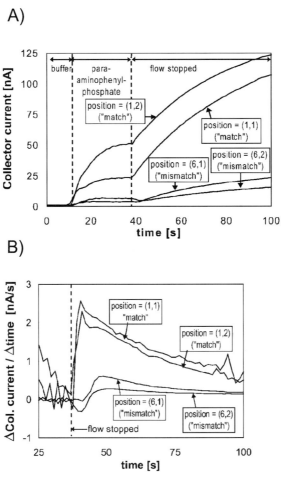

Fig. 9.4 Current data acquired on four electrodes during hybridization. (A) A current increase at the electrode pairs (1,1) and (1,2) indicates binding of a matched DNA molecule, while the current at positions (6,1) and (6,2) remains low. (B) The time derivative of the current clearly shows the difference between matched and mismatched analyte. Reprinted from [1], with permission

amplifies the signal 100-fold by two cascaded current mirrors at a dynamic range of 10^{-12}–10^{-7} A. To achieve this, the transistors are operated in the subthreshold region and pixel-by-pixel compensation of the transistor parameter-related gain variations is possible.

Fig. 9.4 shows data obtained by presenting matching and mismatching analyte molecules to the functionalized electrodes while monitoring the current. A clear difference between the bound matching molecule and negative sample (mismatch) can be seen.

9.2.2
Optical Read-out

An alternative to direct biosensor integration is the combination of a conventional fluorescence or chemiluminescence assay format with a light-sensitive CMOS sensor or sensor array [26–30]. In this case, the CMOS system does not take part in the actual assay, but detects photons emitted from a separate biosensing layer deposited on the chip. To give an example, Davenport et al. coupled a xerogel-based thin-film sensor with a 30-site active pixel sensor for chemiluminescence assays [30]. Compared with other imaging techniques (photomultiplier tubes and charge-coupled devices), the CMOS device consumed 1000 times less power while providing equivalent analytical data. Recently, a single-channel microluminometer for high-sensitivity luminescence detection has been presented, which is capable of detecting as few as 5000 *Pseudomonas fluorescens* 5RL bacterial cells by reducing on-chip leakage currents at the amplifier input to the sub-femtoampere range [31, 32]. The 1.47 mm^2 photodiode is formed by the n-well/p-substrate junction of a standard CMOS process and a current-to-frequency converter circuit accomplishes on-chip signal processing.

Dill et al. [122] have proposed a CMOS-based microelectrode array as substrate for patterned immobilization of probe molecules for binding assays. The 1 cm^2 chips feature 1024 electrodes of 100 µm diameter, which can be individually switched to four independent electrical channels. These serve as current sources for electrochemical immobilization of biotin probe molecules in a porous matrix layer deposited on top of the chip. By adjusting the currents accordingly, the amount of bound biotin can be precisely controlled. For the actual assay, an external scanner is used to measure the resulting fluorescence intensity with local resolution.

9.3
Cell-based Assays

If chemical information about the impact of an analyte on a whole cell or organism is desired, conventional strategies involving pure chemical analysis permit only limited insight. Cell-based assays, on the other hand, allow one to monitor cell responses including ion-channel activities or spontaneous electric signaling, recorded from a population of living cells confined in the analysis device. In fact, living organisms have been used for a long time as indicators for environmental hazards. Since the Middle Ages, canaries served as methane detectors in mines and, even today, water analysis is sometimes performed by monitoring the swimming behavior of small crabs [33].

Owing to their capability to form array-type sensing structures, CMOS-based cell assays allow real-time monitoring of a comparatively large cell population. This permits cells to be perfused with different sample solutions on the same chip, to include reference cells for calibration or even to study the intra-cellular communication in grown neuronal networks. Applications for these systems include pharmacological analysis, neuroscience and environmental monitoring, including warfare agent detection.

9.3.1
Cell Handling

As a prerequisite for most cell-based biosensors, cells have to be selected from a suspension and directed towards the sensing element. In some cases, cells remain suspended and the analysis is performed while passing a suitable detector in a moving stream of buffer solution, and in others it is desirable to capture and immobilize the cell at a specific location on the chip.

9.3.1.1 Flow Systems for Cells and Particles

Fluorescence-activated cell sorting (FACS) and related flow-cytometric analysis techniques are the main tools for high-throughput screening of individual cells. With throughput rates reaching 10^4 cells/s, biological information, such as a specific ion content or the presence of a receptor on the cell surface, are determined by multi-wavelength fluorescent interrogation of analyte-specific fluorescent markers incorporated into the cell beforehand. Additional information, for instance the cell size, is acquired by the analysis of the light-scattering patterns generated when the cell passes the detector.

The miniaturization of flow cytometers aims at reducing the overall instrument complexity and cost, at reducing the sample volume and at including additional functions such as sorting facilities controlled by the detector readout. Miniaturized cytometry flow cells for focusing suspended cells into a well-confined stream have been proposed based on microfluidic effects, such as sheath flow and hydrodynamic focusing [34] or dielectrophoresis [35, 36] (see the next section for details). A functional integration of such a flow system micromachined in silicon with an integrated photodiode array for near-field optical measurement of the cell size and shape has been presented recently [37]. Readout of electrical cell parameters by impedance spectroscopic techniques using integrated microelectrodes has been proposed as an alternative to optical methods [36, 38]. In order to increase the sample throughput of miniaturized cytometers, an array architecture has been proposed relying on dielectrophoretic cell trapping [39].

A complex microfluidic structure based on a soft-molded, two-layer elastomer structure has been used to extract and capture a single cell from a suspension stream for subsequent perfusion with analyte and indicator solutions [40]. Finally, the cell can be resuspended again and transferred into a growth chamber integrated on the chip for culturing. Flow and cell handling are achieved by integrated valves and peristaltic pumps, which are actuated by a layer of pneumatic control channels laminated on to the fluidic channel system. Fabrication techniques such as elastomer molding generally lend themselves to post-process integration of a fluidic architecture with CMOS devices, as the fluidic and CMOS chips can be fabricated separately followed by a final low-temperature bonding step.

9.3.1.2 Dielectrophoresis

Dielectrophoresis, introduced by Pohl in 1951 [41], describes the force acting on a dielectric particle in a spatially non-homogenous electric field. Depending on the physical properties of both the particle and the surrounding medium, the object might be attracted to regions of increased field intensity (positive DEP) or to those with reduced fields (negative DEP). The technique can thus be used to capture, move and immobilize objects such as cells, polymer beads and macromolecules (e.g. DNA). Owing to the relationship between cell properties and the resulting dielectric net force acting on it, the technique can also be used to sort a cell suspension depending on the cell type or state [35].

The force acting on a dielectric particle in a non-homogeneous field can be described by introducing a gradient operator such as [42]

$$\vec{F} = (\vec{M} \cdot \vec{\nabla})\vec{E}$$

where $\vec{M}$ denotes the induced dipole moment of the particle and $\vec{E}$ the electric field. Assuming a spherical particle surrounded by a single, sinusoidal electric field, $\vec{F}$ can be written as

$$\vec{F} = 2\pi\varepsilon_0\varepsilon_p r^3 \text{Re}[K(\omega)]\vec{\nabla} E^2$$

where $\varepsilon_0\varepsilon_p$ is the absolute electric permittivity of the particle and $K(\omega)$ denotes the Clausius–Mossotti factor, which is related to the frequency-dependent, complex permittivity factor of the particle (ε_p^*) and that of the surrounding medium (ε_m^*):

$$K(\omega) = \frac{\varepsilon_p^* - \varepsilon_m^*}{\varepsilon_p^* + 2\varepsilon_m^*}$$

$K(\omega)$ thus defines the magnitude and the sign of the resulting dielectrophoretic force, which is dependent on the particle and medium properties in addition to the frequency of the electric field, ω. The analysis of the Clausius–Mossotti factor allows a dielectrophoretic system to be tuned to separate a mixture of different

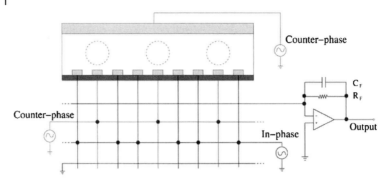

Fig. 9.5 Schematic setup for dielectrophoretic cell transport on a microelectrode substrate. During the actuation, each electrode can be connected to an in-phase or a counter-phase stimulus. Reprinted from [48], with permission

cell types or to attract and repel cells from a target region by switching the frequency of the electric field.

Early DEP devices for cell handling and manipulation relied on capturing particles from a suspension by applying a non-homogeneous electric field between electrodes deposited on two parallel substrates [35, 36, 43–45]. To allow optical inspection of the device, indium tin oxide (ITO) is often used as a transparent conducting film on glass or fused-silica substrates. While simple coplanar electrode pairs allow capture and release of particles by applying positive and negative DEP, the objects cannot be actively moved through the device and an additional transport flow is necessary. If instead a poly-phase electrical signal is applied to an array of electrodes as a traveling wave, particles can actually be propelled along the propagation vector of the wave [46, 47].

Still, control of the cell movement is difficult owing to the dependence of the DEP force on physical cell parameters. To ensure a defined transfer of a particle from one position to the next, Medoro et al. developed a one-dimensional microelectrode array that allows one to build and move around dielectrophoretic cages capable of capturing and releasing particles (Fig. 9.5) [48]. A cylindrical cage is formed along a strip electrode by applying a counter-phase sinusoidal stimulus, while the two neighboring electrodes are connected to an in-phase signal. By shifting the pattern from one electrode to the next, the cage and its occupants can be moved accordingly. Although a first device prototype with 39 electrodes was successfully fabricated by low-cost printed circuit-board techniques, it is obvious that the number of electrodes is limited for this approach owing to the interconnects required.

For systems with higher electrode density and to extend the concept to the second dimension, circuitry for addressing and driving of the electrodes has to be integrated into the chip. Recently, the group presented a device featuring 320×320 individually controllable electrodes at a 20 µm pitch, which is based on a standard CMOS fabrication process (two-poly, three-metal, 0.35 µm) [49]. Again, the device is completed by attaching a transparent counter electrode as cover to the

8×8 mm² chip, thus forming a 85 μm high fluidic chamber containing a volume of less than 3.5 μL. The chip allows one to form up to 12 800 independent DEP cages for cell manipulation, the pattern of which can be programmed in real time. Additionally, each electrode has an integrated silicon-based photodiode that allows optical monitoring of the volume above the electrode.

Fig. 9.6 illustrates the chip architecture including the microelectrode array and two 9-bit static column–row decoders for random access to the electrode sites. Electrodes can either be operated in actuation or in sensing mode. In actuation mode, each electrode is driven by one of the two non-overlapping input signals V_{pip} (in-phase) or V_{phim} (counter-phase) depending on the programmed pattern. Programming is achieved by addressing the electrode and storing the desired signal phase (in-phase or counter-phase) into a memory element contained in the electrode site circuitry, which is implemented for each electrode in the array as shown in Fig. 9.7. The setting remains active until a new value is written to the memory. During the readout phase, a fully differential charge integrator based on a switched-capacitor operational amplifier reads out an n-well junction photodiode integrated in the electrode site (similar to active pixel sensors). Fig. 9.8 shows the collection and clustering of yeast cells in a regular pattern on the chip. More complex manipulations, including positioning, mating and re-separating cells, are also possible by applying suitable programming patterns to the chip.

Dielectrophoretic handling of biological and biochemical entities ranging from larger cells down to DNA molecules has a wide spectrum of applications in the field of integrated biosensing devices. Electrode arrays can be comparatively easily

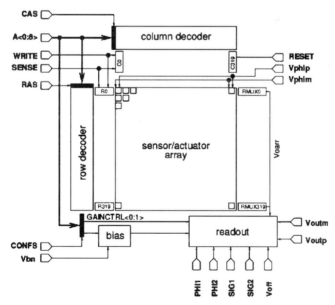

Fig. 9.6 Architecture of the DEP chip. Reprinted from [49], with permission

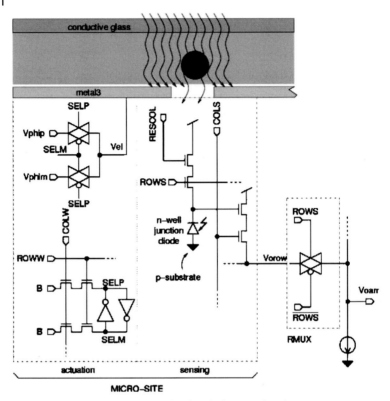

Fig. 9.7 Schematic of a pixel circuit block including switches for DEP actuation, a photodiode with amplifier for optical sensing and a transmission gate for the multiplexed readout scheme. Reprinted from [49], with permission

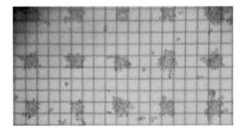

Fig. 9.8 Dielectrophoretic formation and manipulation of clusters of *Saccharomyces cerevisiae* in 280 mM mannitol buffer at a voltage of 3.3 V_{pp} (1 MHz). The picture shows a microscope image after cluster formation. Reprinted from [49], with permission

integrated with standard CMOS technology, which allows the fabrication of massively parallel particle manipulation platforms. When integrated with a sensor array, for instance based on optical interrogation or impedance spectroscopy, these devices can be powerful tools for the life sciences.

9.3.2
Interfacing Electrogenic Cells and ICs

Interfacing microelectronic circuitry with biological cells has applications in medical diagnostics to monitor cell activity, in analytics to relate a cell response to a specific stimulus and in therapeutics, where microelectronic implants are being developed to replace dysfunctional tissue. Due to the nature of the microelectronic signal transfer, research is targeted towards electrogenic cell lines, which are capable of generating electric potentials across their cell membrane by generator proteins in the lipid bilayer. Like many other cells, they also respond to electric stimuli. Among the most important cell types with electrogenic capability are neurons (central and peripheral nervous system), cardiac cells (heart) and muscle cells.

Cellular activity involving electric signaling causes a weak current flow in the extracellular fluid surrounding the cells, which are between 5 and 25 µm in diameter. The resulting voltage drop between two distinct locations on a planar substrate can be monitored as indicator for cell-to-cell signal transmission. Owing to the capacitive transfer characteristics of the cell–electrolyte–electrode interfaces, these action potentials (APs) can be detected as extracellular spikes, which typically have a duration of between 0.2 and 20 ms at a bandwidth of 10 Hz–1 kHz for most mammalian neurons. Signal amplitudes, depending on the cell line and the coupling efficiency between cell and electrode, range between a few hundred µV and several mV.

Electrophysiological assays of living cells or tissue cultures are a major analytical technique used in biochemistry and biology. Beginning with Galvani's electrical excitation of muscle contraction in frogs' legs in 1791, the response of biological tissue and later also of isolated cells or simple cell networks has been studied to understand the underlying simuli–response relationship. In the early 1970s, Thomas *et al.* [50] introduced microelectrode arrays (MEAs) to perform electrophysiological measurements in a non-invasive and parallel fashion to achieve locally resolved mapping of the electric cell activity. As a consequence, MEAs allow the study of cells, although *in vitro*, in their intact cellular environment, which is an advantage over analysis techniques for isolated, single cells (e.g. patch-clamping [51]). Today, MEA chips are used in a variety of implementations including devices with three-dimensional electrodes [52, 53] and perforated MEA substrates for improved cell culturing [54]. Commercial devices based on this technology are available from Multichannel Systems [55], Ayanda Biosystems [56] and Panasonic [57]. An alternative technique, developed by Fromherz and co-workers, employs silicon-based field-effect transistors (FETs) as sensing elements for cellular signals. The cell under investigation is directly placed on the gate oxide of an oxide semiconductor FET (OSFET), so that the cell polarization can affect the inversion channel between source and drain [58–60]. This method allows similar measurements as metal electrode-based MEAs and device implementations in array formats have also been developed. MEAs and FET-based monitoring chips have found widespread applications in drug discovery and basic research [61]. For instance, safety tests for new drugs include the measurement of the drug's impact on the cardiac QT interval (the interval between the characteristic Q and T waves of a heart beat), because a drug-related prolongation of the interval can cause cardiac arrhythmia in sensitive patients (the QT test has been required since 2002 by the safety pharmacology guideline ICH S7B [62]).

For practical applications, passive MEAs are limited in the number of electrode sites (typically < 100) as each electrode needs to be electrically connected to the off-chip signal processing circuitry. The associated production cost and yield problems become prohibitive for larger arrays. Recently, a number of research groups have developed CMOS-integrated MEA devices comprising up to thousands of electrodes with on-chip multiplexing and signal processing, thus allowing one to address and to read out specific electrodes via very few analog or digital connections [63–65].

Aside from action potential measurements, a number of additional factors are important for cell monitoring, including the temperature, pH and oxygen concentration of the culture medium. For parallel and non-invasive measurement of these parameters, Baumann and co-workers have developed CMOS-based microsensor devices featuring a number of ion-sensitive field effect transistors (ISFET) for concentration measurements [123]. These versatile devices have, for instance, been employed to study cellular respiration and acidification in real-time for monitoring of the global cellular metabolism [124]. Furthermore, these authors have developed integrated sensors to quantify additional parameters, for instance interdigitated metal electrodes to assess cell adhesion and growth.

9.3.2.1 Special Fabrication and Design Requirements

MEA devices with integrated circuitry are typically fabricated on standard silicon substrates by means of a commercial CMOS process followed by dedicated post-processing steps. The latter can be divided in three areas: (1) deposition of suitable electrode materials and lithographic structuring, (2) biocompatible surface coating for cell adhesion and viability and (3) placement and immobilization of cells on the array and guidance of the neurite outgrowth.

Electrode Fabrication

The impedance of the metal–electrolyte interface at the electrode plays a key role in the performance of a MEA system. This becomes especially important for CMOS-integrated devices, where the reduction of the electrode area results in higher interface impedance. Also important are the biocompatibility of the electrode material and its stability in physiological buffer solutions.

The aluminum or aluminum–silicon alloy layers used in CMOS processes for metal interconnects are not suitable for cell monitoring applications as the material is easily corroded even by mild buffer solutions, resulting in electrode dissolution and buffer contamination. Therefore, electrodes for cell recording are typically made from better suited materials by post-CMOS thin-film deposition and patterning of the electrode material. If required, a passivation layer (silicon nitride or photopatternable resists such as EPON SU-8 and polyimide) with openings at the measurement sites is added on top of the electrode layer. As electrode material, noble metals (Au, Pt) or compound materials [iridium oxide, indium tin oxide (ITO), titanium nitride] are used. Whereas film deposition is most often done by sputtering or evaporation, electroless plating of gold has the interesting advantage that the electrode material can be directly grown on the CMOS aluminum without additional patterning steps [66, 67]. In order to reduce the electrode impedance, its surface can be roughened to increase the specific surface area, either by dedicated plating techniques [67] or by electrochemically depositing a platinum black coating [68–70].

Biocompatible Surface Coating
To ensure good adhesion of the cells under study to the chip substrate, a surface treatment with an adhesion-promoting layer is generally required. One choice is the deposition of a layer of self-assembling amine derivatives, which bind covalently to silicon-based substrates [e.g. trimethoxysilylpropyldiethylenetriamine (DETA)] [70–72]. The other involves the physisorption of extracellular matrix proteins, which have been shown to stimulate neurite outgrowth and promote cell attachment, chemotaxis and cell differentiation. For instance, the synthetic polyamino acid polylysine (in its D- or L-form) can be non-covalently bound to negatively charged surfaces to promote cell adhesion. The glycoprotein laminin, typically isolated from the mouse tumor cells, is also widely used. For electrophysiological studies, it should be noted that the adhesion layer also affects the electric coupling of the cell membrane to the electrode surface, which makes thin layers preferable [60].

Cell immobilization and Neurite Guidance
The majority of MEA experiments are done using undefined neuronal nets growing on a substrate completely covered by an adhesion layer. However, for reproducible and systematic cell-based experiments involving the assessment of the network dynamics, the growth of defined neuronal nets on an MEA substrate is required. To achieve these, two conditions must be met: neurons have to be placed at defined locations on the MEA grid and neurite outgrowth has to be guided along defined paths on the substrate.

Placing of neurons is typically done manually by pipetting single cells into micromechanical holding structures fabricated on the MEA device. Picket fences to hold cells have been fabricated by photopatterned polyimide [73] and cavities etched into the substrate have been used [74]. In order to achieve cell placement in a more controlled way, Greve *et al.* suggested pneumatic attraction of cells by means of an array of small pores (~ 3 µm diameter) etched by reactive-ion etching into the substrate [75]. Also, automated dielectrophoretic registration of cells on an array of adhesive regions has been proposed [76].

Guidance can be imposed by selectively patterning the substrate chemically using certain amine derivatives or extracellular proteins as discussed in the previous section. These can be structured by photolithography [71, 77], lift-off techniques [78] or microcontact printing [79–81], to name a few. However, the interaction between the patterned protein and the cell membrane is limited, leading to uncontrolled deviation of the growing neurite if the desired path is not straight. To improve the patterning stability, two-level techniques have been developed, where the surface area remaining around the adhesion patterns is modified with a cell-repulsive molecule [80, 82].

Since the early days of cell culture, it was observed that the substrate topography directly affects cell orientation, differentiation and cell function. Patterned, microfabricated substrates with shallow (<1 µm deep) grooves with lateral dimensions in the micrometer range have been used to guide fibroblast cells [83, 84]. Especially interesting for post-CMOS processing, 15 µm wide grooves can also be microfabricated in additive technology by patterning a 15–30 µm thick layer of the epoxy-based photoresist SU-8 on the substrate [85]. Cell attachment was also studied on chaotic, nanometer-scale columnar structures fabricated in silicon by a modified reactive-ion etching process [86] and on lithographically patterned arrays of silicon pillars with sub-µm dimensions [87].

9.3.2.2 Device Examples

The readout electronics required for fully integrated MEAs consists of three subgroups: (1) preamplifier circuitry directly integrated with the measurement electrode or OSFET, (2) multiplexers for addressing the rows or columns of the arrays and (3) additional post-multiplexer condition circuitry, including buffers or analog-to-digital conversion.

Early examples towards the integration of MEAs with microelectronic components were presented by Pancrazio et al., who connected a silicon-based MEA chip and a CMOS signal-processing chip on the same carrier by a miniature flat-ribbon connector [70]. As this hybrid approach does not solve the interconnect problem for larger arrays, a monolithically integrated system was developed by De-Busschere and Kovacs, which features two separated and sealed 10 µL measurement chambers to study two cell populations in parallel [88]. Each chamber includes four arrays of 16 gold microelectrodes, pseudo-reference and stimulation electrodes, multiple temperature sensors and nine 16-to-1 multiplexer circuits. Microelectrode signals can be routed to the off-chip data acquisition system by two multiplexer paths, either directly for unbuffered AP measurements or via a low-noise PMOS source follower to provide impedance conversion (not studied in detail by the authors). An on-chip temperature regulation loop allows the center of the silicon die to be maintained at 37 °C by Joule heating through a NMOSFET transistor acting as an adjustable resistor. The authors also addressed packaging issues for the first time, proposing a cartridge platform consisting of a printed-circuit board for electrical interconnects and a polydimethylsiloxane (PDMS) cover with microfluidic elements such as chambers and septum seals.

Thewes and co-workers recently developed a CMOS microchip featuring 128×128 recording sites at a pitch of 7.8 µm based on a modified OSFET structure with extended gate [1, 63]. The 16 384 electrodes can be read out at a frame

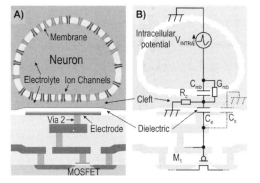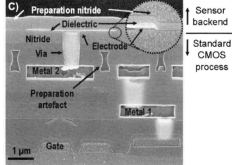

Fig. 9.9 Schematic cross-section through a microelectrode pixel. (A) The MOSFET gate is extended to a post-CMOS metal electrode deposited on the passivation nitride. (B) The cell signal is capacitively coupled to the FET via the cell membrane capacitance C_{mb}, the cleft resistance R_c and the sensor capacitance C_e. (C) Photograph of the cross-section of a fully processed device with the electrode metal stack and dielectric coating. Reprinted from [63], with permission

rate of 2 kiloframes per second using on-chip analog multiplexers and buffer amplifiers. Analog-to-digital conversion and data acquisition are done off-chip using a PC-based setup.

While typical OSFET devices only provide one layer (diffusion lines) for electrical connections to the transistors, a two-dimensional array architecture requires additional interconnect metallization. As a result, the neurons under study are not directly cultured on the gate oxide, but on an extended-gate electrode fabricated by post-CMOS deposition after a standard 5 V, 0.5 µm, two metal-layer, n-well epi-CMOS process. The electrodes are deposited on the passivation nitride after a CMP planarization step and consist of a 50 nm thick Ti/Pt stack connected to the polysilicon MOSFET gate by the CMOS metallization and vias. To ensure purely capacitive coupling to the electrolyte solution, the 4.5 µm diameter electrodes are finally covered with a sputtered, 40–50 nm thick dielectric layer of TiO_2 or TiO_2/ZrO_2. Fig. 9.9 compares the schematic cross-section of the sensing arrangement with a microscope picture of the cross-section of a pixel.

The system architecture of the recording chip and its external instrumentation is depicted in Fig. 9.10. During measurement, the desired column is selected by the column decoder and the signals of the 128 rows are fed into individual readout amplifiers. Subsequently, 16 8-to-1 multiplexers are used to connect the amplifier outputs either to an array of 16 off-chip I/V amplifiers or to a dummy load. Finally, the buffered output voltages of these converters are recorded by a PC-based data acquisition system at more than 8 bit resolution and a data rate of 32 MS/s. Recording sites are selected by digital address lines, which are interfaced by magneto couplers to a PC-based pattern generator. The latter allows one either to read full frames including all electrodes or to focus on a specific area of interest, which can be monitored at higher frame rates. The chip is thermostated by an on-chip temperature sensor and an off-chip control loop combined with a heating ele-

468 | *9 CMOS-based Biochemical Sensing Systems*

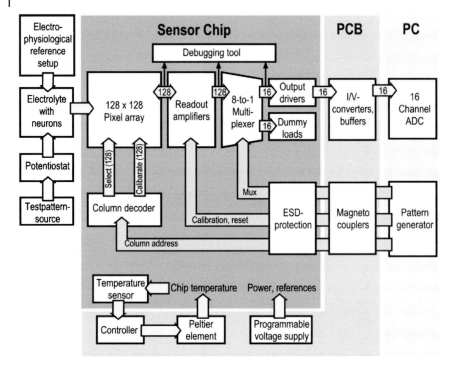

Fig. 9.10 System architecture of the 128×128 pixel cell recording system, including the actual chip, a readout printed-circuit board, a PC-based data acquisition system and other external instrumentation. Reprinted from [63], with permission

ment. The electrolyte solution is biased by an off-chip potentiostat. To insure the signal integrity over the whole array, Thewes and co-workers have developed a pixel-based calibration scheme which is repeatedly carried out after 100 recording cycles to compensate for transistor mismatch and settling behavior [1, 63]. Fig. 9.11 compares action potentials recorded from a snail neuron by intracellular measurement techniques with data acquired by the CMOS chip.

Our laboratory at ETH Zürich has recently presented a highly integrated, CMOS-based cell recording device, which provides full on-chip signal processing including filters and analog-to-digital conversion [64]. Additionally, the device is capable of applying stimulation signals generated at a frequency of 120 kHz by an on-chip 8-bit DAC.

Fig. 9.12 shows a photograph of the chip, which features 16 metal electrodes (50 nm TiW as adhesion layer and 270 nm Pt) 40×40 µm in size at a 250 µm pitch. Different from the system described before, these are not covered by an additional dielectric layer, but are in direct contact with the electrolyte. Owing to the roughness of the aluminum metallization in the CMOS process used (from Austriamicrosystems, Underpremstätten, Austria), Pt deposited directly on the Al layer did not provide sufficient protection of the underlying aluminum. Probably

9.3 Cell-based Assays | 469

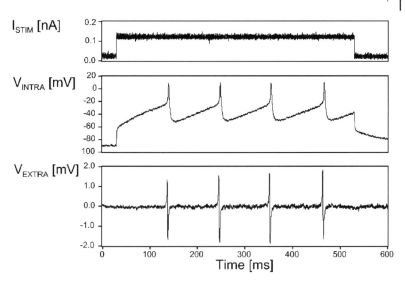

Fig. 9.11 Experimental data from a snail neuron recorded by the CMOS-integrated OSFET array. Top graph: stimulation current I_{STIM} applied by an external potentiostat. Middle graph: intracellular potential V_{INTRA} recorded using an external microelectrode needle inserted into the cell for reference. Bottom graph: V_{EXTRA}, the signal recorded by the OSFET array, is the derivative of the intracellular potential due to the capacitive coupling of the cell and the sensing electrode. Reprinted from [63], with permission

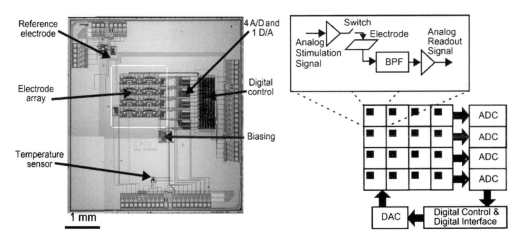

Fig. 9.12 Photograph of the fully integrated CMOS microelectrode array with on-chip signal processing and conversion. Reprinted from [64], with permission

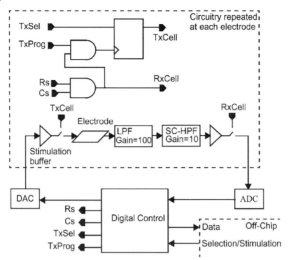

Fig. 9.13 Block diagram of the signal repeating circuitry including an amplifying bandpass filter and transmission gates for stimulation and recording. Column addressing and four ADCs are provided for signal readout and data conversion. A DAC in conjunction with a power amplifier allows to apply arbitrary stimuli to a selected pixel. A digital core handles the interfacing to the host computer. Reprinted from [64], with permission

due to pinholes in the layer, the underlying Al was attacked by the electrolyte after short periods of operation. To solve this problem, the platinum electrode was shifted laterally from the Al pad and the stack of Al and Pt is covered by a stack of six alternating PECVD SiO_2 and Si_3N_4 layers for protection [64].

A signal repeating unit (Fig. 9.13) is designed around each electrode, providing a low-pass filter (cut-off 50 kHz, gain 100), a switched-capacitor high-pass filter (cut-off 100 Hz, gain 10) and a buffer amplifier, resulting in a total signal gain of 1000. During the measurement, the column to be read is selected by a decoder block and the outputs of the buffers are routed to the 8-bit, successive-approximation ADCs. The resulting digital values are sent to the host PC through a parallel, 8-bit bus.

Aside from pure recording, the chip is also capable of applying an arbitrary waveform to any subset of electrodes for stimulation. The signal is generated by an on-chip 8-bit DAC at a sampling frequency of 120 kHz from data delivered by the host computer. To provide sufficient current (up to 10 mA) to drive the low-resistive electrolyte load, a class AB amplifier is integrated on the chip.

9.3.2.3 Electrode Needles for *In Vivo* Electrophysiological Measurements

Whereas MEA devices allow the study of electric cell signaling of planar *in vitro* cell cultures, *in vivo* measurements require a different device topology. Najafi, Wise and co-workers at the University of Michigan have developed a number of micromachined, needle-like neural probes, some of which also feature integrated low-noise

Fig. 9.14 Micromachined neural probe with on-chip CMOS electronics for stimulation and recording. The measurement sites are separated by 400 μm and several chips can be extended to 3D arrays. Image kindly provided by Professor K. D. Wise, University of Michigan

amplifiers and multiplexer circuits [89–91]. Whereas the device examples in the previous sections require only post-CMOS electrode fabrication, these neural probes are fabricated by true micromachining techniques with bulk etching.

To allow insertion of the recording and stimulating probes into biological tissue, needle structures are fabricated from a silicon substrate by frontside bulk micromachining using the p^{++} etch stop technique [92] (see also Chapter 1). This allows the production of 1–3 mm long shanks with a thickness of 10–15 μm and a width of 10–100 μm, which can contain up to 32 probe channels. Due to the minute electrical signals recorded and to allow for the arrangement of probes in one, two or even three dimensions, integrated circuitry was used for amplification and multiplexing. Fig. 9.14 shows a photograph of a 16-channel stimulating microprobe with 64 electrode sites. Eight of these can be driven simultaneously over a current range from –127 to +127 μA by integrated 8-bit D/A converters for stimulation of a 4 mm^3 volume of neural tissue [90].

Probes with additional functions have also been developed, including integrated microchannels for controlled drug infusion at the cellular level [93] or integrated heaters for thermal marking and monitoring of neural tissue [94].

9.4
Future Trends

A number of recent biosensor developments have been targeted towards integration with CMOS circuitry in the near future. This final section discusses these trends in the field.

Fig. 9.15 Photograph of a CMOS chip with cantilever array and fully autonomous on-chip PID controller circuitry. (Photograph kindly provided by Sadik Hafizovic, PEL, ETH Zurich)

9.4.1
Scanning Probe Techniques

Scanning probe techniques have evolved into a powerful branch of nanoscale instrumentation for the life sciences, for instance for imaging of biomolecules [95–97]. The majority of scanning probe instruments today operate using probing cantilevers fabricated in silicon microtechnology and the integration of cantilevers and cantilever arrays with CMOS circuitry is already on the way [98–101]. For full integration, the cantilever deflection is typically measured during the scanning by monitoring the mechanical stress using a piezoresistive Wheatstone bridge. Additionally, cantilevers can be bent by integrated bimorph actuators to follow the surface topography during scanning. The fabrication of tips for high-resolution scanning remains, however, a challenge within the limits of commercial CMOS processes. Wet etching of silicon tips for CMOS processes with deep n-wells [102] and also a post-CMOS transfer of separately fabricated tips [103] have been proposed as a solution. When integrated with on-chip electronics, multiple cantilevers can be combined into linear arrays of up to 10 probes [100], which improves the throughput for imaging larger areas significantly. Recently, a fully integrated, CMOS-based AFM has been presented, which allows continuous monitoring of 10 independent cantilevers controlled by on-chip PID loops (Fig. 9.15) [101].

Aside from imaging, cantilever probes can also serve as sensors for surface binding forces between the cantilever and an underlying substrate. This has been accomplished by CMOS-integrated systems for a variety of surfaces including hydrophobic and hydrophilic self-assembled monolayers [104]. An extension of this concept towards atomic force spectroscopy of biomolecules and cells has been pioneered by Gaub and co-workers [105–107] and chip applications are under development [108].

9.4.2
Cantilever-based Sensors

Micromachined cantilever structures have been shown to work as extraordinarily sensitive detectors for inorganic and organic chemical compounds in both the liquid and gas phase [109–111] (see also Chapter 7). If forced to mechanical oscillation, changes in the cantilever mass, e.g. due to analyte molecules bound by suitable probes immobilized on its surface, result in a change in the resonance frequency [112]. Alternatively, physical or chemical changes in a sensing layer deposited on one side of the cantilever often result in a change of surface stress, which can be monitored by measuring the resulting deflection of the structure [111, 112]. Cantilever-based detection of analytes of biochemical interest has been demonstrated for proteins [113] and bacteria [114]. The integration of cantilever sensors and cantilever arrays with CMOS circuitry for readout and excitation has been described for gas-phase analysis [98, 115–117]. Resonating operation of a CMOS cantilever device in a liquid environment has been achieved recently by electromagnetic actuation with on-chip feedback circuitry [118].

9.4.3
Biochemical Sample Pretreatment

As underlined in several sections of this chapter, the diverse sample pretreatment steps needed for a successful chemical analysis are of key importance for integrated sensing systems. Over the last decade, a number of techniques have been implemented on microfluidic chips, including sample filtration, dialysis, analyte preconcentration and preseparation, labeling, cell lysis and biochemical reactions such as enzymatic digestion or the polymerase chain reaction [6]. The latter has also been integrated on silicon chips with microelectrodes [1] or other sensing principles [119]. Full integration of a PCR device or other sample pretreatment techniques with on-chip circuitry has not yet been presented, probably owing to the considerable space needed on the chip for microfluidic elements. At present, this functionality is most often integrated into low-cost polymer cartridges, which house the actual sensing chip.

9.4.4
Sensor Packaging

Typically, biochemical sensors are in direct contact with the generally liquid sample under study. As a consequence, the microchip packaging solution not only needs to provide good mechanical, physical and chemical protection for the chip, but also has to be compatible with the sample solution. Additionally, it is beneficial for the sensing system if additional functions can be implemented into the package as discussed in the previous section. Efforts are generally directed towards cartridge concepts with microchips embedded in a considerably larger polymer substrate with microfluidic networks, reagent reservoirs and reaction chambers [120]. Along the same lines, an integrated, low-cost pump for liquid sample transport for a CMOS-based cantilever sensor has been presented recently [121].

9.5
Conclusions

The examples presented in this chapter illustrate the impressive scalability of CMOS-integrated sensing systems for biochemical applications, which allows thousands of sensor pixels to be monitored in parallel. Despite the minute signals recorded at each sensing site, on-chip amplification combined with filters and offset compensation circuitry enables precise biochemical measurements. Finally, in combination with microfluidic sample handling, complete assay procedures can be integrated into a single device, which is, in our opinion, the key to successful market penetration of biochemical CMOS sensors.

However, technological issues have also emerged including material compatibility of the typical CMOS layers and complex sensor packaging. These, combined with the tough market requirements for the fabrication and testing cost per device, make it important to choose the areas of activity for CMOS biosensor development precisely. However, for these applications the authors expect a considerable market and a great research potential for high-volume, array-type biosensing systems with on-chip circuitry.

9.6
Acknowledgments

The authors thank the editors, especially Professor Oliver Brand, for his invaluable help during the writing of this chapter.

9.7
References

1 M. Tartagni, L. Altomare, R. Guerrieri, A. Fuchs, N. Manaresi, G. Medoro, R. Thewes, in: *Sensors Update 13*, H. Baltes, G.K. Fedder, J.G. Korvink (eds.); Weinheim: Wiley-VCH, **2004**, 155–200.

2 A. Witvrouw, F. Van Steenkiste, D. Maes, L. Haspeslagh, P. Van Gerwen, P. De Moor, S. Sedky, C. Van Hoof, A.C. de Vries, A. Verbist, A. De Caus- semaeker, B. Parmentier, K. Baert, *Microsyst. Technol.* **2000**, 6, 192–199.

3 E. Verpoorte, N.F. De Rooij, *Proc. IEEE* **2003**, *91*, 930–953.

4 A. Manz, N. Graber, H.M. Widmer, *Sens. Actuators B* **1990**, *1*, 244–248.

5 J. Wang, *J. Pharm. Biomed. Anal.* **1999**, *19*, 47–53.

6 J. Lichtenberg, N.F. de Rooij, E. Verpoorte, *Talanta* **2002**, *56*, 233–266.

7 www.istat.com
8 K. A. ERICKSON, P. WILDING, Clin. Chem. 1993, 39, 283–287.
9 L. JOOS, E. ERYUKSEL, M. H. BRUTSCHE, Swiss Med. Wkly. 2003, 133, 31–38.
10 www.scienion.com
11 C. HAGLEITNER, A. HIERLEMANN, D. LANGE, A. KUMMER, N. KERNESS, O. BRAND, H. BALTES, Nature 2001, 414, 293–296.
12 D. J. HARRISON, P. G. GLAVINA, A. MANZ, Sens. Actuators B 1993, 10, 107–116.
13 A. P. F. TURNER, I. KARUBE, G. S. WILSON (eds.), Biosensors: Fundamentals and Applications; Oxford: Oxford University Press, 1987.
14 J. JANATA, Principles of Chemical Sensors; New York: Plenum Press, 1990.
15 E. GIZELI, C. R. LOWE (eds.), Biomolecular Sensors; London: Taylor and Francis, 2002.
16 J. M. COOPER (ed.), Biosensors: a Practical Approach; Oxford: Oxford University Press, 2004.
17 P. ARQUINT, A. VAN DEN BERG, B. H. VAN DER SCHOOT, N. F. DE ROOIJ, H. BUHLER, W. E. MORF, L. F. J. DURSELEN, Sens. Actuators B 1993, 13, 340–344.
18 E. LAUWERS, J. SULS, W. GUMBRECHT, D. MAES, G. GIELEN, W. SANSEN, IEEE J. Solid-State Circuits 2001, 36, 2030–2038.
19 L. LORENZELLI, B. MARGESIN, S. MARTINOIA, M. T. TEDESCO, M. VALLE, Biosens. Bioelectron. 2003, 18, 621–626.
20 H. MEYER, H. DREWER, B. GRUNDIG, K. CAMMANN, R. KAKEROW, Y. MANOLI, W. MOKWA, M. ROSPERT, Anal. Chem. 1995, 67, 1164–1170.
21 R. KAKEROW, Y. MANOLI, W. MOKWA, M. ROSPERT, H. MEYER, H. DREWER, J. KRAUSE, K. CAMMANN, Sens. Actuators A 1994, 43, 296–301.
22 R. THEWES, F. HOFMANN, A. FREY, B. HOLZAPFL, M. SCHIENLE, C. PAULUS, P. SCHINDLER, G. ECKSTEIN, C. KASSEL, M. STANZEL, R. HINTSCHE, E. NEBLING, J. ALBERS, J. HASSMAN, J. SCHULEIN, W. GOEMANN, W. GUMBRECHT, in: Proceedings of the 2002 IEEE International Solid State Circuits Conference; Piscataway, NJ: IEEE, 2002, p. 350.
23 A. FREY, M. JENKNER, M. SCHIENLE, C. PAULUS, B. HOLZAPFL, P. SCHINDLER-BAUER, F. HOFMANN, D. KUHLMEIER, J. KRAUSE, J. ALBERS, W. GUMBRECHT, D. SCHMITT-LANDSIEDEL, R. THEWES, in: Proceedings of the ISCAS 2003 International Symposium on Circuits and Systems; Piscataway, NJ: IEEE, 2003, pp. V-9–12.
24 M. SCHIENLE, A. FREY, F. HOFMANN, B. HOLZAPFL, C. PAULUS, P. SCHINDLER-BAUER, R. THEWES, in: Proceedings of the 2004 IEEE International Solid State Circuits Conference; Piscataway, NJ: IEEE, 2004, p. 1320.
25 F. HOFMANN, A. FREY, B. HOLZAPFL, M. SCHIENLE, C. PAULUS, P. SCHINDLER-BAUER, D. KUHLMEIER, J. KRAUSE, R. HINTSCHE, E. NEBLING, J. ALBERS, W. GUMBRECHT, K. PLEHNERT, G. ECKSTEIN, R. THEWES, in: Proceedings of the IEEE International Electron Devices Meeting 2002; Piscataway, NJ: IEEE, 2002, p. 957.
26 J. P. GOLDEN, F. S. LIGLER, Biosens. Bioelectron. 2002, 17, 719–725.
27 M. C. MORENO-BONDI, J. P. ALARIE, T. VO-DINH, Anal. Bioanal. Chem. 2003, 375, 120–124.
28 U. LU, B. C. P. HU, Y. C. SHIH, Y. S. YANG, C. Y. WU, C. J. YUAN, M. D. KER, T. K. WU, Y. K. LI, Y. Z. HSIEH, W. Y. HSU, C. T. LIN, IEEE Sens. J. 2003, 3, 310–316.
29 U. LU, B. C. P. HU, Y. C. SHIH, C. Y. WU, Y. S. YANG, Biosens. Bioelectron. 2004, 19, 1185–1191.
30 M. DAVENPORT, A. H. TITUS, E. C. TEHAN, Z. Y. TAO, Y. TANG, R. M. BUKOWSKI, F. V. BRIGHT, IEEE Sens. J. 2004, 4, 180–188.
31 E. K. BOLTON, G. S. SAYLER, D. E. NIVENS, J. M. ROCHELLE, S. RIPP, M. L. SIMPSON, Sens. Actuators B 2002, 85, 179–185.
32 M. L. SIMPSON, G. S. SAYLER, G. PATTERSON, D. E. NIVENS, E. K. BOLTON, J. M. ROCHELLE, J. C. ARNOTT, B. M. APPLEGATE, S. RIPP, M. A. GUILLORN, Sens. Actuators B 2001, 72, 134–140.
33 S. KOOIJMAN, J. J. M. BEDAUX, Water Res. 1996, 30, 1711–1723.
34 P. S. DITTRICH, P. SCHWILLE, Anal. Chem. 2003, 75, 5767–5774.
35 S. FIEDLER, S. G. SHIRLEY, T. SCHNELLE, G. FUHR, Anal. Chem. 1998, 70, 1909–1915.
36 S. GAWAD, L. SCHILD, P. RENAUD, Lab Chip 2001, 1, 76–82.

37 J. H. Nieuwenhuis, J. Bastemeijer, A. Bossche, M. J. Vellekoop, *IEEE Sens. J.* **2003**, *3*, 646–651.
38 H. E. Ayliffe, A. B. Frazier, R. D. Rabbitt, *J. Microelectromech. Syst.* **1999**, *8*, 50–57.
39 J. Voldman, M. L. Gray, M. Toner, M. A. Schmidt, *Anal. Chem.* **2002**, *74*, 3984–3990.
40 A. R. Wheeler, W. R. Throndset, R. J. Whelan, A. M. Leach, R. N. Zare, Y. H. Liao, K. Farrell, I. D. Manger, A. Daridon, *Anal. Chem.* **2003**, *75*, 3581–3586.
41 H. A. Pohl, *J. Appl. Phys.* **1951**, *22*, 869–871.
42 M. P. Hughes, *Nanoelectromechanics in Engineering and Biology*, Boca Raton, FL: CRC Press, **2003**.
43 J. Suehiro, R. Pethig, *J. Phys. D* **1998**, *31*, 3298–3305.
44 M. Dürr, J. Kentsch, T. Muller, T. Schnelle, M. Stelzle, *Electrophoresis* **2003**, *24*, 722–731.
45 P. R. C. Gascoyne, J. V. Vykoukal, *Proc. IEEE* **2004**, *92*, 22–42.
46 S. Masuda, M. Washizu, I. Kawabata, *IEEE Trans. Ind. Appl.* **1988**, *24*, 217–222.
47 R. Hagedorn, G. Fuhr, T. Muller, J. Gimsa, *Electrophoresis* **1992**, *13*, 49–54.
48 G. Medoro, N. Manaresi, A. Leonardi, L. Altomare, M. Tartagni, R. Guerrieri, *IEEE Sens. J.* **2003**, *3*, 317–325.
49 N. Manaresi, A. Romani, G. Medoro, L. Altomare, A. Leonardi, M. Tartagni, R. Guerrieri, *IEEE J. Solid-State Circuits* **2003**, *38*, 2297–2305.
50 C. A. Thomas, P. A. Springer, L. M. Okun, Y. Berwaldn, G. E. Loeb, *Exp. Cell Res.* **1972**, *74*, 61–66.
51 A. Molleman, *Patch Clamping: an Introductory Guide to Patch Clamp Electrophysiology*, New York: Wiley, **2003**.
52 P. Thiebaud, C. Beuret, N. F. de Rooij, M. Koudelka-Hep, *Sens. Actuators B* **2000**, *70*, 51–56.
53 M. O. Heuschkel, M. Fejtl, M. Raggenbass, D. Bertrand, P. Renaud, *J. Neurosci. Methods* **2002**, *114*, 135–148.
54 B. W. Kristensen, J. Noraberg, P. Thiebaud, M. Koudelka-Hep, J. Zimmer, *Brain Res.* **2001**, *896*, 1–17.
55 www.multichannelsystems.com
56 www.ayanda-biosys.com
57 www.med64.com
58 P. Fromherz, A. Offenhausser, T. Vetter, J. Weis, *Science* **1991**, *252*, 1290–1293.
59 P. Fromherz, A. Stett, *Phys. Rev. Lett.* **1995**, *75*, 1670–1673.
60 P. Fromherz, in: *Nanoelectronics and Information Technology*, R. Waser (ed.); Weinheim: Wiley-VCH, **2003**, pp. 781–810.
61 A. Stett, U. Egert, E. Guenther, F. Hofmann, T. Meyer, W. Nisch, H. Haemmerle, *Anal. Bioanal. Chem.* **2003**, *377*, 486–495.
62 www.ich.org
63 B. Eversmann, M. Jenkner, F. Hofmann, C. Paulus, R. Brederlow, B. Holzapfl, P. Fromherz, M. Merz, M. Brenner, M. Schreiter, R. Gabl, K. Plehnert, M. Steinhauser, G. Eckstein, D. Schmitt-Landsiedel, R. Thewes, *IEEE J. Solid-State Circuits* **2003**, *38*, 2306–2317.
64 F. Heer, W. Franks, A. Blau, S. Taschini, C. Ziegler, A. Hierlemann, H. Baltes, *Biosens. Bioelectron.* **2004**, *20*, 358–366.
65 L. Berdondini, P. D. van der Wal, O. Guenat, N. F. de Rooij, M. Koudelka-Hep, P. Seitz, R. Kaufmann, P. Metzler, N. Blanc, S. Rohr, *Biosens. Bioelectron.* **2004**, in press. DOI: 10.106/j.bios.2004.08.011.
66 H. Honma, *Electrochim. Acta* **2001**, *47*, 75–84.
67 L. Berdondini, P. D. van der Wal, N. F. de Rooij, A. Koudelka-Hep, *Sens. Actuators B* **2004**, *99*, 505–510.
68 R. C. Gesteland, B. Howland, J. Y. Lettvin, W. H. Pitts, *Proceedings of the Institute of Radio Engineers* **1959**, *47*, 1856–1862.
69 C. A. Marrese, *Anal. Chem.* **1987**, *59*, 217–218.
70 J. J. Pancrazio, P. P. Bey, A. Loloee, S. R. Manne, H. C. Chao, L. L. Howard, W. M. Gosney, D. A. Borkholder, G. T. A. Kovacs, P. Manos, D. S. Cuttino, D. A. Stenger, *Biosens. Bioelectron.* **1998**, *13*, 971–979.
71 D. Kleinfeld, K. H. Kahler, P. E. Hockberger, *J. Neurosci.* **1988**, *8*, 4098–4120.
72 D. A. Stenger, C. J. Pike, J. J. Hickman, C. W. Cotman, *Brain Res.* **1993**, *630*, 136–147.

73 G. Zeck, P. Fromherz, *Proc. Natl. Acad. Sci. USA* **2001**, *98*, 10457–10462.
74 M. P. Maher, J. Pine, J. Wright, Y. C. Tai, *J. Neurosci. Methods* **1999**, *87*, 45–56.
75 F. Greve, J. Lichtenberg, H. Hall, A. Hierlemann, H. Baltes, in: *Proceedings of the µTAS 2003 Symposium, Squaw Valley, CA, October 5–9, 2003*, M. A. Northrup, K. F. Jensen, D. J. Harrison (eds.); Cleveland, OH: Transducers Research Foundation, **2003**, pp. 327–330.
76 D. S. Gray, J. L. Tan, J. Voldman, C. S. Chen, *Biosens. Bioelectron.* **2004**, *19*, 771–780.
77 P. Fromherz, H. Schaden, *Eur. J. Neurosci.* **1994**, *6*, 1500–1504.
78 B. Lom, K. E. Healy, P. E. Hockberger, *J. Neurosci. Methods* **1993**, *50*, 385–397.
79 C. D. James, R. C. Davis, L. Kam, H. G. Craighead, M. Isaacson, J. N. Turner, W. Shain, *Langmuir* **1998**, *14*, 741–744.
80 B. C. Wheeler, J. M. Corey, G. J. Brewer, D. W. Branch, *J. Biomech. Eng. – Trans. ASME* **1999**, *121*, 73–78.
81 C. D. James, R. Davis, M. Meyer, A. Turner, S. Turner, G. Withers, L. Kam, G. Banker, H. Craighead, M. Isaacson, J. Turner, W. Shain, *IEEE Trans. Bio-Med. Eng.* **2000**, *47*, 17–21.
82 Y. Nam, J. C. Chang, B. C. Wheeler, G. J. Brewer, *IEEE Trans. Bio-Med. Eng.* **2004**, *51*, 158–165.
83 G. A. Dunn, A. F. Brown, *J. Cell. Sci.* **1986**, *83*, 313–340.
84 D. M. Brunette, *Exp. Cell Res.* **1986**, *164*, 11–26.
85 M. Merz, P. Fromherz, *Adv. Mater.* **2002**, *14*, 141–144.
86 S. Turner, L. Kam, M. Isaacson, H. G. Craighead, W. Shain, J. Turner, *J. Vac. Sci. Technol. B* **1997**, *15*, 2848–2854.
87 A. M. P. Turner, N. Dowell, S. W. P. Turner, L. Kam, M. Isaacson, J. N. Turner, H. G. Craighead, W. Shain, *J. Biomed. Mater. Res.* **2000**, *51*, 430–441.
88 B. D. DeBusschere, G. T. A. Kovacs, *Biosens. Bioelectron.* **2001**, *16*, 543–556.
89 K. Najafi, K. D. Wise, *IEEE J. Solid-State Circuits* **1986**, *21*, 1035–1044.
90 C. Kim, K. D. Wise, *IEEE J. Solid-State Circuits* **1996**, *31*, 1230–1238.
91 K. Najafi, in: *Handbook of Microlithography, Micromachining and Microfabrication*, P. Rai-Choudhury (ed.); Bellingham, MA: SPIE, **1997**, Vol. 2, pp. 517–616.
92 K. Najafi, K. D. Wise, T. Mochizuki, *IEEE Trans. Electron Devices* **1985**, *32*, 1206–1211.
93 J. K. Chen, K. D. Wise, J. F. Hetke, S. C. Bledsoe, *IEEE Trans. Bio.-Med. Eng.* **1997**, *44*, 760–769.
94 J. K. Chen, K. D. Wise, *IEEE Trans. Bio.-Med. Eng.* **1997**, *44*, 770–774.
95 S. Scheuring, D. Fotiadis, C. Moller, S. A. Muller, A. Engel, D. J. Muller, *Single Mol.* **2001**, *2*, 59–67.
96 P. L. T. M. Frederix, T. Akiyama, U. Staufer, C. Gerber, D. Fotiadis, D. J. Muller, A. Engel, *Curr. Opin. Chem. Biol.* **2003**, *7*, 641–647.
97 M. B. Viani, L. I. Pietrasanta, J. B. Thompson, A. Chand, I. C. Gebeshuber, J. H. Kindt, M. Richter, H. G. Hansma, P. K. Hansma, *Nat. Struct. Biol.* **2000**, *7*, 644–647.
98 D. Lange, O. Brand, H. Baltes, *CMOS Cantilever Sensor Systems*; Berlin: Springer, **2002**.
99 T. Akiyama, U. Staufer, N. F. de Rooij, D. Lange, C. Hagleitner, O. Brand, H. Baltes, A. Tonin, H. R. Hidber, *J. Vac. Sci. Technol. B* **2000**, *18*, 2669–2675.
100 T. Volden, M. Zimmermann, D. Lange, O. Brand, H. Baltes, *Sens. Actuators A* **2004**, *115* (2–3), 516–522.
101 D. Barrettino, S. Hafizovic, T. Volden, J. Sedivy, K. Kirstein, A. Hierlemann, H. Baltes, in: *Digest of Technical Papers of the 2004 Symposium on VLSI Circuits*; **2004**, 461.
102 M. Ono, D. Lange, O. Brand, C. Hagleitner, H. Baltes, *Ultramicroscopy* **2002**, *91*, 9–20.
103 T. Akiyama, U. Staufer, N. F. de Rooij, *J. Microelectromech. Syst.* **1999**, *8*, 65–70.
104 W. Franks, D. Lange, S. Lee, A. Hierlemann, N. Spencer, H. Baltes, *Ultramicroscopy* **2002**, *91*, 21–27.
105 M. Grandbois, M. Beyer, M. Rief, H. Clausen-Schaumann, H. E. Gaub, *Science* **1999**, *283*, 1727–1730.
106 H. Clausen-Schaumann, M. Seitz, R. Krautbauer, H. E. Gaub, *Curr. Opin. Chem. Biol.* **2000**, *4*, 524–530.
107 M. Benoit, D. Gabriel, G. Gerisch, H. E. Gaub, *Nat. Cell Biol.* **2000**, *2*, 313–317.

108 G. Villanueva, J. Montserrat, F. Perez-Murano, G. Rius, J. Bausells, *Microelectron. Eng.* **2004**, *73–74*, 480–486.

109 T. Thundat, R. J. Warmack, G. Y. Chen, D. P. Allison, *Appl. Phys. Lett.* **1994**, *64*, 2894–2896.

110 J. R. Barnes, R. J. Stephenson, M. E. Welland, C. Gerber, J. K. Gimzewski, *Nature* **1994**, *372*, 79–81.

111 J. Fritz, M. K. Baller, H. P. Lang, H. Rothuizen, P. Vettiger, E. Meyer, H. J. Güntherodt, C. Gerber, J. K. Gimzewski, *Science* **2000**, *288*, 316–318.

112 M. Sepaniak, P. Datskos, N. Lavrik, C. Tipple, *Anal. Chem.* **2002**, *74*, 568A–575A.

113 G. H. Wu, R. H. Datar, K. M. Hansen, T. Thundat, R. J. Cote, A. Majumdar, *Nat. Biotechnol.* **2001**, *19*, 856–860.

114 B. Ilic, D. Czaplewski, H. G. Craighead, P. Neuzil, C. Campagnolo, C. Batt, *Appl. Phys. Lett.* **2000**, *77*, 450–452.

115 D. Lange, C. Hagleitner, A. Hierlemann, O. Brand, H. Baltes, *Anal. Chem.* **2002**, *74*, 3084–3095.

116 D. Lange, C. Hagleitner, C. Herzog, O. Brand, H. Baltes, *Sens. Actuators A* **2003**, *103*, 150–155.

117 Z. J. Davis, G. Abadal, B. Helbo, O. Hansen, F. Campabadal, F. Perez-Murano, J. Esteve, E. Figueras, J. Verd, N. Barniol, A. Boisen, *Sens. Actuators A* **2003**, *105*, 311–319.

118 Y. Li, C. Vancura, C. Hagleitner, J. Lichtenberg, O. Brand, H. Baltes, in: *Proceedings of IEEE Sensors 2003;* **2003**, pp. 809–813.

119 M. A. Burns, B. N. Johnson, S. N. Brahmasandra, K. Handique, J. R. Webster, M. Krishnan, T. S. Sammarco, P. M. Man, D. Jones, D. Heldsinger, C. H. Mastrangelo, D. T. Burke, *Science* **1998**, *282*, 484–487.

120 R. H. Liu, J. N. Yang, R. Lenigk, J. Bonanno, P. Grodzinski, *Anal. Chem.* **2004**, *76*, 1824–1831.

121 W. H. Song, H. Baltes, J. Lichtenberg, in: *Proceedings of the Symposium on Micro Total Analysis Systems 2004,* Malmö, Sweden, T. Laurell et al. (eds.); London, UK: Royal Society of Chemistry, **2004**, 336–338.

122 K. Dill, D. D. Montgomery, W. Wang, J. C. Tsai, *Anal. Chim. Acta* **2001**, *444*, 69–78.

123 W. Baumann, M. Lehmann, M. Bitzenhofer, A. Schwinde, M. Brischwein, R. Ehret, B. Wolf, *Sens. Actuators B* **1999**, *55*, 77–89.

124 M. Lehmann, W. Baumann, M. Brischwein, H. J. Gahle, I. Freund, R. Ehret, S. Drechsler, H. Palzer, M. Kleintges, U. Sieben, B. Wolf, *Biosens. Bioelectron.* **2001**, *16 (3)*, 195–203.

10
CMOS-based Thermal Sensors

*T. Akin, Department of Electrical and Electronics Engineering,
Middle East Technical University, Ankara, Turkey*

Abstract

This chapter presents various CMOS-based thermal sensors, including thermal radiation sensors, thermal converters, and thermal flow sensors. Two thermal radiation sensor approaches are described in detail: thermopiles and microbolometers. Thermopile-based uncooled infrared imaging arrays are relatively simple to implement and are low cost; however, their performances are limited. They typically have small responsivity (5–15 V/W), moderate noise equivalent temperature difference (NETD) values ($\sim$ 500 mK), large pixel sizes (pixel pitch of 100–400 µm), and small array sizes (typically 16×16 and 32×32). Microbolometer type uncooled infrared detectors have shown impressive developments in recent years. They are more expensive than thermopiles, however, much cheaper than cooled photon detectors, while approaching to their performances. Currently, there are microbolometer infrared cameras in the market with array sizes of 320×240, pixel sizes of 25 µm×25 µm, and NETD values smaller than 30 mK. There are also 640×480 format microbolometer array demonstrations with NETD values smaller than 50 mK. Efforts are continuing to reduce the prices of uncooled infrared detectors to widen their use in many commercial applications. CMOS-based thermal converters are used for true root mean square (RMS) voltage and ac signal measurements, independent of signal waveform. There are sensors reported to operate up to 1.2 GHz with a linearity error of less than 1%. Thermal flow sensors are used to measure the movement of a fluid (liquid or gas) by measuring either physical deflection, or heat loss, or pressure variation. Demonstrated sensors include a mass flow sensor that operates over the 1cm/s to 5m/s range and a wind sensor that can measure wind speeds up to 38 m/s.

Advanced Micro and Nanosystems. Vol. 2. CMOS – MEMS.
Edited by H. Baltes, O. Brand, G. K. Fedder, C. Hierold, J. Korvink, O. Tabata
Copyright © 2005 WILEY-VCH Verlag GmbH & Co. KGaA, Weinheim
ISBN: 3-527-31080-0

10 CMOS-based Thermal Sensors

Keywords

Thermal sensors; uncooled infrared detectors; microbolometers; thermopiles, infrared detectors; radiation sensors; flow sensors; thermal converters; thermal flow sensors.

10.1	Introduction	480
10.2	Thermal Radiation Sensors	480
10.2.1	Thermal (Uncooled) Infrared Detectors	481
10.2.2	Types of Thermal (Uncooled) Infrared Detectors	483
10.2.2.1	Thermopile-based Radiation Sensors	483
10.2.2.2	Microbolometers	491
10.3	Thermal Converters	499
10.4	Thermal Flow Sensors	502
10.5	Acknowledgements	506
10.6	References	507

10.1 Introduction

There are a number of thermal sensors based on the CMOS process, including thermal radiation sensors, thermal converters, thermal flow sensors, thermal pressure sensors, thermal acceleration sensors, resonant position sensors, chemical sensors, and sensors for determining thermal properties of materials [1]. Some of these sensors have already been covered in the previous chapters, including pressure sensors (see Chapter 6), acceleration sensors (see Chapter 3), chemical sensors (see Chapter 7), and sensors for determining thermal properties of materials (see Chapter 2). This chapter will summarize the state of the art in three CMOS-based thermal sensors, namely thermal radiation sensors, thermal converters, and thermal flow sensors. Section 10.2 will introduce thermal radiation sensors, explaining the differences between the cooled photon detectors and uncooled thermal infrared detectors. Two uncooled infrared detector approaches will be described in detail: thermopile-based radiation sensors and microbolometers. State-of-the-art in these approaches will be presented. Section 10.3 will summarize the work on thermal converters implemented with CMOS technology together with post-CMOS micromachining. Finally, Section 10.4 will summarize the state-of-the-art on CMOS thermal flow sensors.

10.2 Thermal Radiation Sensors

Infrared radiation is part of the electromagnetic spectrum with wavelengths above the visible spectrum, ranging from 1 µm to several tens of µm [2]. Detectors that can sense infrared radiation are called infrared detectors, while ensembles of infrared detectors in two-dimensional arrays are often called focal plane arrays (FPAs). Infrared detectors are used in many military and commercial applications such as night vision, mine detection, reconnaissance, fire fighting, medical imaging, and industrial control. Detectors for infrared imaging are sensitive in two regions of the infrared spectrum where the infrared transmission is allowed by atmosphere: the 3–5 µm wavelength region (mid wave infrared, MWIR) and the 8–12 µm wavelength region (long wave infrared, LWIR).

There are basically two types of detectors that can sense infrared radiation. The first type is photon detectors [3, 4], where the absorbed infrared photons generate free electron–hole (E–H) pairs, which are then collected by the application of an electric field for electronic processing. The second type of infrared detectors is known as thermal detectors [2], where the energy of the absorbed infrared photon raises the temperature of the detector, and the temperature-induced change in an electrical parameter is measured with the help of suitable circuitry.

Photon infrared detectors are fast, and their sensitivities are much higher than thermal detectors, with noise equivalent temperature difference (NETD) values as low as 7–20 mK. However, the number of thermally generated E–H pairs at room temperature is much larger than the infrared-induced E–H pairs, which makes their use for infrared imaging impossible, unless they are cooled to cryogenic temperatures, i.e., 77 K or below. For this purpose, special and expensive coolers are used, increasing the size, cost, and operating power of the detector systems or cameras. Commonly used cooled (or photon) infrared detectors are fabricated using indium antimonide (InSb) [5], mercury cadmium telluride (HgCdTe or MCT) [6, 7], or quantum well infrared photodetector (QWIP) [8–10] technologies. InSb detectors are used in the 3–5 µm wavelength range, while MCT and QWIP can be used in both 3–5 and 8–12 µm wavelength ranges. The fabrication of these detectors involves complicated processing steps, owing the known difficulties of handling low-bandgap materials required for the detection of low-energy infrared photons. Therefore, the cost of the photonic detectors and infrared cameras using these detectors is very high (typically $50000–150000), finding application areas only in expensive weapon platforms, in astronomical observation instruments, or in special medical instruments, where the performance is the primary issue. On the other hand, infrared cameras using thermal detectors are small in size, consume less power and are less expensive (typically $8000–15000 for 320×240 arrays), making them the ideal choice for applications which require high unit numbers with relatively lower performance. There are continuing efforts to decrease the cost of uncooled detectors to much lower values (such as $50–500) while achieving reasonable performance. The next section explains the principles of thermal detectors.

10.2.1
Thermal (Uncooled) Infrared Detectors

Thermal or uncooled infrared detectors sense the change in an electrical parameter with change in the device temperature related to the amount of absorbed infrared energy. Therefore, the thermal detection mechanism is an indirect means of infrared detection, and the response times of these detectors are longer than those of photon detectors. In most cases, the signal-to-noise ratio and detectivity of the uncooled thermal detectors are lower than those of cooled photon detectors. Therefore, the performance of thermal detectors is in general lower than that of cooled photon detectors. It should be mentioned that photon detectors are usually implemented as linear arrays (such as 240×1 or 240×4) that are scanned optically, while thermal detectors are implemented as 2-D staring arrays (such as 320×240). Since the electrical bandwidth of staring arrays is much lower than that of scanned arrays, it is possible to improve the signal-to-noise ratio of the thermal detectors when operated in staring arrays. Further, it is easier to fabricate staring arrays using thermal detectors than arrays that use cooled photon detectors. Furthermore, at scanning speeds close to the TV frame rate (30 frames/s), the performance degradation of the thermal detectors due to their relatively longer thermal time constants can be minimized by proper detector design. Considering these factors, although cooled detector arrays still provide better performance, the performance difference between thermal and cooled photonic detectors becomes smaller than expected by just comparing them on a pixel basis [2].

The most important advantage of thermal detectors is that they can operate at room temperature without requiring any complex and expensive cooling equipment. The resulting infrared imaging systems utilizing uncooled detector technology have much smaller size, lower cost, lower power consumption, and extended operation durations. Owing to these advantages, uncooled detectors are used in many military and commercial applications, such as night vision, mine detection, driver night vision enhancement, fire fighting, and industrial control applications. Wide application areas have made uncooled infrared technology a highly requested technology, and there is a worldwide effort to implement high-performance and low-cost uncooled infrared detector arrays. There are already infrared cameras on the market using 320×240 format detector arrays with 50 µm×50 µm and 25 µm×25 µm pixel sizes and with noise equivalent temperature differences (NETD) better than 30 mK [11–13]. The uncooled technology has also demonstrated uncooled FPAs with a 640×480 array format with 25×25 and 28×28 µm pixel sizes and NETD values lower than 50 mK [14–17]. The target of uncooled technology is to fabricate detectors with less than 25×25 µm pixel size and with NETD values better than 10 mK [2], i.e., performances similar to that of cooled detectors. Therefore, as the uncooled technology develops, many other infrared imaging systems that currently use cooled infrared detector arrays may start using uncooled detector arrays. The cooled photonic detectors, on the other hand, may find future application areas in more sophisticated and expensive infrared imaging platforms requiring even higher performance with added new features such as multi-spectral and/or multi-color infrared radiation sensing capability, possibly achieved using the rapidly developing QWIP technology.

10.2.2
Types of Thermal (Uncooled) Infrared Detectors

There are two major types of thermal radiation sensors based on post-CMOS fabrication approaches (see Chapter 1): thermopile sensors and microbolometers. Thermopile sensors are mostly based on bulk micromachining of the CMOS-processed wafers, whereas microbolometers are mostly based on surface micromachining of the CMOS-processed wafers. There are also recent examples of microbolometers which are implemented with post-CMOS bulk micromachining. Additional types of thermal detectors include electro-junction devices, where the absorbed infrared energy causes bending of a surface micromachined cantilever beam, changing the capacitance of the pixel [18, 19].

Most of the advances in thermal or uncooled infrared technology have been achieved with the microbolometer technology. Although this technology is already much cheaper than the photon detector technology, the fabrication and system costs are still high for many commercial applications, resulting in a worldwide effort to decrease further the cost of uncooled technology to enter the low-cost, high-volume markets where the system costs of $50–500 are required. Apart from decreasing cost of detector fabrication, there are other key issues to achieve these system costs, including wafer level vacuum packaging of the detectors, their wafer level testing, and the use of low-cost infrared optics. The following sections provide further details on the CMOS-compatible thermal radiation sensors.

10.2.2.1 Thermopile-based Radiation Sensors

Thermoelectric detectors are formed using thermocouples, whose operation principle is based on the Seebeck effect. Fig. 10.1 shows a schematic of a thermocouple. When two pieces of different materials A and B are joined and heated at one end, there will be a self-generated potential difference between the two other ends of the structures depending on the difference of the two material's Seebeck coefficients and the temperature difference between the hot and cold end points.

The output voltage of a thermocouple, V_{AB}, is given as [20–23]

$$V_{AB} = a_{AB} \Delta T_{hot-cold} = (a_A - a_B)(T_{hot} - T_{cold}) \tag{1}$$

where a_A and a_B are the Seebeck coefficients of the two different materials, a_{AB} is the Seebeck coefficient of the thermocouple, T_{hot} and T_{cold} are the temperatures of the hot and cold junctions, respectively, and $\Delta T_{hot-cold}$ is the temperature difference between the hot and cold junctions. The output voltage of a single thermocouple is usually not sufficient; therefore, a number of thermocouples are connected in series to form a so-called thermopile. Fig. 10.2 shows a thermopile that is constructed by series connection of five thermocouples. The thermopile can be used as an infrared detector if the thermocouples are placed on a suspended dielectric layer and if an absorber layer is placed close to or on top of the hot contacts of the thermopile. Fig. 10.3 shows a perspective view of this arrangement.

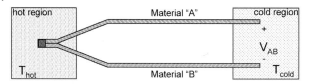

Fig. 10.1 Schematic of a thermocouple. The self-generated voltage V_{AB} at the open ends of the thermocouple is called the Seebeck voltage

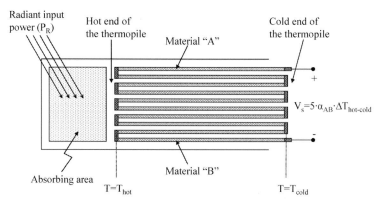

Fig. 10.2 Thermopile consisting of five series-connected thermocouples. The thermopile can be used as an infrared detector if the thermocouples are placed on a suspended and thermally insulating dielectric layer and if an absorber layer is placed close to the hot contacts of the thermopile

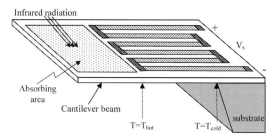

Fig. 10.3 Perspective view of a thermopile that can be used as an infrared detector. Thermocouples are placed on a dielectric cantilever beam and an absorber layer is placed on the tip next to the hot junction

An important factor for obtaining a large output voltage from a thermopile is to obtain a high thermal isolation in order to maximize the temperature difference between hot and cold junctions, $\Delta T_{hot-cold}$, for a specific absorbed power. To achieve this thermal isolation between hot and cold junctions, the thermocouples are often placed on top of dielectric diaphragms where the silicon underneath is removed to increase the thermal resistance. Of course, the layers that are used for

Tab. 10.1 Seebeck coefficients and thermal conductance values of various materials. After [21]

Material	Seebeck coefficient (at 273 K) (µV/K)	Thermal conductance (W/K m)
Aluminum	−1.7[a]	237
Chromium	18.8	
Gold	1.79	318
Copper	1.70	
Platinum	−4.45	
Nickel	−18.0	90
Bismuth	−79[b]	
Antimony	43[b]	
p-Type silicon	300 to 1000[a]	149
n-Type polysilicon	−200 to −500[a]	

[a] At 300 K.
[b] Averaged over 0 to 100 °C.

the thermocouples also contribute to the thermal conduction between the hot and cold junctions, and their contribution to the thermal conduction should be minimized for better performance (see figure of merit).

Thermopiles can be constructed using semiconductors and/or metals. Tab. 10.1 lists Seebeck coefficients and thermal conductance values of various materials. Since the Seebeck coefficients of semiconductors are larger than that of metals, semiconductor thermopiles are more responsive than their metal counterparts [20–30]. The Seebeck coefficient of semiconductor materials depends on the variation of the Fermi level of the semiconductor with respect to temperature; therefore, for semiconductor thermopiles, the magnitude and sign of the Seebeck coefficient can be adjusted by adjusting the doping type and doping level. The following equation can be used to estimate the Seebeck coefficient of silicon as a function of its electrical resistivity at room temperature [21]:

$$a_s = \pm 0.216 \times 10^{-3} \ln(\rho/\rho_0) \qquad (2)$$

where ρ is the resistivity of silicon in ohm cm and $\rho_0 \approx 5 \times 10^{-4}$ Ω cm. Here, a_s is given in V/K, and a positive sign is selected for p-type silicon, whereas a negative sign is selected for n-type silicon. Equation 10.2 suggests that the Seebeck coefficient of a semiconductor increases in magnitude with increased resistivity, and therefore, with decreased doping level. However, a thermopile material with very low electrical resistivity is not necessarily the best choice for a particular infrared detector, as the Seebeck coefficient is only one of the parameters influencing its overall performance. The next section explains the parameters that are important for selection of a material for thermopiles.

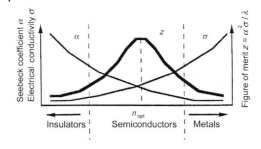

Fig. 10.4 Thermoelectric properties of metals, semiconductors and insulators [33]

Figure of Merit for Thermocouples

An ideal thermocouple material for infrared detector applications should have a very high Seebeck coefficient, a very low thermal conductance, and very low noise. These parameters are usually conflicting, and a figure of merit, z, is defined for the performance of a particular thermocouple material [31, 32]:

$$z = \frac{a^2}{\rho\kappa} \qquad (3)$$

where a is the Seebeck coefficient, κ is the thermal conductivity, and ρ is the electrical resistivity. The figure of merit for a thermocouple increases if materials with high Seebeck coefficients (a) and low thermal conductivities (κ) are used. Other than these two parameters, the electrical resistivity (ρ) should also be taken into account when choosing a thermocouple material, since Johnson noise is generated by the finite electrical resistance of the thermocouple structure. The Johnson noise, which is also called thermal noise, limits the minimum detectible temperature difference with the thermocouple. The Johnson noise voltage (V_N) for a resistor is defined as follows:

$$V_N = \sqrt{4kT\Delta f R_{el}} \qquad (4)$$

where k is Boltzmann's constant (1.381×10^{-23} J/K), T is the temperature in K, Δf is the bandwidth in Hz, and R_{el} is the electrical resistance in Ω. For a thermocouple, a low amount of Johnson noise voltage is required. To decrease the Johnson noise, the electrical resistance, R_{el}, should be decreased. Since R_{el} is proportional to ρ, the thermocouple materials should be chosen from low-resistivity materials. However, as mentioned previously, a lower ρ value also gives a lower Seebeck coefficient. Therefore, an optimum point needs to be determined considering all these parameters based on the figure of merit defined above. Fig. 10.4 shows a graph for the visual interpretation of the figure of merit concept [33]. As can be seen, an optimum value for the figure of merit is achieved for semiconductors at a specific doping level n_{opt}. This doping value is around 10^{19} cm^{-3} for both single-crystal silicon and polysilicon materials.

Tab. 10.1 also lists the thermal conductance values of some of the materials used in CMOS thermopiles. It should be noted that when the number of thermocouples is increased to obtain a high output voltage, it also increases the thermal conduction between the hot and cold junctions and the series electrical resistance (and therefore the Johnson or thermal noise). This means that increasing the number of thermocouples does not necessarily increase the performance after a certain number of thermocouples and that care should be taken when determining the optimum number of thermocouples for a thermopile.

Equation 10.3 is valid for a single thermocouple material. The figure of merit for a thermocouple constructed with two different materials, A and B, having a relative Seebeck coefficient a_{AB}, is defined as follows [31–34]:

$$z = \frac{a_{AB}^2}{\left(\sqrt{\rho_A \kappa_A} + \sqrt{\rho_B \kappa_B}\right)^2} \tag{5}$$

where ρ_A and ρ_B are electrical resistivity of the materials A and B, respectively, in Ω m, and κ_A and κ_B are the thermal conductivity of the materials A and B, respectively, in W/m K.

There are a number of materials that provide very high figures of merit, such as Bi–Te, Bi–Sb–Te, and Pb–Te, and there are thermopile arrays that use some of these materials [35, 36]; however, these materials are not readily available in a CMOS technology. In a CMOS technology, polysilicon is readily available, but polysilicon has a small figure of merit compared with the above materials. Furthermore, in a standard CMOS process, there is only one type of polysilicon (usually n-type), and the other material used to make a thermopile is usually aluminum, which has a very low Seebeck coefficient. Although n-polysilicon/aluminum thermopiles are widely used to implement thermopiles in a standard CMOS process, there are also other options, as described in the next section.

Thermocouple Options in CMOS

There are limited number of layers available as thermocouple materials in CMOS technology. If a standard CMOS process is used for making thermocouples, doping levels, thickness values, and thermal conductivity values of these layers cannot be changed. One should consider the available layers and their thermocouple parameters and decide among various thermocouple options. Fig. 10.5 shows perspective views of four different thermocouple options and approaches used for implementing thermopiles in standard CMOS processes. The most widely used approach is n-poly/aluminum thermopiles [27, 37–39], as shown in Fig. 10.5a. Although aluminum has a very low Seebeck coefficient, this approach is used widely, as it is easy to implement with post-CMOS processes. The approach in Fig. 10.5 (b) is better in terms of Seebeck coefficient, as the p^+-active layer, i.e., the p^+-source/drain implantation, provides a higher Seebeck coefficient than n-poly [40, 41]. However, the p^+-active layer is protected inside an n-well during etching, and this reduces the thermal isolation between the hot and the cold junction, reducing, the overall performance of the thermopile. The third approach, shown in Fig. 10.5 (c), is very attractive, as it provides relatively very high thermocouple coefficients, a_{AB}, with the use of p-poly/n-poly thermo-

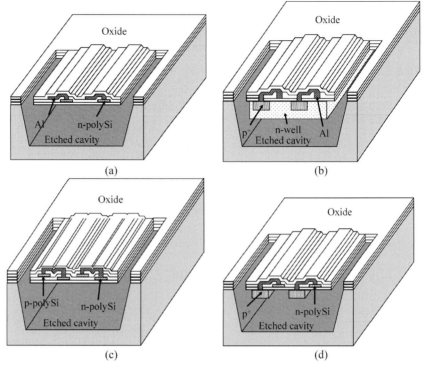

Fig. 10.5 Perspective views of four different approaches used for implementing thermopiles in standard CMOS processes: (a) n-poly/aluminum; (b) p^+-active/aluminum; (c) p-poly/n-poly; (d) n-poly/p^+-active

couples [22, 26, 28, 29, 42] and is easy to implement during post-CMOS processes [similar to Fig. 10.5 (a)]. However, p-poly is not available in most of the standard CMOS processes, and therefore, implementing p-poly/n-poly thermocouples is not possible in most of the standard CMOS processes. The fourth approach, shown in Fig. 10.5 (d), uses n-poly/p^+-active thermocouples [30], maximizing the overall thermocouple Seebeck coefficient. Post-CMOS fabrication steps are slightly more complicated than (a) and (c); however, it can be implemented in any standard n-well CMOS process. Fig. 10.6 shows an SEM photograph of a semiconductor thermopile structure implemented using 20 n-poly/p^+-active thermocouples in a standard n-well CMOS process [30].

Absorber layers are necessary for thermal detectors, as silicon and most of the detector materials are transparent to infrared radiation. There are various layers in literature that are used as absorber layers, such as Gold Black, Titanium, and Titanium Nitride. However, it is also possible to use the standard layers in CMOS as absorbers [43, 44], where it is possible to obtain absorbance of 30–70% in 8–12 μm wavelength regions.

Fig. 10.6 SEM photograph of a semiconductor thermopile structure implemented using 20 n-poly/p$^+$-active thermocouples (10 on each arm) in a standard n-well CMOS process [30]. The structure measures 325 μm × 180 μm in a 1.2 μm CMOS process.

Thermopile Imaging Arrays

Thermopiles are widely used as single elements in a number of applications, mostly as single-point detectors, but array implementations of thermopiles are limited, mainly owing to the large pixel size required for implementing each thermopile pixel. There are some successful focal plane array (FPA) implementations merged with CMOS readout electronics [26–29, 42].

Researchers at the University of Michigan demonstrated a 32×32 FPA where each pixel is implemented with 32 n-poly/p-poly thermocouples on dielectric diaphragms [26]. The pixel size is 375 μm × 375 μm with an active area of 300 μm × 300 μm, i.e., with a fill factor of 64%. The FPA has a responsivity of 15 V/W and a D^* of 1.6×10^7 cm/Hz$^{1/2}$/W. The etching is mainly done from the back side of the wafers, but small etch-cavities are also placed on the front-site of the wafers to achieve heat sinks between the pixels, to prevent the heating of the cold junction and to achieve thermal isolation between adjacent pixels. The process is based on an in-house 3 μm CMOS process, but it is customized to accommodate dielectric window deposition and to implement n-poly and p-poly thermocouples.

Researchers at ETH Zurich also developed thermal imager FPAs using bulk micromachining of CMOS processed wafers [27, 28]. The group first demonstrated a 10×10 FPA with 12 n-poly/aluminum thermocouples that provide a Seebeck coefficient of 108 μV/K. The pixel size is 250 μm × 250 μm. The FPA has a measured responsivity of 5.83 V/W and a D^* of 1.5×10^7 cm/Hz$^{1/2}$/W. The noise-equivalent-temperature difference (NETD) of the FPA is reported to be 530 mK with a low-cost polyethylene Fresnel lens. The FPA is fabricated by back-side etching of the CMOS wafers and stopping at large dielectric diaphragms. The thermal isolation between pixels is achieved by using 25 μm thick and 80 μm wide gold lines on top of the membrane. The group also achieved a 16×16 FPA with the same approach [22] and demonstrated thermal images, as shown in Fig. 10.7.

An interesting array implementation uses post-CCD surface micromachining to implement a 128×128 thermopile array [29]. Each thermopile pixel in the array has 32 pairs of p-type polysilicon and n-type polysilicon thermocouples, and the pixel size is 100 μm × 100 μm with a fill factor of 67%. The reported NETD for this array is 0.5 K with $f=1$ optics. Although these performances are very good for

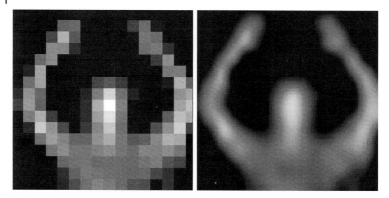

Fig. 10.7 Thermal image of a person, acquired with a 16×16 thermal imager developed at ETH Zurich: (a) unprocessed image; (b) the same image after cubic spline interpolation between pixels. After [22] with permission of the author

a thermopile FPA, the FPA requires vacuum packaging for operation, increasing its cost. Further, CCD technology is not as widespread a technology as CMOS.

Another example of large-format FPA with thermopiles was recently presented by the Nissan Research Center [42]. The FPA has a 120×90 format, and it is monolithically integrated with a CMOS process. Each detector consists of two pairs of n-poly/p-poly thermocouples with a pixel size of 100 μm×100 μm, and front-end bulk-etching is used to thermally isolate the detectors. With vacuum packaging and a precisely patterned Au-black absorbing layer, the detectors provide a responsivity of 3900 V/W after amplification of 2000 V/W. The thermopiles are monolithically integrated with a 0.8 μm CMOS process, but seven special processes are added to the CMOS process, making it a custom-made CMOS process.

The uncooled infrared detectors implemented with thermopiles have an important advantage, as they do not require temperature stabilizers, owing to their inherent differential operation between hot and cold junctions [23]. However, the temperature gradient in the thermopile array may cause significant offsets; therefore, spatial variation in the array temperature should be minimized by careful array design. In one application it has been reported that the heat generated by the on-chip preamplifiers causes noticeable offset in the thermopile outputs [22]. Furthermore, the responsivity of the thermopiles is very low, of the order of 5–15 V/W, and the pixel sizes in these devices are large, such as 250 μm×250 μm, limiting their use for large format detector arrays. Furthermore, these detectors require extra processing for thermal isolation between pixels, such as processing steps to obtain silicon islands [22, 26] or electroplated gold lines [22, 23]. Although there are successful implementations of CMOS thermopile arrays, their low responsivity values and large pixel sizes limit their performance and application areas. These limitations prevent the use of thermopiles for infrared imagers that require large FPAs, and the attention on uncooled infrared detectors has shifted to microbolometers.

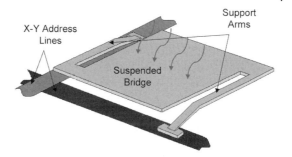

Fig. 10.8 Simplified perspective view of a microbolometer structure obtained using surface micromachining techniques. After [2]

10.2.2.2 Microbolometers

One of the most widely used approaches for uncooled infrared imaging is to use resistive microbolometers implemented using surface micromachined bridges on CMOS-processed wafers [14, 46–55]. Fig. 10.8 shows a simplified perspective view of a microbolometer structure obtained using surface micromachining techniques [2]. Infrared radiation increases the temperature of a material on the thermally isolated and suspended bridge, causing a change in its resistance related to its TCR value. The performance of resistive microbolometers depends on both the temperature sensitive layer along with its thermal isolation and the quality of the readout circuit.

The surface micromachining technique allows the deposition of temperature-sensitive layers with very small thickness, very low mass, and very good thermal isolation on top of bridges over the readout circuit chips. With the removal of the sacrificial layers between the bridge structures and readout circuit chip, suspended and thermally isolated detector structures are obtained, assuming vacuum packaging. There are a number of uncooled infrared cameras on the market with pixel sizes of 50 µm×50 µm and array formats of 320×240, and these cameras provide a noise equivalent temperature difference (NETD) better than 100 mK. At the research level, there are uncooled detector FPAs with pixel sizes as small as 25 µm×25 µm allowing very large format FPAs such as 640×480 to be implemented. The goal is to reach NETD values lower than 10 mK for these large-format FPAs. Therefore, as the uncooled technology develops, many other infrared imaging systems that currently use cooled infrared detector arrays may start to use uncooled detector arrays.

Today's achievements on uncooled infrared technology were initiated by the US Department of Defense in the 1980s, when it gave large classified contracts to both Honeywell and Texas Instruments (TI) to develop two different uncooled infrared technologies [2]. Texas Instruments concentrated on pyroelectric technology [45], whereas Honeywell concentrated on microbolometer technology [46, 47], and they both successfully developed uncooled infrared 320×240 format FPAs. These technologies were unclassified in 1992, and since then many other companies have started working in this technology. These companies include Raytheon [11, 15, 48], BAE (formerly Honeywell) [14], DRS (formerly Boeing) [17, 49], Sarcon [18, 50], Indigo Systems [51, 52], and InfraredVision Technologies [53] in the USA, INO in Canada [54], ULIS in France [55, 56], NEC [57, 58] and Mitsubishi [59–61] in Japan, QuinetiQ [62], in the UK, and XenICs in Belgium [63]. There are also many research institutions working on microbolometer-based uncooled infrared arrays, such as LETI LIR in France [64, 65], IMEC in Belgium [63, 66], University of Texas, Arlington [67–69] and University of Michigan [70] in the USA, KAIST in Korea [71, 72], and METU in Turkey [73–77].

The main advantage of microbolometers over pyroelectric infrared detectors is that they can be monolithically integrated with a CMOS readout circuit. Most of the microbolometers use resistive approaches combined with surface micromachining technology, i.e., the microbolometers are formed using a material with high temperature coefficient of resistivity (TCR) deposited on a suspended and thermally isolated bridge, as shown in Fig. 10.8.

There are efforts to implement resistive surface micromachined microbolometers using many different materials, such as vanadium oxide (VO_x) [11–17, 46–49, 51–54, 57, 58, 80], amorphous silicon (a-Si) [55, 56, 64, 65, 81], polycrystalline silicon-germanium (poly-SiGe) [63, 66], yttrium barium copper oxide (YBa-CuO) [59, 67–69], and metal films [62, 71, 72, 78, 79].

Vanadium oxide is the most widely used material, and it has a high TCR value of 2–3%/K. The main drawback of VO_x is that it is not compatible with a CMOS line and requires a separate fabrication line after the CMOS process to prevent contamination of the CMOS line. In addition, VO_x exhibits large $1/f$ noise due to its non-crystalline structure, limiting its performance. Nevertheless, there are a number of companies fabricating large-format and very high-performance FPAs based on VO_x under license from Honeywell, including Raytheon (formerly Hughes), DRS (formerly Boeing and Rockwell), BAE (formerly Lockheed Martin and Loral), Indigo Systems, InfraredVision Technology, and NEC. Currently, there are large-format FPAs such as 640×480 available with a pixel pitch of 28 or 25 μm using a conventional simple deck structure. Some companies have managed to reduce the pixel pitch to 25 μm for their 640×480 and 320×240 FPAs with the use of double deck structures, first presented by researchers at KAIST [71, 72] in 1998. Fig. 10.9 shows a cross-sectional view of a double deck structure, which is used to obtain a high fill factor in a smaller pixel area [71, 72].

Outstanding performance has been demonstrated with the uncooled technology, with performances close to those of cooled infrared detectors at much lower cost. Fig. 10.10 shows an image taken by the camera developed by Raytheon based on their 640×480 pixel VO_2 FPA, showing excellent resolution and image quality [80]. However, the cost of detectors are still very high for many commercial applications.

Another material that is used successfully to implement high-format FPAs is a-Si, which is a CMOS line compatible material. Detectors with a-Si have been developed by CEA/LETI and transferred to ULIS to implement large-format FPAs such as 320×240 with a pixel pitch down to 35 μm while providing an NETD of 36 mK [55]. Fig. 10.11(a) shows SEM photographs of an a-Si surface micromachined microbolometer detector array from the top, and Fig. 10.11(b) shows a zoomed view of the pixel support arm structure [55]. This figure shows that it is possible to implement very thin layers with a-Si, allowing short arms to be made. Raytheon is also developing a-Si-based uncooled infrared FPAs [81].

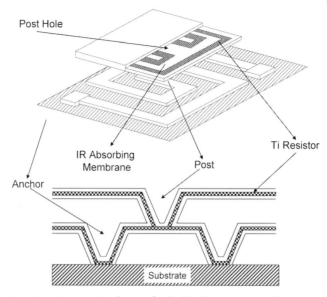

Fig. 10.9 Cross-sectional view of a double deck structure, which is used to obtain a high fill factor in a smaller pixel area. After [72]

Fig. 10.10 An image taken by the camera developed by Raytheon based on their 640 × 480 pixel VO$_2$ FPA, showing excellent resolution and image quality. After [80]

Poly-SiGe is also used for uncooled resistive microbolometers, as it provides a TCR of 2–3%/K. This material was developed by IMEC, Belgium, and subsequently the technology was transferred to XenICs, Belgium. This material needs to be deposited at high temperatures; therefore, it is not easy to integrate with CMOS. A large effort

Fig. 10.11 SEM photographs of a-Si surface micromachined microbolometer detectors: (a) top view; (b) zoomed view of the pixel support arm structure [55]

was made to decrease the deposition temperature to a level that allows its monolithic integration with CMOS. Another problem is the internal stress of the deposited films, making the deposited films buckle. Nevertheless, XenICs has managed to develop 200×1 and 14×14 detector arrays [63]. However, this material exhibits high $1/f$ noise owing to its non-crystalline structure, and it also requires complicated post-CMOS processing to reduce the effects of residual stress.

Another high-TCR material that is used for resistive microbolometer development is YBaCuO [59, 67–69]. This material is attractive as it is deposited at room temperature, and it has low $1/f$ noise. There are efforts at the research level to implement YBaCuO detectors in various substrates, mainly at the University of Texas, Arlington [69], and formerly at Southern Methodist University [67, 68]. A large-format FPA implementation using YBaCuO has been demonstrated by Mitsubishi. A 320×240 format FPA with a 50 µm pitch was demonstrated, and this FPA provides an NETD of 80 mK [59].

There are also efforts to implement resistive microbolometers using metals with the motivation that they are CMOS compatible and their deposition does not require any high-temperature process. However, metal microbolometers have low performance due to the low TCR values of metal films. The most widely used metal for resistive microbolometers is titanium, which has a reported TCR of 0.26%/K when deposited as a thin film [72]. QinetiQ (UK) is currently developing resistive microbolometers [62] based on titanium, where the array size is selected as 64×64 for commercial applications with a pixel size of 75×75 µm. Since the production is done in a commercial CMOS foundry (XFAB), the production costs for these detectors are predicted to be below $100; however, the performance of the detectors is relatively low and the detectors heat up by about 10–20 °C when biasing them to read out their resistance change, decreasing their dynamic range. The use of metals does not bring much advantage compared with post-CMOS processes, and therefore, it is very difficult for metal microbolometers to compete with high-TCR and CMOS line compatible materials, such as a-Si.

All of the surface micromachined resistive bolometers require post-CMOS material deposition and etching steps, as well as a number of high-precision lithography

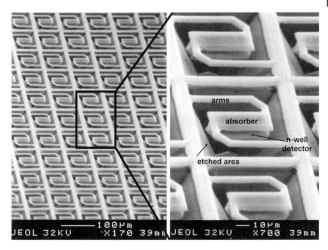

Fig. 10.12 SEM photograph of a fabricated and post-processed array die verifying that all n-well structures are suspended and none of the support arms are broken. The suspended structures remain flat without any extra stress-reducing process steps after CMOS fabrication

steps. An alternative to these approaches was proposed by METU [73], where a resistive microbolometer FPA was demonstrated using the CMOS n-well as the bolometer material, which has a moderate TCR of 0.5–0.7%/K. Fig. 10.12 shows an SEM photograph of a fabricated and post-processed array die verifying that all n-well structures are suspended, and none of the support arms are broken. The suspended structures remain flat without any extra stress-reducing process steps after CMOS fabrication. The pixel size is 80 µm×80 µm, and the fill factor is only 13% owing to large openings required to access silicon for etching after the CMOS process. In addition, the effective pixel TCR is reduced to 0.26%/K owing to the diode implemented within each pixel for easy readout and to the use of polysilicon in the arms to achieve larger thermal isolation. These limitations prevent the implementation of larger FPAs with this approach. The same group demonstrated later that it is possible to reduce the pixel size to 40 µm×40 µm while increasing the fill factor to 44% [74]. The performance of the detectors is increased by implementing the microbolometers using suspended diodes, instead of resistors.

Fig. 10.13 shows the post-processing steps used for implementing (a) resistive-type and (b) diode-type n-well microbolometers. In resistive n-well microbolometers, a number of layers of the CMOS process are removed by proper design in order to obtain openings to the silicon substrate for subsequent post-CMOS bulk etchings while using electrochemical etch-stop in TMAH. This approach requires the openings in the arms to be more than 10 µm to allow etching of the extra metals in these openings, especially in sub-micron CMOS technologies. A different approach is used for diode microbolometers, as shown in Fig. 10.13(b), to reach silicon for post-CMOS anisotropic wet etching. Here, an additional RIE (reactive ion etching) step is included before the bulk micromachining. During RIE,

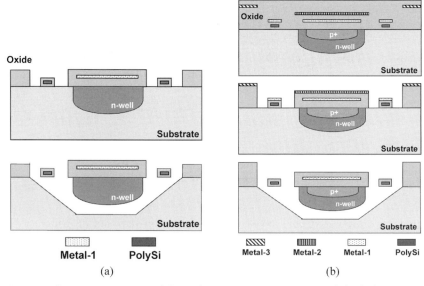

Fig. 10.13 The post-processing used for implementing (a) resistive-type and (b) diode-type n-well microbolometers [74]

CMOS metal layers are used as etch masks to eliminate any critical lithography step after CMOS [101] while achieving narrow etch openings of about 1.2 μm between the support arms. After RIE, the masking metal layers are removed, and then, the bulk silicon underneath the detector is removed while protecting the n-well layer using an electrochemical etch-stop technique in TMAH. This approach was used to implement 64×64 [75] and 128×128 [77] FPAs using a commercial 0.35 μm CMOS process. Fig. 10.14 shows SEM photographs of the fabricated de-

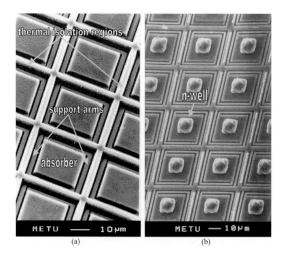

Fig. 10.14 SEM photographs of the fabricated detector array pixels after post-CMOS processing: (a) top view; (b) bottom view after removing pixels from the substrate using a sticky tape [75, 76]

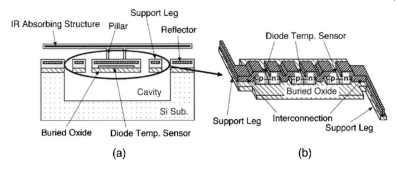

Fig. 10.15 SOI diode microbolometer: (a) schematic of the detector cross-section and (b) schematic of SOI diode-array pixels [28]. Reproduced with permission from the author

tector array pixels after post-CMOS processing: (a) top view and (b) bottom view after removing pixels from the substrate using a sticky tape [75, 77]. The expected NETD values for these arrays are 0.8 and 1 K for 64×64 and 128×128 arrays, respectively, but it is expected to reduce the NETD values to 0.3 K, which is sufficient for a number of low-cost infrared imagers. In comparison with CMOS thermopiles, this approach seems attractive as it allows the implementation of a much smaller pixel and much higher FPA format, while providing similar NETD performances. The performance of diode-based CMOS microbolometers can be increased by connecting a number of suspended diodes in series.

Series-connected diode microbolometers were successfully developed by Mitsubishi using a custom SOI CMOS technology [60]. Fig. 10.15 (a) and (b) show a schematic of the detector cross-section and schematic of SOI diode-array pixels, respectively. Arrays consisting of 320×240 FPA pixels are based on suspended multiple series diodes with 40 μm×40 μm pixel sizes on silicon on insulator (SOI) wafers. The reported NETD value is 120 mK for $f/1$ optics at a scanning rate of 30 frames per second (fps). The pixel structure and the fabrication process of the diode detector FPAs have been changed to increase the fabrication yield. Fig. 10.16 shows the modified fabrication steps and final cross-sectional view of the diode microbolometer, where deep trench etching and trench filling steps are added to remove the silicon underneath the detectors using XeF_2 gas, while still keeping the silicon between the pixels to prevent thermal cross talk. Fig. 10.17 shows an SEM view of the recent SOI diode pixels developed by Mitsubishi [61]. These pixels also utilize high fill factor absorbing structures and this allows the pixel size to be reduced without compromising the performance. The recent work of this group shows that it is possible to obtain an NETD of 87 mK with a 320×240 FPA, while the pixel size is reduced to 28 μm×28 μm [61]. Although this approach provides very uniform arrays with very good potential for low-cost and high-performance uncooled detectors, its fabrication is based on a dedicated in-house SOI CMOS process. Since these detectors cannot be implemented in a standard CMOS process, it would be difficult to reduce their costs to the limits that ultra-low-cost applications require. For ultra-low-cost applications, the best approach would be to implement the detector arrays

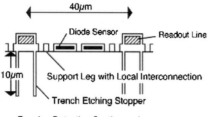
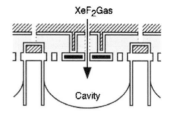

a. Forming Detection Section and Thermally Isolated Structure.

c. Dry Etching of Si Substrate using XeF$_2$ Gas.

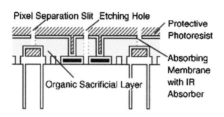
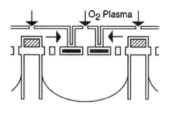

b. Forming Absorbing Membrane.

d. Dry Etching of Organic Sacrificial Layer using O$_2$ Plasma.

Fig. 10.16 Modified fabrication steps and final cross-sectional view of the SOI diode microbolometer, where deep trench etching and trench filling steps are added to remove the silicon underneath the detectors using XeF$_2$ gas, while still keeping the silicon between the pixels to prevent thermal cross talk. Reproduced with permission from the author

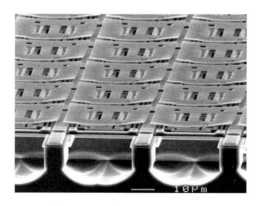

Fig. 10.17 SEM view of the recent SOI diode pixels developed by Mitsubishi. Reproduced with permission from the author

together with their readout circuitry fully in standard CMOS, with some simple post-CMOS micromachining steps [75, 77].

Other than detector fabrication costs, there are other factors that affect the cost of uncooled or thermal infrared detector FPAs, including the need for thermoelectric cooler (TEC) use, vacuum packaging, and optics. Uncooled infrared detectors based on thermopiles do not need the use of TEC, as the detectors respond to the difference between the hot and cold junctions, where the cold junction is usually at ambient temperature. However, microbolometers require the use of TEC, as they also respond to ambient temperature variations. TEC stabilizes the detector temperature to a fixed level around room temperature, and the bolometer resistance only increases with the absorbed heat. This increases the cost of microbolometers. There are recent approaches to eliminate the use of TEC in microbolometers by changing the bias used during the readout of the microbolometer resistance [82, 83]. Differential readout techniques also reduce the need for TEC use [77].

Another issue with microbolometers is self-heating. For reading of the microbolometer resistor or diode, a current is passed through the pixel, and this causes the pixel to heat. There are various approaches to reduce this effect, including differential reading. Vacuum operation is another issue that increases costs. Vacuum operation is very important, especially for surface micromachined microbolometers, as the separation between the substrate and the detector is only 2–2.5 µm. A vacuum environment for the detectors is usually created on a package level, i.e., by vacuuming the package through a small hole and then sealing the hole that is used for pumping air out. This is a costly approach, and recently efforts have been concentrated on vacuum packaging the detectors with wafer level vacuum packaging approaches [84, 85]. Nevertheless, no detector FPA that is vacuum packaged at the wafer level has yet been reported. A satisfactory solution to this bottleneck will be one of the key elements for cost reduction of uncooled infrared detectors. The cost of infrared optics is a further major issue for low-cost infrared imaging applications, and there are continuing efforts to develop low-cost infrared optics with the use of plastics and molding [86], along with the investigation of other low-cost optic materials [87]. There are also efforts to develop microoptical components that can be used together with FPAs to increase their performance [88].

10.3
Thermal Converters

Thermal converters are used for true root mean square (r.m.s.) voltage and AC signal measurements [89–97]. The basic measurement principle of thermal converters is to convert the electrical signal to a heat power and measure the dissipated power as the temperature elevation of a (micromachined) structure with a temperature sensor. This technique allows the true r.m.s. measurement of the signal, independent of the signal waveform. Implementation of such a sensor using CMOS and MEMS technologies provides a number of advantages. First, MEMS techniques allow the heat-generating resistors to be placed on thermally isolated regions, which allows the achievable temperature elevation per dissipated heating power to be maximized. On the other hand, the use of microstructures enables the thermal time constants to be reduced, yielding increased signal bandwidth. The ability to implement these structures in a CMOS process enables one to put readout circuitry together with the sensor, allowing the implementation of precision measurement techniques. Implementing thermal converters with CMOS and MEMS technologies also allows their cost to be reduced.

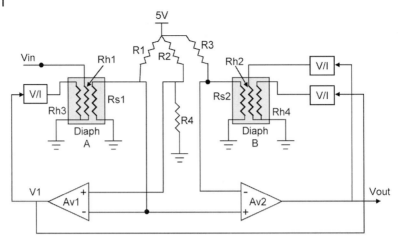

Fig. 10.18 Circuit diagram of the thermal r.m.s. converter developed at the University of Michigan. After [90]

There are various approaches to co-integrate thermal r.m.s. converters with CMOS technologies [89–97]. Devices developed by the US National Institute for Standards and Technology (NIST) use a polysilicon resistor as the heater and thermocouples as the sensor element, with both elements placed on a dielectric diaphragm suspended using front-side bulk micromachining of CMOS-fabricated chips [89]. The bandwidth of the implemented devices is limited to about 100 kHz owing to the relatively large diaphragms of 2×4 mm [89].

A different approach pursued at the University of Michigan utilizes a differential readout mechanism together with a feedback circuit to obtain higher performance in terms of sensitivity and bandwidth and to decrease the influence of ambient temperature variations [90]. Fig. 10.18 shows the circuit diagram of the device. There are two thermally isolated diaphragms and a feedback circuit to keep the temperature of both diaphragms the same and about 20 °C above the ambient temperature. In each diaphragm, there are three resistors, two of which are made with polysilicon and used as heating resistors and one made with gold and used as a temperature sensor. One of the heating resistors (Rh1) on diaphragm A is connected to the unknown AC input signal, while the other heating resistor (Rh3) is driven by the feedback circuit (V_1). Similarly, one of the heating resistors (Rh2) in diaphragm B is connected to the output voltage (V_{out}) generated by the circuit, and the other (Rh4) is connected to the V_1 signal. The V_1 signal is used to keep the temperatures of each diaphragm at a constant value of about 20 °C above ambient regardless of the input signal. The output voltage of the second feedback loop, V_{out}, is the true r.m.s. value of the unknown input signal. If the input signal V_{AC} is zero, then V_{out} is zero. If there is an AC signal on V_{in}, a DC signal V_{out} that provides the same heating power is generated by the circuit, as the feedback circuit forces the temperatures of diaphragms A and B to be the same. Meanwhile, the voltage V_1 is decreased to compensate for the extra heatings from V_{in}

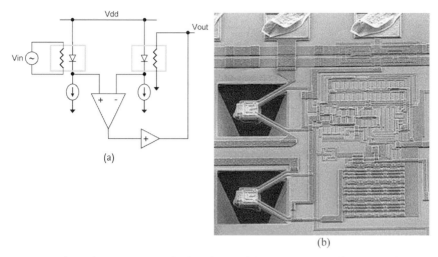

Fig. 10.19 Thermal r.m.s. converter developed at Stanford University: (a) circuit diagram; (b) SEM photograph showing the CMOS circuitry and the suspended n-well structures holding the heater resistors and sensor diodes [91, 92]

and V_{out} on each diaphragm. This approach allows one to obtain true r.m.s. values of input signals up to 20 MHz signal bandwidths with a non-linearity less than 1%. The device measures 3×3.5 mm and is fabricated using a custom 3 µm CMOS process in combination with back-side bulk micromachining, releasing a dielectric membrane with a thermal resistance of 7000 K/W.

The thermal converter developed at Stanford University uses a similar concept compared to the University of Michigan device in terms of keeping the temperatures of two thermally isolated regions the same, where one region is heated with an unknown input AC signal and the other region is heated with a known DC output signal [91, 92]. However, there are a number of differences in the implementation. Fig. 10.19(a) shows the circuit diagram of the thermal r.m.s. converter developed by Stanford University. The temperature sensors are implemented with diodes that are suspended and thermally isolated from the substrate. These suspended microstructures also have heating resistors. When an unknown AC signal is applied to the heating resistor of the microstructure A, the temperature of that region is increased by an amount proportional to the r.m.s. power of the AC signal. The feedback circuitry now generates a DC output voltage, V_{out}, that provides a similar power to the microstructure B to raise its temperature to a value similar to that of microstructure A. Therefore, the r.m.s. value of V_{in} is given by V_{out}. Fig. 10.19(b) shows the fabricated device, where the suspended diode structures are implemented by post-CMOS anisotropic etching from the front of the wafer using TMAH (see also Chapter 1) in combination with an etch-stop technique at the CMOS n-wells. The suspended diode structures provide a high thermal resistance of 37 000 K/W and a low thermal time constant of 5 ms, resulting in a good overall performance. The device has a packaging limited 415 MHz bandwidth, a

60 dB dynamic range, 1% nonlinearity, and 1 mW power dissipation, while occupying a small area of only 400 µm×400 µm.

Substantial work on thermal converters has also been conducted at ETH Zurich [93–96]. A number of devices have been developed with the same basic idea, i.e., a polysilicon resistor on a dielectric diaphragm is used as the heating resistor connected to the unknown AC input signal and thermopiles are used as temperature-sensing elements. The dielectric diaphragms are obtained using either front-side bulk micromachining or back-side bulk micromachining of standard CMOS-fabricated wafers. One of the fabricated devices is reported to provide a linearity error of less than 0.1% below 400 MHz and less than 1% up to 1.2 GHz [93].

Thermal converters to measure even higher frequency microwave signals have also been implemented with post-CMOS bulk micromachining approaches. One such device is implemented using thermocouples in a standard CMOS process. The aluminum/polysilicon thermocouples are suspended using post-CMOS etching with XeF_2 dry etching first, followed by an EDP etch [97]. The fabricated device is reported to operate up to 20 GHz with a non-linearity of ±0.16%, while having a dynamic range of 40 dB and a sensitivity of 5.32 V/W.

10.4
Thermal Flow Sensors

Flow sensors are used to measure the movement of a fluid. They have a wide variety of application areas including biomedical, environmental monitoring, automotive, and process control. There are different techniques to realize a flow sensor depending on the application [1, 98–100], including measuring physical deflection, heat loss, and pressure variation. This section will summarize the state of the art on CMOS-based thermal flow sensors.

Thermal flow sensors measure the fluid flow by measuring the convective heat transport due to the liquid/gas flow [1]. There are three main categories of thermal flow sensors, depending on their operation principles [98]: anemometers, calorimetric flow sensors, and time-of-flight flow sensors.

An anemometer is a device measuring the heat loss due to liquid flow. Generally, anemometers have a single element, which is heated, while its temperature is measured. Owing to the external flow, the heater element temperature, and hence its resistance, change. By monitoring this change, the flow rate can be measured. MEMS-based anemometers employ thermally isolated thin films as the heating element. This heating element is located on a membrane or microbridge to achieve good thermal isolation and high sensitivity. Anemometers can operate in three different modes: constant current, constant temperature, or constant power.

Calorimetric flow sensors and time-of-flight flow sensors require two or more elements. Fig. 10.20 shows the operation principle of a calorimetric flow sensor. In this type of flow sensor, a heating element, such as a polysilicon resistor, heats the fluid. There are two resistive thermometers located on both sides of the heater. In MEMS implementations, these thermometers can be made of thermopile,

10.4 Thermal Flow Sensors

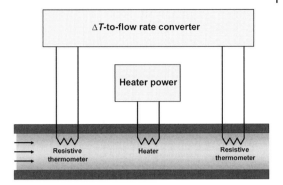

Fig. 10.20 Schematic view of a calorimetric flow sensor. After [98]

polysilicon resistors or p–n junction diodes. Depending on the flow rate and direction, hence, a temperature difference between the thermometer readings develops, giving a measure of the flow rate.

In time-of-flight sensors, a heat pulse is applied to the fluid, and its delay is measured by a temperature sensor located at a different position. This type of flow sensors is more accurate in determining the velocity of the flow [99].

Honeywell developed a microfabricated (but not CMOS integrated) air-flow sensor in 1987 [102] to measure gas velocity, mass flow or differential pressure using the constant-heater-power-mode calorimetric approach. This device measures the temperature difference between upstream and downstream resistors as shown in Fig. 10.20. The sensor is commercially available in a package with a flow inlet employing a dust filter.

Another thermal flow sensor operating in the constant-heater-power-mode was developed at the University of Michigan in 1992 [103, 104]. In this work, an integrated mass-flow sensor with on-chip CMOS interface electronics is presented. This device is an intelligent sensor which is capable of measuring flow velocity, direction, gas type, temperature, and pressure. Fig. 10.21 shows the sensor structure. The sensor operates over the 1 cm/s to 5 m/s range.

Another packaged, miniaturized flow sensor microsystem was reported by Mayer et al. [105] in 1997. The reported system includes a thermal CMOS flow sensor with on-chip power management, signal conditioning, and A/D conversion. Fig. 10.22 shows a schematic view of the system. The flow sensor is based on a membrane made of dielectric layers of the CMOS process. Gate polysilicon is used as the heating resistor and a temperature difference due to heat loss is measured by integrated polysilicon-based thermopiles. The chip consumes 3 mW of power and can measure wind speeds up to 38 m/s with a dynamic range of 65 dB. Fig. 10.23 shows different gas sensors with a similar idea developed by the same group [105–111].

Fig. 10.24 shows a commercial gas flow meter developed by ABB [112], which utilizes the CMOS-based flow sensor developed by Sensirion in Switzerland [113]. The flow sensor is composed of a closed membrane with temperature sensors sandwiched between stress-free dielectric layers. An integrated resistor provides

504 | 10 CMOS-based Thermal Sensors

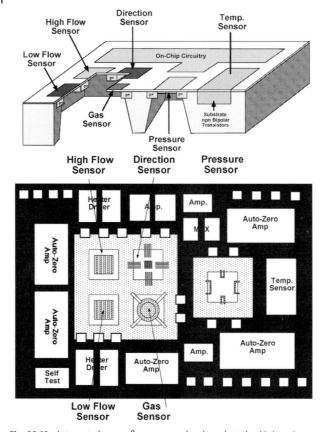

Fig. 10.21 Integrated mass-flow sensor developed at the University of Michigan. After [103]

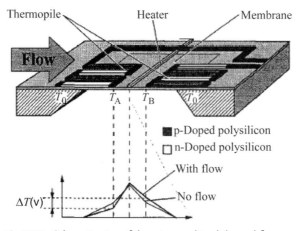

Fig. 10.22 Schematic view of the micromachined thermal flow sensor developed by ETH Zurich. After [105]

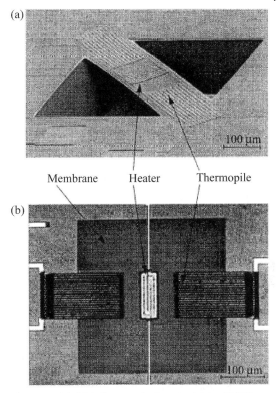

Fig. 10.23 (a) SEM photograph of a microbridge CMOS gas flow sensor. (b) Optical micrograph of a membrane CMOS gas flow sensor developed by ETH Zurich. After [107]

Fig. 10.24 A commercial gas flow meter developed by ABB, which utilizes the CMOS-based flow sensor developed by Sensirion. After [112]

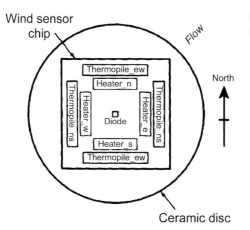

Fig. 10.25 Schematic view of the two-dimensional wind sensor. After [114]

the heating power. The sensors measure the temperature of the gas before and after it flows over the heater and this gives a measure for the gas flow.

A smart wind sensor was reported by Delft University, combining thermal flow sensor and sigma–delta interface electronics on a single chip [114–116]. This chip is capable of measuring wind speed and direction with an accuracy of ±4% and ±2°, respectively, over the range 2–18 m/s. Fig. 10.25 shows the schematic layout of the CMOS wind sensor. It consists of a square silicon substrate on which four heaters, four thermopiles and a diode have been integrated. Each thermopile consists of 12 p^+-diffusion/Al thermocouples and has an estimated sensitivity of 6 mV/K. Heaters are made of polysilicon resistors.

10.5
Acknowledgements

Author would like to thank Dr. Selim Eminoglu for allowing to partially use his Ph.D. thesis for the uncooled infrared detector part and Mr. Zeynel Olgun for allowing to partially use his M.S. thesis for the thermopile detector part. Author also would like to thank Dr. Haluk Kulah for his helps, especially for the section on Thermal Flow Sensors. Author also would like to thank Mr. M. Yusuf Tanrikulu for his helps on drawing of some of the figures and collecting some of the references.

10.6
References

1. H. Baltes, O. Paul, O. Brand, *Micromachined thermally based CMOS microsensors*, Proc. IEEE **1998**, 86, 1660–1678.
2. P. W. Kruse, *Uncooled Thermal Imaging Arrays, Systems and Applications*; SPIE Press, Bellingham, Washington USA, **2001**.
3. A. Rogalski, *Infrared Photon Detectors*, SPIE Press, Bellingham, Washington USA, **1995**.
4. A. Rogalski, *Infrared detectors: status and trends*, Prog. Quantum Electron. **2003**, 27, 59–210.
5. M. E. Greiner, M. Davis, J. W. Devitt, R. Rawe, D. R. Wade, J. Voelker, *State of the art in large-format IR FPA development at CMC Electronics, Cincinnati*, Proc. SPIE **2003**, 5074, 60–71.
6. J. T. Montroy, J. D. Garnett, S. A. Cabelli, M. Loose, A. Joshi, G. Hughes, L. Kozlowski, A. Haas, S. Wong, M. Zandian, A. Chen, J. G. Pasko, M. Farris, C. A. Cabelli, D. E. Cooper, J. M. Arias, J. Bajaj, K. Vural, *Advanced imaging sensors at Rockwell Scientific Company*, Proc. SPIE **2002**, 4721, 212–226.
7. J. Ziegler, M. Bruder, W. Cabanski, H. Figgemeier, M. Finck, P. Menger, Th. Simon, R. Wollrab, *Improved HgCdTe technology for high performance infrared detectors*, Proc. SPIE **2002**, 4721, 242–251.
8. S. D. Gunapala, S. V. Bandara, J. K. Liu, S. B. Rafol, C. A. Shott, R. Jones, S. Laband, J. Woolaway, J. M. Fastenau, A. K. Liu, *9 μm cutoff 640×512 pixel GaAs/ $Al_xGA_{1-x}As$ quantum well infrared photodetector hand-held camera*, Proc. SPIE **2002**, 4721, 144–150.
9. S. Bandara, S. Gunapala, S. Rafol, D. Ting, J. Liu, J. Mumolo, T. Trinh, *Quantum well infrared photodetectors for low background applications*, Proc. SPIE **2002**, 4721, 159–164.
10. W. Cabanski, R. Breiter, W. Rode, J. Ziegler, H. Schneider, M. Walter, M. Fauci, *QWIP LWIR cameras with NETD < 10 mK and improved low frequency drift for long observation time in medicine and research*, Proc. SPIE **2002**, 4721, 165–173.
11. D. Murphy, A. Kennedy, M. Ray, R. Wyles, J. Wyles, J. Asbrock, C. Hewitt, D. Van Lue, T. Sessler, *Resolution and sensitivity improvements for VOx microbolometer FPAs*, Proc. SPIE **2003**, 5074, 402–413.
12. W. Radford, D. Murphy, A. Finch, K. Hay, A. Kennedy, M. Ray, J. Wyles, R. Wyles, J. Varesi, E. Moody, F. Cheung, *Sensitivity Improvements in Uncooled Microbolometer FPAs*, Proc. SPIE **1999**, 3698, 119–130.
13. B. Backer, M. Kohin, A. Leary, R. Blackwell, R. Rumbaugh, *Advances in Uncooled Technology at BAE Systems*, Proc. SPIE **2003**, 5074, 548–556.
14. M. N. Gurnee, M. Kohin, R. Blackwell, N. Butler, J. Whitwam, B. Backer, A. Leary, T. Nielsen, *Developments in uncooled IR technology at BAE Systems*, Proc. SPIE **2001**, 4369, 287–296.
15. D. Murphy, M. Ray, J. Wyles, J. Asbrock, C. Hewitt, R. Wyles, E. Gordon, T. Sessler, A. Kennedy, S. Baur, D. Van Lue, *Performance improvements for VOx microbolometer FPAs*, Proc. SPIE **2004**, 5406, 531–540.
16. R. Blackwell, S. Geldart, M. Kohin, A. Leary, R. Murphy *Recent technology advancements and applications of advanced uncooled imagers*, Proc. SPIE **2004**, 5406, 422–427.
17. P. E. Howard, J. E. Clarke, A. C. Ionescu, C. Li, A. Frankenberger, *Advances in uncooled 1-mil pixel size focal plane products at DRS*, Proc. SPIE **2004**, 5406, 512–520.
18. S. R. Hunter, R. A. Amantea, L. A. Goodman, D. B. Kharas, S. Gershtein, J. R. Matey, S. N. Perna, Y. Yu, N. Maley, L. K. White, *High-sensitivity uncooled microcantilever infrared imaging arrays*, Proc. SPIE **2003**, 5074, 469–480.
19. H. Lakdawala, G. K. Fedder, *CMOS micromachined infrared imager pixel*, in: *11th Int. Conf. on Solid-State Sensors and Actuators (Transducers'01 and Eurosensors XV)*; **2001**, pp. 1548–1551.
20. S. Middelhoek, S. A. Audet, *Silicon Sensors*; Department of Electrical Engineer-

ing, Laboratory of Electronic Instrumentation, TU Delft, **1994**.

21 S. M. Sze, *Semiconductor Sensors*, New York: Wiley, **1994**.

22 A. Schaufelbuhl, *Thermal imagers in CMOS technology*, PhD Thesis, ETH Zurich, **2001**.

23 N. Schneeberger, *CMOS microsystems for thermal presence detection*, PhD Thesis; ETH Zurich, **1998**.

24 N. Schneeberger, O. Paul, H. Baltes, *Optimization of CMOS infrared detector microsystems*, Proc. SPIE **1996**, *2882*, 122–131.

25 W. G. Baer, K. Najafi, K. D. Wise, R. S. Toth, *A 32-element micromachined thermal imager with on-chip multiplexing*, Sens. Actuators A **1995**, *48*, 47–54.

26 A. D. Oliver, K. D. Wise, *A 1024-element bulk-micromachined thermopile infrared imaging array*, Sens. Actuators A **1999**, *73*, 222–231.

27 A. Schaufelbuhl, N. Schneeberger, U. Munch, M. Waelti, O. Paul, O. Brand, H. Baltes, C. Menolfi, Q. Huang, E. Doering, M. Loepfe, *Uncooled low-cost thermal imager based on micromachined CMOS integrated sensor array*, J. Microelectromech. Syst. **2001**, *10*, 503–510.

28 A. Schaufelbuhl, U. Munch, C. Menfoli, O. Brand, O. Paul, Q. Huang, H. Baltes, *256-pixel CMOS-integrated thermoelectric infrared sensor array*, in: *14th IEEE Int. MicroElectroMechanical Systems Conf. (MEMS 2001)*; **2001**, pp. 200–203.

29 T. Kanno, M. Saga, S. Matsumoto, M. Uchida, N. Tsukamoto, A. Tanaka, S. Itoh, A. Nakazato, T. Endoh, S. Tohyama, Y. Yamamoto, S. Murashima, N. Fujimoto, N. Teranishi, *Uncooled infrared focal plane array having 128×128 thermopile detector elements*, Proc. SPIE **1994**, *2269*, 450–459.

30 T. Akin, Z. Olgun, O. Akar, H. Kulah, *An integrated thermopile structure with high responsivity using any standard CMOS technology*, Sens. Actuators A **1998**, *66*, 218–224.

31 P. M. Sarro, *Integrated silicon thermopile infrared detectors*, PhD Thesis; Delft Technical University, **1987**.

32 R. Lenggenhager, *CMOS thermoelectric infrared sensors*, PhD Thesis; ETH Zurich, **1994**.

33 F. Voelklein, *Review of the thermoelectric efficiency of bulk and thin-film materials*, Sens. Mater. **1996**, *8*, 389–408.

34 D. Jaeggi, *Thermal converters by CMOS technology*, PhD Thesis; ETH Zurich, **1996**.

35 M. C. Foote, E. W. Jones, T. Caillat, *Uncooled thermopile infrared detector linear arrays with detectivity greater than 10^9 cm $Hz^{1/2}$/W*, IEEE Trans. Electron Devices **1998**, *45*, 1896–1902.

36 M. C. Foote, S. Gaalema, *Progress towards high-performance thermopile imaging arrays*, Proc. SPIE **2001**, *4369*, 350–354.

37 E. Socher, O. Bochobza-Degani, Y. Nemirovsky, *A novel spiral CMOS compatible micromachined thermoelectric IR microsensor*, J. Micromech. Microeng. **2001**, *11*, 574–576.

38 E. Socher, Y. Sinai, O. Bochobza-Degani, Y. Nemirovsky, *Modeling, design and fabrication of uncooled IR CMOS compatible thermoelectric sensors*, Proc. SPIE **2003**, *4820*, 736–743.

39 C.-H. Du, Z. Lin, C. Lee, *Two-level thermoelectric structures based on CMOS process*, in: *14th European Conference on Solid-State Transducers, Copenhagen*; 33–36, **2000**.

40 A. W. van Herwaarden, P. M. Sarro, *Thermal sensors based on the Seebeck effect*, Sens. Actuators A **1986**, *10*, 321–346.

41 A. W. van Herwaarden, D. C. van Duyn, B. W. van Oudheusden, P. M. Sarro, *Integrated thermopile sensors*, Sens. Actuators A **1989**, *21–23*, 621–630.

42 M. Hirota, Y. Nakajima, M. Saito, F. Satou, M. Uchiyama, *120×90 element thermopile array fabricated with CMOS technology*, Proc. SPIE **2003**, *4820*, 239–249.

43 N. Schneeberger, O. Paul, H. Baltes, *Optimized structured absorbers for CMOS infrared detectors*, in: *Solid-State Sensors and Actuators (Transducers '95, Eurosensors IX)*; **1995**, pp. 648–651.

44 N. Schneeberger, *CMOS microsystems for thermal presence detection*, PhD Thesis; ETH Zurich, **1998**.

45 C. M. Hanson, H. E. Beratan, R. A. Owen, M. Corbin, S. McKenny, *Un-*

cooled thermal imaging at Texas Instruments, Proc. SPIE **1992**, *1735*, 17–26.

46 R. A. Wood, C. J. Han, P. W. Kruse, *Integrated uncooled IR detector imaging arrays*, in: *IEEE Solid State Sensor and Actuator Workshop*; **1992**, pp. 132–135.

47 R. A. Wood, *Uncooled thermal imaging with monolithic silicon focal arrays*, Proc. SPIE **1993**, *2002*, 322–329.

48 D. Murphy, M. Ray, R. Wyles, J. Asbrock, N. Lum, J. Wyles, C. Hewitt, A. Kennedy, D. V. Lue, *High sensitivity 25 μm microbolometer FPAs*, Proc. SPIE **2002**, *4721*, 99–110.

49 P. E. Howard, J. E. Clarke, A. C. Ionescu, C. Li, *DRS U6000 640× 480 VO$_x$ uncooled IR focal plane*, Proc. SPIE **2002**, *4721*, 48–55.

50 R. Amantea, L. A. Goodman, F. Pantuso, D. J. Sauer, M. Varghese, T. S. Villani, L. K. White, *Progress towards an uncooled IR imager with 5 mK NEΔT*, Proc. SPIE **1998**, *3436*, 647–659.

51 W. Parish, J. T. Woolaway, G. Kincaid, J. L. Heath, J. D. Frank, *Low cost 160× 128 uncooled infrared sensor array*, Proc. SPIE **1998**, *3360*, 111–119.

52 W. A. Terre, R. F. Cannata, P. Franklin, A. Gonzalez, E. Kurth, W. Parrish, K. Peters, T. Romeo, D. Salazar, R. Van Ysseldyk, *Microbolometer production at Indigo Systems*, Proc. SPIE **2004**, *5406*, 557–565.

53 K. A. Hay, D. Van Deusen, T. Y. Liu, W. A. Kleinhans, *Uncooled focal plane array detector development at InfraredVision Technology Corp.*, Proc. SPIE **2003**, *5074*, 491–499.

54 T. D. Pope, H. Jeronimek, C. Alain, F. Cayer, B. Tremblay, C. Grenier, P. Topart, S. LeClair, F. Picard, C. Larouche, B. Boulanger, A. Martel, Y. Desroches, *Commercial and custom 160× 120, 256× 1 and 512× 3 pixel bolometric FPAs*, Proc. SPIE **2002**, *4721*, 64–74.

55 E. Mottin, A. Bain, J. Martin, J. Ouvrier-Buffet, S. Bisotto, J. J. Yon, J. L. Tissot, *Uncooled amorphous silicon technology enhancement for 25 μm pixel pitch achievement*, Proc. SPIE **2003**, *4820*, 200–207.

56 E. Mottin, J. Martin, J. Ouvrier-Buffet, M. Vilain, A. Bain, J. Yon, J. L. Tissot, J. P. Chatard, *Enhanced amorphous silicon technology for 320× 240 microbolometer arrays with a pitch of 35 μm*, Proc. SPIE **2001**, *4369*, 250–256.

57 Y. Tanaka, A. Tanaka, K. Iida, T. Sasaki, S. Tohyama, A. Ajisawa, A. Kawahara, S. Kurashina, T. Endoh, K. Kawano, K. Okuyama, K. Egashira, H. Aoki, N. Oda, *Performance of 320× 240 uncooled bolometer-type infrared focal plane arrays*, Proc. SPIE **2003**, *5074*, 414–424.

58 S. Tohyama, M. Miyoshi, S. Kurashina, N. Ito, T. Sasaki, A. Ajisawa, N. Oda, *New thermally isolated pixel structure for high-resolution uncooled infrared FPAs*, Proc. SPIE **2004**, *5406*, 428–436.

59 H. Wada, T. Sone, H. Hata, Y. Nakaki, O. Kaneda, Y. Ohta, M. Ueno, M. Kimata, *YBaCuO uncooled microbolometer IR FPA*, Proc. SPIE **2001**, *4369*, 297–304.

60 T. Ishikawa, M. Ueno, Y. Nakaki, K. Endo, Y. Ohta, J. Nakanishi, Y. Kosasayama, H. Yagi, T. Sone, M. Kimata, *Performance of 320× 240 uncooled IRFPA with SOI diode detectors*, Proc. SPIE **2000**, *4130*, 1–8.

61 Y. Kosasayama, T. Sugino, Y. Nakaki, Y. Fujii, H. Inoue, H. Yagi, H. Hata, M. Ueno, M. Takeda, M. Kimata, *Pixel scaling for SOI-diode uncooled infrared focal plane arrays*, Proc. SPIE **2004**, *5406*, 504–511.

62 P. A. Manning, J. P. Gillham, N. J. Parkinson, T. P. Kaushal, *Silicon foundry microbolometers: the route to the mass-market thermal imager*, Proc. SPIE **2004**, *5406*, 465–472.

63 V. N. Leonov, Y. Creten, P. De Moor, B. Du Bois, C. Goessens, B. Grietens, P. Merken, N. A. Perova, G. Ruttens, C. A. Van Hoof, A. Verbist, J. P. Vermeiren, *Small two-dimensional and linear arrays of polycrystalline SiGe microbolometers at IMEC-XenICs*, Proc. SPIE **2003**, *5074*, 446–457.

64 E. Mottin, A. Bain, J. Martin, J. Ouvrier-Buffet, S. Bisotto, J. J. Yon, J. L. Tissot, *Uncooled amorphous silicon technology enhancement for 25 μm pixel pitch achievement*, Proc. SPIE **2002**, *4820*, 200–207.

65 E. Mottin, J. Martin, J. Ouvrier-Buffet, M. Vilain, A. Bain, J. Yon, J. L. Tis-

SOT, J. P. CHATARD, *Enhanced amorphous silicon technology for 320×240 microbolometer arrays with a pitch of 35 μm*, Proc. SPIE **2001**, *4369*, 250–256.

66 S. SEDKY, P. FIORINI, K. BAERT, L. HERMANS, R. MERTENS, *Characterization and optimization of infrared poly-SiGe bolometers*, IEEE Trans. Electron Devices **1999**, *46*, 675–682.

67 M. ALMASRI, D. P. BUTLER, Z. C. BUTLER, *Semiconducting YBCO bolometers for uncooled IR detection*, Proc. SPIE **2000**, *4028*, 17–26.

68 A. JAHANZEB, C. M. TRAVERS, Z. CELIK-BUTLER, D. P. BUTLER, S. G. TAN, *A semiconductor YBaCuO microbolometer for room temperature IR imaging*, IEEE Trans. Electron Devices **1997**, *44*, 1795–1801.

69 A. YILDIZ, Z. CELIK-BUTLER, D. P. BUTLER, *Microbolometers on a flexible substrate for infrared detection*, IEEE Sens. J. **2004**, *4*, 112–117.

70 C. C. LIU, C. H. MASTERANGELO, *A CMOS uncooled heat-balancing infrared imager*, IEEE J. Solid State Circuits **2000**, *35*, 527–535.

71 H. K. LEE, J.-B. YOON, E. YOON, S.-B. JU, Y.-J. YONG, W. LEE, S.-G. KIM, *A high fill-factor IR bolometer using multi-level electrothermal structures*, in: *Int. Electron Devices Meeting (IEDM)*; **1998**, p. 463.

72 H.-K. LEE, J.-B. YOON, E. YOON, S.-B. JU, Y.-J. YONG, W. LEE, S.-G. KIM, *A high fill-factor infrared bolometer using micromachined multilevel electrothermal structures*, IEEE Trans. Electron Devices **1999**, *46*, 1489–1491.

73 D. SABUNCUOGLU TEZCAN, S. EMINOGLU, T. AKIN, *A low cost uncooled infrared microbolometer detector in standard CMOS technology*, IEEE Trans. Electron Devices **2003**, *50*, 494–502.

74 S. EMINOGLU, D. SABUNCUOGLU TEZCAN, M. Y. TANRIKULU, T. AKIN, *Low-cost uncooled infrared detectors in CMOS process*, Sens. Actuators A **2003**, *109*, 102–113.

75 S. EMINOGLU, M. Y. TANRIKULU, T. AKIN, *A low-cost 64×64 uncooled infrared detector in standard CMOS*, in: *Int. Conf. on Solid-State Sensors, Actuators and Microsystems (TRANSDUCERS'03)*; **2003**, pp. 316–319.

76 S. EMINOGLU, M. Y. TANRIKULU, D. SABUNCUOGLU TEZCAN, T. AKIN, *A low-cost, small pixel uncooled infrared detector for large focal plane arrays using a standard CMOS process*, Proc. SPIE **2002**, *4721*, 111–121.

77 S. EMINOGLU, M. Y. TANRIKULU, T. AKIN, *Low-cost uncooled infrared detector arrays in standard CMOS*, Proc. SPIE **2003**, *5074*, 425–436.

78 A. TANAKA, S. MATSUMOTO, N. TSUKAMOTO, S. ITOH, T. ENDOH, A. NAKAZATO, Y. KUMAZAWA, M. HIJIKAWA, H. GOTOH, T. TANAKA, N. TERANISHI, *Silicon IC process compatible bolometer infrared focal plane array*, in: *Int. Conf. on Solid-State Sensors and Actuators (TRANSDUCERS'95)*; **1995**, pp. 632–635.

79 J. S. SHIE, Y. M. CHEN, M. O. YANG, B. C. S. CHOU, *Characterization and modelling of metal-film microbolometer*, J. Microelectromech. Syst. **1996**, *5*, 298–306.

80 J. S. ANDERSON, D. BRADLEY, C. W. CHEN, R. CHIN, H. GONZALEZ, R. G. HEGG, K. KOSTRZEWA, C. LE PERE, S. TON, A. KENNEDY, D. F. MURPHY, D. RAY, R. WYLES, J. E. MILLER, G. W. NEWSOME, *Advances in uncooled systems applications*, Proc. SPIE **2003**, *5074*, 557–563.

81 T. R. SCHIMERT, N. CUNNINGHAM, G. L. FRANCISCO, R. W. GOOCH, J. GOODEN, P. MCCARDEL, B. E. NEAL, B. RITCHEY, J. RIFE, A. J. SYLLAIOS, J. H. TREGILGAS, J. F. BRADY III, J. GILSTRAP, S. J. ROPSON, *Low-cost low-power uncooled 120×160 a-Si-based microinfrared camera for law enforcement applications*, Proc. SPIE **2001**, *4232*, 187–194.

82 W. J. PARRISH, J. T. WOOLAWAY, *Improvements in uncooled systems using bias equalization*, Proc. SPIE **1999**, *3698*, 748–755.

83 P. E. HOWARD, J. E. CLARKE, C. LI, J. W. YANG, W. Y. WONG, A. BOGOSYAN, *Recent advances in TEC-less uncooled FPA sensor operation*, Proc. SPIE **2003**, *5074*, 527–536.

84 B. E. COLE, R. E. HIGASHI, J. A. RIDLEY, R. A. WOOD, *Integrated vacuum packaging for low-cost light-weight uncooled microbolometer arrays*, Proc. SPIE **2003**, *4369*, 235–239.

85 A. ASTIER, A. ARNAUD, J.-L. OUVRIER-BUFFET, J.-J. YON, E. MOTTIN, *Advanced packaging development for very low cost uncooled IRFPA*, Proc. SPIE **2004**, *5406*, 412–421.

86 N. E. Claytor, R. N. Claytor, *Polymer imaging optics for the thermal infrared*, Proc. SPIE **2004**, *5406*, 107–113.

87 Y. M. Guimond, J. Franks, Y. Bellec, *Comparison of performances between GASIR molded optics and existing IR optics*, Proc. SPIE **2004**, *5406*, 114–120.

88 D. Krogmann, H. D. Tholl, *Infrared microoptics technologies*, Proc. SPIE **2004**, *5406*, 121–132.

89 J. R. Kinard, D. B. Novotny, T. E. Lipe, D.-X. Huang, *Development of thin-film multijunction thermal converters at NIST*, IEEE Trans. Instrum. Meas. **1997**, *46*, 347–351.

90 E. Yoon, K. D. Wise, *A wide band monolithic rms-dc converter using micromachined diaphragm structures*, IEEE Trans. Electron Devices **1994**, *41*, 1666–1668.

91 E. H. Klaassen, R. J. Reay, G. T. A. Kovacs, *Diode-based thermal rms converter with on-chip circuitry fabricated using standard CMOS technology*, in: *8th Int. Conf. of Solid-State Sensors and Actuators (Transducers'95 and Eurosensors IX), Stockholm*; **1995**, pp. 154–157.

92 E. H. Klaassen, R. J. Reay, G. T. A. Kovacs, *Diode-based thermal r.m.s. converter with on-chip circuitry fabricated using CMOS technology*, Sens. Actuators A **1996**, *52*, 33–40.

93 D. Jaeggi, H. Baltes, D. Moser, *Thermoelectric AC power sensor by CMOS technology*, IEEE Electron Device Lett. **1992**, *13*, 336–368.

94 D. Jaeggi, C. AzeredoLeme, P. O'Leary, H. Baltes, *Improved CMOS AC power sensor*, in: *7th Int. Conf. on Solid-State Sensors and Actuators (Transducers'95), Yokohama*; **1993**, pp. 462–465.

95 D. Jaeggi, J. Funk, A. Haberli, H. Baltes, *Overall system analysis of a CMOS thermal converter*, in: *8th Int. Conf. on Solid-State Sensors and Actuators (Transducers'95 and Eurosensors IX), Stockholm*; **1995**, pp. 112–115.

96 D. Jaeggi, *Thermal converters by CMOS technology*, PhD Thesis; ETH Zurich, **1996**.

97 V. Milanovic, M. Gaitan, E. D. Bowen, N. H. Tea, M. E. Zaghloul, *Thermoelectric power sensor for microwave applications by commercial CMOS fabrication*, IEEE Electron Device Let. **1997**, *18*, 450–452.

98 A. Rasmussen, M. E. Zaghloul, *In the flow with MEMS*, IEEE Circuits Devices Mag. **1998**, *14*, 12–24.

99 M. Elwenspoek, *Thermal flow microsensors*, in: *Semiconductor Conference, 1999. Int. CAS '99 Proc.* **1999**, Vol. 2, 423–435.

100 B. W. van Oudheusden, *Silicon flow sensors*, Control Theory Applic., IEE Proc. D **1988**, *135*, 373–380.

101 G. K. Fedder, S. Santhanam, M. L. Reed, S. C. Eagle, D. F. Guillou, M. S.-C. Lu, L. R. Carley, *Laminated high-aspect ratio microstructures in a conventional CMOS process*, Sens. Actuators A **1996**, *57*, 103–110.

102 R. G. Johnson, R. E. Higashi, *A highly sensitive silicon chip microtransducer for air flow and differential pressure sensing applications*, Sens. Actuators **1987**, *11*, 63–72.

103 E. Yoon, K. D. Wise, *An integrated mass flow sensor with on-chip CMOS interface circuitry*, IEEE Trans. Electron Devices **1992**, *39*, 1376–1386.

104 E. Yoon, K. D. Wise, *A multi-element monolithic mass flowmeter with on-chip CMOS readout electronics*, in: *Solid-State Sensors and Actuator Workshop*; **1990**, pp. 161–164.

105 F. Mayer, A. Haberli, H. Jacobs, G. Ofner, O. Paul, H. Baltes, *Single-chip CMOS anemometer*, in: *Int. Electron Devices Meeting*; **1997**, pp. 895–898.

106 F. Mayer, O. Paul, H. Baltes, *Flip-chip packaging for thermal CMOS anemometers*, in: *10th IEEE Int. MicroElectroMechanical Systems Conf. (MEMS '97)*; **1997**, pp. 203–206.

107 F. Mayer, G. Salis, J. Funk, O. Paul, H. Baltes, *Scaling of thermal CMOS gas flow microsensors: experiment and simulation*, in: *9th IEEE Int. MicroElectroMechanical Systems Conf. (MEMS '96)*; **1996**, pp. 116–121.

108 F. Mayer, O. Paul, H. Baltes, *Influence of design geometry and packaging on the response of thermal CMOS flow sensors*, in: *8th Int. Conf. on Solid-State Sensors and Actuators (TRANSDUCERS '95)*; **1995**, pp. 528–531.

109 D. Moser, H. Baltes, *A high sensitivity CMOS gas flow sensor on a thin dielectric membrane*, Sens. Actuators A **1993**, *37/38*, 33–37.

110 D. Moser, R. Lenggenhager, G. Wachutka, H. Baltes, *Fabrication and modelling of CMOS microbridge gas-flow sensors,* Sens. Actuators B **1992**, *6*, 165–169.

111 D. Moser, R. Lenggenhager, H. Baltes, *Silicon gas flow sensor using industrial CMOS and bipolar IC technology,* Sens. Actuators A **1991**, *25–27*, 577–581.

112 D. Matter, T. Kleiner, B. Kramer, B. Sabbattini, *Microsensor-based gas flow meter wins innovation prize,* ABB Rev. **2003**, March, 49–50.

113 www.sensirion.com

114 K. A. A. Makinwa, J. H. Huijsing, *A smart wind sensor using thermal sigma-delta modulation techniques,* Sens. Actuators A **2002**, *97/98*, 15–20.

115 K. A. A. Makinwa, J. H. Huijsing, *A smart CMOS wind sensor,* in: *IEEE Int. Solid-State Circuits Conf. (ISSCC 2002);* **2002**, pp. 352–544.

116 K. A. A. Makinwa, J. H. Huijsing, *A wind-sensor with integrated interface electronics,* in: *IEEE Int. Symp. on Circuits and Systems (ISCAS 2001);* **2001**, pp. 356–359.

11
Circuit and System Integration

C. Hagleitner, IBM Research – Zurich Research Laboratory, Ruschlikon, Switzerland
K.-U. Kirstein, Physical Electronics Laboratory, ETH Zurich, Zurich, Switzerland

Abstract
CMOS-integrated micro- and nanosystems offer the possibility to co-integrate sensors and actuators together with the read-out/driving circuitry on a single chip. This chapter provides some insight into the design of integrated CMOS electronics (analog frontend, analog-to-digital converters, digital interfaces, calibration circuitry) for the sensors and actuators described in this book.

Keywords
Sensor interface circuitry; analog frontend; serial interface; resolution; offset; signal-to-noise ratio (SNR); linearity; bandwidth; sensitivity; drift; Wheatstone bridge; instrumentation amplifier; capacitive read-out; analog filters; switched-capacitor (SC) filters; digital-to-analog converters (DAC); analog-to-digital converters (ADC); digital filter; decimation filter; parameter sensitivity; 6 sigma design; power consumption; design for testability; physical design verification

11.1	Introduction 515
11.1.1	Fabrication Technology for the Interface Circuitry 515
11.1.1.1	CMOS Technology and Device-fabrication Approach 515
11.1.1.2	Technology Node 517
11.1.1.3	Analog Options 518
11.1.2	Specification of the Interface and System Architecture 520
11.2	**Analog Frontend 521**
11.2.1	Key Properties of Analog Frontends 522
11.2.1.1	Static Properties 522
11.2.1.2	Dynamic Properties 523

Advanced Micro and Nanosystems. Vol. 2. CMOS – MEMS.
Edited by H. Baltes, O. Brand, G. K. Fedder, C. Hierold, J. Korvink, O. Tabata
Copyright © 2005 WILEY-VCH Verlag GmbH & Co. KGaA, Weinheim
ISBN: 3-527-31080-0

11.2.1.3	System Properties	525
11.2.1.4	Fully Differential Design	526
11.2.2	Resistive Sensors	527
11.2.3	Capacitive Sensors	530
11.2.3.1	Sensing with AC Modulation	530
11.2.3.2	Switched-capacitor Amplifiers	531
11.2.4	Thermoelectric, Piezoelectric and Pyroelectric Devices	534
11.2.4.1	Thermoelectric Devices	534
11.2.4.2	Piezo- and Pyroelectric Sensors	538
11.3	**Circuitry Building Blocks for CMOS MEMS/NEMS**	**539**
11.3.1	Analog Filters	539
11.3.1.1	Biquad Realization	540
11.3.1.2	SC Filter Architecture	541
11.3.1.3	GmC Filter Architecture	543
11.3.2	Analog-to-Digital Converters (ADCs)	544
11.3.2.1	ADCs for Individual Sensors	544
11.3.2.2	ADCs for Sensor Arrays	545
11.3.2.3	Sigma–Delta ADCs	546
11.3.3	Digital-to-Analog Converters	549
11.3.4	Digital Filters	551
11.3.5	Calibration and Data Interfaces	554
11.3.6	Current and Voltage References	554
11.3.6.1	Temperature Dependence of the Base-emitter Voltage	555
11.3.6.2	Bandgap References in Standard CMOS Technologies	557
11.3.6.3	Current References in Standard CMOS Technologies	558
11.4	**System Integration**	**559**
11.4.1	Parameter Sensitivity/Power Consumption at the System Level	559
11.4.2	Design for Testability	561
11.4.3	Microsystem Verification	564
11.5	**Example: a Single-Chip Gas-detection System**	**566**
11.5.1	Design Flow and Layout Verification	567
11.5.2	Capacitive Chemical Sensor	568
11.5.3	Calorimetric Sensor	570
11.5.4	Mass-sensitive Resonant Cantilever	570
11.5.4.1	Resonant-beam Oscillator	571
11.5.5	Controller and Digital Interface	572
11.5.5.1	Serial Interface	572
11.5.5.2	Digital Controller	573
11.6	**References**	**573**

11.1
Introduction

One of the most important advantages of CMOS-integrated micro- and nanosystems over hybrid solutions is the possibility of co-integrating sensors and actuators together with the necessary read-out/driving circuitry on a single chip. The purpose of this chapter is to provide some insight into the system architecture of the CMOS electronics and to give an overview of the most important building blocks needed to interface with the sensors and actuators described in this book.

The chapter starts with a brief discussion of the IC technology, which is used to fabricate the interface circuitry. Next, a general system architecture is introduced and the two chip interfaces (chip ↔ external signal-processing unit and chip ↔ micromachined devices) are defined. In Section 11.2 we define the key requirements for the analog frontend and frontend circuitry for various sensing principles. In Section 11.3, the most important building blocks for the overall system, e.g. filters and analog-to-digital converters (ADC) are presented. This is followed in Section 11.4 by a discussion of the various requirements to simulate, design and verify a complete system. An exemplary implementation of a CMOS single-chip gas detection system concludes the chapter (Section 11.5).

11.1.1
Fabrication Technology for the Interface Circuitry

The main focus in the initial phase of most micro- and nanosystem projects is on the development of the device technology. The circuit technology is chosen at a later stage. In the following, we analyze the limitations of this choice imposed by the device-fabrication approach.

11.1.1.1 CMOS Technology and Device-fabrication Approach
The three different approaches that can be used to fabricate the sensors and actuators in CMOS-based micro- and nanosystems (pre-CMOS micromachining, intra-CMOS micromachining and post-CMOS micromachining) were discussed in Chapter 1. Each approach has some implications for the choice of IC technology:

Intra-CMOS Micromachining
The intra-CMOS micromachining approach leads to the most stringent constraints for the choice of the CMOS technology. The CMOS technology is chosen at an early stage and must be adapted and further developed together with the MEMS device. Major development work is required to switch to a more advanced CMOS technology. The interruption of the CMOS process module makes it difficult to have CMOS fabrication and micromachining in different locations. Furthermore, the high-temperature steps required for the fabrication of the micromachined devices affect the

thermal budget of the CMOS process, which makes it almost impossible to combine them with modern deep sub-micron CMOS technologies [1].

Pre-CMOS Micromachining

Pre-CMOS micromachining provides more flexibility in the choice of the CMOS technology because the CMOS process module is not interrupted and the fabrication of the micromachined devices does not affect the thermal budget of the CMOS process [1]. Despite the greater flexibility in the choice of the CMOS process, major development work is still required to prepare the micromachined wafers before they can be introduced as starting material into a standard CMOS line and to contact the devices. Furthermore, the CMOS process must be re-qualified for the micromachined wafers by the CMOS foundry. Finally, most pre-CMOS micromachining modules also require post-CMOS processing to release the devices after completion of the CMOS process.

Post-CMOS Micromachining

Post-CMOS micromachining offers the greatest flexibility in the choice of the underlying IC fabrication technology. Any CMOS technology can be used as long as the post-processing steps do not degrade the performance of the integrated circuitry and no expensive re-qualification of the CMOS process by the foundry is required. This drastically reduces the cost for fabless design houses, because they only need to guarantee that their process steps do not affect the characteristics of the transistors and of the passive devices in the CMOS process. There are only few restrictions that limit the range of available technologies. For devices based on bulk-CMOS micromachining, the choice of the substrate material is critical. Some of them use, e.g., a p^{++} diffusion in an n-type substrate (instead of the standard p-type substrate, which is used for most CMOS technologies) as an etch stop [2] or require a defined oxygen content to achieve a sufficient quality of back-side etching [3, 4]. Sub-micron CMOS technologies are very sensitive to the choice of the substrate material. An expensive and time-consuming qualification procedure is required if a new substrate material needs to be introduced. Therefore, a qualified CMOS technology should be chosen that is based on the substrate material required. Another issue in post-CMOS micromachining is the planarity of the area used to fabricate the MEMS devices. For technologies on a scale larger than 0.6 µm, typically a maximum of three metallization layers are available and the surface of the wafer might exhibit topography features of a few micrometers. On the other hand, modern CMOS processes with up to eight metallization layers make extensive use of planarization steps, which reduces the topography to the sub-micrometer regime (without accounting for the passivation opening).

In most technologies based on pre-CMOS and intra-CMOS micromachining, no circuitry can be placed in the area where the MEMS devices are located. Furthermore, the size of analog readout circuitry cannot be reduced according to the scaling laws that apply to digital circuitry. This reduces the savings in area usually achieved when a more advanced technology (smaller minimum feature size) is

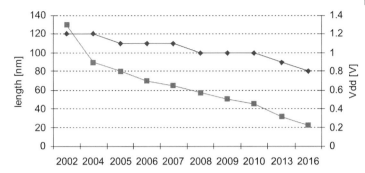

Fig. 11.1 ITRS roadmap [5] for the scaling of the minimum feature size and the digital supply voltage.

used. Therefore, there is less (cost-) pressure to follow the scaling trend of the digital CMOS technologies.

11.1.1.2 Technology Node

The scaling of CMOS technology and DRAM devices has been following the well-known Moore's law for almost three decades. Fig. 11.1 shows the projections for future technology nodes according to the ITRS road map [5]. ASIC technology always lags behind this trend by one or two generations. The majority of current CMOS integrated micro- and nanosystems are based on significantly older technologies. The choice of CMOS technology which is used for a MEMS device is determined by the amount of digital on-chip circuitry, the power-consumption specifications and the costs of mask fabrication.

Digital Circuitry

The percentage of the chip area that is occupied by digital circuitry chiefly determines the savings in area that can be achieved by using a more advanced CMOS technology. The size of the analog circuitry cannot be reduced significantly because the transistor size is mainly determined by the transconductance, noise and matching specifications. Similar arguments apply to the micromachined devices. Therefore, only a few applications with a large amount of on-chip digital circuitry (e.g. large sensor arrays for imaging or probe storage [6, 7]) will benefit from technologies with feature sizes below 0.5 µm. An exception to this rule is RF design, where the increase in transit frequency (f_t) and the reduction in parasitic capacitance of advanced CMOS technologies permit designs in a frequency range that until a few years ago was reserved for bipolar and BiCMOS technologies.

Power Consumption

The dynamic power consumption of the digital circuitry exhibits a quadratic dependence on the supply voltage. In most analog building blocks, a fixed current is required to obtain the specified transconductance and slew rate and therefore only a linear reduction of the power consumption can be achieved through supply-voltage scaling. As a result, the reduced threshold voltage of the analog transistors and small digital supply voltages of sub-micron CMOS technologies (see Fig. 11.1) allow a significant reduction of the overall power consumption at the circuit level for advanced CMOS technologies.

Mask-fabrication Cost

As a consequence of the exponential increase of the mask-fabrication costs for advanced CMOS technologies, these costs have become a major obstacle for the scaling of MEMS technologies. A mask set for a typical 130 nm CMOS process costs approx. 600 000 $ [8] and will roughly double for every technology node (more expensive equipment for mask fabrication and larger number of masks). There are – at best – a few applications of MEMS devices that offer the production quantities required to justify the large initial investment for a 90 nm CMOS technology. Furthermore, the development cycle of CMOS-integrated devices usually requires several re-spins and mask sets because many of the device characteristics (e.g. Young's modulus of CMOS layers, residual stress of released structures) can only be determined through experiments and the simulation capabilities for CMOS MEMS systems are not as advanced as for modern circuit simulation tools (post-layout extraction, etc.).

11.1.1.3 Analog Options

Most ASIC technologies offer some additional options that enable or simplify the design of high-performance analog circuitry. The most popular extensions and their applications for the interface circuitry in CMOS-based micro- and nanosystems are discussed in this section. Each of the following extensions requires extra masks and therefore increases the fabrication cost. Many analog functions can also be realized in standard CMOS technologies, if advanced analog design techniques and careful system design are applied (e.g. transistor-only implementations of ADCs or filters [9–11]).

BiCMOS Technologies

BiCMOS technologies offer (isolated) bipolar transistors in addition to standard CMOS devices. This extension typically requires approximately five additional masks and some major changes in the processing of the original CMOS technology (doping, thermal budget). The bipolar transistors are advantageous for low-noise amplifiers, bandgap references and high-frequency design (some process technologies offer transistors based on SiGe, a material that exhibits a larger bandgap). The larger values and the linear dependence of the bipolar transistor transconductance g_m on the bias current over up to seven orders of magnitude are other attractive features for analog designers.

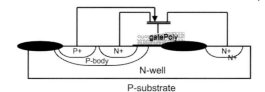

Fig. 11.2 Cross-section of a lateral DMOS transistor.

High-voltage Devices

A large number of MEMS devices require bias voltages that exceed the specifications for core CMOS technology. Most CMOS processes with feature sizes below 0.5 μm offer more than one gate thickness. The increased breakdown voltages of thick oxide devices are needed for IO devices (e.g. 3.3 or 5 V IO elements for a 2.5 V, 0.25 μm CMOS technology). For technologies with feature sizes below 0.18 μm, the thick oxide devices are also needed to reduce the leakage currents in critical regions. At least two additional masks are required.

For applications that require higher voltages, dedicated high-voltage transistors (e.g. lateral or vertical DMOS transistors) can be used. Fig. 11.2 shows the cross-section of a lateral DMOS transistor.

There are several foundries (e.g. [12, 13]) that offer technologies that can withstand several tens to a few hundred volts. Reviews on high-voltage options can be found in [14–16].

Triple-well Technologies for Isolated Transistors

Most standard CMOS technologies use a *p*-type wafer as starting material and, therefore, the bulk-node of the NMOS transistor is connected to the common substrate for all devices.

The research on rf applications of standard CMOS technologies has led to the development of a new class of devices, which are isolated from the substrate by means of a deep *n*-well and trenches. This way, electrical isolation of the sensitive NMOS devices from the noisy substrate can be achieved. This isolation can be used in MEMS interface circuitry to achieve a larger voltage range or handle negative voltages on the chip [17].

Analog Resistors and Capacitors

Many OpAmp-based designs require feedback resistors. Moreover, resistor values of several tens to a few hundred kΩ are needed to meet the desired power-consumption specifications. Standard digital CMOS technologies do not have a well-defined high-resistivity layer available. The polysilicon layer for the gate is well defined but the resistivity is of the order of only 10 Ω/□. The *n*-well (or *p*-diffusion) resistors exhibit higher resistivity but are non-linear owing to diode effects. At the price of one or two extra masks, the doping of the polysilicon layer can be selectively blocked and a linear resistor with 1–10 kΩ/□ is obtained (there are other options for high-resistivity layers but this is probably the most common one). The only capacitors available in digital CMOS technologies are MOS capacitors (i.e.

transistors with their drain and source connected together), which are non-linear and require complex biasing circuitry. Linear capacitors with a reasonable capacitance per area can be obtained by adding an extra polysilicon layer on top of the gate polysilicon (poly/poly capacitors) or by inserting thin metal layers in between two regular metallization layers (MIM capacitors or stacked MIM capacitors). Again, approximately two additional masks are required for each of these devices. The capacitance per area is of the order of 1 fF/μm^2.

Thick Metal Layers
Many advanced CMOS technologies offer one or two thick metal layers for low-ohmic supply voltage distribution. This is also increasingly being used for RF applications to fabricate on-chip coils and can be beneficial for many analog readout circuits. This option usually does not require extra masks.

Some of these analog options also offer interesting opportunities for post-CMOS micromachining of CMOS layers as shown in Chapter 1. Examples include [18], where the second polysilicon layer has been used as a sacrificial layer to release a micromachined membrane and where the deep n-well (of a triple-well process) is used to define different thicknesses for parts of an AFM cantilever.

11.1.2
Specification of the Interface and System Architecture

The next step after choosing the IC technology is to partition the system into on-chip and off-chip components [19]. The number of external components (e.g. off-chip capacitors) and precision reference elements (e.g. voltage references, quartz oscillators) must be minimized to achieve a competitive system solution. There are three possible interfaces between the CMOS MEMS chip and an off-chip data processing unit:

- *Analog output:*
  For systems that have only one or a small number of devices and are based on an old CMOS technology, the co-integration of digital circuitry and ADCs consumes too much area. In this case, the most economic solution is to include only the necessary analog biasing and signal-conditioning circuitry on the micromachined CMOS chip and use external components for the complex functions. This approach is used, e.g., by Analog Devices for their acceleration sensors and gyroscope, which are based on a 3 μm BiCMOS technology [20]. As the signal-to-noise ratio (SNR) and resolution are determined by the first stages in the signal processing chain (preamplifier, filter, demodulation), there is no loss of performance compared with solutions with higher degrees of integration. The main disadvantage of this approach is that no digital circuitry to perform calibration, program gains, correct for temperature drifts and transmit status information is available. There are only a few companies (e.g. Analog Devices) which have sufficient precision analog devices, analog trimming capabilities on wafer level (e.g. laser trimming) and analog design expertise (e.g. for analog temperature compensation) to do without digital compensation options.

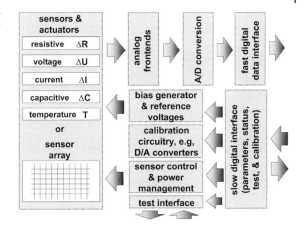

Fig. 11.3 Template for the system architecture of a micro- or nanosystem chip with a digital interface to the signal processing unit.

- *Digital output:*
  Most systems include an on-chip ADC to avoid transmission of sensitive analog signals through an unknown, noisy, external data channel and some memory (registers, RAM or non-volatile flash-memory) to store calibration information, compensate non-idealities of the micromachined devices and adapt the read-out circuitry to the current environment. An external microcontroller or FPGA is used to acquire and process the data and display or transmit the results via a standard interface to a PC or distant control unit. Fig. 11.3 shows a template for a system based on this approach.
- *Complete system on-chip:*
  For systems based on advanced CMOS technologies that are designed for high-volume applications or include a large number of devices (sensor arrays), it may be advantageous to include the microcontroller, the standard digital interface and the display on the same die together with the micromachined devices and the analog read-out circuitry. There is no reason to believe that the general trend towards systems-on-chip should not apply – on a longer time scale – to MEMS/NEMS devices.

For many MEMS devices, standard IC packages cannot be used because they require access of an external measurand (e.g. analyte-loaded air for a chemical sensor) to the micromachined devices. Other devices require hermetic sealing, e.g. high-Q resonators in order to avoid air damping. Therefore, a high degree of on-chip integration (e.g. bandgap references, bias generators, digital interfaces with a small number of connections) is desirable to minimize the number of IO connections from the chip to the data-processing unit and to simplify the packaging.

11.2 Analog Frontend

The analog frontend includes the circuitry that interfaces directly with the sensor elements and is specific to each application. Therefore, we start by defining the key properties that are shared by all analog frontends and that to a large extent define the performance of the overall system. Then, frontend circuitry for the most popular sensing devices is presented. The measurement options and basic frontend circuitry for electrochemical sensors have already been described in Chapter 7. Interface circuitry for RF applications can be found in Chapter 5. Many building blocks for analog frontends use operational amplifiers (OpAmps). A detailed discussion of OpAmp implementations would exceed the scope of this chapter, but can be found in [21, 22].

11.2.1
Key Properties of Analog Frontends

There are several properties and parameters that are adequate for the characterization and comparison of CMOS-based microsystems. In the following section, key properties are introduced that are relevant in the field of microsystems and instrumentation in general. Some of them apply only to the signal-conditioning circuitry, whereas others are characteristic properties of the entire transducer system. A common classification distinguishes between static and dynamic properties.

11.2.1.1 Static Properties

Range

The range describes the interval of signal values that can be handled by the signal-conditioning unit. It is often referred to as input or output range, describing the acceptable signals at the input and output of a building block, respectively. A transconductance amplifier, for example, will be characterized by a voltage input range and a current output range. The signal range also indicates the operating condition that underlies other given parameters of the system such as linearity and resolution measures.

Offset

The offset is the signal that is obtained at the output of the system for a zero input signal. While calibration is necessary to eliminate the sensor offset, the offset of the read-out circuitry can be reduced by using chopping or auto-zero techniques. The offset is by definition a DC signal but it can vary owing to temperature changes or aging.

Resolution

The resolution is the smallest signal change that can be recognized by the system. In a discrete value (digital) system, this value is given by the signal range divided by the number of possible values. In digital systems, the resolution therefore is equal to the size of the least significant bit (LSB):

$$\text{LSB} = \frac{X_{\max} - X_{\min}}{2^{\text{numBits}}} \qquad (1)$$

In the analog domain, the resolution is limited by distortion, interference and noise signals and is therefore determined by the sum of all these 'unwanted' signals. It can vary over the signal range.

Signal-to-Noise Ratio
As stated above, signal interference and noise are the main limiting factors for the signal resolution. The ratio between the actual noise and the signal is another important figure in describing signal processing and instrumentation systems. The signal-to-noise ratio (SNR) gives the ratio of the power of signal and noise and is usually expressed in decibels (dB):

$$\text{SNR}_{\text{dB}} = 10 \cdot \log\left(\frac{P_{\text{signal}}}{P_{\text{noise}}}\right) \qquad (2)$$

If electronic noise is dominant over other interference sources, the signal resolution is given by a signal that has the same power as the noise:

$$\text{SNR}_{\text{dB}} = 0 \Rightarrow P_{\text{signal}} = P_{\text{resolution}} \qquad (3)$$

Linearity
The linearity of a system is a measure of the distortion of the signal transfer characteristic. It can be applied to both static and dynamic signals, therefore linearity definitions are given in both sections. In the case of a static transfer characteristic of a building block, the non-linearity describes the deviation of the output signal from the output of an ideal linear system. There are different approaches for the ideal linear characteristic. One of the most common is a simple linear fit of the actual output range of the system. Sometimes not the actual output range but the ideal output range is taken for the linear fit. This approach also takes into account gain and range deviations in the non-linearity measure. Moreover, a least-squares fit approximation can also be used to calculate the non-linearity. This non-linearity is usually expressed relative to the actual or ideal signal range. Fig. 11.4 shows an example transfer characteristic and the calculated linear fit (a) and the resulting non-linearity plot (b). Instead of the entire plot, sometimes only the maximum values are given, which often occur at the minimum or maximum signal level. In the case of analog–digital and digital–analog conversion systems, special definitions of linearity measures have been developed, namely the differential non-linearity (DNL) and the integral non-linearity (INL), which are described in [69].

11.2.1.2 Dynamic Properties

Bandwidth
The bandwidth is similar to the range specification of a signal and describes the frequency range that can be processed by the actual system block. In instrumentation systems the bandwidth is always a matter of optimization, because in order

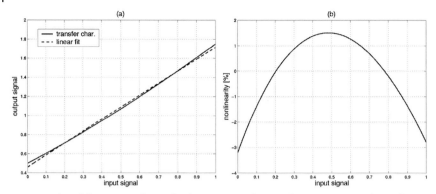

Fig. 11.4 Plot of the measured transfer characteristic of a transducer system together with its linear fit (a) and the calculated non-linearity (b).

to maximize the SNR the bandwidth has to be chosen as narrow as possible to limit the contribution of noise and other interferences, but must still be large enough to process the measurement signal. Several different definitions of bandwidth exist. Here the two most important ones for integrated sensor systems will be given. The 3 dB bandwidth, also known as the 3 dB corner frequency, is the frequency of the transfer characteristic at which an input signal is damped to half of its power (−3 dB). In the case of a band-pass system, the signal bandwidth is given by the difference between the upper and the lower corner frequencies. The second bandwidth definition is the (equivalent) noise bandwidth, which describes the bandwidth of a rectangular-shaped filter (rectangular window) yielding the same output power as the actual system transfer characteristic and is very useful for noise calculations:

$$B_N = \frac{1}{|H_0|^2} \int_{-\infty}^{\infty} |H(j\omega)|^2 d\omega \qquad (4)$$

where H_0 is the static transfer characteristic and $H(j\omega)$ the frequency response function of the described system. For of a first-order low-pass filter, there is a simple correlation between the 3 dB bandwidth and the noise bandwidth:

$$B_N = \frac{\pi}{2} f_{3dB} \qquad (5)$$

Harmonic Distortion
Another measure of (non-)linearity is the harmonic distortion. This is a dynamic characteristic and takes into account the fact that non-linear distortions in electronic circuits usually increase with increasing signal frequency. The distortion factor adds the amplitudes of the harmonic tones relative to the fundamental tone; the total harmonic distortion (THD) is the ratio of the signal powers of the harmonics, t_k, divided by the power of the fundamental tone, t_0:

$$\text{THD}_{\text{dB}} = 10 \cdot \log\left(\frac{\sum_{k=1}^{n} t_k}{t_0}\right) \tag{6}$$

11.2.1.3 System Properties

Many of the above-mentioned parameters are not only used to describe individual building blocks, but also characterize the overall system behavior. This is especially true for parameters such as range and resolution and also for definitions of the linearity such as the distortion factor. Additionally, there are special parameters that describe and quantize important properties of the entire transducer system.

Sensitivity

The sensitivity describes the change of the system output with a change of the system input signal. It corresponds to the gain of an electronic amplifier stage, but usually covers signals of different physical disciplines such as acceleration as the input signal and a digital word at the output for an integrated acceleration sensor. The sensitivity usually depends on the operating conditions, so that additional measures such as signal range, signal bandwidth and other environmental conditions are important when a sensitivity is given:

$$S_0 = \left.\frac{\partial y(x_0, x_1, \ldots, x_n)}{\partial x_0}\right|_{x_1,\ldots,x_n=\text{const.}} \tag{7}$$

Limit of Detection

The limit of detection (LOD) is the smallest recognizable input signal and is similar to the signal resolution given by the influence of noise, distortion and interference by other signals. Especially when characterizing the entire system, the sensitivity to signals other than the measurement signal is an important parasitic effect and is often specified by the next property.

Cross-sensitivity

The cross-sensitivity basically has the same definition as the sensitivity except that it is the system response to a change of signals other than the desired measurement signal. The selectivity describes the robustness of a system to cross-sensitivity and is the ratio between sensitivity and cross-sensitivity:

$$\text{Sel}_1 = \frac{S_0}{S_1} = \left.\frac{\frac{\partial y(x_0, x_1, \ldots, x_n)}{\partial x_0}}{\frac{\partial y(x_0, x_1, \ldots, x_n)}{\partial x_1}}\right|_{x_2,\ldots,x_n=\text{const.}} \tag{8}$$

Accuracy and Precision

Accuracy and precision have two different definitions that are sometimes mixed up. Accuracy is a measure of how accurately the true value is determined by a measurement. Accuracy can be improved by averaging several measurements to eliminate random measurement errors. Precision, on the other hand, is a measure of the reproducibility of a measurement. It describes the distribution of several measurements.

Drift

Whereas the precision is an indicator for the short-time reproducibility, drift is a measure of the long-time stability of the instrumentation system. The time scale of drift specification can vary from hours to several years depending on the actual application and the intended lifetime of the system. In general, for every previously described parameter, a drift specification can be given and is commonly done for key properties such as resolution, sensitivity, accuracy and offset.

11.2.1.4 Fully Differential Design

The majority of analog frontends made from discrete components use a single-ended design, where all signals are measured relative to a well-defined ground-potential (see Fig. 11.5). Even differential sensing signals from, e.g., a resistive Wheatstone bridge are converted into single-ended signals at an early stage. If analog frontend circuitry is combined with ADCs and digital logic in a system-on-chip device, the ground and supply lines are noisy because of crosstalk from the digital part and external interference. Moreover, signals from integrated sensors are often small and the supply voltage of modern CMOS technologies approaches 1 V. As a result, fully differential circuitry, in which the signal is defined as the difference between a positive output node V^+ and a negative output node V^- has to be employed (see Fig. 11.5). If the design is symmetric, noise from the supply lines affects both voltages in

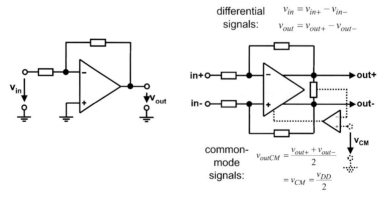

Fig. 11.5 Single-ended versus fully differential circuit design.

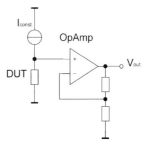

Fig. 11.6 Measurement setup to monitor the resistance of a device with a constant-current source and an operational amplifier.

the same way and cancels out. This suppression of noise and interference comes at the price of an additional amplifier (inside the main amplifier, see Fig. 11.5) that is required to keep the common-mode voltage $V_{CM} = (V_{out+} + V_{out-})/2$ at a predefined value. This voltage is usually chosen to be equal to half the supply voltage V_{dd}.

The performance of most integrated micro- and nanosystems can be significantly improved by using fully differential frontend circuitry. Therefore, most of the analog frontends discussed in this section are fully differential designs.

11.2.2
Resistive Sensors

As has been shown in the preceding chapters, many transducers are based on the change of the electric resistance upon a change of parameters such as temperature or mechanical stress. In general, there are two possible configurations to measure an electrical resistance: a defined voltage is applied to the device under test (DUT) and the current is measured, or a defined electrical current is led through the DUT while the voltage is monitored. For the latter approach, a voltage meter with high input resistance is crucial for accurate measurements, which can be realized by an instrumentation amplifier with a MOS transistor-based input stage. Accordingly this approach is mostly realized in CMOS-based technologies. Fig. 11.6 shows the application of a MOS-based operational amplifier to monitor the voltage drop over a resistor under test. This approach is used, for instance, to measure the temperature of microhotplates used for environmental gas sensing and material characterization [23].

Especially for temperature sensing, a constant current source is important. Section 11.3.6.3 describes possible realizations of current references that are temperature insensitive.

For sensing mechanical signals, such as force or pressure, resistive transducers are usually arranged in a Wheatstone bridge configuration (see Fig. 11.7). The advantages of this bridge configuration are a reduced cross-sensitivity to the temperature if all four resistors are subject to the same thermal influence and an increased sensitivity when more than one device acts as a transducer. The sensitivity is increased by a factor of 2 when, for example, R_1 and R_4 are piezoresistors and R_2 and R_3 are constant and by a factor of 4 if in addition R_2 and R_3 act as sensors with an opposite sign of their sensitivity. The drawback of this approach is that

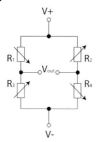

Fig. 11.7 Wheatstone bridge configuration for resistive transducer elements.

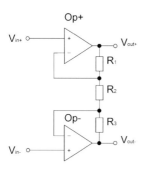

Fig. 11.8 Pseudo-differential realization of an instrumentation amplifier.

additional transducer elements are needed, which have to be arranged appropriately so that the correct sensitivities are achieved. Fig. 11.8 shows a pseudo-differential instrumentation amplifier with high input impedance that keeps the differential signal path at its output. A more elegant way to implement such an amplifier comprises of a differential difference amplifier (DDA) [24, 25].

The main limiting factor for the resolution of resistive Wheatstone bridges and also for single resistive devices is the thermal noise contribution of the resistors themselves. The thermal noise power when monitoring the voltage drop with a high input impedance device is given by Eq. 9, whereas the signal power is given by Eq. 10. The resulting SNR is then proportional to the sensitivity S_{res} and the square of current $I_{const.}^2$ and inversely proportional to the signal bandwidth Δf, which shows the importance of optimizing the bandwidth for the actual application (see also Section 11.2.1.2).

$$P_N = 4kTR\Delta f \tag{9}$$

$$P_S = S_{res} I_{const.}^2 R \tag{10}$$

Eq. (11) shows that the SNR can be increased by reducing the signal bandwidth and by increasing the current flowing through the resistor.

$$\text{SNR} = \frac{P_S}{P_N} = \frac{S_{res} I_{const.}^2}{4kT\Delta f} \tag{11}$$

Fig. 11.9 Wheatstone bridge with four p-channel MOS transistors.

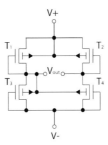

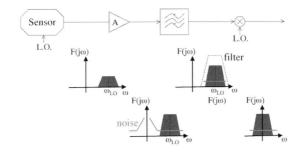

Fig. 11.10 A lock-in architecture to read out a Wheatstone bridge, showing the corresponding frequency graphs.

In the case of a Wheatstone bridge configuration, the current through the individual resistors depends on the bridge supply voltage, which therefore should be increased. The maximum current is limited by the power dissipation in the DUT, which often results in cross-sensitivities due to thermally induced stress and the heating of the transducer itself. For many applications a large resistor value is desirable, because the sensitivity of the transducer can be increased, e.g. in the case of a temperature sensor, or the power dissipation has to be reduced, e.g. in the case of a stress sensor for chemical sensing [26]. In a CMOS technology, high-resistive devices can also be realized by p- or n-channel MOS transistors, biased in the linear region. Fig. 11.9 shows a Wheatstone bridge comprising of four p-channel MOS transistors [27]. This approach permits the realization of a high-resistive measurement bridge in a CMOS technology without special high-resistive process options.

The drawback of using MOS transistor devices instead of passive resistors is the additional flicker noise that is introduced. As flicker noise is only predominant for low frequencies, MOS-based Wheatstone bridges are better suited for resonant applications, where the low-frequency noise can be filtered out without sacrificing the measurement signal. In addition, the sensor signal can be transferred to higher frequencies by using lock-in techniques, as shown in Fig. 11.10. The concept of readout modulated sensor signals can also be found in capacitive sensor applications (see Section 11.2.3.1) and a more detailed description of general lock-in amplification architectures is given in [28].

11.2.3
Capacitive Sensors

Capacitive detection has become a method of choice for the read-out of sensor signals [29], e.g. for accelerometers [20, 30–33] or chemical sensors [34–36]. Whereas in hybrid designs the large parasitic capacitances and the bond wires make capacitive read-out a difficult task, monolithic designs with their small parasitics and small distances between sensor and interface circuitry are ideally suited for capacitive detection. Furthermore, capacitive detection offers low power consumption and usually fewer drift problems than, e.g., piezoresistive sensors. In [20], Analog Devices claims a resolution of 12×10^{-21} F for their integrated gyroscope and numerous other publications report detection levels in the attofarad regime [30–33]. As a result, the majority of the high-volume MEMS devices to sense pressure and acceleration are based on capacitive detection. There are basically two methods to realize the frontend circuitry: sensing with AC modulation (continuous-time approach) and switched-capacitor interfaces.

11.2.3.1 Sensing with AC Modulation

Fig. 11.11 shows a schematic of two possible continuous-time detection schemes for capacitive sensors in a full-bridge configuration. An AC modulation voltage (sine-wave or square-wave) is applied to the capacitive bridge. The sensor signal can then be read out as a current (left-hand side of Fig. 11.11) or as a voltage (right-hand side) and is then amplified by the OpAmp. The best SNR is achieved for a full capacitive bridge, in which all four capacitors change based on a measured signal (e.g. deflection for an accelerometer) and C_1/C_3 have opposite signs to C_2/C_4 for the sensitivity. The sensor design often does not allow different signs of the sensitivity. In this case, C_1 and C_3 are used as the sensor elements, whereas C_2 and C_4 are constant reference capacitors. The values and temperature

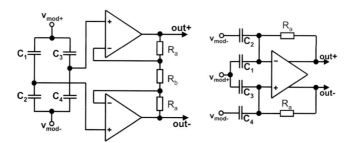

Fig. 11.11 Continuous-time detection schemes for capacitive sensors in full-bridge configuration. The modulated sensor signal can be read out as either a voltage (left-hand side) or as a current (right-hand side). In both cases, the output signal is proportional to the capacitance change after demodulation (not shown). Additional circuitry is needed to define the DC operating point at the input of the amplifier.

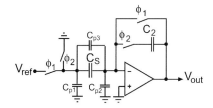

Fig. 11.12 Basic switched-capacitor amplifier for capacitive sensors. The voltage at the output of the amplifier is $V_{out}=V_{ref}C_S/C_2$. The two clock-phases must be non-overlapping.

coefficients of the reference capacitors should either be matched to the sensing elements or a tuning capability must be provided.

For designs in which area or other constraints permit only one sensing element, a capacitive half-bridge (C_1/C_2 in Fig. 11.11) can be used in combination with single-ended read-out circuitry.

A detailed discussion of the advantages and disadvantages of current versus voltage sensing and a noise analysis for the interface circuitry together with the sensor element can be found in Chapter 3.

11.2.3.2 Switched-capacitor Amplifiers

Switched-capacitor interface circuitry is probably the most popular approach to reading out capacitive sensors. Often switched-capacitor circuits are also used to design filters. A more detailed introduction to switched-capacitor design is given in Section 11.3.1.2 and in [29, 37, 38]. Fig. 11.12 illustrates the basic principle of a switched-capacitor amplifier. During phase Φ_1, the capacitive sensing element C_S is charged up to the voltage V_{ref} (switch S_1), whereas the charge on capacitor C_2 is reset to zero. In the next phase, Φ_2, the charge on the sensing element is transferred to capacitor C_2 and the voltage at the output of the amplifier equals

$$V_{out} = \frac{C_S}{C_2} V_{ref} \tag{12}$$

This basic implementation already reveals two advantages of the switched-capacitor approach. First, the parasitic capacitors to ground do not contribute to the sensor signal. The charge on C_{P1} is discharged into the ground node during phase Φ_2 and the charge on C_{P2} is constant because it is connected to the virtual ground at the negative input of the OpAmp. Therefore, only the parasitic capacitance C_{P3} in parallel to the sensing element has to be taken into account. The second advantage is that the DC offset of the OpAmp is cancelled owing to the correlated double sampling (CDS). During phase Φ_1, the offset of the OpAmp is stored on the sensing capacitor C_s and subsequently cancelled during the charge-transfer phase Φ_2.

As in the continuous-time approach, the CMRR and PSRR can be improved by a fully differential design and the sensitivity and dynamic range are increased through the use of reference elements or elements with opposite signs for the sensitivity. Furthermore, some parasitic effects, such as charge injection, are greatly reduced by a fully differential design. Fig. 11.13 shows a fully differential

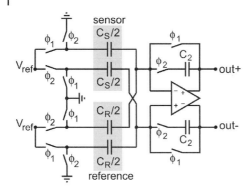

Fig. 11.13 Fully differential frontend amplifier for capacitive sensors based on switched-capacitor techniques. The differential voltage at the output of the amplifier is $V_{out} = V_{ref}(C_S - C_R)/C_2$. The clock phases with subscript d are delayed to reduce charge injection.

design of a switched-capacitor amplifier with two sensing and two reference capacitors. The phases are identical to the single-ended implementation, but now the voltage at the amplifier output at the end of phase Φ_2 is given by

$$V_{out} = \frac{C_S - C_R}{C_2} V_{ref}, \tag{13}$$

where C_S is the capacitance of the sensing element and C_R is the reference capacitance. Note that to improve the charge-transfer efficiency, both capacitors are split into two parts. If only one of the capacitors C_S and C_R were connected to the positive and negative terminals of the OpAmp, the common-mode voltage at the input of the amplifier would shift away from the (regulated) common-mode output voltage and the sensitivity would be reduced to

$$S = \frac{V_{out}}{V_{ref}} = \left(\frac{C_p}{C_S + C_p}\right)\left(\frac{C_S - C_R}{C_2}\right), \tag{14}$$

where C_p is the parasitic capacitance at the input of the OpAmp.

Noise Analysis of Switched-capacitor Amplifiers

In general, the theoretical noise performance of a switched-capacitor circuit is inferior to that of continuous-time implementations because the switching will undersample the high-frequency noise of the switch transistors and the wide-band OpAmp. In practice, a switched-capacitor implementation often allows significant savings in area and power consumption, which can be used to improve noise performance. Furthermore, the superior noise performance of continuous-time implementations is only realized if high-order band-limiting filters are employed before the ADC or if an oversampled ADC in combination with digital filtering is used.

The equivalent circuit model for the noise analysis of the single-ended switched-capacitor amplifier is shown in Fig. 11.14. For reasons of symmetry, the same circuit can be used to analyze the fully differential design. First, the contribution of the equivalent input noise-source of OpAmp will be analyzed. The low-

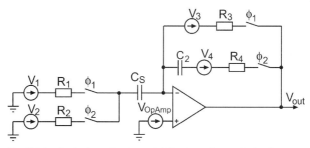

Fig. 11.14 Equivalent circuit model for the noise analysis of a switched-capacitor amplifier. For reasons of symmetry, the fully differential design can be analyzed in the same way.

frequency (e.g. $1/f$) noise of the OpAmp is cancelled by the CDS scheme in the same way as the equivalent input offset of the OpAmp. To achieve full settling of the OpAmp in one clock phase (half the clock period), the bandwidth of the OpAmp must be about 5–10 times larger than the sampling frequency, depending on the accuracy requirements. Therefore, the noise contribution of the OpAmp will be 5–10 times the thermal noise-floor of the OpAmp. For a detailed analysis, the overall transfer function from the noise-source to the output of the amplifier has to be derived:

$$H_{\text{CDS}}[\omega] = H_{\text{transfer}}[\omega] - H_{\text{reset}}[\omega] e^{-j\omega \frac{T_s}{2}}, \tag{15}$$

where T_s is the clock period and H_{transfer} and H_{reset} are the transfer functions from the noise source at the input of the OpAmp to capacitor C_2 during the corresponding clock phases Φ_1 (reset) and Φ_2 (charge transfer). The noise power after the CDS is then given by

$$S_{\text{CDS}}[\omega] = \tau^2 \left[\frac{\sin\left(\pi \tau \frac{\omega}{\omega_s}\right)}{\pi \tau \frac{\omega}{\omega_s}} \right]^2 \sum_{k=-\infty}^{\infty} (S_{\text{OpAmp}}[\omega - k\omega_s] \cdot |H_{\text{CDS}}[\omega - k\omega_s]|^2), \tag{16}$$

where S_{OpAmp} is the input-referred noise power density of the OpAmp and τ is the hold time of the sampled signals (typically $\tau = T_s/2$). The switches are 'on' only during either the reset or the charge-transfer phase. Therefore, there is no cancellation of their low-frequency noise due to the CDS scheme and their noise contributions can be calculated by using the transfer function from the corresponding noise source to output capacitor C_2. More information on noise analysis of switched-capacitor circuits can be found in [39].

11.2.4
Thermoelectric, Piezoelectric and Pyroelectric Devices

The common feature of the three signal-detection methods that are addressed in this section is that they do not require any external bias. The signal is directly related to some physical effect (e.g. displacement of charge due to mechanical stress or heat). Therefore, these devices usually exhibit less offset and reduced drift than biased sensor devices.

11.2.4.1 Thermoelectric Devices

Thermoelectric devices measure the voltage difference at the output of a thermocouple consisting of two different conductors, which is proportional to the temperature difference between the hot junction and the cold junction of the thermocouple (Seebeck effect). Owing to the small difference in the Seebeck coefficient of the materials available in CMOS technologies [~110 µV/K for an aluminum/polysilicon thermocouple (see Chapter 7)], a large number of thermocouples are usually connected in series to form a thermopile and achieve reasonable signal amplitude. The best performance is achieved when the overall SNR, including the transduction mechanism, the thermopile and its source resistance and the amplifier, is optimized. An example of a detailed SNR analysis can be found in [40, 41].

For the calorimetric chemical sensor described in chapter 7, the center of the thermopile is connected to analog ground and a differential signal in the microvolt regime is obtained. A low-noise (thermal noise ~10 nV/Hz^{½}, no $1/f$ noise), low-offset (~1 µV) amplifier with a large amplification factor is required because of the small signal amplitudes. Standard CMOS amplifiers cannot be used because they show an offset in the millivolt range and large $1/f$ noise. Fig. 11.15 shows the input-referred noise of a standard CMOS instrumentation amplifier. The corner frequency, where the $1/f$ noise contribution is equal to the thermal noise floor, is typically in the range 1–100 kHz.

There are two categories of amplifiers that offer the desired performance: chopper amplifiers and auto-zero amplifiers. The chopper amplifier modulates the input signal to a frequency above the corner frequency in order to avoid the $1/f$ noise contribution in the signal band. At the output of the chopper amplifier, the signal is demodulated. In this way, the DC offset of the amplifier is also elimi-

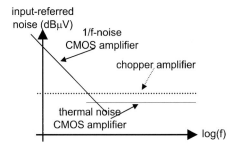

Fig. 11.15 Input-referred noise of standard CMOS amplifier and chopper amplifier.

nated (modulated to the chopping frequency and then filtered). The auto-zero amplifier first samples the amplifiers offset and then subtracts it from the subsequent sampling of the signal. This also eliminates $1/f$ noise for frequencies well below the sampling frequency. Overall, the chopper amplifier offers better noise performance and a smaller residual offset than the auto-zero amplifier.

A comparison of the two techniques and examples of integrated chopper amplifiers can be found in [31, 42–46]. In the following, an implementation of a chopper amplifier will be explained in detail.

Chopper Amplifier
Fig. 11.16 shows a block diagram of a chopper amplifier, as first published by Menolfi and Huang for infrared sensing applications [45–47] and later also developed for calorimetric chemical sensors [48].

The specifications for the amplifier are given in Table 11.1. The signal is modulated to the chopping frequency by the cross-coupled pair of switches at the input of the amplifier. The first amplification stage is a low-noise amplifier with a gain of ~20 and defines the overall noise performance of the amplifier. The band-pass filter eliminates the DC offset of the first amplification stage, reduces the residual offset at the output of the demodulator [46] and further amplifies the signal by a factor of 20. The final amplification stage again has a gain of 20 and the cross-coupled pair of switches at the output of the amplifier demodulates the chopped

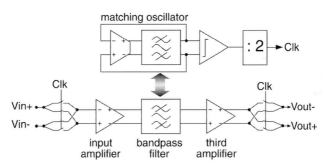

Fig. 11.16 Block diagram of a low-noise chopper amplifier for a calorimetric chemical sensor.

Tab. 11.1 Specifications and measured performance of a CMOS chopper amplifier for a calorimetric chemical sensor [48] (the amplifier was realized in 0.8 µm, 5 V CMOS technology)

	gain	bandwidth	f_{chop}	eq. input noise	residual offset
designed	6400	500	5000	< 12 nV/Hz$^{1/2}$	< 10 µV
measured	6130	477	4948	7 nV/Hz$^{1/2}$	0.24 µV
std. dev.	150	26	88		0.60 µV

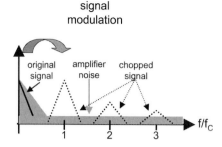
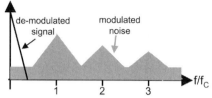

Fig. 11.17 Principle of a chopper amplifier explained in the frequency domain. The signal is first modulated to the chopping frequency f_C. Then, the signal is amplified and the noise of the first amplification stage is added. After band-pass filtering, the amplified signal is de-modulated, while the large $1/f$ noise of the CMOS amplifier is modulated to f_C. Finally, a low-pass filter is needed to remove the noise and chopping residuals (not shown).

signal to the baseband. At the output of the amplifier, a low-pass filter is required to remove the modulated DC offset of the filter and the third amplification stage. As an alternative, an oversampled ADC together with a synchronous sampling scheme can be used to avoid the need for additional analog filtering [47]. Fig. 11.17 explains the function of the chopper amplifier in the frequency domain. While the input-offset and low-frequency noise of the first amplification stage is removed by the band-pass filter and the remaining components are modulated to the chopping frequency at the output of the amplifier, the signal is modulated at the input of the amplifier and demodulated at the output after passing three amplification stages. The overall gain of the amplifier is less than the multiplied gain of the three amplification stages, because the high-frequency components of the modulated signal are removed by the band-pass filter. The gain is given by

$$A_V = \frac{8}{\pi^2} A_{V1} A_{V2} A_{V3} \cong 6400, \tag{17}$$

where A_{V1-3} are the gains of the three amplification stages. The chopping frequency should be chosen as low as possible because the residual DC offset is proportional to it. The lower limit of the chopping frequency is determined by the bandwidth of the signal, the required passband flatness, the order and the Q-factor of the band-pass filter. Furthermore, the chopping frequency should exceed the $1/f$ corner frequency of the input-referred noise in the first amplification stage. For the calorimetric chemical sensor with a signal bandwidth of ~500 Hz and a second-order band-pass filter with a Q-factor of 5, a chopping frequency of 5 kHz was chosen. The size of the switch transistors for the input chopper should be the minimum, again because the residual offset is proportional to it. Here, the lower limit is determined by the maximum tolerable on-resistance of the switch transistors.

A standard low-noise instrumentation amplifier, as described in Section 11.2.2, could be used to realize the first amplification stage of the chopper amplifier, but

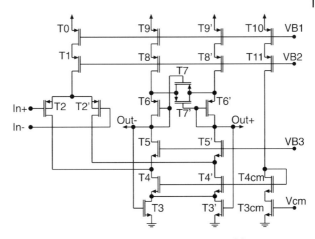

Fig. 11.18 Schematic of a low-noise gm-ratio amplifier.

considerable power is required for the two OpAmps and to drive the feedback resistors. For typical thermopile configurations, the signal swing from the sensor is of the order of only a few millivolts down to the microvolt range. Therefore, a gm-ratio amplifier as shown in Fig. 11.18 was used to realize the first amplification stage. The gain is defined by the ratio of the transconductance gm of transistor T_2 and the degenerate transconductance of the transistors T6 and T7 [49]:

$$A_V = \frac{gm_2}{\frac{gm_6}{\left(1+\frac{\beta_6}{\beta_7}\right)}} = 20 \tag{18}$$

The size of the transistors is determined by the scaling factors needed to realize the desired gain and by the noise requirements of the amplifier. A factor of 2.5 can be achieved by the degeneration transistor T7 at the output of the amplifier without degrading linearity. The remaining factor of 8 (to the overall gain of 20) comes from the transconductance ratio of transistors T2 and T6. Good matching is required to keep the spread of the gain due to process variations and tolerances as small as possible (<3% for this example). Therefore, it is better to scale the width and bias current of the transistors than to scale the length. Careful layout and the use of multiples of unit transistors are important factors for the final spread.

The noise performance of the amplifier is mostly determined by the differential input pair and the NMOS transistors T4/T4'. For the transistor sizes shown in Fig. 11.18, an input-referred noise of 7 nV/Hz$^{1/2}$ was obtained. The power consumption of the complete amplifier is 1.95 mW. Further reduction of the power consumption at the price of a reduced input swing can be achieved by using transistors in weak inversion [48].

A gm-C filter was used to implement the band-pass filter for the chopper amplifier. The main challenge in this filter design is to realize the low center frequency of 5 kHz with the small integrated capacitors available in CMOS technologies. A tuning capability can be useful to match the center frequency of the filter to the chopping frequency. The third amplification stage is similar to the input amplifier. Degeneration transistors have to be used for the input differential pair in order to handle the large signal swing at the output of the band-pass filter.

11.2.4.2 Piezo- and Pyroelectric Sensors

These types of sensors show a similar electrical behavior, so readout electronics for them will have the same properties and the same architectures can used for both piezo- and pyroelectric devices. As the physical function of these sensors has been already described in Chapter 2, only the electrical interface will be described in more detail here.

With a change of the environmental signal, which is the mechanical tension for the piezoelectric and the temperature for the pyroelectric sensor, an electric voltage is generated. As these devices are sensitive to changes in the input signal, the output voltage is ideally proportional to the time derivative of mechanical tension and of temperature for the piezo- and pyroelectric sensors, respectively.

This voltage is small compared with the output voltage that is provided by the bridge configurations of the resistive and capacitive sensors already described (see Sections 11.2.2 and 11.2.3). Additionally, the sensor materials are used to have a high specific resistance, resulting in a high output impedance of the sensing device.

For most applications it is sufficient to know the amount of temperature or pressure change; the signal dynamics are not required. In that case, the readout circuit records the amount of charge generated at the interface of the transducer (actually, no charge is generated, but rather separated and 'visible' to the interfaces of the transducer element).

Looking at the equivalent circuit model of the piezo- and pyroelectric sensors (see Fig. 11.19), we recognize mainly capacitive behavior with a large parasitic resistor and capacitor R_{par} and C_{par}, respectively. The charge moved upon a change of stress or pressure is modeled by a current pulse I_{ch}.

A charge-sensitive amplifier (CSA) acts as a transimpedance amplifier to convert the current pulse into a voltage and has a frequency characteristic of an integrator to make the output voltage proportional to the charge at the CSA input. Fig. 11.20 shows an implementation of a CSA comprising an operational amplifier together with an integrating capacitor and a reset circuit.

A reset circuit is necessary to prevent the integrator from saturation if the input signal has a DC component, which is usually the case. This reset can be applied periodically by a switch that short-circuits and discharges the feedback capacitor. This is a very convenient approach, because the integrator remains (almost) ideal and no charge is lost. The time interval has to be chosen according to the output signal range of the CSA, to prevent saturation reliably. The drawback of this approach is the need for a switching clock, which might cause additional interference to the weak charge signals. To overcome this drawback, a continuous time reset can by realized with a simple resistor in parallel to the feedback capacitor. This architecture will introduce another pole to the transfer characteristic. This

Fig. 11.19 Equivalent circuit of a piezo- or pyroelectric element with parasitic capacitance and conductance.

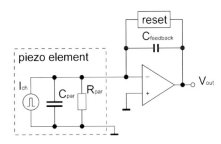

Fig. 11.20 Piezo- or pyroelectric element connected to a charge-sensitive amplifier.

pole must be lower than the sensor signal bandwidth, in order to keep the integration operation of the current pulses, but must also be large enough to prevent saturation of the amplifier.

In designing a charge amplifier for the readout of piezo- or pyroelectric signals, special attention has to be given to the electronic noise contribution, because of the weak electrical signals provided by those transducers. Examples of low-noise realizations of charge-sensitive amplifiers can be found in [50–52]. Usually the charge-sensitive amplifier is followed by an analog filter to maximize the SNR (see Section 11.3.1).

11.3
Circuitry Building Blocks for CMOS MEMS/NEMS

11.3.1
Analog Filters

One of the main functions of signal conditioning circuitry besides amplification is filtering. The main goal of filters is to improve the SNR of a sensor signal. In this context, noise stands not only for an undetermined statistical process, such as electronic white or flicker noise, but also for any interference signal that disturbs the sensor signal. In addition to the electronic noise mentioned, the origin of such signals can be the electromagnetic emission of other circuits, such as digital processors or radio signals, present in the environment of the microsystem.

Filters improve the SNR by selectively damping frequencies containing noise or amplifying the frequencies of interest that contain the sensor signal.

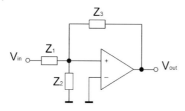

Fig. 11.21 Schematic of a biquad filter, which can be realized as a discrete or integrated circuit.

11.3.1.1 Biquad Realization

The most common analog filter realization is the biquad filter. Fig. 11.21 shows a typical circuit realization with an operational amplifier and three passive components [53]. Such a circuit can be realized both with discrete devices on a circuit-board level and as an integrated circuit. However, the realization as an integrated circuit has some limitations for certain applications, which will be explained later.

By choosing different configurations of the passive components (resistors, capacitors or inductors) one can realize low-pass, high-pass, band-pass or band-reject filter characteristics.

In a common CMOS technology, only resistors and capacitors are available, which restricts the number of possible configurations. Eq. 19 shows the general transfer function of a biquad filter:

$$H(j\omega) = \frac{a_0 + a_1 j\omega + a_2 (j\omega)^2}{b_0 + b_1 j\omega + b_2 (j\omega)^2} \tag{19}$$

A good overview of the theory of biquad filters and the different filter characteristics that can be derived is given in [54].

One biquad architecture that is well suited to be realized in CMOS technology is the Sallen and Key configuration [55]. Usually, Sallen-Key filters are used to realize second-order low- or high-pass filters, but a band-pass configuration is also possible. Fig. 11.22 shows a schematic of a Sallen-Key second-order low-pass filter and Eq. 20 is the corresponding transfer function:

$$H_{LP}(j\omega) = \frac{\omega_0^2}{(j\omega)^2 + \frac{\omega_0}{Q} j\omega + \omega_0^2} \tag{20}$$

where

$$\omega_0 = \frac{1}{\sqrt{R_1 R_2 C_1 C_2}} \text{ and } Q = \frac{\omega_0}{\frac{1}{R_1 C_1} + \frac{1}{R_2 C_2}}.$$

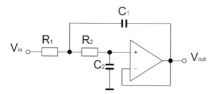

Fig. 11.22 Schematic of a Sallen-Key second-order low-pass filter.

A characteristic property of many sensor signals is their narrow bandwidth. When the CMOS-based microsystem consists of chemical or biological transducers, the processed information only shows slow changes, leading to a narrow bandwidth of the electrical signals. As the main goal of filters in the signal-conditioning chain is the maximization of the SNR, also the analog filters have to incorporate low-cutoff frequencies. This reveals one of the main drawbacks of the 'traditional' filter implementation with operational amplifiers in an integrated circuit technology: As can be seen from Eq. 20, the cutoff frequency of this second-order low-pass filter is inversely proportional to the RC product of the passive devices. Hence low values of f_c lead to high values of R and C. That this is a serious limitation for the application of integrated filters can be seen from the following example:

Assuming a sensor signal bandwidth of 1 kHz, a filter would be used to maximize the SNR by damping signals beyond the 1 kHz bandwidth. Choosing a second-order low-pass implementation for this filter, such as the Sallen-Key biquad from Fig. 11.22 with a 3 dB cutoff frequency of 1 kHz leads to $\omega_0 = 6.3 \times 10^3$ rad/s, which can be realized by values of 500 kΩ and 320 pF for $R_{1,2}$ and $C_{1,2}$ respectively. In a standard CMOS technology with a minimum feature size of 0.8 µm, the passive devices alone would occupy an area of nearly 0.18 mm^2 [12].

This limitation can be improved by the use of CMOS technologies that offer special passive devices, such as high-resistive layers and high-density capacitors, but this leads to more expensive solutions, as these technologies are notably more expensive and can have a lower production yield [56].

Another drawback when using passive components, such as resistors and capacitors, in an integrated circuit technology is the lack of accuracy. Integrated resistors and capacitors show good relative matching, but poor accuracy for the absolute values. In a standard CMOS technology, a polysilicon layer is used to form resistors, which usually has a spread of ±15% for the specific sheet resistance. This is due to the process deviations of the doping of the polysilicon layer and the geometric variations during lithography and etching. For integrated capacitors these deviations are usually less critical because these devices are only affected by geometric deviations of the CMOS technology.

Together with the need for high RC time constants, this fact was the main driving force for the development of a filter architecture called switched-capacitor filters (SC filters).

11.3.1.2 SC Filter Architecture

This filter architecture is very suitable to implement precise gain and filter circuits in integrated circuit technologies, especially in CMOS technologies. Many signal-conditioning circuits of integrated CMOS-based MEMS with their narrow signal band requirements are realized in this architecture, whereas it is not very common for PCB-level realizations.

CMOS SC circuits were first realized about 40 years ago, but they are still one of the best solutions for integrated circuits when signal conditioning for instrumentation is required. A good overview of the different SC amplifier and -filter structures can be found in [37].

The main idea of the SC technique is to replace each resistor of a conventional active filter circuit with a capacitor, connected by a pair of switches. When using a CMOS technology, these switches can be realized by a single MOS transistor or a CMOS pair, depending on the voltage range. This leads to very compact layouts and has the advantage of reduced power consumption, as during the static phase CMOS switches do not consume any power. The emulated resistor value is proportional to the reciprocal of the capacitor multiplied by the switching frequency. Therefore, there are two main advantages of the SC circuit architecture:

1. The use of capacitors instead of resistors usually improves the accuracy of time constants and therefore the accuracy of the filter characteristic, compared with the use of both resistors and capacitors to form the filter characteristics. The RC time constants are now formed by a ratio of capacitor values, so first-order process variations cancel out.
2. With an appropriate selection of switching frequency, much higher time constants can be realized for the same amount of area than with a conventional RC approach. That is why SC circuits are so often used for low-bandwidth signal conditioning as is required for many MEMS applications.

The concept of realizing the resistor element by switching a current through a capacitor leads to a discrete-time system. The input signal is sampled during one clock phase and the output signal is available during the same or another clock phase depending on the circuit architecture. The drawbacks of discrete-time systems and the influence of sampling on the signal properties are described more detailed in [37] and [69]. To fulfil the sampling theorem, an anti-aliasing filter has to precede the SC filter. This continuous-time low-pass filter usually is implemented as a standard second-order filter (see Section 11.3.1.1). As the actual signal conditioning is performed by the SC filter, the requirements of the anti-aliasing filter are more relaxed than if the entire signal conditioning function were implemented as a continuous-time filter. Assuming again a signal bandwidth of 1 kHz, the necessary filter can now be implemented as a second-order SC filter with $f_c \approx 1$ kHz. When a switching frequency of 100 kHz is used, the area consumption of the passive elements of this filter implementation can be reduced to about 0.11 mm^2 compared with that of the continuous time approach (see Section 11.3.1.1) accompanied by an increased accuracy of the filter characteristic. But now an additional anti-aliasing filter would be needed: This additional anti-aliasing filter can be implemented again as a second-order continuous-time low-pass filter with sufficient damping at 100 kHz. A rule of thumb suggests a damping of 40 dB at the switching frequency for an anti-aliasing filter, which would lead to a 3 dB cutoff frequency of 10 kHz for a second-order low-pass filter. Compared with the Sallen-Key example in Section 11.3.1.1, the time constant is reduced by a factor of 10, so the area consumption of the passive elements is 10 times lower.

The use of CMOS transmission gates as switches for the SC filters has a general impact on the performance of SC circuits. When such switches are opened, charge is injected from the transistor gates to the drain and source nodes. This charge is integrated on the connected capacitors and causes an error voltage on these capacitors. This effect has been well known since the beginning of the use of SC circuits. A good analytical description of this clock feed-through effect can be found in [57], which also describes the common method to decrease it: the so-called dummy compensation. Other general means of reducing the clock feed-through are the use of a differential signal path, where a symmetric feed-through cancels out, and the use of the smallest possible switching transistors.

New trends in the field of SC filter architectures are the design for higher signal bandwidth [58–60] and the optimization for low-power applications [61–63]. Recent publications [64, 65] also show that one of the advantages of digital filters, namely the possibility to reconfigure and adapt the filter characteristic, can also be realized in analog SC architectures.

11.3.1.3 GmC Filter Architecture

Another architecture to realize even large time constants in a CMOS technology are the GmC filters. Here a first-order low-pass stage is formed by the transconductance of an operational transconductance amplifier (OTA) connected to a capacitive load as shown in Fig. 11.23. The name GmC filter is derived from the typical CMOS implementation, where the gain of an OTA is defined by the MOS parameter Gm. These filters are therefore also known as OTA-C filters.

To realize large time constants or low cut-off frequencies, a small Gm together with a large C has to be realized. As already pointed out in the preceding sections, the realization of large capacitors in conventional CMOS technologies is limited, but small values of Gm can be realized and makes the GmC architecture very well suited for low-bandwidth signal conditioning such as is required for many microsensor applications.

The requirements on the OTAs for GmC filters differ from the requirements for the previously described SC filters: In contrast to these, the OTAs for GmC filters must have a smaller transconductance Gm together with a large linear input range. Today several advanced OTA architectures exist that are used to implement OTAs in CMOS technology for GmC filters [66].

In order to build up more advanced analog filters such as low- or high-pass filters of orders greater than one or band-pass or band-reject filters, several stages can be combined. The necessary zeros of the desired transfer functions are realized by using feedback structures in the signal path, which causes the pole of the simple low-pass structure to be transformed into a zero. Fig. 11.24 shows a typical implementation of a general-purpose biquad filter [54].

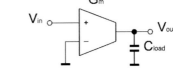

Fig. 11.23 A single transconductance amplifier with a capacitor load form a first-order low-pass filter.

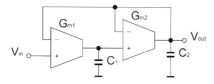

Fig. 11.24 Biquad filter realized with a GmC architecture, comprising two OTAs and capacitor loads.

One of the main advantages of the GmC architecture over the SC architecture is the property of continuous time processing. These filters do not need any anti-aliasing measures and are also suited for higher frequency signal processing. However, in contrast to the SC approach, the GmC filters lack accuracy because is the actual filter characteristic depends on the transconductance Gm of the integrated MOS transistors. This parameter is not only affected by process tolerances during the doping and structuring steps but also depends strongly on the operating conditions of the transistor, such as biasing and the operating temperature.

Therefore, GmC filters are usually accomplished in a tunable form and calibration methods at power-on or even during the signal processing have to be introduced [67, 68].

11.3.2
Analog-to-Digital Converters (ADCs)

A general discussion of ADCs would exceed the scope of this chapter. A good overview of integrated ADCs can be found in [69]. Various ADC architectures and their capabilities are compared in [70, 71]. This section focuses on frequently used ADC architectures for MEMS interfaces. The ADCs for CMOS integrated micro/nanosystems can be divided into two categories with different general requirements: ADCs for individual devices and ADCs for device arrays. After a general discussion of the two categories, the sigma–delta ($\Sigma\Delta$) ADC, which is the most popular and versatile architecture for sensors, will be detailed.

11.3.2.1 ADCs for Individual Sensors

The most important specifications for single-sensor ADCs are accuracy and linearity. The conversion rate is usually less critical because most sensors (except for, e.g., particle detectors) monitor slowly changing environmental signals. Besides $\Sigma\Delta$ modulators, which will be described in Section 11.3.2.3, single-slope/dual-slope converters, successive approximation converters and algorithmic converters are frequently used.

Fig. 11.25 shows a possible implementation of a dual-slope ADC. The input voltage is integrated for a known duration t_1. Then, the time that is needed to return back to zero with a known reference signal V_{ref} is measured. The digital output code is calculated as the ratio of the measured time t_2 and the known time t_1 (see Fig. 11.25). Single-slope and dual-slope converters are popular because they are simple, require only a minimum of circuitry (integrator, comparator, some dig-

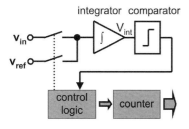 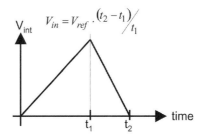

Fig. 11.25 Schematic of a dual-slope ADC.

ital circuitry) and require no precision elements. Furthermore, they are inherently linear. Single-slope and dual-slope ADCs are mostly used for low-speed (<100 ks/s)#Q2#, low-accuracy (~8 bit) applications. For medium speed and accuracy (up to a few Ms/s, up to 12 bits), successive approximation ADCs [72–74] and algorithmic ADCs [75–77] can be used. Both types of ADCs require a conversion time of n cycles to achieve n-bit accuracy. The successive approximation ADC uses an n-bit DAC to match the unknown value at the input of the ADC. Starting from the most significant bit, one DAC bit is set in each cycle and the resulting DAC voltage is compared with the input voltage. If the DAC voltage exceeds the input voltage then bit m of the DAC is reset before bit $m+1$ of the DAC is set. In this way, the remaining error after n cycles is less than half of the least-significant bit (LSB). The simplest version of the algorithmic ADC, which is sometimes also referred to as cyclic ADC, subtracts the reference voltage from the input voltage if the reference voltage exceeds the input voltage. Then the remaining error is multiplied by a factor of 2 and the result compared with the input voltage again, thus yielding an error of less than half an LSB after n conversion cycles.

11.3.2.2 ADCs for Sensor Arrays

There are two possible approaches for sensor arrays with a large number of elements: either one fast ADC is used to convert the multiplexed signals from all elements or each individual element employs a slow ADC. In the latter case of a cell-level ADC, small area and low power consumption are the most important features. This makes single-slope ADCs and dual-slope ADCs an attractive choice [7].

The counter and the ramp generator can be shared between all cells, thus reducing the size of the ADC. Yang et al. [78] developed a modified conversion algorithm (novel ramp generator) that also eliminates the need for a latch inside the cell. As a result, four detectors and the ADC of their 320×256 pixel CMOS imager (0.35 µm technology) fit into a 20×20 µm² cell. An 8-bit first-order $\Sigma\Delta$ ADC (cell area 30×30 µm²) [79] and a 13-bit two-step ADC (230×230 µm²) [80] have been presented for applications in which the area is not so critical.

The use of cell-level ADCs allows data to be processed already in the sensor array. This reduces the amount of data that has to be transmitted off-chip and can eliminate the need for additional data-processing hardware.

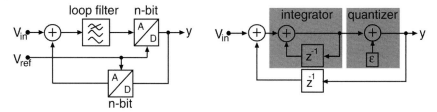

Fig. 11.26 Block diagram and signal-flow graph of first-order $\Sigma\Delta$ modulator.

If a single ADC is multiplexed between all elements (or shared for one row), the size of the converter is not critical. For most sensors, the sampling rate is rather low (between a few samples per second and 100 ks/s), which leads to an overall sampling rate below 100 Ms/s for an array with 1000 elements. Therefore, in most cases well-known high-speed ADC architectures [69], such as flash converters, pipelined converters and folding converters, are used.

11.3.2.3 Sigma–Delta ADCs

$\Sigma\Delta$ modulators with a single-bit quantizer are the most popular choice because of the following features:

- High accuracy can be achieved without reference-voltage scaling; no accurate matching of resistors or capacitors is required.
- No calibration is required.
- They are faster than single-slope and dual-slope converters.
- They are strictly monotonic.
- They contain only a small number of elements (small area).
- The specifications of the anti-aliasing filter are relaxed because of the oversampling.

The basic idea of the $\Sigma\Delta$ ADC is to use a very coarse quantizer to convert the signal. In most cases, a simple comparator (1-bit ADC) is employed. Instead of sampling the signal at the Nyquist frequency f_N (twice the bandwidth of the signal, $f_N = 2f_{BW}$), oversampling is used in combination with a feedback mechanism (noise shaping) to reduce the large quantization error of the 1-bit ADC.

Fig. 11.26 shows a block diagram and the signal-flow graph of a first-order $\Sigma\Delta$ modulator. If we assume that initially the output of the integrator is equal to the (positive) input voltage, then the output of the comparator is logic '1' and the DAC output is set to $-V_{ref}$ for the following conversion cycle. At the end of the next cycle, the output of the comparator is equivalent to

$$V_{out} = V_{in} + (V_{in} - V_{ref}), \tag{21}$$

where $(V_{in}-V_{ref})$ is the quantization error from the first comparison. Therefore, the integrator filters the quantization error. The resulting transfer function (in the z-domain) of the $\Sigma\Delta$-modulator is given by

$$Y = [X + (1 - z^{-1})Q], \tag{22}$$

where Q is the quantization error. The quantization noise is concentrated at high frequencies outside the signal band (noise shaping). At the output of the modulator, a digital decimation filter is needed to remove the high-frequency noise and reduce the sampling-rate from the oversampling frequency f_S to the Nyquist frequency f_N. The noise floor at low frequencies is given by the thermal noise of the integrators. If we neglect this thermal noise and assume an ideal low pass for the decimation filter, the SNR at the output of the modulator is given by

$$\text{SNR[dB]} \cong 30 \log(\text{OSR}) - 3.4, \tag{23}$$

where OSR is the oversampling ratio. From the SNR, the number of bits N for the ADC can be calculated by

$$N = \frac{\text{SNR[dB]} - 1.76}{6.02} \tag{24}$$

A higher SNR for a given OSR is obtained if the order of the loop filter or the accuracy of the quantizer and DAC is increased. The majority of the designs for micro- and nanosystem applications use single-bit quantizers and first- or second-order noise-shaping, because multi-bit quantizers require precision elements and high-order loop filters are complex owing to stability issues. Detailed information about multi-bit and high-order $\Sigma\Delta$ ADCs and band-pass $\Sigma\Delta$ ADCs can be found in [81–84].

Second-order single-bit $\Sigma\Delta$ modulators are frequently used because they require a smaller OSR to achieve a given SNR. Furthermore, they experience fewer problems with tones and dead zones than first-order ADCs do [83]. The SNR of a second-order $\Sigma\Delta$ modulator for an OSR >10 is given by

$$\text{SNR[dB]} \cong 50 \log(\text{OSR}) - 11.1 \tag{25}$$

Fig. 11.27 shows the signal-flow diagram of a second-order $\Sigma\Delta$ modulator. The transfer function is now given by

$$Y = z^{-\frac{3}{2}}[X + (1 - z^{-1})^2 Q] \tag{26}$$

The second-order filtering of the quantization error leads to a 20 dB gain in SNR for a 10-fold increase in OSR over the first-order integrator (see Eqs. 23 and 25).

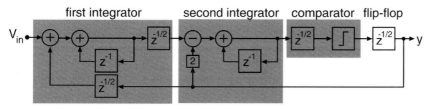

Fig. 11.27 Signal-flow diagram of a second-order $\Sigma\Delta$ modulator.

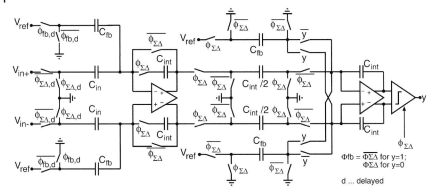

Fig. 11.28 Fully differential second-order switched-capacitor $\Sigma\Delta$.

Fig. 11.28 shows a fully differential switched-capacitor implementation of a second-order single-bit $\Sigma\Delta$ modulator. The first integrator strongly resembles the switched-capacitor amplifier discussed in Section 11.2.3. This $\Sigma\Delta$ modulator requires only a single-ended reference voltage V_{ref}, which is defined relative to the common-mode voltage (analog ground in Fig. 11.28). The negative reference voltage $-V_{ref}$ is generated by interchanging the clock phases on the switches of the negative input of the OpAmp with those of the switches on the positive input. A single-stage folded-cascode operational transconductance amplifier with a switched-capacitor common-mode feedback can be used to implement the OpAmps in Fig. 11.28 [48]. An example including specifications can be found in Section 11.5. The most important specifications for the switches are the on-resistance and the charge injection. To minimize the charge injection, the transistors should be designed as small as possible without violating the on-resistance specification. The effects of charge injection are greatly reduced by a fully differential design because the (signal-independent) charge-injection spikes appear as a common-mode signal. Half-size dummy switches can be used to reduce the charge injection further wherever single transistors are used for the switches. In the case of CMOS switches (NMOS transistor in parallel with PMOS transistor), the clock phases must be carefully selected to allow charge-injection compensation [48].

The output of the $\Sigma\Delta$ modulator is a bit stream at the oversampled frequency f_S. A digital decimation filter is required to remove the high-frequency quantization noise and decimate the sampling frequency to the Nyquist frequency f_N. The simplest form of a decimation filter is a digital counter, which is equivalent to a first-order low-pass filter (or moving-average filter plus decimator). The most popular decimation filter is a combination of a comb filter that performs the coarse filtering and reduces the sampling rate to $8 \times f_N$ and a FIR filter that reaches the specifications and decimates by a factor of 8. The comb filter should be of order $n+1$ for an n-th order $\Sigma\Delta$ modulator. More information on the design of decimation filters for $\Sigma\Delta$ ADCs can be found in [82, 85] and Section 11.3.4 of this chapter.

$\Sigma\Delta$ modulators are very versatile building blocks for micro- and nanosystems applications. The basic architecture of the $\Sigma\Delta$ modulator can easily be adapted to perform additional tasks:

- By changing the oversampling factor and/or the decimation filter, resolution can be traded for conversion rate and power consumption without modification of the converter.
- The sensor element can be incorporated into the modulator, e.g. for capacitive sensors the sensing capacitor replaces the input capacitor C_{in}. The same principle has also been applied to resistive and inductive sensors.
- Reference elements that compensate sensor offsets can also be incorporated.
- The generation of the reference voltage or compensation signals can be performed inside the modulator. This approach has been used [86] to design a temperature sensor that generates the PTAT voltage in addition to the temperature-independent bandgap reference voltage internally.

11.3.3
Digital-to-Analog Converters

There are two main application areas of digital-to-analog converters (DACs) in CMOS microsystems:

1. Static applications, i.e. for tuning voltages and automatic calibration schemes;
2. Dynamic applications, i.e. for actuation.

Static DACs have to be used when calibration or compensation of the measured signal delivered by the integrated transducer is necessary. Micro-scale transducers usually provide weak electric signals, which need high amplification. Mismatch and offsets would lead to saturation of these gain stages, therefore compensation has to be performed at the beginning of the signal conditioning path in the analog domain. The requirements for such DACs are medium resolution (for most cases a resolution of 8 bits has been proved to be sufficient), low power consumption and often also low area consumption. When a closed-loop system for automatic control of the baseline signal is implemented, additionally a monotonic transfer characteristic and the corresponding linearity requirements are important for stable operation. Fig. 11.29 shows an example of such an offset compensation system. Conversion speed and signal bandwidth are low compared with those of the sensor signal, as only static errors or slow drift signals have to be processed.

The second type of applications will be called dynamic applications and are driven by the increasing use of digital signal processing in sensor/actuator systems (see Section 11.3.4). A typical signal flow in a controlled microsystem is that a transducer signal is fed to a digital signal-processing unit via an ADC, in which

Fig. 11.29 Block diagram showing the use of a DAC in an offset compensation system.

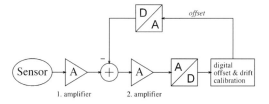

an actuation signal is generated that has to be converted to the analog domain to be applied to an integrated actuator. A good example of such a microsystem is the hotplate-based chemical sensor [23, 87], where a digital controller keeps the temperature of the microhotplate constant for accurate measurements. In contrast to the static application of DACs, in dynamic applications the performance of the converter affects the resolution and accuracy of the whole sensor system. The bandwidth of such converters extends from a few hertz, in the case of the chemical sensor mentioned before, to several kilohertz for resonant mechanical sensors such as an inertial sensor (see Chapter 3).

The requirements of DACs for offset compensation and calibration signals can often be fulfilled by using current- or voltage-weighting signals. The weighting signals can be generated by a matching network of passive components such as resistors or capacitors or by matching MOS transistors acting as current sources. Especially MOS transistor-based converters are also suited for dynamic applications, where a certain signal bandwidth is required. A brief summary of different circuit architectures can be found in [69]. Without special means, a resolution of 8 bits with reasonable accuracy can be achieved in standard CMOS technologies. The actual SNR and transfer linearity are limited by the matching properties of the CMOS technology and can be improved to up to 10 bits with certain layout measures, e.g. implementing a thermometer coding scheme [88].

If DACs with higher resolution are required, different circuit architectures have to be considered: The accuracy can be improved by applying a switching scheme called dynamic element matching. Here the matching elements (this architecture is usually used in conjunction with matching MOS-based converters) are exchanged by a switching scheme. The error due to unmatched devices is supposed to have a statistical distribution, which can be averaged by using different elements for one code. The realizations of dynamic element matching architectures vary from a periodic switching between two elements to the implementation of random generators and permutation of all matching elements. There is a trade-off between achieved resolution and area and power consumption. Other methods use calibration schemes to improve the conversion accuracy.

Other architectures for high-accuracy converters use algorithmic schemes, which are only suitable for low-bandwidth signals, but typically have lower area and power consumption. Such architectures include the use of integrating converters, which are also used for the corresponding ADC type, pulse-width modulation schemes and oversampling schemes (see Section 11.3.2). A good starting point for a literature search on the different converter architectures is again [69].

A common drawback of the mentioned converters with dynamic element matching and the algorithmic converters is that they introduce additional signal frequencies at the analog output, caused by the switching clocks and their intermodulation with the converted signal. These signals usually have to be filtered out by additional analog circuits which makes these DAC architectures less suitable for static applications such as offset compensation.

11.3.4
Digital Filters

In the past few years of integrated circuit development (and therefore also in the field of integrated microsystem development), an increased number of realizations of digital signal processing on-chip could be observed [89].

The combination of analog building blocks with their full-custom design style and a semi-custom or universal digital building block is often termed system-on-chip (SOC) in order to refer to the challenges in the verification of such a complex monolithic system (see also Section 11.4.3). The SOC design approach in general is not new to microsystem designers, as they are used to combine functional blocks with different design approaches together in one system, namely electronic circuitry and transducers.

A main driving force for digital signal processing is the availability of high-end CMOS technologies, which on the one hand also permit the realization of complex digital system at reasonable cost of area and power and on the other hand presume a reliable design verification, owing to large development costs, which can so far only be fulfilled by a digital design flow with reuse of tested and characterized intellectual property (IP). The realization of analog blocks in advanced CMOS technologies with feature sizes beyond 0.25 µm incur a higher risk of redesigns leading to higher development costs.

A general advantage of digital signal processing is the possibility of reconfiguring the processing function merely by reprogramming the system. As stated in [89], an increasing number of signal-processing functions are implemented in software on increasingly standardized digital hardware, in order to cover applications with one (hardware) design, because of the increased (hardware) development costs.

The main advantage of implementing a signal-conditioning function in the digital domain for microsensor applications is the increased accuracy that can be achieved. As described in Section 11.3.1, analog filters are subject to parameter deviations of the CMOS technology, which affect the transfer characteristics. These deviations are totally absent in digital designs. Moreover, limitations in the realization of filter coefficients and the order of the filter function, as described in Section 11.3.1, are eliminated when using a digital implementation. However, general limitations of the filter theory, such as stability, still have to be taken into account.

The resolution and to a certain extent the accuracy of a digital filter are determined by the word length of the processed data [90]. Whereas a high resolution is easy to achieve by increasing the number of bits to be processed, the main bottleneck in a digital signal-processing approach for a sensor system is usually caused by the necessary analog-to-digital conversion (and of course by the digital-to-analog conversion, if an actuator system is realized).

The error induced by the finite bit length of the processed data causes distortion and intermodulation of the signal rather than increasing the noise, although the quantization error is often also called quantization noise.

In contrast to an analog filter, a digital filter adds no electronic noise to the signal, but usually inhibits a worse dynamic range than its analog counterpart,

$$F_{SINC^3}(z) = \frac{1}{N^3} \cdot \left(\frac{1-z^{-N}}{1-z^{-1}}\right)^3 = \frac{1}{N^3} \cdot \frac{1}{1-z^{-1}} \cdot \frac{1}{1-z^{-1}} \cdot \frac{1}{1-z^{-1}} \cdot (1-z^{-N}) \cdot (1-z^{-N}) \cdot (1-z^{-N})$$

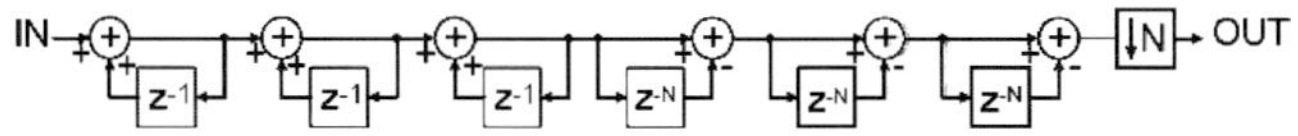

Fig. 11.30 Transfer function and block diagram of a third-order comb filter.

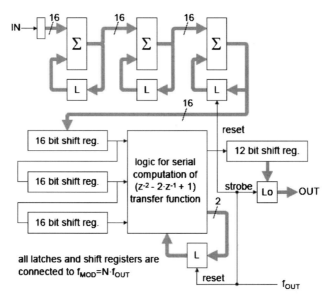

Fig. 11.31 Area-optimized implementation of a third-order sinc filter.

which is responsible for increased harmonic distortion and intermodulation. For low-order filtering having time constants that are not overly long (i.e. corner frequencies of more than 1 Hz), an analog implementation in a CMOS technology is often better suited, because it consumes less area and power than its digital counterpart.

The following example shows a digital decimation filter for a 12-bit $\Sigma\Delta$ AD converter, which is used in a capacitive tactile sensor to monitor the human blood pressure [91]. The modulator is a second-order $\Sigma\Delta$ modulator with an integrated capacitive sensor interface (see Section 11.2.3.2 and 11.3.2.3) and is operated at an oversampling ratio of 128. The specifications of 12-bit resolution and a signal bandwidth of 500 Hz are achieved by a two-stage filter architecture. The first stage is a third-order comb filter, providing a $sinc^3$ transfer function. Fig. 11.30 shows the block diagram of the $sinc^3$ filter and Fig. 11.31 the area-optimized implementation.

The second stage is a 64-tap finite-impulse response (FIR) filter with a linear-phase transfer characteristic, which reduces the number of coefficients to 32 [90]. The actual implementation of this second filter stage is shown in Fig. 11.32. A

11.3 Circuitry Building Blocks for CMOS MEMS/NEMS

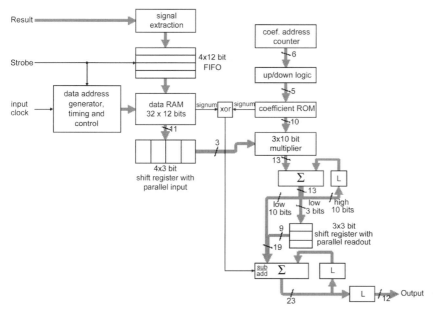

Fig. 11.32 Implementation of a 64-tap linear-phase FIR filter for a medium-scale CMOS technology.

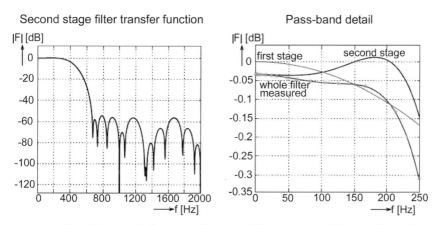

Fig. 11.33 Plot of the transfer functions of the second filter stage and of the overall pass-band characteristics.

data RAM is used to realize an area-efficient implementation of the 32-step-long shift register with a sequential address counter. The coefficients themselves are stored in a ROM block and the multiplication is done partly sequentially in 3-bit pieces.

Fig. 11.33 shows the transfer functions of the second stage and the complete filter. The second stage shows an overshoot around 200 Hz, which increases the flatness of the pass-band transfer function together with the transfer characteristic of the first filter stage. The entire two-stage filter occupies an area of 3.25 mm² including the RAM block in a 0.8 µm CMOS technology.

11.3.5
Calibration and Data Interfaces

Serial interfaces are a common choice because they require only a small number of IO connections. Numerous different serial interfaces have been described in the literature. Over the past few years, some standards for sensor interfaces have been established and some interfaces are widely used in certain areas of application:

- IEEE 1451.1-4 [92]: This standard was established only a few years ago and is not yet widely used. The large number of connections (10 IO ports), the lack of bus capability and the need for on-chip memory to store the 'Transducer Electronic Data Sheet' (TEDS) render it difficult to use for integrated sensor interfaces.
- SPI bus: Widely used in the USA; included with many microcontrollers.
- I2C bus: Developed by Philips, parts of it are proprietary; widely used in Europe (with small modifications).
- CAN bus: The standard bus interface used in the automotive industry.

11.3.6
Current and Voltage References

The on-chip integration of ADCs and DACs has led to a demand for accurate and temperature-independent on-chip voltage references, which are needed to avoid extra IO pins and a large number of external components. The same reasons also triggered the design of on-chip bias-current generators to supply the various analog circuitry building blocks, e.g. amplifiers, filters and ADCs. There are several approaches to generate accurate and temperature-independent voltages or currents, but in this section only the most popular approach, which is based on the approximately linear temperature dependence of a diode, will be described in detail. The weighted sum of a diode voltage and a voltage that is proportional to the absolute temperature (PTAT) results in a temperature-independent reference voltage, which is approximately identical with the bandgap voltage of silicon (see Fig. 11.35). Therefore, the term bandgap reference is used for circuits based on this principle. Most designs use the base-emitter junction of a bipolar transistor instead of a simple p–n junction because the base-emitter junction has fewer parasitic effects.

In addition to diode-based designs, also the breakdown voltage of Zener diodes can be used to design precision voltage references [93], but only very few CMOS technologies offer this option. There are also a number of designs based on the threshold voltage of MOS transistors [94, 95] but they suffer from large fabrication tolerances and the parasitic effects of MOS transistors.

11.3.6.1 Temperature Dependence of the Base-emitter Voltage

The base-emitter voltage V_{BE} of a bipolar transistor as a function of the collector current I_C is accurately described by

$$V_{BE}[T] = \frac{kT}{q} \ln\left(\frac{I_C[T]}{I_S[T]}\right), \tag{27}$$

where T denotes the absolute temperature, q the electron charge and k the Boltzmann constant. The current $I_S[T]$ is given by

$$I_S[T] = \frac{kT}{N_B} A n_i^2[T] \bar{\mu}[T], \tag{28}$$

where N_B is the Gummel number, A the base-emitter junction area, $n_i[T]$ the intrinsic carrier concentration and $\mu[T]$ the effective mobility of minority carriers in the base. The temperature dependence of $n_i[T]$ is described by

$$n_i^2[T] = KT^3 e^{\frac{qV_G[T]}{kT}}, \tag{29}$$

where K is a constant with respect to temperature and $V_G[T]$ is the silicon bandgap voltage. The effective mobility $\mu[T]$ is given by

$$\bar{\mu}[T] = LT^{-n}, \tag{30}$$

where L and n are constants. Unfortunately, it is not possible to calculate all the constants from process data with sufficient accuracy. Therefore, the parameters A, K, L and N_B are eliminated from the equation by using the base-emitter voltage $V_{BE}[T_R]$ measured at a reference current I_R and a reference temperature T_R ($T_R = 300$ K). Only the temperature-dependent terms are retained and they are all relative to T_R. This leads to the equation

$$V_{BE}[T] = V_G[T] - \frac{T}{T_R}(V_G[T_R] - V_{BE}[T_R]) + V_{tR} \ln\left(\frac{I_C[T]}{I_C[T_R]}\right)$$
$$- (4-n)\frac{T}{T_R} V_{tR} \ln\left(\frac{T}{T_R}\right), \tag{31}$$

where $V_{tR} = \frac{kT_R}{q} = 25.85$ mV.

The main contribution to the temperature dependence comes from the second, linear term in Eq. 31. The last term is the curvature term, which exhibits a logarithmic dependence on temperature and is smaller than the linear contribution. If a PTAT current is used to bias the bipolar transistor, then the third term has the same form as the curvature term and the factor $(4-n)$ in the curvature term reduces to $(3-n)$.

Eq. 30 is only an approximation of the temperature dependence of the mobility and the temperature dependence of the bandgap voltage is non-linear [96]. The

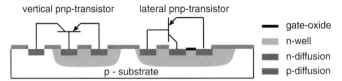

Fig. 11.34 Bipolar transistors available in a standard *n*-well CMOS process. Vertical substrate transistor (left) and lateral *pnp* transistor with *n*-well as base (right).

majority of publications [97–99] uses a constant, extrapolated bandgap V_{G0} and a further parameter η is introduced to fit the measured data:

$$V_{BE}[T] = V_{G0} - \frac{T}{T_R}(V_G[T_R] - V_{BE}[T_R]) - \eta \frac{T}{T_R} V_{tR} \ln\left(\frac{T}{T_R}\right) \qquad (32)$$

The disadvantage of this approach is that the fit parameter η exhibits large variations for different processes. Better results are obtained if the full equation together with the correct temperature dependence of the bandgap voltage $V_G[T]$ is used [86].

Vertical pnp Transistors in *n*-Well CMOS Technology

Two parasitic bipolar transistors are available in standard *n*-well CMOS technology. The lateral *pnp* transistor shown in Fig. 11.34 has a large process spread and exhibits non-idealities as a gate is used to separate emitter and collector. It is rarely used in bandgap references. The vertical *pnp* transistor (see Fig. 11.34) has better performance and less process spread but has a major disadvantage: The collector, which is formed by the *p*-substrate, is tied to the negative supply voltage and hence the collector current is inaccessible. As only the emitter current of the vertical *pnp* transistor can be controlled, the base current has to be subtracted before Eq. 32 can be applied. Furthermore, the current amplification factor $\beta[T]$ is small and depends strongly on temperature and current. Depending on the depth of the well implants and the drain-source diffusions, the current amplification factor $\beta[T]$ can be less than unity. Most of the CMOS references rely on this transistor, but compared with bipolar designs some additional parameters have to be considered: the base current, the base resistance, the Early effect and high- and low-level injection.

PTAT Voltages

A PTAT voltage can be generated as the difference between the base-emitter voltages of two bipolar transistors T_1 and T_2, which are either biased with different currents or have different areas:

$$\Delta V_{BE} = \frac{kT}{q} \ln N, \qquad (33)$$

where N is either the ratio of the bias currents or the ratio of the transistor areas.

11.3.6.2 Bandgap References in Standard CMOS Technologies

The principle and schematic of a CMOS bandgap reference based on vertical *pnp* transistors are shown in Fig. 11.35. Transistor T_1 consists of N unit transistors T_2. The current mirror formed by transistors T_3 and T_4 can either be omitted or used to generate a PTAT current for the OpAmp or other circuits. Because of the OpAmp, V_{ref} is the sum of the base-emitter voltage of transistor T_1 and the difference ΔV_{BE} of T_1 and T_2 scaled with the resistor ratio R_2/R_1 (see Fig. 11.35). In this way, only the linear terms in Eq. 32 are compensated. The non-linear curvature term is linearized around T_R and the current through the transistors T_1 and T_2 is assumed to be PTAT. V_{BE1} is then described by

$$V_{BE}[T] = V_{G0} - \frac{T}{T_R}\{V_G[T_R] - V_{BE}[T_R] - V_{tR}(\eta - 1)\}. \tag{34}$$

To compensate the linear term in Eq. 32, the resistor ratio must be

$$\frac{R_2}{R_1} = \frac{V_G[T_R] - V_{BE}[T_R] - V_{tR}(\eta - 1)}{V_{tR} \ln N} \tag{35}$$

The value of N is typically chosen around 15 to achieve a ΔV_{BE} that is significantly larger than the offset voltage of the OpAmp. Depending on the technology and the reference temperature T_R chosen, this leads to a resistor ratio between 5 and 10. The bias current is usually chosen around 10 mA to avoid high- and low-level injection effects. The remaining error for a design with a careful layout is of the order of a few millivolts over a temperature range from –40 to 120 °C.

For technologies with a large $\beta[T]$, the base voltage can be connected to an analog common-mode voltage (VCM). In this way, the reference voltage V_{ref} is generated relative to VCM.

In addition to the non-linearities, the major sources of error in this approach include the offset voltage of the operational amplifier, the non-ideal PTAT current and the mismatch of the two bipolar transistors.

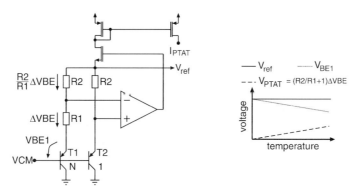

Fig. 11.35 Principle and schematic of a standard bandgap reference in CMOS technology.

If higher accuracies are required, the offset of the operational amplifier must be reduced [42] and the curvature must be corrected with additional circuitry. Furthermore, the temperature dependence of the resistors has to be taken into account. A different approach is to use switched-capacitor techniques to sum up the different contributions [86, 98]. Together with dynamic-element matching techniques and a simple calibration procedure, an accuracy of better than 1 °C over a temperature range from –40 to 120 °C can be achieved in a digital CMOS technology.

11.3.6.3 Current References in Standard CMOS Technologies

The simplest form of current reference is a resistor to V_{dd} in combination with a current mirror, but in many cases the resulting variations due to changes in V_{dd}, temperature and process tolerances cannot be tolerated.

PTAT Current Source

The same circuit that is used for the bandgap reference can also be used to generate a PTAT current. This is done simply by adding a current mirror on top of the source-follower transistor T_5, as shown in Fig. 11.35. If the bandgap reference is used simultaneously to generate a reference voltage and a PTAT bias current, the output of the bandgap reference must be buffered to avoid changes of the PTAT current due to load current.

Constant-current Source

A temperature-independent current results from adding a PTAT current to a current with negative temperature coefficient, which can be obtained by biasing a poly-resistor with constant voltage as shown in Fig. 11.36.

The calculation of the resistor values is similar to the bandgap reference and only first-order contributions are compensated. The detailed calculation can be found in [86], where an error of less than 1.0% has been obtained for a temperature range between –40 and 120 °C.

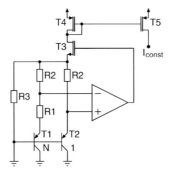

Fig. 11.36 Schematic of the constant-current generator.

11.4
System Integration

11.4.1
Parameter Sensitivity/Power Consumption at the System Level

Parameter Sensitivity

As already mentioned in preceding sections, most of the specifications of integrated circuits, e.g. the frequency characteristic of analog filters (see Section 11.3.1), depend on specifications of the CMOS technology used. These technology parameters are subject to deviations. Referring again to Section 11.3.1.1, a typical value for this deviation would be ±15% for the 3 sigma spread of the polysilicon sheet resistance. Using technologies with smaller feature size (deep sub-micron), the parameter spread of the passive devices and sometimes of the analog properties of the MOS transistors can be increased significantly, which makes these technologies less suitable for analog signal processing, as needed in instrumentation systems. This is tolerated, because these CMOS technologies are usually optimized for high-level digital systems, such as processors, memory and digital signal processing. The circuit designers have to cope with the given parameter deviations. They can choose circuit architectures such as continuous-time analog filters that permit power- and/or area-efficient designs, but are more sensitive to parameter deviations or they use switched-capacitor circuits or a complete digital filter approach for the sake of simplicity, area and sometimes also power but obtain a more robust design to technology variations. Already in the design phase of a new product, the production yield targeted must be a specification. In quality management and the related business science, an expression for high yield and customer satisfaction is the '6 sigma' design. The technology deviation and therefore also the yield spread of the product are supposed to have normal (Gaussian) distribution. To achieve a 6 sigma design means that 99% of the manufactured devices must meet the product specification and failures are only caused by technological defects. In order to achieve this goal the circuit has to be tuned during the design phase to meet its specification with a 6 sigma spread of the technology parameters. One common method to verify this is to provide so-called corner parameters of the devices offered by the CMOS technology, which include the 6 sigma spread, meaning, the minimum and the maximum deviation of the technology parameters, such as sheet resistance and effective geometries. Applying these parameter sets to the circuit simulations often lead to a waste in area or power because the worst-case combination of two or more parameters has a smaller probability than 6 sigma, but will be included in the design specification with this verification approach. New verification approaches use Monte Carlo simulations to obtain a better estimation of the final production yield, than just applying the worst-case scenarios, but they are limited to small building blocks, because of the necessary computing power. Recent research activities have been aimed at another solution, namely the mapping of the technology statistics to the device models and parameters. This is not trivial, especially for MOS transistor devices, because there are many technology parameters that influence the transistor behavior in different ways, e.g. geometries, implantation and diffusion profiles or layer thicknesses. Some of these deviations add to each other, others cancel out. The most promising approach to generate such device models are statistical methods such as principle component analysis [100].

Power Consumption

Another important factor that has to be taken into account in the early design phase is the power dissipation of the device. In addition to the technology parameters, the device temperature is important for proper operation of the electronic circuit (and of course also for the integrated transducers). Usually a temperature range is specified, for which the given specifications are valid. The power dissipated in the device together with the package technology and the ambient temperature range determine the operating temperature of the microsystem (see Chapter 4 in Volume 1). The power dissipation of electronic circuits can be divided into static and dynamic power dissipation. Analog circuits mostly exhibit static power dissipation: the operating points of the electronic devices define the currents and voltages across them and therefore the electric power, independent of the applied signals. The power dissipation is constant over time and easy to calculate for these building blocks

$$P_{static} = V_{supply} \cdot I_{supply} \tag{36}$$

A suitable method to reduce the static power is to downscale the supply voltage. However, this approach is mainly limited by the needed signal range and SNR and will yield only a proportional improvement to the power dissipation. The minimum current consumption is given by the dynamic specifications of the building block, such as signal bandwidth and harmonic distortion.

In addition to the static power dissipation, electronic circuits also exhibit dynamic power dissipation. This mainly occurs in time-discrete systems and is caused by the dynamic currents to charge and discharge capacitive loads and the electric losses of these dynamic currents. This dynamic power dissipation is proportional to the square of the supply voltage, to the applied frequency and to the sum of capacitive load that has to be (dis)charged:

$$P_{dynamic} \propto V_{supply}^2 \cdot f \cdot C_\Sigma \tag{37}$$

This dynamic power dissipation is one of the main driving forces to scale modern CMOS technologies for digital systems down to smaller supply voltages, although this scaling is limited by the occurrence of another power dissipation source: the leakage currents [101].

The above-mentioned methods to reduce the power consumption of integrated circuits are of limited use for instrumentation systems, because the reduction in supply voltage always limits the usable signal range and therefore limits the SNR [102].

There are other methods to reduce the power dissipation of such circuits, which can be summarized by the term 'power management'. One is to use more intelligent amplifier stages that reduce the power dissipation, when no driving current is needed, such as the class AB architecture for large resistive amplifier loads [103]. There has also been extensive research devoted to the development of power efficient output stages for large capacitive loads [104]. However, the term power management includes also methods to reduce the power dissipation of blocks that are not active at certain times. These methods can include temporary reductions in the clock frequency and supply voltage (mainly for digital blocks) [105], and also the complete shutdown of building blocks and the defined waking up, which is more suitable for analog blocks.

11.4.2
Design for Testability

This topic is often crucial for a successful product development. Thus in industrial companies, designers are often instructed by internal guidelines to 'design for testability'.

When selling a product, one must be able to measure all specified parameters of the product, in order to avoid customer dissatisfaction. Many industrial branches force their suppliers to furnish proof of the given specifications by measurements. This applies especially to the automotive and military industries, where quality documentation is standardized.

Design Guidelines

When looking at conventional CMOS products, a good rule of thumb for the origin of costs is that one-third of them are caused by the CMOS wafer fabrication, one-third by the chip packaging and the last third by the testing of the product. For most CMOS-based MEMS devices this rule still applies, but sometimes with a slight increase in the cost share for packaging and testing. That is why reducing the costs of testing is a good means of reducing the overall system costs. 'Design for testability' describes a design strategy that includes the test capabilities of the entire system from the first steps on.

So far guidelines for this design approach can hardly be found in the literature. Many companies have internal guidelines including testability issues, but they are considered to be important intellectual property and therefore kept confidential.

Considering system tests in the early design phase and including reasonable test features can greatly reduce the effort needed for the final production test of the microsystem. This affects both the test equipment required and the time required to test the system functions, both of which are key parameters in determining the testing costs.

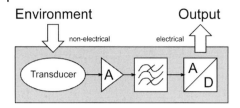

Fig. 11.37 Block diagram of a monolithic sensor system, showing input and output signals.

Test Strategy

Objectives of a test strategy are:

- increase test coverage (ideal, 100%; real, 70–95%);
- reduce costs of testing (equipment and time);
- localize occuring failures in the design and allow further analysis.

Fig. 11.37 shows the signal flow of a monolithic integrated sensor system. The input signal for such a system is in general a non-electrical stimulus, e.g. a change of the ambient temperature in the case of a temperature sensor. A 'straightforward' production test would be to stimulate the sensor device with a defined change of temperature and to record the system response at the digital interface. Such a test would require a large effort compared with the test of a conventional CMOS circuit.

The device under test (DUT) has to be placed in a test environment that has a controlled temperature. Then a temperature stimulus is fed to the device and the response of the DUT is recorded. The advantages of such a test procedure are that most of the specified product parameters, such as sensor resolution, linearity and bandwidth, can be obtained from this test setup.

The disadvantage of such a test strategy is usually the costs. Even for a temperature measurement a thermo-chamber or a precise thermo-chuck for tests on a wafer level are needed. In addition, the bandwidth of the entire measurement setup determines the time needed for the tests. Especially in the case of environmental sensors such as temperature sensors, the system bandwidth is limited by the application of the stimulus rather than by the microsystem itself. A possible failure in the electronic signal processing, for instance, can only be detected after the electrical output has been recorded, which means that a large amount of expensive testing time is wasted on testing a faulty device.

It also makes a further yield improvement of the final product very difficult, because the building block with the highest failure rate cannot be identified. The process of yield improvement is usually an iterative procedure. How much effort should be spent on this procedure depends strongly on the production count and the market price of the product and usually has to be estimated for every design individually.

First, the part of the microsystem with the highest failure rate has to be identified. This can by done be carefully analyzing the manufacturing tolerances and failures together with the system architecture chosen early in the design process if possible. These assumptions have to be confirmed by the failure statistics of the

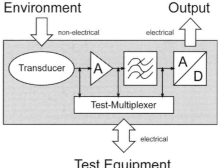

Fig. 11.38 Signal flow of a sensor readout system with additional access to single building blocks.

first product tests and can lead to design corrections in order to improve the test strategy and finally also the production yield.

In order to perform these steps, an individual test of each building block is necessary. Fig. 11.38 shows the block diagram of Fig. 11.37 now with additional test inputs and outputs to access the input and output signals of each functional block.

For electronic circuits and building blocks there are two different approaches to generate suitable tests: the automatic generation of test patterns and the development of application-oriented tests. In the first approach, test stimuli and the expected test responses are generated by the design tools automatically. This has the advantage that a well defined test coverage can be achieved with a reduced effort, because these test patterns are already generated during the design process. The disadvantage is that in the case of a test failure, direct identification of the origin of failure is difficult. Also, there is usually no relationship between a test and the product specification, so in the case of a low production yield, there is no possibility of selling devices with reduced product specifications (binning). This drawback can be circumvented by the second test approach, the development of application-oriented tests, which are based on the product specification. These tests give detailed information on the product specifications that are responsible for the desired production yield. However, they usually need more effort to be developed and mostly require a more complex test environment and longer testing time, thus making them more expensive. The first approach is applied mainly to digital systems and system blocks, whereas the second one is better suited for analog and mixed-signal circuits.

As already mentioned, the time needed for device testing is an important cost factor. Therefore, special care has to be taken when setting up the test order. Tests with the highest failure rate should be performed at the beginning of the production tests and then together with a 'break on failure', no testing time is wasted on faulty devices. During production development and production ramp-up, this rule is often violated, because additional data also on broken devices are collected in order to improve the production yield or the test procedure.

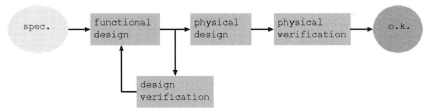

Fig. 11.39 Design flow of integrated circuits development.

11.4.3
Microsystem Verification

Physical Design Verification
After the logical and parametric functionality of the design has been verified, the physical design verification has to be performed.

Whereas in VLSI designs the generation of the layout for analog and digital building blocks is quite different, the necessary verification of the physical design uses the same methods and CAD tools. Fig. 11.39 shows the general design flow of integrated circuits.

The tasks of the physical design verification can be divided into two main objectives:

1. Verification of process rules and guidelines and that the physical design matches with the earlier verified netlist (physical design verification).
2. Verification of functional design including physical parasitics (post-layout functional verification).

The physical design verification is usually split into three different procedures, which can be accomplished interactively or in a batch process:

1. The so-called design rule check (DRC) analyzes the drawn geometries of the different masks layers and checks for violations of the geometric rules for the target technology. In the case of a full-custom physical design, which is still common for analog and key digital building blocks, this check is usually applied interactively during the design phase. Errors are reported graphically, so that the user can fix them interactively. The DRC itself comprises a successive application of geometric operations, in order to calculate sizes, distances and overlapping of the geometric shapes of the mask layers [106, 107].
2. The next step in physical verification is the extraction of the electronic devices and their connectivity from the layout. The algorithms for this task are similar to the DRC and are also mainly based on geometric operations on the layout data to identify electronic devices of the target technology and electric connections between them [106]. The devices and their connectivity are listed in a tabular format, which is called netlist and gives a unique representation of the circuit.
3. The final step to verify that the layout represents the correct devices and connectivity is to compare the netlist from the functional design with the pre-

viously extracted netlist from the physical extraction [layout versus schematic verification (LVS)]. Most software tools use algorithms from the mathematical graph theory to accomplish this task [108].

The physical realization of an integrated circuit in a CMOS technology can greatly influence the electric behavior of the circuit, up to rendering the entire functionality. Some of these effects can already be taken into account in the functional design phase, others can not.

The main effects are caused by parasitic capacitances between the connections layers (polysilicon and metal lines). These lead to a limited bandwidth of the transmitted signal and increase the delay or are responsible for crosstalk of higher frequencies to neighboring lines. The first drawback can be taken into account in the functional design phase. However, the estimation of parasitic delays often leads to suboptimal designs, because a safety margin of the estimated signal delay has to be incorporated to ensure proper functionality. As there are only algorithms for delay estimation during the semi-custom digital design, in the full-custom analog design this has to done by hand by an experienced designer. In critical designs, which exhaust the possibilities of a technology chosen and especially with decreasing feature sizes, the impact of the physical realization on the system functionality increases. In these situations post-layout verification is necessary in order to verify the design functionality. For a post-layout verification the following additional steps of the physical verification have to be accomplished:

1. During the netlist extraction not only the usable devices and their interconnectivity are generated, but also parasitic devices such as interconnection resistances and interlayer capacitances are generated and added to the extracted netlist. These parasitic devices can often be added to the functional circuit representation, which is called back-annotation [109]. Whereas in MEMS design three-dimensional extraction tools are used, such as finite-element or finite-differences solvers, only two-dimensional tools are available for the circuit extraction. This is because the layout of the mask data is generated and stored only for two dimensions. Modern CMOS technologies show aspect ratios, in which the height of the structured layer has the same magnitude as or is even larger than the lateral sizes. Here two-dimensional extractions cannot produce results of sufficient precision for a reasonable verification. As the generation and analysis of three-dimensional data is not accomplishable with currently available computer power, the so-called 2.5-dimensional layout extraction has been developed [110].
2. CMOS-based MEMS products show a similar increase in circuit complexity to conventional CMOS circuits. Starting from 10–100 integrated devices, CMOS-based MEMS today integrate full mixed-signal circuits with analog signal processing, AD/DA converters, processors and sometimes even memory, adding up to tens of thousands of devices. When a large number of parasitic devices such as resistors and capacitors are added to these netlists (back-annotation), verification with the currently available computing power becomes impossible. Therefore, the resulting netlist for the post-layout verification has to be reduced. One straightforward method for netlist reduction is to remove from the netlist parasitic devices

whose values are below a given threshold. This threshold has to be chosen as a trade-off between verification accuracy and simulation time. A more advanced approach is the use of model order reduction techniques [111]. Here the number of circuit elements and nodes is reduced by transformation of the resulting matrix to a matching subspace with a controlled reduction of accuracy. Currently this method is applied only for the passive devices of the circuits [112]. Actual research extends this method also to the non-linear parts of the simulation matrix.

11.5
Example: a Single-Chip Gas-detection System

In this section, a versatile single-chip gas-detection system for the detection of volatile organic compounds (VOCs, e.g. alcohols, benzenes) is described [48, 86]. The chip comprises three polymer-coated micromachined transducers that respond to fundamentally different analyte properties and hence permits quantitative or qualitative analysis of gas mixtures.

Fig. 11.40 shows a chip micrograph with the three transducers and an additional temperature sensor. The temperature sensor accounts for the strong temperature dependence of the bulk physisorption of volatiles in polymers. The three different sensors are all coated with the same sensitive polymer layer. Additional discrimination capabilities can be added to the system by creating or combining an array of identical chips in a multi-chip module and coat each chip with a different sensitive layer.

Fig. 11.41 shows a block diagram of the on-chip interface electronics. The weak transducer signals are first amplified by the analog frontend circuitry and then

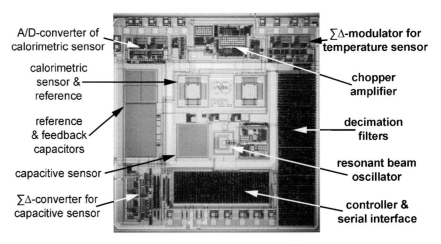

Fig. 11.40 Micrograph of the single-chip gas-detection microsystem. The three micromachined transducers in the center are surrounded by a metallic frame, which shields the interface circuitry and the IO connections in the periphery of the chip from the measurands after flip-chip packaging. The size of the chip is approximately 6×6 mm^2.

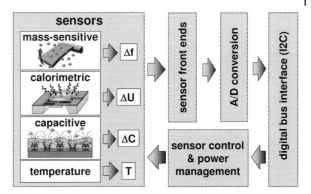

Fig. 11.41 Block diagram of the single-chip gas-detection system. No accurate external references whatsoever are required to operate the chip.

digitized by dedicated ADCs. The digital values are then transmitted to an off-chip data-processing unit via a standard I2C serial interface. The chip also includes a digital controller and all the circuitry to generate the bias-current reference voltages for the four transducers.

11.5.1
Design Flow and Layout Verification

Fig. 11.42 depicts the design flow used during the development of the gas-detection system. Dedicated software packages were employed in the initial design phase: Finite Element Modeling with ANSYS [113] for the transducers, Spectre [114] to simulate the analog interface circuitry and Modelsim [115] together with Synopsys [116] to simulate and synthesize the digital circuitry. So far, there is no single software package that supports all aspects of CMOS MEMS design.

Top-level simulation of the entire system and layout-versus-schematic (LVS) verification are key to avoiding expensive redesigns, which are caused by errors during the layout assembly of complex microsystems. SpectreVerilog was used to perform mixed-signal simulations of the top-level schematics, including analog circuitry, digital components and lumped circuit models of the transducer elements. In this way, the overall functionality of the system and the correct behavior of the various interfaces (transducers/circuitry, analog circuitry/digital components) are verified. The standard design-rule check (DRC) and LVS of the CMOS process produces about 10 000 design-rule violations and extraction errors caused by the transducer elements. A custom extension module that includes the design rules and extraction parameters of the transducers was developed to achieve an error-free top-level DRC and LVS.

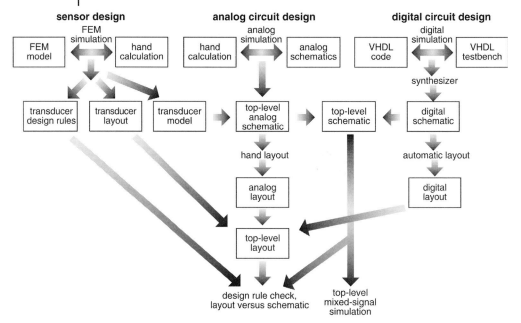

Fig. 11.42 Design flow for the single-chip gas-detection system. The use of a single design environment for sensors and interface circuitry allows top-level simulation and layout verification of the complete system.

11.5.2
Capacitive Chemical Sensor

The sensing principle and structure of the chemocapacitor have already been described in Chapter 7. Fig. 11.43 shows a block diagram of the second-order switched-capacitor ADC, which has been used to read out the capacitive information. The sensing element is incorporated into the first integrator. A tunable reference capacitor that has the same structure as the sensing element but without sensitive coating is used. To reduce thermal drift, the feedback capacitors are also implemented as interdigital capacitors (see Fig. 11.43).

The schematic of the $\Sigma\Delta$ modulator is almost identical with that of the $\Sigma\Delta$ modulator shown in Fig. 11.28. Only the capacitors in the input stage have been replaced by the split-sensing and reference capacitors, as described for the capacitive amplifier in Section 11.2.3. The $\Sigma\Delta$ modulator can be operated at clock frequencies of up to 4 MHz. The most important specifications for the OTA are summarized in Table 11.2. Fig. 11.44 shows a schematic of the folded-cascode OTA. The NMOS current source formed by T_2 has been split to allow independent optimization of the noise performance (to reduce $1/f$ noise, T_2 should be large) and of the gain bandwidth of the CMFB (small, fast transistors) [48]. A switched-capacitor CMFB has been used, because of its low power consumption and superior input range compared with other configurations [21]. The $\Sigma\Delta$ modulator achieves 19-bit resolution for a bandwidth of 1 Hz. Owing to the small bandwidth and the high accuracy of the sensor, an on-chip digital decimation filter would consume too much area. Therefore, a simple counter (20 bits) has been used to decimate the bitstream from the $\Sigma\Delta$ modulator.

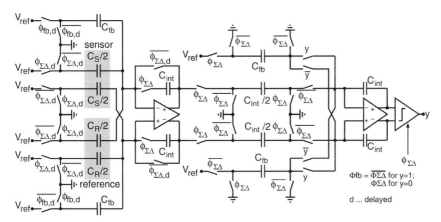

Fig. 11.43 Schematic of the second-order switched-capacitor $\Sigma\Delta$ modulator for the capacitive chemical sensor.

Tab. 11.2 Specifications of the OTA for the first switched-capacitor integrator in the $\Sigma\Delta$ modulator

specification	DC-gain 90 dB	GBW 30 MHz	therm. noise 10 nV/Hz$^{1/2}$	1/f-corner 30 kHz	slew rate 5 V/s	output swing +/− 3 V
affects....	integrator leakage	max. sampling frequency	SNR	SNR	settling	input range

GBW... gain-bandwidth product

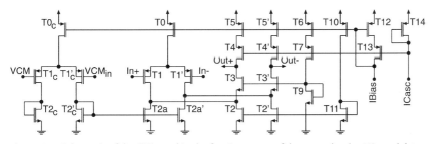

Fig. 11.44 Schematic of the OTA used in the first integrator of the second-order $\Sigma\Delta$ modulator.

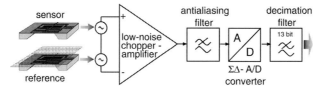

Fig. 11.45 Block diagram of the read-out circuitry for the calorimetric sensor.

11.5.3
Calorimetric Sensor

The polymer-coated thermopile, which is used for the calorimetric chemical sensor, has already been described in Chapter 7. Fig. 11.45 shows a block diagram of the read-out circuitry for the calorimetric sensor. The chopper amplifier has already been described in Section 11.2.4.1. The anti-aliasing filter at the output of the chopper amplifier is a second-order gm-C filter with a cut-off frequency of 5 kHz. The $\Sigma\Delta$ modulator, which has been described in Section 11.3.2, is used in combination with a third-order comb filter and a 13-tap FIR filter (13 bits) to achieve an overall accuracy of 12 bits at the output of the ADC. The reference voltage V_{ref} (relative to VCM) is provided by an on-chip bandgap reference, as described in Section 11.3.6.

11.5.4
Mass-sensitive Resonant Cantilever

There are basically three ways to determine the resonance frequency of a cantilever:

1. Noise spectrum: The frequency spectrum of the thermal noise is measured and analyzed. The peak frequency is equivalent to the resonance frequency. This method requires no excitation of the cantilever, but a low-noise spectrum analyzer is needed for the measurement.
2. Frequency sweep of excitation signal: The cantilever is excited at different frequencies and the response is evaluated to determine the resonance frequency. A gain-phase meter is needed for the measurement.
3. Oscillator: The cantilever is used as the loop filter in an oscillator. A fast feedback is needed in this case, but a simple counter can be used to determine the resonance frequency.

The third option leads to the most accurate results because the Q-factor of the cantilever is enhanced by the feedback [27]. An actuation mechanism and a deflection sensor are needed to realize a monolithically integrated design of an oscillator. Furthermore, only the third option can be realized in CMOS technology with reasonable effort because integrated spectrum analyzers and gain-phase meters are difficult to design.

11.5.4.1 Resonant-beam Oscillator

The circuitry for the resonant-beam oscillator is shown in Fig. 11.46. The output signal of the Wheatstone bridge is first amplified by a low-noise differential difference amplifier (DDA). The amplification factor can have a maximum value of 35 in order to prevent saturation of the amplifier by the DC-offset of the Wheatstone bridge. The signal is then high-pass filtered to remove the offset voltages of the bridge and the first amplifier. The high-pass filter also prevents up-conversion of the amplifier's $1/f$ noise. Finally, the filter eliminates the low-frequency thermal crosstalk and therefore avoids resonances due to this mechanism. AC coupling at the input of the first amplifier would allow a higher gain in the first amplification stage, but would entail the disadvantages of added noise and of an additional path for switching interference and substrate noise to the most critical point in the circuit.

The signal is then amplified and high-pass filtered a second time before it is converted into a square-wave signal by the comparator. This second amplification stage is needed to achieve a sufficiently large amplitude at the input of the comparator. The minimum amplitude at the input of the comparator is defined by the following:

- The input offset of the comparator: An amplitude at least 10 times larger than the offset-voltage (~1 mV) is needed for the desired duty cycle of 45–55%.
- The noise and crosstalk at the input of the comparator: Switching of the comparator because of noise coupling from the digital circuitry and the large switching transistor prevents stable oscillation. A hysteresis at the input of the comparator cannot be used as this would prevent the onset of the oscillation owing to thermal noise in the Wheatstone bridge and mechanical noise from the cantilever.

After the comparator, a digital delay line is used to adjust the phase in order to achieve positive feedback at the resonance frequency. A source follower at the output of the delay line drives the small heating resistor. The short-term stability of the oscillator was characterized using the Allen variance [27]. Using a measurement interval of 1 s and a dynamic heating power of 10 mW, the Allen variance was determined to be 9×10^{-8}. This corresponds to a short-term frequency stability of 0.04 Hz.

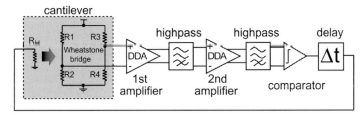

Fig. 11.46 Schematic of the resonant-beam oscillator.

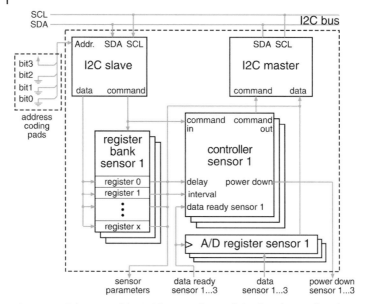

Fig. 11.47 Schematic of the I2C bus interface and the digital controller. The controller, register bank and the register for the digital output values have been replicated for each sensor.

11.5.5
Controller and Digital Interface

Fig. 11.47 shows a schematic of the controller and the I2C-interface, which is included on the multisensor chip. Both have been designed and simulated using VHDL.

11.5.5.1 Serial Interface

The multisensor chip was designed for a gas-detection system that combines several identical chips with different sensitive coatings. The I2C bus offers some advantages for this task compared with other serial interfaces:

- Simple, serial protocol: The I2C protocol clearly defines the most important features (addressing, word length, master/slave communication, start/stop condition) without adding expensive overheads that require large amounts of on-chip memory and digital circuitry.
- Small number of connections: Only two wires for clock and data are needed.
- Bus-enabled: Up to 127 devices can be connected on the same bus. A simple arbitration algorithm solves the problem of data collision.

The I2C-bus offers two possible modes of operation:
- Single-master: An external unit polls the multisensor chips and collects the data from the chips.
- Multi-master: Each individual chip can initiate a data transfer.

The multi-master mode was chosen in order to reduce the data traffic on the bus. The data-recording unit only initializes the multisensor chips and sets their timing parameters. The multisensor chips then send their data at regular intervals without being polled by the data-recording unit. Besides reducing the traffic on the bus, this also provides more flexibility for future extensions, e.g. threshold-limiting values where one sensor triggers the system when its signal exceeds a certain threshold.

11.5.5.2 Digital Controller

The digital controller (Fig. 11.47) interprets the commands coming from the data-recording unit, stores the sensor parameters and manages the access of the individual sensors to the I2C bus. Each sensor on the chip can be addressed individually via the serial interface. A minimal set of commands was implemented for the controller: power on/off, read data (with programmable intervals), set/read parameters.

11.6 References

1 M.W. JUDY, 'Evolution of integrated inertial MEMS technology,' in: *Solid-State Sensor, Actuator and Microsystem Workshop, Hilton Head*; **2004**, 27–32.
2 J. JI, K. NAJAFI, K.D. WISE, 'A scaled electronically-configurable multichannel recording array,' *Sens. Actuators A* **1990**, *22*, 589–591.
3 T. MULLER, *An Industrial CMOS Process Family for Integrated Silicon Sensors*; PhD Thesis, ETH 13463: ETH Zurich, **1999**.
4 T. MULLER, M. BRANDL, O. BRAND, H. BALTES, 'An industrial CMOS process family adapted for the fabrication of smart silicon sensors,' *Sens. Actuators A* **2000**, *84*, 126–133.
5 ITRS, 'International technology roadmap for semiconductors,' http://public.itrs.net, **2003**.
6 P. VETTIGER, G. BINNIG, 'The nanodrive project,' *Sci. Am.* **2003**, *288*, 34–41.
7 S. KLEINFELDER, L. SUKHWAN, L. XINQIAO, A. EL GAMAL, 'A 10000 frames/s CMOS digital pixel sensor,' *IEEE J. Solid-State Circuits* **2001**, *36*, 2049–2059.
8 C.R. HELMS, 'Semiconductor technology research, development, and manufacturing: status, challenges, and solutions,' in: *International Conference on Characterization and Metrology for ULSI Technology, University of Texas, Austin, TX*; **2003**.
9 C.M. HAMMERSCHMIED, Q. HUANG, 'Design and implementation of an untrimmed MOSFET-only 10-bit A/D converter with –79-dB THD,' *IEEE J. Solid-State Circuits* **1998**, *33*, 1148–1157.
10 L.J. PU, Y.P. TSIVIDIS, 'Transistor-only frequency-selective circuits,' *IEEE J. Solid State Circuits* **1990**, *25*, 821–832.
11 J. SAUERBREY, T. TILLE, D. SCHMITT-LANDSIEDEL, R. THEWES, 'A 0.7-V MOSFET-only switched-opamp Sigma Delta modulator in standard digital CMOS technology,' *IEEE J. Solid-State Circuits* **2002**, *37*, 1662–1669.
12 Austriamicrosystems, www.austriamicrosystems.com.
13 B. MURARI, 'Bridging the gap between the digital and real worlds: the expanding role of analog-interface technologies,' in: *IEEE International Solid-State Circuits Conference, San Francisco*; **2003**, 30–35.
14 H. BALLAN, M. DECLERCQ, *High Voltage Devices and Circuits in Standard CMOS Technologies*; Dordrecht: Kluwer, **1999**.
15 B.J. BALIGA, *Modern Power Devices*; New York: Wiley, **1987**.

16 B. J. BALIGA, *High Voltage Integrated Circuits*; New York: IEEE Press, **1988**.
17 C. HAGLEITNER, D. LANGE, T. AKIYAMA, A. TONIN, R. VOGT, H. BALTES, 'On-chip circuitry for a CMOS parallel scanning AFM,' *Proc. SPIE* **1999**, *3673*, 240–248.
18 T. SALO, T. VANCURA, O. BRAND, H. BALTES, 'CMOS-based sealed membranes for medical tactile sensor arrays,' *IEEE Micro Electro Mech. Syst.* **2003**, 590–593.
19 K. NAJAFI, 'Sensor–system interface: the influence of on-chip electronics,' in: *International Symposium on Circuits and Systems, Philadelphia, PA*; **1987**, 233–236.
20 J. A. GEEN, S. J. SHERMAN, J. F. CHANG, S. R. LEWIS, 'Single-chip surface micromachined integrated gyroscope with 50/h Allan deviation,' *IEEE J. Solid-State Circuits* **2002**, *37*, 1860–1866.
21 K. R. LAKER, W. SANSEN, *Design of Analog Integrated Circuits and Systems*; New York: McGraw-Hill, **1994**.
22 J. H. HUIJSING, *Operational Amplifiers, Theory and Design*; Boston: Kluwer, **2000**.
23 D. BARRETTINO, M. GRAF, W. H. SONG, K.-U. KIRSTEIN, A. HIERLEMANN, H. BALTES, 'Hotplate-based monolithic CMOS microsystems for gas detection and material characterization for operating temperatures up to 500 °C,' *IEEE J. Solid-State Circuits* **2004**, *39*, 1202–1207.
24 E. SÄCKINGER, W. GUGGENBÜHL, 'A versatile building block: the CMOS differential difference amplifier,' *IEEE J. Solid-State Circuits* **1987**, *SC 22*, 287–294.
25 A. J. GANO, J. E. FRANCA, 'New fully differential variable gain instrumentation amplifier based on a DDA topology,' in: *Instrumentation and Measurement Technology Conference, IMTC*; **1999**, 60–64.
26 C. VANCURA, M. RÜEGG, Y. LI, D. LANGE, C. HAGLEITNER, O. BRAND, A. HIERLEMANN, H. BALTES, 'Magnetically actuated CMOS resonant cantilever gas sensor for volatile organic compounds,' in: *International Conference on Solid-State Sensors, Actuators and Microsystems*; **2003**, 1355–1358.
27 D. LANGE, O. BRAND, H. BALTES, *CMOS Cantilever Sensor Systems*; Heidelberg: Springer, **2002**.
28 M. L. MEADE, *Lock-in Amplifiers: Principles and Applications*; London: Peregrinus, **1983**.
29 V. P. PETKOV, B. E. BOSER, 'Capacitive interfaces for MEMS,' in: *Enabling Technology for MEMS and Nanodevices*, vol. 1, H. BALTES, O. BRAND, G. K. FEDDER, C. HIEROLD, J. G. KORVINK, O. TABATA (eds.); **2004**, 49–92.
30 N. YAZDI, K. NAJAFI, A. S. SALIAN, 'A high-sensitivity silicon accelerometer with a folded-electrode structure,' *J. Microelectromech. Syst.* **2003**, *12*, 479–486.
31 J. WU, G. K. FEDDER, L. R. CARLEY, 'A low-noise low-offset chopper-stabilized capacitive-readout amplifier for CMOS MEMS accelerometers,' in: *IEEE International Solid-State Circuits Conference*; **2002**, 428–429.
32 M. LEMKIN, B. E. BOSER, 'A three-axis micromachined accelerometer with a CMOS position-sense interface and digital offset-trim electronics,' *IEEE J. Solid-State Circuits* **1999**, *34*, 456–468.
33 A. A. SESHIA, M. PALANIAPAN, T. A. ROESSIG, R. T. HOWE, R. W. GOOCH, T. R. SCHIMERT, S. MONTAGUE, 'A vacuum packaged surface micromachined resonant accelerometer,' *J. Microelectromech. Syst.* **2002**, *11*, 784–793.
34 C. HAGLEITNER, D. LANGE, A. HIERLEMANN, O. BRAND, H. BALTES, 'CMOS single-chip gas detection system comprising capacitive, calorimetric and mass-sensitive microsensors,' *IEEE J. Solid-State Circuits* **2002**, *37*, 1867–1878.
35 A. KOLL, *CMOS Capacitive Chemical Microsystems for Volatile Organic Compounds*; PhD Thesis, ETH 13460, ETH Zurich, **1999**.
36 P. P. L. REGTIEN, 'Solid-state humidity sensors,' *Sens. Actuators* **1981**, *2*, 85–95.
37 R. UNBEHAUEN, A. CICHOCKI, *MOS Switched-capacitor and Continuous-time Integrated Circuits and Systems: Analysis and Design*; Berlin: Springer, **1989**.
38 V. S. L. CHEUNG, L. H. C., *Design of Low-voltage CMOS Switched-opamp Switched-capacitor Systems*; Boston: Kluwer, **2003**.
39 J. GOETTE, *Contributions to the Noise Analysis of Switched-capacitor Circuits*; PhD Thesis, ETH 10214, ETH Zurich, **1994**.

40 A. SCHAUFELBUEHL, N. SCHNEEBERGER, U. MUNCH, M. WAELTI, O. PAUL, O. BRAND, H. BALTES, C. MENOLFI, H. QIUTING, E. DOERING, M. LOEPFE, 'Uncooled low-cost thermal imager based on micromachined CMOS integrated sensor array,' *J. Microelectromech. Syst.* **2001**, *10*, 503–510.

41 A. SCHAUFELBUEHL, *Thermal Imagers in CMOS Technology*, PhD Thesis, ETH 14484, ETH Zurich, **2002**.

42 A. BAKKER, K. THIELE, J.H. HUIJSING, 'A CMOS nested-chopper instrumentation amplifier with 100-nV offset,' *IEEE J. Solid-State Circuits* **2000**, *35*, 1877–1883.

43 C.C. ENZ, E.A. VITTOZ, F. KRUMMENACHER, 'A CMOS chopper amplifier,' *IEEE J. Solid-State Circuits* **1987**, *22*, 335–342.

44 H.W. KLEIN, W.L. ENGL, 'Design techniques for low noise CMOS operational amplifiers,' in: *European Solid-State Circuits Conference*; **1984**, 27–30.

45 C. MENOLFI, Q. HUANG, 'A low-noise CMOS instrumentation amplifier for thermoelectric infrared detectors,' *IEEE J. Solid-State Circuits* **1997**, *32*, 968–976.

46 C. MENOLFI, Q. HUANG, 'A fully integrated, untrimmed CMOS instrumentation amplifier with submicrovolt offset,' *IEEE J. Solid-State Circuits* **1999**, *34*, 415–420.

47 C. MENOLFI, *Low-noise CMOS Chopper Instrumentation Amplifiers for Thermoelectric Microsensors*; PhD Thesis, ETH 13583, ETH Zurich, **2000**.

48 C. HAGLEITNER, A. HIERLEMANN, H. BALTES, 'CMOS single-chip gas detection systems: Part II,' in *Sensors Update*, vol. 12, H. BALTES, G.K. FEDDER, J.G. KORVINK (eds.) Weinheim: Wiley-VCH, **2003**, 51–120.

49 F. KRUMMENACHER, N. JOEHL, 'A 4-MHz CMOS continuous-time filter with on-chip automatic tuning,' *IEEE J. Solid-State Circuits* **1988**, *23*, 750–758.

50 S. TEDJA, J. VAN DER SPIEGEL, H.H. WILLIAMS, 'A CMOS low-noise and low-power charge sampling integrated circuit for capacitive detector/sensor interfaces,' *IEEE J. Solid-State Circuits* **1995**, *30*, 110–119.

51 W.M.C. SANSEN, Z.Y. CHANG, 'Limits of Low noise performance of detector readout front ends in CMOS technology,' *IEEE Trans. Circuits Syst.* **1990**, *37*, 1375–1382.

52 Y. HU, E. NYGARD, 'A new design of a low noise, low power consumption CMOS charge amplifier,' *Nucl. Instrum. Methods Phys. Res. A* **1995**, *365*, 193–197.

53 U. TIETZE, C. SCHENK, *Electronic Circuits*; New York: Springer, **2003**.

54 K. SU, *Analog Filters*, 2nd edn; Boston: Kluwer, **2002**.

55 R.P. SALLEN, E.L. KEY, 'A practical method of design in RC-active filters,' *IEEE Trans. Circuit Theory* **1955**, *2*, 74–85.

56 Y. NISHI, R. DOERING, *Handbook of Semiconductor Manufacturing Technology*, New York: Marcel Dekker, **2000**.

57 B.J. SHEU, C. HU, 'Switch-induced error voltage on a switched capacitor,' *IEEE J. Solid-State Circuits* **1984**, *19*, 911–913.

58 W. ALOISI, G. GIUSTOLISI, G. PALUMBO, 'Exploiting the high-frequency performance of low-voltage low-power SC filters,' *IEEE Trans. Circuits Systems II: Express Briefs* **2004**, *51*, 77–84.

59 A. BASCHIROTTO, F. SEVERI, R. CASTELLO, 'A 200-Ms/s 10 mW switched-capacitor filter in 0.5-µm CMOS technology,' *IEEE J. Solid-State Circuits* **2000**, *147*, 196–200.

60 L. LENTOLA, G.M. CORELAZZI, E. MALAVASI, A. BASCHIROTTO, 'Design of SC filters for video applications,' *IEEE Trans. Circuits Syst. Video Technol.* **2000**, *10*, 14–22.

61 P. FILORAMO, G. GIUSTOLISI, G. PALMISANO, G. PALUMBO, 'Approach to the design of low-voltage SC filters,' *IEE Proc. Circuits Devices Syst.* **2000**, *147*, 196–200.

62 V.S.L. CHEUNG, H.C. LUONG, K. WING-HUNG, 'A 1-V 10.7-MHz switched-opamp band-pass sigma-delta modulator using double-sampling finite-gain-compensation technique,' *IEEE J. Solid-State Circuits* **2002**, *37*, 1215–1225.

63 L. LENTOLA, A. MOZZI, A. NEVIANI, A. BASCHIROTTO, 'A 1 µA front end for pace maker atrial sensing channels with early sensing capability,' *IEEE Trans. Circuits Syst. II: Analog Digital Signal Process.* **2003**, *50*, 397–403.

64 A. CARUSONE, D.A. JOHNS, 'Analogue adaptive filters: past and present,' *IEE Proc. Circuits Devices Syst.* **2000**, *147*, 82–90.

65 T. Ndjountche, R. Unbehauen, 'Analogue discrete-time basic structures for adaptive IIR filters,' *IEE Proc. Circuits Devices Syst.* **2000**, *147*, 250–256.

66 E. Sanchez-Sinencio, J. Silva-Martinez, 'CMOS transconductance amplifiers, architectures and active filters: a tutorial,' *IEE Proc. Circuits Devices Syst.* **2000**, *147*, 3–12.

67 J. Silva-Martinez, M. J. Steyaert, W. Sansen, *High Performance CMOS Continuous Time Filters*; Boston: Kluwer, **1993**.

68 J. Silva-Martinez, M. J. Steyaert, W. Sansen, 'A 10.7-MHz 68-dB SNR CMOS continuous-time filter with on-chip automatic tuning,' *IEEE J. Solid-State Circuits* **1992**, *27*, 1843–1853.

69 R. van de Plassche, *CMOS Integrated Analog-to-Digital and Digital-to-Analog Converters*; Boston: Kluwer, **2003**.

70 J. Sevenhans, Z. Y. Chang, 'A/D and D/A conversion for telecommunication,' *IEEE Circuits Devices Mag.* **1998**, *14*, 32–42.

71 R. H. Walden, 'Analog-to-digital converter survey and analysis,' *IEEE J. Sel. Areas Commun.* **1999**, *17*, 539–550.

72 C. S. Lin, B. D. Liu, 'A new successive approximation architecture for low-power low-cost CMOS A/D converter,' *IEEE J. Solid-State Circuits* **2003**, *38*, 54–62.

73 T. P. Redfern, J. J. Connolly, Jr., S. W. Chin, T. M. Frederiksen, 'A monolithic charge-balancing successive approximation A/D technique,' *IEEE J. Solid-State Circuits* **1979**, *14*, 912–920.

74 J. Yuan, C. Svensson, 'A 10-bit 5-MS/s successive approximation ADC cell used in a 70-MS/s ADC array in 1.2 µm CMOS,' *IEEE J. Solid-State Circuits* **1994**, *29*, 866–872.

75 S. Chin, C. Wu, 'A CMOS ratio-independent and gain-insensitive algorithmic analog-to-digital converter,' *IEEE J. Solid-State Circuits* **1996**, *31*, 1201–1207.

76 A. Kitagawa, M. Kokubo, T. Tsukada, T. Matsuura, M. Hotta, K. Maio, E. Yamamoto, E. Imaizumi, 'A 10b 3MSample/s CMOS cyclic ADC,' in: *IEEE International Solid-State Circuits Conference*; **1995**, 280–281.

77 H. Onodera, T. Tateishi, K. Tamaru, 'A cyclic A/D converter that does not require ratio-matched components,' *IEEE J. Solid-State Circuits* **1988**, *23*, 152–158.

78 D. X. D. Yang, B. Fowler, A. El Gamal, 'A Nyquist-rate pixel-level ADC for CMOS image sensors,' *IEEE J. Solid-State Circuits* **1999**, *34*, 348–356.

79 B. Fowler, *CMOS Area Image Sensors with Pixel Level A/D Conversion*; PhD Thesis, Stanford University, **1995**.

80 H. Eltoukhy, K. Salama, A. El Gamal, M. Ronaghi, R. Davis, 'A 0.18 µm CMOS 10^{-6} lux bioluminescence detection system-on-chip,' in: *IEEE International Solid-State Circuits Conference, San Francisco*; **2004**, 222–223.

81 Y. Geerts, M. Steyaert, W. Sansen, *Design of Multi-bit Delta–Sigma A/D Converters*; Boston: Kluwer, **2002**.

82 J. C. Candy, G. C. Temes, *Oversampling Delta–Sigma Data Converters: Theory, Design and Simulation*; New York: IEEE Press, **1992**.

83 S. R. Norsworthy, R. Schreier, *Delta–Sigma Data Converters: Theory, Design and Simulation*; New York: IEEE Press, **1997**.

84 O. Bajdechi, J. H. Huijsing, *Systematic Design of Sigma–Delta Analog-to-Digital Converters*; Boston: Kluwer, **2004**.

85 T. Saramaki, H. Tenhunen, 'Efficient VLSI-realizable decimators for sigma–delta analog-to-digital converters,' in: *International Symposium on Circuits and Systems, Espoo, Finland*; **1988**, 1525–1528.

86 C. Hagleitner, O. Brand, A. Hierlemann, H. Baltes, 'CMOS single-chip gas detection systems: Part I,' in *Sensors Update*, vol. 11, J. G. Korvink (ed.); Weinheim: Wiley-VCH, **2003**, 51–120.

87 D. Barrettino, M. Graf, S. Taschini, C. Hagleitner, A. Hierlemann, H. Baltes, 'A single-chip CMOS micro-hotplate array for gas detection and material characterization,' in: *IEEE International Solid-State Circuits Conference, San Francisco*; **2004**, 314–315.

88 R. J. Baker, H. W. Li, D. E. Boyce, *CMOS: Circuit Design, Layout and Simulation*; Piscataway, NJ: IEEE Press, **1998**.

89 N. M. Donofrio, 'Processors and memory: the drivers of embedded systems toward the networked world,' in: *IEEE International Solid-State Circuits Conference, San Francisco*; **2004**, 20–23.

90 D. SCHICHTHÄRLE, *Digital Filters, Basics and Design*; Berlin: Springer, **2000**.

91 K.-U. KIRSTEIN, J. ŠEDIVÝ, T. SALO, C. HAGLEITNER, T. VANCURA, H. BALTES, 'A CMOS-based tactile sensor for continuous blood pressure monitoring,' in: *European Solid-States Circuits Conference, Leuven*; **2004**, 463–466.

92 R. ALLAN, 'The IEEE family of transducer interface standards,' *www.elecdesign.com/Articles/Index.cfm?ArticleID=2990*, **2003**.

93 B. GILBERT, 'Monolithic voltage and current references: theme and variations,' in: *Analog Integrated Circuits*, J. H. HUIJSING, W. SANSEN, R. J. VAN DE PLASSCHE (eds.); Boston: Kluwer, **1995**, 269–353.

94 K. N. LEUNG, P. K. T. MOK, 'A CMOS voltage reference based on weighted difference between PMOS and NMOS transistors for low-dropout regulators,' in: *European Solid-State Circuits Conference*; **2001**, 88–91.

95 R. A. BLAUSCHILD, P. TUCCI, R. S. MULLER, R. G. MEYER, 'A NMOS voltage reference,' in: *IEEE International Solid-State Circuits Conference, San Francisco*; **1978**, 50–51.

96 Y. P. TSIVIDIS, 'Accurate analysis of temperature effects in I/sub c/V/sub BE/ characteristics with application to bandgap reference sources,' *IEEE J. Solid-State Circuits* **1980**, *SC 15*, 1076–1084.

97 R. C. S. FREIRE, S. DAHER, G. S. DEEP, 'A highly linear single p–n junction temperature sensor,' *IEEE Trans. Instrum. Meas.* **1994**, *43*, 127–132.

98 M. TUTHILL, 'A switched-current, switched-capacitor temperature sensor in 0.6-µm CMOS,' *IEEE J. Solid-State Circuits* **1998**, *33*, 1117–1122.

99 B. S. SONG, P. R. GRAY, 'A precision curvature-compensated CMOS bandgap reference,' *IEEE J. Solid-State Circuits* **1983**, *18*, 634–643.

100 T. B. TARIM, M. ISMAIL, H. H. KUNTMAN, 'Robust design and yield enhancement of low-voltage CMOS analog integrated circuits,' *IEEE Trans. Circuits Syst. I: Fundam. Theory Applic.* **2001**, *48*, 474–486.

101 T. SAKURAI, 'Perspectives on power-aware electronics,' in: *IEEE International Solid-State Circuits Conference, San Francisco*; **2003**, 26–29.

102 V. GOPINATHAN, D. RO, 'Does Moore's law apply to analog? Past, present and future implications of technology progress and higher levels of integration for mixed-signal circuits,' in: *IEEE International Solid-State Circuits Conference, San Francisco*; **2002**, 118–119.

103 P. E. ALLEN, D. R. HOLBERG, *CMOS Analog Circuit Design*, 2nd edn.; New York: Oxford University Press, **2002**.

104 R. HOGERVORST, J. H. HUIJSING, *Design of Low-voltage, Low-power Operational Amplifier Cells*; Boston: Kluwer, **1996**.

105 D. M. MONTICELLI, 'Taking a system approach to energy management,' in: *European Solid-State Circuits Conference, Estoril, Portugal*; **2003**, 15–19.

106 R. BRÜCK, *Entwurfswerkzeuge für VLSI-Layout*; Munich: Carl Hanser, **1993**.

107 T. OHTSUKI, *Layout Design and Verification*; New York: Elsevier Science, **1986**.

108 E. BARKE, 'A network comparison algorithm for layout verification of integrated circuits,' *IEEE Trans. Comput.-Aided Des. Integrated Circuits Syst.* **1984**, *3*, 135–141.

109 E. SICARD, T. DEMONCHAUX, J. L. NOULLET, A. RUBIO, 'Cross-talk extraction from mask layout,' in: *European Conference on Design Automation*; **1993**, 414–418.

110 Cadence, 'Diva Reference,' *www.cadence.com*, Product Version 4.4.6 ed.

111 E. B. RUDNYI, J. G. KORVINK, *Review: Automatic Model Reduction for Transient Simulation of MEMS-based Devices*, vol. 11; Weinheim: Wiley-VCH, **2002**.

112 R. W. FREUND, 'Krylov-subspace methods for reduced-order modeling in circuit simulation,' *J. Comput. Appl. Math.* **2000**, *123*, 395–421.

113 ANSYS, *www.ansys.com*.

114 Spectre, *www.cadence.com*.

115 ModelSim, *www.model.com*.

116 Synopsys, 'Design Compiler,' *www.synopsys.com*.

Index

absolute pressure sensors   300, 309 ff
absorber layers   488
absorption   335 ff, 352 f
AC modulation   530
accelerometers   140 ff, 160–184
accuracy
– analog frontends   526
– digital-to-analog converters (DACs)   550
– filters   541
– fingerprint images   425
– operational transconductance amplifiers (OTAs)   569
acetic acid etching   18, 302
acidity sensors   364
acoustic devices   193–224
action potentials   463
add-on layers   43 ff
adsorption   335 ff
Agilent 0.5 μm process   218
aging   150
air damping   170
air discharge   412 ff
alkaline phosphatase   455
Allen variance   570
ALP2LV/MV   72
aluminum
– bulk material   97 ff
– Seebeck coefficients   485
– thin films   83
aluminum bridge   233
aluminum capacitors   247
aluminum metallization   3 ff, 14 ff, 32 f, 43, 182
– elastic properties   126
aluminum-silicon alloy layers   464
ammonium hydroxide compound etching   20
amorphous silicon   492
amperometric sensors   359 ff, 449
amplifiers   196, 217 f, 227
– chemical sensors   347, 359
– pressure sensors   259
– system integration   513 ff
analog capacitors   519
analog devices (ADI) SOI MEMS   146, 183
analog filters   513 ff, 535 ff
analog frontend   513, 521 ff
analog resistors   519

analog-to-digital converters (ADCs)   513 ff, 523, 544 ff
analysis quality   451
anemometers   502 f
anisotropic etching   8 f, 18 ff, 241
anisotropy factor   287
annealing   34, 43
anodically bonded glass encapsulation   309 ff
anti-aliasing filters   542
anti-fraud techniques   440
antiresonant reflecting optical waveguide (ARROW)   355
application-oriented tests   563
architectures see: block diagrams
array
– four-cantilever   377 ff
– phased   221
artificial fingerprints   441
ASIC technology   517
Atmel's AT77C103A FingerChip   403
atmospheric pressure vapor deposition (APCVD)   5
atomic force microscopy (AFM)   345
Austria Micro System (AMS) process   215 f
authentication   397
auto-zero techniques   522, 534
automatic fingerprint identification system (AFIS)   395 f, 417 f
automatic generation of test pattern   563
Avogadro's number   75
axisymmetric diaphragms   301
axisymmetric plates   274

back annotation   565
back-end metallization   10
bandgap voltages   555 ff
bandpass filters   535 ff
bandwidth
– analog frontends   523
– filters   541
– inertial sensors   179
– operational transconductance amplifiers (OTAs)   569
– switched capacitor amplifier   513, 535
base emitter voltage   555 ff

basicity sensors 364
batch fabrication, biochemical sensors 449 f
beam test structures 115 ff
bending stresses, thin plates 274, 277
benzocyclobutene (BCB) switch 233
BEOL accelerometer 169 ff
Bessel filter 153
bias stability 144
BiCMOS technology 2 ff, 72, 234
– interface circuitry 518, 521
bifurcation 416, 422
BioAPI 436
biochemical sensing systems 378 ff, 447–478
biocompatible materials 14, 464
biological entities 338
biometric capacitive fingerprint sensor systems 391–446
biometric engine level error rate 426
biosensor arrays 454 ff
biquad filters 540 ff
blank reticle 27
blister test 114 f
block diagrams
– ADI XL150 gyroscope 158
– ADI XL50 148
– ADI XL76 152
– calorimetric sensor 503 f, 570
– chopper amplifier 535
– constant current generator 558
– DEP chip 461
– dielectrophoretic cell 460
– digital-to-analog converters 549
– DMOS transistor 519
– DNA hybridization 455
– double check structure 493
– dual slope ACD 545
– earphone 210
– fingerprint sensor 405
– gas detection system 567
– gas flow meter 505
– hotplates 369
– interdigitated capacitive sensor 373
– i-Stat system 450
– microelectrode pixel 467
– microphone 199
– neural probe 471
– operational transconductance amplifiers (OTAs) 569
– oscillator feedback circuitry 347
– pixel recording 468
– pressure sensors 304, 312, 316
– RF switch 225 ff, 232

– Sallen-Key filter 540
– Siemens accelerometer 168
– Sigma–Delta modulator 373, 547
– switched capacitor amplifier 531
– thermal RMS converter 500
– thermocouple 484
– transceiver 229
– Wheatstone bridge 529
blood gas analysis 454
Boltzmann constant 91, 143, 198, 486
boron etch-stop 303 ff
borophosphosilicate glass (BPSG) 38
boundary elements simulation (BEM) 28
breakage test 415
breakdown, dielectric 453
bridge bending 100, 122 f
brittle layers 119
Brownian noise 143 ff, 152 ff, 170 f, 197 f
Bubnov–Galerkin variation 273, 276
buckling 107, 111, 115
– acoustic devices 202 ff
buffered oxide etching (BOE) 8
building blocks 522, 539 ff, 559
bulge test 100 f
bulk material properties 83 ff, 267
bulk micromachining 18 ff, 29, 42
– pressure sensors 257, 300, 305 ff
bulk silicon 368
– accelerometer 181
– inertial sensors 180 ff
bus-enabled interfaces 572

CAE (1.3 μm) 72
cage compounds 338
calibration
– analog frontends 522
– building blocks 554 ff
– interface circuitry 521
– pressure sensors 257, 259
calixarenes 338
calorimetric sensors 353, 502 f
– gas detection system 570 f
– switched capacitor amplifier 535
CAN bus 554
cantilever based sensors
cantilevers 13, 19, 56, 90 f
– biochemical sensors 472 ff
– chemical sensors 345 ff, 377 ff
– gas detection system 570 f
capacitive acoustic devices 196
capacitive bridges 143, 161, 171
capacitive chemical sensors 15, 359, 568 f
capacitive detection mode 151

capacitive fingerprint sensor systems 391–446
capacitive forces 73 f
capacitive MEMS ultrasonic transducers (cMUTs) 217
capacitive microphones 195
capacitive pressure sensors 257, 301–314
capacitive read-out 513
capacitive sensors 527 f
capacitive transducers 401 f
capacitor inductor combination 210
capacitor layers 80
capacitors 142, 229 ff, 235
– chemical sensors 372
– interface circuitry 519
– tunable 227
capture conditions, fingerprint images 424
carbon black-loaded polymers 367
Carnegie Mellon series 169 ff, 186 f
cartridges 450
catalytic thermal sensors 349 ff
cavities 300
CBT (2 μm) 72
cell-based assays 457 ff
cell–electrolyte electrode interfaces 463
cell handling 458 f, 464
cell level ACD 545
cell trapping, dielectrophoretic 458
cellular activity 463
charge carrier transport 74 f
charge-sensitive amplifier (CSA) 538
charge transfer 317, 340 ff
chemical mechanical polishing (CMP) 11
chemical mechanical rubbing test 414
chemical reactions 340 ff
chemical robustness 416 f
chemical sensors 335–390, 568 f
chemical vapor deposition (CVD) 3, 44, 202, 210, 220
chemiluminescence 353, 457
chemisorption 337
chemocapacitors 372 f, 568
chemoresistors 365
chemotransistors 361 ff
chopper amplifier 522, 534 f
circuit integration 27 ff, 513–577
circular diaphragms 293 f, 301
circular plates 277 ff
clamped clamped beams 115
Clark cell 361
Clausius–Mossotti factor 459
clock period
– filters 542

– pressure sensors 319
– switched capacitor amplifier 533
closed couplers, acoustics 194
closed/open loop architectures 150
coatings 336, 401
comb fingers 157 f, 162, 170, 237
common biometric exchange file format (CBEFF) 437
common mode feed back (CMFB) 317
comperator 570
compliance
– acoustic devices 198
– coefficients 73 f, 76
– inertial sensors 147 ff
– pressure sensors 265 ff
– tensors 261, 266
component-based integration 437 ff
compressibility 262, 267
computer-aided engineering (CAE) 6
condensation heat 352
condenser microphones 195
conducting polymers 367
conducting thermocouple materials 85
conducting thin films 69
conductivity
– electrical 82 f
– thermal 69, 73, 86 ff, 96 ff, 323
conductometric sensors 358, 365 ff
constant-current sources 558
constant-flow modes 502
contact opening reticle 27
contact area shifts 424
contact electrode 454
contact oxides 12, 96
contact resistance 80
contamination 453
continuous-time detection 530
control layer, fingerprint sensors 435
controllers, gas detection system 572 f
conversion radio architectures 227
converters
– analog-to-digital (ADCs) 315
– digital-to-analog 513 ff, 523, 549 ff
– thermal 480, 499 ff
cooled photon detectors 482
coplanar waveguides (CWG) 231
copper
– bulk material 97 ff
– metallization 71, 126
– Seebeck coefficients 485
– spiral inductor 242
– thin films 83, 87
core-cladding interface 355
Coriolis acceleration 144 ff, 156 ff, 164 f

correlated double sampling (CDS) 317, 531 f
Cottrell equation 359
Cr–Al–Cr spacer 180
cracking 114
critical pressure 107
cross sensitivity 525
crosstalk 570
crowding effects 240
crown ethers 338
cryptosmartcards 441
current mode 151
current references 554 ff, 558 ff
cut-frame structures 119
cyclodextrins 338
CYE (0.8 μm) 72
cytometers 458

Daimler-Benz accelerometer 181
damascence copper process 234
damping 170, 182
– acoustic devices 198
– coefficients 140, 143
DARPA 165, 183
data interfaces 554 ff
databases, fingerprint sensors 429 f, 435
decimation filter 513, 552
deep reactive ion etching (DRIE) 21, 35, 53, 201 f
defects 24
defence evaluation 167
deflection 103 ff
– acoustic devices 211
– diaphragms 257, 301 ff
deformation functions 105
delay time 570
Delta–Sigma converters 313, 319, 321 ff, 513
demagnificating projection 399 f
demodulation 225, 520
demultiplexer 379
deployment 438 ff
deposition techniques 4 ff, 25, 301
design guidelines
– accelerometers 141 ff
– biochemical sensors 464 ff
– gas detection system 566 f
– system integration 561 f
design-rule-check (DRC) 26 ff, 564, 567 ff
desorption 352
device-under-test (DUT) 205, 527, 562
Dialog Semiconductor accelerometer 181
diamond shape structures 115

diaphragms 106 f
– acoustic devices 195
– pressure sensors 257
dielectric constant 73 f
– chemical sensors 374
– transducers 401
dielectric layers 5, 53, 126 f, 368
– biochemical sensors 452
– fingerprint sensors 407
dielectric thin films 69
dielectrometers 372 f
dielectrophoresis (DEP) 458 ff
diffential nonlinearity (DNL) 523
differential circuit design 526
differential difference amplifier (DDA) 347, 528, 570
differential switched capacitor Sigma–Delta modulator 548
diffraction effects 197
diffusion barrier metallization 71
diffusion coefficients 359
diffusion effects 370
diffusion layers 80, 283 ff
diffusivity 73 f
digital circuitry 517
digital controllers, gas detection system 573 f
digital filters 551 ff
digital light processing (DLP) 46
digital micromirror device (DMD) 22, 46
digital output 521
digital serial protocol controller 452
digital signal processor (DSP) 50
digital sound reconstruction (DSR) 193, 214 ff
digital-to-analog converters (DACs) 513 ff, 523, 549 ff
dilatation 268
diode array pixels 497
diodes, varactor 249
dirt problems, fingerprint sensors 440
discrete time system 542
distance transducers 401
distortions
– analog frontends 524
– filters 551
– fingerprint images 424
DNA analysis 14
doping 4, 24 f
– piezoresistivity 281 ff
– pressure sensors 303
double polysilicon–double metal (DPDM) 184
drain/source implantation 12, 17, 40

drift 526
dry etching 18 ff
dual-band transceiver 229, 248
dual-slope analog-to-digital converters 544
duality principle 353
Dulong–Petit law 75
dynamic mode
– chemical sensors 345
– digital-to-analog converters (DACs) 549
dynamic properties, analog frontend 523 f
dynamic range, filters 551
dynamic signature verification (DSV) 396

earphone construction 208 ff
ECPD10 process 72
elastic moduli 69 ff, 73 f, 126 f
– silicon 266
elastic solids 261
elastic stiffness 73 ff, 76
elastoresistivity tensor 280
electret microphones 195
electrical conductivity 82 f
electrical–electrical transduction 74 ff
electrical material data 80 ff
electrical resistivity 69, 73 ff, 80, 485
electrical–thermal transduction 74 ff
electrically programmable read-only memory (EPROM) 378 f
electrochemical reactions 340 ff
electrochemical read-out 454 f
electrochemical sensors 342, 358 ff
electrode fabrication 454, 464, 470 ff
electrogenic cells 463 ff
electromagnetic resonators 243 ff
electromagnetic waves 353
electromechanical drives 244
electromechanical interfaces 407 f
electron–hole pairs 353, 481
electronic design automation (EDA) 27
electronic noise 551
electrophysiological assays 463
electroplated gold lines 490
electroplating 5 ff, 42, 47
electrostatic deflection 100
electrostatic/thermal actuation 117 f, 235
electrothermomechanical material data 73 ff
encoding 394 f, 403 ff
endothermic reactions 348
energy minimization method 113
engine level error rate 426
enrichment factor 339

enzyme label 455
epipolysilicon 128, 160 f
equal error rate 427
equivalent input noise 197
error rates
– analog frontends 526 f
– filters 551
– fingerprint sensors 426 ff, 432
– piezoresistivity 284
ESD air discharge 412 ff
etch-stop techniques 20, 42, 50
– electrochemical (ECE) 20, 51, 304 ff
– gate polysilicon 167 f
– microbolometers 495
– pressure sensors 259, 300 ff
etching 4, 18 ff, 38 ff, 241
– chemical sensors 346, 372
– interface circuitry 516
ethylcellulose (EC) 380
ethylenediamine-pyrocatechol (EDP) etching 20, 53, 84, 241
– chemical sensors 350 ff
– pressure sensors 303
Euler buckling 115
evanescent waves 354
excitation signal 570
exothermic reactions 348
expansion coefficients 273
   see also: thermal expansion coefficient
externally polarized microphones 195

fabrication 1–68
– biochemical sensors 452, 464 ff
– infrared detectors 481
– interface circuitry 515 ff
– microphones 197 f, 201 f
– SOI diode microbolometer 498
– speakers 209
– system integration 515 ff
– ultrasonic sensors 218 f
Fabry–Perot devices 354 ff
face recognition 395
failure statistics 562
failure-to-capture (FTC) rate 428
failure-to-enrol (FTE) rate 428
false acception rate (FAR) 397, 426
false rejection rate (FRR) 397, 426
faradaic currents 359
Faraday constant 340
feature extraction 416 ff, 424
feedback
– chemical sensors 345
– fingerprint sensors 418, 439

Fermi–Dirac integrals 281 f
Fermi levels 282, 485
ferromagnetic films 16
field-effect transistors (FETs)
   122, 230, 361, 463
field-oxide layers 12, 96
field tests, fingerprint sensors 411
figures-of-merit, thermocouples 486
film bulk acoustic resonator (FBAR) 246
filters 244 ff
– building blocks 539 ff
– digital 551 ff
– fingerprint sensors 421
– interface circuitry 520 ff
– system integration 513 ff
fingerprint sensor systems 16, 391–446
finite element modeling (FEM) 27
– ANSYS 567
– gas detection system 567
– material properties 79 f
finite impulse response (FIR) 313, 552
flexural plate wave devices (FPWs) 343
flexural rigidity 103, 271
flicker noise 163
flow cytometric analysis 454, 458
flow diagram, IC fabrication 4
flow sensors, thermal 480 ff
fluorescence 353
fluorescence-activated cell sorting (FACS)
   458
focal plane arrays (FPA) 481 ff
folded cascode operational transconductance amplifiers (OTAs) 568
force balance equation 140
force feedback 147, 150 f
fork-type rate sensors 157 f, 164
four-terminal transverse voltage strain
   gauge (X-ducer) 260, 297
fractional resistances 294 ff
fracture mechanical properties, silicon 123
fracture strength/toughness
   69 ff, 73 f, 76, 120
frame-stabilized structures 119
fraud, fingerprint sensors 440
frequency conversion 225
frequency response 212, 216
frequency shift 336, 343
frequency sweep 570
Fresnel lenses 489
friction 73
front-side processing, wafers 301
fully differential analog frontend 526 f
fused silica 452

Gabor filter 421
gain stability 144
gallium arsenide 9
– chemical sensors 353
– RF CMOS MEMS 226
Galvanielectrics 463
gas detection system 379 ff, 566 ff
gas flow meters 505
gate capacitance 144
gates 3, 10
– layers 80
– materials 6 f, 167
– metal electrode 362
– polysilicon 86, 167
gauge factors
– capacitive 300
– polysilicon films 284 f
gauge sections 119
Gaussian distribution
– piezoresistivity 283
– system integration 559
Gaussian plane wave filters 421
geometric effects 109, 353
geometric rule violations 564
geometry–material property relations 79 f
germanium 45, 187
glass encapsulation 309 ff
glucose 449
– oxidase 454
GmC filters 537, 543 ff
   see also: transconductance
gold
– black 488
– electrodes 465
– electroplating 181
– layers 6
– Seebeck coefficients 485
gold–tin solder frame 307
graph matching 425
gravimetric mass detection 217
Gray counter 379
gray scale dynamics 408, 425
Greek cross device 81 ff, 92 ff
Gummel number 555
gyroscopes 144 ff

half-bridge topology 170
Hall plates 15, 37
hardware description language (HDL) 27
harmonic distortion 524
heat capacity 69, 73, 95 ff
heat loss 85, 502
heat transfer 75 ff

heating elements 502
Hermann–Schönflies notation 265
high-g protptype 183
high-impedance amplifiers 196
high-temperature capacitive pressure
 sensors 323 ff
high-temperature chemoresistors 367
high-temperature polysilicon 167
high-voltage devices 519
Hooke's law 261–269
hot junctions 350
hotplates 368
 see also: microhotplates
housing, acoustic devices 211
human serum albumin (HSA) 355
hybrid systems 454, 515
hydrazine water etching 49
hydrocarbons 337, 344, 352
hydrofluoric acid (HF) etching 8, 18 f
– pressure sensors 260, 302
hydrogen monoxide 337
hydrogen peroxide 360

I2C bus interface 571
identification 391, 416 ff, 439 ff
identifier on card (IoC) 441
IEEE1451.1-4 interfaces 554
image capture, fingerprint sensors
 408–419
imaging arrays, thermopiles 489
immobilization 464
impedance, acoustic devices 203
impedance spectroscopy 461
implantation techniques 12
impurity concentration, piezoresistors 298
in vivo electrophysical measurements
 470 ff
indentation 121 f
indium phosphide 353
indium tin oxide (ITO) 460, 465
indiumantimonide 481
inductors 229 ff, 238 ff, 250
inertial sensors 137–192, 204
infinite impulse response (IIR) 313
information check, fingerprint sensors 420
infrared detectors, uncooled 480, 483 ff
inkjet printing 455
input common mode feedback (ICMFB)
 amplifier 163
input noise 197 f
input signal domain 73
insertion loss 225
instrumentation amplifier 513 ff, 536

integral nonlinearity (INL) 523
integrated circuits (ICs) 4, 79 ff, 341
integrated electronics 450 ff
integrated inertial sensors 137–192
integrated optics 354 ff
interconnects 3, 38, 452
interface circuitry
– gas detection system 567 ff
– system integration 513 ff
interfaces
– electrogenic cells 463 ff
– electronics 566
– fingerprint sensors 407 ff, 435
interfacial cracking 114
interferences, analog frontends 527
interferometers 354
intermediate CMOS integration 257
intermetal insulators 71
intermetal oxides 12, 96
intermodulation filters 551
internal reflection transducer 400
intra CMOS micromachining 29, 37 ff
– interface circuitry 515 f
– pressure sensors 314 ff
inversion layers 78, 129
ion implantation 5
ion-selective field-effect transistor (ISFET)
 362, 378, 464
ionic compounds 338
iris recognition 394 f
irradiation 353
islands, silicon layers 368
isolated transistors 519
isopropyl alcohol (IPA) 303
isotropic etching 8 f, 18 f, 21, 302
isotropic materials, Hooke's law 269 ff
i-Stat Portable Clinical Analyzer (PCA)
 449
ITRS road map 517

Johnson noise 198 ff, 206 f, 486
junctions 9 ff, 37 ff, 180 f, 234
– chemical sensors 350
– pressure sensors 260, 283, 322

K band frequencies 243
Kanda effect 281, 299
Kelvin relation 74 f
kernel layer, fingerprint sensors 435
Kirchoff condition 102
KPY60 high pressure sensor 298
Krautkramer transducer 220

Lamé's constants 269
Langmuir–Blodgett films 338
Laplacian operator 272, 278
laser trimming 149
lateral accelerometers 169 ff
lattice defects 370
layers
– acoustic devices 196 ff
– add-on 43 ff
– biochemical sensors 453, 464
– brittle 119
– chemical sensors 336 ff, 352
– dielectric 368
– fabrication 3 ff, 25
– fingerprint sensors 407
– interface circuitry 516, 520
– inversion 78
– material characterization
 80 ff, 89 ff, 96, 234 f
– piezoresistivity 129, 283 ff
– polysilicon 167
– post CMOS micromachining 43 ff
– pressure sensors 259 ff, 300, 305
– system integration 565
– thermocouples 488
 see also: passivation-, sacrificial-, oxide-layers
layout-versus-schematic (LVS) verification
 26, 565 ff
lead zirconoate titanate (PZT) tranducers
 219 f
least significant bit (LSB) 522, 545
lift-off technique 7, 15
LIGA 178 f
light-emitting diode (LED) 357
limit-of-detection (LOD) 525
linearity
– analog frontends 523
– pressure sensors 276
– system integration 513
lipids 338
liquid flow 502
lithographic structuring 27, 464
living cells 463
load deflection models 103 ff
local oxidation of silicon (LOCOS) 9
long diaphragms 109 f
long-term stability 72
– analog frontends 526
long-wave infrared thermal sensors 481
Lorentz forces 346
loudspeaker 205
low-doped drain (LDD) technology 3, 10
low-noise amplifiers (LNA) 227, 536

low-pressure CVD (LPCVD) 5, 25, 38
– chemical sensors 350, 355
low-pressure sensors 298
low-temperature chemoresistors 366
low-temperature oxide (LTO)–spin on
 glass–LTO sandwich (LTO–SOG–LTO)
 47, 180
lower explosive limit (LEL) 336, 349

M test (mechanical) 118
Mach–Zehnder interferometers 354
magnetic actuation 346
magnetic modes 354
MAMOS transistors 182
manifold air pressure (MAP) sensing
 260 f
masking 6, 18, 27
– interface circuitry 518
– layers 564
– materials 18
mass-sensitive resonant cantilever 570 f
mass-spring damper system 140
matcher-on-card (MoC) 441
matching, fingerprint sensors
 394 f, 419 f, 424
material characterization 11 ff, 69–136,
 338, 452
Mauguin notation 265
measurement techniques 69 ff, 79
mechanical–electrical transduction 77
mechanical energy domain 69
mechanical material data 123 ff
mechanical–mechanical transduction 76
mechanical properties 100 ff, 125
mechanical sensitivity 140
mechanical shock resistance 414 ff
mechanical switch 420
medium temperature deposition 25
membrane-based test structures 89 f
membranes 105 f
mercury cadmium telluride 481
metal bulk materials 97 ff
metal dielectric structures 172
metal–electrolyte interface 464
metal films 492, 296
metal inertial sensors 176 ff
metal layers 6 ff, 53, 169, 235
– acoustic devices 196
– biochemical sensors 453
– interface circuitry 520
– system integration 565
metal oxide semiconductor field effect
 transistor (MOSFET) 3, 361

metal oxide semiconductor implementation system (MOSIS)   215 f, 218
metal oxide–polymer diaphragms   196
metal-plated accelerometers   177 ff
metal silicon contacts   44
metal thin films   83
metallic sensor materials   338
metallization
– elastic properties   124 ff
– fabrication   2–16, 32–40
– interface circuitry   516
– material properties   71 ff, 80 ff
metrics, inertial sensors   140 f
microbolometers   479–512
microcontact printing   7, 455
microelecrode arrays (MAs)   463
microfabriation   1–68
micro-g accelerometers   139 f
microhotplates   367 ff
micromachining
– cantilevers   345 ff
– pressure sensors   261 ff
– switches   229 ff
micromechanical structuring   79
microphones   193 ff, 201 ff
microsensors   335
microspectrometers   356 ff
microstripline filters   243
microtensile test   100, 119 f
mid-wave infrared thermal sensors   481
miniaturization   195
– biochemical sensors   448 ff
– fingerprint sensors   394, 409
– RF CMOS MEMS   225
minutiae encoding   403 f, 417 f, 422
mixed-signal processing   315 ff, 245
mobility   144
– data interfaces   555
– RF CMOS MEMS   225
Modelsim   567
modular integrated CMOS and microstructures (MICS)   161
modular monolithic microelectromechanical systems (M'EMS)   33
modulation, RF CMOS MEMS   225
  see also: Delta–Sigma
moisture, fingerprint sensors   412
molding   7
molecular crystals   338
monolithic integration
– capacitive sensors   530
– inertial sensors   137–192
– pressure sensors   259, 299
– RF CMOS MEMS   226 ff

Monte Carlo simulations   559
Moore's law   517
multiband transceivers   228 ff
multibit quantizers   547
multichip solutions   372
multilevel metallization   12
multimaster interfaces   572
multiparameter biochemical microsensor systems   378 ff
multiple component processing   194
multiplexers
– biochemical sensors   451, 466
– chemical sensors   379
– system integration   545
multisensor systems   376 ff
multitransducer gas sensor systems   379 ff

n-wells   9, 12, 357
nanoindentation   100, 121 f
Nasicon   338
near the edge (E) sensor   219
Nernst equation   340, 361
neural activity   14
neural probe   471
neurite guidance   464
nickel   338
– bulk material   97 ff
– ring gyroscope   47
– Seebeck coefficients   485
– structure materials   180
– thin films   83
nitric acid etching   18, 302
nitride layers   38
noble metals   338
– coatings   369
– doping   372
– electrodes   465
– layers   6
noise
– acoustic devices   193 ff, 197 ff
– analog frontends   525 ff
– filters   551
– inertial sensors   140 ff, 152 ff, 163 ff, 170
– power spectrum   207
– resonant devices   570
– RF CMOS MEMS   225
– switched capacitor amplifier   532, 537
  see also: signal-to-noise ratio
noise-equivalent temperature difference (NETD)   479–512
normal stresses   262, 280 f
normalization, fingerprint images   422
Nyquist frequency   315, 546

offset
- analog frontends 513, 522
- switched capacitor amplifier 535
ohmic conductors 291
ohmic switches 231
on-chip amplifier 217
on-chip multiplexer 451
on-chip signal processing 195, 340
Onsager thermodynamics 75
OpAmp-based interface circuits 519 f, 530 f, 548
operating temperatures, chemical sensors 368
operation modes, acoustic devices 195
operational amplifiers 359
operational transconductance amplifiers (OTAs) 310, 543 f, 568
optical interrogation 461
optical path 399
optical read-out 457 f
optical sensors 337, 342, 353 ff
optical transducer 399 ff
orientation dependence, piezoresistivity 288
oscillating mechanical surface deformation 343
oscillators
- circuits 164
- frequencies 228
- gas detection system 571 f
- resonant cantilevers 570
output, interface circuitry 520
output signal domain 73
oxide nitride layer sandwich 89
oxide semiconductor field effect transistor (OSFET) 463, 466
oxygen measurements 360
oxygen precipitation 24
oxygen vacancies 370

p-wells 9, 12
packaging 231
- acoustic devices 197
- biochemical sensors 453, 473 ff
- inertial sensors 150
- interface circuitry 521
- pressure sensors 300
- RF CMOS MEMS 226
- system integration 560
palladium 338, 362
  see also: noble metals
parallel-plate capacitors 142, 162
parameter deviation 559

parameter sensitivity 513, 559 ff
parametric adjustments, inertial sensors 149 f
parasitic capacitance 240
- inertial sensors 144, 152 f, 172
- interface circuitry 517
- pyroelectric elements 539
- system integration 565
parasitic resistance 80
partition coefficient 337 ff, 343, 348
passivation 5, 12 f
- thermal properties 96, 99
passivation layers 25, 44
- biochemical sensors 453, 464
- fingerprint sensors 407
passive devices 516
passive test structures 115 f
patterning 6, 25, 455
Pauw test structure 80, 92 f
pellistors 349 ff
Peltier coefficient 73 f
pencil-drop test 415
pendulous accelerometer 140
performance, biometric system 425 ff
permittivity factor 459
person's verification 391, 416 ff, 439 ff
personal identification number (PIN) 393, 398 f
- diodes 230 f
personal rights protection 396
pH sensors 364, 454
phantom powered microphones 196
phased arrays 221 f
phosphate buffered saline (PBS) 360
phospholipids 338
phosphorescence 353
phosphorus implantation 38
phosphosilicate glass (PSG) 10, 44
photolithography 4 ff, 153
photons 353, 482
photopatterned silicone gasket 454
photoresists 39, 167
- layers 6, 234
- plating mold 47
photostructurable layers 7
phototransistors 357
phthalocyanines 338
physical design verification 513, 564
physical properties 69 ff
physical vapor deposition (PVD) 5, 44
physisorption 337, 566
Pierce oscillator circuits 164
piezoelectric elements 534 ff, 539
piezoelectric materials 196

piezoresistive accelerometers 169, 180 f
piezoresistive coefficients 69 ff, 73 ff, 102, 122 ff, 129
piezoresistive pressure sensors 49, 257 ff, 278 ff
piezoresistors
– analog frontends 527
– diaphrams 297 ff
Pirani pressure gauges 56, 261, 323 f
pixel arrays, fingerprint sensors 406 f
pixel-by-pixel compensation 457
plane stress approximation 270 ff
plane wave filters 421
plasma-enhanced CVD (PECVD) 3, 71
plastic properties 119 ff
plating mold 47
platinum 6, 14, 338, 349
– bulk material 97
– electrodes 454, 465
– Seebeck coefficients 485
– thin films 83
plug-up process sequences 36
pn-junctions 180, 234
– fabrication 11 ff, 37 ff, 50 ff
– pressure sensors 260, 283, 322
pnp-transistors 556
pointer structures 115
Poisson ratio 69 ff, 73 ff, 101, 110, 119, 126 f
– pressure sensors 267, 272, 284
polarization 354
polyether urethane (PEUT) 375 ff, 380
polyaniline 367
polybenzoxarole 456
polycrystalline silicon *see:* polysilicon
polydimethylsiloxane (PDMS) 7, 219, 352, 380
polygermanium 45, 187
polyimides 196, 465
polymer coatings 336, 453
polymer layers 6 f, 566
polymers 338
– conducting 367
polypyrroles 367, 454
polysilicon 3 ff, 10, 32, 43, 53
– accelerometer 167
– acoustic devices 196
– chemical sensors 349
– elastic properties 126
– inertial sensors 146 ff
– layers 85 ff, 167 ff, 305
– material properties 83
– piezoresistors 284 f, 305 ff
– pressure sensors 259, 305, 312 f

– refill 37
– residual stress 124
– resistors 502
– Seebeck coefficients 485
– system integration 559, 565
– tensile properties 128
– thermal properties 97
polysilicon–germanium (polySi-Ge) materials 187, 492
polysilicon–polysilixon interconnects 44
polythiophene 367
polyurethanes 338, 347, 375 ff, 380
post-CMOS micromachining 43 ff
– chemical sensors 344, 372
– interface circuitry 515 f
– passivation layers 453
– pressure sensors 257
potassium hydroxide (KOH) etching 5, 19 ff, 24 f, 41, 50 f
– chemical sensors 346, 350 ff, 372
– inductors 241
– inertial sensors 180
– pressure sensors 260, 303
potentiometric sensors 358, 361 ff
power amplifier 228
power consumption 151
– interface circuitry 517
– RF CMOS MEMS 226
– system integration 513, 559 ff
pre-CMOS micromachining 29, 33 ff
– pressure sensors 257
– thermal sensors 515 f
preamplifiers
– biochemical sensors 466
– interface circuitry 520
precipitation defects 24
precision, analog frontends 526
preprocessing, fingerprint sensors 417 f, 421 f
pressure sensors 49, 56, 257–334
Privium 394
processing 9 ff, 17 ff, 24 ff, 69
– biochemical sensors 452
– diaphragms 201
– inertial sensors 145 f
– pressure sensors 300–313
processor smartcards 441
proportional-integral-differential (PID) controller 369
proportional-to-the-absolute-temperature (PTAT) 554 ff, 558 ff
protocols, interfaces 572
pseudo-differential amplifiers 528
Pseudomas fluorescense 450

pyroelectric analog frontend 534 ff
pyroelectric elements 539
pyroelectric infrared detectors 492
PZT tranducers 219 f

quadrature signals 228
quality factor
– biochemical sensors 451
– chopper amplifier 536
– fingerprint sensors 409 ff
– resonant cantilevers 570
– RF CMOS MEMS 225
– spiral inductors 240
quantization error 546
quantization noise 162 f
quantum efficiency 353
quantum-well infrared photodetectors 481
quartz microbalance (QMB) 343

radiation resistance 198, 209
radiation sensors 480 ff
radio receivers 196
raw image enhancement 421
Rayleigh–Ritz variation 273
Rayleigh surface acoustic waves 343 ff
reactive ion etching (RIE) 8, 11, 169
– chemical sensors 346, 355
– microbolometers 495
receiver–operator curve (ROC) 427
receivers 227 ff
receptors 338, 348
recognition processes 394 ff, 411
redox cycling 455
reduced Fermi level 282
reference electrodes 359, 454
reference fingerprints 425
reflection optical waveguides 355
refractive index 354
regeneration transistors 538
reliability 72, 397
request fingerprints 425
residual stress 73 f, 109, 236
residual stress/strain 101, 123
resistive sensors 526 f
resistivity
– electrical 69, 73 ff, 80
– fingerprint sensors 412 ff
resistors 149, 502
resolution
– analog frontends 513, 522
– digital-to-analog converters (DACs) 549

– filters 551
– fingerprint images 425
resonance frequencies 77, 144
– chemical sensors 336, 343
resonant-beam oscillator 571 f
resonant cantilever 345, 570 f
resonators
– inertial sensors 156 ff
– RF CMOS MEMS 229 ff, 243 ff, 343
– thin film bulk acoustic 246
reticles 27
reversibility, chemical sensors 338
RF CMOS MEMS 225–256
RF filtering 206
ridges, fingerprint sensors 416, 430
rigidity 103, 267, 271
ring gyroscope 47, 179
ring plates 277 ff
Robert Bosch accelerometer 160 f
robustness 79, 414 ff
Root Allan deviation 160
root mean square (RMS) voltage 479, 499 f
rotated coordinate systems 284 f
rotating pointer structures 115
rotation, fingerprint images 424
round robin test 128
rub-off test 414

Saccharomyces cerevisae 461
sacrificial aluminum etching (SALE) 22, 55
sacrificial layers 36 ff, 89
– pressure sensors 259 ff, 300, 305
sacrificial oxide etching 38
sacrificial photoresist layer 234
sacrificial spacer 180
Sallen–Key filters 540
sample handling 120, 473 ff
sample-to-answer-path 450
sandwich technology 3 ff, 47, 54, 89, 96
scalability 451
scaling 283, 451
scanning area limits 425
scanning probe techniques 472 ff
scattering 353
scenario evaluation, fingerprint sensors 429
schematics *see:* block diagrams
scratch resistivity 415
security, fingerprint sensors 438 ff
Seebeck coefficients 69 ff, 73, 84 ff, 87 ff
– thermocouple materials 350 ff, 483 ff
segmentation, fingerprint sensors 421
selectivity 338, 375
selfalignement 305

selfheating 88, 499
semiconducting substrates 9
semiconducting thin films 69
semiconductor switches 230
sense nodes 172 f
sensitivity
– acoustic devices 193 ff
– analog frontends 525
– biochemical sensors 451
– fingerprint sensors 407
– inertial sensors 140 ff, 160, 177 f, 183
– system integration 513, 559 ff
sensor arrays 12
sensor device integration 434 ff
separation-by-implantation-of-oxygen (SIMOX) 40, 128, 322, 372
serial interfaces 513, 572 f
serpentine metal-oxide mesh pattern 201 f, 211 f, 218 f
Severinghaus pH FET 378
shear mode resonator 343
shear modulus 76, 267
shear stress 262, 280 f
sheet resistance 73, 80, 94, 559
shock tolerance 140
short-term stability 570
sidewall implantation 37
Siemens gate polysilicon accelerometer 167
Sigma–Delta converter
– analog-to-digital 546 ff
– chemical sensors 351, 373, 380
Sigma–Delta electrostatic force 161 ff
Sigma–Delta modulators 544, 568
signal processing
– biochemical sensors 452
– fingerprint sensors 399 ff
– pressure sensors 259, 315 ff
signal-to-noise ratio (SNR) 152, 340, 513 577
signature recognition 396
silicides metallization 71
silicon 3 ff
– acoustic devices 196
– bulk material 97 ff
– chemical sensors 368
– data interfaces 555
– diaphragms 257
– elastic properties 126
– material properties 74 ff, 287 ff
– microbolometers 492
– pressure sensors 259, 304 ff
– RF CMOS MEMS 226

– Seebeck coefficients 485
– wafers 287 ff, 452
silicon dioxide
– layers 3 ff
– material properties 74 ff
– thermal properties 96 ff
silicon germanium BiCMOS devices 234
silicon nitride
– chemical sensors 355
– diaphragms 260, 305 ff
– elastic properties 127
– layers 3 ff, 38, 52 f, 305
– material properties 74 ff, 78
– residual stress 125
– tensile properties 128
silicon-on-glass (SOG) 47, 55, 180
– accelerometers 184
– coatings 71
silicon-on-insulator (SOI) 2 ff, 35 f, 88
– chemical sensors 372
– inertial sensors 183 f
– microbolometers 497
– pressure sensors 322
silicon oxide
– elastic properties 127
– residual stress 124
– tensile properties 128
silicon trench etching 37
simulation results, acoustic devices 212
sinc filters 552
single-chip gas detection system 566 ff
single-chip integration 250 f
single-chip radios 225
single-master interfaces 572
skin effects 240
small-size design 197 ff
smart cart integration 440 ff
software development kit (SDK) 434 ff, 438
software solutions, fingerprint sensors 416 ff, 419 ff
sol-gel process 372
solder-bonded capacitive pressure sensors 307 ff
sound pressure displacement 200
sound pressure level (SPL) 205
sound waves 195
source drain currents 361
source drain implantation 12, 17, 40
space charge region 361
spacers 180
speaker recognition 396
speakers 193 f, 208 ff

specific heat 69 ff
specifications
- interface circuitry 520 ff
- operational transconductance amplifiers (OTAs) 569
- switched capacitor amplifier 535
Spectre Synopsys 567
spiral inductor 240
spot noise 160, 455
spring constant 140, 153
springs, acoustic devices 202 f
sputtering 6, 42, 45
- chemical sensors 344
square diaphragms 106 f
squeeze film damping 182, 186, 198 f, 201 ff
ST's TouchStrip TCS3B capacitive transducer 403
stability 72
- analog frontends 526
- fingerprint sensors 432
- resonant beam oscillator 570
standard biometric templates 437
standard software, fingerprint sensors 435 ff
static mode, chemical sensors 345
static properties, analog frontend 522 f
static random access memory (SRAM) 46
stepper based lithography 27
stiction, inertial sensors 147, 149 ff
stiffness 73 f, 76
- tensors 261, 265 ff
storage smartcards 441
strain gauges
- diaphragms 304 ff
- pressure sensors 259
strain tensors 261
stress 150
- diaphragms 257
- relaxation 85
stress–strain relationship 102
SU-8 substrates 7, 465
substrate materials 5 ff
- biochemical sensors 452
- interface circuitry 516 f
substrate temperatures 88
super-heterodyne radio architectures 227
surface acoustic waves (SAW) 343 ff
surface coating 464
surface deformation 343
surface doping 372
surface energy, fingerprint sensors 412

surface micromachining 22 ff, 29
- microbolometers 491
- pressure sensors 257, 300 ff, 344 ff
- test structures 89 f
surface properties 7, 78, 115
- fingerprint sensors 399
suspended inductors 238 ff
suspended rings 115
suspension 151
- spring constant 140
swipe sensors 403 f, 440
switched capacitor
- Delta–Sigma modulator 317 ff
- filters 541 ff
- pressure sensors 257
- system integration 513 ff, 531 ff, 568
switching 229 ff
- fingerprint sensors 420 ff
symmetry transition 113
synthetic fingerprints 430
system integration 513–577
- fingerprint sensors 433 ff
system level error rates 428
system-on-card (SoC) 441
system-on-chip
- filters 551
- interface circuitry 521
system properties, analog frontend 525 f
system scalabililty 451

temperature coefficient of resistance (TCR) 74, 82 ff, 95 ff
- microbolometers 492
temperature dependence 150, 350
- base-emitter voltage 555 ff
- layers 491
- piezoresistivity 281 ff
- pressure sensors 259
- sensitivity 207
- transducers 402
template-on-card (ToC) 441
templates
- fingerprint sensors 425, 437
- interface circuitry 521
tensile stress 147, 262
tensile test 119 f
tensor notation 261 f
tented arches 418
termination 416
test structures 88 ff, 115 ff, 150, 513, 561 ff
tetramethyl ammoniumhydroxide (TMAH) 20, 24, 53, 303, 495

thermal accelerometer 175 f
thermal actuation 345
thermal conductivity 69, 73, 86 ff
– chemical sensors 368
– pressure sensors 323
thermal–electrical transduction 74 ff
thermal energy domain 69
thermal expansion coefficient (TCE) 69, 73 ff, 101, 114 ff, 126 f, 236
thermal–mechanical transduction 77
thermal pressure sensors 324 ff
thermal properties 86 ff
thermal sensors 337, 342, 348 f, 479–512
thermal–thermal transduction 77 ff
thermal transducers 402 ff
thermal van der Pauw test structure 80, 92 f
thermally stabilized accelerometers 173
thermistors 6
thermochuck test 562
thermocouples 350, 483 ff
thermoelastic friction 73
thermoelectric coolers 498
thermoelectric analog frontend 534 ff
thermoelectric material data 80 ff, 86 ff
thermoelectric sensors 350 ff
thermomechanical noise 142
– acoustic devices 193, 197 ff
  see also: Brownian noise
thermopiles 12, 479–512
– chemical sensors 350
– imaging arrays 489
thermopower 69, 82 ff
thermovoltage 350
thickness control, diaphragms 302 ff
thickness-shear-mode resonator 343
thin diaphragms 300
thin films 69–136, 149
thin plate deflection 271 ff
thinned fingerprint images 423
Thomson coefficient 73 f, 76
three-electrode system 359
threshold limited value (TLV) 336
time constants
– filters 541
– thermal 100
time etch-stop 303 ff
  see also: etch-stop
time-of-flight flow sensors 502 f
tin dioxide 337
tin-lead solder frame 307
tin oxides 338, 370
tissue cultures 463

titanium
– bulk material 97
– bolometer structures 41
– thin films 83
titanium–gold seed layer 47
titanium nitride thermocouples 488
total harmonic distortion (THD) 524
total internal reflection transducer 400
TouchStrip TCS3B capacitive transducer 403
toughness 69 ff, 73 f, 76, 101
transceivers, multiband 228 ff
transconductance 144
– switched capacitor amplifier 537, 543 ff, 548
transducer electronic data sheet (TEDS) 554
transducers
– acoustic 194, 214 ff, 219 ff
– biochemical sensors 451
– chemical sensors 336 ff, 376 f, 379 ff
– fingerprint sensors 399 ff
– gas detection system 567
– pressure sensors 259 ff
– system integration 560
transduction effects 70 ff, 203 ff
transfer functions
– analog frontends 523
– biquad filter 540
– Delta–Sigma modulators 257, 321 ff
– FIR filter 553
– switched capacitor amplifier 538
transformation tensors 263 ff, 287 ff
transients, chemical sensors 352
transistors 167, 182, 361
– interface circuitry 516 f, 556
– RF CMOS MEMS 226
– switched capacitor amplifier 537
transmission gates 542
transmission lines 228, 243
transverse modes 354
trench isolation 35, 184
tri-axial piezoresistive accelerometer 182
trimming capability 149
triple-well technologies 519
tunable capacitors 227
tungsten
– bulk material 97
– elastic properties 126
– plug 71
– thin films 83
tuning fork resonant sensors 164
tuning voltage 234

twin-well technology 3
two-beam interferometer 355

ultrasonic sensors 193 f, 217 ff
uncooled infrared detectors 480, 483 ff
user interaction, fingerprint sensors 439 ff

vacancies 370
vacuum cavity 259
vacuum operations 499
valleys, fingerprints 416
van der Pauw test 80, 92 f
vanadium oxide 492
vaporization heat 352
varactor diodes 248
variable gain amplifier (VGA) 227
verification
– fingerprint sensors 391, 416 ff, 439 ff
– system integration 513, 564 ff
vertical axis accelerometer 170
vertical axis gyroscope 163
vertical parallel-plate capacitors 235
vertical pnp transistors 556
vibrating metal-ring gyroscope 178 ff
vibration rejection 193 ff, 200 f
vibratory rate gyroscope 144 ff
voice scan 396
volatile organic compounds (VOCs) 338, 344, 379, 566
voltage-controlled charge source (VCQS) 319
voltage-controlled oscillator (VCO) 234, 247
voltage modes 151, 554 ff
voltammetric sensors 337, 358 ff
volume reduction 226

wafer bonding 300, 310 ff
wafer curvature 114 f
wafer materials 9 f
wafer technology 46 f, 452, 516 f

water acoustics 219 ff
wave guides 355
weight reduction 226
well formation 9, 12
wet etching 18 f, 24 f, 50
– pressure sensors 259, 302
wet skin, fingerprint sensors 421, 439
Wheatstone bridge 11, 50, 181
– biochemical sensors 472
– chemical sensors 365
– circuit integration 513 ff, 526 ff
– piezoresistors 261, 295 ff, 300 f
– resonant beam oscillator 570
whorl, fingerprint class 418
wire bonding 16
wireless communication 225 ff
working electrodes 359
working scheme
– biometric system 318 ff
– fingerprint sensors 431
worst case combinations 559

X-ducer 260, 297
xenon difluoride etching 169, 241
xerogels 457
XL series inertial sensors 146, 151 ff, 155, 158 f

yield stress 73 f, 76
Young's modulus 69 ff, 73 ff, 101 ff, 119, 126 f
– acoustic devices 199, 204
– inertial sensors 146 f
– pressure sensors 267, 272, 284
yttrium barium copper oxide 492 ff

zeolites 338
zero-input signal 522
zero-rates 427
zinc oxide 343
zirconium dioxide 337

Errata / AMN 1 / Chapter 7 / References

7
Scanning Micro- and Nanoprobes for Electrochemical Imaging

C. Kranz, A. Kueng, B. Mizaikoff, School of Chemistry and Biochemistry,
Georgia Institute of Technology, Atlanta, GA, USA

246 J. ZHANG, Q. CHI, A. M. KUZNETSOV, A. G. HANSEN, H. WACKERBARTH, H. E. M. CHRISTENSEN, J. E. T. ANDERSEN, J. ULSTRUP, J. Phys. Chem. B **2002**, 106, 1131–1152.

247 A. G. HANSEN, H. WACKERBARTH, J. U. NIELSEN, J. ZHANG, A. M. KUZNETSOV, J. ULSTRUP, Russ. J. Electrochem. **2003**, 39, 108–117.

248 J. ZHANG, M. GRUBB, A. G. HANSEN, A. M. KUZNETSOV, A. BOISEN, H. WACKERBARTH, J. ULSTRUP, J. Phys.: Condens. Matter **2003**, 15, S1873–S1890.

249 E. P. FRIIS, J. E. T. ANDERSEN, P. MOLLER, J. ULSTRUP, J. Electroanal. Chem. **1997**, 431, 35–38.

250 Q. CHI, J. ZHANG, E. P. FRIIS, J. E. T. ANDERSEN, J. ULSTRUP, Electrochem. Comm. **1999**, 1, 91–96.

251 E. P. FRIIS, J. E. T. ANDERSEN, Y. I. KHARKATS, A. M. KUZNETSOV, R. J. NICHOLS, J.-D. ZHANG, Proc. Natl. Acad. Sci. USA **1999**, 96, 1379–1384.

252 P. FACCI, D. ALLIATA, S. CANNISTRARO, Ultramicroscopy **2001**, 89, 291–298.

253 Q. CHI, J. ZHANG, J. U. NIELSEN, E. P. FRIIS, I. CHORKENDORFF, G. W. CANTERS, J. E. T. ANDERSEN, J. ULSTRUP, J. Am. Chem. Soc. **2000**, 122, 4047–4055.

254 A. G. HANSEN, A. BOISEN, J. U. NIELSEN, H. WACKERBARTH, I. CHORKENDORFF, J. E. T. ANDERSEN, J. ZHANG, J. ULSTRUP, Langmuir **2002**, 19, 3419–3427.

255 L. ADOLFINI, B. BONANNI, G. W. CANTERS, M. PH. VERBEET, S. CANNISTRARO, Surf. Sci. **2003**, 530, 181–194.

256 J. BRASK, H. WACKERBARTH, K. J. JENSEN, J. ZHANG, I. CHORKENDORFF, J. ULSTRUP, J. Am. Chem. Soc. **2003**, 125, 94–104.

257 R. GUCKENBERGER, M. HEIM, G. CEVC, H. F. KNAPP, W. WIEGRAEBE, A. HILLEBRAND, Science **1995**, 266, 1538–1540.

258 F.-R. F. FAN, A. J. BARD, Science **1995**, 270, 1849–1851.

259 R. GUCKENBERGER, M. HEIM, Science **1995**, 270, 1851–1852.

260 D. P. ALLISON, P. HINTERDORFER, W. HAN, Curr. Opinion Biotech. **2002**, 13, 47–51.

261 Y. E. KORCHEV, C. L. BASHFORD, M. MILOVANOVIC, I. VODYANOY, M. J. LAB, Biophys. J. **1997**, 73, 653–658.

262 Y. E. KORCHEV, J. GORELIK, M. J. LAB, E. V. SVIDERSKAYA, C. L. JOHNSTON, C. R. COOMBES, I. VODYANOY, C. R. EDWARDS, Biophys. J. **2000**, 78, 451–457.

263 Y. E. KORCHEV, Y. A. NEGULYAEV, C. R. EDWARDS, I. VODYANOY, M. J. LAB, Nat. Cell Biol. **2000**, 2, 616–619.

264 A. I. SHEVCHUK, J. GORELIK, S. E. HARDING, M. J. LAB, D. KLENERMAN, Y. E. KORCHEV, Biophys. J. **2001**, 81, 1759–1764.

265 J. GORELIK, A. I. SHEVCHUK, G. I. FROLENKOV, I. A. DIAKONOV, M. J. LAB, C. J. KROS, G. P. RICHARDSON, I. VODYANOY, C. R. W. EDWARDS, D. KLENERMAN, Y. E.

Advanced Micro and Nanosystems. Vol. 1
Edited by H. Baltes, O. Brand, G. K. Fedder, C. Hierold, J. Korvink, O. Tabata
Copyright © 2004 WILEY-VCH Verlag GmbH & Co. KGaA, Weinheim
ISBN: 3-527-30746-X

Korchev, *Proc. Natl. Acad. Sci.* **2003**, *100*, 5819–5822.

266 Y. Takii, K. Takoh, M. Nishizawa, T. Matsue, *Electrochim. Acta* **2003**, *48*, 3381–3385.

267 T. Kaya, Y. Torisawa, D. Oyamatsu, M. Nishizawa, T. Matsue, *Biosens. Bioelectron.* **2003**, *18*, 1379–1383.

268 M. Nishizawa, K. Takoh, T. Matsue, *Langmuir* **2002**, *18*, 3645–3649.

269 C. Kranz, T. Lotzbeyer, H.-L. Schmidt, W. Schuhmann, *Biosens. Bioelectron.* **1997**, *12*, 257–266.

270 D. T. Pierce, P. R. Unwin, A. J. Bard, *Anal. Chem.* **1992**, *64*, 1795–1804.

271 B. D. Bath, H. S. White, E. R. Scott, in *Scanning Electrochemical Microscopy*, A. J. Bard, M. V. Mirkin (eds.); New York: Marcel Dekker, **2001**, Chapter 9.

272 S. Amemiya, A. J. Bard, *Anal. Chem.* **2000**, *72*, 4940–4948.

273 D. T. Pierce, A. J. Bard, *Anal Chem.* **1993**, *65*, 3598–3604.

274 C. Kranz, G. Wittstock, H. Wohlschlager, W. Schuhmann, *Electrochim. Acta* **1997**, *42*, 3105–3111.

275 G. Wittstock, T. Wilhelm, *Electroanalysis* **2001**, *13*, 669–675.

276 H. Shiku, Y. Hara, T. Matsue, I. Uchida, T. Yamauchi, *J. Electroanal. Chem.* **1997**, *438*, 187–190.

277 G. Wittstock, *Fresenius J. Anal. Chem.* **2001**, *370*, 303–315.

278 G. Wittstock, W. Schuhmann, *Anal. Chem.* **1997**, *69*, 5059–5066.

279 T. Wilhelm, G. Wittstock, *Angew. Chem. Int. Ed.* **2003**, *42*, 2248–2250.

280 G. Denuault, M. H. T. Frank, L. M. Peter, *Faraday Discuss.* **1992**, *94*, 23–35.

281 B. R. Horrocks, M. V. Mirkin, D. T. Pierce, A. J. Bard, G. Nagy, K. Toth, *Anal. Chem.* **1993**, *65*, 1213–1224.

282 C. Wei, A. J. Bard, G. Nagy, K. Toth, *Anal. Chem.* **1995**, *67*, 1346–1356.

283 B. R. Horrocks, M. V. Mirkin, *J. Chem. Soc. Faraday Trans.* **1998**, *94*, 1115–1118.

284 B. M. Quinn, P. Liljeroth, K. Kontturi, *J. Am. Chem. Soc.* **2002**, *124*, 12915–12921.

285 M. H. Troise-Frank, G. Denuault, L. M. Peter, *Faraday Discuss.* **1992**, *94*, 23–35.

286 S. Daniele, C. Bragato, I. M. Ciani, A. Baldo, *Electroanalysis* **2003**, *15*, 621–628.

287 D. Oyamatsu, Y. Hirano, N. Kanaya, Y. Mase, M. Nishizawa, T. Matsue, *Bioelectrochem.* **2003**, *60*, 115–121.

288 A. Bruckbauer, L. Ying, A. M. Rothery, Y. E. Korchev, D. Klenerman, *Anal. Chem.* **2002**, *74*, 2612–2616.

289 C. Kranz, A. Kueng, A. Lugstein, E. Bertagnolli, B. Mizaikoff, *Ultramicroscopy* **2003**, in press.

290 J. V. Macpherson, C. E. Jones, A. L. Barker, P. R. Unwin, *Anal. Chem.* **2002**, *74*, 1841–1848.

291 W. Schuhmann, *Mikrochim. Acta* **1995**, *121*, 1–29.

292 W. Schuhmann, *Rev. Mol. Biotech.* **2002**, *82*, 425–441.

293 T. Wilhelm, G. Wittstock, R. Szargan, *Fres. J. Anal. Chem.* **1999**, *365*, 163–167.

294 C. Kranz, A. Kueng, A. Lugstein, E. Bertagnolli, B. Mizaikoff, *unpublished results*.

295 Y. Matsumura, K. Kajino, M. Fujimoto, *Membr. Biochem.* **1980**, *3*, 99–129.

296 C. Giaume, R. T. Kado, *Biochim. Biophys. Acta* **1983**, *762*, 337–343.

297 S. Glab, A. Hulanicki, G. Edwall, F. Ingmann, *Crit. Rev. Anal. Chem.* **1989**, *21*, 29–47.

298 A. Kueng, C. Kranz, B. Mizaikoff, *Sens. Lett.* **2003**, *1*, 2–15.